Grzimek's
Animal Life Encyclopedia

Second Edition

●●●●

Grzimek's Animal Life Encyclopedia

Second Edition

●●●●

Volume 5
Fishes II

Dennis A. Thoney, Advisory Editor
Paul V. Loiselle, Advisory Editor
Neil Schlager, Editor

Joseph E. Trumpey, Chief Scientific Illustrator

Michael Hutchins, Series Editor
In association with the American Zoo and Aquarium Association

Detroit • New York • San Diego • San Francisco • Cleveland • New Haven, Conn. • Waterville, Maine • London • Munich

THOMSON

GALE

Grzimek's Animal Life Encyclopedia, Second Edition

Volume 5: Fishes II
Produced by Schlager Group Inc.
Neil Schlager, Editor
Vanessa Torrado-Caputo, Assistant Editor

Project Editor
Melissa C. McDade

Editorial
Stacey Blachford, Deirdre S. Blanchfield, Madeline Harris, Christine Jeryan, Kate Kretschmann, Jacqueline L. Longe, Mark Springer

Permissions
Margaret Chamberlain

Imaging and Multimedia
Randy Bassett, Mary K. Grimes, Lezlie Light, Christine O'Bryan, Barbara Yarrow, Robyn V. Young

Product Design
Tracey Rowens, Jennifer Wahi

Manufacturing
Wendy Blurton, Dorothy Maki, Evi Seoud, Mary Beth Trimper

For more information contact
The Gale Group, Inc.
27500 Drake Rd.
Farmington Hills, MI 48331–3535
Or you can visit our Internet site at
http://www.gale.com

ISBN 0-7876-5362-4 (vols. 1–17 set)
0-7876-6572-X (vols. 4–5 set)
0-7876-5780-8 (vol. 4)
0-7876-5781-6 (vol. 5)

LIBRARY OF CONGRESS CATALOGING-IN-PUBLICATION DATA

Grzimek, Bernhard.
[Tierleben. English]
Grzimek's animal life encyclopedia.— 2nd ed.
v. cm.
Includes bibliographical references.
Contents: v. 1. Lower metazoans and lesser deuterosomes / Neil Schlager, editor — v. 2. Protostomes / Neil Schlager, editor — v. 3. Insects / Neil Schlager, editor — v. 4-5. Fishes I-II / Neil Schlager, editor — v. 6. Amphibians / Neil Schlager, editor — v. 7. Reptiles / Neil Schlager, editor — v. 8-11. Birds I-IV / Donna Olendorf, editor — v. 12-16. Mammals I-V / Melissa C. McDade, editor — v. 17. Cumulative index / Melissa C. McDade, editor.
ISBN 0-7876-5362-4 (set hardcover : alk. paper)
1. Zoology—Encyclopedias. I. Title: Animal life encyclopedia. II. Schlager, Neil, 1966- III. Olendorf, Donna IV. McDade, Melissa C. V. American Zoo and Aquarium Association. VI. Title.
QL7 .G7813 2004
590′.3—dc21
2002003351

Printed in Canada
10 9 8 7 6 5 4 3 2 1

Recommended citation: *Grzimek's Animal Life Encyclopedia*, 2nd edition. Volumes 4–5, Fishes I–II, edited by Michael Hutchins, Dennis A. Thoney, Paul V. Loiselle, and Neil Schlager. Farmington Hills, MI: Gale Group, 2003.

Contents

• • • • •

Foreword

Earth is teeming with life. No one knows exactly how many distinct organisms inhabit our planet, but more than 5 million different species of animals and plants could exist, ranging from microscopic algae and bacteria to gigantic elephants, redwood trees and blue whales. Yet, throughout this wonderful tapestry of living creatures, there runs a single thread: Deoxyribonucleic acid or DNA. The existence of DNA, an elegant, twisted organic molecule that is the building block of all life, is perhaps the best evidence that all living organisms on this planet share a common ancestry. Our ancient connection to the living world may drive our curiosity, and perhaps also explain our seemingly insatiable desire for information about animals and nature. Noted zoologist, E.O. Wilson, recently coined the term "biophilia" to describe this phenomenon. The term is derived from the Greek *bios* meaning "life" and *philos* meaning "love." Wilson argues that we are human because of our innate affinity to and interest in the other organisms with which we share our planet. They are, as he says, "the matrix in which the human mind originated and is permanently rooted." To put it simply and metaphorically, our love for nature flows in our blood and is deeply engrained in both our psyche and cultural traditions.

Our own personal awakenings to the natural world are as diverse as humanity itself. I spent my early childhood in rural Iowa where nature was an integral part of my life. My father and I spent many hours collecting, identifying and studying local insects, amphibians and reptiles. These experiences had a significant impact on my early intellectual and even spiritual development. One event I can recall most vividly. I had collected a cocoon in a field near my home in early spring. The large, silky capsule was attached to a stick. I brought the cocoon back to my room and placed it in a jar on top of my dresser. I remember waking one morning and, there, perched on the tip of the stick was a large moth, slowly moving its delicate, light green wings in the early morning sunlight. It took my breath away. To my inexperienced eyes, it was one of the most beautiful things I had ever seen. I knew it was a moth, but did not know which species. Upon closer examination, I noticed two moon-like markings on the wings and also noted that the wings had long "tails", much like the ubiquitous tiger swallow-tail butterflies that visited the lilac bush in our backyard. Not wanting to suffer my ignorance any longer, I reached immediately for my *Golden Guide to North American Insects* and searched through the section on moths and butterflies. It was a luna moth! My heart was pounding with the excitement of new knowledge as I ran to share the discovery with my parents.

I consider myself very fortunate to have made a living as a professional biologist and conservationist for the past 20 years. I've traveled to over 30 countries and six continents to study and photograph wildlife or to attend related conferences and meetings. Yet, each time I encounter a new and unusual animal or habitat my heart still races with the same excitement of my youth. If this is biophilia, then I certainly possess it, and it is my hope that others will experience it too. I am therefore extremely proud to have served as the series editor for the Gale Group's rewrite of *Grzimek's Animal Life Encyclopedia*, one of the best known and widely used reference works on the animal world. *Grzimek's* is a celebration of animals, a snapshot of our current knowledge of the Earth's incredible range of biological diversity. Although many other animal encyclopedias exist, *Grzimek's Animal Life Encyclopedia* remains unparalleled in its size and in the breadth of topics and organisms it covers.

The revision of these volumes could not come at a more opportune time. In fact, there is a desperate need for a deeper understanding and appreciation of our natural world. Many species are classified as threatened or endangered, and the situation is expected to get much worse before it gets better. Species extinction has always been part of the evolutionary history of life; some organisms adapt to changing circumstances and some do not. However, the current rate of species loss is now estimated to be 1,000–10,000 times the normal "background" rate of extinction since life began on Earth some 4 billion years ago. The primary factor responsible for this decline in biological diversity is the exponential growth of human populations, combined with peoples' unsustainable appetite for natural resources, such as land, water, minerals, oil, and timber. The world's human population now exceeds 6 billion, and even though the average birth rate has begun to decline, most demographers believe that the global human population will reach 8–10 billion in the next 50 years. Much of this projected growth will occur in developing countries in Central and South America, Asia and Africa-regions that are rich in unique biological diversity.

Finding solutions to conservation challenges will not be easy in today's human-dominated world. A growing number of people live in urban settings and are becoming increasingly isolated from nature. They "hunt" in super markets and malls, live in apartments and houses, spend their time watching television and searching the World Wide Web. Children and adults must be taught to value biological diversity and the habitats that support it. Education is of prime importance now while we still have time to respond to the impending crisis. There still exist in many parts of the world large numbers of biological "hotspots"-places that are relatively unaffected by humans and which still contain a rich store of their original animal and plant life. These living repositories, along with selected populations of animals and plants held in professionally managed zoos, aquariums and botanical gardens, could provide the basis for restoring the planet's biological wealth and ecological health. This encyclopedia and the collective knowledge it represents can assist in educating people about animals and their ecological and cultural significance. Perhaps it will also assist others in making deeper connections to nature and spreading biophilia. Information on the conservation status, threats and efforts to preserve various species have been integrated into this revision. We have also included information on the cultural significance of animals, including their roles in art and religion.

It was over 30 years ago that Dr. Bernhard Grzimek, then director of the Frankfurt Zoo in Frankfurt, Germany, edited the first edition of *Grzimek's Animal Life Encyclopedia*. Dr. Grzimek was among the world's best known zoo directors and conservationists. He was a prolific author, publishing nine books. Among his contributions were: *Serengeti Shall Not Die*, *Rhinos Belong to Everybody* and *He and I and the Elephants*. Dr. Grzimek's career was remarkable. He was one of the first modern zoo or aquarium directors to understand the importance of zoo involvement in *in situ* conservation, that is, of their role in preserving wildlife in nature. During his tenure, Frankfurt Zoo became one of the leading western advocates and supporters of wildlife conservation in East Africa. Dr. Grzimek served as a Trustee of the National Parks Board of Uganda and Tanzania and assisted in the development of several protected areas. The film he made with his son Michael, *Serengeti Shall Not Die*, won the 1959 Oscar for best documentary.

Professor Grzimek has recently been criticized by some for his failure to consider the human element in wildlife conservation. He once wrote: "A national park must remain a primordial wilderness to be effective. No men, not even native ones, should live inside its borders." Such ideas, although considered politically incorrect by many, may in retrospect actually prove to be true. Human populations throughout Africa continue to grow exponentially, forcing wildlife into small islands of natural habitat surrounded by a sea of humanity. The illegal commercial bushmeat trade-the hunting of endangered wild animals for large scale human consumption-is pushing many species, including our closest relatives, the gorillas, bonobos and chimpanzees, to the brink of extinction. The trade is driven by widespread poverty and lack of economic alternatives. In order for some species to survive it will be necessary, as Grzimek suggested, to establish and enforce a system of protected areas where wildlife can roam free from exploitation of any kind.

While it is clear that modern conservation must take the needs of both wildlife and people into consideration, what will the quality of human life be if the collective impact of short-term economic decisions is allowed to drive wildlife populations into irreversible extinction? Many rural populations living in areas of high biodiversity are dependent on wild animals as their major source of protein. In addition, wildlife tourism is the primary source of foreign currency in many developing countries and is critical to their financial and social stability. When this source of protein and income is gone, what will become of the local people? The loss of species is not only a conservation disaster; it also has the potential to be a human tragedy of immense proportions. Protected areas, such as national parks, and regulated hunting in areas outside of parks are the only solutions. What critics do not realize is that the fate of wildlife and people in developing countries is closely intertwined. Forests and savannas emptied of wildlife will result in hungry, desperate people, and will, in the long-term lead to extreme poverty and social instability. Dr. Grzimek's early contributions to conservation should be recognized, not only as benefiting wildlife, but as benefiting local people as well.

Dr. Grzimek's hope in publishing his *Animal Life Encyclopedia* was that it would "...disseminate knowledge of the animals and love for them", so that future generations would "...have an opportunity to live together with the great diversity of these magnificent creatures." As stated above, our goals in producing this updated and revised edition are similar. However, our challenges in producing this encyclopedia were more formidable. The volume of knowledge to be summarized is certainly much greater in the twenty-first century than it was in the 1970's and 80's. Scientists, both professional and amateur, have learned and published a great deal about the animal kingdom in the past three decades, and our understanding of biological and ecological theory has also progressed. Perhaps our greatest hurdle in producing this revision was to include the new information, while at the same time retaining some of the characteristics that have made *Grzimek's Animal Life Encyclopedia* so popular. We have therefore strived to retain the series' narrative style, while giving the information more organizational structure. Unlike the original *Grzimek's*, this updated version organizes information under specific topic areas, such as reproduction, behavior, ecology and so forth. In addition, the basic organizational structure is generally consistent from one volume to the next, regardless of the animal groups covered. This should make it easier for users to locate information more quickly and efficiently. Like the original Grzimek's, we have done our best to avoid any overly technical language that would make the work difficult to understand by non-biologists. When certain technical expressions were necessary, we have included explanations or clarifications.

Considering the vast array of knowledge that such a work represents, it would be impossible for any one zoologist to have completed these volumes. We have therefore sought specialists from various disciplines to write the sections with which they are most familiar. As with the original *Grzimek's*, we have engaged the best scholars available to serve as topic editors, writers, and consultants. There were some complaints about inaccuracies in the original English version that may have been due to mistakes or misinterpretation during the complicated translation process. However, unlike the original *Grzimek's*, which was translated from German, this revision has been completely re-written by English-speaking scientists. This work was truly a cooperative endeavor, and I thank all of those dedicated individuals who have written, edited, consulted, drawn, photographed, or contributed to its production in any way. The names of the topic editors, authors, and illustrators are presented in the list of contributors in each individual volume.

The overall structure of this reference work is based on the classification of animals into naturally related groups, a discipline known as taxonomy or biosystematics. Taxonomy is the science through which various organisms are discovered, identified, described, named, classified and catalogued. It should be noted that in preparing this volume we adopted what might be termed a conservative approach, relying primarily on traditional animal classification schemes. Taxonomy has always been a volatile field, with frequent arguments over the naming of or evolutionary relationships between various organisms. The advent of DNA fingerprinting and other advanced biochemical techniques has revolutionized the field and, not unexpectedly, has produced both advances and confusion. In producing these volumes, we have consulted with specialists to obtain the most up-to-date information possible, but knowing that new findings may result in changes at any time. When scientific controversy over the classification of a particular animal or group of animals existed, we did our best to point this out in the text.

Readers should note that it was impossible to include as much detail on some animal groups as was provided on others. For example, the marine and freshwater fish, with vast numbers of orders, families, and species, did not receive as detailed a treatment as did the birds and mammals. Due to practical and financial considerations, the publishers could provide only so much space for each animal group. In such cases, it was impossible to provide more than a broad overview and to feature a few selected examples for the purposes of illustration. To help compensate, we have provided a few key bibliographic references in each section to aid those interested in learning more. This is a common limitation in all reference works, but *Grzimek's Encyclopedia of Animal Life* is still the most comprehensive work of its kind.

I am indebted to the Gale Group, Inc. and Senior Editor Donna Olendorf for selecting me as Series Editor for this project. It was an honor to follow in the footsteps of Dr. Grzimek and to play a key role in the revision that still bears his name. *Grzimek's Animal Life Encyclopedia* is being published by the Gale Group, Inc. in affiliation with my employer, the American Zoo and Aquarium Association (AZA), and I would like to thank AZA Executive Director, Sydney J. Butler; AZA Past-President Ted Beattie (John G. Shedd Aquarium, Chicago, IL); and current AZA President, John Lewis (John Ball Zoological Garden, Grand Rapids, MI), for approving my participation. I would also like to thank AZA Conservation and Science Department Program Assistant, Michael Souza, for his assistance during the project. The AZA is a professional membership association, representing 205 accredited zoological parks and aquariums in North America. As Director/William Conway Chair, AZA Department of Conservation and Science, I feel that I am a philosophical descendant of Dr. Grzimek, whose many works I have collected and read. The zoo and aquarium profession has come a long way since the 1970s, due, in part, to innovative thinkers such as Dr. Grzimek. I hope this latest revision of his work will continue his extraordinary legacy.

Silver Spring, Maryland, 2001
Michael Hutchins
Series Editor

How to use this book

Grzimek's Animal Life Encyclopedia is an internationally prominent scientific reference compilation, first published in German in the late 1960s, under the editorship of zoologist Bernhard Grzimek (1909–1987). In a cooperative effort between Gale and the American Zoo and Aquarium Association, the series has been completely revised and updated for the first time in over 30 years. Gale expanded the series from 13 to 17 volumes, commissioned new color paintings, and updated the information so as to make the set easier to use. The order of revisions is:

Volumes 8–11: Birds I–IV
Volume 6: Amphibians
Volume 7: Reptiles
Volumes 4–5: Fishes I–II
Volumes 12–16: Mammals I–V
Volume 3: Insects
Volume 2: Protostomes
Volume 1: Lower Metazoans and Lesser Deuterostomes
Volume 17: Cumulative Index

Organized by taxonomy

The overall structure of this reference work is based on the classification of animals into naturally related groups, a discipline known as taxonomy—the science in which various organisms are discovered, identified, described, named, classified, and catalogued. Starting with the simplest life forms, the lower metazoans and lesser deuterostomes, in volume 1, the series progresses through the more advanced classes, culminating with the mammals in volumes 12–16. Volume 17 is a stand-alone cumulative index.

Organization of chapters within each volume reinforces the taxonomic hierarchy. In the case of the volumes on Fishes, introductory chapters describe general characteristics of fishes, followed by taxonomic chapters dedicated to order and, in a few cases, suborder. Readers should note that in a few instances, taxonomic groups have been split among more than one chapter. For example, the order Cypriniformes is split among two chapters, each covering particular families. Species accounts appear at the end of the taxonomic chapters. To help the reader grasp the scientific arrangement, order and suborder chapters have distinctive symbols:

● = Order Chapter
○ = Suborder Chapter

The order Perciformes, which has the greatest number of species by far of any fishes order—and in fact is the largest order of vertebrates—has been split into separate chapters based on suborder. Some of these suborder chapters are again divided into multiple chapters in an attempt to showcase the diversity of species within the group. For instance, the suborder Percoidei has been split among four chapters. Readers should note that here, as elsewhere, the text does not necessarily discuss every single family within the group; in the case of Percoidei, there are more than 70 families. Instead, the text highlights the best-known and most significant families and species within the group. Readers can find the complete list of families for every order in the "Fishes family list" in the back of each Fishes volume.

As chapters narrow in focus, they become more tightly formatted. Introductory chapters have a loose structure, reminiscent of the first edition. Chapters on orders and suborders are more tightly structured, following a prescribed format of standard rubrics that make information easy to find. These taxonomic chapters typically include:

- Scientific name of order or suborder
- Common name of order or suborder
- Class
- Order
- Number of families

Main chapter

- Evolution and systematics
- Physical characteristics
- Distribution
- Habitat
- Feeding ecology and diet
- Behavior
- Reproductive biology
- Conservation status
- Significance to humans

Species accounts

- Common name
- Scientific name

- Family
- Taxonomy
- Other common names
- Physical characteristics
- Distribution
- Habitat
- Feeding ecology and diet
- Behavior
- Reproductive biology
- Conservation status
- Significance to humans

Resources
- Books
- Periodicals
- Organizations
- Other

Color graphics enhance understanding

Grzimek's features approximately 3,500 color photos, including nearly 250 in the Fishes volumes; 3,500 total color maps, including more than 200 in the Fishes volumes; and approximately 5,500 total color illustrations, including nearly 700 in the Fishes volumes. Each featured species of animal is accompanied by both a distribution map and an illustration.

All maps in *Grzimek's* were created specifically for the project by XNR Productions. Distribution information was provided by expert contributors and, if necessary, further researched at the University of Michigan Zoological Museum library. Maps are intended to show broad distribution, not definitive ranges.

All the color illustrations in *Grzimek's* were created specifically for the project by Michigan Science Art. Expert contributors recommended the species to be illustrated and provided feedback to the artists, who supplemented this information with authoritative references and animal specimens from the University of Michigan Zoological Museum library. In addition to illustrations of species, *Grzimek's* features drawings that illustrate characteristic traits and behaviors.

About the contributors

All of the chapters were written by ichthyologists who are specialists on specific subjects and/or families. The volumes' subject advisors, Dennis A. Thoney and Paul V. Loiselle, reviewed the completed chapters to insure consistency and accuracy.

Standards employed

In preparing the volumes on Fishes, the editors relied primarily on the taxonomic structure outlined in *Fishes of the World*, 3rd edition, by Joseph S. Nelson (1994), with some modifications suggested by expert contributors for certain taxonomic groups based on more recent data. Systematics is a dynamic discipline in that new species are being discovered continuously, and new techniques (e.g., DNA sequencing) frequently result in changes in the hypothesized evolutionary relationships among various organisms. Consequently, controversy often exists regarding classification of a particular animal or group of animals; such differences are mentioned in the text.

Grzimek's has been designed with ready reference in mind, and the editors have standardized information wherever feasible. For ***Conservation Status***, *Grzimek's* follows the IUCN Red List system, developed by its Species Survival Commission. The Red List provides the world's most comprehensive inventory of the global conservation status of plants and animals. Using a set of criteria to evaluate extinction risk, the IUCN recognizes the following categories: Extinct, Extinct in the Wild, Critically Endangered, Endangered, Vulnerable, Conservation Dependent, Near Threatened, Least Concern, and Data Deficient. For a complete explanation of each category, visit the IUCN web page at <http://www.iucn.org/themes/ssc/redlists/categor.htm>.

In addition to IUCN ratings, chapters may contain other conservation information, such as a species' inclusion on one of three Convention on International Trade in Endangered Species (CITES) appendices. Adopted in 1975, CITES is a global treaty whose focus is the protection of plant and animal species from unregulated international trade.

In the Species accounts throughout the volume, the editors have attempted to provide common names not only in English but also in French, German, Spanish, and local dialects. Readers can find additional information on fishes species on the Fishbase Web site: <http://www.fishbase.org>.

Grzimek's provides the following standard information on lineage in the ***Taxonomy*** rubric of each Species account: [First described as] *Acipenser brevirostrum* [by] LeSueur, [in] 1818, [based on a specimen from] Delaware River, United States. The person's name and date refer to earliest identification of a species, although the species name may have changed since first identification. However, the entity of fish is the same.

Readers should note that within chapters, species accounts are organized alphabetically by family name and then alphabetically by scientific name. In each chapter, the list of species to be highlighted was chosen by the contributor in consultation with the appropriate subject advisor: Dennis A. Thoney, who specializes in marine fishes; and Paul V. Loiselle, who specializes in freshwater fishes.

Anatomical illustrations

While the encyclopedia attempts to minimize scientific jargon, readers will encounter numerous technical terms related to anatomy and physiology throughout the volume. To assist readers in placing physiological terms in their proper context, we have created a number of detailed anatomical drawings. These can be found on pages 6 and 7, and 15–27 in the "Structure and function" chapter. Readers are urged to make heavy use of these drawings. In addition, selected terms are defined in the ***Glossary*** at the back of the book.

Appendices and index

In addition to the main text and the aforementioned ***Glossary,*** the volume contains numerous other elements. ***For Further Reading*** directs readers to additional sources of information about fishes. Valuable contact information for ***Organizations*** is also included in an appendix. An exhaustive ***Fishes family list*** records all recognized families of fishes according to *Fishes of the World*, 3rd edition, by Joseph S. Nelson (1994). And a full-color ***Geologic time scale*** helps readers understand prehistoric time periods. Additionally, the volume contains a ***Subject index.***

Acknowledgements

Gale would like to thank several individuals for their important contributions to the volume. Dr. Dennis A. Thoney, subject advisor specializing in marine fishes, created the overall topic list for the volumes and suggested writers and reviewed chapters related to marine fishes. Dr. Paul V. Loiselle, subject advisor specializing in freshwater fishes, suggested writers and reviewed chapters related to freshwater fishes. Neil Schlager, project manager for the Fishes volumes, coordinated the writing and editing of the text. Finally, Dr. Michael Hutchins, chief consulting editor for the series, and Michael Souza, program assistant, Department of Conservation and Science at the American Zoo and Aquarium Association, provided valuable input and research support.

Advisory boards

Library advisors

Contributing writers

Fishes I–II

Arturo Acero, PhD
INVEMAR
Santa Marta, Colombia

M. Eric Anderson, PhD
J. L. B. Smith Institute of Ichthyology
Grahmstown, South Africa

Eugene K. Balon, PhD
University of Guelph
Guelph, Ontario, Canada

George Benz, PhD
Tennessee Aquarium Research
Institute and Tennessee Aquarium
Chattanooga, Tennessee

Tim Berra, PhD
The Ohio State University
Mansfield, Ohio

Ralf Britz, PhD
Smithsonian Institution
Washington, D.C.

John H. Caruso, PhD
University of New Orleans, Lakefront
New Orleans, Louisiana

Marcelo Carvalho, PhD
American Museum of Natural History
New York, New York

José I. Castro, PhD
Mote Marine Laboratory
Sarasota, Florida

Bruce B. Collette, PhD
National Marine Fisheries Systematics
Laboratory and National Museum of
Natural History
Washington, D.C.

Roy Crabtree, PhD
Florida Fish and Wildlife
Conservation Commission
Tallahasee, Florida

Dominique Didier Dagit, PhD
The Academy of Natural Sciences
Philadelphia, Pennsylvania

Terry Donaldson, PhD
University of Guam Marine
Laboratory, UOG Station
Mangliao, Guam

Michael P. Fahay, PhD
NOAA National Marine
Fisheries Service,
Sandy Hook Marine Laboratory
Highlands, New Jersey

John. V. Gartner, Jr., PhD
St. Petersburg College
St. Petersburg, Florida

Howard Gill, PhD
Murdoch University
Murdoch, Australia

Lance Grande, PhD
Field Museum of Natural History
Chicago, Illinois

Terry Grande, PhD
Loyola University Chicago
Chicago, Illinois

David W. Greenfield, PhD
University of Hawaii
Honolulu, Hawaii

Melina Hale, PhD
University of Chicago
Chicago, Illinois

Ian J. Harrison, PhD
American Museum of Natural History
New York, New York

Phil Heemstra, PhD
South African Institute for
Aquatic Biodiversity
Grahamstown, South Africa

Jeffrey C. Howe, MA
Freelance Writer
Mobile, Alabama

Liu Huanzhang, PhD
Chinese Academy of Sciences
Hubei Wuhan,
People's Republic of China

G. David Johnson, PhD
Smithsonian Institution
Washington, D.C.

Scott I. Kavanaugh, BS
University of New Hampshire
Durham, New Hampshire

Frank Kirschbaum, PhD
Institute of Freshwater Ecology
Berlin, Germany

Kenneth J. Lazara, PhD
American Museum of Natural History
New York, New York

Andrés López, PhD
Iowa State University
Ames, Iowa

John A. MacDonald, PhD
The University of Auckland
Auckland, New Zealand

Jeff Marliave, PhD
Institute of Freshwater Ecology
Vancouver, Canada

John McEachran, PhD
Texas A&M University
College Station, Texas

Leslie Mertz, PhD
Wayne State University
Detroit, Michigan

Elizabeth Mills, MS
Washington, D. C.

Katherine E. Mills, MS
Cornell University
Ithaca, New York

Randall D. Mooi, PhD
Milwaukee Public Museum
Milwaukee, Wisconsin

Thomas A. Munroe, PhD
National Systematics Laboratory
Smithsonian Institution
Washington, D.C.

Prachya Musikasinthorn, PhD
Kasetsart University
Bangkok, Thailand

John E. Olney, PhD
College of William and Mary
Gloucester Point, Virginia

Frank Pezold, PhD
University of Louisiana at Monroe
Monroe, Louisiana

Mickie L. Powell, PhD
University of New Hampshire
Durham, New Hampshire

Aldemaro Romero, PhD
Macalester College
St. Paul, Minnesota

Robert Schelly, MA
American Museum of Natural History
New York, New York

Matthew R. Silver, BS
University of New Hampshire
Durham, New Hampshire

William Leo Smith, PhD
American Museum of Natural History
and Columbia University
New York, New York

Stacia A. Sower, PhD
University of New Hampshire
Durham, New Hampshire

Melanie Stiassny, PhD
American Museum of Natural History
New York, New York

Tracey Sutton, PhD
Woods Hole
Oceanographic Institution
Woods Hole, Massachusetts

Gus Thiesfeld, PhD
Humboldt State University
Arcata, California

Jeffrey T. Williams, PhD
Smithsonian Institution
Washington, D.C.

Contributing illustrators

Drawings by Michigan Science Art

Joseph E. Trumpey, Director, AB, MFA
Science Illustration, School of Art and Design, University of Michigan

Wendy Baker, ADN, BFA

Brian Cressman, BFA, MFA

Emily S. Damstra, BFA, MFA

Maggie Dongvillo, BFA

Barbara Duperron, BFA, MFA

Dan Erickson, BA, MS

Patricia Ferrer, AB, BFA, MFA

Gillian Harris, BA

Jonathan Higgins, BFA, MFA

Amanda Humphrey, BFA

Jacqueline Mahannah, BFA, MFA

John Megahan, BA, BS, MS

Michelle L. Meneghini, BFA, MFA

Bruce D. Worden, BFA

Thanks are due to the University of Michigan, Museum of Zoology, which provided specimens that served as models for the images.

Maps by XNR Productions

Paul Exner, Chief cartographer
XNR Productions, Madison, WI

Tanya Buckingham

Jon Daugherity

Laura Exner

Andy Grosvold

Cory Johnson

Paula Robbins

Polymixiiformes

(Beardfishes)

Class Actinopterygii
Order Polymixiiformes
Number of families 1

Photo: Beardfish (*Polymixia berndti*) live in deep marine waters, at 60–1,706 ft (18–520 m). They are rarely seen alive. This specimen was found near Oahu. (Photo by John E. Randall. Reproduced by permission.)

Evolution and systematics

Beardfishes of the order Polymixiiformes comprise a single living family, Polymixiidae, containing one genus, *Polymixia*, and, according to the revisional study of Kotlyar (1993), ten species. These seemingly nondescript fishes have a remarkably checkered taxonomic history, and few fishes have been shifted back and forth in different phylogenetic schemes as have these poorly understood animals. Although most systematists agree that Polymixiiformes are basal acanthomorphs, their precise placement at the base of this huge radiation of spiny-rayed fishes remains a mystery. In view of the longstanding confusion regarding the phylogenetic position of beardfishes, it should be clear that the small size of the order is no indication of their evolutionary importance. Although the group is represented today by a single genus, the Polymixiiformes have a much more diverse fossil record, and at least two families, the Aipichthyidae and Polymixiidae, containing some six genera, are currently recognized.

Polymixiiforms appear first in the fossil record in the late Cretaceous period, some 95 million years ago, and have been viewed as being of considerable importance to our understanding of the spiny-rayed fishes and their evolution. In 1964 the noted British paleoichthyologist Colin Patterson reviewed polymixiiform fossil diversity and discussed the striking similarities between them and certain families of living perciforms such as carangids (jacks) and monodactylids (moonfishes). Whether these similarities reflect the common phenomenon of convergent evolution or are the result of recent common ancestry remains an open question. Whatever the case, such an ongoing confusion serves to highlight the importance of Polymixiiformes to our understanding of the evolution of spiny-rayed fishes.

Physical characteristics

Polymixia are rather deep-bodied fishes, with prominent blunt snouts and large eyes. Their dorsal and anal fins bear well-developed spines, but a spine is lacking in the subabdominally positioned pelvic fins, which have a single, segmented, leading ray and six branched rays. All beardfishes are characterized by the possession of a pair of hyoid barbels. These "chin" barbels are internally supported by three branchiostegal rays of the hyoid arch, and their characteristic appearance lends their bearers the name "beardfishes." A 2001 study by Kim and his colleagues highlights the unique nature of the anatomy and muscular control of beardfish barbels.

Distribution

Beardfishes are known to occur in the Atlantic, Indian and Western Pacific oceans, as well as in tropical and subtropical waters. Worldwide collection data for *Polymixia* species are poor, and sampling in many regions is limited or nonexistent. When caught, most species are taken at depths ranging from 492–2,132.5 ft (150–650 m), but catches from as shallow as 164 ft (50 m) have been recorded.

Habitat

Very little is known of the habitat preferences of *Polymixia* species, but the presence of chin barbels suggests a bottom-dwelling mode of life, probably over sandy or muddy substrates.

Behavior

Nothing is known.

Feeding ecology and diet

The stomach of *Polymixia* is thick walled and muscular, often with over 100 pyloric ceca (pouches). Analyses of the gut contents indicate an opportunistic diet of crustaceans, squid, and small fishes. Interestingly, beardfishes are recorded among the stomach contents of the Indian Ocean coelacanths, as well as in those of a wide range of other fish predators.

Reproductive biology

Polymixia eggs are unknown, and the smallest individual so far identified is a 0.15 in (0.39 cm) postflexion larva, which was described in some detail by Konishi and Okiyama, who noted the presence of moderately developed head spination similar to that found in many beryciform fishes. Reproductive biology is virtually unknown, but sexual dimorphism in coloration has been noted in *P. lowei*, where an intense black marking on the anal fin and both lobes of the tail fin are present only in male fish.

Conservation status

None of the *Polymixia* species are included on the IUCN Red List. Although catches of some species are relatively high in some regions, there are no indications of overfishing.

Significance to humans

Beardfishes are marketed for human consumption in most regions; however, commercial catches are limited but growing. As catches of other species have declined, a number of Pacific beardfish species have been tagged as having unexploited fisheries potential.

Species accounts

Stout beardfish
Polymixia nobilis

FAMILY
Polymixiidae

TAXONOMY
Polymixia nobilis Lowe, 1836, off Madeira.

OTHER COMMON NAMES
Portuguese: Salmonete do Alto.

PHYSICAL CHARACTERISTICS
Maximum length 19.6 in (50 cm). Deep-bodied, with a prominent blunt snout and large eyes. The ctenii on the scales are arranged in wedgelike rows, and the hyoid barbels are long and filamentous. Dark bronzy-grey dorsally and silvery ventrally.

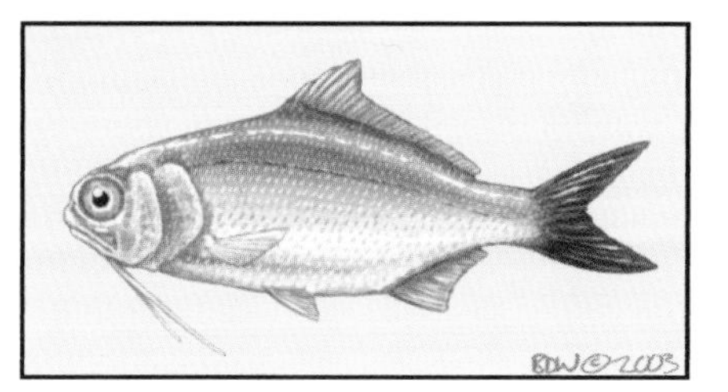

Polymixia nobilis

DISTRIBUTION
Found only in Atlantic Ocean. In western Atlantic, they occur from Norfolk south to Lesser Antilles, Cuba, and Bahamas, where they are sympatric with *Polymixia lowei*. The stout beardfish is the only beardfish in the eastern Atlantic, where it occurs from the Azores and Canary Islands south to St. Helena. Records of *Polymixia nobilis* from the Pacific are now known to be misidentifications.

HABITAT
Semihard and soft bottoms on the continental shelf and slope.

BEHAVIOR
Nothing is known.

FEEDING ECOLOGY AND DIET
Feeds on crustaceans, squid, and small fishes.

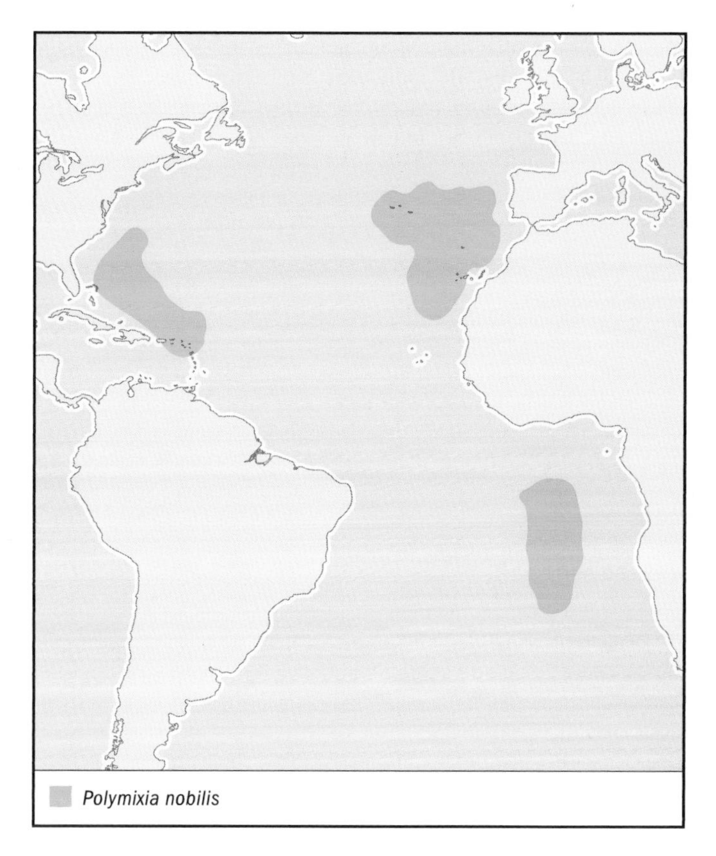

Polymixia nobilis

REPRODUCTIVE BIOLOGY
Eggs are unknown and reproductive biology is virtually unknown.

CONSERVATION STATUS
Not listed by the IUCN.

SIGNIFICANCE TO HUMANS
Stout beardfishes are of some commercial importance, particularly in Madeira, where they are marketed fresh and frozen. ◆

Resources

Periodicals

Kim, B. J., M. Yabe, and K. Nakaya. "Barbels and Related Muscles in Mullidae (Perciformes) and Polymixiidae (Polymixiiformes)." *Ichthyological Research* 117 (2001): 409–413.

Konishi, Y., and M. Okiyama. "Morphological Development of Four Trachichthyoid Larvae (Pisces: Beryciformes), with Comments on Trachichthyoid Relationships." *Bulletin of Marine Science* 60 (1997): 66–88.

Kotlyar, A. N. "A New Species of the Genus *Polymixia* from the Kyushu-Palau Submarine Ridge and Notes on the Other Members of the Genus." *Journal of Ichthyology* 2 (1993): 30–49.Patterson, C. "A Review of Mesozoic Acanthopterygian Fishes, with Special Reference to Those of the English Chalk." *Philosophical Transactions of the Royal Society of London* 3 (1994): 213–482.

Other

"Coelacanths: Coelacanth Fact Sheet." South African Institute for Aquatic Biodiversity. Ichthos, 2001. (March 20, 2003). <http://www.saiab.ru.ac.za/educoel1.htm>

Melanie Stiassny, PhD

Percopsiformes
(Troutperches and relatives)

Class Actinopterygii
Order Percopsiformes
Number of families 3

Photo: Troutperch (*Percopsis omiscomaycus*) combine the characteristics of spiny-rayed and soft-rayed fishes, resembling both trout and perch. (Photo by Animals Animals ©Raymond A. Mendez. Reproduced by permission.)

Evolution and systematics

The oldest fossils date back to the Paleocene Paskapoo Formation (between 60 and 62 million years ago) in western Canada. The Percopsiformes may be related remotely to the codfishes (Gadiformes) and toadfishes (Batrachoidiformes). One point of controversy about their phylogeny is that they show primitive conditions, such as the presence of an adipose fin, which suggests character reversal in their evolutionary history. The monophyly of this group has been questioned. Murray and Wilson proposed removal of the amblyopsid family from the group and created a new order: Amblyopsiformes.

There are two recognized suborders. The first, the Percopsoidei, is characterized by the presence of an adipose fin and a complete lateral line. The suborder is represented by one family: Percopsidae, or troutperches, with one genus and two species. The second, the Aphredoderoidei, is characterized by the absence of an adipose fin and includes two families: Aphredoderidae (pirate perch, one species) and Amblyopsidae (swampfishes and cavefishes, four genera and six species).

Physical characteristics

These are small fishes (less than 8 in, or 20 cm) with a mosaic of primitive characters, such as an adipose fin, and advanced characters, such as a pelvic girdle located farther back from the cranium compared with most other fishes. They also have fewer fin spines and ray-supported dorsal and anal fins, each usually with one to four anterior soft spines. If pelvic fins are present, they are located in a position below the abdomen and behind the pectorals, with three to eight soft rays. The body is covered with cycloid or ctenoid scales.

Distribution

The troutperches are distributed in North America from Alaska and the Great Lakes drainage to the southern and eastern United States.

Habitat

All species are freshwater, with two species found in swamps, one as a facultative cave dweller. Four species are obligatory cavernicoles (cave dwellers).

Behavior

Besides the fact that all species are solitary, little is known about their behavior. The exception is certain types of behavior studied among cavefishes. At least two of the noncavernicolous species are nocturnal.

Feeding ecology and diet

Members of this order are opportunistic predators that eat a variety of food items; at least one species is cannibalistic. Percopsiformes are preyed upon by other fishes, water snakes, and fish-eating birds. Fish larvae may be preyed upon by aquatic insects. Cavefishes are not generally preyed upon since they are the top predators in their habitats.

Reproductive biology

They are oviparous, but nothing else is known at the family level. Spawning (at least for the noncavernicolous species) takes place in the spring. Fecundity tends to be low.

Conservation status

The IUCN Red List includes four species from this order, all of which are cave-dwelling species from the family Amblyopsidae. *Speoplatyrhinus poulsoni* is listed as Critically Endangered, while *Amblyopsis rosae*, *A. spelaea*, and *Typhlichthys subterraneus* are listed as Vulnerable.

Significance to humans

Some species can be found in both the commercial trade and public aquaria. Cave species have been important in understanding evolutionary issues.

1. Alabama cavefish (*Speoplatyrhinus poulsoni*); 2. Ozark cavefish (*Amblyopsis rosae*); 3. Swampfish (*Chologaster cornuta*); 4. Spring cavefish (*Forbesichthys agassizii*); 5. Southern cavefish (*Typhlichthys subterraneus*); 6. Northern cavefish (*Amblyopsis spelaea*); 7. Sand roller (*Percopsis transmontana*); 8. Troutperch (*Percopsis omiscomaycus*); 9. Pirate perch (*Aphredoderus sayanus*). (Illustration by Emily Damstra)

Species accounts

Ozark cavefish
Amblyopsis rosae

FAMILY
Amblyopsidae

TAXONOMY
Typhlichthys rosae Eigenmann, 1897, "caves of Missouri."

OTHER COMMON NAMES
None known.

PHYSICAL CHARACTERISTICS
Grows to 2.56 in (6.5 cm). Pinkish-white in coloration. The eyes are not externally visible because they have only vestigial tissue under the skin. This fish also lacks pelvic fins.

DISTRIBUTION
This species can be found at 41 sites on the Springfield Plateau, over seven counties in three states: southwest Missouri (20 sites), northwest Arkansas (10 sites), and northeast Oklahoma (11 sites). (The verified historic range was larger.)

HABITAT
Individuals of this species are found mostly in small cave streams with a chert or rubble bottom, in pools over a silt and sand bottom, or in karst windows or wells, but never too deep.

BEHAVIOR
Almost nothing is known about their behavior.

FEEDING ECOLOGY AND DIET
Stomach contents have been found to contain copepods, which constituted about 70–90% of the contents by volume; the balance was primarily small salamanders, crayfish, isopods, amphipods, and young of their own species. Most individuals grow between April and October. Cannibalism does not always occur in this species.

REPRODUCTIVE BIOLOGY
Breeding habits are not well understood. They have an extended spawning season, with a peak in late summer. The maximum life span is four to five years. Growth is sporadic.

CONSERVATION STATUS
Classified as Vulnerable by the IUCN and as Threatened by the U.S. Fish and Wildlife Service.

SIGNIFICANCE TO HUMANS
Of particular scientific interest because it is a cave species. ◆

Northern cavefish
Amblyopsis spelaea

FAMILY
Amblyopsidae

TAXONOMY
Amblyopsis spelaeus DeKay, 1842, Mammoth Cave, Kentucky, United States.

OTHER COMMON NAMES
German: Nördlicher Höhlenfisch.

PHYSICAL CHARACTERISTICS
The species grows to 4.33 in (11.0 cm). They are pink-white in coloration. The eyes are not externally visible because they have only vestigial tissue under the skin. The pelvic fins are rarely absent; when present, they are always very small. They have a large, broad head.

DISTRIBUTION
Individuals of this species are found in about 100 caves in Kentucky and southern Indiana. Based on field observations, Keith suggested that the species distribution may be limited by competition with the southern cavefish, *Typhlichthys subterraneus.*

HABITAT
Their typical habitats are caves and subterranean passages of well-developed karst terrain. Can be found on consolidated mud-rock substrates in shoals and silt-sand substrates in pools but more often in caves with uniform silt-sand substrates.

BEHAVIOR
They respond to light by moving away (scotophilia).

FEEDING ECOLOGY AND DIET
They feed on benthic crustaceans and worms but can live for two years without food because of their low metabolic rate. They are considered a top predator.

REPRODUCTIVE BIOLOGY
These fish have external fertilization, and spawning takes place during high water between February and April. They have a low reproduction rate. The females brood eggs in the gill cav-

ity for about two to five months. The young appear in late summer and early fall.

CONSERVATION STATUS
This species is classified as Vulnerable by the IUCN. It occupies a highly restricted habitat and is susceptible to any disturbance in the water, such as groundwater pollution, sedimentation, runoff, impoundment, quarrying, and overcollecting.

SIGNIFICANCE TO HUMANS
Of particular scientific value as a cave species. ◆

Swampfish
Chologaster cornuta

FAMILY
Amblyopsidae

TAXONOMY
Chologaster cornutus Agassiz, 1853, "ditches of the rice fields in South Carolina."

OTHER COMMON NAMES
None known.

PHYSICAL CHARACTERISTICS
Grows to 2.86 in (6.8 cm). These fish are strongly bicolored—dark brown above and white to yellow (creamy white) below. They also have three narrow black stripes on each side and an orange or yellow cast to the head. The head is depressed, with small eyes. Pink gills are visible through the unpigmented gill covers. The cycloid scales are embedded, and the fish lack pelvic fins.

DISTRIBUTION
This species is found in North America on the Atlantic Coastal Plain from the Roanoke River drainage in Virginia to the Altamaha River drainage in Georgia (United States).

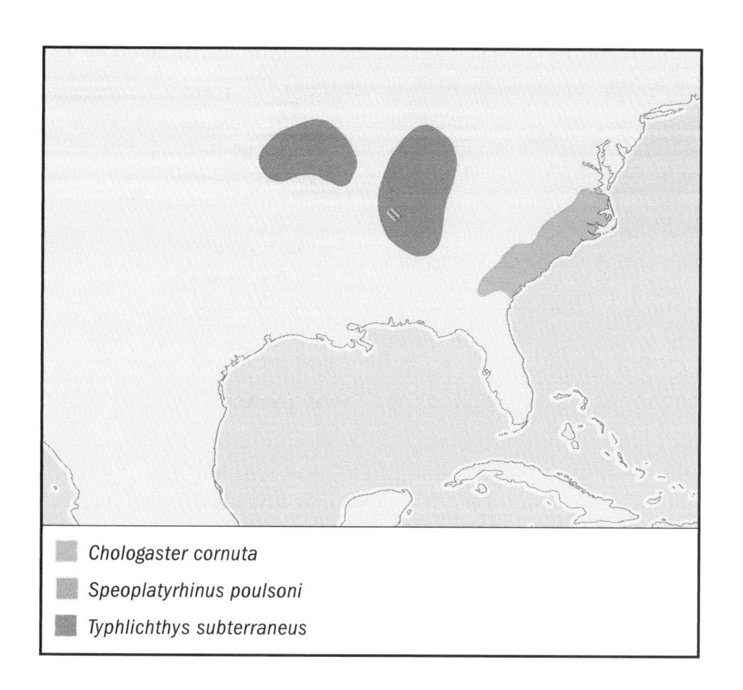

HABITAT
They occur year-round on vegetation and debris in low-lying swamps, ponds, ditches, sloughs, and quiet pools and backwaters of streams, usually in well-shaded, small bodies of waters. The chemical nature of these waters is acidic and boglike. Often this species does not show in many faunal surveys; its sensitive response to touch (thigmotaxis) makes it difficult to find in the roots and debris of its preferred habitat along the edges of submerged weed banks.

BEHAVIOR
This species is largely nocturnal.

FEEDING ECOLOGY AND DIET
Feeds on midge larvae, ostracods, and copepods. Vulnerable to dragonfly nymphs, larger fishes, water snakes, and fish-eating birds.

REPRODUCTIVE BIOLOGY
They spawn between early March and mid-April and usually die after spawning. They lay up to 430 eggs and may live as long as two years.

CONSERVATION STATUS
Not listed by the IUCN.

SIGNIFICANCE TO HUMANS
Sometimes found in the aquarium trade. ◆

Spring cavefish
Forbesichthys agassizii

FAMILY
Amblyopsidae

TAXONOMY
This species originally was described under two names: *Chologaster agassizii* and *Forbesella agassizi* Putnam, 1872, a well in Lebanon, Wilson County, Tennessee, United States. The genus *Forbesichthys* was eventually adopted since *Forbesella* was already in use for marine animals known as tunicates.

OTHER COMMON NAMES
None known.

PHYSICAL CHARACTERISTICS
Grows to 3.54 in (9.0 cm). This species is dark brown to nearly black on the dorsum, grading to lighter brown laterally; it is cream yellow ventrally, often with a thin yellow stripe along each side. These fish have minute scales embedded under the skin. They lack pelvic fins and have triads of sensory papillae midlaterally and scattered clusters called neuromasts on the head.

DISTRIBUTION
Found in central and western Kentucky (west to the Tennessee River) to southern central Tennessee and west across southern Illinois to southeastern Missouri. When the Mississippi River changed its course, the Missouri population may have been isolated from the others for about 2,000 years. It was intentionally stocked from southern Illinois sites to establish a population near Quincy College. This population was intended to serve as a nearby source of fish for research.

HABITAT
Individuals of this species are active in springs at night, almost always near the surface; they usually retreat underground during the day.

BEHAVIOR
The few individuals that venture into the spring portions of their habitat may have a strong tendency to move against the current (rheotaxist) for periods of half a minute to one minute, but they typically show strong thigmotaxis and hide under rocks or debris. They prefer highly oxygenated over less oxygenated water but respond to light by moving away (scotophilia). They tolerate a wide range of temperatures.

FEEDING ECOLOGY AND DIET
They feed at night on amphipods, midge larvae, tiny worms, and microcrustaceans.

REPRODUCTIVE BIOLOGY
Spawning takes place in February. The life span is about three years.

CONSERVATION STATUS
Not listed by the IUCN.

SIGNIFICANCE TO HUMANS
It has no particular significance except for its ecological and scientific value in researching the process of cave colonization. ◆

Alabama cavefish
Speoplatyrhinus poulsoni

FAMILY
Amblyopsidae

TAXONOMY
Speoplatyrhinus poulsoni Cooper and Kuehne, 1974, Key Cave, Alabama, United States.

OTHER COMMON NAMES
None known.

PHYSICAL CHARACTERISTICS
Grows to 2.83 in (7.2 cm). This species has an extremely elongated and anteriorly depressed head that makes up one-third of the standard length in adults. The snout is laterally compressed, with a terminal mouth. The species is white in coloration and lacks externally visible eyes as well as pelvic fins.

DISTRIBUTION
Its distribution is restricted to Key Cave, Lauderdale County, on the north bank of the Tennessee River.

HABITAT
They are found mostly in still waters of cave systems.

BEHAVIOR
Nothing is known about this very rarely studied species.

FEEDING ECOLOGY AND DIET
They feed on copepods, isopods, amphipods, and other small cavefish.

REPRODUCTIVE BIOLOGY
Little is known about the reproductive biology of this species, and what is known is not encouraging about its future potential for survival. It appears that only a small percentage of females reproduce, producing only a few eggs, and reproduction does not take place every year.

CONSERVATION STATUS
Its total population size is estimated to be less than 100 individuals, which would make it one of the most endangered fish species in the world. It is classified as Critically Endangered by the IUCN and Endangered by the U.S. Fish and Wildlife Service. It is being threatened by groundwater pollution from agricultural runoff.

SIGNIFICANCE TO HUMANS
Of particular scientific value as a cave species. ◆

Southern cavefish
Typhlichthys subterraneus

FAMILY
Amblyopsidae

TAXONOMY
Typhlichthys subterraneus Girard, 1859, a well near Bowling Green, Warren County, Kentucky, United States.

OTHER COMMON NAMES
None known.

PHYSICAL CHARACTERISTICS
Grows to 3.54 in (9.0 cm). They are pinkish in coloration and have a large, broad head. The eyes are not visible, being only vestigial in nature and covered by skin. Other defining characters include seven to 10 dorsal soft rays, seven to 10 anal soft rays, 10 to 15 caudal rays, and 28 to 29 vertebrae.

DISTRIBUTION
This species is found in the subterranean waters of two major disjunct ranges separated by the Mississippi River: the Ozark Plateau of central and southeastern Missouri and northeastern Arkansas and the Cumberland and Interior Low Plateaus of northwestern Alabama, northwestern Georgia, central Tennessee and Kentucky, and southern Indiana.

HABITAT
They are found in caves near the water table.

BEHAVIOR
They do not respond to light.

FEEDING ECOLOGY AND DIET
Feeds mainly on copepods, amphipods, isopods, insects, and worms.

REPRODUCTIVE BIOLOGY
Breeding probably occurs in late spring in association with rising water levels, and spawning takes place between April and May. The females lay fewer than 50 eggs each. They grow slowly and can live up to four years.

CONSERVATION STATUS
Classified as Vulnerable by the IUCN.

SIGNIFICANCE TO HUMANS
They are of particular scientific value as a cave species. ◆

Pirate perch
Aphredoderus sayanus

FAMILY
Aphredoderidae

TAXONOMY
Scolopsis sayanus (Gilliams, 1824), fishponds, Harrowgate, "near Philadelphia." Two subspecies have been proposed.

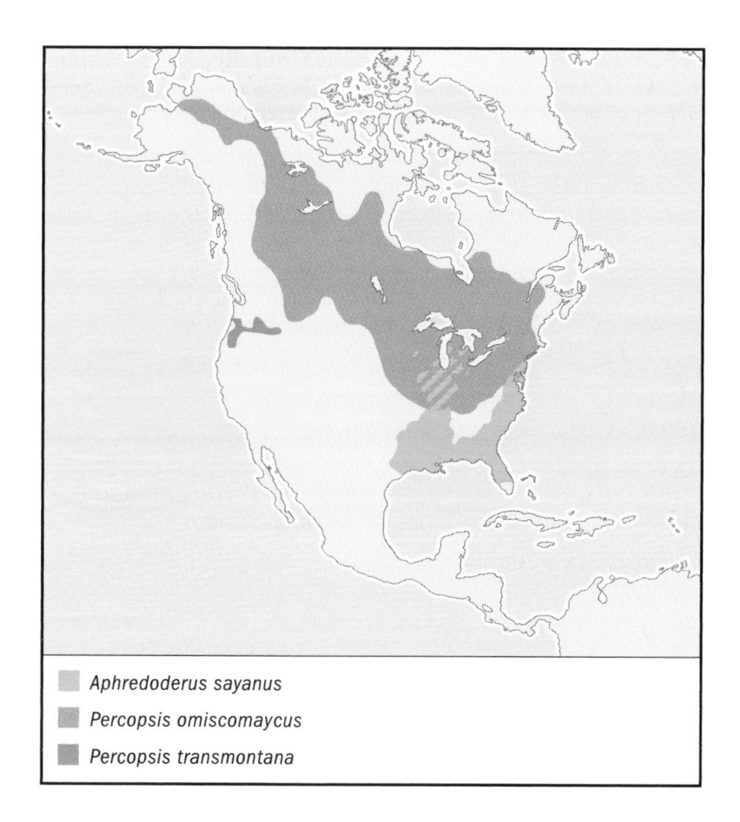

OTHER COMMON NAMES
German: Piratenbarsch.

PHYSICAL CHARACTERISTICS
Grows to 5.51 in (14.0 cm). A short, deep body, with a large head and mouth and a protruding jaw. They lack an adipose fin, and the lateral line is either absent or incomplete. The head is covered by ctenoid scales on the sides.

DISTRIBUTION
This species is found in waters of the Atlantic and Gulf slopes, the Mississippi Valley, and scattered parts of the eastern Great Lakes Basin in the United States from Minnesota south through the Mississippi Valley across the Gulf coast to Florida and north along the Atlantic coast to New York. It also can be found in the southeastern corner of Oklahoma, in the easternmost tributaries of the Red River, and throughout the Coastal Plain of Arkansas (but not in the Ozark Mountains). There are isolated populations in the Lake Ontario and Lake Erie drainages in New York, and the species has been reported in Wisconsin outside what is considered their native range, which suggests introduction. Populations on the Atlantic slope have been considered a subspecies (*Aphredoderus sayanus sayanus*) distinct from the subpopulation of the Mississippi Valley (*A. s. gibbosus*). The populations from the Gulf of Mexico drainage have been termed intermediate.

HABITAT
They usually occur over mud in quiet bodies of water, such as swamps, vegetated sloughs, ponds, oxbow lakes, ditches, backwaters, and pools of creeks and in small to large rivers on mud and silt bottoms. Adults most frequently are found at sites whose bottoms are overlain with leaf litter. The larvae of this species can be quite abundant in some areas.

BEHAVIOR
This is a solitary and crepuscular species.

FEEDING ECOLOGY AND DIET
They feed on insects, blue-green algae, and small crustaceans and fishes, which suggests that, like other members of this order, they are an opportunistic species that goes after almost any food item. Vulnerable to dragonfly nymphs, larger fishes, water snakes, and fish-eating birds.

REPRODUCTIVE BIOLOGY
The major spawning period for pirate perch in the Atchafalaya River Basin, Louisiana, is February through March. It appears that adult pirate perch are not branchial brooders but rather release their adhesive eggs over leaf litter and woody debris. They can live up to four years or longer.

CONSERVATION STATUS
Not listed by the IUCN.

SIGNIFICANCE TO HUMANS
This species is considered a water quality indicator species by the Arkansas Department of Environmental Quality for the Gulf Coastal Ecoregion. ◆

Troutperch
Percopsis omiscomaycus

FAMILY
Percopsidae

TAXONOMY
Salmo omiscomaycus Walbaum, 1792, Hudson Bay.

OTHER COMMON NAMES
English: Silver chub; Finnish: Lohiahven; French (Canada): Omisco.

PHYSICAL CHARACTERISTICS
Grows to 7.87 in (20 cm). Coloration can vary from yellowish to silvery to almost transparent, depending on the sexual state. There is a row of about 10 dark spots along the midline of the back and 10 or 11 spots along the lateral line, with another row of spots high on the sides and above the lateral line. The fins are always transparent. The most distinguishing characteristic is an adipose fin with small, weak spines on the dorsal and anal spines. Other characters include short gill rakers and rough ctenoid scales. The lateral line is nearly straight.

DISTRIBUTION
The original distribution was the Atlantic and Arctic basins throughout most of Canada, from Quebec to the Yukon and British Columbia, and south to the Potomac River drainage in Virginia; the Yukon River drainage, the Yukon and Alaska; and the Great Lakes and Mississippi River basins south to West Virginia, eastern Kentucky, southern Illinois, central Missouri, North Dakota, and northern Montana. It has been introduced in the Housatonic River drainage of Connecticut and Massachusetts and into Willard Bay Reservoir and Utah Lake, Utah.

HABITAT
They occur in lakes, deep-flowing pools of creeks, and rivers and usually are found over sand.

BEHAVIOR
Individuals of this species move into the shallows of lakes at night to feed and then move back to deeper water as dawn approaches. Some populations spawn exclusively at night.

FEEDING ECOLOGY AND DIET
Feeds on smaller fish, benthic crustaceans, insects, and phytoplankton. Vulnerable to larger fish, water snakes, and fish-eating birds.

REPRODUCTIVE BIOLOGY
Spawning takes place between April and August. Two or more males compete for a single female by chasing her near the surface, often breaking the surface of the water. Eggs and milt then are released. Death has been recorded after spawning. They can live up to four years.

CONSERVATION STATUS
Not listed by the IUCN.

SIGNIFICANCE TO HUMANS
Occasionally used as bait. ◆

Sand roller
Percopsis transmontana

FAMILY
Percopsidae

TAXONOMY
Columbia transmontana Eigenmann and Eigenmann, 1892, near the mouth of the Umatilla River, Umatilla County, Oregon, United States.

OTHER COMMON NAMES
None known.

PHYSICAL CHARACTERISTICS
Grows to 3.78 in (9.6 cm). Like the troutperch, this species has a large and naked head and chambers in the lower jaws and cheeks known as "pearl organs."

DISTRIBUTION
This species is found in the Columbia River system and some tributaries from the middle Columbia River in Washington downstream to within 25 mi (40 km) above its mouth, including western Idaho, southern Washington, and northern and western Oregon, United States.

HABITAT
They occur in slow-moving portions of streams and rivers, such as backwaters and marginal pools. They prefer mud-sand bottoms, although they have been reported over rubble substrate with considerable aquatic vegetation.

BEHAVIOR
Nothing is known.

FEEDING ECOLOGY AND DIET
Feeds on small aquatic invertebrates. Vulnerable to dragonfly nymphs, larger fishes, water snakes, and fish-eating birds.

REPRODUCTIVE BIOLOGY
Little is known, except that they can live up to six years.

CONSERVATION STATUS
Not listed by the IUCN, but the species may have disappeared from Idaho waters.

SIGNIFICANCE TO HUMANS
They have no significant economic or cultural importance to humans. ◆

Resources

Books

Etnier, D. A., and W. C. Starnes. *The Fishes of Tennessee.* Knoxville: University of Tennessee Press, 1993.

Lee, D. S., C. R. Gilbert, C. H. Hocutt, R. E. Jenkins, D. E. McAllister, and J. R. Stauffer, Jr. *Atlas of North American Freshwater Fishes.* Raleigh: North Carolina State Museum of Natural History, 1980.

Morrow, James E. *The Freshwater Fishes of Alaska.* Anchorage: Alaska Northwest Publishing, 1980.

Murray, A. M., and M. V. H. Wilson. "Contributions of Fossils to the Phylogenetic Relationships of the Percopsiform Fishes (Teleostei: Paracanthopterygii): Order Restored." In *Mesozoic Fishes. 2. Systematics and Fossil Record,* edited by G. Arratia and H.-P. Schultze. Munich: Dr. Friedrich Pfeil Verlag. 1999.

Page, Lawrence M., and Brooks M. Burr. *A Field Guide to Freshwater Fishes of North America North of Mexico.* Boston: Houghton Mifflin Company, 1997.

Patterson, C., and D. E. Rosen "The Paracanthopterygii Revisited: Order and Disorder." In *Papers on the Systematics of Gadiform Fishes,* edited by D. M. Cohen. Science Series no. 32. Los Angeles: Natural History Museum of Los Angeles City, 1989.

Romero, Aldemaro, ed. *The Biology of Hypogean Fishes.* Dordrecht: Kluwer Academic Publishers, 2001.

Whitworth, W. R. *Freshwater Fishes of Connecticut.* State Geological and Natural History Survey of Connecticut Bulletin 114. Hartford: Connecticut Department of Environmental Protection, 1996.

Periodicals

Adams, Ginny L., and James E. Johnson. "Metabolic Rate and Natural History of Ozark Cavefish, *Amblyopsis rosae,* in Logan Cave, Arkansas." *Environmental Biology of Fishes* 62 (2001): 97–105.

Boltz, J. M., and J. R. Stauffer. "Systematics of *Aphredoderus sayanus* (Teleostei, Aphredoderidae)." *Copeia* 1993, no. 1 (1993): 81–98.

Boschung, Herbert T. "Catalogue of Freshwater and Marine Fishes of Alabama." *Bulletin of the Alabama Museum of Natural History* 14 (1992): 1–266.

Brown, J. Z., and J. E. Johnson. "Population Biology and Growth of Ozark Cavefish in Logan Cave National Wildlife Refuge, Arkansas." *Environmental Biology of Fishes* 62 (2001): 161–169.

Fontenot, Q. C., and D. A. Rutherford. "Observations on the Reproductive Ecology of Pirate Perch *Aphredoderus sayanus.*" *Journal of Freshwater Ecology* 14 (1999): 545–549.

Resources

Green, S. M., and A. Romero. "Responses to Light in Two Blind Cave Fishes (*Amblyopsis spelaea* and *Typhlichthys subterraneus*) (Pisces: Amblyopsidae)." *Environmental Biology of Fishes* 50 (1997): 167–174.

Keith, J. H. "Distribution of Northern Cavefish, *Amblyopsis spelaea* DeKay, in Indiana and Kentucky and Recommendation for Its Protection." *Natural Areas Journal* 8 (1988): 69–79.

Killgore, K. J., and J. A. Baker. "Patterns of Larval Fish Abundance in a Bottomland Hardwood Wetland." *Wetlands* 16 (1996): 288–295.

Kuhajda, B. R., and R. L. Mayden. "Status of the Federally Endangered Alabama Cavefish, *Speoplatyrhinus poulsoni* (Amblyopsidae), in Key Cave and Surrounding Caves, Alabama." *Environmental Biology of Fishes* 62 (2001): 215–222.

Monzyk, F. R., W. E. Kelso, and D. A. Rutherford. "Characteristics of Woody Cover Used by Brown Madtoms and Pirate Perch in Coastal Plain Streams." *Transactions of the American Fisheries Society* 125, no. 4 (1997): 665–675.

Murray, A. M., and M. V. H. Wilson. "New Paleocene Genus and Species of Percopsiform (Teleostei: Paracanthopterygii) from the Paskapoo Formation, Smoky Tower, Alberta." *Canadian Journal of Earth Sciences* 33, no. 3 (1996): 429–438.

Poulson, T. L. "Cave Adaptation in Amblyopsid Fishes." *American Midland Naturalist* 70 (1963): 257–290.

Romero, A. "Threatened Fishes of the World: *Amblyopsis rosae* (Eigenmann, 1897) (Amblyopsidae)." *Environmental Biology of Fishes* 52 (1998): 434.

———. "Threatened Fishes of the World: *Typhlichthys subterraneus* (Girard, 1860) (Amblyopsidae)." *Environmental Biology of Fishes* 53 (1998): 74.

———. "Threatened Fishes of the World: *Speoplatyrhinus poulsoni* Cooper and Kuehne, 1974 (Amblyopsidae)." *Environmental Biology of Fishes* 53 (1998): 293–294.

Romero, A., and L. Bennis. "Threatened Fishes of the World: *Amblyopsis spelaea* DeKay, 1842 (Amblyopsidae)." *Environmental Biology of Fishes* 51, no. 4 (1998): 420.

Rosen, D. E. "An Essay on Euteleostean Classification." *American Museum Novitates*, no. 2782 (1985): 1–57.

Scott, W. B., and E. J. Crossman. "Freshwater Fishes of Canada." *Bulletin of the Fisheries Research Board of Canada* 184 (1973): 1–966.

Other

Fuller, Pam. *Percopsis omiscomaycus.* Nonindigenous Aquatic Species. 17 April 2000 (20 March 2003). <http://nas.er.usgs.gov/fishes/accounts/percopsi/pe_omisc.html>

Romero, Aldemaro. Guide to Hypogean Fishes. (20 March 2003). <http://www.macalester.edu/environmentalstudies/ARLab/HypogeanFishes/hypogeanfishesguide.htm>

Aldemaro Romero, PhD

Ophidiiformes

(Cusk-eels and relatives)

Class Actinopterygii
Order Ophidiiformes
Number of families 4 or 5

Photo: A pearlfish (*Carapus bermudensis*) about to enter a sea cucumber. (Photo by Chesher/Photo Researchers, Inc. Reproduced by permission.)

Evolution and systematics

The order Ophidiiformes is a group of slender, elongate, eel-like, and mostly bottom-dwelling fishes that is relatively unremarkable in external morphological features but exhibits highly evolved behavioral and reproductive traits. The order contains 355 named species, but there are numerous undescribed forms in scientific collections that have not been formally named. Collectively, there are probably 380–400 ophidiiform species in 90 genera, living mostly in marine environments but also including some freshwater and estuarine species. The order has the deepest-dwelling fish known to science, *Abyssobrotula galatheae*, captured at a depth of approximately 5 mi (8,370 m) below the ocean surface! The ophidiiform family Carapidae (the pearlfishes) comprises about 31 species in seven genera, including some species that are parasitic in the body cavities of marine invertebrates, such as starfish and sea cucumbers. The family Ophidiidae, the cusk-eels, is the most diverse family (about 218 species in 48 genera) and includes several large benthic species that are fished commercially. The family Bythitidae contains 96 species in 32 genera and includes live-bearing species that reside in shallow waters. The family Aphyonidae contains 22 live-bearing species (in six genera) that occur in deeper waters. Another family, the Parabrotulidae (three species in two genera), often is included in the order Ophidiiformes, but its placement there is the subject of controversy. Some researchers consider parabrotulids to be close relatives of the perciform family Zoarcidae, the eelpouts. Parabrotulids, however, share with Ophidiiformes many skeletal and soft anatomical features that are lacking in zoarcids. Thus, the taxonomic limits of the order Ophidiiformes remain uncertain and require further research.

The cusk-eels and their allies are classified in the superorder Paracanthopterygii, a large and morphologically diverse assemblage that includes the freshwater trout-perch and its relatives (Percopsiformes), the codfish and its relatives (Gadiformes), and the goosefish and its relatives (Lophiiformes). Some ichthyologists doubt that the Paracanthopterygii is a natural grouping (that is, monophyletic, or derived from a common ancestor), and the superorder has been reorganized several times. At present the superorder is defined by the common possession by its members of certain skeletal traits of the caudal fin and the cranium as well as the presence of a small bone (termed a supraneural) that lies in the dorsal musculature above the anterior vertebrae. Some paracanthopterygian fishes also possess certain morphological traits that are characteristic of the higher perchlike fishes (Acanthomorpha). These traits include the presence of true fin spines, particular patterns of small bones and ligaments associated with anterior vertebrae, and patterns of ossification of the pelvic bones. Thus, it is generally accepted that the Paracanthopterygii are acanthomorph fishes, currently classified within the Holacanthopterygii, a newly proposed name for a higher category that includes stephanoberyciform, zeiform, beryciform, and all other perchlike fishes. Within the Paracanthopterygii, ophidiiform fishes may be aligned most closely with gadiformes. Ophidiiforms and gadiforms share certain structural and developmental features of the caudal skeleton and gut anatomy. The phylogenetic relationships among paracanthopterygian orders remain speculative, however.

Extensive morphological analysis has not produced a satisfactory hypothesis of the evolutionary history of the ophidiiform fishes, and the phyletic relationships among ophidiiform families are largely unknown. Within the order, ichthyologists have shown natural groupings at family, subfamily, and generic levels, and there are solid proposals for sister group relationships among some families. Evidence for monophyly of the entire ophidiiform lineage is lacking, however. A number of characters have been proposed to classify ophidiiform families, including the position of the pelvic fins, but most of these proposals have proved unsatisfactory. Ichthyologists currently subdivide ophidiiform fishes into two groups, based on the presence or absence of viviparity and the anatomical features that are associated with live bearing.

The suborder Bythitoidei (containing the live-bearing families Bythitidae and Aphyonidae) is considered to be a natural grouping, since it is apparent that these fishes share highly specialized reproductive traits. Within the suborder, aphyonids and bythitids are sister taxa, each possessing uniquely specialized morphological characters (termed "synapomorphies"). Ichthyologists have been unable to specify synapomorphies supporting monophyly of the suborder Ophidioidei (containing the oviparous Carapidae and Ophidiidae), however. Within the Ophidioidei, the Carapidae has been shown to be monophyletic and is considered the closest relative of all other ophidiiforms. The Ophidiidae has not been shown to be monophyletic, but there are natural groupings within the family. The ophidioid subfamily Ophidiinae (the true cusk-eels) is considered monophyletic, because all its members share certain characteristics of the pelvic girdle. The ophidioid subfamilies Brotulinae and Brotulotaeniinae each contain a single, well-defined genus. The ophidioid subfamily Neobythitinae, containing 38 genera and about 170 species, is probably not a natural group and requires further research.

Fossil evidence of paracanthopterygian fishes is not extensive. A freshwater Paleocene fossil taxon, *Mcconichthys*, may be associated with the ophidiiform lineage, but this conclusion is disputed by some researchers. Fossil ophidiiform fishes, especially fossilized otoliths, or ear bones, are abundant in some Tertiary deposits. To date, several ophidiiform taxa have been described from the Paleocene and Eocene and from more recent deposits, including species of *Hoplobrotula* and *Ampheristus*.

Physical characteristics

Most ophidiiforms are long, relatively slender fishes with big heads, often small eyes, long dorsal and anal fins, and a caudal portion that tapers to a point. Sizes range from the tiny *Microbrotula*, which matures at 1.5 in (38 mm) in length, to commercially exploited species such as *Genypterus blacodes* that attain lengths of 3.3–6.6 ft (1–2 m). The cusk-eels often are strongly pigmented along the dorsal or anal midline, with lateral black or brown markings or bands extending the length of the body. In other ophidiiforms, the body is covered uniformly with small pigment spots, or pigment is entirely lacking. The mouth usually is large, and the upper jaw (maxilla) reaches a point at or beyond a vertical drawn through the eye.

A cusk-eel (*Ophidion scrippsae*) swimming near southern California. (Photo by Kerstitch. Bruce Coleman, Inc. Reproduced by permission.)

The caudal fin is small, sometimes reduced to a bony point, never forked, and often inconspicuous when it is confluent with the dorsal and anal fins. The pectoral fins often are long, sometimes exceeding the length of the head. The pelvic rays sometimes are absent. When present, the pelvic rays usually are long and conspicuous. Scales can be absent, but when present, they generally are small. The order is defined by the following set of external characteristics. (1) The pelvic fins (when present) have only one or two rays. (Some species have an additional spinelike splint.) (2) The pelvic fins are inserted anteriorly, just below the opercular margin or sometimes farther forward. (3) The pelvic fin bases typically are close together. (4) The dorsal and anal fin bases are long, reaching the caudal fins in most species. (5) There are no dorsal or anal fin spines. (6) Bones (termed "pterygiophores") supporting the anal- and dorsal-fin rays outnumber total vertebrae. (The ratio of dorsal and anal rays to total vertebrae is about 1.5:1.) (7) The nostrils are paired on each side of the head.

In the pearlfishes (Carapidae), the anal-fin rays are longer than the dorsal-fin rays, and the upper jaw lacks a supramaxillary bone. In ophidiids the supramaxillary bone is present, and the anal-fin rays are either equal in length or shorter than the dorsal-fin rays. Within these families, subfamilies are distinguished on the basis of the presence or absence of pelvic fins, scales, barbels, and other internal traits. Carapids and ophidiids are egg-laying fishes, and males lack the external intromittent organ that is characteristic of bythytoid fishes. In addition, the nostrils of ophidioid fishes are positioned higher on the snout than those of bythytoid fishes. Bythytids possess a swim bladder, and most species have scales; aphyonids lack both scales and a swim bladder. Bythitid subfamilies are distinguished by caudal-fin morphological features. Aphyonids are fragile, transparent fishes that have weakly developed skeletal systems and possess other traits characteristic of larval ophidiiforms. Thus, many ichthyologists believe that aphyonids are neotenic. Neotenic organisms are those that retain larval features as adults, and the process of neoteny is thought to be important in the evolution of many animal species, including humans.

Distribution

Most ophidiiform fishes are distributed broadly in all oceans, sometimes to abyssal depths and extending to shallow seas and estuaries. Ophiidiforms occur from Greenland south to the Weddell Sea, but most species are found in warmer waters of the tropics and subtropics. Carapids, ophidiids, and aphyonids are strictly marine, whereas some bythitid species are estuarine or reside in freshwater. Some bythitids have extremely restricted home ranges in caves and sinkholes.

A pearlfish (*Carapus bermudensis*) living inside a sea cucumber. (Illustration by Patricia Ferrer)

Habitat

Ophidiiform fishes generally are secretive and tend to associate with structures or with other animals. Most opidiiforms are free-living benthic species hovering close to the bottom and residing in mucus-lined mud or sand burrows, rock or coral crevices, or sea caves or associated with bottom-dwelling invertebrate communities, including deep sea vent fauna. Some species are pelagic or benthopelagic. One mesopelagic genus is associated uniquely with a species of deep-sea jellyfish. Pearlfishes are either free-living or inquiline species. Free-living species (the pyramondontimes and some echiodontines) are pelagic or bottom dwelling in deep oceanic water or on the continental shelf. Some echiodontines are believed to be associated with tube worm communities. The commensal carapids (*Onuxodon*, *Carapus*, and *Encheliophis*) are all shallow-dwelling species that reside within the body cavity of invertebrate hosts, such as pearl oysters, giant clams, tunicates, sea stars, and sea cucumbers. Species in the bythitid genera *Lucifugia* and *Ogilbia* reside in freshwater caves and sinkholes.

Behavior

The air bladder, anterior vertebrae, and associated ligaments and muscles are modified in many ophidiiform fishes to produce sound. In some species, the air bladder is partly ossified and serves as a resonating chamber. Sound production is thought to be associated with reproduction, as some species have been observed to produce sound just before mating. Hydroacoustic surveys often show large assemblages of vocal cusk-eels. Ophidiiform fishes generally have a close association with the bottom, hovering near holes, ledges, and drop-offs or hiding in mud and sand. Burrowing is accomplished by tail-first entry into soft and movable sediments. Ophidiiform fishes usually hide in burrows or crevices or within or around invertebrate hosts during daylight hours and then exit at night to forage. One bythitidae species has been seen living inside a hot thermal vent at great depths. Many ophidiiform fishes have highly evolved commensal associations with invertebrates. In the pearlfishes, some species are obligatory inquilines that never leave their hosts and feed on a host's internal organs.

Feeding ecology and diet

Ophidiiform fishes consume a wide variety of invertebrate and fish prey. Most are bottom dwellers that are strongly nocturnal, suggesting that they forage for benthic organisms during evening hours. This feeding behavior is facilitated by well-developed sensory pores on the head; a large, inferior mouth; and long, fleshy barbels on the chin in some species that are believed to aid in locating prey. Food consists of worms, crustaceans, echinoderms, and small bottom fishes such as gobies and small flatfishes. In turn, ophidiiform fishes are important food sources for many larger fish predators including skates, rays, sharks, eels, cod, hakes, goosefishes, and flounders. In addition, cusk-eels and their relatives living in shallow waters are prey to wading birds.

Reproductive biology

Ophidiiform fishes either deposit eggs or bear live young. The oviparous (egg-bearing) species include the pearlfishes (Carapidae) and the cusk-eels (Ophidiidae). Eggs of most carapids and ophidiids are unknown. Those that have been identified either are spawned in open water as individual, free-floating eggs or are deposited in a mucilaginous raft or gummy matrix, much like eggs of the goosefish and its allies (Lophiiformes). The egg rafts float at the ocean surface until they hatch, usually within several days. Larvae of carapids and ophidiids are pelagic, typically floating near the surface and sometimes traveling great distances from their hatching locality. Scientists believe that larvae of these species have the ability to regulate growth and metamorphosis (the transformation to juvenile and adult form). This developmental strategy allows species to disperse over great distances into habitats that are underutilized, thus reducing competition for limited resources.

Pearlfish larvae are unique in their possession of a long, highly ornamented predorsal filament (actually, the first dorsal fin ray), known as the vexillum. The vexillum may have a sensory function, because it contains stout cranial nerve fibers and its position around the head and mouth can be controlled by the larva. Vexillifer larvae have been identified for almost all pearlfish genera. Larvae are elongate, and their bodies typically are sparsely pigmented. After a long pelagic period, vexillifer larvae of species that are commensal or parasitic in invertebrates as adults lose the vexillum and shrink in length to transform abruptly to a so-called tenuis stage. The tenuis larvae seek out their benthic invertebrate hosts before maturing into adults.

Most cusk-eel larvae do not undergo an abrupt transition in morphological features in their development. One remarkable exception is a strange larva captured near South Africa. The larva has an enormous appendage that is highly pigmented and resembles the tentacles of a jellyfish. The appendage contains the larval gut. This so-called exterilium larva ("the larva with an outside gut") is tentatively identified as an unknown ophidiid species. Larvae of the ophidiid subfamily Neobythitinae have elongate dorsal and pelvic rays, but all other larval cusk-eels lack elongate rays. These larvae generally resemble adults, with a conspicuous coiled gut, pelvic fins placed far forward on the throat, long dorsal and anal rays, and big heads and mouths.

Larvae of the live-bearing bythytoid fishes are poorly known, and only a few have been described. Larvae of *Brosmophysis* (Bythitidae) and *Barathronus* (Aphyonidae) generally resemble ophidiid larvae, in that they lack elongate rays and possess coiled guts and long dorsal and anal fin bases. Aphyonid larvae hatch at larger sizes than do bythitid larvae. Some bythytoid embryos have specialized feeding appendages termed "trophotaena," through which they gain maternal nourishment. Some bythytid embryos consume or suckle bulbs of ovigerous tissue during development to supplement embryonic nutrition through the yolk sac.

Conservation status

The IUCN lists seven ophidiiform species as Vulnerable due to their rarity and limited habitats. All are live-bearing bythytids: *Lucifugia* (*Stygicola*) *dentata*, *L. simile*, *L. spelaeotes*, *L. subterranea*, *L. teresianarum*, *Ogilbia pearsei*, and *Saccogaster melanomycter*. One species, *O. galapagosensis*, is listed as Data Deficient.

Significance to humans

Most ophidiiform fishes are unknown to the public and are rare, deep-water species taken only in research cruises or commensal and parasitic forms that hide inside the bodies of invertebrate hosts or otherwise secretive species living in caves or small crevices in the reef. Four species in the genus *Genypterus* are large; their flesh is tasty, and they are fished commercially. In Chile and New Zealand, landings of *G. blacodes* reach 33,000 tons (30,000 metric tonnes) annually. *Brotula barbata* is a food source of growing importance to the peoples of several west African countries.

1. Key brotula (*Ogilbia cayorum*); 2. Pearlfish (*Carapus bermudensis*); 3. Band cusk-eel (*Ophidion holbrooki*). (Illustration by Patricia Ferrer)

Species accounts

Key brotula
Ogilbia cayorum

FAMILY
Bythytidae

TAXONOMY
Ogilbia cayorum Evermann and Kendall, 1898, Key West, Florida, United States. The holotype is a female. It was unknown at the time of the original description that the species is sexually dimorphic. Thus, considerable taxonomic confusion occurred when the male of the species was described separately. Taxonomic confusion persists, since it is believed currently that numerous undescribed *Ogilbia* species reside in the western tropical Atlantic Ocean. The extent to which their distributions overlap with the Key brotula is unknown. Thus, care must be taken in identification.

OTHER COMMON NAMES
English: West Indies brotula.

PHYSICAL CHARACTERISTICS
A small and moderately elongate ophidiiform fish with long anal and dorsal fins. The body and fins are uniformly pigmented, appearing yellow to olive-brown. Adults can attain about 3.9 in (10 cm) in length. The caudal fin is distinct and separate, not confluent with the dorsal- and anal-fin bases. The pelvic fins are positioned at the isthmus, anterior to the pectoral-fin bases. Each pelvic fin has one fleshy ray that is longer than the length of the head. The eye is small and the mouth large, with the upper jaw extending well posterior to the eye. The body is covered with small overlapping scales that do not appear on the cheeks (opercular bones).

DISTRIBUTION
The species is known from the Florida Keys, the Bahamas, and other localities in the Caribbean, including Belize, Cuba, Puerto Rico, and Jamaica south to Venezuela. Many of these distributional records may be for undescribed and similar species with overlapping distributions in this region.

HABITAT
The holotype of the Key brotula was collected by seine on an algae-covered shoal over coral rubble. The coral fragments were encrusted with organisms. The species usually is collected at depths of less than 9.8 ft (3 m), often by uprooting and shaking clumps of *Halimeda.* This secretive bythitid also may reside in reef crevices, rocky holes, and mangrove roots in shallow waters and out to the edge of the reef. It is rarely seen by divers and typically is collected only in research surveys at poison stations, sometimes locally abundant. For example, more than 15 individuals were collected at one station in Belize. The species probably is tolerant of large fluctuations in temperature and salinity.

BEHAVIOR
Specimens kept in aquaria constantly hide under shells or rubble. The species probably is nocturnal, judging by its small eyes and cryptic nature. There are few data on its natural habits.

FEEDING ECOLOGY AND DIET
There are no detailed studies of the diet or feeding behavior of this species. Juvenile and adult Key brotulas probably eat small crustaceans and are preyed upon by larger fishes and wading birds. Embryonic nutrition is remarkable. To supplement energy reserves in yolk sacs, embryos of Key brotula "suckle" on fluid-filled bulbs of ovigerous tissue. This highly unusual developmental strategy may be an evolutionary response to the poor feeding evironment of larvae after hatching.

REPRODUCTIVE BIOLOGY
Males and females have delicate, interlocking copulatory organs. Genital morphological features have been described in detail, but mating has never been observed. The copulatory apparatus of the male is partly calcified and is thought to be derived from the first anal ray. The penis rests on a pedestal surrounded by muscular pseudoclaspers and covered by a fleshy urogenital hood. The urogenital pore of the female is similarly protected. Most bythitids have the capability of sperm storage in the oviduct or ovary. Fertilization is internal, and embryos grow rapidly. Late embryos are well developed, with pigmented eyes and fully developed and functional digestive system, musculature, gills, and caudal fin. Newborn young are about 0.5 in (12 mm) in length, and brood size is about 14 individuals.

CONSERVATION STATUS
Not listed by the IUCN.

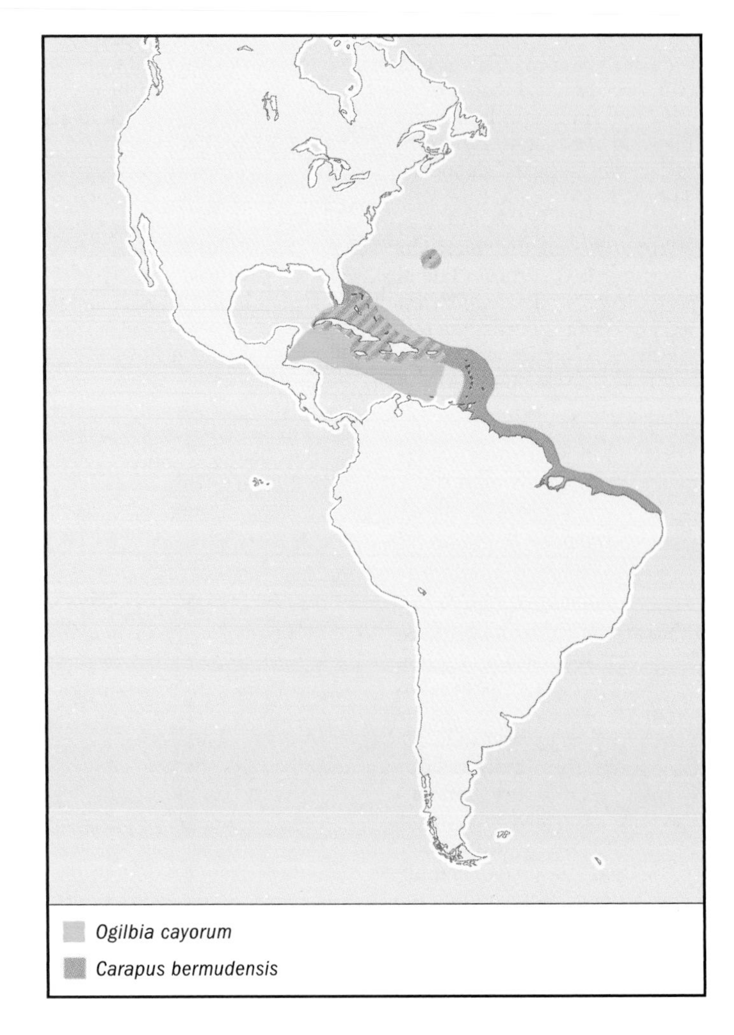

SIGNIFICANCE TO HUMANS
This species is rarely observed and is not fished commercially. It has limited appeal as an aquarium species, since it rarely shows itself to the viewer. ◆

Pearlfish
Carapus bermudensis

FAMILY
Carapidae

TAXONOMY
Lefroyia bermundensis Jones, 1874, Bermuda. The holotype was named in honor of Lefroy, a former governor of Bermuda. The species is now valid as *Carapus bermudensis* (Jones) following several taxonomic revisions. Numerous names are probably referable to *C. bermudensis*, including *C. recifensis*, *C. chavesi*, and *Fierasfer dubius*. *C. bermudensis* may be related most closely to *C. acus*, its eastern Atlantic relative. The species-level relationships of the genus, however, are poorly known.

OTHER COMMON NAMES
None known.

PHYSICAL CHARACTERISTICS
Carapus bermudensis is long, slender, and eel-like, with a large head and relatively large eyes. It is translucent, with silvery bands along the flanks, black internal pigment visible along the vertebral column, a silver cheek patch, and large pigment blotches along the bases of the dorsal and anal fins and head. The anal fin origin is anterior to the dorsal fin origin. There are 13–18 anal rays anterior to the first dorsal ray. This number varies among pearlfish species and is useful in identification. There are no pelvic fins, and the caudal fin usually is absent. The pectoral fin has 17–20 rays. The teeth on the upper jaw are small, and some are heart-shaped. The teeth on the lower jaw are larger and conical. The air bladder is separated into two parts by an internal constriction under vertebrae 11 and 12. This feature of the internal anatomy is characteristic of all species in the genus *Carapus*, and the position of the constriction relative to the vertebrae allows for separation of species.

DISTRIBUTION
Distributed in shallow waters along the shores of the western Atlantic, Bermuda, and the Caribbean Sea south to Brazil. Its larvae sometimes are collected far north and east of this range in plankton samples taken by research cruises.

HABITAT
All species of the genus *Carapus* have obligatory commensal relationships with sea cucumbers (Holothuria), starfishes (Asterioidea), or sea squirts (Ascidiacea). Many species exhibit host specificity. *Carapus bermudensis* has been collected in the body cavity of nine holothurian species in the genera *Actinopyga*, *Isostichopus*, *Thone*, *Astichopus*, *Holothuria*, and *Theelothuria*. These host species generally reside in shallow waters, to about 98.4 ft (30 m) on sandy bottoms or grass beds in tropical and subtropical lagoons near reefs. In one study in the Bahamas, pearlfishes were found in relatively few restricted areas, although more than 1,000 sea cucumbers were surveyed.

BEHAVIOR
Resides within the body of its host during daylight and is believed to exit at night to forage and perhaps spawn. This strategy limits the probability of predation by larger fishes. This species has been observed in aquaria as it rapidly enters its primary host, *Actinopyga agassizi*. The species first locates the anal opening of the sea cucumber with its snout, presumably through olfaction. As the fish holds its head in the proper position at the anal opening, the body curves and the tip of the tail tracks along the mid-lateral line until it reaches the anus. Once the tail tip is aligned and pointed into the opening, the fish abruptly turns, forcing its way tail first into the host by body undulations. There are no observations of living pearlfishes in the wild and little data on its habits and behaviors.

FEEDING ECOLOGY AND DIET
Some inquiline pearlfishes are parasitic, dining on the internal organs of the invertebrates they occupy. This species of pearlfish is not parasitic and feeds outside the holothurian host, probably at night. There have been no detailed studies of the food habits of this species, but gut contents of individuals are mostly crustacean invertebrates, such as amphipods, small shrimps, crabs, and mysids. Rarely, a pearlfish is found in the stomach of larger, predaceous fishes.

REPRODUCTIVE BIOLOGY
Spawning of pearlfishes has not been observed, and carapid reproductive behavior is poorly known. Some investigators have identified eggs collected in plankton samples by subsequent incubation in the laboratory. There also are a few reports of pearlfish species spawning in aquaria. In these cases, the scientists did not observe spawning directly but found eggs in tanks after periods of darkness. The eggs of pearlfishes are ellipsoid, usually containing an oil droplet and deposited into a jellylike, mucous matrix that floats at the surface. The egg mass has been described as oval, spherical, or somewhat flattened. Eggs hatch in one to two days. Early larvae are easily identified, since they possess a vexillum that first appears as a small protuberance but rapidly grows in length. Older pearlfish larvae have a long pigmented and ornamented vexillum that often is damaged in collection. Pearlfish larvae are extremely elongate, reaching about 7.1 in (180 mm) in length, and possess a distinct small ring of melanophores on the snout. Larvae are remarkable, in that they undergo two separate growth phases: the first as vexillifer larvae that become very elongate and the second as tenuis larvae that shrink to about half their original length.

CONSERVATION STATUS
Not listed by the IUCN.

SIGNIFICANCE TO HUMANS
This species is rarely observed and is not fished commercially. ◆

Band cusk-eel
Ophidion holbrooki

FAMILY
Ophidiidae

TAXONOMY
Ophidium holbrooki Putnam, 1874, Key West, Florida, United States.

OTHER COMMON NAMES
None known.

PHYSICAL CHARACTERISTICS
Variations in squamation (scale pattern), counts of fin rays, vertebrae, and other skeletal features, and body coloration are

often useful in identifying ophidiid fishes. The band cusk-eel lacks scales on the top of the head and has 66–69 total vertebrae, 117–132 dorsal rays, 977–109 anal rays, and 19–21 pectoral rays. The dorsal and anal fins are continuous with the caudal fins. The pelvic fins, each consisting of two rays, are located far forward on the chin. The head and the body are tan in color, with no mottled patterns, blotches, or bands of pigment. The dorsal and anal fins are edged in black. This species is larger, attaining about 11.8 in (30 cm) in total length, and somewhat deeper bodied than other cusk-eels (genera *Ophidion*, *Lepophidium*, *Otophidion*, and *Parophidion*) found in its home range.

DISTRIBUTION

Along the Atlantic coast of the United States from North Carolina south to the Gulf of Mexico and extending along the coast of Brazil to Lagoa dos Patos. The species is reported to be absent from the Bahamas.

HABITAT

Little is known of the specific habitats of the band cusk-eel. It has been collected with research and commercial trawls on soft muddy to sandy bottoms from near shore to about 246 ft (75 m).

BEHAVIOR

There have been no studies of the behavior of this species. Ophidiid fishes are bottom dwellers, and many reside in burrows dug into soft mud and sand. Observations by sumersibles suggest that some ophidiid species are nocturnal and abundant in some areas. Many, perhaps all, cusk-eels produce sound, and recent acoustic surveys have found large and vocal assemblages of ophidiid fishes in some areas. Sound production most likely is related to spawning behavior.

FEEDING ECOLOGY AND DIET

There have been no studies of the feeding habits of the band cusk-eel. Ophidiid fishes consume benthic invertebrates, primarily small crustaceans (shrimps, amphipods, mysids, and crabs) and worms. Small fishes, such as anchovies, gobies, and tonguefish, also are consumed. In turn, the band cusk-eel is preyed upon by larger fishes, especially dogfish, skates, conger eels, and flounders.

REPRODUCTIVE BIOLOGY

Unlike bythytids, male band cusk-eels do not have a copulatory organ, and fertilization occurs externally. The eggs and larvae have not been described, and the early life stages of this and most cusk-eel species are unknown.

CONSERVATION STATUS

Not listed by the IUCN.

SIGNIFICANCE TO HUMANS

This species is landed as by-catch in trawl fisheries for shrimps and bottom fishes and may appear in fish markets in some countries of South America. It is relatively small and has limited value in commercial markets, although its flesh is considered good. ◆

Resources

Books

Bohlke, James E., and Charles C. G. Chaplin. *Fishes of the Bahamas and Adjacent Tropical Waters.* 2nd edition. Austin: University of Texas Press, 1993.

Collette, Bruce B., and Grace Klein-MacPhee. *Bigelow and Schroeder's Fishes of the Gulf of Maine.* 3rd edition. Washington, DC: Smithsonian Institution Press, 2002.

Gordon, D. J., D. F. Markle, and J. E. Olney. "Ophidiiformes: Development and Relationships." In *Ontogeny and Systematics of Fishes*, edited by H. G. Moser. Special Publication no. 1. American Society of Ichthyologists and Herpetologists, 1984.

Nelson, J. S. *Fishes of the World.* 3rd edition. New York: John Wiley & Sons. 1994.

Nielsen, Jørgen G., Daniel M. Cohen, Douglas F. Markle, and C. Richard Robins. *FAO Species Catalogue: Ophidiiform Fishes of the World (Order Ophidiiformes).* FAO Fisheries Synopsis, vol. 18, no. 125. Rome: Food and Agriculture Organization of the United Nations, 1999.

Robins, C. Richard, and Carleton R. Ray. *A Field Guide to Atlantic Coast Fishes of North America.* Boston: Houghton Mifflin Co., 1999.

Periodicals

Fahay, M. P. "Development and Distribution of Cusk-eel Eggs and Larvae in the Middle Atlantic Bight with a Description of *Ophidion robinsi* n. sp. (Teleostei: Ophidiidae)" *Copeia* 1992 (1992): 799–819.

Resources

Johnson, G. D., and C. Patterson. "Percomorph Phylogeny: A Survey of Acanthomorphs and a New Proposal." *Bulletin of Marine Science* 52 (1993): 554–626.

Markle, D. F., and J. E. Olney. "Systematics of the Pearlfishes (Pisces: Carapidae)." *Bulletin of Marine Science* 47 (1990): 269–410.

Suarez, S. S. "The Reproductive Biology of *Ogilbia cayorum*, a Viviparous Brotulid Fish." *Bulletin of Marine Science* 25 (1975): 143–173.

John E. Olney, PhD

Gadiformes

(Grenadiers, hakes, cods, and relatives)

Class Actinopterygii
Order Gadiformes
Number of families 11

Photo: A hake, or forkbeard (*Urophycis blennoides*), in the Mediterranean Sea. (Photo by Sophie de Wilde/Jacana/Photo Researchers, Inc. Reproduced by permission.)

Evolution and systematics

It is important to note that agreement is lacking among ichthyologists as to the composition, origins, hierarchy, or relationships within the Gadiformes. Nor is there agreement concerning external relationships of members of this rather arbitrary group of fishes. There are no groupings of characteristics that can be assigned to the Gadiformes alone; therefore, synapomorphies (derived characters shared by all members of the group in question) have not been identified for all members of the order to the exclusion of nonmembers. Important recent contributions concerning the phylogeny and systematics of the order Gadiformes include Marshall and Cohen 1973; Markle 1982; Cohen 1984; Fahay and Markle 1984; Cohen (Ed.) 1989 (and several contributions therein); and Cohen et al. 1990. To quote Cohen et al. "The assignment to the order of many species is presently as much a matter of ichthyological convention as it is a result of logic." For the purposes of the present volume, we ground our classification on a taxonomic model proposed by Cohen et al. 1990, wherein taxa are listed alphabetically, in order not to suggest phylogenetic relationships. However, we elevate several genera or groups of genera to family status, not arbitrarily, but based on a consensus of several studies. Those 11 families are listed below (alphabetically) with a tally of genera and species presently understood to be contained in each.

- Bregmacerotidae (1 genus, 15 species)
- Euclichthyidae (1 genus, 1 species)
- Gadidae (12 genera, 22 species)
- Lotidae (3 genera, 4 species)
- Macrouridae (19 genera, 300+ species)
- Melanonidae (1 genus, 2 species)
- Merlucciidae (3 genera, 18 species)
- Moridae (18 genera, 89 species)
- Muraenolepididae (1 genus, 4 species)
- Phycidae (6 genera, 27 species)
- Steindachneriidae (1 genus, 1 species)

This listing departs from certain other studies presently available in the following ways: 1) The Gadidae, Lotidae, and Phycidae are usually considered subfamilies (Gadinae, Lotinae, and Phycinae) of the family Gadidae in those studies; 2) the genus *Steindachneria* is included in the Merlucciidae in some other studies rather than being elevated to full family status; and, 3) the Merlucciidae is included in an expanded family Gadidae in some other studies. The listing above is used here largely because it is the result when ontogenetic evidence is considered. It also better emphasizes the diversity present within the order Gadiformes.

Physical characteristics

The families listed above are diagnosed as follows:

- Bregmacerotidae: These species are sometimes called unicorn cods. The first dorsal fin is a single, elongate ray arising from a position over the head. The second dorsal and anal fins are long mirrors of each other; there are longer rays at the anterior and posterior ends forming rounded lobes. The elongate, trailing pelvic fin rays arise from the throat position and extend well beyond the anus. The lateral line is high on the body and is parallel to the dorsal margin of the body. There are no chin barbels.
- Euclichthyidae: The single species is called the eucla cod. The body is long and tapering to a very narrow caudal peduncle. It has a large mouth and no chin barbel. The first dorsal fin is short and high,

nearly touching the second dorsal fin, which is shorter and extends the length of the body to the caudal fin. The anal fin is long, with a greatly enlarged anterior lobe. The caudal fin is small and asymmetrical, with the lower rays being longest. Each pelvic fin is comprised of four separate, filamentous rays.

- Gadidae: This family includes cods, haddock, pollock, tomcod, and others. There are three separate dorsal fins and two separate anal fins. The dorsal and anal fins are either touching at their bases or separated by gaps. A chin barbel is usually present. There are pelvic fins, sometimes with one or more elongate rays.
- Lotidae: This includes the tusk (or cusk), burbot, and lings. The dorsal fin is single (in *Brosme*), or there is a short first dorsal followed by a long second dorsal in *Lota* and *Molva*. The single anal fin has a long base and includes many fin-rays. There is a well-developed chin barbel. The pelvic fins are normal and are not modified into elongate rays.
- Macrouridae: This includes rattails and grenadiers. The head and trunk are short, and the tail is compressed and greatly elongate, tapering to a point and lacking a caudal fin (with one exception). The chin barbel is usually present. The head and mouth shape and size vary. Eye size varies, but eyes are usually very large. There are two dorsal fins; the first is high, often including spinous anterior rays. The second dorsal and anal fins are long, meeting at the tail tip. The pectoral fins are narrow based and positioned high on the trunk. The pelvic fins are narrow based, thoracic to jugular in position, and comprised of 5–17 rays. Some species have a light organ on the mid-ventral line of the trunk. The benthic species have a well-developed air bladder, and the bathypelagic species lack an air bladder.
- Melanonidae: This includes the pelagic cod. It is small, not exceeding 5.9 in (15 cm). The body is long, tapering to a very narrow caudal peduncle. The head has numerous fleshy ridges. There is no chin barbel. The dorsal and anal fins are single and long-based. A slight gap separates the caudal fin from the dorsal and anal fins.
- Merlucciidae: This includes hakes and grenadiers. The genus *Merluccius* (hakes) has a large head (1/3 to 1/4 of body length), with a large, oblique mouth. The lower jaw is longer than the upper. There are two separate dorsal fins, with the first being short-based, high, and triangular. The second is long and partially divided by a notch in midsection. The single anal fin is similar in shape to the second dorsal. There are pelvic fins with seven rays. The genera *Lyconus* and *Macruronus* are characterized by long, tapering bodies lacking caudal fins. The dorsal fin is single with elevated anterior portion in the former, double with short first dorsal followed by very long second dorsal in the latter. Both have large, oblique mouths. Both have normal pelvic fin without elongate rays. The pectoral fin includes elongate rays in *Lyconus*.
- Moridae: This includes the moras. The body tapers to a very narrow peduncle. There are two or three dorsal fins, and one or two anal fins. The pelvic fins are thoracic and are wide apart at the bases. The caudal fin is symmetrical and separated from the dorsal and anal fins by a gap.
- Muraenolepididae: This includes moray cods. The body is long and compressed. The head is small with a chin barbel. The gill openings are restricted and do not open above the pectoral fin bases. There are two dorsal fins, the first comprised of a single, slim ray, and the second long-based and merging with the caudal fin. The anal fin is single, also merging with the caudal fin. The pelvic fins are thoracic with five rays, 2–3 of which are elongate and not attached to others. The lateral line ends at mid-body.
- Phycidae: This includes hakes and rocklings. There is a single anal fin. There are two dorsal fins, the first either short-based and moderate in height, or comprised of a single elongate, filamentous ray followed by many, very short hair-like rays. There is a single chin barbel, or 2–4 barbels on the snout as well as on the chin. The pelvic fin is normal in shape and length, or with two very elongate rays, often reaching the level of the anus.
- Steindachneriidae: This includes the luminous hake. The body is long, compressed laterally, and tapers to a point. The head is compressed laterally, and the mouth is very large. There are two dorsal fins, the first with one spine and 7–9 rays, the second with 123 or more rays. The anal fin is comprised of 123 to more than 125, very short rays. The first ray of the pelvic fin is elongate and filamentous. The anus is between the pelvic fin rays and is separated from the urogenital opening just anterior to the anal fin. A purplish, striated light organ covers the lower body and sides of head.

Distribution

The distribution of gadiform fishes varies by family, but in general, gadiforms are residents of cool water, therefore occurring throughout the water column in high latitudes, but mainly in deeper layers of tropical waters, where temperatures are lower (a phenomenon known as "tropical submergence"). Members of the gadiforms are primarily marine, but a few freshwater or estuarine species occur. The distribution of each family was summarized in Marshall and Cohen (1973). Bregmacerotids differ somewhat from the rest of the order in their distribution in tropical and subtropical seas. The single euclichthyid species occurs in Australian and New Zealand waters, where it lives near the bottom in deep water.

Gadids are centered in continental shelf waters of the temperate and boreal North Atlantic, although a few members

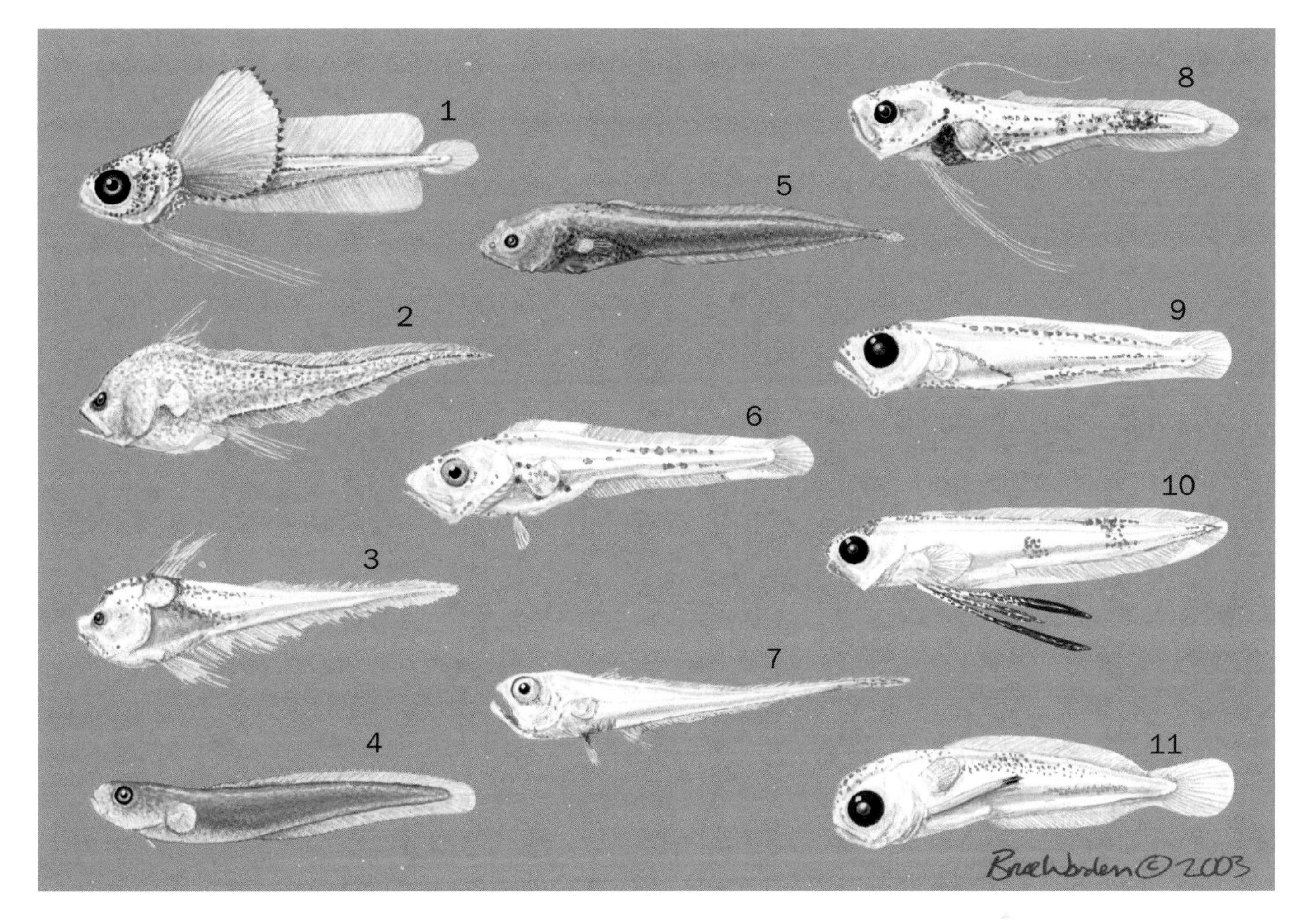

Examples of larval stages of gadiform fishes: 1. Moridae, *Gadella maraldi*; 2. Macrouridae, *Gadomus* sp.; 3. Macrouridae, *Coryphaenoides* sp.; 4. Muraenolepididae, *Muraenolepis* sp.; 5. Melanonidae, *Melanonus* sp.; 6. Merlucciidae, *Merluccius productus*; 7. Steindachneriidae, *Steindachneria argentea*; 8. Bregmacerotidae, *Bregmaceros mcclellandi*; 9. Gadidae, *Gadus morhua*; 10. Lotidae, *Brosme brosme*; 11. Phycidae, *Urophycis chuss*. (Illustration by Bruce Worden)

have wandered into cool waters of the North Pacific and some species are circumpolar. Three species in *Gadus*, including the Atlantic cod, occur circumboreally, extending into Arctic waters north of Europe, where they are found as deep as 1,640 ft (500 m). *Melanogrammus aeglefinus*, the economically important haddock, is restricted to the North Atlantic where it occurs off the northeastern United States and northern coasts of Europe. *Microgadus* includes the tomcod, a brackish and freshwater species found along the east coast of Canada and the northeastern United States. *Pollachius* contains two species, one of which is restricted to the eastern North Atlantic, the other occurring on both sides of the North Atlantic. *Theragra chalcogramma*, the Alaska pollock, is a North Pacific species that is widely distributed in temperate to boreal waters. Fishes in the family Lotidae are also centered in the North Atlantic, and one genus (*Lota*) has successfully invaded fresh waters of northern Europe and northern North America.

Members of the speciose family Macrouridae (rattails and grenadiers) occur primarily in deep-water habitats throughout the world's oceans (except for the Arctic). Almost all deep oceanic basins contain a macrourid fauna, and it has been estimated that in the Pacific, members of this family comprise the greatest vertebrate biomass between certain depth strata (Cohen et al. 1990). More macrourid species occur in tropical waters than at high latitudes. Several have very restricted ranges, but the deeper-living species are more widely distributed.

The two species in the family Melanonidae are not well-known or often collected. One species is circumantarctic, the other circumglobal in tropical/subtropical waters. Merlucciids occur in continental slope and deep-shelf habitats along coastlines throughout the world. The several species in *Merluccius* are found on both sides of the Atlantic Ocean, the eastern Pacific Ocean, and off southern New Zealand. The enigmatic genus *Lyconus* has been found (rarely) in both the North and South Atlantic oceans. One genus (*Macruronus*) is restricted to subantarctic waters.

The family Moridae is found in all oceans. Most species have very restricted ranges, although *Antimora microlepis* occurs in much of the entire North Pacific. Four species in the family Muraenolepididae all occur in the Southern Ocean, near the bottom in cold-temperate waters. Each species is restricted to waters around capes or groups of islands surrounding Antarctica.

The Phycidae are bottom-living fishes with a center around the coastlines of the North Atlantic, but with a few Southern Hemisphere species. *Gaidropsarus* exhibits a center of abundance in the northeast Atlantic Ocean, but also extends to Japan, New Zealand, and South Africa. *Phycis* contains two species, which are benthopelagic along the eastern North Atlantic Ocean coast. The several species in *Urophycis* are distributed in the western Atlantic Ocean from Canada through Argentina. *Enchelyopus cimbrius* is found on both sides of the North Atlantic, where it occurs along the east coasts of the United States and Canada, the Gulf of Mexico, southern Greenland, Iceland, and the north coasts of Europe. Finally, the monotypic family Steindachneriidae (*Steindachneria argentea*) is restricted to fairly deep waters of the Gulf of Mexico, Caribbean Sea, and continental slope waters off the eastern United States.

Habitat

Habitats occupied by gadiforms vary by family and by genera. Bregmacerotids occur epi- or mesopelagically in open-oceanic waters, extending at times to shallow, coastal habitats. Some have occasionally been found in estuaries. *Euclichthys polynemus*, the sole species in the family Euclichthyidae, occurs benthopelagically in depths of 820–2,625 ft (250–800 m).

Most species in the Gadidae occur demersally (on the bottom) or benthopelagically. Very few (e.g. *Gadiculus argenteus*) are pelagic (live in the water column off the bottom). The demersal species occur over a variety of substrates (rock, sand, mud, gravel, or shell debris), most prefer one over the others, and some undertake seasonal migrations between habitat types. The two species in *Pollachius* are pelagic, and sometimes form large wandering schools that migrate seasonally. The many species in the Macrouridae are found in all oceans, where they occur over very deep bottoms, including the deep ocean basins. The deepest-occurring species occur at greater than 3.7 mi (6,000 m), and few occur shallower than 328 ft (100 m). Melanonids are widely distributed in open ocean meso- and bathypelagic depths between tropical and subantarctic waters.

Most species in the genera *Merluccius* and *Melanonus* occur over continental shelves or upper continental slopes. Favored bottom types occupied by species in the genus *Merluccius* range from sandy to muddy. The ill-known genus *Lyconus* is pelagic and occurs in open waters of the Atlantic Ocean.

Morids are pelagic to benthopelagic, and occur from shallow coastal habitats (rarely including estuarine habitats) to deep oceanic waters. Favored bottom habitats range from soft to hard bottoms, and from sand to mud. Muraenolepidids live near bottom in moderate depths in waters surrounding Antarctica. Phycids are demersal fishes living on a variety of substrates ranging from mud to sand to shell debris. Finally, *Steindachneria argentea* lives in the lower water column over soft bottoms on the deeper parts of the continental shelf and upper continental slope.

Habitat requirements of most gadiforms vary according to their life-history stage or age. It is impossible to propose a generalized model of those varying habitat requirements because each family and each species exhibits its own pattern. For example, within the Phycidae, the white hake, *Urophycis tenuis*, spawns beyond the continental shelf off the northeastern United States, and the fertilized eggs rise into open ocean layers near the surface in an area known as the "Slope Sea." The larvae hatch and develop into a pelagic-juvenile stage, strongly associated with the surface. As they grow they migrate across the entire breadth of the continental shelf toward shore. They arrive in estuaries as small juveniles and spend one season there growing rapidly, finally leaving the estuary at sizes of about 6 in (15.2 cm) at the end of the summer. One- and two-year-olds may remain segregated from older fish by virtue of their occupation of shallower bays and near coastal waters. Subsequently they mingle with the rest of the adult population on deeper parts of the continental shelf, with seasonal migrations into shallower waters. The white hake's range of habitat requirements, therefore, extends from the estuary, across the breadth of the continental shelf, to the upper part of the continental slope, and also includes the entire water column, from surface to bottom (Fahay and Able 1989).

Another phycid, *Urophycis chuss*, the red hake, has very specialized habitat requirements. The fertilized eggs occur in near-surface layers of oceanic waters off the northeastern United States. After hatching, the developing larvae occupy the same general layers. The larvae gradually acquire silvery coloration and go through a pelagic-juvenile stage when they occur epipelagically or neustonically, within a few inches of the surface, often associated with floating weed or debris. After about two months of this pelagic existence, red hakes settle to the ocean bottom and seek shelter in a variety of structured habitats including beds of clam shells, anemone or polychaete tubes, depressions made by fishes or crustaceans, or most frequently, beds of scallops, *Placopecten magellanicus*. The young fishes have been found hiding under scallops, but more frequently they enter into an inquiline association with them whereby they live within the scallops' mantle cavities. The young fish remain in this association through their first winter, finally emerging when they are about 4 in (10.2 cm) long in the spring, and then they remain in coastal or estuarine waters through the next summer. When waters cool in the fall, they join older fish in an offshore migration toward the edge of the continental shelf, where they spend their second winter. Their required habitats, therefore, include the entire water column from surface to bottom, specific beds of invertebrate hosts on the substrate, and the entire breadth of the continental shelf from estuary to shelf edge depending on the season (Able and Fahay 1998).

The Atlantic cod also has habitat requirements specific to its life history stages. The fertilized eggs are pelagic and occur in waters overlying bays, the continental shelf, and important banks and shoals, (e.g. eastern Georges Bank, Grand Banks). Larvae are also pelagic and drift slowly away from spawning areas as they develop. The early juveniles descend to the bottom when they are about 2 in (5.1 cm) long and settle on pebble-gravel deposits on important banks, such as the northeast peak of Georges Bank. After settlement, young fishes appear to favor vegetated habitats (such as eelgrass) in coastal embayments, where they avoid predation by older cod, as well as other predators. After two or three years of segregation from adults, these one- and two-year-olds finally join the adult population. Adults exhibit seasonal movements associated with depth and temperature fluctuations, generally moving into shallower waters during summer and retreating to deeper waters for the winter. Data are available demonstrating habitat preferences based on depth, temperature, and salinity, but surprisingly little is known about bottom types favored. Rocky, pebbly, sandy, and gravelly have all been used to describe cod haunts.

Behavior

In general, gadiform fishes are demersal and highly piscivorus, but there are exceptions to these characterizations. These fishes occur primarily in colder waters, but many exhibit seasonal migrations associated with reproduction or the quest for important prey items. Observations on actual spawning behavior are few, and data on day–night differences in their behavior are also few. Feeding behavior also varies seasonally for many species, with a characteristic pattern involving cessation of feeding activity during spawning seasons.

Feeding ecology and diet

Many gadiform fishes feed on prey items occurring in the substrate, and they are assisted in their search for food by the presence of tactile barbels on their chins. Barbels are well developed in almost all of the gadids, phycids, lotids, macrourids, morids, and muraenolepidids, all of whom feed actively on benthic items. Conversely, the diet of one of the most pelagic of the gadids, the pollock (*Pollachius virens*), consists primarily of euphausiids and Atlantic herring, and its much-reduced barbel reflects this focus on pelagic prey. Certain other pelagic gadiforms, such as the melanonids, bregmacerotids, and *Euclichthys polynemus*, lack chin barbels, and although their food habits are not well studied, it can be assumed that their important diet items are also pelagic.

The phycid hakes *Urophycis chuss* and *U. tenuis* (red hake and white hake, respectively) have similar food habits, although they do not always occur in the same habitats. Both species focus on crustaceans and eat other fishes only secondarily. Annelids, molluscs, and all other prey constitute a minor fraction of their diets. Their feeding behavior also takes advantage of sensitive pelvic fin rays, which are deployed in advance of the fish and aid in the search for prey through their tactile abilities. The Atlantic cod is primarily a piscivore. Among the fishes it consumes, the herring, *Clupea harengus*, is perhaps the most important, but redfish, mackerel, and smaller cod are also important prey items, and diet components are likely to vary between different study sites. Certain Newfoundland studies have found the capelin to be critically important. Cod also eat crabs, and in some studies crustaceans have been identified as more important than fishes. The haddock, *Melanogrammus aeglefinus*, has slightly different food habits than its relative, the cod. Although it focuses on crustaceans as a major dietary component, fishes are unimportant in the remainder of its diet, while polychaetes and echinoderms are secondarily important. The diets of macrourids are highly variable and consist of a wide range of fishes and benthic and pelagic invertebrates. Merlucciids are voracious predators, and the several species of *Merluccius* are highly piscivorous.

Specific information concerning the species that prey on gadiform fishes is lacking for most species. Adults of larger, commercially important species (cod, haddock, etc.) are probably only preyed upon by sharks, billfishes, and other large predators, and their most important predator is undoubtedly man. Young stages of all species, however, face predation by a large number of species, and this plays a large part in determining year-class strength. Some of their predators are larger members of the same species.

Reproductive biology

More is known about reproduction and egg and larval development in the families Gadidae and Merlucciidae, for those two contain commercially important species that have received the most attention. Gadiform fishes, in general, release masses of eggs that are then fertilized externally. Almost all of those eggs are pelagic (although little is known about reproduction in the deep-water species). The Atlantic cod is one of the world's most fecund fishes. A female of 11.0 lb (5 kg) is capable of producing 2.5 million eggs, and larger females can produce more. The haddock is not far behind. A female just under a meter in length can produce close to 2 million eggs. Eggs of gadiforms range from about 0.02 in (0.5 mm) in diameter in some morids and phycids, to about 0.08 in (2.0 mm) in certain gadids and macrourids. The chorion (outer shell) is smooth in most, but it may have a hexagonal pattern in the Macrouridae. Most gadiform eggs have a single, small oil globule, although eggs of the gadids lack an oil globule. Early life history stages are known for fewer than a third of the species described in the Gadiformes, but some descriptions are available for each family. Gadiform larvae exhibit a diverse array of shapes and specializations. According to Fahay and Markle (1984), "There does not seem to be any character unique or diagnostic for young gadiforms. The features of body shape, anus morphology, and pelvic fin development in combination with specific familial characters appear to be the most useful for initial identification. Transformation is gradual and direct with no striking changes in ontogeny." The gut of most gadiform larvae coils early in ontogeny, and combined with a tapering postanal region and rounded head, contributes to an overall tadpole-like appearance. It has not been documented in all gadiform families and is not always easily observed, but very young gadiform larvae have an anus that exits laterally through the finfold rather than at its edge as in most fish larvae. Another characteristic of gadiform larvae is that some secondary caudal rays develop before some primary rays (in forms that have a caudal fin).

Conservation status

The IUCN lists three gadiform species: *Physiculus helenaensis* is categorized as Critically Endangered, and *Gadus morhua* and *Melanogrammus aeglefinus* are categorized as Vulnerable. There are no gadiform fishes listed by the United States as endangered or threatened. However, among the factors threatening the sustainability of viable populations of gadiform fishes, overfishing figures high, especially regarding the Atlantic cod, *Gadus morhua*. In some important areas, for example eastern Canadian provinces, the cod is commercially extinct, meaning its population levels are so low that it can no longer sustain a fishery. The loss of various marine habitats, critically important to the survival of young fishes (as well as older stages), is often cited as contributory to fish populations' declines. For almost all fishes, our knowledge of the critical function of these habitats is lacking or superficial, and increased research in these areas is often cited as necessary for proper management of marine resources.

Significance to humans

Certain gadiforms are among the world's most commercially important fishes. In the late 1980s, for example, some 15,101,665 tn (13,700,000 metric tons [t]), representing fully 17% of the world's landings of marine fishes, were comprised of gadiforms. Of this total, 95% was contributed by the Gadidae (cods and their relatives), followed by the merlucciids, the macrourids, and morids.

The Atlantic cod has been an important fishery for centuries, and this fishery has actually influenced the development of western civilization in countries around the perimeter of the North Atlantic Ocean (Kurlansky 1997). It is said that when John Cabot arrived in Newfoundland waters, supposedly the first European explorer to do so, he was greeted by a well-established fleet of Basque fishermen, who in turn had been fishing for cod in Grand Banks waters for centuries before that. In the days before refrigeration, during the time when the Catholic Church mandated the eating of fish on Fridays and holy days, a good-tasting, lean fish that dried well was in a position to dominate the European markets. The Basques were pioneers and masters both in catching and processing cod and therefore enjoyed a commanding position in the world's economy. Despite the heavy fishing pressure exerted on the cod stocks during past centuries, it sustained a huge fishery until the late 1900s, when increased exploitation, based on increasingly efficient fishing methods (possibly combined with changing environmental trends), finally contributed to the collapse of the cod population. The collapse of the stocks off Labrador and Newfoundland has had particularly devastating and tragic economic consequences in eastern Canada, where the effects have been compared to the Great Depression during the 1930s in the United States.

Other gadiforms are also the basis for valuable fisheries. The Alaska (or walleye) pollock, *Theragra chalcogramma*, contributes more to the world's fisheries than any other demersal fish species, of any family. The total annual landings of this fish in the late 1980s reached 7,389,750 tn (6,703,868 t). The merlucciids (hakes of the genus *Merluccius* and two species of *Macruronus*) were once considered trash fish, but now 11 species are being exploited. In the late 1980s, 2,180,192.1 tn (1,977,837 t) were harvested, making them the second most commercially important family of gadiforms after the Gadidae.

1. Roundnose grenadier (*Coryphaenoides rupestris*); 2. Luminous hake (*Steindachneria argentea*); 3. Atlantic tomcod (*Microgadus tomcod*); 4. Alaska pollock (*Theragra chalcogramma*); 5. Burbot (*Lota lota*); 6. Haddock (*Melanogrammus aeglefinus*); 7. Silver hake (*Merluccius bilinearis*); 8. White hake (*Urophycis tenuis*); 9. Atlantic cod (*Gadus morhua*); 10. Red hake (*Urophycis chuss*); 11. Pollock (*Pollachius virens*). (Illustration by Emily Damstra)

Species accounts

Atlantic cod
Gadus morhua

FAMILY
Gadidae

TAXONOMY
Gadus morhua Linnaeus, 1758, Atlantic Ocean and the coasts of Europe.

OTHER COMMON NAMES
None known.

PHYSICAL CHARACTERISTICS
Three separate dorsal fins, two separate anal fins. Dorsal and anal fins touching at their bases or separated by very narrow gaps. Chin barbel present. Pelvic fins sometimes with one elongate ray. Head relatively narrow and long. Snout to base of first dorsal fin length <33% of total length. Overall brownish to greenish gray on upper sides, paler ventrally. Body covered with spots, sometimes vague.

DISTRIBUTION
East coast of North America, north of Cape Hatteras, North Carolina, Hudson Bay, both coasts of southern Greenland, Iceland, coast of Europe from Bay of Biscay to Barents Sea.

HABITAT
Widely distributed in a variety of habitats from close to shore to depths >1,968 ft (600 m), but most common over continental shelf between 492 and 656 ft (150 and 200 m). Mostly demersal, although incursions into water column may coincide with feeding or reproduction. Also found in river mouths from late fall to early winter. Tolerates a wide range of temperatures and salinities, but larger fish generally occur in colder water 32–41°F (0–5°C).

BEHAVIOR
The Atlantic cod is a highly migratory fish. Patterns of migrations differ somewhat between regions. This pattern is associated with reproduction and seasonal temperature change in the Newfoundland stock (Rose 1993). Here, huge schools of cod leave their wintering areas in deep, oceanic waters, and follow tongues of deep, relatively warm, oceanic waters (or highways) across the continental shelf to summer feeding areas nearer to the coast. Spawning occurs in dense concentrations (>1 fish/m^3) as they begin this mass movement, with multiple pairs of spawning fish observed in columns above the mass. As this huge mass migrates inshore, it periodically encounters important prey aggregations (such as capelin or shrimp) and disperses in order to feed. The mass is led by the largest fish (or scouts), and the smallest bring up the rear. After reaching nearshore waters, they turn and move northward along the Newfoundland coast in late summer, then eventually return to their deep-water wintering areas.

Off New England, Atlantic cod typically move into coastal waters during the fall, and then retreat into deeper waters during spring. A slightly different pattern occurs in the Great

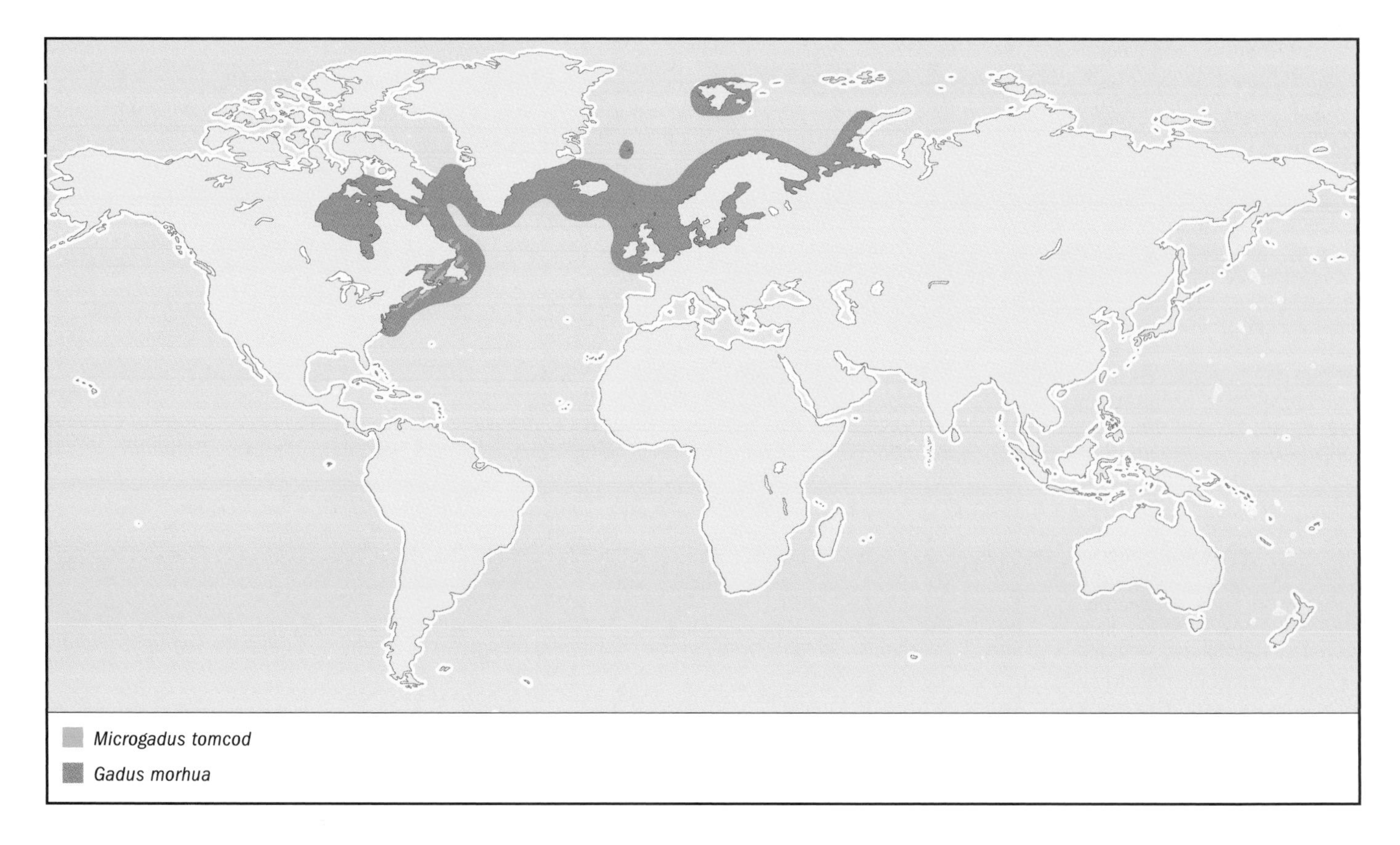

South Channel area where they move southwesterly during the fall, spend the winter off southern New England and the Middle Atlantic coast, and then reverse this movement during the spring.

FEEDING ECOLOGY AND DIET

Diet changes with life history stage. The cod is generally omnivorous and voracious. For most ages, feeding occurs in twilight (dawn and dusk), but young fish feed almost continuously. Larvae feed on plankton; juveniles feed on invertebrates, especially small crustaceans; older fish feed on invertebrates and fishes, including young cod. Important diet items are likely to vary between study areas. Herring and capelin are important items in some areas.

REPRODUCTIVE BIOLOGY

Size and age at maturity have declined in recent years, most likely as a response to the fishery harvesting older and larger fish, or to a general decline in the stock biomass due to intense exploitation. A Scotian Shelf study (Beacham 1983) found that median age at maturity declined about 50% from 1959 (when age at 50% maturity was 5.4 years in males; 6.3 years in females) to 1979 (when age at 50% maturity was 2.8 years in both sexes). Median lengths at maturity declined from 20.1 to 15.4 in (51 to 39 cm) in males; 21.3 to 16.5 in (54 to 42 cm) in females. This smaller-and-younger-at-maturity trend continued between 1972 and 1995 in all zones between Georges Bank and Labrador, until presently in United States waters, maturity is reached between 1.7 and 2.3 years (median age) and 12.6 and 16.1 in (32 and 41 cm) (average length). Off the northeastern United States, the distribution of eggs indicates that important spawning occurs over the northeast peak of Georges Bank and around the perimeter of the Gulf of Maine. Reproduction peaks in winter and spring, but continues weakly throughout the year. The North Sea is a major center for reproduction in the eastern Atlantic, where spawning peaks between December and May. Eggs and larvae are pelagic, and juveniles begin descending to the bottom at sizes between 1.0 and 2.4 in (2.5 and 6.0 cm).

CONSERVATION STATUS

The Atlantic cod is listed as Vulnerable by the IUCN. Populations are heavily overexploited by fisheries and are at reduced levels of abundance. Both commercial landings and estimates of spawning stock size are at their lowest levels since 1960. Catch limits are strictly managed, and several important fishing grounds, e.g. portions of Georges Bank, have been closed to all fishing, largely in response to these low levels.

SIGNIFICANCE TO HUMANS

The importance of the Atlantic cod through history can hardly be overemphasized. For the past 1,000 years, the capture, preparation, and distribution of the cod has influenced the development of Western Civilization, especially around the perimeter of the North Atlantic Ocean. The Vikings crossed the Atlantic in pursuit of the cod. The Basques turned the cod into a commercial product in medieval times. Cape Cod was named in honor of the cod in 1602. The cod has actually been the cause of wars between countries, from American colonial times to recent conflicts between Iceland and Great Britain in the twentieth century. Newfoundland was settled by Irish and English natives in the early eighteenth century, largely because of opportunities in the cod fishery. Throughout most of the nineteenth century, this fishery was the most important source of employment and income for people in Newfoundland and much of Eastern Canada. In 1992, the cod population nearly reached a point of commercial extinction in waters off eastern Canada and Newfoundland, and a fishing moratorium was imposed. This moratorium has removed the main source of employment and income for thousands of fishermen from hundreds of small fishing communities and has truly devastated the Canadian economy. ◆

Haddock

Melanogrammus aeglefinus

FAMILY

Gadidae

TAXONOMY

Melanogrammus aeglefinus Linnaeus, 1758, Oceano Europeo.

OTHER COMMON NAMES

None known.

PHYSICAL CHARACTERISTICS

Three separate dorsal fins, two separate anal fins. Dorsal and anal fins separated by narrow gaps. Small chin barbel present. Pelvic fins sometimes with one elongate ray. Lateral line dark. A prominent blotch on side over the pectoral fin.

DISTRIBUTION

Eastern North Atlantic from Bay of Biscay to Spitzbergen; Barents Sea; around Iceland and southern tip of Greenland; Western North Atlantic from Labrador to Cape Charles, Virginia. In the western Atlantic, highest abundance occurs over Georges Bank, Scotian Shelf, and southern Grand Bank. The highest concentrations off the United States are associated with the two major stocks located on Georges Bank and in the southwestern Gulf of Maine.

HABITAT

Haddock are most common at depths of 148–443 ft (45–135 m) and temperatures of 36–50°F (2.2–10°C). Substrates preferred include rock, sand, gravel, or broken shell. Gravelly sand and gravel are preferred in the Western Atlantic. Haddock exhibit age-dependent shifts in habitat use, with juveniles occupying shallower water on bank and shoal areas, and larger adults associated with deeper water.

BEHAVIOR

Adult haddock do not undertake long migrations, but seasonal movements occur in the western Gulf of Maine, the Great South Channel, and on the northeast peak of Georges Bank.

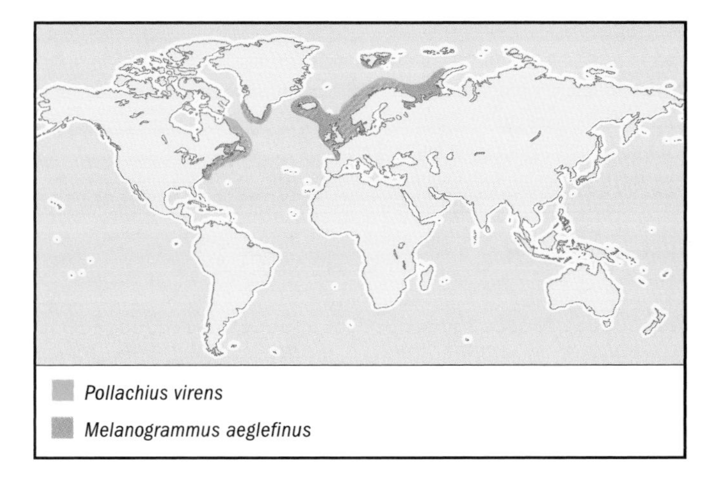

FEEDING ECOLOGY AND DIET
Crustaceans, echinoderms, polychaetes, and mollusks are the most important prey items for juveniles and adults combined. Juveniles prey mostly on crustaceans. Other fishes are of minor importance in the haddock's diet.

REPRODUCTIVE BIOLOGY
Spawning occurs between January and June, with peak activity during late March and early April. An average-sized 21.7 in (55 cm) female produces approximately 850,000 eggs, and larger females are capable of producing up to three million eggs annually. Spawning concentrations occur on eastern Georges Bank, to the east of Nantucket Shoals, and along the Maine coast. Growth and maturation rates of haddock have changed significantly over the past 30 to 40 years. During the early 1960s, all females age four and older were fully mature, and approximately 75% of age three females were mature. Presently, growth is more rapid, with haddock reaching 18.9 to 19.7 in (48 to 50 cm) at age three. Nearly all age three and 35% of age two females are mature. Although early maturing fish increase spawning stock biomass, the degree to which these younger fish contribute to reproductive success of the population is uncertain.

CONSERVATION STATUS
Listed as Critically Endangered by the IUCN. The spawning stock biomass of Georges Bank haddock declined from 76,000 tn (69,000 t) in 1978 to 12,125 tn (11,000 t) by 1993, and has since increased to 41,900 tn (38,000 t) in 1998. However, spawning stock biomass is presently below the minimum threshold level of 58,400 tn (53,000 t), indicating the stock is in an overfished condition. Observed increases in spawning stock biomass of Georges Bank haddock have resulted from conservation of existing year classes. This is a necessary first step in the stock rebuilding process. Recent research vessel surveys provide indications that the 1998 year class may be the strongest in two decades. If this recruitment is realized, there is a potential for significant stock rebuilding.

SIGNIFICANCE TO HUMANS
An extremely valuable fishery on both sides of the North Atlantic. In 1987, the FAO reported that landings of this species amounted to 439,295 tn (398,522 t), of which total most (400,530 tn; 363,353 t) was taken in the northeastern Atlantic. Leading fishing countries are United Kingdom, Russia, Norway, and Iceland, followed by France, Denmark, and others. Northwest Atlantic landings are dominated by Canada, followed by the United States. ◆

Atlantic tomcod
Microgadus tomcod

FAMILY
Gadidae

TAXONOMY
Microgadus tomcod Walbaum, 1792, Artedi.

OTHER COMMON NAMES
English: Frostfish.

PHYSICAL CHARACTERISTICS
Three separate dorsal fins, two separate anal fins. Dorsal and anal fins separated by gaps and rounded in outline. Chin barbel present. Pelvic fins with one elongate ray. Caudal fin rounded. Olive brown to green dorsally, paler ventrally, with darker mottling on sides.

DISTRIBUTION
Once reported to occur along the coast of North America from Labrador to North Carolina, the tomcod now occurs only as far south as the Hudson River in New York, where it is common.

HABITAT
The tomcod is a coastal fish, ascending rivers into habitats with very low salinities, and living near or on the bottom. It is strictly riverine in the Hudson River, but also can survive landlocked in lakes. Young stages are found in estuaries throughout its range, but not those with limited (or no) freshwater input.

BEHAVIOR
No migrations to offshore waters, but tomcod prefer colder temperatures and move into deeper waters during summer, returning to shallow waters during fall and winter.

FEEDING ECOLOGY AND DIET
Feeds mostly on small crustaceans (especially shrimps and amphipods), worms, small molluscs, squid, and very young fishes.

REPRODUCTIVE BIOLOGY
Tomcod spawning is accomplished during winter, and involves elaborate courtship behavior of small groups. Most egg deposition occurs well upstream, in the lowest salinities available, and eggs are weakly adhesive. In the Hudson River, 93–99% of spawners are young-of-the-year fish approaching their first birthday. Larvae hatch at 0.2–0.24 in (5–6 mm) total length and grow rapidly during their first spring. Growth is then depressed during summer and resumes in the fall.

CONSERVATION STATUS
Not threatened.

SIGNIFICANCE TO HUMANS
Tomcod is a popular sport and food fish, mostly in Canada and New England states. In addition to a limited hook and line fishery, tomcod are taken with bag nets, pocket nets, and weirs. ◆

Pollock
Pollachius virens

FAMILY
Gadidae

TAXONOMY
Pollachius virens Linnaeus, 1758, Oceano Europeo.

OTHER COMMON NAMES
English: Coalfish, saithe.

PHYSICAL CHARACTERISTICS
Three separate dorsal fins, two separate anal fins. Dorsal and anal fins separated by narrow gaps. A very small chin barbel present. Lateral line pale. Brownish-green dorsally, slightly paler ventrally. Fins same color as body.

DISTRIBUTION
The pollock occurs on both sides of the North Atlantic. In the western North Atlantic, it is found from the Hudson Strait to Cape Hatteras, North Carolina. In the eastern North Atlantic, from Spitzbergen to Bay of Biscay. Also found in Barents Sea and around Iceland.

HABITAT

The pollock is strongly pelagic and occurs most frequently over depths of 361–590 ft (110–180 m), although its range can vary with food supply and season. Adult fishes occur in temperatures as low as 32°F (0°C), and they do not tolerate temperatures >52°F (11°C). Young stages are known as harbor pollock and are commonly found in bays and estuaries throughout their range.

BEHAVIOR

The pollock is a schooling species and is found throughout the water column, not just near bottom. Pollock engage in short migrations associated with temperature changes or for spawning, but otherwise remain fairly stationary within their range.

FEEDING ECOLOGY AND DIET

The pollock feeds most actively on pelagic prey. Important prey items include euphausiids (especially *Meganyctiphanes norvegica*), fishes, and molluscs (especially the squid *Loligo*). Crustaceans are most important in juveniles' diets. Fishes comprise only 12% of juveniles' diets and 28% of adults' diets.

REPRODUCTIVE BIOLOGY

In the western Atlantic, spawning occurs from September to April with peaks between December and February. Both sexes reach sexual maturity during their third year, at lengths of 19.9 and 18.9 in (50.5 cm and 47.9 cm) in males and females, respectively. Spawners occasionally form huge aggregations. Spawning occurs over hard, rocky bottoms, and activity peaks when temperatures are between 40.1 and 42.8°F (4.5 and 6.0°C). Eggs and larvae develop pelagically, and small pelagic juveniles begin to enter inlets in February and March, when they are <2.0 in (50 mm) long.

CONSERVATION STATUS

Not listed by the IUCN. Although there are recognized western Atlantic centers of abundance of pollock on the Scotian Shelf, Georges Bank, and Gulf of Maine, tagging studies suggest considerable movement of pollock between these centers and accordingly, pollock from Cape Breton, Nova Scotia, and south continue to be assessed as a single, unit stock by United States scientists. The total nominal catch from this stock, including commercial and recreational, has been steadily declining since 1986, and the 1996 total represents an 82% reduction from 1986 landings. Spawning biomass is increasing, but within the Gulf of Maine, stock abundance and biomass remain low. Overall the stock is considered to be fully exploited, but not yet in an overfished condition.

SIGNIFICANCE TO HUMANS

Pollock is an important commercial species, and it is marketed in several ways: fresh, as chilled fillets, frozen, canned, dried and salted, and in brine. A large percentage of the 1987 total landings 524,680 tn (475,981 t) was landed in the northeast Atlantic by Norway, Iceland, France, Germany, the United Kingdom, and Denmark. Most of the catch in the northwest Atlantic is landed by Canada, the United Kingdom, and France. ◆

Alaska pollock

Theragra chalcogramma

FAMILY

Gadidae

TAXONOMY

Theragra chalcogramma Pallas, 1811, sea of Okhotsk and the shores of Kamchatka.

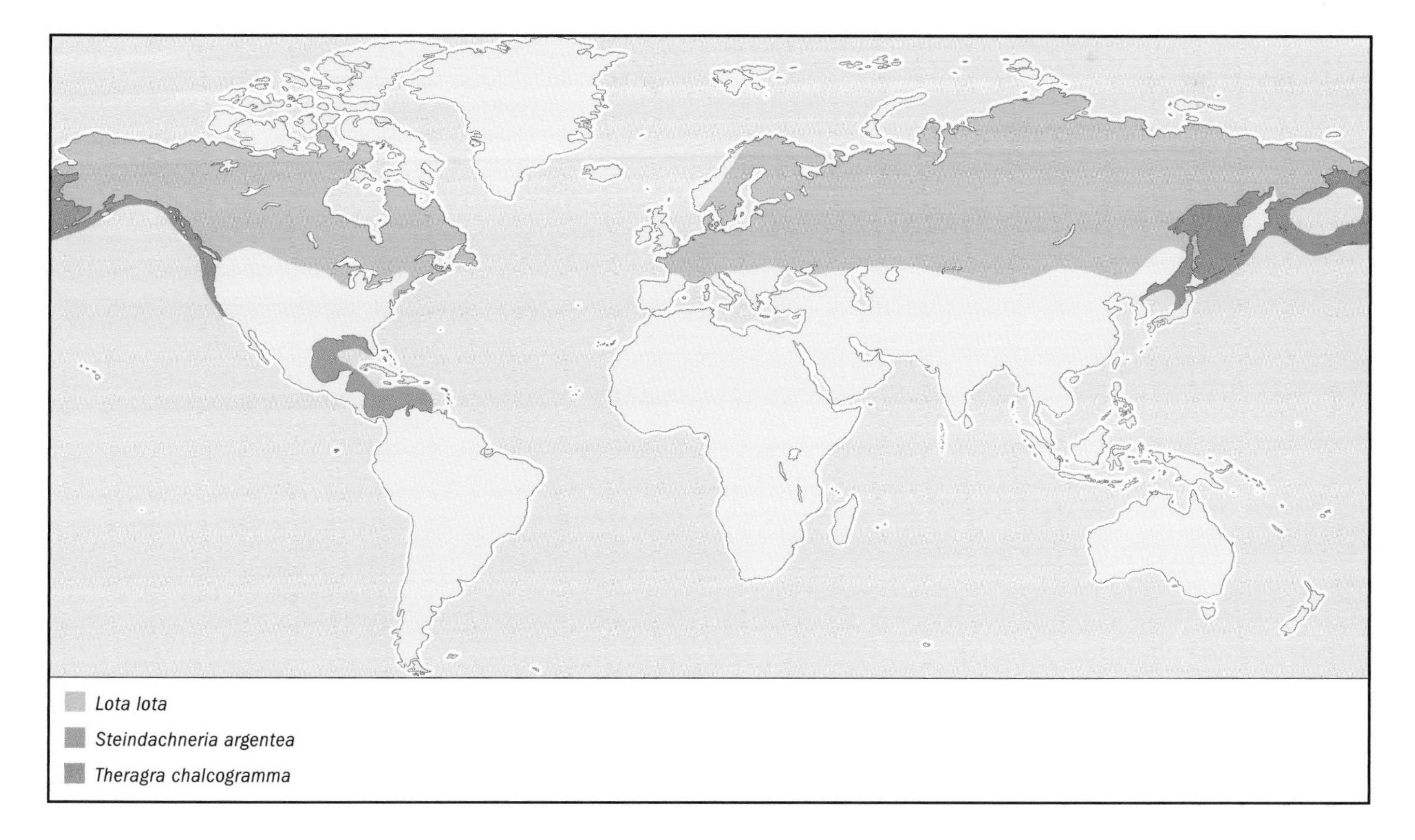

OTHER COMMON NAMES
English: Walleye pollock.

PHYSICAL CHARACTERISTICS
Three separate dorsal fins, two separate anal fins. Dorsal and anal fins separated by gaps. A very tiny chin barbel present. Pelvic fins sometimes with one ray elongate. Olive green to brown dorsally, silvery on sides, pale ventrally. Often mottled or blotchy.

DISTRIBUTION
Found in temperate and subarctic waters of the Northern Pacific Ocean from Sea of Japan, through Okhotsk Sea, Bering Sea, and Gulf of Alaska, then south to waters off central California. Separate stocks (as many as 12) occur in the North Pacific, including those in the Gulf of Alaska, Aleutian Islands, and Bering Sea.

HABITAT
A schooling fish found on or near the bottom, but also in midwater to near-surface depths. Heaviest catches made at depths between 164 and 984 ft (50 and 300 m).

BEHAVIOR
Alaska pollock perform vertical migrations on a daily basis. Juveniles ascend at night to feed on zooplankton near the surface. Migrations of the entire population associated with spawning and feeding. Alaska pollock follow a circular pattern of migrations in the Bering Sea, moving inshore in the spring to spawn and feed, and offshore to warmer, deeper waters in winter.

FEEDING ECOLOGY AND DIET
Young fish feed on copepod adults and eggs. Adults prey on shrimps, sand lance, and herring off British Columbia; on pink, chum, and coho salmon in Alaskan waters; on mysids, euphausiids, silver smelt, and capelin in Asian waters.

REPRODUCTIVE BIOLOGY
Alaska pollock begin spawning at age two, but ages four and five contribute most to reproduction. Most spawning begins late February in the Bering Sea, March or April in the Gulf of Alaska. Major spawning aggregations are found in Shelikof Strait, Straits of Georgia, Aleutian Basin, and off the Pribilof Islands. Spawning fish form dense schools high in the water column. Fecundity ranges from 37,000 eggs per female to nearly one million off the coast of Canada. In the western Bering Sea, an 11-year-old female can produce 15 million eggs. Eggs are pelagic and occur mostly within 98 ft (30 m) of the surface.

CONSERVATION STATUS
Not listed by the IUCN. Based on estimates of spawning stock biomass and projections into the near future, the Gulf of Alaska stock is not in an overfished condition, nor is it predicted to be.

SIGNIFICANCE TO HUMANS
The Alaska pollock has become an increasingly important human food resource. In the early 1990s, landings of this species were the largest of any demersal fish. The FAO Yearbook of Fishery Statistics for 1987 reported 7,389,750 tn (6,703,868 t) were landed, primarily in the western part of the North Pacific by USSR, Japan, Poland, and Republic of Korea. The largest catches are made over the outer shelf and slope of the eastern Bering Sea, between the Aleutian Islands and the Pribilofs. ◆

Burbot
Lota lota

FAMILY
Lotidae

TAXONOMY
Lota lota Linnaeus, 1758, Europe.

OTHER COMMON NAMES
English: American burbot (Canada), lush (Alaska), lawyer, ling (Canada), eelpout.

PHYSICAL CHARACTERISTICS
Short first dorsal fin followed by long second dorsal fin. Anal fin single, nearly as long-based as second dorsal. Pelvic fins normal, not modified into elongate rays. Well-developed chin barbel. Anterior nostril has barbel-like flap. Color yellow, light tan, to brown, overlain with a blotchy pattern of darker brown or black.

DISTRIBUTION
The burbot occurs in freshwaters of northern North America and Europe and Asia. Occurs farther north than 40° N (to nearly 80° N).

HABITAT
Occurs on the bottoms of lakes and rivers, from depths of 1.6 ft (0.5 m) to more than 755 ft (230 m).

BEHAVIOR
The burbot moves into shallower waters during summer nights. They also move into shallower water to spawn in some parts of their range.

FEEDING ECOLOGY AND DIET
The burbot has been characterized as a voracious predator and night feeder. Young fish feed on insect larvae, crayfish, molluscs and other invertebrates, whereas adults >19.7 in (50 cm) feed almost exclusively on other fishes.

REPRODUCTIVE BIOLOGY
Burbot spawning occurs from November to May, but primarily between January and March in Canada, and December in parts of Russia. Spawning usually occurs under the ice, over sand or gravel substrates, at night, and in shallow water (<9.8 ft [3 m] depth). Eggs are semi buoyant. Fecundity ranges from 45,600 eggs per 13.4-in (34-cm) female to 1,362,077 eggs in a 25.2-in (64-cm) female.

CONSERVATION STATUS
Not listed by the IUCN. The burbot may occur in considerable numbers in many inland lakes, but has declined over past levels in the Great Lakes, where it had been considered a nuisance species.

SIGNIFICANCE TO HUMANS
The burbot is an important competitor for food of other species, such as lake trout and whitefish. It is fished commercially in Finland, Sweden, and the European part of Russia, but it is only moderately important as a commercial species in Canada and Alaska. Often marketed salted or as pet food. ◆

Roundnose grenadier
Coryphaenoides rupestris

FAMILY
Macrouridae

TAXONOMY
Coryphaenoides rupestris Gunnerus, 1765, Trondheim.

OTHER COMMON NAMES
English: Black grenadier, rock grenadier.

PHYSICAL CHARACTERISTICS
Abdominal region (tip of snout to beginning of anal fin) short. Snout broad and rounded. Head broad, soft and deep, with tiny chin barbel. First dorsal fin with two spines and 8–11 rays. Pelvic fin has 7–8 rays, with one elongate. A modified scute-like scale at tip of snout. A wide gap between first and second dorsal fins. Anal fin rays much longer than dorsal fin rays. Color brownish gray, with blackish fins. Reaches greater than a meter in length.

DISTRIBUTION
Occurs in the North Atlantic north of Cape Hatteras, North Carolina to Baffin Island, Greenland, off Iceland and Norway, south to Spain in the Eastern Atlantic. Isolated occurrences off North Africa and Bahamas.

HABITAT
Found between 590 and 7,217 ft (180 and 2,200 m) depth, but concentrated between 1,312 and 3,937 ft (400 and 1,200 m). Adults may be distributed in shallower waters than younger fishes.

BEHAVIOR
Undertakes a post-spawning migration during the winter, back into relatively shallow waters.

FEEDING ECOLOGY AND DIET
Undertakes diurnal, vertical feeding migrations that may take it well off the bottom >3,280 ft (1,000 m). Consumes a variety of fishes and invertebrates, primarily pelagic crustaceans such as shrimps, amphipods, and cumaceans. Also feeds on lanternfishes and cephalopods.

REPRODUCTIVE BIOLOGY
Undertakes spawning migrations during the summer into deeper waters, particularly near Iceland. Females mature at about 23.6 in (60 cm), males at about 15.8 in (40 cm). Fecundity estimates range from 12,000 to 35,500 eggs per female. In some areas, may spawn year-round. The eggs are pelagic, spherical, and 0.091–0.095 in (2.3–2.4 mm) in diameter. Their shell is honeycombed and they contain a single oil globule. The larva has been illustrated by Merrett (1978).

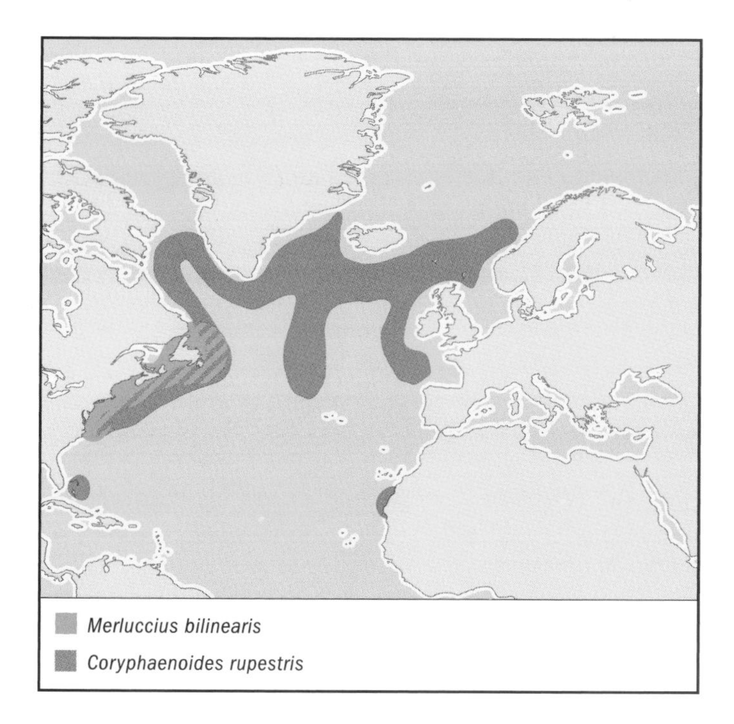

CONSERVATION STATUS
Not listed by the IUCN. A valuable commercial fish species currently facing overexploitation in the North Atlantic.

SIGNIFICANCE TO HUMANS
The roundnose grenadier is a fish with excellent taste and texture. Some 22,000 tn (20,000 t) are landed annually, primarily by fishing fleets from Russia, Germany, and Poland. Its liver is rich in fats and vitamins. ◆

Silver hake
Merluccius bilinearis

FAMILY
Merlucciidae

TAXONOMY
Merluccius bilinearis Mitchill, 1814, New York.

OTHER COMMON NAMES
English: Atlantic whiting.

PHYSICAL CHARACTERISTICS
Genus (hakes) characterized by large head (1/3 to 1/4 of body length), with large, oblique mouth. Lower jaw longer than upper. Two separate dorsal fins, the first short based, high, and triangular, separated from the second, which is long and partially divided by a notch in the midsection. Single anal fin similar in shape to second dorsal. Pelvic fins with seven rays. Silver hake has 16–20 gill rakers on the first arch (cf. 8–11 in closely related *Merluccius albidus*). Color purplish gray dorsally, silvery white lower on sides.

DISTRIBUTION
Occurs on the continental shelf of the northwest Atlantic Ocean from Gulf of St. Lawrence and Grand Banks to offings of North Carolina. Silver hake are most abundant between Nova Scotia and New Jersey. Two stocks have been identified in American waters. One occurs in the Gulf of Maine and northern edge of Georges Bank, the other from the southern edge of Georges Bank to Cape Hatteras, North Carolina.

HABITAT
Silver hake apparently prefer temperatures between 44.6 and 51.8°F (7 and 11°C) in the summer and fall, and 44.6 and 55.4°F (7 and 13°C) in the spring. They occur in depths between 33 and 4,100 ft (10 and 1,250 m), moving into deeper waters during the coldest time of year. Little is known about substrates they prefer, although juveniles 0.6–2.0 in (1.5–5.0 cm) are most abundant on silt-sand bottoms with concentrations of amphipod tubes. During the colder times of year, silver hake become concentrated in deep basins of the Gulf of Maine and along the upper continental slope.

BEHAVIOR
All stocks of silver hake exhibit inshore-offshore migrations associated with temperature changes and availability of important diet items. They are primarily demersal fish, but often move upward in the water column, especially at night, presumably following prey items.

FEEDING ECOLOGY AND DIET
Young silver hake <7.9 in (20 cm) eat mostly crustaceans, such as euphausiids and shrimps. As they grow, they consume increasing proportions of fishes, and adults >13.8 in (35 cm) feed almost exclusively on fishes. Said to be a voracious predator, their diet sometimes includes smaller silver hake.

REPRODUCTIVE BIOLOGY
Major spawning areas include coastal Gulf of Maine, southern Georges Bank, and waters south of Rhode Island. In these areas spawning reaches a peak in June and July. Spawning occurs during late summer off Sable Island Bank in Canadian waters. Females are asynchronous spawners and lay their eggs in several batches through the season. Age and length at maturity have both declined in recent years. In the early 1960s, silver hake reached maturity at two to three years of age and lengths between 11.4 and 13.0 in (29 and 33 cm). By 1989, these figures had declined to 1.6–1.7 years old and 8.8–9.1 in (22.3–23.2 cm), respectively. Another study found that 20% of two-year-olds had reached maturity in 1973, while 80% of two-year-olds had reached maturity in 1990.

CONSERVATION STATUS
Not listed by the IUCN. The decline in size and age at maturity are usually the result of a stock that has been overfished. In both the northern and southern stocks, significant mortality of juvenile silver hake has occurred through discarding in the large mesh and small mesh otter trawl fisheries directed at other species. Annual discard estimates over 1989–1992 ranged from 1,430–11,020 tn (1,300–10,000 t [10 million to 81 million fish]) per year. Excessive discard mortality on juveniles may severely limit opportunities to rebuild either silver hake stock. The southern stock is considered to be in an overfished condition.

SIGNIFICANCE TO HUMANS
The silver hake fishery is important to the United States, Russia, and Cuba. Centers of the fishery are Nova Scotia, Gulf of Maine, and Georges Bank. The total catch in 1987 was 85,950 tn (77,975 t), down from 479,500 tn (435,000 t) in 1973. Silver hake flesh is flaky and good tasting. It is marketed frozen or filleted, and preparation methods include smoking, boiling, and frying. ◆

Red hake
Urophycis chuss

FAMILY
Phycidae

TAXONOMY
Urophycis chuss Walbaum, 1792, Artedi.

OTHER COMMON NAMES
English: Squirrel hake.

PHYSICAL CHARACTERISTICS
Single anal fin. Two dorsal fins, the first short based and moderate in height, with one elongate ray. A single chin barbel. Pelvic fin comprised of two very elongate rays, tip of the longest reaching level of anus. Upper limb of gill raker with three gill rakers; caudal fin with 28–34 rays.

DISTRIBUTION
Western North Atlantic Ocean from Cape Hatteras, North Carolina to Nova Scotia, rarely to Gulf of St. Lawrence. Two stocks occur off the northeastern United States: a northern stock from the northern slopes of Georges Bank and Gulf of Maine, and a southern stock, from the southern slopes of Georges Bank to North Carolina.

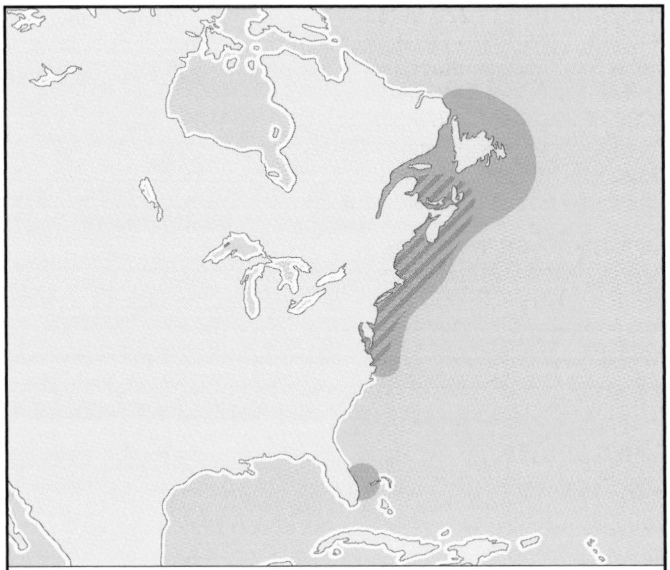

Urophycis tenuis
Urophycis chuss

HABITAT
Found on muddy or sandy bottoms, less common on gravelly or hard bottoms. Adults found between 16.4 and >984 ft (5 and >300 m), but some seasonal migrations take place. Early settled juveniles live in an inquiline association with sea scallops, *Placopecten magellanicus*, or other structured habitats, after which they remain in relatively shallow coastal waters until their second year. Thereafter, they are most common in depths <328 ft (100 m) in warmer months, >328 ft (100 m) in colder months.

BEHAVIOR
Red hake migrate seasonally in reaction to changing temperatures. During summer, they are quite common in nearshore bays and estuaries in New England. During the winter, they migrate into deeper waters.

FEEDING ECOLOGY AND DIET
Hakes use their pelvic fin rays as sensory organs to find food (Bardach and Case 1965). Juveniles leave their shelters at night and prey on small benthic organisms such as crustaceans. Adults also prey on crustaceans, but consume a wide variety of fish and squid as well.

REPRODUCTIVE BIOLOGY
Spawning occurs spring through fall off the coast of northeastern United States, but may be restricted to mid-summer in the Gulf of Maine. Eggs and larvae develop pelagically, and the larvae transform into a specialized pelagic–juvenile stage that is highly neustonic (lives very near the surface), often gathering around floating debris. They settle to the bottom at sizes of 1.4–1.6 in (35–40 mm).

CONSERVATION STATUS
Not listed by the IUCN. During the early 1960s, total landings from both stocks (northern and southern) peaked at 125,220 tn (113,600 t) in 1966. Annual landings then declined

sharply to only 14,220 tn (12,900 t) in 1970, increased again to 84,220 tn (76,400 t) in 1972, and then have declined steadily since. Red hake landings averaged only 1,870 tn (1,700 t) per year during 1990–1999, a decline of over 40% from the 1980–1989 average. Red hake landings in 1999 were well below historic levels. Despite these declines, neither stock is presently considered to be in an overfished condition, and recruitment of younger fish appears to be strong.

SIGNIFICANCE TO HUMANS
A variable constituent of the United States and Canadian trawl fisheries. Large fish marketed fresh or frozen, and small fishes sometimes sold for animal feeds. ◆

White hake
Urophycis tenuis

FAMILY
Phycidae

TAXONOMY
Urophycis tenuis Mitchill, 1814, New York.

OTHER COMMON NAMES
English: Mud hake (Canada).

PHYSICAL CHARACTERISTICS
Single anal fin. Two dorsal fins, the first short based and moderate in height, with one elongate ray. A single-chin barbel. Pelvic fin comprised of two very elongate rays, not quite reaching level of anus. Upper limb of gill raker with two gill rakers; caudal fin with 33–39 rays.

DISTRIBUTION
Most commonly from Newfoundland to Cape Hatteras, North Carolina, as far south as Florida in deeper water and around the coasts of Iceland.

HABITAT
Soft, muddy bottoms of the outer continental shelf and upper continental slope. Most occur deeper than 656 ft (200 m). Also found in deeper basins in the Gulf of Maine and submarine canyons along the edge of the continental shelf. Juveniles depend on estuaries during their first spring and summer, where they are common in eel grass beds and other structured habitats.

BEHAVIOR
Young white hake engage in sand-hiding behavior, whereby they bury themselves in sand with only their heads protruding. The entire population typically moves into deeper waters in the fall, although the largest sizes occur in depths of 656 ft (200 m) and deeper and move very little, if at all.

FEEDING ECOLOGY AND DIET
Juveniles feed on polychaetes, shrimps, and other crustaceans. Adults also feed on crustaceans, but augment their diet with fishes, including juvenile white hake.

REPRODUCTIVE BIOLOGY
Spawning occurs in early spring off the northeastern United States, with a separate spawning event during summer over the Scotian Shelf and Gulf of St. Lawrence (Able and Fahay 1998). Larvae develop into a silvery pelagic–juvenile stage that is strongly associated with the ocean surface. These soon migrate into estuaries, where they settle to their first bottom stage, and remain through the summer.

CONSERVATION STATUS
Not listed by the IUCN. Total landings of white hake increased from about 1,102 tn (1,000 t) during the late 1960s to 9,150 tn (8,300 t) in 1985. Landings then declined to 5,622 tn (5,100 t) in 1989, rose sharply to 10,582 tn (9,600 t) in 1992, and have since steadily declined to levels not seen since the early 1970s. Total landings in 1998 were 2,866 tn (2,600 t), a 30% decline from 1996. Results of the most recent assessment indicate the Gulf of Maine–Georges Bank white hake stock is in an overfished condition.

SIGNIFICANCE TO HUMANS
Since 1968, the United States fishery has accounted for approximately 90% of the Gulf of Maine–Georges Bank white hake catch, but Canadian fishermen also land significant amounts from Newfoundland and northern Gulf of Maine. Larger fish are marketed fresh or as frozen fillets. Smaller fish are sometimes used for animal feed. ◆

Luminous hake
Steindachneria argentea

FAMILY
Steindachneriidae

TAXONOMY
Steindachneria argentea Goode and Bean, 1896, off Mississippi River delta.

OTHER COMMON NAMES
None known.

PHYSICAL CHARACTERISTICS
Body long, compressed, tapering to a fine point and tiny caudal fin. Anus separated from urogenital opening, the former situated between the pelvic fin bases, the latter just ahead of the anal fin. A light organ present on the ventral part of body and sides of head, appearing as purplish, striated area. First dorsal fin has one spine and 7–9 rays; second dorsal and anal fins each has more than 123 rays. Anterior portion of anal fin elevated, containing 10–12 rays. Pectoral fin has 14–17 rays. First ray of pelvic fin filamentous. Body silvery, upper part somewhat brownish, belly purplish.

DISTRIBUTION
Found in the central western Atlantic Ocean and off the East Coast of the United States, Gulf of Mexico, and Caribbean Sea as far as Venezuela.

HABITAT
Occurs on outer part of continental shelf and upper continental slope, usually over soft bottoms.

BEHAVIOR
Unknown.

FEEDING ECOLOGY AND DIET
Unknown.

REPRODUCTIVE BIOLOGY
Eggs of this species are undescribed. The larvae are pelagic and are uncommonly collected in the Gulf of Mexico. They have large heads and large eyes, and the pectoral fins are somewhat stalked as they develop. The striated luminous organ begins to develop in larvae as small as 0.9 in (24.0 mm), when fin

rays are completely formed. The anus and urogenital opening initially are found together, but the anus migrates forward with development.

CONSERVATION STATUS
Not listed by the IUCN.

SIGNIFICANCE TO HUMANS
No fishery is directed at this species, although large quantities are sometimes landed between depths of 1,300 and 1,640 ft (400 and 500 m) in the northern Gulf of Mexico. ◆

Resources

Books

Able, K. W., and M. P. Fahay. *The First Year in the Life of Estuarine Fishes in the Middle Atlantic Bight.* Piscataway, NJ: Rutgers University Press, 1998.

Kurlansky, M. *Cod. A Biography of the Fish That Changed the World.* New York: Walker and Company, 1997.

Periodicals

Bardach, J. E., and J. Case. "Sensory Capabilities of the Modified Fins of Squirrel Hake (*Urophycis chuss*) and Searobins (*Prionotus carolinus* and *P. evolans*)." *Copeia* 2 (1965): 194–206.

Beacham, T. D. "Variability in Size or Age at Sexual Maturity of White Hake, Pollock, Longfin Hake, and Silver Hake in the Canadian Maritimes Area of the Northwest Atlantic Ocean." *Canadian Technical Report of Fisheries and Aquatic Sciences* 1157: iv+43p.

Cohen, D. M. "Gadiformes: Overview." In *Ontogeny and Systematics of Fishes*, edited by H. G. Moser, W. J. Richards, D. M. Cohen, M. P. Fahay, A. W. Kendall, Jr. and S. L. Richardson. *American Society of Ichthyology and Herpetology, Special Publication* no. 1 (1984): 259–265.

———, ed. "Papers on the Systematics of Gadiform Fishes." *Science Series No. 32; Los Angeles County Museum of Natural History* (1989): 143–158.

Cohen, D. M., T. Inada, T. Iwamoto, and N. Scialabba. "Gadiform Fishes of the World (Order Gadiformes)." *An Annotated and Illustrated Catalogue of Cods, Hakes, Grenadiers, and Other Gadiform Fishes Known to Date. FAO Fisheries Synopsis* 10, no. 125 (1990): 1–442.

Fahay, M. P. "The Ontogeny of *Steindachneria argentea* Goode and Bean with Comments on Its Relationships." In *Papers on the Systematics of Gadiform Fishes. Science Series No. 32*, edited by D. M. Cohen. *Los Angeles County Museum of Natural History* (1989): 143–158.

Fahay, M. P., and K. W. Able. "The White Hake, *Urophycis tenuis* in the Gulf of Maine: Spawning Seasonality, Habitat Use, and Growth in Young-of-the-Year, and Relationships to the Scotian Shelf Population." *Canadian Journal of Zoology* 67 (1989): 1715–1724.

Fahay, M. P., and D. F. Markle. "Gadiformes: Development and Relationships." In *Ontogeny and Systematics of Fishes*, edited by H. G. Moser, W. J. Richards, D. M. Cohen, M. P. Fahay, A. W. Kendall, Jr., and S. L. Richardson. *American Society of Ichthyplogy and Herpetology, Special Publication* no. 1 (1984): 265–283.

Fahay, M. P., P. L. Berrien, D. L. Johnson, and W. W. Morse. "Essential Fish Habitat Source Document: Materials for Determining Habitat Requirements of Atlantic Cod, *Gadus morhua* Linnaeus." *NOAA Technical Memorandum* (1999) NMFS-F/NEC: 41 p.

Markle, D. F. "Identification of Larval and Juvenile Canadian Atlantic Gadoids with Comments on the Systematics of Gadid Subfamilies." *Canadian Journal of Zoology* 60, no. 12 (1982): 3420–3438.

Marshall, N. B., and D. M. Cohen. "Order Anacanthini (Gadiformes): Characters and Synopsis of Families." *Memoirs of the Sears Foundation for Marine Research* 1, no. 6 (1973): 479–495.

Merrett, N. R. "On the Identity and Pelagic Occurrence of Larval and Juvenile Stages of Rattail Fishes (Family Macrouridae) from 60 N, 20 W, and 53 N, 20 W." *Deep-Sea Research* 25 (1978): 147–60.

———. "The Elusive Macrourid Alevin and Its Seeming Lack of Potential in Contributing to Intrafamilial Systematics." In *Papers on the Systematics of Gadiform Fishes. Science Series No. 32*, edited by D. M. Cohen. *Los Angeles County Museum of Natural History.* (1989): 175–185.

Rose, G. A. "Cod Spawning on a Migration Highway in the North-west Atlantic." *Nature* 366 (1993):458–461.

Scott, W. B., and E. J. Crossman. "Freshwater Fishes of Canada." *Fisheries Research Board of Canada Bulletin.* 184 (1973): 1–966.

Michael P. Fahay

Batrachoidiformes

(Toadfishes)

Class Actinopterygii
Order Batrachoidiformes
Number of families 1

Photo: A splendid coral toadfish (*Sanopus splendidus*) near Cozumel, Mexico. Its striking appearance sets it apart from the other toadfishes. (Photo by J. W. Mowbray/Photo Researchers, Inc. Reproduced by permission.)

Evolution and systematics

The family Batrachoididae is the only family in the order Batrachoidiformes, and is thought to be most closely related to the fishes in the order Lophiiformes containing the goosefishes, frogfishes, and deepsea anglers. It has been divided into three subfamilies: Porichthyinae (*Porichthys*, 14 species and *Aphos*, 1 species) and Thalassophryninae (*Daector*, 5 species and *Thalassophryne*, 6 species) are the most derived and are restricted to the New World; Batrachoidinae containing at least 16 genera (including *Opsanus* and *Sanopus*) and about 51 species occur worldwide.

Physical characteristics

Toadfishes are small- to medium-sized fishes with a broad, flattened head and a wide mouth that usually has barbels and/or fleshy flaps around it. The eyes are on top of the head and directed upwards. The pelvic fins are forward, in front of the pectoral fins, and have one spine and three soft rays. There are two separate dorsal fins, the first with two or three spines, and the second is long with 15 to 25 soft rays. The anal fin is somewhat shorter than the second dorsal fin. The pectoral fins are large with a broad base. The gill openings are small and are restricted to the sides of the body. Species of *Porichthys* have photophores (light organs) along their sides and ventral surface. Species in the subfamily Thalassophryninae have hollow, venomous spines in their first dorsal fin and opercles. *Bifax lacinia* has a flap with an eye spot at the end of the maxilla on each side of the mouth.

Toadfishes usually are rather drab colored, often brownish with darker saddles, bars, or spots; however, some species in the Atlantic genus *Sanopus* are brightly colored. Maximum size of species ranges from 2.2 in (56 mm) to at least 20.1 in (510 mm) standard length.

Distribution

Worldwide between about 51° N and 45° S along continents in marine and brackish waters, occasionally entering rivers. Several freshwater species in South America.

In the New World Pacific Ocean: the genus *Porichthys* occurs from southeast Alaska south to Ecuador; *Aphos* from Peru south to Chile; *Daector* from Costa Rica to Peru; and *Batrachoides* from Mexico to Peru. In the New World Atlantic Ocean: *Porichthys* occurs from Virginia south to Argentina; *Opsanus* from Massachusetts south to Belize; *Sanopus* from Yucatan, Mexico to Panama; *Triathalassothia* in Belize, Honduras, and Argentina; *Amphichthys* from Panama to Brazil; *Thalassophryne* from Panama to Uruguay. Species of *Daector*,

An oyster toadfish (*Opsanus tau*) resting on the ocean floor. It makes a "grunting" sound when caught. (Photo by Tom McHugh/Photo Researchers, Inc. Reproduced by permission.)

The oyster toadfish (*Opsanus tau*) has been used in studies of insulin and diabetes, drug metabolism, hearing, dizziness and motion sickness. (Photo by David Hall/Photo Researchers, Inc. Reproduced by permission.)

Thalassophryne, and *Potamobatrachus* occur in freshwater in South America.

In the eastern Atlantic Ocean *Batrachoides*, *Halobatrachus*, *Perulibatrachus*, and *Chatrabus* occur along the African coast. South Africa has several genera: *Chatrabus*, *Batrichthys*, and *Austrobatrachus*. *Barchatus* is found in the Red Sea and adjacent areas, *Bifax* in the Gulf of Oman as well as *Austrobatrachus*, which ranges to India. *Allenbatrachus* ranges from India through the Indo-Australian archipelago to the Philippines and north to Thailand. Toadfishes have not been recorded from Taiwan or Japan. The genus *Halophryne* occurs on the west, north, and east costs of Australia, and north to the Philippine Islands, and *Batrachomoeus* has the same range in Australia and ranges north through the Indo-Australian archipelago to Vietnam.

Habitat

Toadfishes rest on and bury in the substrata and are found from the shoreline down to deep water, at least to 1,200 ft (366 m). They occur in full-strength sea water, brackish water, and also freshwater. Their usual cryptic coloration allows them to blend with the subtrata where they can function as ambush predators. The species in the genus *Sanopus* typically live in sand depressions under coral heads.

Behavior

Toadfishes are known for their sound production resulting from the contraction of muscles on the swim bladder. Both males and females produce agonistic grunts, whereas only males make longer courtship calls, "boat whistles," or "hums."

Feeding ecology and diet

In addition to being ambush predators, toadfishes also move about feeding on invertebrates, mostly crabs, shrimps, mollusks, sea urchins, and fishes, and others take planktonic organisms from the water column.

Reproductive biology

Males prepare nests, usually in a cavity under a rock or shell, including objects discarded by humans such as cans or bottles. Males attract females by vocalizations, and then females lay large, adhesive eggs and leave the area. Males guard and fan the eggs until after hatching. The young may remain in the nest after hatching, still attached to the nest surface and even after free swimming. The plainfin midshipman has two types of males, larger nest holding ones and smaller sneaker males that dart into nests attempting to fertilize eggs of a nesting pair.

Conservation status

The IUCN lists five species of toadfishes as Vulnerable—the Cotuero toadfish (*Batrachoides manglae*), the whitespotted toadfish (*Sanopus astrifer*), the whitelined toadfish (*S. greenfieldorum*), the reticulated toadfish (*S. reticulatus*), and the splendid coral toadfish (*S. splendidus*). A number of species appear to have very limited distributions.

Significance to humans

Larger toadfish species are eaten, although there is no specific fishery for them. Species in the genus *Allenbatrachus* are occasionally collected and sold in the aquarium trade as freshwater fishes. Species in the genera *Opsanus* and *Porichthys* are used in laboratory studies. Venomous toadfishes in the subfamily Thalassophryninae can inflict pain if handled.

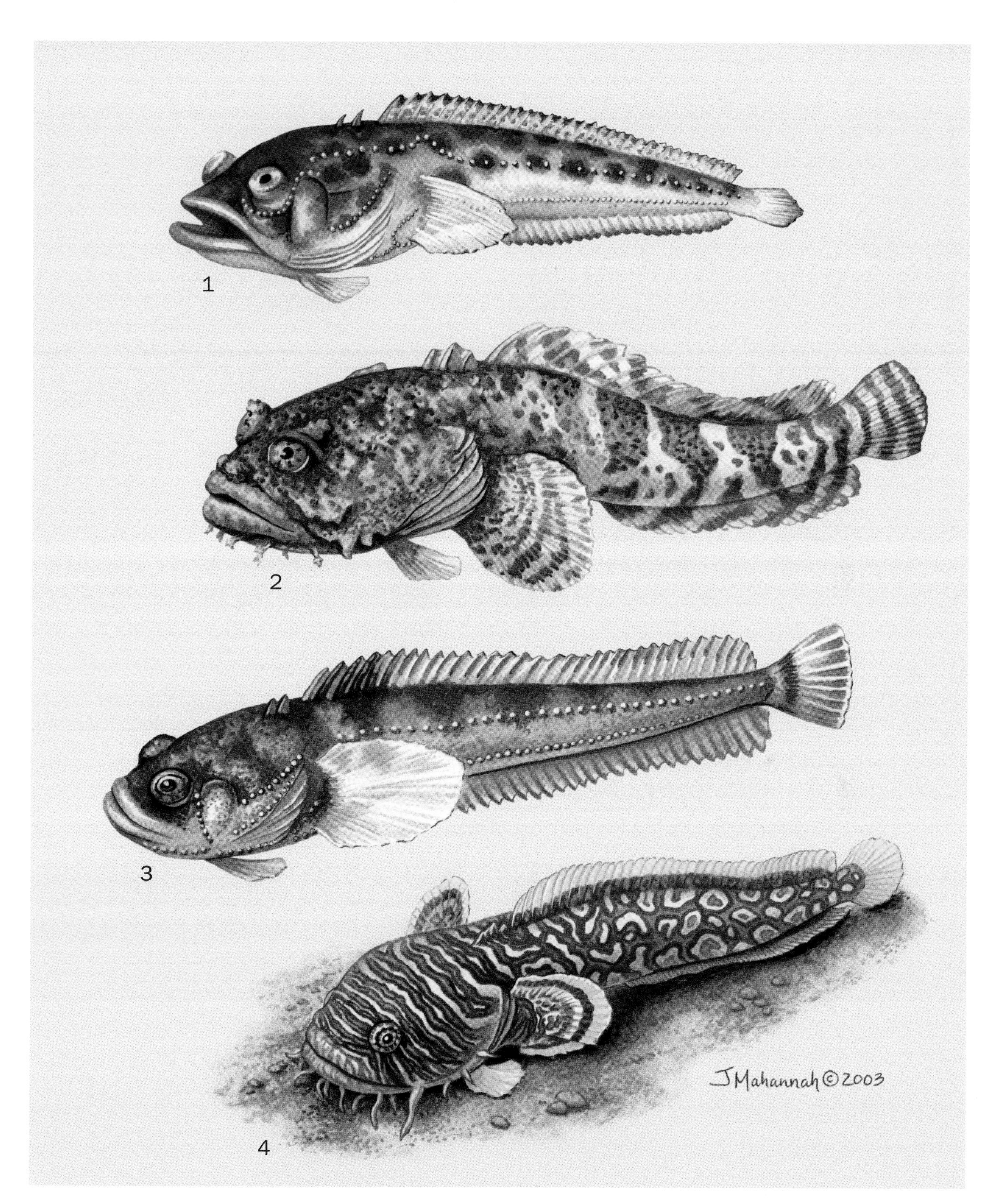

1. Atlantic midshipman (*Porichthys plectrodon*); 2. Oyster toadfish (*Opsanus tau*); 3. Plainfin midshipman (*Porichthys notatus*); 4. Splendid coral toadfish (*Sanopus splendidus*). (Illustration by Jacqueline Mahannah)

Species accounts

Oyster toadfish
Opsanus tau

FAMILY
Batrachoididae

TAXONOMY
Gadus tau Linnaeus, 1766, Carolina, United States.

OTHER COMMON NAMES
None known.

PHYSICAL CHARACTERISTICS
The maximum size is 15 in (381 mm) standard length. There are three dorsal-fin spines, a body lacking scales, and a single subopercular spine. The inner surface of the pectoral fins has discrete glands between the upper rays, and the pectoral fin with an axillary pore behind it. The second dorsal fin has 23–27 soft rays, and the anal fin has 19–23 soft rays. The tongue, gill arches, roof of mouth and inner surface of the gill covers are light, not black, and the background color of the body is dark with no spots. The pectoral fin has definite cross bars.

DISTRIBUTION
Atlantic coast of the United States, from Maine to Miami, Florida.

HABITAT
Usually found over rock, sand, or mud, and oyster shell bottoms, often most abundant in estuaries.

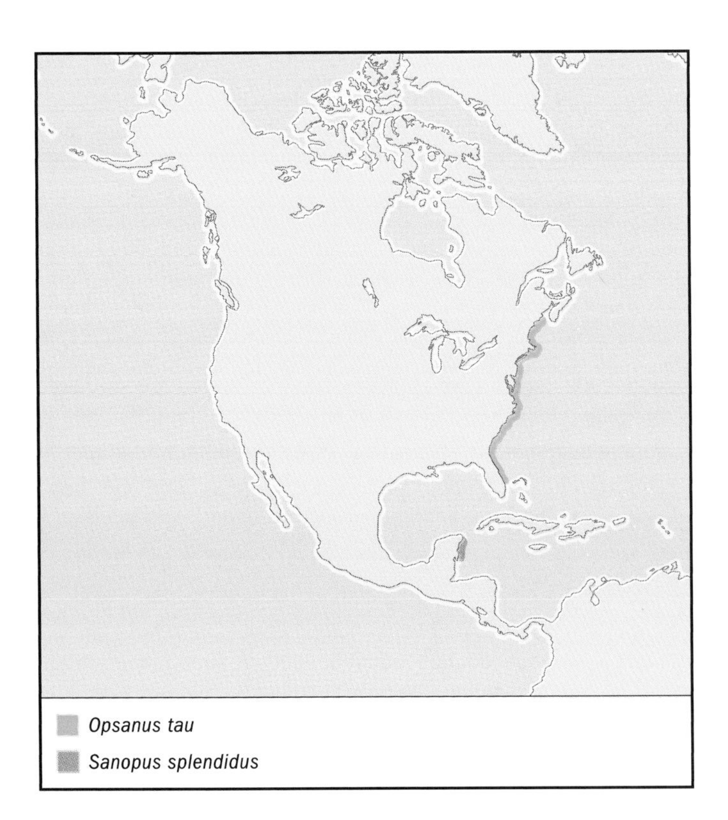

BEHAVIOR
This species makes a grunting noise by rubbing muscles across the swim bladder. Males use a boat-whistle call to attract females to nesting sites.

FEEDING ECOLOGY AND DIET
Feeds mainly on small crabs and other crustaceans.

REPRODUCTIVE BIOLOGY
Males establish nesting sites from April through October. Reproductive behavior typical of the family.

CONSERVATION STATUS
Not threatened.

SIGNIFICANCE TO HUMANS
Generally considered a nuisance when caught by fishers, but used as an experimental animal for studies involving insulin and diabetes, drug metabolism, hearing, dizziness, and motion sickness. ◆

Plainfin midshipman
Porichthys notatus

FAMILY
Batrachoididae

TAXONOMY
Porichthys notatus Girard, 1854, San Francisco, California, United States.

OTHER COMMON NAMES
English: Singing fish, canary bird fish.

PHYSICAL CHARACTERISTICS
The maximum size is 15 in (380 mm) standard length. There are two spines in the first dorsal fin, and the second dorsal and anal fins are long. There are rows of photophores on the head and body, and those on the underside of the head are V-shaped.

DISTRIBUTION
Southeast Alaska to Baja, California, Mexico.

HABITAT
Usually buried in soft bottom, mud, or sand, but can be found in rocky intertidal areas during breeding season.

BEHAVIOR
Like other family members it produces sound, making grunts and a hum that is a mating call. It has been demonstrated that the photophores can be used as a countershading mechanism to hide the midshipman from predators while it is foraging in the water column at night.

FEEDING ECOLOGY AND DIET
A nocturnal species that is partially buried during the day but at night rises into the water column to feed on planktonic organisms. It has been suggested that its photophores help illuminate food items. It also can function as an ambush predator, taking crustaceans and fishes while buried.

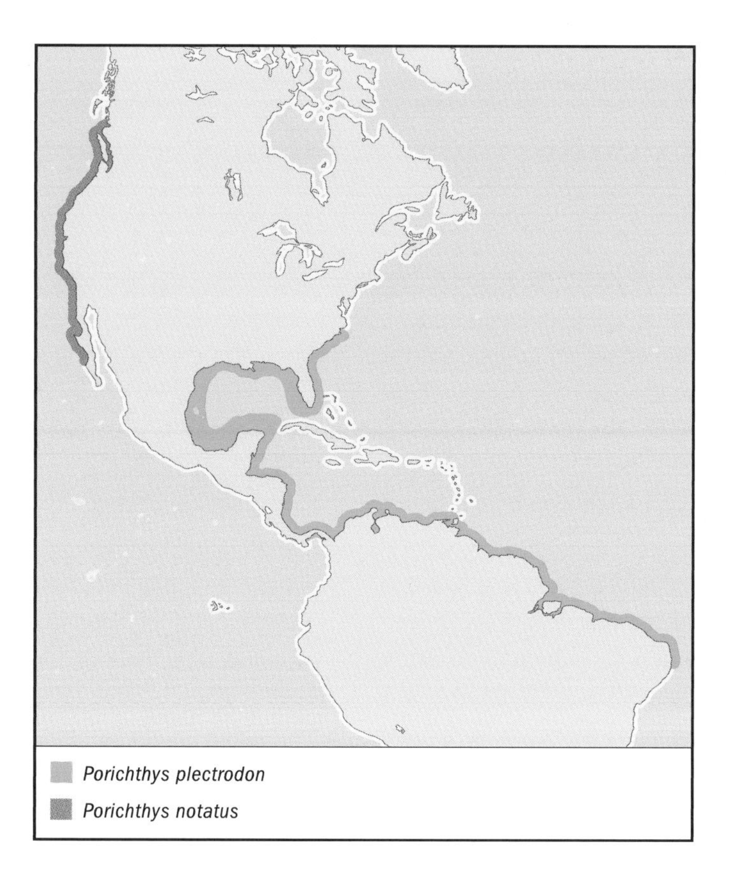

REPRODUCTIVE BIOLOGY
Males establish nests under objects and attract females by humming. Two kinds of males are present, larger nest-holding ones and smaller sneaker males that dart into nests attempting to fertilize eggs of a nesting pair. There have been reports that the photophores also are involved in mating behavior.

CONSERVATION STATUS
Not threatened.

SIGNIFICANCE TO HUMANS
The humming of breeding males bothers persons living in houseboats. It is used as a laboratory animal and is an important food item for seals and sea lions. ◆

Atlantic midshipman
Porichthys plectrodon

FAMILY
Batrachoididae

TAXONOMY
Porichthys plectrodon Jordan and Gilbert, 1882, Galveston, Texas, United States.

OTHER COMMON NAMES
None known.

PHYSICAL CHARACTERISTICS
The maximum size is 8.6 in (218 mm) standard length. There are two spines in the first dorsal fin, and the second dorsal and anal fins are long. There are rows of photophores on the head and body, and those on the underside of the head are U-shaped.

DISTRIBUTION
Cape Henry, Virginia south to northern Brazil, including the Gulf of Mexico and the Caribbean Sea.

HABITAT
Prefers mud bottoms where it buries itself during the day

BEHAVIOR
It produces both aggressive and mating vocalizations. Photophores may be used as a countershading mechanism to avoid predators while foraging in the water column at night.

FEEDING ECOLOGY AND DIET.
Although some benthic invertebrates and fishes have been found in their stomachs, most of the food consists of planktonic crustaceans and larval fishes taken at night while foraging in the water column.

REPRODUCTIVE BIOLOGY
It moves from deeper water into shallow bays and spawns in the spring and summer. The reproductive behavior is typical of the family.

CONSERVATION STATUS
Not threatened.

SIGNIFICANCE TO HUMANS
Used as an experimental animal. ◆

Splendid coral toadfish
Sanopus splendidus

FAMILY
Batrachoididae

TAXONOMY
Sanopus splendidus Collette, Starck, and Phillips, 1974, Cozumel Island, Mexico.

OTHER COMMON NAMES
None known.

PHYSICAL CHARACTERISTICS
The maximum size is 9.9 in (252 mm) standard length. There are three dorsal-fin spines; a single subopercular spine; no scales; no glands between the upper rays of the inner surface of pectoral fins; and an axillary pore present. All fins are bordered with black and orange-yellow.

DISTRIBUTION
Known only from off Cozumel Island, Quintana Roo, Mexico.

HABITAT
Usually found in sand depressions under coral heads.

BEHAVIOR
Not known.

FEEDING ECOLOGY AND DIET
An ambush predator feeding on small fishes and gastropods.

REPRODUCTIVE BIOLOGY
Not known.

CONSERVATION STATUS
Listed as Vulnerable by the IUCN. It is known only from off Cozumel Island and thus is potentially at risk.

SIGNIFICANCE TO HUMANS
Commonly photographed by scuba divers. ◆

Resources

Books

Breder, C. M., Jr., and D. E. Rosen. *Modes of Reproduction in Fishes.* Garden City, NY: Natural History Press, 1966.

Collette, B. B. "Order Batrachoidiformes, Batrachoididae, Toadfishes." In *FAO Western Central Atlantic Identification Guide to Living Marine Resources of the Western Central Atlantic*, edited by K. E. Carpenter. Rome: FAO, in press.

Collette, B. B., and G. Klein-MacPhee, eds. *Bigelow and Schroeder's Fishes of the Gulf of Maine.* 3rd edition. Washington, DC: Smithsonian Institution Press, 2002.

Eschmeyer, W. N., E. S. Herald, and H. Hammann. *A Field Guide to Pacific Coast Fishes of North America.* Boston: Houghton Mifflin Company, 1983.

Greenfield, D. W. "Batrachoididae." In *FAO Species Identification Guide for Fishery Purposes. The Living Marine Resources of the Western Central Pacific*, edited by K. E. Carpenter and V. H. Niem. Rome: FAO, 1999.

Hart, J. L. *Pacific Fishes of Canada.* Ottawa: Fisheries Research Board of Canada, Bulletin 180, 1973.

Hoese, H. D., and R. H. Moore. *Fishes of the Gulf of Mexico, Texas, Louisiana, and Adjacent Waters.* 2nd edition. College Station, TX: Texas A & M University Press, 1998.

Murdy, E. O., R. S. Birdsong, and J. A. Musick. *Fishes of Chesapeake Bay.* Washington, DC: Smithsonian Institution Press, 1997.

Potts, D. T., and J. S. Ramsey. *A Preliminary Guide to Demersal Fishes of the Gulf of Mexico Continental Slope (100 to 600 fathoms).* Mobile, AL: Alabama Sea Grant Extension, 1987.

Robins, C. R., and G. C. Ray. *A Field Guide to Atlantic Coast Fishes of North America.* Boston: Houghton Mifflin Company, 1986.

Periodicals

Bass, A. H. "Shaping Brain Sexuality." *American Scientist* 84 (1996): 352–363.

Collette, B. B. "A Review of the Venomous Toadfishes, Subfamily Thalassophryninae." *Copeia* 1966, no. 4 (1966): 846–864.

———. "A Review of the Coral Toadfishes of the Genus *Sanopus* with Descriptions of Two New Species from Cozumel Island, Mexico." *Proceedings of the Biological Society of Washington* 87, no. 18 (1974): 185–204.

———. "Two New Species of Coral Toadfishes, Family Batrachoididae, Genus *Sanopus*, from Yucatan, Mexico, and Belize." *Proceeding of the Biological Society of Washington* 96, no. 4 (1983): 719–724.

———. "*Potamobatrachus trispinosus*, a New Freshwater Toadfish (Batrachoididae) from the Rio Tocantins, Brazil." *Ichthyological Explorations of Freshwaters* 6, no. 4 (1995): 333–336.

Collette, B.B., and J. L. Russo. "A Revision of the Scaly Toadfishes, Genus *Batrachoides*, with Descriptions of Two New Species from the Eastern Pacific." *Bulletin of Marine Science* 31, no. 2 (1981): 197–233.

Crane, J. M., Jr. "Bioluminescent Courtship Display in the Teleost *Porichthys notatus.*" *Copeia* 1965, no 2 (1965): 239–241.

Gilbert, Carter R. "Western Atlantic Batrachoidid Fishes of the Genus *Porichthys*, Including Three New Species." *Bulletin of Marine Science* 18, no. 3 (1968): 671–730.

Greenfield, D. W. "*Perulibatrachus kilburni*, a New Toadfish from East Africa (Teleostei: Batrachoididae)." *Copeia* 1996, no. 4 (1996): 901–904.

———. "*Allenbatrachus*, a New Genus of Indo-Pacific Toadfish (Batrachoididae)." *Pacific Science* 51, no. 3 (1997): 306–313.

———. "*Halophryne hutchinsi*: a New Toadfish (Batrachoididae) from the Philippine Islands and Pulau Waigeo, Indonesia." *Copeia* 1998, no. 3 (1998): 696–701.

Greenfield, D. W., and T. Greenfield. "*Triathalassothia gloverensis*, a New Species of Toadfish from Belize (=British Honduras) with Remarks on the Genus." *Copeia* 1973, no. 3 (1973): 560–565.

Greenfield, D. W., J. K. L. Mee, and J. E. Randall. "*Bifax lacinia*, a New Genus and Species of Toadfish (Batrachoididae) from the South Coast of Oman." *Fauna Saudia Arabia* 14 (1994): 276–281.

Hutchins, J. B. "A Revision of the Australian Frogfishes (Batrachoididae)." *Records of the Western Australian Museum* 4, no. 1 (1976): 3–43.

Lane, E. D. "A Study of the Atlantic Midshipmen, *Porichthys porosissimus*, in the Vicinity of Port Arkansas, Texas." *Contributions in Marine Science* 12 (1967): 1–53.

Roux, Ch. "Révision des Poissons Marins de la Famille des Batrachoididae de la Côte Occidentale Africaine." *Bulletin du Muséum National D'Historie Naturelle 2nd series* 42, no. 4 (1971): 626–643.

Smith, J. L. B. "The Fishes of the Family Batrachoididae from South and East Africa." *Annals and Magazine of Natural History 12th series* no. 52 (1952): 313–339.

Walker, H. J., and R. H. Rosenblatt. "Pacific Toadfishes of the Genus *Porichthys* (Batrachoididae) with Descriptions of Three New Species." *Copeia* 1988, no. 4 (1988): 887–904.

David W. Greenfield, PhD

Lophiiformes
(Anglerfishes)

Class Actinopterygii
Order Lophiiformes
Number of families 18

Photo: A frogfish (*Antennarius* sp.) eating a fish. (Photo by Tom McHugh/Steinhart Aquarium/Photo Researchers, Inc. Reproduced by permission.)

Evolution and systematics

The anglerfishes (order Lophiiformes) are a remarkable assemblage of approximately 65 genera and nearly 300 species of curious and wonderfully bizarre fishes. Lophiiformes is one of six orders that make up the euteleostean superorder Paracanthopterygii. Of the six paracanthopterygian orders, the toadfishes (order Batrachoidiformes) are considered to be most closely related to the anglerfishes. The fossil record extends back to the Eocene, with representatives from the families Antennariidae and Lophiidae having been found in Eocene formations of Italy.

Pietsch and Grobecker recognized 18 lophiiform families and divided them among five suborders. In phylogenetic order and from most primitive to most dervied, they are: Lophioidei (containing only one family, Lophiidae, the monkfishes), Antennarioidei (four families, the frogfishes), Chaunacoidei (one family, Chaunacidae, the sea toads or coffinfishes), Ogcocephaloidei (one family, the batfishes), and Ceratioidei (11 families, the bathypelagic anglerfishes).

Physical characteristics

The anglerfishes share two very conspicuous derived characters (synapomorphies). (1) The anterior portion of the spinous dorsal fin has migrated forward to a position on the anterior surface of the head, just behind the snout. The first dorsal fin spine has been modified further to serve as an angling apparatus (illicium), by loosing its membranous connection to the other cephalic dorsal fin spines and acquiring a distal bait or lure (esca) to help attract prey. (2) The bases of the pectoral fins (radials) have become greatly elongated so that the pectoral fins appear to be at the end of long, jointed arms. The number of pectoral radials has been reduced to between two and five. Additional lophiiform synapomorphies are gill openings that are reduced in size and, in all but the most primitive lophioids, located posterior to the pectoral fin base; five or six branchiostegal rays; absent ribs; and first vertebra fused to the neurocranium. Anglerfishes are very diverse in both color and size; this order ranges from a few inches to approximately 6.6 ft (2 m) in length.

Distribution

Anglerfishes are worldwide in distribution. With the exception of one antennariid (frogfish) species that has been known to enter freshwater, they are exclusively marine, and most species occur in deep water.

The batfish (*Ogcocephalus* sp.) "walks" along the bottom on limb-like fins. A "lure" is located below the base of the rostrum. (Photo by David M. Schleser/Nature's Images, Inc/Photo Researchers, Inc. Reproduced by permission.)

A longlure frogfish (*Antennarius multiocellatus*) extending its lure to attract a fish. (Photo by Andrew J. Martinez/Photo Researchers, Inc. Reproduced by permission.)

Habitat

Most anglerfishes occur in deep oceanic waters. The ceratioid anglerfishes are midwater fishes, inhabiting primarily the mesopelagic and bathypelagic zones. Most other anglerfishes are benthic and occur in the waters of the outer continental shelf and continental slope. A few lophioids enter the shallows at high latitudes, and many antennarioids occur in coral reef or other shallow habitats in tropical and warm temperate waters.

Behavior

Most knowledge of anglerfish behavior is associated with the specialized adaptations that have made them such effective and successful ambush predators. Most important among them are the angling apparatus and the functional morphological features of the feeding mechanism. These adaptations, along with certain reproductive adaptations, have made the anglerfish among the most successful predators in the deep sea.

The angling apparatus (illicium) is derived from the anteriormost spine in the dorsal fin. The illicium and second dorsal fin spine (if present) articulate with a single, enlarged pterygiophore. The illicium has a rather elaborate set of associated muscles, which allows the angler to move the spine rapidly, thus thrashing the bait (esca) vigorously above or in front of the mouth. In some anglerfishes the bait may be a simple bulb or clublike structure, but in others it can be quite

An angler (*Lophius piscatorius*) resting on the sandy bed of the Atlantic Ocean near Brittany, France. (Photo by Jeff Rotman/Photo Researchers, Inc. Reproduced by permission.)

An anglerfish "angling" with lure for catching prey. (Illustration by Joseph E. Trumpey)

elaborate. In many deep-sea species, the esca is bioluminescent, while in forms that live in sunlit regions, it may be an elaborate fleshy structure, occasionally resembling a small, shrimplike crustacean in some lophiid species or even a small fish in one species of antennariid.

The mouth in most anglerfishes is cavernous, and in many species it is lined with long, sharp, thin, depressible teeth. The functional morphological features associated with the anglerfish feeding apparatus are quite fascinating. Like many derived teleostean fishes, anglerfishes are "gape-and-suck" feeders. Many anglerfishes, however, have a greatly elongated and highly mobile suspensorium, or hyopalatine arch (the portion of the skull that supports the lower jaw). This feature serves to permit swallowing of very large prey and allows the volume of the buccal cavity to increase to a much greater extent than in other fishes, which results in much greater negative pressure or suction when the mouth is opened rapidly.

Not only can anglerfishes create much greater negative pressure when opening the mouth and expanding the buccal cavity, but they appear to be capable of doing so much more rapidly than other fishes. Pietsch and Grobecker used high-speed cinematography (800 and 1,000 frames/second) to study the functional morphological characteristics of the feeding apparatus in antennariid anglerfishes. They showed that maximum oral expansion and subsequent prey engulfment may occur in as little as four milliseconds, and the average speed for three species of *Antennarius* was seven milliseconds. For comparison, this takes 40 milliseconds in the European perch (*Perca fluviatilis*), 16 milliseconds in the freshwater butterflyfish (*Pantodon buchholzi*), and 15 milliseconds in the stonefish (*Synanceia verrucosa*).

During typical ambush feeding, the anglerfish remains motionless (either on the bottom or in the water column, depending on the family) until suitable prey is detected. Many benthic anglerfishes (e.g., antennariids and lophiids) are capable of remarkable cryptic coloration, whereas the midwater forms are usually very dark brown or black (cryptic in a pelagic environment with little or no light). Upon sensing prey, the angler brings the illicium into play, attempting to attract the prey to within reach (aggressive mimicry). When

A sargassumfish (*Histrio histrio*) floating in a sargassum mat at the ocean surface. (Illustration by Joseph E. Trumpey)

the prey reaches the strike zone, the mouth of the angler opens rapidly, and the buccal cavity expands greatly, creating a strong suction that draws in the prey.

Feeding ecology and diet

Anglerfishes feed primarily upon nekton, especially fishes and other nektonic organisms that probably are attracted by angling apparatus. Batfishes are one of the few groups of anglerfishes that feed primarily on benthic invertebrates. There is evidence suggesting that various kinds of anglerfishes—including large species—are consumed by larger predatory fishes such as sharks.

Reproductive biology

Very little is known about the reproductive biology and early life history of most anglerfishes, but it would appear that the early larval stages of all anglerfishes are pelagic, and thus meroplanktonic. One adaptation that does appear to be common to all anglerfishes is the acquisition of sexual dimorphism in the size and structure of the olfactory organs. This appears to be associated with the production of species-specific pheromones by the females, which attract the males during the breeding season. What little else is known about the reproductive biology of these fishes is restricted primarily to the families Lophiidae, Antennariidae, and a few ceratioid families; within these families, knowledge is likewise restricted to very few species.

A warty frogfish (*Antennarius maculatus*) near a brown sponge and the lava sand of the Lembeh Strait, near Indonesia. (Photo by Fred McConnaughey/Photo Researchers, Inc. Reproduced by permission.)

Early life-history stages have been described for a few of the 25 species of lophiid anglerfishes (*Lophius americanus*, *L. budegassa*, and *L. piscatorius*). Larvae of only four of the 41 antennariid species have been described, but only *Antennarius striatus* and *Histrio histrio* are reasonably well known.

In the case of all anglerfishes from which eggs have been identified, eggs are released from the female embedded in a long, bouyant, ribbonlike mucoid veil. This is known to reach a length of 39.4 ft (12 m) and a width of 5 ft (1.5 m) and has been estimated to contain more than 1.3 million eggs in one large lophiid anglerfish species (*L. americanus*). The pigmentation of the embryos imparts a dark hue to the veil, and they are so conspicuous as they float at the surface of the sea that they have long been known to commercial fishermen. In anglerfishes in which the larvae have been identified, the larvae pass through a planktonic stage before becoming either benthic or nektonic (in the case of the midwater ceratioid anglerfishes).

Conservation status

Only one of the anglerfish families, Lophiidae, contains species of economic importance. These are harvested commercially in U.S., European, and Japanese waters, and the species that occurs off the Atlantic Coast of the United States, *L. americanus*, has received protection from the U. S. National Marine Fisheries Service owing to overfishing both for the meat, which is marketed in the United States as monkfish, and the liver, which is highly prized in Japan. The IUCN lists the spotted handfish (*Brachionichthys hirsutus*) of Tasmania as Critically Endangered. The population of this species has been greatly reduced by an introduced starfish that preys on its egg clusters.

Significance to humans

Other than the lophiid species that are harvested commercially, lophiiform fishes have little importance to humans.

A painted frogfish (*Antennarius pictus*) near the Russell Islands. (Photo by Fred McConnaughey/Photo Researchers, Inc. Reproduced by permission.)

1. Male seadevil (*Cryptopsaras couesii*); 2. Sargassumfish (*Histrio histrio*); 3. Big-eye frogfish (*Antennarius radiosus*); 4. Longnose batfish (*Ogcocephalus corniger*); 5. Monkfish (*Lophius americanus*). (Illustration by Joseph E. Trumpey)

Species accounts

Big-eye frogfish

Antennarius radiosus

FAMILY
Antennariidae

TAXONOMY
Antennarius radiosus Garman, 1896.

OTHER COMMON NAMES
English: Singlespot frogfish.

PHYSICAL CHARACTERISTICS
The big-eye frogfish has skin with closely set, forked dermal spinules. The illicium is about as long as the second dorsal fin spine and has no dermal spinules. There is a single dark spot (ocellus) surrounded by a lightly colored ring below the posterior portion of the soft dorsal fin.

DISTRIBUTION
The big-eye frogfish occurs in the western Atlantic Ocean from Long Island, New York, to Florida and across the northern Gulf of Mexico. It has been reported from the eastern Atlantic, but it is apparently very rare there. It has been taken at depths ranging from 20 to 900 ft (6–275 m).

HABITAT
Like other members of its genus, the big-eye frogfish is benthic. This species is primarily an inhabitant of mud bottoms. It is commonly taken by shrimp trawlers in the northern Gulf of Mexico, and it is frequently found in deeper portions of higher-salinity estuaries.

BEHAVIOR
Unknown.

FEEDING ECOLOGY AND DIET
Like other frogfishes and most other anglerfishes, the big-eye frogfish is a voracious ambush predator and will take any prey it can capture.

REPRODUCTIVE BIOLOGY
Unknown.

CONSERVATION STATUS
Not threatened. A significant death rate may result from those that are taken as bycatch by shrimp trawlers.

SIGNIFICANCE TO HUMANS
None known. ◆

Sargassumfish

Histrio histrio

FAMILY
Antennariidae

TAXONOMY
Histrio histrio (Linnaeus, 1758).

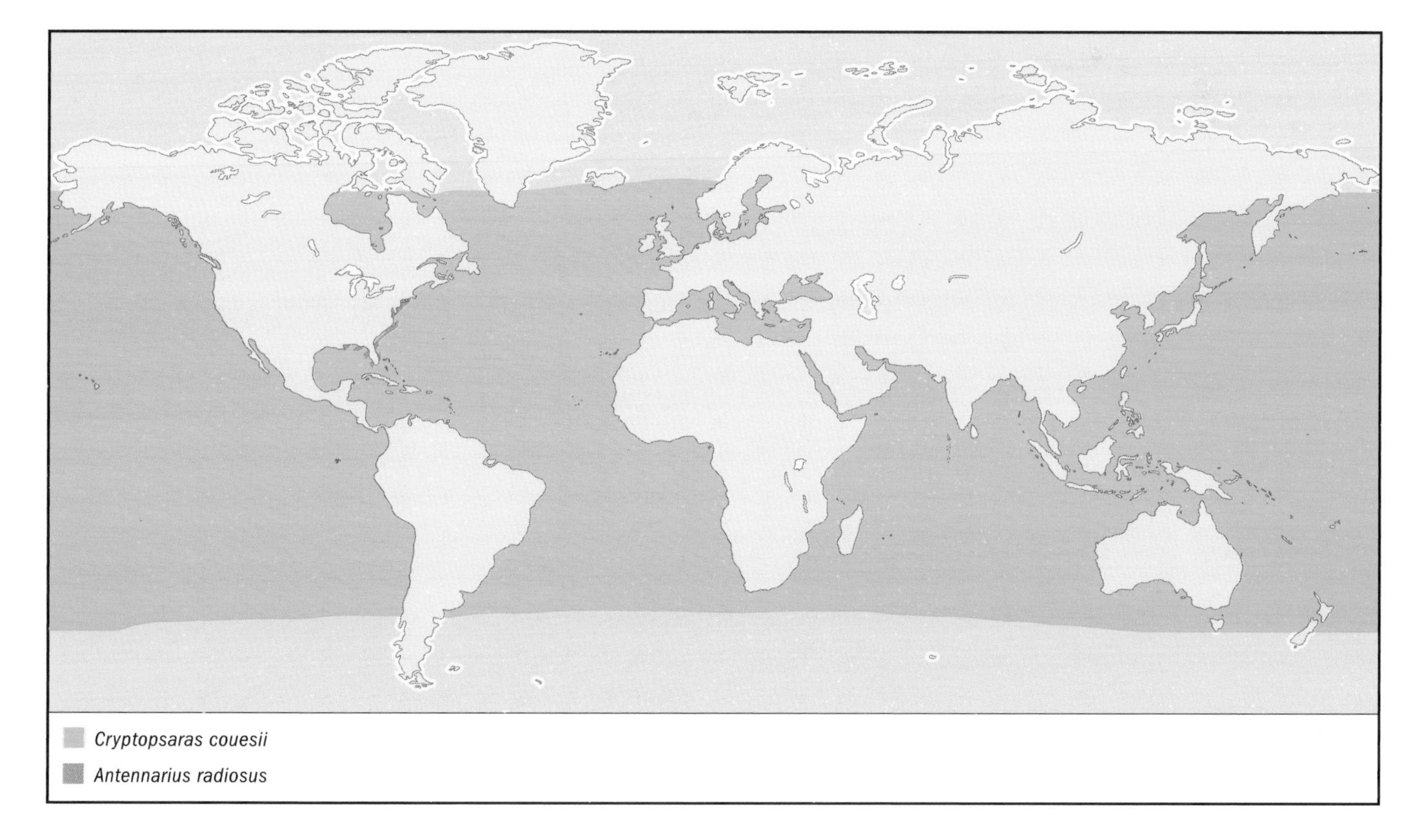

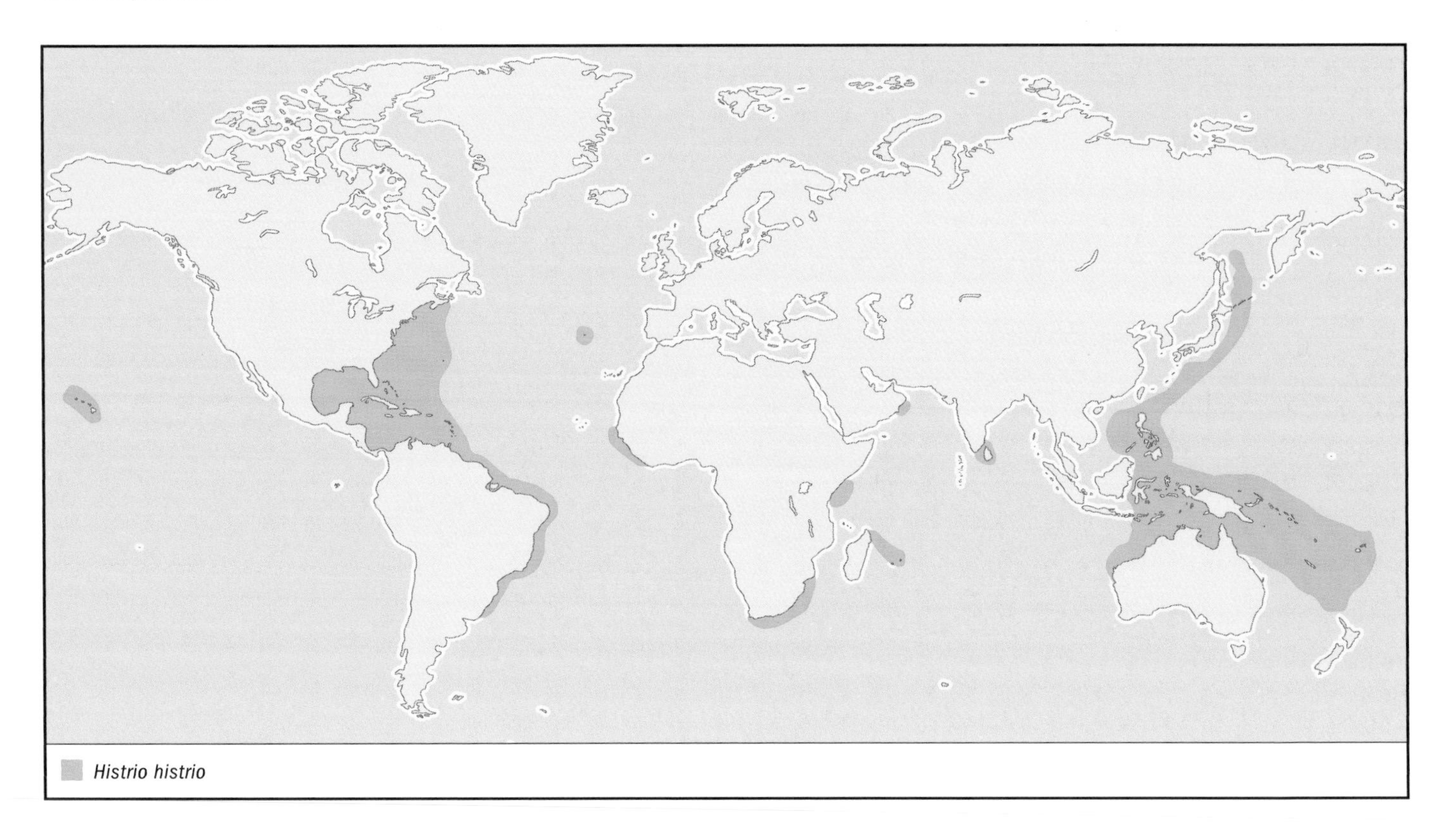

OTHER COMMON NAMES
English: Sea mouse, mousefish.

PHYSICAL CHARACTERISTICS
Frogfishes (family Antennariidae) are somewhat similar to their monkfish relatives in that they have a large head on which are located several isolated dorsal fin spines; the pectoral fins appear armlike. They are strikingly different in that the head is globose and the body is flattened laterally (compressed) rather than dorsally (depressed). In addition, the second and third dorsal fin spines usually are surrounded by so much fleshy skin that it may be difficult to recognize them as fin spines. The second, or soft, dorsal and anal fins are much more conspicuous, and the soft dorsal is usually much longer than the anal fin. The sargassumfish is characterized by its very short angling apparatus and its smooth skin, which lacks the small, spiny scales or dermal spinules typical of most other frogfishes. Its color pattern is changeable and highly variable, but it is usually cryptic, consisting of streaks, spots, and mottling of brown, olive, and yellow, making it nearly impossible to detect among the sargassum weed in which it hides.

DISTRIBUTION
The sargassumfish has the widest geographic distribution of any frogfish. Being an obligate associate of sargassum weed, they are found wherever this species occurs, that is, in tropical, warm temperate, and occasionally cool temperate waters of the Atlantic, Pacific, and Indian Oceans. They appear to be very rare in the eastern Atlantic, and they are absent from the eastern Pacific.

HABITAT
The sargassumfish is the only frogfish species that is not benthic. It is only found associated with the epipelagic brown algae, *Sargassum*, in tropical and warm temperate environments. Although *Sargassum* is usually pelagic, members of the community associated with it, including the sargassumfish, are technically pseudopelagic.

BEHAVIOR
Most of what is known about the behavior of sargassumfish is associated with their feeding. This behavior is described in the introductory section.

FEEDING ECOLOGY AND DIET
Like most other anglerfishes, sargassumfish are ambush predators that use aggressive mimicry to attract prey. Most observers describe them as particularly voracious, feeding readily and indiscriminately on anything they can swallow, including fishes as long as or longer than themselves. They even consume other sargassumfish. They are one of the major predators of the sargassum community.

REPRODUCTIVE BIOLOGY
Reproduction in the sargassumfish is similar to the known reproductive behavior of other anglerfishes. The eggs are released in the typical egg veil or raft, which in sargassumfish may be 35–47 in (90–120 cm) long and 2–4 in (5–10 cm) wide.

CONSERVATION STATUS
Not threatened.

SIGNIFICANCE TO HUMANS
None known. ◆

Seadevil
Cryptopsaras couesii

FAMILY
Ceratiidae

TAXONOMY
Cryptosaras couesii Gill, 1883.

OTHER COMMON NAMES
None known.

PHYSICAL CHARACTERISTICS
Like other midwater anglerfishes, the seadevil is characterized by the absence of pelvic fins, pseudobranchs, and scales or other dermal derivatives. Moderate to extreme degrees of sexual dimorphism characterize members of this suborder. Ceratiids exhibit extreme sexual dimorphism, with the adult males being reduced to ectoparasites of the females. As with most species, the lure (esca) appears to be bioluminescent in the seadevil. Female seadevils have a vertical to strongly oblique mouth, and the first two or three rays of the soft dorsal fin are modified as caruncles (low, fleshy appendages).

DISTRIBUTION
Their distribution is worldwide between 63° N and 43° S latitudes.

HABITAT
The species inhabits the lower mesopelagic and bathypelagic zones, between 246 and 13,123 ft (75–4,000 m), but is most commonly found between 1,640 and 4,101 ft (500–1,250 m).

BEHAVIOR
Little is known besides feeding ecology and reproductive biology.

FEEDING ECOLOGY AND DIET
Ceratioid anglerfishes, like the seadevil, are one of the most important groups of predatory fishes in the lower mesopelagic and bathypelagic zones of the world oceans. Feeding is by aggressive mimicry using the angling apparatus. They feed upon zooplankton and nekton, with fishes and cephalopods probably forming a major component of the diet.

REPRODUCTIVE BIOLOGY
The reproductive adaptations found among the ceratioid anglerfishes are among the most fascinating in the animal kingdom. As mentioned earlier, all lophiiform anglerfishes appear to have some degree of sexual dimorphism in the structure and size of the olfactory organs. It is assumed that this is associated with the females' production of species-specific pheromones, which attract the males when the female is ready to breed. In ceratioid anglerfishes this sexual dimorphism has reached its extreme, not only among the vertebrates but also perhaps the entire animal kingdom. In most of the 11 ceratioid families, the female is considerably larger than the male (females range from three to 13 times the length of conspecific males). Furthermore, in at least four ceratioid families the males have been so drastically reduced that they have become essentially a pair of swimming testes directed by a huge pair of olfactory organs. Once these "minimized" males find a female, they attach themselves to her with their jaws, eventually achieving histological fusion and leading the remainder of their lives as essentially a pair of "ectoparasitic" testes.

CONSERVATION STATUS
Not threatened.

SIGNIFICANCE TO HUMANS
None known. ◆

Monkfish
Lophius americanus

FAMILY
Lophiidae

TAXONOMY
Lophius americanus Curvier and Valenciennes in Valenciennes, 1837, North America.

OTHER COMMON NAMES
English: Goosefish, anglerfish.

PHYSICAL CHARACTERISTICS
Like most lophiid anglerfishes, the monkfish is distinguished easily from other anglerfishes by its very large, wide, flattened head. Lophiids also have an enormous mouth armed with long, sharp, conical teeth, and the angling apparatus and second dorsal fin spine arise very close to each other at the anterior end of the snout. The monkfish has a number of derived character states, among which are increased numbers of dorsal and anal fin rays and vertebrae, gill openings restricted to a location completely behind the pectoral fin base, and a variety of cranial osteological features. The monkfish is distinguished from its congeners by several morphological characters.

DISTRIBUTION
The monkfish is known to occur from the Grand Banks off Newfoundland south to the east coast of Florida (about 29° N latitude) and from just below the tide line (only at high latitudes) to depths of more than 2,625 ft (800 m). However, few large individuals are taken below 1,312 ft (400 m). This is the only member of the genus found in the western Atlantic north of Cape Hatteras.

HABITAT
Adult monkfish have been found on soft and hard substrates including soft mud, soft sand, hard sand, pebbles, gravel, and broken shells.

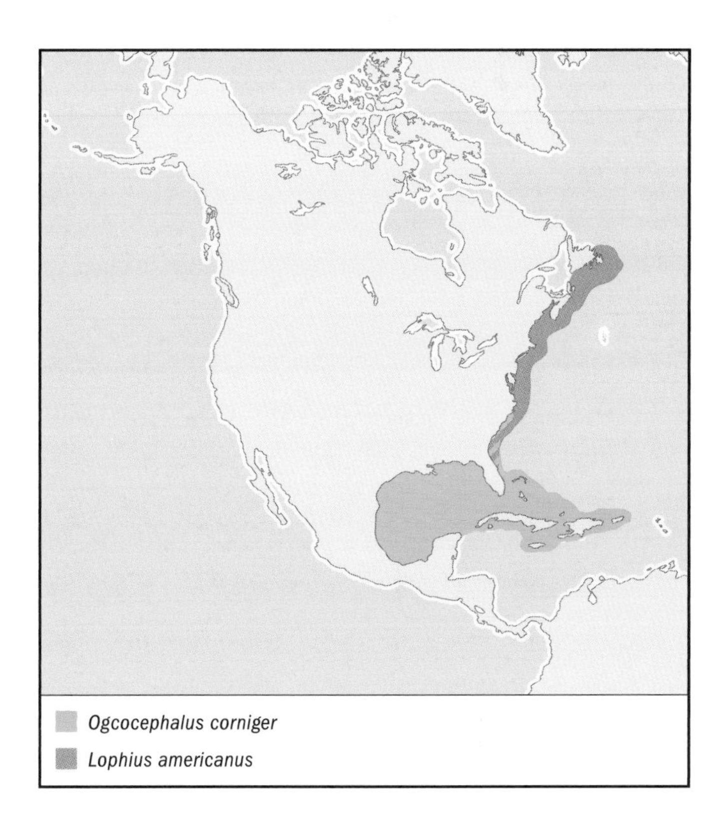

BEHAVIOR

Most of what is known about lophiid behavior comes from observations of the monkfish and another species, *Lophius piscatorius*; general aspects of this behavior are described earlier, in the ordinal account. Aristotle made the very first observations of angling behavior. He described them as having hairlike filaments hung before their eyes, with knobs attached like bait to the end of each filament. When little fishes come in range of the filaments and strike at them, they are led down with the filaments into the monkfish's mouth. Such observations were not recorded again until 1925, when Bigelow and Welsh reported the observations of W. F. Clapp, who observed monkfish catching young tomcod (*Microgadus tomcod*) in the eelgrass beds of Duxbury Bay, Massachusetts.

FEEDING ECOLOGY AND DIET

Like all lophiiform anglerfishes, monkfishes are voracious ambush predators. Their diet consists primarily of fishes, and when given a choice, they prefer soft-rayed fishes to spiny-rayed species. When the opportunity presents itself, however, they take any prey large enough to engulf. This includes fishes nearly as long as itself. The alternate name goosefish alludes to the fact that the monkfish has been known to engulf marine birds when it enters shallow coastal waters at high latitudes during the winter months. (Cormorants, herring gulls, widgeons, scoters, loons, guillemots, razor-billed auks, grebes, and other diving fowl, such as scaup ducks and mergansers, have all been recorded as stomach contents.) Smaller monkfishes are known to feed on a wide variety of invertebrates, including small lobsters, crabs, squid, and polychaetes.

REPRODUCTIVE BIOLOGY

As mentioned earlier, the monkfish is one of the few lophiid species for which detailed life history information is available. Long, mucoid egg veils are characteristic of this species.

CONSERVATION STATUS

During the latter few decades of the twentieth century, the popularity of monkfish increased steadily in U. S. markets. Because the liver also is highly sought by the Japanese market, fishing pressure on this species has resulted in overexploitation. Fortunately, this species has received some protection from the U. S. National Marine Fisheries Service. This species is not listed by the IUCN.

SIGNIFICANCE TO HUMANS

As *Lophius americanus* has gained popularity with consumers in the United States and there is strong demand for it in the Japanese market, this species has tremendous economic importance. ◆

Longnose batfish

Ogcocephalus corniger

FAMILY

Ogcocephalidae

TAXONOMY

Ogocephalus corniger Bradbury, 1980.

OTHER COMMON NAMES

None known.

PHYSICAL CHARACTERISTICS

Most batfishes are dorsoventrally flattened. They have only a single, very short dorsal fin spine (the illicium), although the vestige of the second spine may be present but embedded. The illicium usually is contained within a deeply concave illicial cavity, and it typically extends forward and downward rather than forward and upward, as in most other anglerfishes. The skin is covered with prominent, tubercle-like scales, which may be very large and conical (bucklers) in some genera. The scales fuse at the anterior end of the snout, forming a moderate to greatly elongated rostrum. The mouth is relatively small compared with other anglerfishes and has small teeth. The longnose batfish has a long, slender rostrum at the end of a triangular head and dark bands of pigment on the distal portions of the pectoral and caudal fins. The overall body color is dark brown or gray, with small, pale spots uniformly distributed on the dorsal surface.

DISTRIBUTION

The longnose batfish inhabits the western Atlantic from Cape Lookout, North Carolina, south to the Yucatan Peninsula, including the Gulf of Mexico and the Caribbean. It has been taken at depths ranging from 95 ft (29 m) to 755 ft (230 m).

HABITAT

All batfishes are benthic, occurring on a wide variety of bottom types.

BEHAVIOR

Unknown.

FEEDING ECOLOGY AND DIET

Batfishes feed using aggressive mimicry, as do most anglerfishes; however, they are unusual among anglerfishes in that a greater proportion of their diet seems to be small, benthic invertebrates.

REPRODUCTIVE BIOLOGY

Unknown.

CONSERVATION STATUS

Not threatened.

SIGNIFICANCE TO HUMANS

None known. ◆

Resources

Books

Collette, Bruce B., and Grace Klein-MacPhee, eds. *Bigelow and Schroeder's Fishes of the Gulf of Maine.* 3rd edition. Washington, DC: Smithsonian Institution Press, 2002.

McEachran, J. D., and J. D. Fechhelm. *Fishes of the Gulf of Mexico.* Vol. 1. Austin: University of Texas Press, 1998.

Nelson, Joseph S. *Fishes of the World.* 3rd edition. New York: John Wiley & Sons, 1994.

Paxton, J. R., and W. Eschmeyer, eds. *Encyclopedia of Fishes.* San Diego: Academic Press, 1995.

Pietsch, T. W., and D. B. Grobecker. *Frogfishes of the World: Systematics, Zoogeography, and Behavioral Ecology.* Stanford: Stanford University Press, 1987.

Robins, C. Richard, and G. Carleton Ray. *A Field Guide to Atlantic Coast Fishes: North America.* Boston: Houghton Mifflin Co., 1986.

Resources

Periodicals

Bigelow, H. B., and W. W. Welsh. "Fishes of the Gulf of Maine." *Bulletin of the U. S. Bureau of Fisheries* 40, no. 1 (1925): 1–567.

Caruso, J. H. "The Systematics and Distribution of the Lophiid Anglerfishes. I. A Revision of the Genus *Lophiodes* with the Description of Two New Species." *Copeia* 1981, no. 3 (1981): 522–549.

———. "The Systematics and Distribution of the Lophiid Anglerfishes. II. Revisions of the Genera *Lophiomus* and *Lophius*." *Copeia* 1983, no. 1 (1983): 11–30.

———. "The Systematics and Distribution of the Lophiid Anglerfishes. III. Intergeneric Relationships." *Copeia* 1985, no. 4 (1985): 870–875.

Hoffman, H. A., and D. S. Jordan. "A Catalog of the Fishes of Greece, with Notes on the Names Now in Use and Those Employed by Classical Authors." *Proceedings of the Academy of Natural Sciences of Philadelphia* (1892): 230–285.

Pietsch, T. W., and D. B. Grobecker. "The Compleat Angler: Aggressive Mimicry in an Antennariid Anglerfish." *Science* 201, no. 4353 (1978): 369–370.

John H. Caruso, PhD

Mugiliformes
(Mullets)

Class Actinopterygii
Order Mugiliformes
Number of families 1

Photo: The white mullet (*Mugil cerema*) is native to the East Coast of the United States, from Massachusetts to Texas. (Photo by Tom McHugh/Shedd Aquarium, Chicago/Photo Researchers, Inc. Reproduced by permission.)

Evolution and systematics

A revision in 1997 recognized 14 genera and 62 species of mullets as valid. Their relationships to other spiny-rayed teleosts are unclear. Skeletal features suggest affinities either to perciforms or to atherinomorphs. Morphological analyses have indicated that mugilids are part of the group Smegmamorpha, with the Synbranchiformes (swamp and spiny eels), Elassomatidae (pygmy sunfishes), Gasterosteiformes (sticklebacks, pipefishes, and relatives), and Atherinomorpha (silversides, livebearers, and relatives). Analyses of complete mitochondrial DNA sequences indicate that mugilids are related most closely to atherinomorphs and that the Synbranchiformes form the next most closely related group to the mugilids and atherinomorphs.

According to anatomical studies of the pharyngobranchial region of the head, *Agonostomus* is the most primitive of the mugilid genera. *Joturus*, *Cestraeus*, and *Aldrichetta* are, respectively, the next most derived lineages. Evolutionary relationships between the remaining, higher mullets, such as *Mugil*, *Myxus*, *Liza*, and *Valamugil* have not been resolved in published work. The oldest known mugilid fossils are skeletal remains of *Mugil princeps*, collected from 30- to 40-million-year-old Menilite beds of Poland and the Ukraine.

Physical characteristics

Most species commonly reach about 7.9 in (20 cm) in total length, but some (e.g., *Mugil cephalus*) may attain 31.5–39.4 in (80–100 cm). The head is broad and flattened dorsally in most species. The snout is short, and the mouth is small. The gill arches of many species are specialized, forming a characteristic pharyngobranchial organ that has an expanded, denticulate pad used for filtration of ingested material. In many (but not all) species of mullets, the teeth are positioned on the lips; this is unlike most species of fishes, in which teeth, if present, are attached directly to the jawbones. In most species of mullets, the teeth are very small or may even be absent.

The eyes may be partially covered by adipose tissue. There are two short, well-separated dorsal fins, the first with four spines and the second with eight to ten segmented rays. The anal fin is short, with two or three spines and seven to twelve segmented rays in adults. The pectoral fins are placed high on the body, and the caudal fin is weakly forked. The lateral line is absent. The scales are moderate to large in size, with one or more longitudinal grooves. There are two or more pyloric caeca associated with the stomach, which also has a thick-walled, muscular gizzard in most species. Mullets usually are grayish green or blue dorsally, and their flanks are silvery, often with dark longitudinal stripes. They are pale or yellowish ventrally.

Distribution

Worldwide through tropical, subtropical, and warm temperate regions; some species inhabit cool temperate waters. Several species and genera are common in the Indo-Pacific (e.g., *Liza* and *Valamugil*). *Mugil* generally is restricted to the Atlantic and eastern Pacific.

Habitat

Most mullets are found in coastal marine and brackish waters. They are nektonic, usually in shallow inshore environments, such as coastal bays, reef flats, tide pools, and around harbor pilings and in brackish water estuaries, lagoons, and

Mullets (*Mugil* sp.) in the Homosassa River, Florida, USA. (Photo by Douglas Faulkner/Photo Researchers, Inc. Reproduced by permission.)

mangroves. They usually swim over sandy-muddy bottoms and sea grass meadows, in relatively still waters. They commonly occur at water depths of 65.6 ft (20 m) but may be found offshore or in deeper waters. Many species are euryhaline and move between marine and freshwater environments of rivers and flooded rice fields. Some species occasionally swim far upriver, and a few species spend their entire adult lives in rivers.

Behavior

Feeding behavior apparently follows daily cycles, which may change through the seasons according to water temperature and prey availability. Several species form schools, particularly at night; schooling adults may show leaping behavior, especially during the evening.

Feeding ecology and diet

Mullet fry are planktivorous; larger specimens browse on submerged surfaces and use their pharyngobranchial organ to filter particulate material, microalgae, microorganisms, and small invertebrates, such as polychaetes, crustaceans, and mollusks. Mullets are subject to predation from larger fishes (e.g., drums and basses), crocodiles, birds (in particular, pelicans), and various aquatic mammals such as seals, sea lions, and dolphins. They are also hunted by humans.

Reproductive biology

Coastal species typically spawn in shallow open areas or offshore, forming large schools before moving out to the spawning grounds. Freshwater species possibly move downstream to spawn in brackish waters; alternatively they might spawn upstream, and the fry are swept downstream for a short period before subsequently migrating back upriver.

Conservation status

One nominal species, *Liza luciae*, known only from the Saint Lucia estuary, Natal, is considered Endangered according to the 2002 IUCN Red List. A taxonomic revision in 1997 included *Liza luciae* as conspecific with the more widespread *Liza melinoptera*.

Significance to humans

Mullets are important food fishes, caught in subsistence and commercial fisheries, and are used in aquaculture in many parts of the world. Food and Agriculture Organization (FAO) statistics give the global commercial capture of mullets for 2000 as 409,892 metric tons (26% from China). Global aquaculture of mullets in 2000 was 100,091 metric tons (80% coming from Egypt), with a value of almost $333.4 million dollars.

1. Large-scale mullet (*Liza grandisquamis*); 2. Shark mullet (*Rhinomugil nasutus*); 3. Flathead mullet (*Mugil cephalus*); 4. Mountain mullet (*Agonostomus monticola*). (Illustration by Barbara Duperron)

Species accounts

Mountain mullet

Agonostomus monticola

FAMILY
Mugilidae

TAXONOMY
Mugil monticola Bancroft, 1834, Jamaica. *Agonostomus monticola* is morphologically variable. Several different species have been recognized from Central America and the West Indies, but a review in 1997 listed them as conspecific.

OTHER COMMON NAMES
French: Mulet de fleuve; Spanish: Lisa de río, tepemechín.

PHYSICAL CHARACTERISTICS
Grows to 14.2 in (36 cm) in total length. Dorsal surface of head is convex between the eyes. Teeth are small and attached directly to the jawbones. Pharyngobranchial organ is rudimentary. There are 16–23 gill rakers on the lower part of first gill arch. Anal fin has two spines and 10 soft rays in adults. There are 38–46 scales in a longitudinal series along the flanks. Body is brownish, sometimes with a silvery band on the flanks from the pectoral to the caudal fin. Dorsal, anal, and caudal fins are dusky yellowish; there is a dark spot at the base of the caudal fin.

DISTRIBUTION
Rivers of the West Indies, Central America, Colombia, Venezuela, and the Galapagos Islands. Occasionally reported from the rivers of Florida and Louisiana.

HABITAT
Freshwaters of fast-flowing hill streams.

BEHAVIOR
Little is known.

FEEDING ECOLOGY AND DIET
Diet varies seasonally. Omnivorous, feeding mainly on insects, prawns, fruits, and algae.

REPRODUCTIVE BIOLOGY
Adults may spawn in the lower reaches of rivers or in the sea. Spawning apparently correlates with peak rainfall. Larvae and juvenile fish spend perhaps six weeks at sea before migrating back upstream.

CONSERVATION STATUS
Not threatened.

SIGNIFICANCE TO HUMANS
There are small commercial and subsistence fisheries in the West Indies and Central America. ◆

Fringelip mullet

Crenimugil crenilabis

FAMILY
Mugilidae

TAXONOMY
Mugil crenilabis Forsskal, 1775, Red Sea.

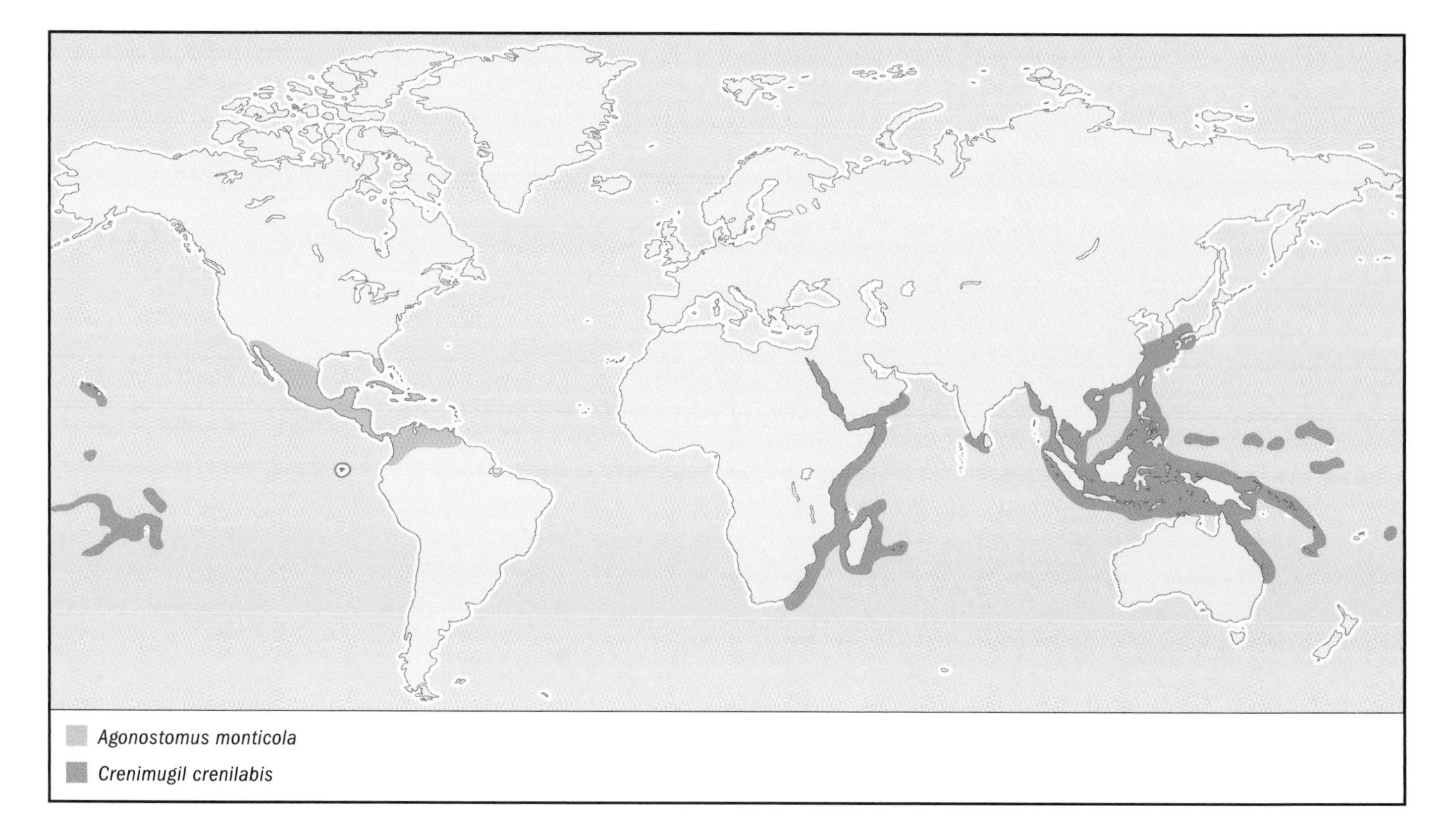

OTHER COMMON NAMES
French: Mulet boxeur; Spanish: Lisa labride.

PHYSICAL CHARACTERISTICS
Commonly reaches 10.2 in (26 cm) in standard length but may reach 19.7 in (50 cm). The upper lip is very thick, with up to 10 rows of small papillae in fish larger than 2.4 in (6 cm). There are no teeth on the lips. The anal fin has three spines and usually nine soft rays in adults. There are 36–42 scales in a longitudinal series along the flanks. Body is olive-green dorsally and silvery on the flanks and abdomen. Pectoral fins are yellowish, with a distinct dark purplish spot at the upper part of the fin base.

DISTRIBUTION
Indo-Pacific, from the Red Sea to Polynesia.

HABITAT
In shallow waters (up to 65.6 ft, or 20 m) of lagoons with sandy and muddy bottoms, reef flats, tide pools, and harbors.

BEHAVIOR
Little is known.

FEEDING ECOLOGY AND DIET
Larvae and juveniles probably feed on plankton. Adults probably feed on particulate organic material, algae, and invertebrates.

REPRODUCTIVE BIOLOGY
Adult fish form schools in shallow waters around lagoons before spawning.

CONSERVATION STATUS
Not threatened.

SIGNIFICANCE TO HUMANS
Small commercial and subsistence fisheries, particularly in Polynesia; some aquaculture. ◆

Large-scale mullet
Liza grandisquamis

FAMILY
Mugilidae

TAXONOMY
Mugil grandisquamis Valenciennes, 1836, Gorée, Senegal.

OTHER COMMON NAMES
French: Mulet écailleux; Spanish: Liza escamuda.

PHYSICAL CHARACTERISTICS
Commonly reaches 9.8 in (25 cm) in standard length but may reach up to 15.7 in (40 cm) in fork length. Teeth on both lips are absent or are very small, well spaced, and positioned in a single row on the upper lip, or rarely, in two rows. The anal fin has three spines and nine soft rays in adults. There are 26–30 scales in a longitudinal series along the flanks. Body is darkish dorsally and slightly paler and silvery laterally and ventrally. The anal fin and lower lobe of the caudal fin may be yellowish.

DISTRIBUTION
West Africa from Senegal to the Republic of Congo.

HABITAT
Usually found in brackish waters covering muddy substrates, such as mangroves, creeks, estuaries, and inundated mudflats.

BEHAVIOR
Little is known.

FEEDING ECOLOGY AND DIET
Adults feed in schools, mostly at night, on particulate organic matter, mud, diatoms, algae, microarthropods, foraminiferans,

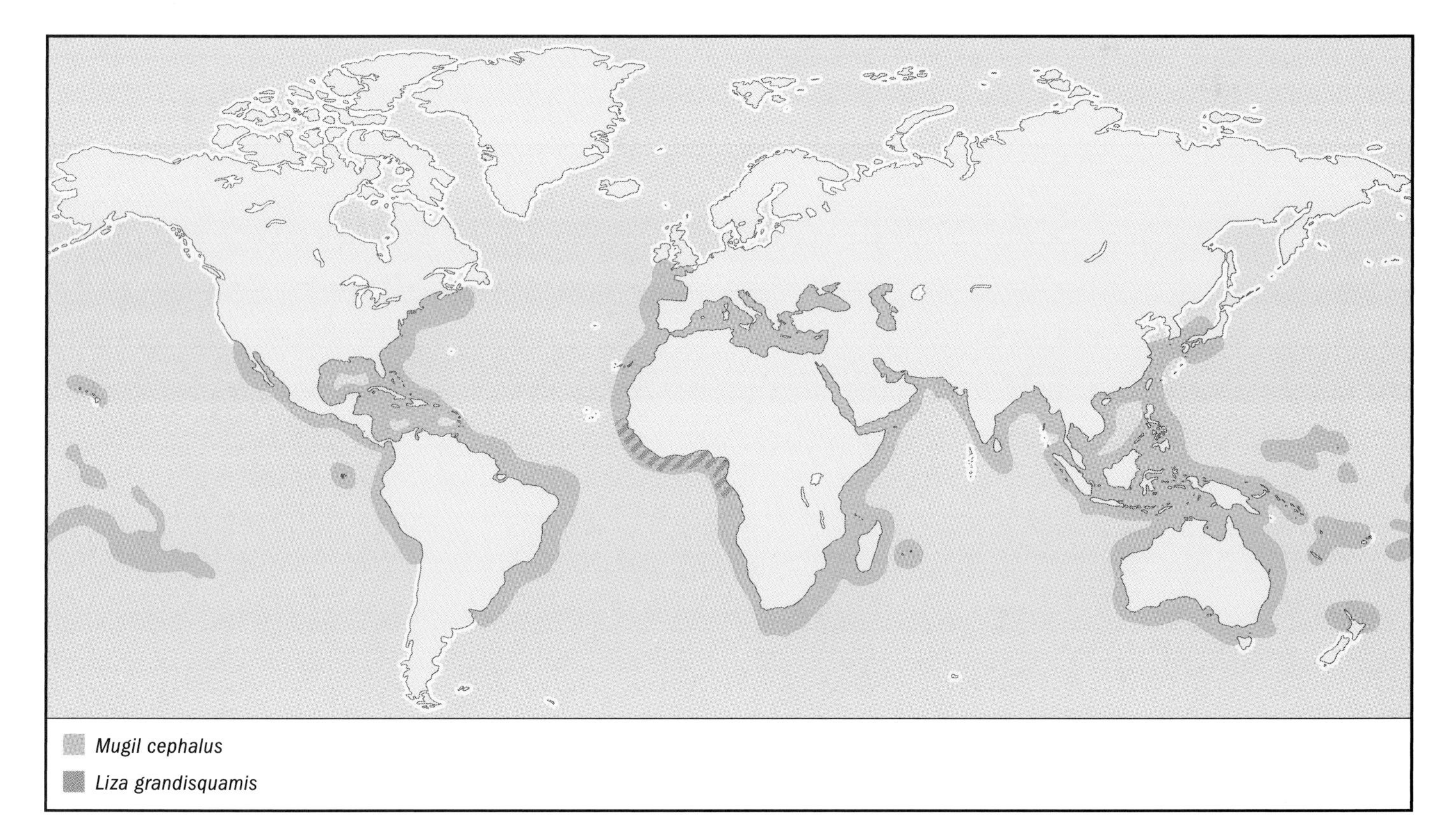

and free-living nematodes. During the wet season fish take a greater proportion of fine organic material than coarser sands, because increased rainfall brings more fine allochthonous material into rivers and "washes" the sand sediments free of fine food material.

REPRODUCTIVE BIOLOGY
Populations from the Ebrie lagoon, Ivory Coast, do not show a seasonal spawning cycle; spawning occurs principally in the lagoon but also may occur in the sea.

CONSERVATION STATUS
Not threatened.

SIGNIFICANCE TO HUMANS
Small commercial and subsistence fisheries. A potentially important species for aquaculture. ◆

Flathead mullet
Mugil cephalus

FAMILY
Mugilidae

TAXONOMY
Mugil cephalus Linnaeus, 1758, European seas. It has been suggested that several species, initially distinguished by slight differences in morphological features, fin ray counts, dentition, and fin coloration, are synonymous with *Mugil cephalus.* Some authors have indicated that the concept of *Mugil cephalus* as a widespread, polymorphic species may be incorrect. According to comparative analyses of the morphometrics, isozymes, and mitochondrial DNA of populations of *Mugil cephalus* from around the world, however, the variation between populations is not large enough to indicate definitively that they are differentiated species.

OTHER COMMON NAMES
English: Striped mullet, black mullet; French: Mulet cabot; Spanish: Lisa pardete.

PHYSICAL CHARACTERISTICS
Commonly reaches 13.8 in (35 cm) in total length but may reach 47.2 in (120 cm) in standard length. Adipose tissue covers most of the eye. There are several rows of very small unicuspid and bicuspid teeth on the edges of the lips. Anal fin has three spines and eight soft rays in adults. There are 36–44 scales in a longitudinal series along the flanks. Body is olive-green dorsally and silvery on the flanks and abdomen, with about seven longitudinal dark stripes along the flanks. The pelvic fin, anal fin, and lower lobe of the caudal fin are yellowish in some populations.

DISTRIBUTION
Worldwide in tropical, subtropical, and warm temperate waters from 42° S latitude to almost 51° N latitude; less abundant in the tropics.

HABITAT
Tolerant of salinity levels from zero to 81 parts per thousand (freshwater to hypersaline) and temperatures from 41°F (5°C) to 98.6°F (37°C). Inhabits inshore marine waters, estuaries, lagoons, and rivers, usually in shallow waters, and rarely moving deeper than 656 ft (200 m). Adults may move far upriver.

BEHAVIOR
The adults form schools and sometimes jump.

FEEDING ECOLOGY AND DIET
Larvae and juveniles feed on plankton. Adults feed on particulate organic material, algae, and invertebrates. Fish may gulp and filter sediment, browse over submerged surfaces, or feed at the surface.

REPRODUCTIVE BIOLOGY
Adult fish move offshore in shoals to spawn, usually at night, before returning to inshore brackish waters and freshwaters. Fry and juveniles remain in sheltered bays, lagoons, and estuaries until they are sexually mature.

CONSERVATION STATUS
Not threatened.

SIGNIFICANCE TO HUMANS
Major commerical and subsistence fisheries in many parts of the world. Food and Agriculture Organization (FAO) of the United Nations statistics give the global commercial capture of flathead mullets for 2000 as 29,335 metric tons (18% and 19% coming from Mexico and Venezuela, respectively). Global aquaculture of flathead mullets in 2000 was 89,078 metric tons (90% coming from Egypt), with a value of almost $312.3 million dollars. ◆

Shark mullet
Rhinomugil nasutus

FAMILY
Mugilidae

TAXONOMY
Mugil nasutus De Vis, 1883, Cardwell, Rockingham Bay, Queensland, Australia.

OTHER COMMON NAMES
English: Mud mullet.

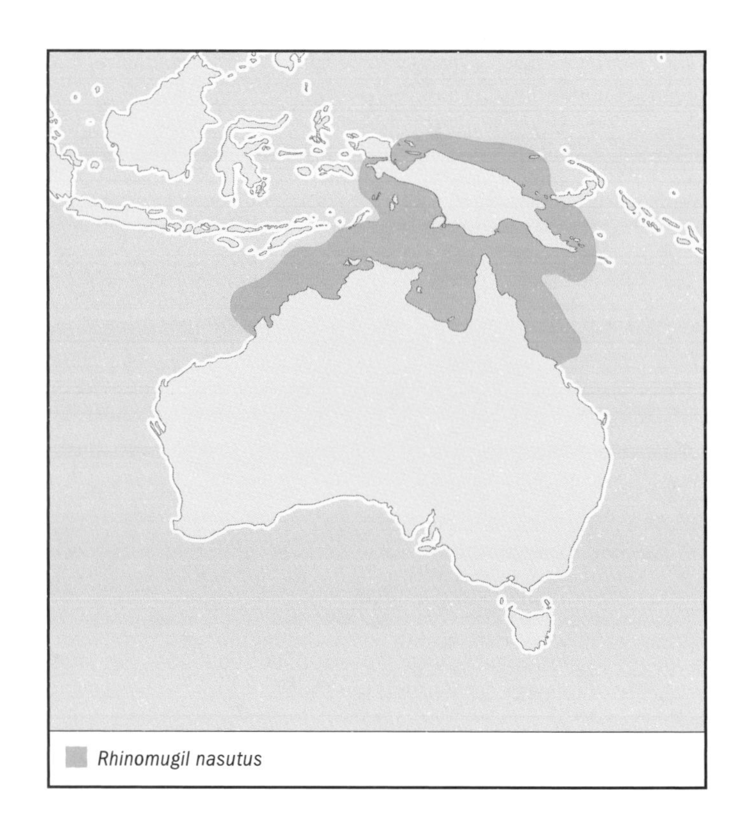
Rhinomugil nasutus

PHYSICAL CHARACTERISTICS
Reaches 12.6 in (32 cm) in fork length. The shark mullet looks unlike other mullets, because the eyes are positioned high, toward the top of the head, which is concave between the eyes. Nostrils are placed low on the snout, which projects beyond the upper lip. Teeth on both lips are small and spatulate. The anal fin has three spines and eight soft rays in adults. There are 28–30 scales in a longitudinal series along flanks. Body is slate gray dorsally and silvery laterally and ventrally; the fins are yellowish.

DISTRIBUTION
Tropical Australia and New Guinea.

HABITAT
Usually occurs in muddy freshwaters and coastal water habitats, such as mangroves.

BEHAVIOR
Swims in schools at the water surface, with the eyes and snout exposed. Apparently capable of breathing air and wriggling for short distances over mud banks.

FEEDING ECOLOGY AND DIET
Feeds on muddy substrates and may feed on algae and insects at the water surface.

REPRODUCTIVE BIOLOGY
Little is known.

CONSERVATION STATUS
Not threatened.

SIGNIFICANCE TO HUMANS
No reported fisheries data. Species of *Rhinomugil* (usually *Rhinomugil corsula*) are sold in the aquarium trade, because of their similarity to species of four-eyed fish (*Anableps*). ◆

Resources

Books

De Sylva, D. P. "Mugiloidei: Development and Relationships." In *Ontogeny and Systematics of Fishes*, edited by H. G. Moser, W. J. Richards, D. M. Cohen, M. P. Fahay, A. W. Kendall, Jr., and S. L. Richardson. Lawrence, KS: American Society of Ichthyologists and Herpetologists, 1984.

Harrison, I. J. "Mugilidae." In *Guía FAO para la Identificacíon de Especies para los Fines de la Pesca Pacifico Centro-Oriental*, edited by W. Fischer, F. Krupp, W. Schneider, C. Sommer, K. E. Carpenter, and V. H. Niem. Vol. 3. Rome: FAO, 1995.

Harrison, I. J., and H. Senou. "Mugilidae." In *FAO Species Identification Guide for Fisheries Purposes. The Living Marine Resources of the Western Central Pacific*, Vol. 4. *Bony Fishes, Part 2, (Mugilidae to Carangidae)*, edited by K. Carpenter and V. H. Niem. Rome: FAO, 1999.

Nelson, J. S. *Fishes of the World.* 3rd edition. New York: John Wiley and Sons, 1994.

Oren, O. H. *Aquaculture of Grey Mullets.* International Biological Programme vol. 26. Cambridge: Cambridge University Press, 1981.

Patterson, C. "Osteichthyes: Teleostei." In *The Fossil Record*, edited by M. J. Benton. Vol. 2. London: Chapman and Hall, 1993.

Smith, M. M., and J. L. B. Smith. "Mugilidae." In *Smiths' Sea Fishes*, edited by M. M. Smith and P. C. Heemstra. Berlin: Springer-Verlag, 1986.

Periodicals

Albaret, J.-J., and M. Legendre. "Biologie et Ecologie des Mugilidae en Lagune Ebrié (Côte d'Ivoire): Intérét Potential pour l'Aquaculture Lagunaire." *Revue d'Hydrobiologie Tropicale* 18, no. 4 (1985): 281–303.

Blaber, S. J. M., and A. K. Whitfield. "The Feeding Ecology of Juvenile Mullet (Mugilidae) in South-east African Estuaries." *Biological Journal of the Linnean Society* 9 (1977): 277–284.

Corti, M., and D. Crosetti. "Geographical Variation in the Grey Mullet: A Geometric Morphometric Analysis Using Partial Warp Scores." *Journal of Fish Biology* 48 (1996): 255–269.

Crosetti, D., W. S. Nelson, and J. C. Avise. "Pronounced Genetic Structure of Mitochondrial DNA Among Populations of the Circumglobally Distributed Grey Mullet (*Mugil cephalus* Linnaeus)." *Journal of Fish Biology* 44 (1994): 47–58.

Cruz, G. A. "Reproductive Biology and Feeding Habits of Cuyamel, *Joturus pichardi*, and Tepemechín, *Agonostomus monticola* (Pisces; Mugilidae), from Río Platano, Mosquitia, Honduras." *Bulletin of Marine Science* 40, no. 1 (1987): 63–72.

De Silva, S. S. "Biology of Juvenile Grey Mullet: A Short Review." *Aquaculture* 19 (1980): 21–36.

Drake, P., A. M. Arias, and L. Gallego. "Biología de los Mugílidos (Osteichthyes, Mugilidae) en los Esteros de las Salinas de San Fernando (Cádiz). III. Hábitos Alimentarios y su Relación con la Morfometría del Aparato Digestivo." *Investigacion Pesquera* 48, no. 2 (1984): 337–367.

Harrison, I. J., and G. J. Howes. "The Pharyngobranchial Organ of Mugilid Fishes: Its Structure, Variability, Ontogeny, Possible Function and Taxonomic Utility." *Bulletin of the British Museum of Natural History (Zoology Series)* 57, no. 2 (1991): 111–132.

Johnson, G. D., and C. Patterson. "Percomorph Phylogeny: A Survey of Acanthomorphs and a New Proposal." *Bulletin of Marine Science* 2, no. 1 (1993): 554–626.

King, R. P. "Observations on *Liza grandisquamis* (Pisces: Mugilidae) in Bonny River, Nigeria." *Revue d'Hydrobiologie Tropicale* 19, no. 1 (1986): 61–66.

———. "New Observations on the Trophic Ecology of *Liza grandisquamis* (Valenciennes, 1836) (Pisces: Mugilidae) in the Bonny River, Niger Delta, Nigeria." *Cybium* 12, no. 1 (1988): 23–36.

Miya, M., A. Kawaguchi, and M. Nishida. "Mitogenomic Exploration of Higher Teleostean Phylogenies: A Case Study for Moderate-Scale Evolutionary Genomics with 38 Newly Determined Complete Mitochondrial DNA Sequences." *Molecular Biology and Evolution* 18, no. 11 (2001): 1993–2009.

Resources

Phillip, D. A. T. "Reproduction and Feeding of the Mountain Mullet, *Agonostomus monticola*, in Trinidad, West Indies." *Environmental Biology of Fishes* 37 (1993): 47–55.

Senou, H. "Redescription of a Mullet, *Chelon melinopterus* (Perciformes: Mugilidae)." *Bulletin of the Kanagawa Prefectural Museum of Natural Sciences* 26 (1997): 51–55.

Stiassny, M. L. J. "What Are Grey Mullets?" *Bulletin of Marine Science* 52, no. 1 (1993): 197–219.

Thomson, J. M. "The Mugilidae of the World." *Memoirs of the Queensland Museum* 41, no. 3 (1997): 457–562.

Torres-Navarro, C. I., and J. Lyons. "Diet of *Agonostomus monticola* (Pisces: Mugilidae) in the Río Ayuquila, Sierra de Manantlán Biosphere Reserve, México." *Revista de Biología Tropical* 47, no. 4 (1999): 1087–1092.

Other

Food and Agriculture Organization of the United Nations. "Fishery Software." (12 Feb. 2003). <http://www.fao.org/fi/statist/FISOFT/FISHPLUS.asp>

Ian J. Harrison, PhD

Atheriniformes

(Rainbowfishes and silversides)

Class Actinopterygii
Order Atheriniformes
Number of families 8

Photo: A salmon rainbowfish (*Glossolepis incisus*). (Photo by Tom McHugh/Photo Researchers, Inc. Reproduced by permission.)

Evolution and systematics

The order Atheriniformes originally was conceived by Donn Rosen in 1964 to include the exocoetoids (flying fishes and allies), scomberesocoids (sauries), adrianichthyoids (ricefishes), cyprinodontoids (killifishes and allies), atherinoids (silversides and allies), and phallostethoids (priapium fishes). Today this disparate assemblage of fishes still is thought to be a natural group, but it has been parceled into three orders. Now called the Atherinomorpha, the group contains the orders Cyprinodontiformes and Beloniformes (put together as Division II atherinomorphs by Rosen and Parenti in 1981) and the Atheriniformes (Division I atherinomorphs according to Rosen and Parenti). Characters of the testis, egg, embryo, rostral cartilage and skull, and dorsal gill arch skeleton, among others, unite the Atherinomorpha. This chapter concerns the Atheriniformes, which at present includes the silversides, phallostethids, rainbowfishes, and related taxa.

At more than 50 million years old, fossils of the genus *Atherinidarum* from the early Eocene of France are the oldest known atheriniforms. Researchers disagree about the identity of the most basal atheriniform lineages and have variously proposed both freshwater (bedotiids and melanotaeniids) and partly or wholly marine (atherinopsids and notocheirids) groups. Through the course of their evolution different groups of atheriniforms have repeatedly breached the marine-freshwater barrier. Such an ability to colonize patchily distributed freshwater habitats has fostered speciation within the order, especially in the Americas and Australia. At present there are about 315 species and 49 genera of atheriniforms known to science.

Ichthyologists disagree on whether the Atheriniformes are a monophyletic group, descended from a single branch on the atherinomorph tree, or a hodgepodge of lineages on multiple branches that are associated closely with the Cyprinodontiformes and Beloniformes (which are considered sister groups). Even the number and constitution of families is in dispute: recent classifications have cited from six to ten families. In light of the disagreement in the literature, this treatment follows Joseph Nelson's book *Fishes of the World*, in which eight families are recognized within the order: Bedotiidae (Malagasy rainbowfishes), Melanotaeniidae (rainbowfishes), Pseudomugilidae (blue-eyes), Telmatherinidae (Sulawesi rainbows), Atherinidae (silversides), Notocheiridae (surf silversides), Dentatherinidae (pygmy or tusked silverside), and Phallostethidae (priapium fishes).

In 1996, the most recent treatment of the issue, Dyer and Chernoff marshaled 10 morphological characters supporting the Atheriniformes as a monophyletic, or natural, group. In addition, they proposed a new classification for the order, in which the number of families was reduced to six. Familial allocations within this new classification are in stark contrast to Nelson's scheme. Differences include an expansion of the family Melanotaeniidae to subsume the families Bedotiidae, Pseudomugilidae, and Telmatherinidae; an expansion of the family Phallostethidae to include Dentatherinidae; and an elevation of the Old World silverside genus *Atherion* to family status (Atherionidae). Furthermore, in agreement with a 1994 article by Saeed and collaborators, the New World silverside subfamilies Menidiinae and Atherinopsinae are shunted from the family Atherinidae into their own family, Atherinopsidae. Dyer and Chernoff's morphology-based hypothesis awaits further testing, especially using DNA sequence data. It is too soon to say whether their proposal will gain wide acceptance. Certainly, though, theirs will not be the last word on atheriniform relationships.

A school of silversides (Atherinidae), near Thailand. (Photo by Animals Animals ©Joyce & Frank Burek. Reproduced by permission.)

Physical characteristics

A reasonable degree of morphological diversity is found among the Atheriniformes, and, in particular, phallostethids exhibit a form that is unique among all fishes. The typical silverside generally is elongate and laterally compressed, with two dorsal fins, a single anal fin spine, usually cycloid scales, and no lateral line. Most are silvery in color and have a prominent mid-lateral stripe along each side of the body. Melanotaeniids are constructed similarly but often are deeper bodied and have sexually dimorphic color patterns. Male rainbowfishes frequently are brilliantly colored, with complex patterns, in various shades of red, yellow, orange, blue, and green. In telmatherinids, as well as some pseudomugilids and melanotaeniids, the anal and dorsal fins sometimes are elongated into filamentous extensions or elaborate fanlike shapes.

Most noteworthy among atheriniforms, however, are the extraordinarily modified phallostethids (priapium fishes). In both male and female phallostethids, the anal and genital openings are shifted anteriorly and are located under the throat. Males have an elaborate, bilaterally asymmetric copulatory structure under the head, called a priapium, for which no parallel exists among other fishes. A suspensory component of the priapium is made up of modified anterior pleural ribs and pelvic bones. Emerging from the posterior end of the priapium and arching forward almost the entire length of the head are the ctenactinia—elongate, curved bones used for clasping the female during mating. In different species, the anus is offset to one or the other side of the priapium, and the seminal papilla is offset opposite the anus. Both open anteriorly. The seminal papilla, which is elaborated into a copulatory organ in some species, is used to transfer bundles of sperm to the female.

Distribution

The Atheriniformes are distributed in mainly coastal areas of tropical and temperate seas throughout the world and also occur in freshwater lakes and streams in many regions. Largely or entirely marine atheriniform families include the Atherinidae, Notocheiridae, and Dentatherinidae. Atherinids occur worldwide in near-shore marine environments and in freshwaters of primarily the Americas and Australia. Members of the atherinid subfamily Menidiinae show a propensity for invasion of freshwater habitats from coastal marine environments. In the lakes and streams of the Mesa Central of southern Mexico, freshwater menidiines have diversified into a species flock of 20 or so closely related species with exceedingly circumscribed ranges (many are found in only a single lake). Notocheirids are exclusively marine: five species of the genus *Iso* are distributed along the coasts of South Africa, India, Japan, Australia, and Hawaii, and the species *Notocheirus hubbsi* is found only along the coast of Chile. Finally, the family Dentatherinidae consists of but a single species, *Dentatherina merceri*, from the inshore seas of the Philippines, New Guinea, and northeastern Australia.

All of the remaining atheriniform families are predominantly freshwater and are wholly restricted to Australia and the Indo-Pacific region, except for the Bedotiidae from the freshwaters of Madagascar. Rainbowfishes of the family Melanotaeniidae are abundant in New Guinea at elevations below 5,249 ft (1,600 m) as well as on numerous islands to the west of New Guinea. Additionally, they are found in Australia, mostly in river drainages in the northern and eastern regions of the continent. Pseudomugilids, from New Guinea, Australia, and Indonesia, and telmatherinids, from Sulawesi and Indonesia, can be found in fresh, brackish, and marine waters. Likewise, phallostethids occur in freshwaters and estuaries of the Southeast Asian mainland, the Philippines, and Sulawesi.

Habitat

Atheriniform habitats are as diverse as the fishes themselves. From the crashing surf zone of a Pacific island, where one might observe the silvery flashes of a school of *Iso*, to the still depths of a rainforest pool in New Guinea, where shoaling rainbowfishes dominate, atheriniforms have adapted to a range of aquatic conditions. Marine silversides can be found in almost all nonpolar, shallow marine habitats, including reefs, estuaries, and lagoons, and in the surf along beaches. Only a few species live in the open water. Freshwater atherini-

Boeseman's rainbowfish (*Melanotaenia boesemani*) are found in New Guinea. Photo by Animals Animals ©M. Gibbs, OSF. Reproduced by permission.)

forms occur in a variety of temperate and tropical lakes and streams, including spring-fed desert waterholes in Australia and mountain lakes in Patagonia.

Behavior

Atheriniforms are gregarious fishes that form schools of various sizes. Some marine silversides aggregate into enormous schools––numbering in the thousands—which cruise just below the surface, continually feeding on plankton. For instance, *Atherinomorus capricornensis* has been observed forming schools more than 328 ft (100 m) long and 65.6 ft (20 m) wide. At night silversides are attracted to bright lights and are caught easily by fishermen for use as bait. Some of the most interesting types of behavior found in atheriniforms, however, relate to reproduction.

Feeding ecology and diet

In general, atheriniforms are omnivorous, sometimes feeding on algae but depending more heavily on the most abundant forms of invertebrate protein in their respective habitats. For marine species, this means a heavy reliance on minute crustaceans and other forms of zooplankton. Fish larvae also are eaten by some species. In freshwaters, terrestrial insects often represent a significant portion of the diet. This is certainly true for the melanotaeniid rainbowfishes, which have a special preference for ants. In addition, freshwater atheriniforms feed on aquatic insects, zooplankton, and algae. Atheriniforms themselves are preyed upon by birds, mammals, and larger fishes, among other organisms. Marine silversides, in particular, which often aggregate in large schools, are sought after by larger fishes of commercial importance.

Reproductive biology

Peculiar reproductive repertoires are numerous within the atheriniformes. Large eggs with adhesive filaments are characteristic of the group; the filaments are used to anchor the eggs to aquatic plants or other forms of substrate. Among the rainbowfishes, spawning occurs day after day for an extended period, each day of which the female attaches a small number of thread-bearing eggs, fertilized externally by the male, to underwater plants. Eggs typically hatch in seven to 18 days. Atheriniform larvae generally are large (0.16–0.35 in; 4–9 mm) at hatching, and although they still possess a yolk sac, they have open mouths and can begin feeding immediately.

A truly exceptional reproductive strategy has evolved in grunion (two species of atherinids). *Leuresthes tenuis* and *L. sardina* both time their spawning to take advantage of tidal effects. Grunion spawn during periods of only a few hours on only six nights a month, when tides are highest around the times of the full and new moons (semilunar tides). The fishes ride in with large waves and are left stranded on the sand as the water recedes. Once on the beach, the female burrows into the sand with her tail to lay the eggs, and an accompanying male deposits his milt simultaneously. The fertilized eggs are left buried in the sand a few inches below the surface; they hatch in about two weeks, when semilunar high tides return to uncover them. Possible advantages for the eggs are decreased predation and increased oxygen levels and incubation temperature.

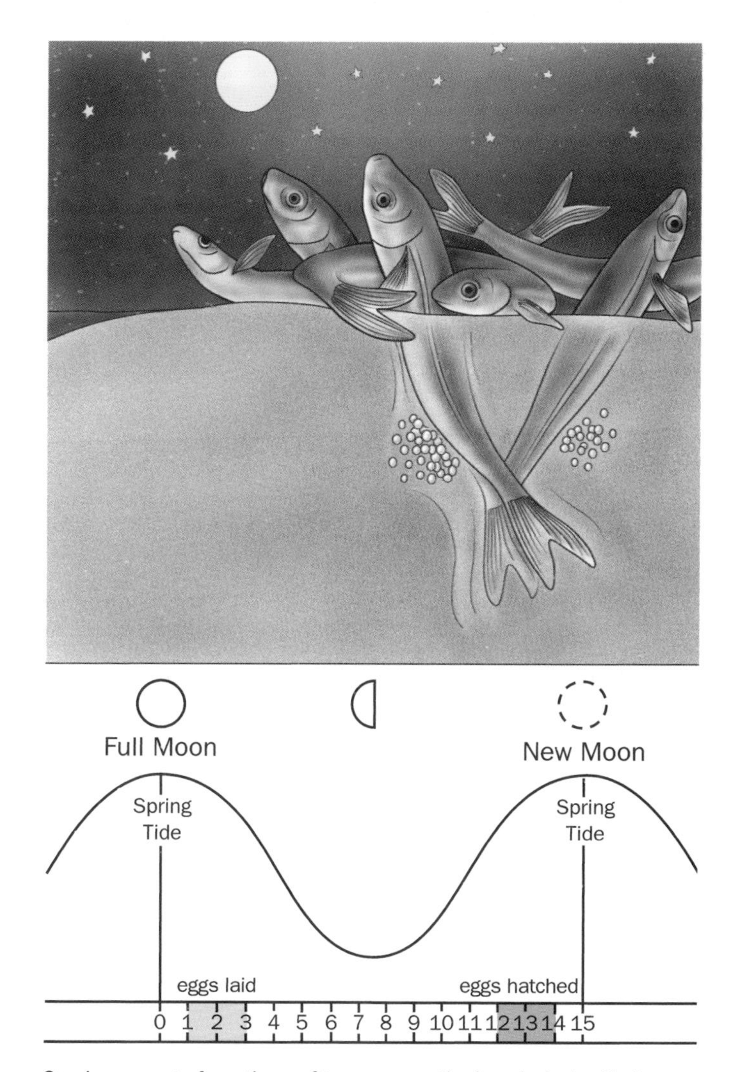

Grunion emerge from the surf to spawn on the beach, laying their eggs soon after the full moon. (Illustration by Patricia Ferrer)

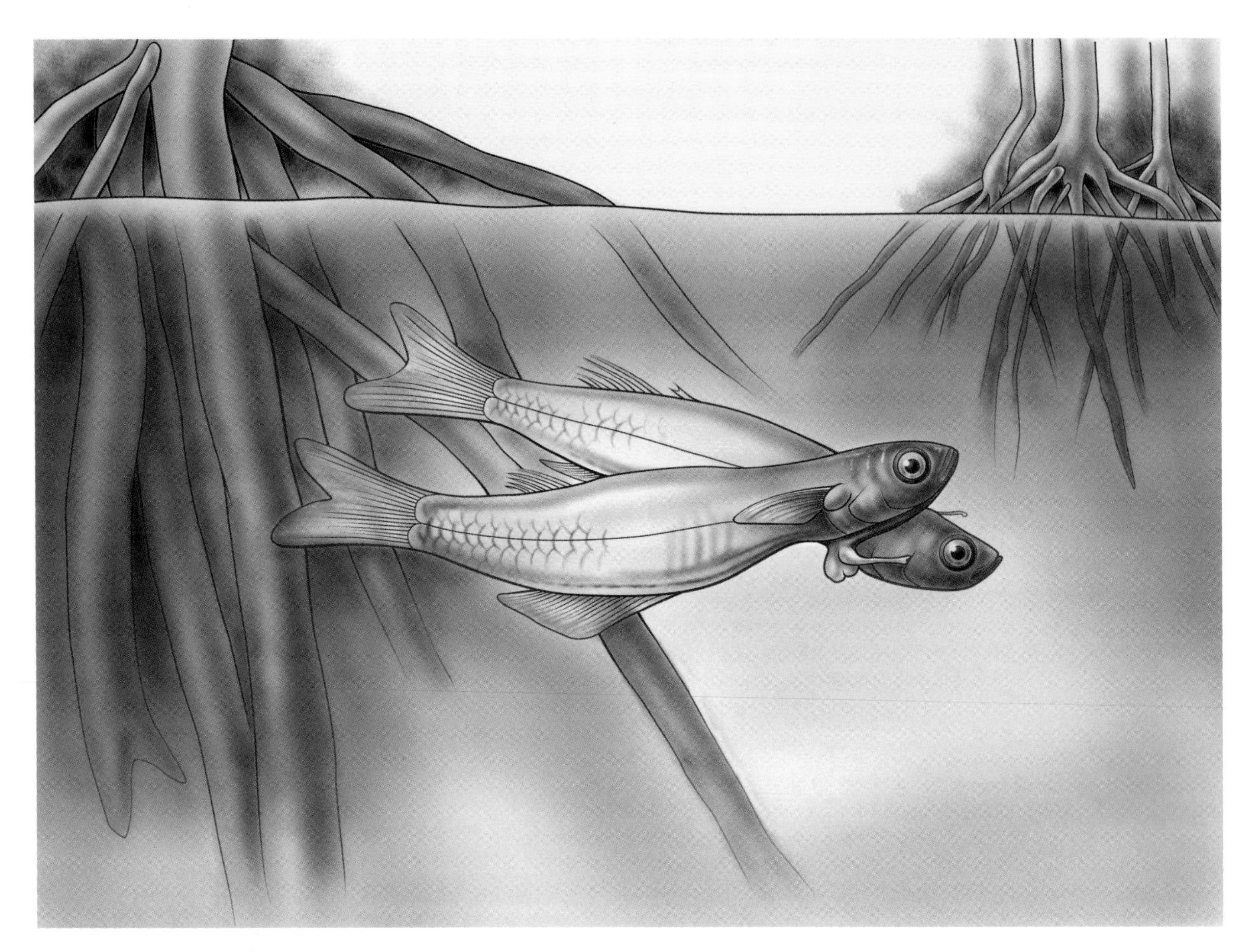

Phallostethids mating; male restrains female with anteriorly situated priapium—a mating organ. (Illustration by Patricia Ferrer)

Male phallostethids, with their bizarre gripping structures (described earlier), ram females, and the two become locked together physically by the priapium during copulation. Frantic bursts of swimming are required for the pair to disengage. Further description of this behavior is provided in the species account of *Neostethis bicornis.*

In a lineage of *Menidia*, sexual reproduction has been forgone entirely. The *Menidia clarkhubbsi* species complex from the Gulf of Mexico is made up of unisexual clonal lineages thought to be the result of hybridization events between *M. beryllina* and an as yet undetermined species, similar to *M. peninsulae.* The unisexuals produce eggs with a complete chromosomal complement that do not require fertilization but must still rely on sperm supplied by one of the bisexual species to stimulate embryo formation. Spawning itself is simple in *Menidia.* A single female—attended by several males—deposits eggs in aquatic vegetation, and the males leave milt; no courtship has been observed.

Conservation status

While no atheriniforms are listed by CITES, 79 species presently are included on the IUCN Red List. Of those, 28 species are listed as Data Deficient, 8 species are listed as Lower Risk/Near Threatened, and 31 species (most of which are rainbowfishes) are listed as Vulnerable. Five species are Endangered: *Poblana letholepis* and *Poblana squamata* from Mexico, *Craterocephalus fluviatilis* and *Pseudomugil mellis* from Australia, and *Melanotaenia boesemani* from New Guinea. Six species are considered Critically Endangered: *Chilatherina sentaniensis* from Irian Jaya, *Glossolepis wanamensis* and *Kiunga ballochi* from Papua New Guinea, *Poblana alchichica* from Mexico, *Rheocles wrightae* from Madagascar, and *Scaturiginichthys vermeilipinnis* from Australia. A single atheriniform, *Rheocles sikorae* from Madagascar, is believed to be Extinct.

A common trend holds true for the threatened atheriniforms: they are restricted to circumscribed freshwater habitats. Such habitats are particularly vulnerable. Water quality often is threatened by pollutants, including nutrients, which are the byproducts of such human activities as deforestation, mining, and waste disposal. Introduced fish species pose another serious problem; they can be detrimental to native fish populations both as resource competitors and as predators. In the United States—where freshwater atheriniforms are few—there are no atheriniforms listed as Endangered by the U. S. Fish and Wildlife Service. Only one species, *Menidia extensa* (the waccamaw silverside) from North Carolina, is listed as Threatened.

Significance to humans

Because of their small size, most atheriniforms are not sought as sources of food for humans, though there are exceptions to this rule. For instance, some large piscivorous *Odontesthes* species support thriving fisheries in lakes and reservoirs in Peru, Argentina, Chile, and southern Brazil. Marine silversides are indirectly important to commercial fisheries, in that they are an significant food source for many larger fish species that are valued as food fishes. When they are directly fished, small silversides typically are used as bait or converted into pet food.

In contrast, the freshwater rainbowfishes, owing to their extraordinary diversity of colors, are valued highly in the aquarium trade. At present, most of the rainbowfishes sold in pet stores are captive bred, but in the recent past some species were fished heavily to satisfy the demands of mostly European and American aquarists. Indeed, consider the case of the Boeseman's rainbow (see species account), which is considered Endangered. During the 1980s, when the fishery was unrestricted, tens of thousands of individuals were being removed from the wild each month for export.

1. Celebes rainbowfish (*Marosatherina ladigesi*); 2. Boeseman's rainbowfish (*Melanotaenia boesemani*); 3. *Rheocles derhami*; 4. Eendracht land silverside (*Atherinomorus endrachtensis*); 5. Flower of the wave (*Iso rhothophilus*); 6. *Neostethus bicornis*; 7. Inland silverside (*Menidia beryllina*); 8. California grunion (*Leuresthes tenuis*). (Illustration by Patricia Ferrer and Michelle Meneghini)

Species accounts

Eendracht land silverside

Atherinomorus endrachtensis

FAMILY
Atherinidae

TAXONOMY
Atherina endrachtensis Quoy and Gaimard, 1825, Shark Bay, Western Australia. Type locality probably is given in error; more likely, it is New Guinea or Waigeo Island.

OTHER COMMON NAMES
English: Endracht hardyhead, striped hardyhead, striped silverside; Misima-Paneati: Galgal.

PHYSICAL CHARACTERISTICS
Maximum length about 3.3 in (8.5 cm). Elongate, with typical silverside shape. Moderately deep body. Large eyes and silvery midlateral band. Swim bladder visible through translucent flesh.

DISTRIBUTION
Northern Australian coast, Vanuatu, Papua New Guinea, Admiralty Islands, New Britain, Solomon Islands, New Caledonia.

HABITAT
Found in shallow coastal waters.

BEHAVIOR
Often schools with other silverside species.

FEEDING ECOLOGY AND DIET
Feeds on zooplankton.

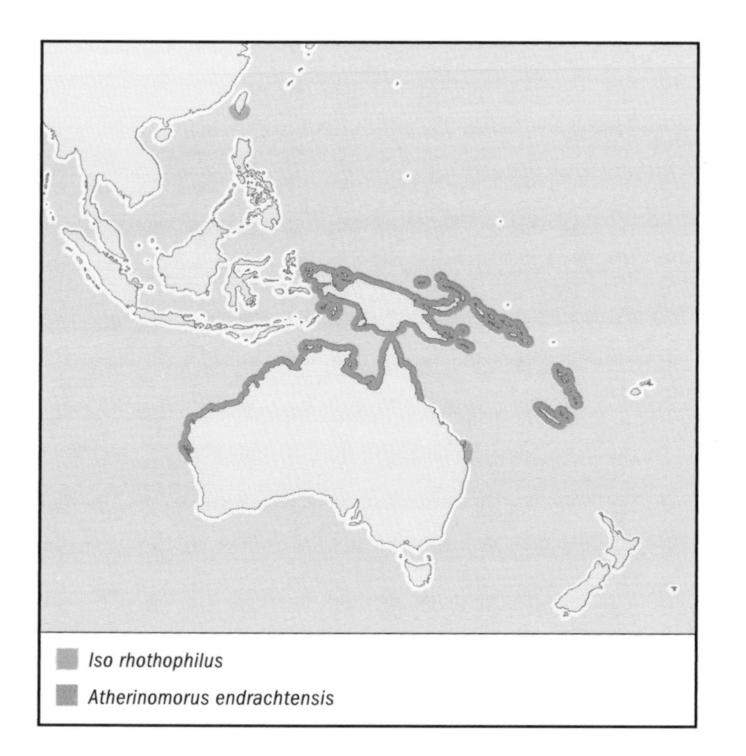

REPRODUCTIVE BIOLOGY
Spawning behavior unknown.

CONSERVATION STATUS
Not threatened.

SIGNIFICANCE TO HUMANS
None known. ◆

California grunion

Leuresthes tenuis

FAMILY
Atherinidae

TAXONOMY
Atherinopsis tenuis Ayres, 1860, San Francisco, California, United States.

OTHER COMMON NAMES
French: Capucette californienne; Spanish: Pejerrey californiano.

PHYSICAL CHARACTERISTICS
Grows to 7.5 in (19.0 cm) maximum length. An elongate silverside with a prominent silvery lateral band and bluish green coloration on the back.

DISTRIBUTION
Monterey Bay, California, to the southern Baja Peninsula.

HABITAT
Coastal marine waters.

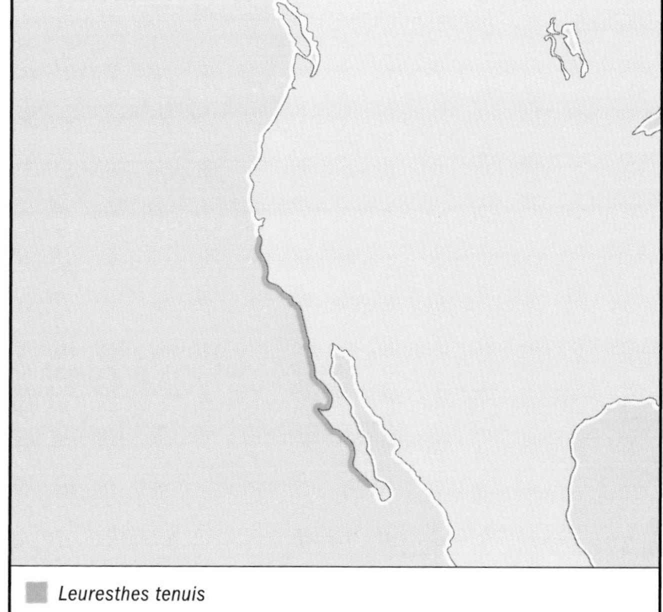

BEHAVIOR
Known for its peculiar spawning behavior. When grunion emerge from the surf to spawn on beaches, they can be collected by hand.

FEEDING ECOLOGY AND DIET
Feeds on zooplankton.

REPRODUCTIVE BIOLOGY
Spawns en masse above the waterline on beaches during highest tides of the spring and summer months. Both males and females ride in with large waves and are left exposed on the sand when the water recedes. Females burrow tail first into the sand to deposit their eggs, and attending males fertilize the eggs as they are released. After spawning, adults return to the sea. They are capable of spawning numerous times. The eggs hatch in about two weeks, upon being agitated by another high tide. Buried grunion eggs provide food for shore birds, crabs, isopods, and beetles.

CONSERVATION STATUS
Not threatened.

SIGNIFICANCE TO HUMANS
Collected by hand during spawning runs. ◆

Inland silverside
Menidia beryllina

FAMILY
Atherinidae

TAXONOMY
Chirostoma beryllinum Cope, 1867, Potomac River, opposite Washington, D.C., at Jackson City, Virginia, United States.

OTHER COMMON NAMES
None known.

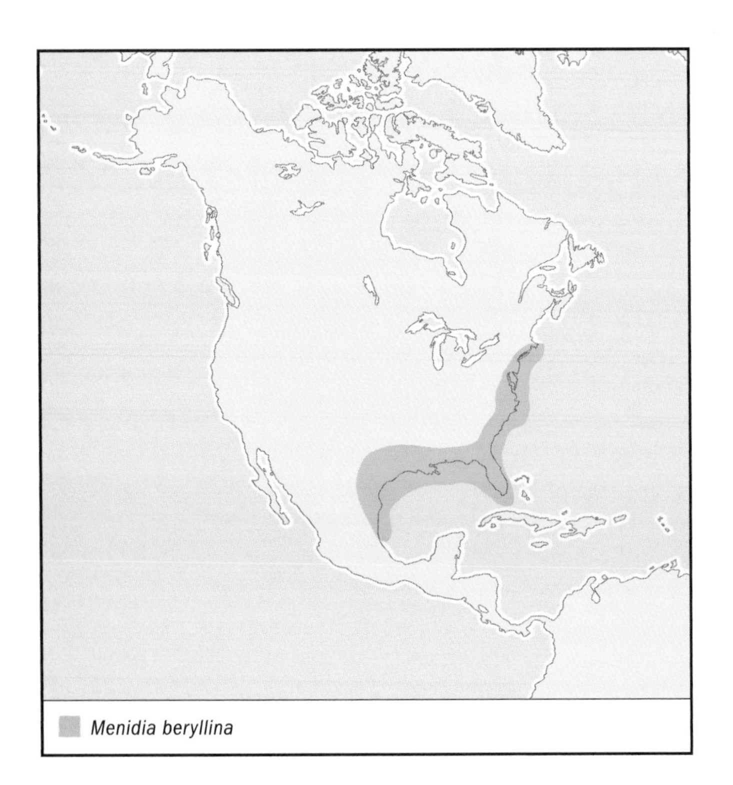

PHYSICAL CHARACTERISTICS
Maximum length of 5.9 in (15.0 cm). Elongate, slender silverside with a fairly compressed body. Silvery lateral band, with a dark line above. Greenish dorsally and whitish ventrally. Scales are smooth to the touch.

DISTRIBUTION
Along the coast of North America from Massachusetts to northern Mexico.

HABITAT
Coastal waters as well as freshwater rivers and streams. Ascends streams and can live entirely in freshwater. Prefers sandy substrate.

BEHAVIOR
Schooling fish that can make the transition between freshwater and saltwater.

FEEDING ECOLOGY AND DIET
Feeds on zooplankton.

REPRODUCTIVE BIOLOGY
Spawns in the spring and summer months.

CONSERVATION STATUS
Not threatened.

SIGNIFICANCE TO HUMANS
Some populations have been established in freshwater impoundments to provide food for sport fishers. ◆

No common name
Rheocles derhami

FAMILY
Bedotiidae

TAXONOMY
Rheocles derhami Stiassny and Rodriguez, 2001, Ambomboa River, Madagascar.

OTHER COMMON NAMES
None known.

PHYSICAL CHARACTERISTICS
Grows to 1.97 in (5 cm) maximum length. Small, moderately deep-bodied fish with relatively large scales. Scales are absent from the nape, chest, and belly. Before spawning, males become darker in color, with blackish blue on the fins and orange-red around the throat; otherwise both males and females are a drab grayish tan. Males have long, filamentous second dorsal, anal, and pectoral fins; this sexual dimorphism is unique within the genus.

DISTRIBUTION
Sofia River drainage in northern Madagascar. Known from the Mangarahara River and one of its tributaries, the Ambomboa River.

HABITAT
Pools in clear streams.

BEHAVIOR
Little is known.

FEEDING ECOLOGY AND DIET
Appears to feed almost exclusively on terrestrial insects.

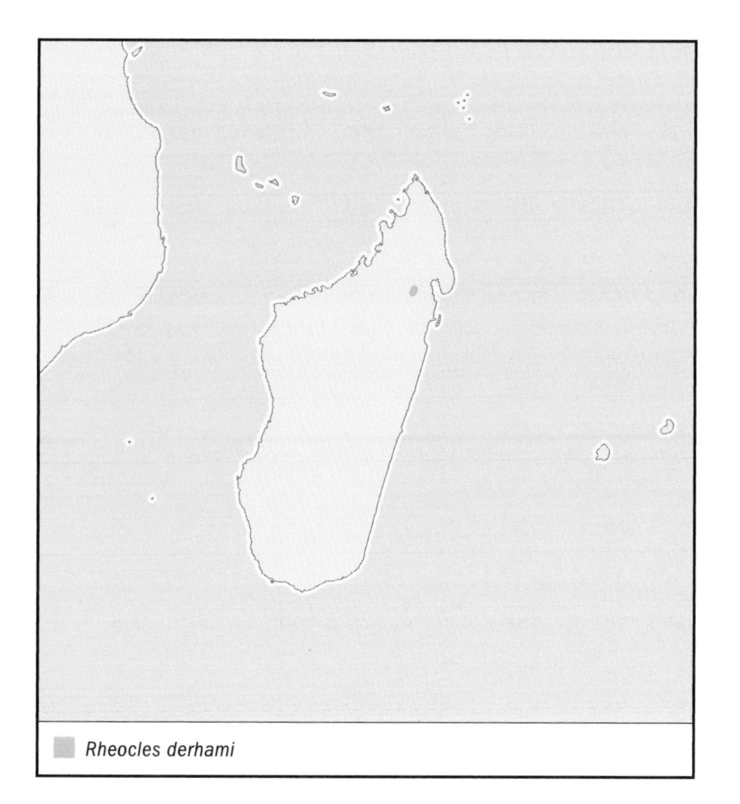

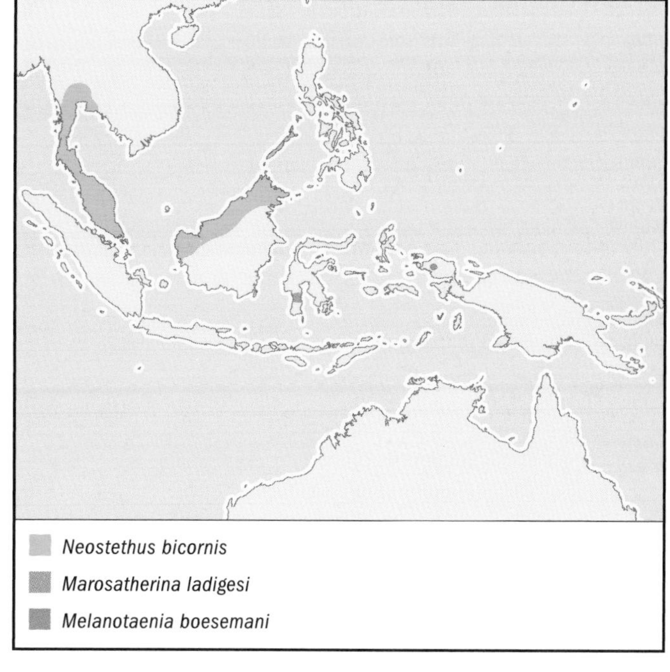

REPRODUCTIVE BIOLOGY
Sexually dimorphic. Spawning behavior is unknown.

CONSERVATION STATUS
Not threatened.

SIGNIFICANCE TO HUMANS
None known. ◆

Boeseman's rainbowfish
Melanotaenia boesemani

FAMILY
Melanotaeniidae

TAXONOMY
Melanotaenia boesemani Allen and Cross, 1980, Ajamaru Lakes, Vogelkop Peninsula, Irian Jaya, Indonesia.

OTHER COMMON NAMES
None known.

PHYSICAL CHARACTERISTICS
Attains 4.5 in (11.5 cm) in maximum length. Laterally compressed and deep bodied, with the head tapering to a point. A striking fish, with blue-gray coloration on the front half of the body and brilliant orange on the back half. Dorsal fins have a white margin.

DISTRIBUTION
Ajamaru Lakes region of Vogelkop Peninsula, New Guinea.

HABITAT
Clear lakes and streams.

BEHAVIOR
Shoaling fishes that feed on insects at the surface.

FEEDING ECOLOGY AND DIET
Omnivorous; prefers terrestrial insects but also feeds on aquatic insects, zooplankton, and algae.

REPRODUCTIVE BIOLOGY
Easy to breed in captivity. Eggs are laid in aquatic plants, and females produce 10–20 eggs per day, spawning daily for an extended period.

CONSERVATION STATUS
Listed as Endangered by the IUCN.

SIGNIFICANCE TO HUMANS
Very popular in the aquarium trade. In the late 1980s, before government restrictions on the fishery, 60,000 males a month were being removed from the wild to be sold as aquarium fish. While rainbows sometimes may be eaten, they are reportedly bony and have a strong formic acid taste, owing perhaps to their preference for ants as a source of food. ◆

Flower of the wave
Iso rhothophilus

FAMILY
Notocheiridae

TAXONOMY
Tropidostethus rhothophilus Ogilby, 1895, Maroubra Bay, N.S.W., Australia.

OTHER COMMON NAMES
English: Surf sardine; French: Surfette commune; Spanish: Rompeolas.

PHYSICAL CHARACTERISTICS
Grows to 2.95 in (7.5 cm) maximum length. Small, laterally compressed fish with a blunt head and a notch in the dorsal

portion of the opercle. Body is deepest around the pectoral fin base, strongly tapering posteriorly and with an abdominal keel. Scales are small, deciduous, and absent anteriorly. Coloration is translucent, with a wide, silvery midlateral band.

DISTRIBUTION
Known from eastern Australia and Taiwan, though likely found throughout Indo-West Pacific.

HABITAT
Typically, rough surf along sandy or rocky coastline; infrequently in estuaries.

BEHAVIOR
Schools in highly oxygenated surf-zone regions around rocky outcroppings or along beaches. Very delicate; does not survive handling.

FEEDING ECOLOGY AND DIET
Likely feeds on zooplankton.

REPRODUCTIVE BIOLOGY
Little is known.

CONSERVATION STATUS
Not threatened.

SIGNIFICANCE TO HUMANS
Rarely caught; not a fish of commercial importance. ◆

No common name
Neostethus bicornis

FAMILY
Phallostethidae

TAXONOMY
Neostethus bicornis Regan, 1916, Kuala Langat, peninsular Malaysia.

OTHER COMMON NAMES
None known.

PHYSICAL CHARACTERISTICS
Small, transparent fish, with urogenital and anal openings placed far anteriorly. Dramatic sexual dimorphism. Males have a complex, asymmetrical structure, the priapium, derived from the pectoral and pelvic girdles and used in reproduction.

DISTRIBUTION
Peninsular Malaysia, Singapore, Thailand, Borneo, Palawan (Philippines).

HABITAT
Fresh and brackish water.

BEHAVIOR
Forms small schools in shallow water near the shore.

FEEDING ECOLOGY AND DIET
Planktivorous.

REPRODUCTIVE BIOLOGY
The male violently swims into the female and knocks her onto her side, gripping her with his ctenactinia, two elongate bones that are part of the priapium. He then attaches a bolus of sperm over her oviduct and swims in rapid spirals and forward bursts that may break the surface of the water, apparently in an effort to break free from the female.

CONSERVATION STATUS
Not listed by the IUCN.

SIGNIFICANCE TO HUMANS
None known. ◆

Celebes rainbowfish
Marosatherina ladigesi

FAMILY
Telmatherinidae

TAXONOMY
Telmatherina ladigesi Ahl, 1936, Makasar, Sulawesi, Indonesia.

OTHER COMMON NAMES
German: Celebes Sonnenstrahlfisch.

PHYSICAL CHARACTERISTICS
Attains maximum length of 3.1 in (8 cm). Laterally compressed and translucent, with bluish midlateral band. Fins bordered in yellow. Elongate second dorsal and anal fins.

DISTRIBUTION
Bantimurung area of south Sulawesi.

HABITAT
Lakes and streams in karst region.

BEHAVIOR
Peaceful community fish in the aquarium.

FEEDING ECOLOGY AND DIET
Omnivorous.

REPRODUCTIVE BIOLOGY
Eggs are attached by a thread to floating plants, in small numbers and over a period of several days. Spawning may last for several months; eggs develop in eight to 11 days. Males behave animatedly during courtship.

CONSERVATION STATUS
Not listed by the IUCN.

SIGNIFICANCE TO HUMANS
A popular aquarium species that is bred commercially in large numbers in Southeast Asia. ◆

Resources

Books

Allen, Gerald R. *Rainbowfishes in Nature and in the Aquarium.* Melle, Germany: Tetra-Verlag, 1995.

Benton, M. J., ed. *The Fossil Record 2.* London: Chapman and Hall, 1993.

Berra, T. M. *Freshwater Fish Distribution.* San Diego: Academic Press, 2001.

Breder, C. M., Jr., and D. E. Rosen. *Modes of Reproduction in Fishes.* Garden City, NY: Natural History Press, 1966.

Resources

Carpenter, K. E., and V. H. Niem, eds. *FAO Species Identification Guide for Fishery Purposes.* Vol. 4, *The Living Marine Resources of the Western Central Pacific.* Rome: FAO, 1999.

Echelle, A. A., and I. Kornfield, eds. *Evolution of Fish Species Flocks.* Orono: University of Maine at Orono Press, 1984.

Fischer, W., F. Krupp, W. Schneider, C. Sommer, K. E. Carpenter, and V. H. Niem, eds. *Guía FAO para Identificacíon de Especies para los Fines de la Pesca Pacifico Centro-Oriental.* Vol. 2. Rome: FAO,1995.

Hieronimus, Harro. *All Rainbows and Related Families.* Mörfelden-Walldorf, Germany: Verlag A.C.S., 2002.

Nelson, J. S. *Fishes of the World.* 3rd edition. New York: John Wiley and Sons, 1994.

Paxton, John R., and William N. Eschmeyer, eds. *Encyclopedia of Fishes.* 2nd edition. San Diego: Academic Press, 1998.

Riehl, Rudiger, and Hans A. Baensch. *Aquarium Atlas,* 2nd edition. Melle, Germany: Baensch, 1989.

Watson, W. "Atherinidae: Silversides." In *The Early Stages of Fishes in the California Current Region,* edited by H. G. Moser. California Cooperative Oceanic Fisheries Investigations (CalCOFI) Atlas no. 33. Lawrence, KS: Allen Press, 1996.

White, B. N., R. J. Lavenberg, and G. E. McGowan. "Atheriniformes: Development and Relationships." In *Ontogeny and Systematics of Fishes,* edited by H. G. Moser. Special Publication of the American Society of Ichthyology and Herpetology. Lawrence, KS: Allen Press, 1984.

Periodicals

Dyer, B. S., and B. Chernoff. "Phylogenetic Relationships Among Atheriniform Fishes (Teleostei: Atherinomorpha)." *Zoological Journal of the Linnean Society* 117, no. 1 (1996): 1–69.

Echelle, A. A., and A. F. Echelle. "Patterns of Abundance and Distribution Among Members of a Unisexual-Bisexual Complex of Fishes (Atherinidae: Menidia)." *Copeia* 1997, no. 2 (1997): 249–259.

Mok, E. Y. M., and A. D. Munro. "Some Anatomical and Behavioural Aspects of Reproduction in Members of an Unusual Teleost Family: The Phallostethidae." *Journal of Natural History* 31, no. 5 (1997): 739–778.

Parenti, L. R. "Relationships of Atherinomorph Fishes (Teleostei)." *Bulletin of Marine Science* 52, no. 1 (January 1993): 170–196.

———. "Phylogenetic Systematics and Biogeography of Phallostethid Fishes (Atherinomorpha, Phallostethidae) of Northwestern Borneo, with Description of a New Species." *Copeia* 1996, no. 3 (1996): 703–712.

Rosen, D. E. "The Relationships and Taxonomic Position of the Halfbeaks, Killifishes, Silversides, and Their Relatives." *Bulletin of the American Museum of Natural History* 127, no. 5 (1964): 217–268.

Rosen, D. E., and L. R. Parenti. "Relationships of *Oryzias,* and the Groups of Atherinomorph Fishes." *American Museum Novitates* 2719 (November 1981): 1–25.

Saeed, B., W. Ivantsoff, and L. E. L. M. Crowley. "Systematic Relationships of Atheriniform Families Within Division I of the Series Atherinomorpha (Acanthopterygii) with Relevant Historical Perspectives." *Journal of Ichthyology* 34 (1994): 27–72.

Stiassny, Melanie L. J. "Notes on the Anatomy and Relationships of the Bedotiid Fishes of Madagascar, with a Taxonomic Revision of the Genus *Rheocles* (Atherinomorpha: Bedotiidae)." *American Museum Novitates* 2979 (August 1990): 1–33.

Stiassny, Melanie L. J. and Damaris M. Rodriguez. "*Rheocles derhami,* a New Species of Freshwater Rainbowfish (Atherinomorpha: Bedotiidae) from the Ambomboa River in Northeastern Madagascar." *Ichthyological Exploration of Freshwaters* 12, no. 2 (2001): 97–104.

Unmack, Peter J. "Biogeography of Australian Freshwater Fishes." *Journal of Biogeography* 28 (2001): 1053–1089.

Robert Schelly, MA

Beloniformes

(Needlefishes and relatives)

Class Actinopterygii
Order Beloniformes
Number of families 5

Photo: A silver needlefish (*Xenentodon cancila*) from Southeast Asia. (Photo by Mark Smith/Photo Researchers, Inc. Reproduced by permission.)

Evolution and systematics

The Beloniformes is one of three orders within the series Atherinomorpha. One of the other two orders, the Atheriniformes, is thought by some ichthyologists to represent an unnatural grouping of several lineages, while others consider it monophyletic (a natural group). The other order, Cyprinodontiformes, is the sister-group of the Beloniformes, as evidenced by numerous internal characters, including modifications to the gill arches and the bones surrounding the eyes. Both Cyprinodontiformes and Beloniformes are agreed to be monophyletic. Beloniformes themselves are united by derived internal characters of the gill arches rather than any conspicuous external morphological characters. Five families of fishes make up the order Beloniformes: Adrianichthyidae (ricefishes), Belonidae (needlefishes), Scomberesocidae (sauries), Exocoetidae (flyingfishes), and Hemiramphidae (halfbeaks). Within these families are 38 genera and about 200 species, 51 of which are either brackish or freshwater, the remainder of which are marine.

The earliest known fossil Beloniformes are just over 50 million years old, and come from two sites: the exocoetids from Monte Bolca in Italy, and the hemiramphids from the Selsey formation in England. Beloniformes are broadly divided into two suborders, the Exocoetoidei (beaked forms: belonids, scomberesocids, exocoetids, and hemiramphids) and the Adrianichthyoidei (lacking a beak; adrianichthyids). Adrianichthyids were traditionally included within the Cyprinodontiformes until Rosen and Parenti argued for their inclusion within the Beloniformes in 1981, based mainly on characters of the gill arches and hyoid apparatus. Adrianichthyids are now considered to be the sister lineage to the rest of the order, within which the sister groups Belonidae-Scomberesocidae and Hemiramphidae-Exocoetidae have been suggested.

Contrary to this scheme, a study based on morphology and molecules, published in 2000 by Lovejoy, places halfbeaks as the ancestral form among the beaked beloniforms. In Lovejoy's tree, some halfbeaks are most closely related to needlefishes and sauries, while the marine halfbeak *Hemiramphus* is sister to flyingfishes. This result refutes an old hypothesis, based on the observation that needlefishes pass through a halfbeak stage during their development, that halfbeaks derive from needlefishes via truncation of the development sequence.

Physical characteristics

Beloniformes are typically elongate fishes, with dorsal and anal fins situated posteriorly on the body and the lateral line situated ventrally. Additional characteristics of the group include fusion of the toothed 5th ceratobranchials into a lower pharyngeal jaw and an open nasal pit. Belonids, aptly called needlefishes, are sleek and garlike piscivores, with very long upper and lower jaws studded with sharp teeth. They can achieve lengths of up to 3.3 ft (1 m). A small number of needlefishes have a reduced upper jaw, and like halfbeaks, feed on plankton and insects. Scomberesocids, of which the largest are about 1.65 ft (0.5 m) long, can be distinguished from belonids by the five or six finlets behind their dorsal and anal fins. The diminutive scomberesocid *Cololabis adocetus*, at 3 in (7.5 cm), is the smallest fish in the surface waters of the open ocean.

In most species of hemiramphids, or halfbeaks, the lower jaw is much longer than the upper. The front margin of the upper jaw is triangular in shape, the scales are large and cycloid, and fin spines are lacking. Exocoetids, the flyingfishes, are torpedo shaped with greatly enlarged pectoral fins, and the lower lobe of the caudal fin is stiffened and much larger than the upper. Interestingly, the most primitive flyingfish genera, *Oxyporhamphus* (once included with hemiramphids),

The four-winged flyingfish, *Cheilopogon pinnatibarbatus* (family Exocoetidae), taxis by rapidly beating its caudal fin in the water. It then achieves free flight, for a distance up to 160 ft (50 m). (Illustration by Emily Damstra)

Fodiator, and *Parexocoetus*, have elongate lower jaws reminiscent of halfbeaks. More derived flyingfishes have acquired oversized pelvic fins in addition to large pectoral fins, and are called four-wingers.

Adrianichthyids, the most basal among the Beloniformes, are superficially unlike other members of the group. Most of the species are in the genus *Oryzias*, and are small, relatively deep-bodied fishes with large eyes, upturned mouths, and a long anal fin base. Noteworthy in the family is the duckbilled buntingi (*Adrianichthys kruyti*), which has a bill-shaped mouth with the upper jaw overhanging the lower. *Xenopoecilus* species also have a bill-shaped mouth, and the carry their eggs at the base of the pelvic fins by way of filamentous attachments.

Distribution

Beloniformes are widely distributed in temperate and tropical marine and fresh waters. Adrianichthyids are found in Asian fresh and brackish waters from India to Japan, and in the Indo-Australian archipelago. Belonids are found in the open ocean in tropical and temperate seas worldwide, with numerous species in the freshwaters of South America and some in Asia. Like marine belonids, scomberesocids are widely distributed in warmer waters of the open ocean. Exocoetids are found in warm waters of the Atlantic, Pacific, and Indian Oceans. Hemiramphids have a similar distribution in marine waters, but have also invaded freshwaters, especially in the Indo-Australian region.

Habitat

Marine beloniforms can be found in the surface waters of the open ocean, as well as in coastal habitats like estuaries and mangrove swamps. Adrianichthyids, belonids, and hemiramphids can be found in a diversity of tropical freshwater habitats, including lakes and rivers.

Behavior

One characteristic of many beloniforms is a strong attraction to lights at night. This behavior is exploited by fishermen, who use lights to capture schools of sauries that cruise just below the ocean surface. Such fishing methods, however, involve a peculiar (although infrequent) hazard: impalement by large needlefishes. In one documented case, a 3.3 ft (1 m) long *Tylosurus* fatally impaled a fisherman after jumping toward the light on board his canoe.

Certainly the most remarkable beloniform behavior is exocoetid flight. (It should be noted that flight is not entirely restricted to exocoetids: Some hemirhamphids are capable of gliding, and *Euleptorhamphus viridis* has been reported to travel 164 ft [50 m] in two jumps.) In the more derived four-winged flyingfish species, flight is achieved as follows. The fish, swimming at a speed of about 33 ft/s (10 m/s), breaks the surface at an oblique angle and taxis for 16.4–82 ft (5–25 m) by rapidly beating the caudal fin in the water. Then a free flight ensues, which may span a distance of 164 ft (50 m) and reach a height of 26.2 ft (8 m). Once the fish loses altitude, caudal fin taxiing can be repeated without returning to the water, so that flights can be stretched to distances of 1,312 ft (400 m). Intriguingly, flyingfishes seem to sense wind direction and take off into the wind, and tantalizing evidence suggests that they can control the direction of flight; *Cypsilurus* appear to successfully seek out patches of seaweed in which to land. So why do flyingfishes fly? One of the most likely hypotheses is that flight has evolved as a tactic for evading predators.

Feeding ecology and diet

Beloniformes utilize a relatively wide spectrum of foods. Most impressive perhaps are the marine needlefishes, which cruise through the surface waters of the open ocean devouring small fishes. However, not all needlefishes are piscivores. In the Amazon, many belonids feed heavily on zooplankton or insects. *Belonion apodion*, which grows only to about 2 in (5 cm), is unusual in that it deftly snaps up individual rotifers, which are less than 0.004 in (0.1 mm) long and usually pass through the gill rakers of filter-feeding planktivorous fishes. *Potamorrhaphis*, which prefers terrestrial insects (especially flying ants), hovers motionlessly and waits for prey to fall to the surface alongside its body. Then it rapidly curls the body and strikes at the prey from the side.

Freshwater halfbeaks also feed on terrestrial insects, and some are particularly well suited to this mode of feeding.

Members of the genus *Hemirhamphodon* are noteworthy for having numerous anteriorly directed teeth on their lower jaws, which ensnare ants and other insects found floating on the surface. Marine halfbeaks, on the other hand, tend to feed on algae, diatoms, and sea grasses, though some species eat small fishes. Planktivorous marine beloniforms include the flyingfishes and sauries. Ricefishes are omnivorous and will eat plankton, small insects, detritus, and plant material.

Beloniformes themselves often fall prey to larger fishes. Flyingfishes in particular are eaten by mackerel, tuna, and marlin, among other predatory fishes, as well as squids and birds.

Reproductive biology

Much of what is known about beloniform reproductive biology involves the eggs and larvae. Typically, eggs develop in one to two weeks, and the larvae are immediately able to feed upon hatching. Many pelagic beloniform eggs have filamentous projections that cause them to stick to floating debris. Needlefish eggs have tendrils that are particularly sticky, and they form egg clusters that stick to other objects in the water. Likewise, sauries produce filamentous eggs that float in open water, but they are less adhesive than needlefish eggs. Flyingfishes lay pelagic eggs that may or may not have filaments, and some species attach their eggs to floating seaweed. Marine halfbeaks lay eggs with tendrils that float about in open water, but some freshwater representatives bear live young, namely *Dermogenys*, *Nomorhamphus*, and *Hemirhamphodon*. In these viviparous forms, long genital papillae are used for internal fertilization, and the male anal fin is modified into an andropodium.

The adrianichthyid *Horaichthys* from India, uniquely among atherinomorphs, produces encapsulated sperm bundles, or spermatophores. In adrianichthyids other than *Oryzias*, fertilization is apparently internal. The eggs of many species of adrianichthyids are retained externally by the female for various lengths of time. Females of the species *Xenopoecilus oophorus*, known as the egg-carrying buntingi, carry a cluster of about 30 fertilized eggs attached by filaments in an external concavity near the vent. The pelvic fins cover and protect this egg mass.

Conservation status

No Beloniformes are CITES listed, or listed as endangered by the U. S. Fish and Wildlife Service. However, 16 species are included on the IUCN Red List. Eight of those, mostly *Oryzias* species, are categorized as Vulnerable, one species is listed as Lower Risk/Near Threatened, and two species are listed as Data Deficient. Three species are listed as Endangered: *Oryzias orthognathus*, *Xenopoecilus oophorus*, and *Xenopoecilus sarasinorum*. Listed as Critically Endangered are *Adrianichthys kruyti* and *Xenopoecilus poptae*, both of which are known only from Lake Poso, Sulawesi. Although no Beloniformes are formally listed as Extinct in the Wild by the U. S. Fish and Wildlife Service or Extinct by the IUCN, *Adrianichthys kruyti* is generally thought to be extinct. Pressure from an introduced species of catfish, in addition to parasites that entered the lake with the catfish, are implicated in the decline of the Lake Poso adrianichthyids.

Significance to humans

Many Beloniformes are fished at night using lights, and some rather creative methods have been devised. Where flyingfishes are abundant, fishermen that leave a light suspended all night over a canoe partially filled with water can return in the morning to a boat full of fresh fish. The fish are drawn to the light and jump into the canoe, but have too little water to jump back out. Flyingfishes are also attracted to leaves or straw scattered about the surface as a place to lay their eggs, and can be fished by using such material.

Some Beloniformes are used by humans as more than just food. Numerous freshwater species, including halfbeaks, ricefishes, and needlefishes, can be found in the aquarium trade. In Thailand, the halfbeak, *Dermogenys pusillus*, is bred in captivity so that males, which will engage rivals by locking jaws, can be used as fighting fish. Members of the genus *Oryzias* are propagated in large numbers in captivity to be used in experimental research.

Finally, it should be mentioned that needlefishes can in rare cases be traumatogenic, causing injury or death by means of impalement. In one such case, a hapless surfer was killed when the snout of a fast-swimming needlefish went through his eye socket and into his brain.

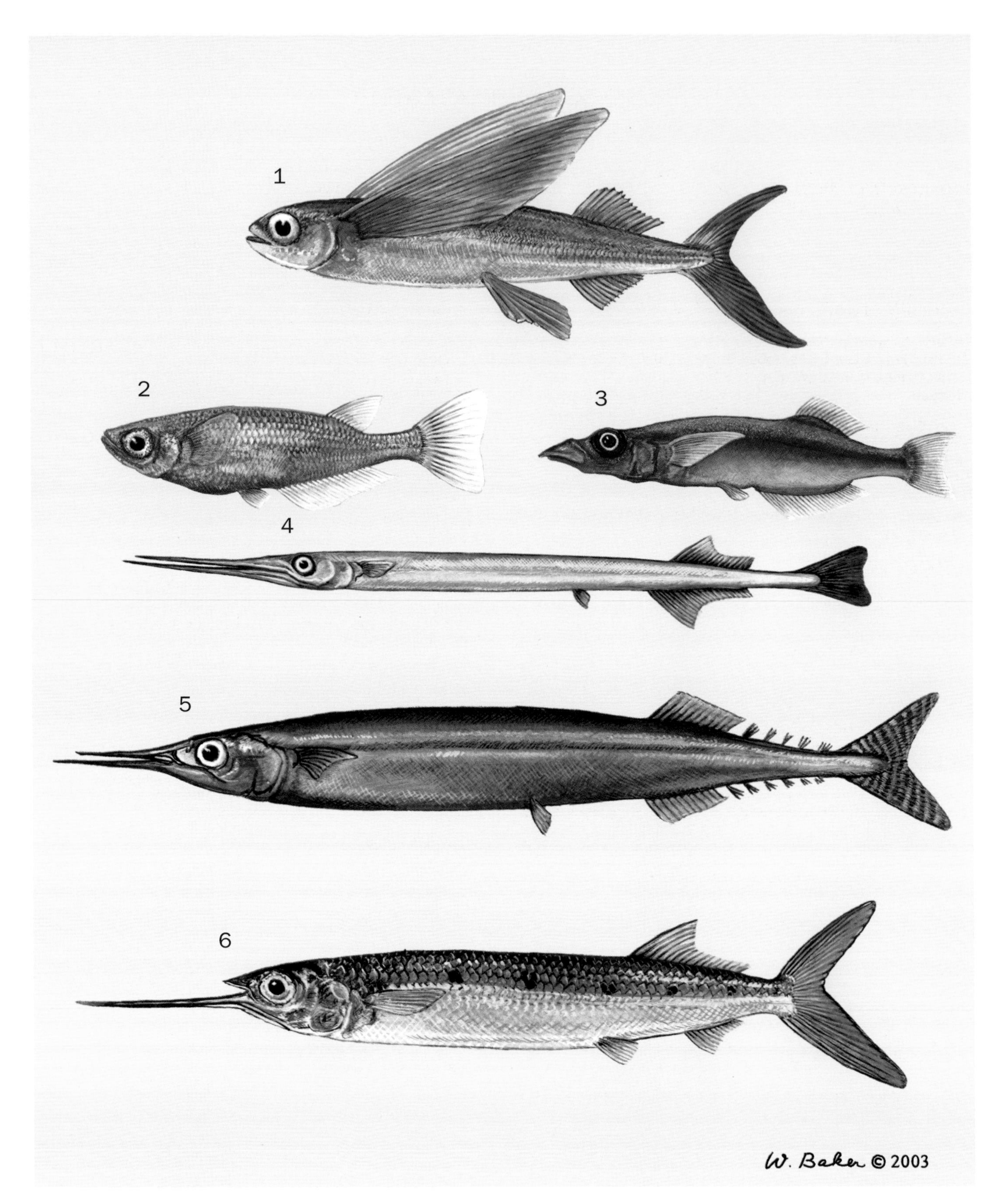

1. California flyingfish (*Cheilopogon pinnatibarbatus californicus*); 2. Japanese rice fish (*Oryzias latipes*); 3. Duckbilled buntingi (*Adrianichthys kruyti*); 4. Californian needlefish (*Strongylura exilis*); 5. Atlantic saury (*Scomberesox saurus saurus*); 6. Blackbarred halfbeak (*Hemiramphus far*). (Illustration by Wendy Baker)

Species accounts

Japanese rice fish
Oryzias latipes

FAMILY
Adrianichthyidae

TAXONOMY
Poecilia latipes Temminck and Schlegel, 1846, Japan.

OTHER COMMON NAMES
English: Japanese medaka, tooth-carp; German: Japan-Reiskärpfling; Cantonese: Fut mei dzeung ue; Japanese: Medaka.

PHYSICAL CHARACTERISTICS
Maximum length 1.6 in (4 cm). Small and shallow bodied, with upturned mouth and silvery olive coloration. No spines in dorsal or anal fins. Many strains of captive-raised Japanese rice fish have been selectively bred for pale yellow color. Strains that appear red or mottled black and gold have also been developed.

DISTRIBUTION
Japan, Korea, China, and Vietnam, as well as the great rivers of Southeast Asia: the Mekong, Red, Irrawaddy, and Salween.

HABITAT
Calm stretches of streams, rice paddies, and wetlands.

BEHAVIOR
Forms schools, generally peaceful.

FEEDING ECOLOGY AND DIET
Feeds on zooplankton and insects, as well as some detritus and plant material.

REPRODUCTIVE BIOLOGY
Fertilization is external, although the eggs are carried for a short time, stuck to the female's abdomen, prior to deposition. Females can produce broods of 10–40 eggs every two days during the breeding season. Eggs are slightly larger than 0.039 in (1 mm) in diameter and usually hatch in 8–14 days.

CONSERVATION STATUS
Not threatened.

SIGNIFICANCE TO HUMANS
Used widely in experimental research, also found in the aquarium trade. ◆

Duckbilled buntingi
Adrianichthys kruyti

FAMILY
Adrianichthyidae

TAXONOMY
Adrianichthys kruyti Weber, 1913, Lake Poso, Sulawesi, Indonesia.

OTHER COMMON NAMES
English: Duckbill Poso minnow; German: Entenschnabelkärpfling.

PHYSICAL CHARACTERISTICS
Maximum length about 4.3 in (11 cm). Large, horizontal, duck-bill shaped mouth, with upper jaw overhanging lower; eyes large, extend beyond dorsal head-profile when viewed from side. Elongate, somewhat compressed.

DISTRIBUTION
Lake Poso, Sulawesi.

HABITAT
Deeper waters of Lake Poso.

BEHAVIOR
Unknown.

FEEDING ECOLOGY AND DIET
Unknown.

REPRODUCTIVE BIOLOGY
Unknown, though one specimen reported to be a hermaphrodite.

CONSERVATION STATUS
Listed as Critically Endangered on the IUCN Red List; Harrison and Stiassny (1999) think it possibly extinct.

SIGNIFICANCE TO HUMANS
Kottelat reported that the adrianichthyids he observed were heavily parasitized by copepods; native fisherman say these parasites became a problem when *Clarias* was introduced into Lake Poso in the early 1980s. Voracious snakeheads have also been introduced into the lake, and may have led to the decline

of the endemic fishes. He argues that the duckbilled buntingi may not be extinct, but just no longer abundant enough for fisherman to expend effort and so never observed. Whatever its status, *Adrianichthys kruyti* is an example of an endemic species important as a fishery declining dramatically following exotic species introductions by humans. ◆

Atlantic saury

Scomberesox saurus saurus

FAMILY
Scomberesocidae

TAXONOMY
Esox saurus Walbaum, 1792, Cornwall and British seas.

OTHER COMMON NAMES
English: Atlantic needlefish; French: Aiguille de mer; German: Echsenhecht; Spanish: Alcrique.

PHYSICAL CHARACTERISTICS
Maximum length 19.7 in (50 cm). Elongate and needlefish-like, but with toothless jaws and finlets following the dorsal and anal fins.

DISTRIBUTION
Mediterranean, North Atlantic, rarely Iceland, Norway, Denmark.

HABITAT
Surface waters of the open ocean.

BEHAVIOR
Schooling fishes that travel long distances through the open ocean. Spawn in warmer waters, migrate to plankton-rich temperate waters to feed. Leap out of the water when pursued by predators.

FEEDING ECOLOGY AND DIET
Zooplankton and fish larvae.

REPRODUCTIVE BIOLOGY
External fertilization, eggs scattered in open water.

CONSERVATION STATUS
Not threatened.

SIGNIFICANCE TO HUMANS
Said to have delicious flesh, but not abundant enough to support a large fishery. ◆

Californian needlefish

Strongylura exilis

FAMILY
Belonidae

TAXONOMY
Belone exilis Girard, 1854, San Diego, California.

OTHER COMMON NAMES
French: Aiguille de Californie; Spanish: Agujón bravo de California.

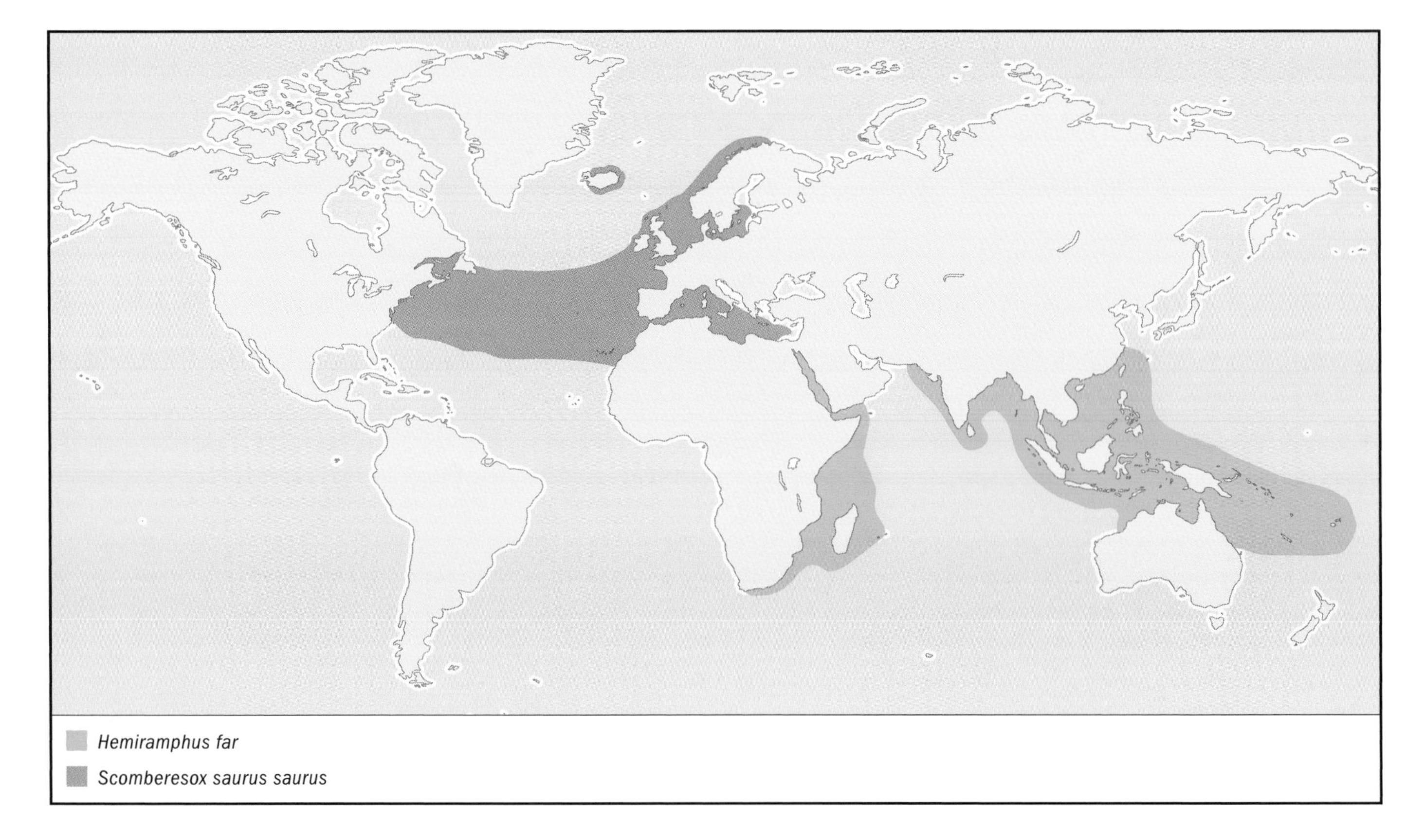

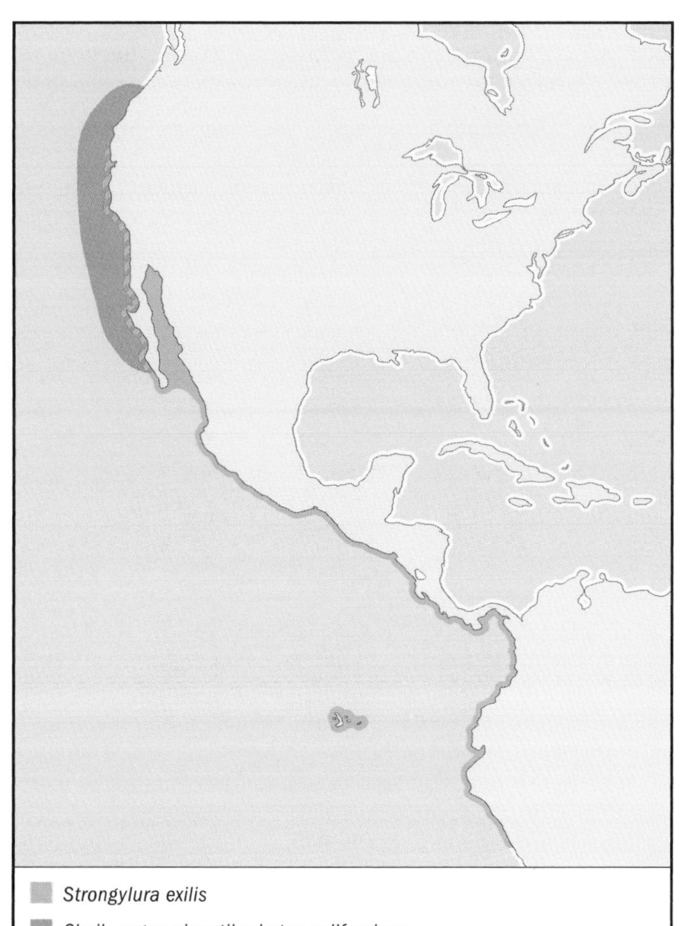

PHYSICAL CHARACTERISTICS
Maximum length 35.8 in (91 cm). Very elongate, with long snout and sharp teeth. Emarginate caudal fin, no dorsal or anal fin spines.

DISTRIBUTION
Coastally from San Francisco to Peru; also in the Galápagos.

HABITAT
Lagoons, harbors, and coastal areas. Frequents mangroves and enters freshwaters.

BEHAVIOR
Sometimes schools in large numbers, leaps out of the water when threatened.

FEEDING ECOLOGY AND DIET
Feeds on small fishes.

REPRODUCTIVE BIOLOGY
Eggs are attached to floating vegetation by means of long filaments; larvae drift in surface waters. The eggs are approximately 0.14 in (3.5 mm) in diameter and hatch in about two weeks. Larvae are 0.35–0.47 in (9–12 mm) at hatching.

CONSERVATION STATUS
Not threatened.

SIGNIFICANCE TO HUMANS
Sold fresh in fish markets. In very rare cases, may cause injury or death by impalement. ◆

Blackbarred halfbeak
Hemiramphus far

FAMILY
Hemiramphidae

TAXONOMY
Esox far Forsskål, 1775, Luhaiya, Yemen, Red Sea.

OTHER COMMON NAMES
English: Blackbarred garfish; French: Aiguillette, demi-bec bagnard.

PHYSICAL CHARACTERISTICS
Maximum length 17.7 in (45 cm). Lower jaw beaklike and dramatically longer than upper jaw; color is bluish dorsally and silvery on sides, with three to nine vertical bars. Dorsal and anal fins posteriorly situated, lower lobe of caudal fin longer than upper.

DISTRIBUTION
Indo-West Pacific: Red Sea and East Africa to Samoa, from northern Australia and New Caledonia to the Ryukyu Islands. Has entered the eastern Mediterranean through the Suez Canal.

HABITAT
Marine; near vegetation in coastal areas.

BEHAVIOR
Forms schools.

FEEDING ECOLOGY AND DIET
Feeds mostly on sea grasses, in addition to green algae and diatoms.

REPRODUCTIVE BIOLOGY
Spawns in estuaries.

CONSERVATION STATUS
Not threatened.

SIGNIFICANCE TO HUMANS
Commercially fished, said to have good-tasting flesh, sometimes used as bait. ◆

California flyingfish
Cheilopogon pinnatibarbatus californicus

FAMILY
Exocoetidae

TAXONOMY
Exocoetus californicus Cooper, 1863, Santa Catalina Island, California.

OTHER COMMON NAMES
French: Exocet californien; Spanish: Volador de California.

PHYSICAL CHARACTERISTICS
Maximum length 15 in (38 cm). One of the "four-winged" flyingfishes; both pectoral and pelvic fins are enlarged. Lower lobe of caudal fin is also considerably larger than the upper. Bluish gray dorsally, silver ventrally.

DISTRIBUTION
Oregon to southern Baja California.

HABITAT
Surface waters of the open ocean.

BEHAVIOR
Schooling fishes, capable of leaping out of the water and gliding for long distances, possibly as a means of evading predators.

FEEDING ECOLOGY AND DIET
Zooplankton and small fishes.

REPRODUCTIVE BIOLOGY
Spawns in the summer months. Eggs are pelagic, and stick to floating seaweed and other debris. The eggs are approximately 0.07 in (1.8 mm) in diameter. Larvae are roughly 0.17 in (4.5 mm) long at hatching.

CONSERVATION STATUS
Not threatened.

SIGNIFICANCE TO HUMANS
Occasionally used as bait. ◆

Resources

Books

Berra, T. M. *Freshwater Fish Distribution.* San Diego: Academic Press, 2001.

Breder, C. M., Jr., and D. E. Rosen. *Modes of Reproduction in Fishes.* Garden City, NY: The Natural History Press, 1966.

Collette, B. B., G. E. McGowen, N.V. Parin, and S. Mito. "Beloniformes: Development and Relationships." In *Ontogeny and Systematics of Fishes,* edited by H. G. Moser. Lawrence, KS: Allen Press, 1984.

Fischer, W., F. Krupp, W. Schneider, C. Sommer, K. E. Carpenter, and V. Niem, eds. *Guia FAO para Identification de Especies para lo Fines de la Pesca. Pacifico Centro-Oriental,* Vol. 2. Rome: FAO, 1995.

Harrison, I. J., and M. L. J. Stiassny. "The Quiet Crisis: A Preliminary Listing of the Freshwater Fishes of the World that Are Extinct or 'Missing in Action.'" In *Extinctions in Near Time,* edited by R. D. E. MacPhee. New York: Kluwer Academic/Plenum Publishers, 1999.

Nelson, J. S. *Fishes of the World.* 3rd edition. New York: John Wiley & Sons, 1994.

Parin, N. V. "Exocoetidae: Flyingfishes." In *FAO Species Identification Guide for Fishery Purposes. The Living Marine Resources of the Western Central Pacific,* Vol. 4. *Bony Fishes, Part 2 (Mugilidae to Carangidae),* edited by K. E. Carpenter and V. H. Niem. Rome: FAO, 1999.

Patterson, C. "Osteichthyes: Teleostei." In *The Fossil Record 2,* edited by M. J. Benton. London: Chapman and Hall, 1993.

Paxton, J. R., and W. N. Eschmeyer, eds. *Encyclopedia of Fishes.* 2nd edition. San Diego: Academic Press, 1998.

Riehl, R., and H. A. Baensch. *Aquarium Atlas.* Melle, Germany: Baensch, 1986.

Periodicals

Anderson, W. D., III, and B. B. Collette. "Revision of the Freshwater Viviparous Halfbeaks of the Genus *Hemirhamphodon* (Teleostei: Hemiramphidae)." *Ichthyological Exploration of Freshwaters* 2 (1991): 151–176.

Boughton, D. A., B. B. Collette, and A. R. McCune. "Heterochrony in Jaw Morphology of Needlefishes (Teleostei: Belonidae)." *Systematic Zoology* 40 (1991): 329–354.

Coates, D., and P. A. M. Van-Zwieten. "Biology of the Freshwater Halfbeak *Zenarchopterus kampeni* (Teleostei: Hemiramphidae) from the Sepik and Ramu River Basin, Northern Papua New Guinea." *Ichthyological Exploration of Freshwaters* 3 (1992): 25–36.

Dasilao, J. C., Jr., and K. Sasaki. "Phylogeny of the Flyingfish Family Exocoetidae (Teleostei, Beloniformes)." *Ichthyological Research* 45 (1998): 347–353.

Davenport, J. "How and Why do Flying Fish Fly?" *Reviews in Fish Biology and Fisheries* 4 (1994): 184–214.

Goulding, M., and M. L. Carvalho. "Ecology of Amazonian Needlefishes (Belonidae)." *Revista Brasileira de Zoologia* 2 (1984): 99–111.

Kottelat, M. "Synopsis of the Endangered Buntingi (Osteichthyes: Adrianichthyidae and Oryziidae) of Lake Poso, Central Sulawesi, Indonesia, with a New Reproductive Guild and Descriptions of Three New Species." *Ichthyological Exploration of Freshwaters* 1 (1990): 49–67.

Lovejoy, N. R. "Reinterpreting Recapitulation: Systematics of Needlefishes and Their Allies (Teleostei: Beloniformes)." *Evolution* 54 (2000): 1,349–1,362.

Lovejoy, N. R., and B. B. Collette. "Phylogenetic Relationships of New World Needlefishes (Teleostei: Belonidae) and the Niogeography of Transitions Between Marine and Freshwater Habitats." *Copeia* 2001, no. 1 (2001): 324–338.

Lovejoy, N. R., and M. L. G. De-Araujo. "Molecular Systematics, Biogeography and Population Structure of Neotropical Freshwater Needlefishes of the Genus *Potamorrhaphis.*" *Molecular Ecology* 9 (2000): 259–268.

Meisner, A. D. "Phylogenetic Systematics of the Viviparous Halfbeak Genera *Dermogenys* and *Nomorhamphus* (Teleostei: Hemiramphidae: Zenarchopterinae)." *Zoological Journal of the Linnaean Society* 133 (2001): 199–283.

Mok, E. Y. M., and A. D. Munro. "Observations on the Food and Feeding Adaptations of Four Species of Small Pelagic Teleosts in Streams of the Sungei Buloh Mangal, Singapore." *Raffles Bulletin of Zoology* 39 (1991): 235–257.

Parenti, L. R. "Relationships of Atherinomorph Fishes (Teleostei)." *Bulletin of Marine Science* 52, no. 1 (Jan. 1993): 170–196.

Parenti, L. R., and D. E. Rosen. "Relationships of *Oryzias,* and the Groups of Atherinomorph Fishes." *American Museum Novitates* 2,719 (Nov. 1981): 1–25.

Parin, N. V., and D. A. Astakhov. "Studies on the Acoustic-Lateralis System of Beloniform Fishes in Connection with Their Systematics." *Copeia* 1982, no. 2 (1982): 276–291.

Resources

Rosen, D. E. "The Relationships and Taxonomic Position of the Halfbeaks, Killifishes, Silversides, and Their Relatives." *Bulletin of the American Museum of Natural History* 127, no. 5 (1964): 217–268.

Whitten, A. J., S. V. Nash, K. D. Bishop, and L. Clayton. "One or More Extinctions from Sulawesi, Indonesia." *Conservation Biology* 1, no. 1 (May 1987): 42–48.

Robert Schelly, MA

Cyprinodontiformes
(Killifishes and live-bearers)

Class Actinopterygii
Order Cyprinodontiformes
Number of families 9

Photo: A male blue platy (*Xiphophorus maculatus*) from Central America. (Photo by Mark Smith/Photo Researchers, Inc. Reproduced by permission.)

Evolution and systematics

As early as 1828 Wagner recognized a suprageneric category for the group known to us today as the Cyprinodontiformes. In 1835 Agassiz erected the Cyprinodontes as the family containing the genera *Cyprinodon*, *Lebias*, *Molinesia* (correctly spelled *Mollienesia*), *Poecilia*, and *Anableps*. By 1883 the term Cyprinodontidae was in general use for 30 genera and 130 species, as reflected in the work of Jordan and Gilbert. In his classic work *The Cyprinodonts*, published in 1895, Garman arranged the many genera of the Cyprinodontes into eight subfamilies, and while he erroneously included the characin genus *Neolebias* and the cyprinid genus *Fundulichthys*, his systematic view of the constituents of the Cyprinodontiformes is about the same today. This is reflected in the work of Berg, who formally erected the group as the order Cyprinodontiformes in 1940.

In 1964 Rosen placed the Cyprinodontiformes in the superfamily Cyprinodontoidea in the order Atheriniformes. As of 2002 the Cyprinodontiformes were considered a natural group (i.e., a monophyletic group) most closely related to the Beloniformes, the order containing halfbeaks, medakas, needlefishes, sauries, and flyingfishes. Rosen took Garman's eight subfamilies and ordered them into five families, the Cyprinodontidae for all the oviparous genera, the Anablepidae for the viviparous genus *Anableps*, the Jenynsiidae for the viviparous genus *Jenynsia*, the Goodeidae for the genera of the viviparous splitfins, and the Poeciliidae for the viviparous genera with gonopodium. In 1924 Hubbs had argued for the placement of *Anableps* and *Jenynsia* into the family Anablepidae, a view not adopted by Rosen. The five families of Rosen correspond today to the order Cyprinodontiformes. In Rosen's work there was a clearly defined family-level separation of the lineages into viviparous and oviparous. Essentially, the five families were thought to be related, but there were no proposals about the details of the relationships among the five families or their genera.

It was not until the iconoclastic work of Parenti in 1981 that relationships for the cyprinodontiform genera and their families were proposed. Shared derived characters (evolutionary novelties), not primitive characters, were used to define the various evolutionary lineages in a systematic phylogenetic analysis of the order. Almost all the known genera were reevaluated, and a comprehensive cladogram was constructed to illustrate the proposed interrelationships of the genera and the families into which they were placed. For the first time the order Cyprinodontiformes was defined using only derived characters, consisting of a suite of various osteological features and, in general, a long developmental period and early breeding habits. In the new ordering the viviparous families did not form a monophyletic group, because some live-bearers turned out to be related more closely to oviparous species than to other live-bearers. There were numerous taxonomic and nomenclature changes as the result of the restructuring of the genera as well as the proposed scheme of their relationships.

Three taxonomic arrangements were proposed, wherein viviparous genera were deemed to be sister taxa to oviparous genera. This kind of relationship had not been contemplated previously and constituted a paradigm shift within cyprinodontiform systematics. The viviparous genera *Anableps* and *Jenynsia* were recognized as sister taxa because of shared characters in their reproductive biology. These, in turn, were considered the sister group of the oviparous *Oxyzygonectes*, which had been aligned with *Fundulus*. These genera constitute the Anablepidae. The viviparous family Goodeidae, with 17 genera, commonly known as splitfins, was realigned with the sister group formed by the oviparous genera *Crenichthys* and *Empetrichthys*. This group now constitutes the family Goodeidae, which may be considered to have two subfamilies, the Goodeinae and the Empetrichthyinae. The Poeciliidae, with 27 genera, was placed in a clade with the oviparous African lampeyes, *Aplocheilichthys*, and its related genera and the

Pacific foureyed fish (*Anableps dowi*) from Central America. (Photo by William E. Townsend, Jr./Steinhart Aquarium/Photo Researchers, Inc. Reproduced by permission.)

oviparous genus *Fluviphylax.* That group, in turn, was hypothesized to be the sister group of the Anablepidae.

As a result of these revisions, the killifishes, that is, the oviparous cyprinodontiform genera, are no longer considered to constitute a monophyletic group in the scientific sense, but in the vernacular the term is used commonly. It also should be noted that the Cyprinodontiformes were subdivided into two suborders, the Aplocheiloidei and Cyprinodontoidei, both of which spanned the continents of South America and Africa, with interesting zoogeographical implications.

Four major revisions of the Cyprinodontiformes have been published since 1981. Using both molecular and osteological methods, overall cyprinodontiform taxonomy was revisited, the phylogenetic relationships of the Old World and New World aplocheiloids were revised sharply, and a major restructuring of the Poeciliidae was undertaken. An additional molecular study of the phylogeny of the family Rivulidae and its two subfamilies, the Rivulinae and Cynolebiatinae, was published in 1999, and this may lead to a radical taxonomic and nomenclatural revision of these two subfamilies.

The fossil record does little to illuminate cyprinodontiform origins. The earliest fossils, the cyprinodontid *Pachylebias* and *Prolebias,* both found in Europe, date to the Oligocene epoch, 25–40 million years ago (mya). Most fossils date to the Miocene, 10–25 mya. The fossil record is relatively recent compared with the inferred history of the Cyprinodontiformes. The phylogenetic relationships of recent killifishes and live-bearers accord well with the realities of plate tectonics and the breaking up of Gondwana, the single supercontinent formed by present-day South America, Africa, Antarctica, Australia, New Zealand, Madagascar, and India plus an assortment of other small plates not part of the continent of Laurasia, which was positioned to the north of Gondwana.

The distribution of the Cyprinodontiformes on today's widely separated pieces of Gondwana argues for a very ancient origin of this order of fishes, more than three times the age of the oldest-known fossil killifishes. The killifishes and live-bearers originated on Gondwana and were contemporaneous with the dinosaurs. Plate tectonics (continental drift) carried these freshwater fishes to their present locations on a journey that began with the breakup of Gondwana 140 mya. The mountain killifishes, genus *Orestias,* found in the Altiplano regions of Peru, Bolivia, and Chile, were in place long before the rising of the Andes lifted them to their lofty positions. In general, one can expect some dispersal within zoogeographical areas, but the distribution of freshwater fish families is a result mainly of continental drift.

Before plate tectonic theory, freshwater fish distributions presented great puzzles, the answers to which were sometimes fanciful speculations. For instance, both the African killifish genus *Aphyosemion* and the South American killifish genus *Rivulus* were placed in the same subfamily, the Rivulinae. If one thinks of the continents as always being in the same positions, how can the freshwater fishes of these two continents possibly be related, since there is no way small freshwater fishes can swim from Africa to South America? One of the more fanciful hypotheses, ludicrous by today's more sophisticated standards, was that a series of islands spanned the Atlantic Ocean from South America to Africa. The freshwater fishes supposedly swam from island to island, thus accounting for the separation of their families. Then the islands conveniently disappeared without a trace. Another theory held that the continents had moved apart because the earth was expanding. (Two marks on a balloon grow farther apart as the balloon is inflated.) That theory suffers from a lack of any plausible mechanism, although in its early days the theory of continental drift was subjected to the same criticism, a criticism that was answered by the now widely accepted hypotheses of seafloor spreading and subduction.

The systematics of the cyprinodontoid family Poeciliidae parallels the biogeographical situation outlined for the aplocheiloid genera *Rivulus* and *Aphyosemion,* since part of the Poeciliidae is South American and part is African. There, too, plate tectonics offers a satisfying explanation of the biogeography of that group.

The taxonomic evaluation of the Cyprinodontiformes is far from complete, and one may reasonably expect many more far-reaching revisions. Many new cyprinodontiform species are being described and will enrich our understanding of this order. In 2000 Lazara pointed out that the number of species described since the first killifish was cited in 1766 by Linnaeus has increased exponentially.

Physical characteristics

The earliest known killifish "description" is a 600-year-old piece of mother-of-pearl jewelry, one inch long, produced by the Native American Mogollon culture in the recognizable shape of *Cyprinodon tularosa,* endemic to New Mexico. As a general rule, killifishes and live-bearers are sexually dimorphic and dichromatic. In 1881 Steindachner described male and female *Cynolebias bellottii* as two different species, the female named appropriately *Cynolebias maculatus.* Cyprinodontiform males and females differ in shape and color and sometimes in numbers of anal and dorsal fin rays, as noted by Steindachner. There is a gestalt to the Cyprinodontiformes that is difficult to describe (because they are so variable) but which makes them instantly recognizable. Few fishes can be mistaken for a killifish or live-bearer, but some of the very few are the mudminnows of the Umbridae. In fact, in 1843 *Umbra pygmaea* was described by Ayres as a killifish (*Fundulus fuscus*). Ayres was perhaps the first person to document his confusion about this resemblance.

The Cyprinodontiformes vary greatly in length, from 0.4 in (1 cm), which meets the formal definition of a miniature fish, to nearly 13 in (33 cm). Many are basically cylindrical in shape, with tapering around the caudal peduncle. Some are sleek, pike-like predators and others elongate and flat-topped (the top minnows of the order), with mouths designed for surface feeding. Some are laterally compressed and elongate for fast movement in streams or in pelagic conditions, and others that occupy benthic ecological niches may or may not be compressed but tend to be deep-bodied. Some *Orestias* have "chunky" body proportions akin to those of various fancy goldfish.

Cyprinodontiformes possess only one dorsal fin, which has its origin anywhere from far forward of the first anal fin ray to a point over the last few anal fin rays. The dorsal fin is never completely ahead and rarely entirely behind the anal fin; there almost always is an overlap. The origin of the anal fin ranges from about the midbody to three-fourths of the way from the snout. Fin rays are soft; Cyprinodontiformes do not have spines. Unpaired fins are rounded, truncated, pointed, elongated, or a combination of these shapes. Caudal fins are sometime lyre-shaped. The unpaired fins may carry very elaborate extensions or filaments, which in some cases extend beyond the caudal fin. Males of most species have contact organs, that is, bony outgrowths along the outer margins of the scales, along the fin rays, or on the snout. These organs help initiate spawning or position the males during spawning. In the live-bearers and some killifishes, the anal fin of the male becomes a gonopodium, which is used as an intromittent organ.

Pelvic fins sometimes are a prominent feature, but mostly they are small, tiny, or absent. Pelvic fin position varies, though usually it is close to the origin of the anal fin; sometimes it is far forward and close to the pectoral fins. There is no lateral line system along the sides, although in some species neuromasts protrude through the scales, running along what normally would be the course of the lateral line. The lateral line system is present around the head, with the cephalic neuromasts either totally exposed or in canals or a combination of the two states. Derived states of the cephalic lateral line system are very useful in taxonomic studies. Some species use the cephalic lateral line system to locate surface prey by its vibrations.

In 1949 Gosline developed an elaborate and very useful classification and numbering system for the sensory canals and pores of the cyprinodontiform head. The anterior naris is tubular in the aplocheiloids and in the cyprinodontoid genera *Cubanichthys* and *Anableps*. Among the cyprinodontoids this is considered to be independently derived. Overall squamation is complete, partial, or absent. There has been some attempt, mainly among those studying aplocheiloids, to use the pattern of scales on the head as a taxonomic tool. Upper and lower jaw teeth are spatulate, unicuspid, bicuspid, or tricuspid or have various combinations of those tooth forms. Sometimes teeth are present on the vomer. Jaw teeth are used to seize food items; teeth on the pharyngeals do the chewing. Mouths are protrusile. In some cases the lower jaw has a marked upward turn, sometimes almost perpendicular to the body axis. Some species have thickened lips to facilitate the eating of algae.

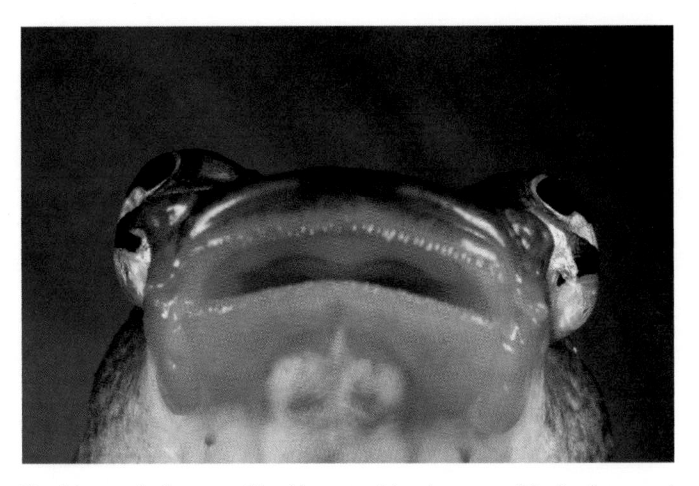

The largescale foureyes (*Anableps anableps*) sees well both above and below the water because its eyes have two regions of retina, one for seeing above water and the other for underwater. It feeds on both insects from the air and small fish from the water. (Photo by Dr. Paul A. Zahl/Photo Researchers, Inc. Reproduced by permission.)

Body proportions and fin lengths and shapes are different in the sexes. Females typically have a more rounded appearance. Aquarists never have problems determining the sex of the Cyprinodontiformes. In other groups, this is not the case. Color differences are always noticeable and, in many cases, dramatic. The females generally are plain—perhaps silvery, olive drab, or brownish—whereas males may be brightly colored in crimson, iridescent greens and blues, bright yellow, bright blue, or a combination constituting a veritable riot of colors.

Distribution

Killifishes have a worldwide distribution, except for Australia, Antarctica, and Europe north of the Pyrenees and the Alps, with the exception of *Aphanius fasciatus* along the Mediterranean coast of France. Live-bearers are found in North, Central, and South America and the Caribbean.

Habitat

The Cyprinodontiformes occupy such diverse habitats that it is impossible to characterize them in a simple way. A small number of species occur in marine environments, some are brackish water species, and others are even found in hypersaline waters. Most species, however, inhabit freshwater. Many species, particularly in the genera *Aphanius* and *Cyprinodon*, are found in hypothermal environments at temperatures close to their upper lethal limit. Many *Rivulus* are semiterrestrial and may occur under leaves or logs or move overland from puddle to puddle, pond to pond, and rivulet to rivulet. In some tropical forest areas they are not to be seen until a light rain fills up tire tracks, forms puddles, or fills in the hoof

A Mombasa killifish (*Nothobranchius guentheri*) from Kenya. (Photo by Tom McHugh/Steinhart Aquarium/Photo Researchers, Inc. Reproduced by permission.)

prints of cows or the footprints of people. Presumably, they are waiting out drier conditions under damp leaf litter until there is enough water in their microhabitats. In small streams they are found along the edges in tiny pockets of water or hidden under the vegetation or stuck on leaves overhanging the water. At least one species, *Rivulus marmoratus*, inhabits land crab burrows.

Killifishes and live-bearers are found in slow-moving to fast-moving streams, tiny rivulets, shallow sheets of flowing water, puddles, ponds, rivers, lakes, swamps, salt marshes, estuaries, tidal flats, marine coastal waters, isolated desert springs, hypersaline lakes, and springheads. Where the habitats are large, they tend to be at the margins–with some notable exceptions, such as the pelagic lacustrine species. Almost all of these habitats are heavy with vegetation. Some fishes are found in areas where there are seasonal torrential conditions, which they manage to survive. Perhaps some of the semiterrestrial species leave the water under these conditions. Pelagic forms, while not common, do occur in the high-altitude lakes of the Andes and in some African lakes, most notably, Lake Tanganyika. In both Africa and South America, aplocheiloid killifishes have successfully colonized habitats with seasonal temporary waters. These species lay eggs in the substrate and die off when the water evaporates. The eggs, protected by the substrate, go into a resting state called "diapause." At times the substrate becomes so dry and cracked that it is difficult to imagine that the eggs can survive. When the rains of the wet season fill the shallow pans (in Africa, some are elephant watering holes), roadside ditches, culverts, meadows, temporary swamps, depressions, and ponds that these species inhabit, most but not all eggs hatch within hours, thereby providing a hedge against the false onset of a rainy season. Ironically, in many areas of South America and Africa human intervention in the form of road construction and its associated culverts and ditches has helped these species. Even though some places have two rainy seasons a year, this seasonal characteristic is termed "annualism."

Behavior

Males of the seasonal fishes are aggressively territorial, defending their breeding sites against other males. In aquaria, where retreat is limited, males may fight until one or both die from their injuries. Aggressive territorial behavior is common in the Cyprinodontidae and is known in the Fundulidae (*Fundulus catenatus*, *F. diaphanus*, and *Lucania goodei*). Aggression sometimes extends to nonbreeding females. Male agonistic behavior is very common among the killifishes and some live-bearers and is not necessarily territorial. In the cyprinodontids aggression is associated with the defense of breeding territories; otherwise they move about in peaceful schools. In one cyprinodontid species, *Jordanella floridae*, the male defends a territory, builds a nest, and fans the eggs—a rare case of cyprinodontiform male parental care. The long-term defense of a breeding territory by most male *Cyprinodon* likewise confers a degree of protection to the eggs deposited there.

The Poeciliinae, Anablepidae, and Goodeidae are active, gregarious, and sometimes scrappy. When they are not occupied by feeding activities, males posture and display as they seek to mate. Female receptivity behavior is complex. Among the poeciliines, a male sometimes rushes in quickly, thrusting his gonopodium, and then beats a hasty retreat, particularly in those species where the female is much larger. Poeciliine females release a pheromone-like substance, thought to be estrogen, which stimulates males into a mating frenzy. Among the goodeids, members of the genus *Allodontichthys* behave much like North American darters. In Africa the nonseasonal killifishes inhabit swamps, trickles, very small streams, and occasionally rivers, but usually they occur in vegetation-choked portions at the edges. Here they are distributed singly in small pockets of open water in the weedy margins, under the vegetation itself, or sometimes under the leaf litter on the bottom but never out in the open. Interestingly, one of these species, *Aphyosemion franzwerneri*, also behaves like a darter. In aquaria, males of all these species range from peaceful to ferociously aggressive toward each other. Those species that exhibit schooling behavior occur exclusively in the suborder Cyprinodontoidei in the families Fundulidae, Cyprinodontidae, Anablepidae, and Poeciliidae. Whether a single species or a mix of species, these schools sometimes are composed of massive numbers of individuals.

Feeding ecology and diet

Cyprinodontiformes are piscivorous, omnivorous, herbivorous, or dedicated to particular food items, such as terrestrial and aquatic invertebrates, zooplankton, detritus, algae, and vascular plants. Some are aggressive feeders and pose a danger to other species when they are introduced outside their natural range. The diet of some species in the Poeciliinae includes a significant cannibalistic component.

Reproductive biology

In the poeciliines, as the male matures, the anal fin is modified into a gonopodium; at the juvenile stage there is no difference between the male and female anal fin. The gonopodium serves as a launching platform for sperm bundles called "spermatozeugmata." In mating the gonopodium is swung forward in a vertical plane and thrust at the female genital opening so as to deposit the sperm bundles either near or inside the opening. The end of the gonopodium has hook-like structures to facilitate the transfer of the spermatozeugmata. It is not known how the bundles are transported to the tip of the gonopodium. Females can store sperm for extended periods of time. Some females are capable of superfetation, that is, they have the ability to carry more than one brood of embryos at different stages of development. There are two gestation extremes: in type 1 the embryo is nourished by the yolk reserves laid down before fertilization ("lecithotrophy"); in type 2 the egg is very small, without much yolk, but the embryo is heavily nourished by maternal fluids transferred by a kind of placenta ("matrotrophy"). There are various intergrade levels of maternal nourishment, depending on the species. In *Tomeurus gracilis*, the egg is expelled, and development is external and dependent on the yolk reserve. This exceptional situation is, in reality, an extreme form of type 1 gestation. For the rest of the species with a gestation period, birth is initiated by muscular contractions that rupture the follicular walls of the embryos, thus initiating their exit into the environment.

There are two unisexual poeciliine "species," the all-female *Poecilia formosa* and *Poeciliopsis monacha-lucida*. Both so-called species are considered to be of hybrid origin, and some taxonomists do not regard them as true species. Matings are needed to cause these "species" to reproduce. In *P. formosa* paternal DNA is excluded, and the mating results in an all-female brood. In *Poeciliopsis*, paternal DNA is not excluded, but the resultant brood is also all female. The origins and relationships of unisexual poeciliines were reviewed by Schultz.

Males of the live-bearing anablepids, the genera *Anableps* and *Jenynsia*, have a tubular gonopodium that is used to transfer free spermatozoa, not sperm bundles, to the female. In *Anableps dowi* sperm bundles are formed but break apart before they make their way down the gonopodium. The gonopodia in these genera swing forward in a horizontal plane but only to one side, the left or the right. The vents of females are accessible only from the left or the right. Thus, left-sided males must mate with right-sided females and vice versa. Embryos are nourished by a pseudo-placenta that transfers nutrients to the enlarged intestine of embryos of *Anableps*, whereas in *Jenynsia* the maternal nutrients are supplied through the mouth and opercular opening of developing embryos.

In the viviparous goodeids, the first several anal fin rays of the male are shorter and offset from the rest of the anal fin by a notch. This is the structure by which sperm bundles are transferred, but the exact mechanism of transfer is unknown. Gestating embryos absorb nutritional ovarian fluids by means of elaborate outgrowths, called "trophotaeniae," which form around the anal region of the embryo.

Excluding *Rivulus marmoratus*, internal fertilization is known in four egg-laying genera, *Campellolebias*, *Cynopoecilus*, *Epiplatys*, and the monotypic *Tomeurus*. *Campellolebias* and *Cynopoecilus* are seasonal genera in the family Rivulidae. Males of *Campellolebias* have a gonopodium, whereas those of *Cynopoecilus* have a modified anal fin in which the first six anal fin rays are bunched together and have prominent contact organs along the rays in the form of papillae. *Epiplatys* is a genus with nearly 50 species, but only one is known to have internal fertilization, a Sudanese population of *Epiplatys bifasciatus*. *Tomeurus gracilis* is a typical gonopodial poeceliine species, except that it expels its eggs. The mechanism of sperm transfer in *Cynopoecilus* and *Epiplatys bifasciatus* is unknown.

The seasonal fishes lay their eggs in the substrate. Typically, a male defends a territory while mostly hovering over and close to the substrate. A receptive female approaches, angling with her head toward the substrate; the male draws near, sometimes wrapping his anal and dorsal fins around the usually smaller female. The pair dives into the substrate, sometimes disappearing from sight. One or more eggs are released by the female and fertilized by the male. After a characteristic rolling motion designed to bury the eggs, the male and female emerge from the substrate. The eggs enter a resting state. The water in these habitats eventually evaporates, and the adults die or, more likely, are eaten by birds before the water is completely gone. Some of the eggs embryonate and hatch during the next rainy season. Even though the next rainy season may be months or, in a drought, years away, in a home aquarium these eggs may hatch after as little as two to three months of storage in damp peat moss. The spawning of seasonal fish continues through the wet season. The eggs of the seasonal killifishes develop asynchronously in the female, not all at once.

Most nonseasonal killifishes are plant spawners. Males court a passing female and, if she is receptive, the pair move to the vegetation and press their bodies together in a characteristic S shape, their bodies quivering as a single egg is released and fertilized. The eggs are adhesive and stick to the vegetation. Aquarists simulate this habitat by providing a faux plant in the form of a small mop made of acrylic yarn. Aquarists often collect the eggs of these killifishes with their fingers. The egg is tough and easily handled without fear of damage, and it has a long developmental period, 10–14 days. As a result, the young hatch without a yolk sac and are fully capable of swimming and feeding, unlike the young of many other fish groups, which pass through a larval stage.

The species *Rivulus marmoratus* is a self-fertilizing hermaphrodite, unique among vertebrates in that respect. Individuals of the species look like typical female *Rivulus*, but they possess both ovaries and testes, with the ability to fertilize their own eggs before laying them. Essentially they are self-cloning. Two types of males are known. Secondary males are hermaphrodites that have become functional males with the color characteristics of a male *Rivulus*, but they retain a small amount of ovarian tissue. Some populations produce primary males, that is, individuals hatching out as males without ovaries. The populations in the Belize Keys have a high percentage of primary males (25%), and there is some evidence that primary males (and possibly secondary males) mate with the hermaphrodites. This might entail a mechanism whereby the presence of a male prompts suppression of self-fertilization in the hermaphrodites. There are killifishes that spawn in cracks and crevices; typically, these are species found in rocky streams and lakes, the procatopodine genera *Procatopus* and *Lamprichthys tanganicanus*, respectively. Some species, such as *Fundulus catenatus*, lay their eggs over shallow gravel beds. The reproductive biology of many killifishes is not known, even though some of them are fairly common.

Conservation status

Approximately 10% of all cyprinodontiform species—92 in all—are cited in the 2000 Red List of the International Union for Conservation of Nature and Natural Resources (IUCN) in the Extinct (10 species), Extinct in the Wild (5 species), Critically Endangered (18 species), Endangered (20 species), Vulnerable (25 species), Lower Risk/Near Threatened (3 species), and Data Deficient (11 species) categories. The cause of all freshwater fish extinctions and the establishment of the other categories of concern are due solely to the harmful effects of human intervention and not to the events of natural history. Sadly, a species of *Cyprinodon* described in 1993 was given the species name *inmemoriam*, since the species was extinct by the time it was described. The Red List is not a static document, and more species may be expected to appear there.

For instance, although huge tracts of the African rainforests are being cleared, not a single killifish from the affected areas is on the 2000 Red List. This is an extreme example of the category Data Deficient, in this case, no data at all. An undescribed *Nothobranchius* species from the Caprivi Strip of Namibia is the only African species listed as Endangered. All *Nothobranchius* habitats, being seasonal, are capable of being severely affected by human activity, so the absence of other *Nothobranchius* from the 2000 list offers small comfort. Numerous species are listed as Extinct in the Wild. Given the history of failure to keep Extinct in the Wild species, such as the monkey spring pupfish, alive as captive animals, unless they are successfully reintroduced into the wild, the future of such species is bleak.

For species listed in any category the reasons why they were listed are cited. Establishing whether a species is extinct is very difficult because of the nature of extinction. The absence of evidence is not necessarily evidence of absence. The difficulties are acute when the actual distribution of a species is imperfectly known or when its taxonomic limits have not been established or when there has been inadequate sampling. Harrison and Stiassny have reviewed this topic. Another 20 species are listed as regionally endangered by various states in the United States. These statistics need to be put into perspective. The percentage (10%) of the Cyprinodontiformes under threat is among the highest in the Actinopterygii, comparable to the carps, of the order Cypriniformes, with 12% of its 2,660 species appearing on the Red List. Live-bearers and killifishes have colonized marginal habitats easily degraded by human intervention or highly vulnerable to introduced exotic species. In addition, their distributions sometimes are highly localized, as, for example, in a single spring or pond. These factors account for the high level of threat occurring in this group.

Significance to humans

Species in the genera *Orestias* in the Andes and *Anableps* in Central and South America are taken in subsistence fishing by local inhabitants. In the bait fish industry in the United States, fundulids, such as *Fundulus heteroclitus*, and cyprinodontids, such as *Cyprinodon variegatus*, are sold routinely as bait fish. In the case of the latter species this sometimes leads to disastrous consequences for the local fauna when end-of-the-day bait-bucket releases occur. Both killifishes and live-bearers are voraciously larvivorous, thus helping to control mosquito populations and their resulting detrimental effects upon humans. In many areas killifishes and live-bearers represent an important forage item for game fishes. In the Everglades, Florida Bay, and the Keys there is an interesting food chain of great economic significance. The salt marsh mosquito is very abundant in southern Florida. The killifishes and live-bearers feed heavily on mosquito larvae, obtaining the energy necessary to produce many offspring. When the killifishes and live-bearers move into the tidal creeks, they provide abundant forage for tarpon, redfish, and snook, which feature heavily in the economy of southern Florida.

A mosquitofish (*Gambusia affinis*) about to feed on mosquitoes. (Photo by Robert Noonan/Photo Researchers, Inc. Reproduced by permission.)

Aquarium hobbyists keep many species of Cyprinodontiformes. Various live-bearers, selectively bred for color and fancy fins, are sold in large numbers. Many specialty hobbyist groups, such as the American Killifish Association, the Deutsche Killifisch Gemeinschaft, the American Livebearer Association, and the Association France Vivipare, have been formed to keep and study these fishes. The cooperation of hobbyists and ichthyologists has had a significant impact on the furtherance of our knowledge of the Cyprinodontiformes. Cyprinodontiform species have been used widely in evolutionary studies, the study of life history patterns, the study of the effects of exotic introductions, and the disciplines of ecology, reproductive biology, genetics, physiology, toxicology, and behavioral psychology. This list is by no means exhaustive. *Xiphophorus hellerii*, *Fundulus heteroclitus*, and *Rivulus marmoratus* alone have accounted for hundreds of articles in professional journals and other publications. The beauty of many of these fishes has led to their being featured on the postage stamps of several countries, which aids in stimulating an interest in conservation efforts.

1. Largescale foureyes (*Anableps anableps*); 2. Redtail splitfin (*Xenotoca eiseni*); 3. Ascotan Mountain killifish (*Orestias ascotanensis*); 4. Northern mummichog (*Fundulus heteroclitus macrolepidotus*); 5. Blackfin pearl killifish (*Austrolebias nigripinnis*); 6. Mangrove rivulus (*Rivulus marmoratus*); 7. Tanganyika pearl lampeye (*Lamprichthys tanganicanus*); 8. Devils Hole pupfish (*Cyprinodon diabolis*); 9. Chocolate lyretail (*Aphyosemion australe*); 10. Green swordtail (*Xiphophorus hellerii* ssp. *hellerii*). (Illustration by Emily Damstra)

Species accounts

Largescale foureyes
Anableps anableps

FAMILY
Anablepidae

TAXONOMY
Anableps anableps Linnaeus, 1758, India (misreported type locality).

OTHER COMMON NAMES
English: Striped foureyed fish, foureyed fish, foureye; Spanish: Cuatro ojos; Portuguese: Quatro-olhos.

PHYSICAL CHARACTERISTICS
Grows to 12 in (30 cm) in total length. Elongate, compressed posteriorly, and flat anteriorly. Blunt snout and toothy horizontal mouth with only a slight curvature. Large, bulging, froglike eyes set far forward, almost on the snout, and divided by a black horizontal band into an upper and lower portion for separately viewing above and below the waterline. Dorsal fin set far back, completely behind the anal fin. Three to five blue to violet horizontal lines of varying lengths run along the sides. Dorsal surface is brownish. A whitish line runs along the back; at the operculum the line divides into a Y shape, each arm of which terminates at an eye. The ventral area is whitish. Sexually dimorphic. Females are larger than males, with rounded fins; males have a tubular, scaled gonopodium.

DISTRIBUTION
Trinidad, Venezuela, Guyana, Suriname, French Guiana, and the Amazon Delta in Brazil.

HABITAT
Found in freshwater, brackish, and saltwater rivers, streams, and estuaries and oceanic saltwater near beaches.

BEHAVIOR
A gregarious schooling fish, sometimes gathering in the hundreds with congeners of equal size. Since they swim with the upper half of their eyes above the water, they have the amusing habit of bobbing their heads up and down so as to keep their eyes wet. This species has been seen riding the breakers near sandy beaches, sometimes getting tossed onto the beach by the waves. Undaunted, foureyes just jumps back in. Because of their acute vision above the water, they are very difficult to capture.

FEEDING ECOLOGY AND DIET
In brackish and marine environments, scores of these fish jump out of the water at low tides to gulp down the mud, which is rich in algae, diatoms, dinoflagellates, amphipods, isopods, and worms. Although they are primarily surface feeders, they leap from the water in pursuit of low-flying insects. Small fishes also are part of their diet, but these fish are not bottom feeders. Not surprisingly, aquarists report that the species will not take food that falls to the bottom of an aquarium.

REPRODUCTIVE BIOLOGY
Foureyes are live-bearers with internal fertilization and matrotrophic (type 2) gestation and development. The male gonopodium swings out in a horizontal plane, but only in one direction, either to the left or to the right. The vent of the female is covered by a hinged scale called a "foricula," which opens to the left or right. Thus, right-handed males must mate with left-handed females and vice versa, an odd situation that does not seem to impede their reproduction. The gestation period is about 20 weeks, with a brood size of 10 to 20 fry. The young are large, up to 2 in (5 cm) at birth. This large size is remarkable, considering that sexual maturity is reached at about 3.5 in (9 cm).

CONSERVATION STATUS
Not listed by the IUCN.

SIGNIFICANCE TO HUMANS
All species are taken for food in subsistence fishing, and, to a limited extent, they are sold in the aquarium trade. ◆

Chocolate lyretail
Aphyosemion australe

FAMILY
Aplocheilidae

TAXONOMY
Aphyosemion australe Rachow, 1921, Port-Gentil (formerly Cape Lopez), Gabon.

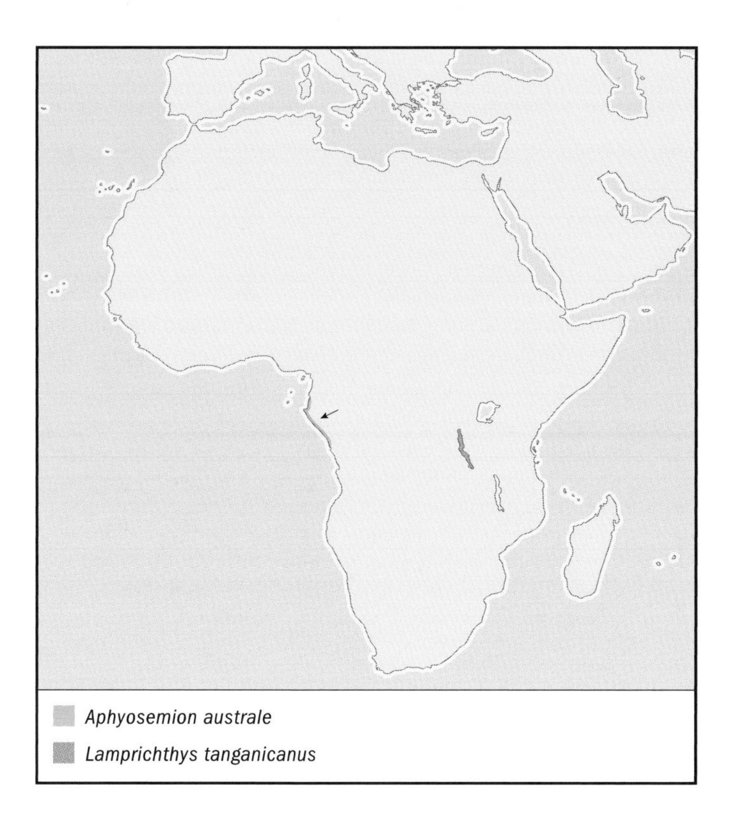

OTHER COMMON NAMES
English: Chocolate australe, australe, Cape Lopez lyretail, lyretail panchax.

PHYSICAL CHARACTERISTICS
Grows to 2–2.4 in (5–6 cm). Sexually dimorphic and dichromatic. The male is slender-bodied and cylindrical in shape; the body has a slight curve to the dorsal surface. The head tapers to a terminal mouth, and the caudal peduncle is compressed and tapering. Dorsal fin is small and set far back, with its origin over the midpoint of the anal fin. Dorsal surface is chocolate brown, and anterior flanks are light metallic blue; red spots and splotches are scattered over the body and on the dorsal and anal fins. All fins show color, the dorsal and anal fins with orange and red margins. Anal fin tapers to a point, with its color grading into white; the upper and lower parts of the caudal fin have curved white extensions, giving this fin its characteristic lyre-tailed shape. The female is smaller, usually without much color. Sometimes golden, gray, or muddy, with rounded fins and iridescent bluish white margins on the pectoral fins. Body and unpaired fins have tiny red dots. All color and color patterns vary widely for both males and females.

DISTRIBUTION
South along the coasts of Gabon starting at the Ogowe River, the Congo, the Cabinda Enclave (Angola), and Zaire.

HABITAT
Found in swamps associated with small streams and rivers, rainforest swamps, and shallow flooded areas—quiet, weed-choked environments.

BEHAVIOR
A peaceful species, easily kept in a heavily planted aquarium or one provided with spawning mops or with a combination of plants and mops.

FEEDING ECOLOGY AND DIET
Aquarists feed the chocolate lyretail brine shrimp—frozen or live—adult or newly hatched, and such live foods as fruit flies, daphnia, and tubificid worms. In nature, it is assumed that aquatic invertebrates and terrestrial insects are the chief component of the diet.

REPRODUCTIVE BIOLOGY
A typical plant spawner, the male courts the female with fins flared. A receptive female moves to the plants or spawning mop, where the pair presses against each other in an S shape, both of them quivering. The female releases an egg, which the male fertilizes. The adhesive egg sticks to the vegetation or mop. These eggs are collected easily and hatch in 14 days. It is estimated that some aquarium populations of the chocolate lyretail have been held and bred in aquaria continuously since 1913, yet differences are small compared with the wild forms from the type locality.

CONSERVATION STATUS
Not listed by the IUCN.

SIGNIFICANCE TO HUMANS
Sold and exchanged among aquarium hobbyists and occasionally available in the aquarium trade. ◆

Devils Hole pupfish
Cyprinodon diabolis

FAMILY
Cyprinodontidae

TAXONOMY
Cyprinodon diabolis Wales, 1930, Devils Hole, Ash Meadows, Nevada, United States.

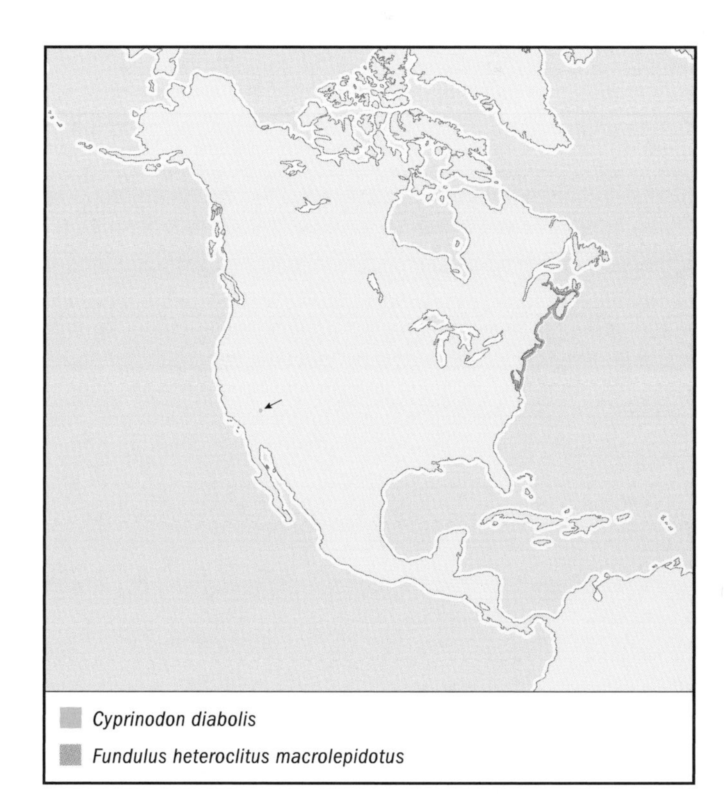

OTHER COMMON NAMES
None known.

PHYSICAL CHARACTERISTICS
A dwarf species seldom exceeding 1 in (2.6 cm) in length. Pelvic fins are absent. Resembles the juvenile pupfish. Sexually dimorphic and dichromatic. Males have iridescent blue body color; the operculum is iridescent with a purple sheen. The unpaired fins are bordered in black, the dorsal fin is iridescent gold, and the anal and caudal fins are whitish toward the base. Prominent genital papilla. Females overall have more yellowish coloration than males. The back and caudal and pectoral fins are yellowish brown, and the dorsal fin is edged in black. The operculum is metallic green, with a hint of a dark bar on the caudal peduncle.

DISTRIBUTION
Ash Meadows, Nevada.

HABITAT
The enclosed pool of an aquifer formed by flooded limestone caverns and reached by a 49 ft (15 m) naturally occurring shaft in the side of a hill, its only source of light. The dimensions of the entire pool are 11.5 ft by 72 ft (3.5 m by 22 m). A boulder divides the surface of the pool into two sections. The breeding and most of the feeding, upon which the survival of the species depends, take place on an algae-covered rocky shelf 11.5 ft by 16.6 ft (3.5 m by 5 m) with 1 ft (0.3 m) of water above it. The shelf is situated at the foot of the shaft. The overall dimensions of this part of the pool are 11.5 ft by 42 ft (3.5 m by 13 m). At the end of the shelf there is an abrupt drop-off of 28 ft (8.5 m). The substrate continues to slope downward below that depth. Water flows in and out of the pool at a year-round temperature of 90°F (32°C). As the lakes and waters of the area disappeared and the desert formed, the species was trapped in the aquifer 10,000 to 25,000 years ago.

BEHAVIOR
The pupfish exhibits a daily cycle of movement to and from the rocky shelf, depending on the time of day and the time of year.

FEEDING ECOLOGY AND DIET
In the summer and fall the species feeds chiefly on the algae growing on the shelf. Algae also grow on the substrate just off the shelf up to a depth of 50 ft (15 m), with only trace amounts below that. In winter and spring, when the algae are much reduced, diatoms are the chief food items. The population varies from about 200 to 500 individuals, depending on the algae growth.

REPRODUCTIVE BIOLOGY
Males do not set up breeding territories, as do other pupfish. Spawning takes place primarily in April and May. Males closely accompany a ripe female to the algae-covered shelf, where they stay together and spawn at irregular time intervals spanning about one hour. The species has spawned in the laboratory, but no eggs were hatched.

CONSERVATION STATUS
Listed as Vulnerable by the IUCN.

SIGNIFICANCE TO HUMANS
The pupfish has endured in its tiny habitat for as many as 25,000 years, providing us with a stimulus to ponder both the durability and the vicarious nature of life. ◆

Ascotan Mountain killifish
Orestias ascotanensis

FAMILY
Cyprinodontidae

TAXONOMY
Orestias ascotanensis Parenti, 1984, Salar de Ascotan, Chile.

OTHER COMMON NAMES
English: Lake Ascotan Mountain killifish

PHYSICAL CHARACTERISTICS
Grows to 2.4 in (6 cm) in length. A member of the *Orestias agassizii* species complex. Basically olive-green in color, with anal and dorsal fins set back and over each other. The lower jaw turns up. Males and females are robust, with a relatively large head compared with the heads of other species in the species complex. Sexually dimorphic and dichromatic. Males are smaller than females and more slender. Breeding males have a bright yellow overlay on the sides and the anal and dorsal fins. Females are larger and more rounded than males. The fins and body are mottled, tending toward more uniform coloration when they are very mature.

DISTRIBUTION
Known only from Salar de Ascotan, a small saline lake in northwestern Chile, the southernmost population of *Orestias.*

HABITAT
The species is found in the slightly brackish water of Lake Ascotan and its associated ponds, where *Ruppia filifolia* is the most common vegetation.

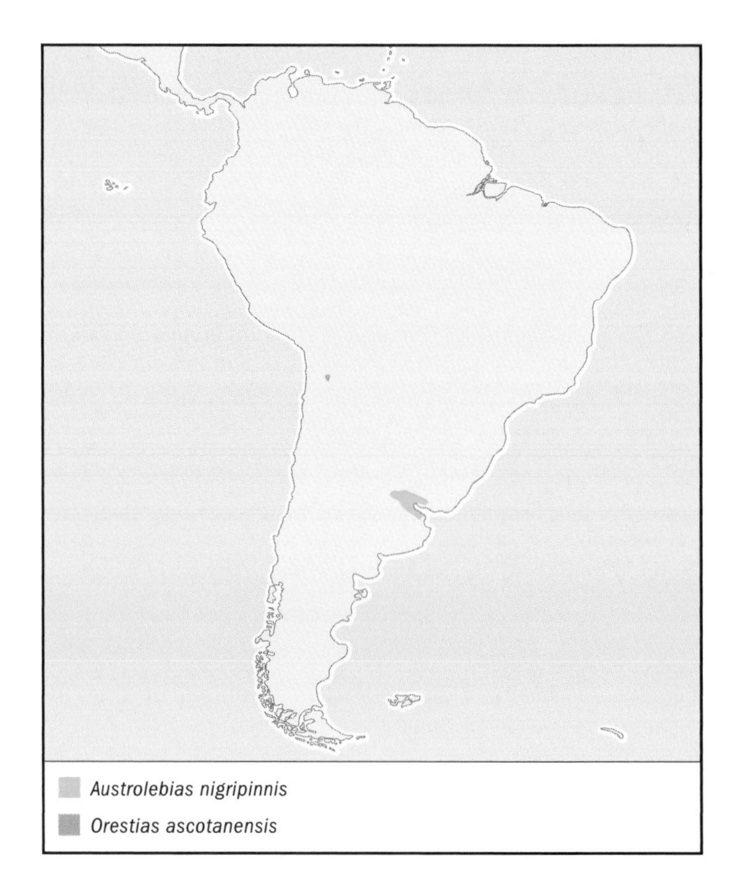

BEHAVIOR
In 1995, for the purpose of studying captive breeding, Jara, Soto, and Palma set up four males and four females in a planted 3 gallon (12 L) aquarium with no reported aggression among the males. The males did not persistently court the females, as happens with many other killifishes; some time was spent apart, feeding and moving about the aquarium.

FEEDING ECOLOGY AND DIET
In aquaria the species readily feeds on insect larva, crustaceans, and commercial flake food, indicating that in the wild it is a benthopelagic feeder on aquatic invertebrates.

REPRODUCTIVE BIOLOGY
This species is a typical plant spawner. At the lake, eggs were found attached to the vegetation. In aquaria the male courts the female by approaching from beneath and nudging her pelvic area. The two then move to the vegetation. With the pair positioned side by side with their vents close together, the male assumes a typical S-shape posture. The female releases two or three adhesive eggs, which are fertilized by the male. The filamentous eggs adhere to the vegetation. The incubation time is about 17 days at 63–68°F (17–20°C).

CONSERVATION STATUS
Not listed by the IUCN.

SIGNIFICANCE TO HUMANS
None known. ◆

Northern mummichog
Fundulus heteroclitus macrolepidotus

FAMILY
Fundulidae

TAXONOMY
Fundulus heteroclitus macrolepidotus Walbaum, 1792, northern America. In the 1980s, Able and Felley recognized the subspecies status of the northern populations of *Fundulus heteroclitus.*

OTHER COMMON NAMES
English: Killie, killifish; Spanish: Fúndulo; Portuguese: Fundulo.

PHYSICAL CHARACTERISTICS
Grows to 5 in (13 cm) in length. Sexually dimorphic and dichromatic. Males and females have up to 15 vertical bars, which tend to be faint or missing in mature adults. The mouth is blunt, with a turned-up lower jaw. The dorsal fin is positioned over the anal fin origin at about the midbody. Males have a dorsal ocellus. Breeding males are light gray to very black in background color. The body and all unpaired fins have bright iridescent white to greenish opalescent spots. The dorsal, anal, and caudal fins have a yellow margin that is less noticeable on the caudal fin in the black body variety. There is a yellowish cast to the abdomen and operculum, the latter with iridescent silver highlights. Color patterns and color intensities vary widely in both sexes. Females are larger than males. Body is chunky with a pale silvery background, brownish color on the back, and whitish color on the abdomen. The fins are clear. Adults rarely are confused with other species in their natural range.

DISTRIBUTION
From Chesapeake Bay, Maryland, United States, to Newfoundland, Canada. Naturalized in the estuary of the Guadalquivir River, southwestern Spain, and possibly established in the salt marshes of the estuary of the Guadiana River, at Castro Marim, southeastern Portugal.

HABITAT
The species is active primarily in tidally influenced coastal salt marshes as well as estuaries, tidal streams and creeks, shallow marine and brackish environments, and back-beach lagoons reached by high tides. They have been reported for freshwater portions of the Bronx River and naturalized in freshwater impoundments, the latter likely due to bait-bucket releases. Most studies of its habitat use have been undertaken in its primary habitat, the salt marshes.

BEHAVIOR
Traveling in huge schools numbering in the hundreds, the mummichog is in constant search of food. As a result, fishers find that they are caught easily in baited minnow traps. As a predator on salt marsh invertebrates and small fish, and as a prey item for wading birds and the blue crab, the northern mummichog has a significant impact on salt marsh trophic dynamics. The annual mortality rate is about 50% for adults and more than 99% for the larval and juvenile class sizes.

FEEDING ECOLOGY AND DIET
Feeds on a variety of marine and freshwater invertebrates, diatoms, mollusks, amphipods, crustaceans, plant material, insects, detritus, worms, and small fishes. It is doubtful that a mummichog would spurn anything edible that fits into its mouth. Mummichogs are preyed upon by larger fish and fish-eating birds.

REPRODUCTIVE BIOLOGY
Judging by its reproductive behavior in aquaria, the mummichog probably spawns continually in the spring and summer, but in nature its peak spawning activity is governed by a lunar cycle. Eggs are laid in clumps on floating algae mats or at the base of marsh grasses and buried in the sand in the high marshes at the very high tides of the new moon or full moon. The eggs are placed where desiccation will be minimized when the tide goes out. The eggs hatch when they are immersed again at the next very high tide. In aquaria water-incubated eggs hatch in about 16 days.

CONSERVATION STATUS
Not listed by the IUCN.

SIGNIFICANCE TO HUMANS
The mummichog is used as a bait fish and as an aquarium animal; it also is one of the most widely used laboratory animals. It is a harmless species unless it is swallowed live. It has been reported that many bored fishers have swallowed live mummichogs. In one such instance, some of the fish were infected with the larva of the nematode *Eustrongyloides ignotus,* causing stomach and intestinal problems for three fishers, two of whom required surgical intervention. Raasch noted, "Though 12 fishermen reported no symptoms from this pastime, the moral is that fishermen should abstain from taking the bait themselves when the fish refuse to do so." ◆

Redtail splitfin
Xenotoca eiseni

FAMILY
Goodeidae

TAXONOMY
Xenotoca eiseni Rutter, 1896, near Tepic, Nayarit, Mexico.

OTHER COMMON NAMES
Spanish: Mexcalpique cola roja.

PHYSICAL CHARACTERISTICS
Males grow to 2.4 in (6 cm) in length and females to 2.8–3 in (7–8 cm). Sexually dimorphic and dichromatic. All fins are rounded. The dorsal fin of males is larger than that of females and is set over the anal fin, far back on the portion of the body that tapers to the caudal peduncle. The body is block-shaped, with a humpbacked appearance in mature fish. The depth of the caudle peduncle is much smaller than the body depth. The head is small and pointed. The basic background color of the body is brownish gray to beige; the degree of dark pigmentation is quite variable. Males are bright orange to red at the base of the caudal peduncle, with iridescent blue highlights on the forward part of peduncle between the anal and dorsal fins. The middle portion of the body has gold highlights, and the abdomen is reddish to yellowish near the operculum. The dorsal fin is sometimes black, the caudal fin is orange to yellow, and the anal fin is orange. Color patterns and intensities vary greatly by location, with colors in alpha males very pronounced. Females have a grayish beige body color, sometimes with silvery highlights; all fins are slightly yellow.

DISTRIBUTION
Mexico in Rio San Leonel, El Sacristan, and Rio Grande de Santiago drainages in Nayarit; also in Rio Tamazula, Jalisco.

HABITAT
Found in rivers, streams, and springs as well as in ditches, which are sometimes highly polluted.

BEHAVIOR
Collects in groups that have an alpha male. Males constantly pursue females and are pugnacious and scrappy toward each other and especially toward other species. In aquaria they have the reputation of being fin nippers.

FEEDING ECOLOGY AND DIET
Omnivorous.

REPRODUCTIVE BIOLOGY
A live-bearing species with internal fertilization, the embryos being nourished by trophotaeniae. The gestation period is about 60 days. Brood size ranges from 10 to 40 fry, with reports of up to 100 fry.

CONSERVATION STATUS
Not listed by the IUCN.

SIGNIFICANCE TO HUMANS
The redtail splitfin is sold in the aquarium trade and is bred and traded by live-bearer aquarium hobbyists. ◆

Tanganyika pearl lampeye
Lamprichthys tanganicanus

FAMILY
Poeciliidae

TAXONOMY
Lamprichthys tanganicanus Boulenger, 1898, Mbiti Rocks, southernmost Lake Tanganyika.

OTHER COMMON NAMES
English: Lake Tanganyika lampeye, Tanganyika pearl killifish, Tanganyika killifish.

PHYSICAL CHARACTERISTICS
Grows to 6 in (15 cm) in length. Sexually dimorphic and dichromatic. Males and females are silvery, sleek, and compressed and taper to the terminal mouth and caudal peduncle. Brightly reflective eyes. They have high-set pectoral fins and dorsal fins set far back, with the origin approximately over the middle of the anal fin. Caudal fins have scales at the base. Recurved teeth noticeable on the outer surface of the upper and lower jaws. Males have an anal fin with a broad base; the rays are shorter than in females. Eight or nine lines of very bright iridescent blue spots along the body, with yellowish spotting on dorsal, anal, and caudal fins and yellow pelvic fins. All observed colors and patterns vary widely, depending on the angle of the lighting. Female are smaller than males, with rounded fins. Shorter dorsal and anal fin base, anal fin rays longer than in the male. The body is silvery or bluish, and the fins are faintly yellow or orange.

DISTRIBUTION
Endemic to Lake Tanganyika.

HABITAT
Rocky shores (not sandy areas) out to a depth of about 30 ft (10 m).

BEHAVIOR
A benthopelagic fish moving in large schools with conspecifics and sometimes in schools of native freshwater herrings. Despite its predilection for schooling, individuals wander about apart from schools. Males are not aggressive toward each other. Fry form huge schools in the hundreds and even in the thousands. In aquaria newly hatched fry form schools at the water surface immediately upon hatching. Fry must be reared apart from the adults.

FEEDING ECOLOGY AND DIET
Feeds on aquatic and terrestrial insects, such as chironomid larvae, termites, and beetles, as well as small crustaceans, ostracods, fish scales, and small fish. The elements of this diet indicate

feeding from the bottom to the surface of the water column. The recurved teeth on the outer surface of the upper and lower jaws facilitate the capture of chironomid larvae and ostracods on rocky substrates. Fish scales and small fishes indicate midwater feeding, while terrestrial insects suggest surface feeding. Aquatic vegetation does not seem to play a part in the diet.

REPRODUCTIVE BIOLOGY
A bottom spawner, utilizing cracks and crevasses for its egg-deposition sites. A male and female meeting in midwater descend to the rocky substrate. The female positions herself close to a crack or crevasse in the rocks, where she ejects an egg, which is fertilized by the male. The eggs are not adhesive but become wedged in place. In aquaria they make use of a similar arrangement of rocky crevasses but also lay their eggs in spawning mops and even on the bare bottom of the aquarium. The eggs are large, 0.1 in (3 mm), but difficult to see. Eggs eye-up in about a week and hatch in two to three weeks.

CONSERVATION STATUS
Not listed by the IUCN.

SIGNIFICANCE TO HUMANS
Occasionally appears in the aquarium trade. ◆

Green swordtail
Xiphophorus hellerii

FAMILY
Poeciliidae

TAXONOMY
Xiphophorus hellerii Heckel, 1848, Orizaba, Mexico.

OTHER COMMON NAMES
English: Swordtail; Spanish: Cola de espada.

PHYSICAL CHARACTERISTICS
Males grow to 5.5 in (14 cm) in total length, with a sword length of 1.6–3 in (4–8 cm). Females grow to 6.3 in (16 cm). Sexually dimorphic. Males and females have basically the same grayish green background color. Males have a gonopodium, a slender body, a long extension on the bottom of the caudal fin forming a "sword," two lines of reddish dots on a squared-off dorsal fin, and a pronounced line of color that is bright to dark red to almost brown running longitudinally along the body from the snout through the eye to the base of the caudal fin. Sometimes paralleled on the flanks by one or two fainter reddish lines above and below. The sword is an iridescent yellowish green bordered in black. Females are robust and rounded, with squared-off anal and dorsal fins. Caudal fin is asymmetrically rounded, with a hint of a protrusion from the lower part of the fin. There is a shadow on the area over the vent (gravid spot).

DISTRIBUTION
Native distribution in Rio Nantla, Veracruz, Mexico, to northwestern Honduras. Naturalized in the continental United States (ten states) and Hawaii; Michoacan, Morelos, Nuevo Leon, and Coahuila, Mexico; Transvaal, South Africa; Sri Lanka; New Caledonia, Australia; Fiji; Guam; Réunion; and Madagascar.

HABITAT
Inhabit rivers, streams, warm springs and their runoffs, canals, and ponds with heavy vegetation.

BEHAVIOR
While they are not territorial, male swordtails form hierarchical groups. A dominant male drives off rivals within a feeding domain or an area where females have congregated. The mating success rate of a dominant male is about 80% within his area of activity. Females form hierarchical groups of several individuals that stay in close proximity to a dominant male.

FEEDING ECOLOGY AND DIET
Plant material and insects form the greatest part of the diet. In aquaria they are voracious, consuming a wide variety of foods, such as fruit flies, frozen or live brine shrimp and chironomid larvae, flake foods, high-protein paste foods, small fish, and algae.

REPRODUCTIVE BIOLOGY
Green swordtails are live-bearers. Males impregnate females by means of a gonopodium. Females can store sperm and may produce, from a single mating, several broods over an eight-month period. Swordtails undergo lecithotrophic (type 1) gestation and development, with fry being produced approximately every 30 days. Brood sizes are fairly large; 100 or more fry may be produced by a large female.

CONSERVATION STATUS
Not listed by the IUCN.

SIGNIFICANCE TO HUMANS
The green swordtail has considerable economic importance in the aquarium trade and frequently is used in genetic and behavioral research. ◆

Blackfin pearl killifish
Austrolebias nigripinnis

FAMILY
Rivulidae

TAXONOMY
Austrolebias nigripinnis Regan, 1912, La Plata, Argentina.

OTHER COMMON NAMES
Vernacular: Nigripinnis.

PHYSICAL CHARACTERISTICS
A small species, 2–2.75 in (5–7 cm) and rarely more than 2 in (5 cm) long. Relatively deep-bodied and sexually dimorphic and dichromatic. Males are larger than females. All fins are blue-black with an iridescent greenish overlay on the pectorals. Fins are rounded, with relatively large anal and dorsal fin bases. Iridescent metallic-green margin on the dorsal fin and similar margins on the anal and caudal fins are made up of a series of separate spots. Opalescent whitish to greenish spots on all fins and the body, forming five to 11 vertical bars. Lines paralleling the fin edges on the unpaired fins. Body color is sometimes pale around the abdomen. All intensities and patterns vary widely. The female has smaller rounded fins and brown mottling on the body, dorsal fin, and anal fin. Clear caudal, ventral, and pectoral fins and whitish abdominal area.

DISTRIBUTION
The La Plata River basin in Uruguay and Argentina.

HABITAT
Inhabit temporary waters, such as flooded meadows, shallow ponds, and roadside ditches.

BEHAVIOR
The male takes over an area, which it defends against other males. A female moves in, signaling her receptivity by tilting toward the substrate. The male joins the female, and they move side by side with their heads together. The pair spawns directly on the substrate or, conditions permitting, by diving into it. In an aquarium with more than one male, a dominant male, obvious by his black body and fins, defends a spawning area of his choice; the color of the other males fades to light gray, with some black or brownish blotching. If the aquarium is heavily planted so that the males cannot see each other, more than one dominant male may appear.

FEEDING ECOLOGY AND DIET
The species feeds on live foods, such as aquatic and terrestrial insects, worms, crustaceans, and other aquatic invertebrates. In aquaria live foods (such as fruit flies, brine shrimp, chironomid larvae, tubificid worms, and daphnia) are accepted eagerly and almost everything else is rejected, with the exception of chopped earthworms and certain paste foods.

REPRODUCTIVE BIOLOGY
Eggs placed in the substrate during spawning go into diapause. When the temporary waters evaporate, the adults die off, but the eggs survive until they hatch in the next rainy season, within hours of being wetted. Aquarists simulate this environment by providing garden peat moss as the substrate, periodically collecting the peat and storing it away in an almost dry state at a temperature of about 70°F (20°C). When the peat is flooded after two to four months of storage, most of the eggs will hatch. If the eggs do not hatch, the peat can be dried again and flooded at a later date.

CONSERVATION STATUS
Not listed by the IUCN.

SIGNIFICANCE TO HUMANS
The blackfin pearl killifish is exchanged among aquarium hobbyists but only rarely becomes available in the aquarium trade. ◆

Mangrove rivulus
Rivulus marmoratus

FAMILY
Rivulidae

TAXONOMY
Rivulus marmoratus Poey, 1880, "from Cuba, if they do not exist in the United States." Poey's description of the type locality has to be one of the strangest in the history of taxonomy. As it turns out, this species is found in Florida and Cuba; the types are presumed to be from Cuba, Poey's homeland.

OTHER COMMON NAMES
English: Rivulus.

PHYSICAL CHARACTERISTICS
Grows to 3 in (75 mm) in length. Hermaphrodites. Cylindrical in shape, with tapering of the head and caudal peduncle and a high scale count. Dorsal and anal fins are set far back (typical for all *Rivulus* species), with the dorsal fin origin set back farther than the anal fin origin. A characteristic splotchy brown pattern, with an irregular dark mark just behind the operculum (humeral blotch) and a rivulus spot (ocellus) on the caudal peduncle. Primary and secondary males have the same body shape, an orange or pink overlay (more often orange) on the body and fins, mottling, and a dark humeral blotch. The rivulus spot sometimes is absent, unpaired fins at times are clear and at times have dark edges. The proportion of primary males in Belize is high (25%), whereas elsewhere the proportion is very low for both primary and secondary males.

DISTRIBUTION
Florida, Mexico, Belize, Nicaragua, Guatemala, Honduras, French Guiana, Venezuela, Brazil as far south as Santos, the Bahamas, and numerous Caribbean locations.

HABITAT
A semiterrestrial species that inhabits coastal mangrove forests in very shallow water or the wet areas of their muddy flats, out of water under detritus and leaf litter, or under or inside rotting logs. Frequently occurs in the burrows of land crabs. Found in marine and brackish water or in hypersaline pools but not in freshwater.

BEHAVIOR
Can be caught in traps set in crab holes but dies if the traps become flooded, possibly indicating that respiration of atmospheric air is a necessity. They also can be caught on a tiny hook baited with a small piece of worm. Captive individuals jump out of the water to snatch a termite or ant held above the water surface. In nature they flip along the water surface one or more times or leave the water altogether to elude danger. Extremely aggressive toward each other in aquaria, yet small aggregations have been caught in crab burrows and under rotting logs.

FEEDING ECOLOGY AND DIET
Feeds on small terrestrial and aquatic insects, mosquito larvae, polychaete worms, mollusks, and gastropods. It leaves the water to seize food items but returns to the water to eat them.

REPRODUCTIVE BIOLOGY
The only known self-fertilizing vertebrate. Although they are highly efficient at self-fertilization, hermaphrodites release some unfertilized eggs. There is evidence that outcrossing takes place, which would seem to indicate that the hermaphrodites can suppress self-fertilization so that males can fertilize the unfertilized eggs released by the hermaphrodites.

CONSERVATION STATUS
Not listed by the IUCN. Regionally are listed as a species of Special Concern in Florida.

SIGNIFICANCE TO HUMANS
Widely used in field and laboratory studies in genetics, toxicology, ecology, and physiology. ◆

Resources

Books
Berra, Tim M. *Freshwater Fish Distribution.* San Diego: Academic Press, 2001.

Brichard, Pierre. *Pierre Brichard's Book of Cichlids and All Other Fishes of Lake Tanganyika.* Neptune City, NJ: T.F.H. Publications, 1989.

Resources

Costa, Wilson J. E. M. *Pearl Killifishes. The Cynolebiatinae: Systematics and Biogeography of the Neotropical Annual Fish Subfamily (Cyprinodontiformes: Rivulidae).* Neptune City, NJ: T.F.H. Publications, 1995.

———. "Phylogeny and Classification of the Cyprinodontiformes (Euteleostei: Atherinomorpha): A Reappraisal." In *Phylogeny and Classification of Neotropical Fishes,* edited by L. R. Malabarba, R. E. Reis, R. P. Vari, Z. M. Lucena, and C. A. S. Lucena. Porto Alegre, Brazil: EDIPUCRS, 1998.

Etnier, David A., and Wayne C. Starnes. *The Fishes of Tennessee.* Knoxville: University of Tennessee Press, 1993.

Frickhinger, Karl Albert. *Fossil Atlas: Fishes.* Blacksburg, VA: Tetra Press, 1996.

Harrison, Ian J., and Melanie L. J. Stiassny. "The Quiet Crisis: A Preliminary Listing of the Freshwater Fishes of the World That Are Extinct or 'Missing in Action.'" In *Extinctions in Near Time: Causes, Contexts, and Consequences,* edited by Ross MacPhee. New York: Kluwer Academic/Plenum Publishers, 1999.

Hilton-Taylor, C., comp. *2000 IUCN Red List of Threatened Species.* Gland, Switzerland, and Cambridge, U.K.: IUCN, 2000.

La Rivers, Ira. *Fish and Fisheries of Nevada.* Reno: University of Nevada Press, 1994.

Lazara, Kenneth J. *The Killifish Master Index: The Killifishes, an Annotated Checklist of the Oviparous Cyprinodontiform Fishes.* 4th edition. Cincinnati: American Killifish Association, 2000.

Mayden, Richard L., ed. *Systematics, Historical Ecology, and North American Freshwater Fishes.* Stanford: Stanford University Press, 1992.

Meffe, Gary K., and Franklin F. Snelson, Jr., eds. *Ecology and Evolution of Livebearing Fishes (Poeciliidae).* Englewood Cliffs: Prentice Hall, 1989.

Minckley W. L., and James E. Deacon, eds. *Battle Against Extinction, Native Fish Management in the American West.* Tucson: University of Arizona Press, 1991.

Page, Lawrence M., and Brooks M. Burr. *A Field Guide to Freshwater Fishes of North America North of Mexico.* Boston: Houghton Mifflin, 1997.

Raasch, Maynard S. *Delaware's Freshwater and Brackish-Water Fishes: A Popular Account.* Neptune City, NJ: T.F.H. Publications, 1996.

Schultz, R. Jack. "Origins and Relationships of Unisexual Poeciliids." In *Ecology and Evolution of Livebearing Fishes (Poeciliidae),* edited by Gary K. Meffe and Franklin F. Snelson, Jr. Englewood Cliffs: Prentice Hall, 1989.

Seegers, Lothar. *Killifishes of the World: Old World Killis I—Aphyosemion, Lampeyes, Ricefishes.* Mörfelden-Walldorf: Verlag A.C.S., 1997.

———. *Killifishes of the World: Old World Killis II—Aplocheilus, Epiplatys, Nothobranchius.* Mörfelden-Walldorf: Verlag A.C.S., 1997.

———. *Killifishes of the World: New World Killis—Cyprinodon, Cynolebias, Rivulus.* Mörfelden-Walldorf: Verlag A.C.S., 2000.

Smith, C. Lavett. *The Inland Fishes of New York State.* Albany: New York State Department of Environmental Conservation, 1985.

Wildekamp, Rudolf H. *A World of Killies: Atlas of the Oviparous Cyprinodontiform Fishes of the World.* Mishawaka, IN: American Killifish Association, 1993.

Wischnath, Lothar. *Atlas of Livebearers of the World.* Neptune City, NJ: T.F.H. Publications, 1993.

Periodicals

Able, Kenneth W., and James D. Felley. "Geographical Variation in *Fundulus heteroclitus*: Tests for Concordance Between Egg and Adult Morphologies." *American Zoologist* 26, no. 1 (1986): 145–157.

Baugh, Thomas. M., and James E. Deacon. "Daily and Yearly Movement of the Devil's Hole Pupfish *Cyprinodon diabolis* Wales in Devil's Hole, Nevada." *Great Basin Naturalist* 43, no. 4 (1983): 592–596.

———. "Maintaining the Devil's Hole Pupfish *Cyprinodon diabolis* Wales in Aquaria." *Journal of Aquariculture and Aquatic Sciences* 3, no. 4 (1983): 73–75.

———. "The Most Endangered Pupfish." *Freshwater and Marine Aquarium* 6, no. 6 (1983): 22–26, 78–79.

Beaugrand, Jacques P., Jean Caron, and Louise Comeau. "Social Organization of Small Heterosexual Groups of Green Swordtails (*Xiphophorus hellerii*, Pisces, Poeciliidae) Under Conditions of Captivity." *Behaviour* 91 (1984): 24–60.

Davis, William P., et al. "Field Observations of the Ecology and Habits of Mangrove Rivulus (*Rivulus marmoratus*) in Belize and Florida (Teleostei: Cyprinodontiformes: Rivulidae)." *Ichthyological Exploration of Freshwaters* 1, no. 2 (1990): 123–134.

Ferdenzi, Joseph. "Aquarium Observations on the Tanganyican Pearl Killifish, *Lamprichthys tanganicanus.*" *Journal of the American Killifish Association* 20, no. 3 (1987): 95–100.

Foster, N. R. "The Tanganyikan Lampeye, *Lamprichthys tanganicanus* (Boulenger)." *Journal of the American Killifish Association* 16, no. 5 (1983): 165–170.

Fuller, Rebecca C., and Joseph Travis. "A Test for Male Parental Care in a Fundulid, the Bluefin Killifish, *Lucania goodei.*" *Environmental Biology of Fishes* 61, no. 4 (2001): 419–426.

Garman, S. "The Cyprinodonts." *Memoirs of the Museum of Comparative Zoology at Harvard College* 19, no. 1 (July 1895): 1–179.

———. "Sexual Lefts and Rights." *American Naturalist* 29 (November 1895): 1012–1014.

Ghedotti, Michael J. "Phylogenetic Analysis and Taxonomy of the Poeciloid Fishes (Teleostei: Cyprinodontiformes)." *Zoological Journal of the Linnean Society* 130, no. 1 (2000): 1–53.

Hrbek, T., and A. Larson. "The Evolution of Diapause in the Killifish Family Rivulidae (Atherinomorpha, Cyprinodontiformes): A Molecular Phylogenetic and Biogeographic Perspective." *Evolution* 53, no. 4 (1999): 1200–1216.

Jara, Fernando, D. Soto, and R. Palma. "Reproduction in Captivity of the Endangered Killifish *Orestias ascotanensis* (Teleostei: Cyprinodontidae)." *Copeia* 1995, no. 1 (1995): 226–228.

Klee, Albert J. "*Anableps*, the Four-Eyed Fish." *Aquarium*, no. 4 (1968): 6–7, 42–48.

Kneib, R. T. "The Role of *Fundulus heteroclitus* in Salt Marsh Trophic Dynamics." *American Zoologist* 26, no. 1 (1986): 259–269.

Lewis, Thomas H. "A Mogollon Description of *Cyprinodon*." *Southwestern Naturalist* 26, no. 1 (1981): 71–72.

Murphy, W. J., and G. E. Collier. "A Molecular Phylogeny for Aplocheiloid Fishes (Atherinomorpha, Cyprinodontiformes): The Role of Vicariance and the Origins of Annualism." *Molecular Biology and Evolution* 14, no. 8 (1997): 790–799.

Murphy, W. J., and G. E. Collier. "Phylogenetic Relationships of African Killifishes in the Genera *Aphyosemion* and *Fundulopanchax* Inferred from Mitochondrial DNA Sequences." *Molecular Phylogenetics and Evolution* 11, no. 3 (1999): 351–360.

Parenti, Lynne R. "A Phylogenetic and Biogeographic Analysis of Cyprinodontiform Fishes (Teleostei, Atherinomorpha)." *Bulletin of the American Museum of Natural History* 168, no. 4 (1981): 335–557.

———. "A Taxonomic Revision of the Andean Killifish Genus *Orestias* (Cyprinodontiformes, Cyprinodontidae)." *Bulletin of the American Museum of Natural History* 178, no. 2 (1984): 107–214.

Rosen, D. E. "The Relationships and Taxonomic Position of the Halfbeaks, Killifishes, Silversides, and Their Relatives." *Bulletin of the American Museum of Natural History* 127, no. 5 (1964): 1–176.

Rosen, D. E., and R. M. Bailey. "The Poeciliid Fishes (Cyprinodontiformes): Their Structure, Zoogeography, and Systematics." *Bulletin of the American Museum of Natural History* 126, no. 1 (1963): 1–176.

Taylor, D. Scott, et al. "Homozygosity and Heterozygosity in Three Populations of *Rivulus marmoratus*." *Environmental Biology of Fishes* 61, no. 4 (2001): 455–459.

Taylor, Edward C. "*Anableps*, the Amphibious Livebearer: Part 1." *Freshwater and Marine Aquarium* 3, no. 11 (1980): 16–19, 85–91.

———. "*Anableps*, the Amphibious Livebearer: Part 2." *Freshwater and Marine Aquarium* 3, no. 12 (1980): 15–19, 88–92.

Wales, J. H. "Biometrical Studies of Some Races of Cyprinodont Fishes from the Death Valley Region, with the Description of *Cyprinodon diabolis* n. sp." *Copeia* 1930, no. 3 (1930): 61–70.

Weedman, David A. "Monkey Spring Pupfish." *Arizona Wildlife Views* 40, no. 11 (1997): 9.

Zahl, Paul A., et al. "Visual Versatility and Feeding of the Four-Eyed Fishes, *Anableps*." *Copeia* 1977, no. 4 (1977): 791–793.

Organizations

American Killifish Association. 280 Cold Springs Drive, Manchester, PA 17345-1243 USA. Web site: <http://www.aka.org>

American Livebearer Association. 5 Zerbe Street, Cressona, PA 17929-1513 USA. Phone: (570) 385-0573. Fax: (570) 385-2781. Web site: <http://livebearers.org>

Desert Fishes Council. 315 East Medlock Drive, Phoenix, AZ 85012 USA. Phone: (602) 274-5544. Web site: <http://www.desertfishes.org/>

Kenneth J. Lazara, PhD

Stephanoberyciformes

(Whalefishes and relatives)

Class Actinopterygii
Order Stephanoberyciformes
Number of families 9

Illustration: Hairyfish (*Mirapinna esau*). (Illustration by Brian Cressman)

Evolution and systematics

The order Stephanoberyciformes comprises deepwater, bathypelagic fishes that are for the most part poorly known anatomically. The order has a checkered systematic history; some of the families currently assigned to this order were treated as part of the formerly larger order Beryciformes (sometimes called Trachichthyiformes), others were previously included in the Lampridiformes. Stephanoberyciforms, as presently constituted, were given the status of an order by G. D. Johnson and C. Patterson in a phylogenetic survey of acanthomorph teleosts in 1993 (which is followed here); parts of the group were previously recognized as separate orders by earlier authors, including the whalefishes (Cetomimiformes) and pricklefishes and allies (Xenoberyces in part, or Stephanoberyciformes *sensu stricto*).

The 9 families that form the Stephanoberyciformes are divided into 28 genera and about 92 species. The families are:

- Stephanoberycidae (pricklefishes; 3 monotypic genera)
- Melamphaidae (bigscales or ridgeheads; 5 genera, about 38 species)
- Gibberichthyidae (gibberfishes, *Gibberichthys*; 2 species)
- Hispidoberycidae (*Hispidoberyx ambagiosus*)
- Cetomimidae (flabby whalefishes; 9 genera, about 35 species)
- Barbourisiidae (*Barbourisia rufa*)
- Rondeletiidae (redmouth whalefishes, *Rondeletia*; 2 species)
- Mirapinnidae (3 genera, 5 species)
- Megalomycteridae (largenoses; 4 genera, 5 species)

The first four families form the superfamily Stephanoberycoidea; the remaining families are united in the Cetomimoidea. Stephanoberyciforms share a specialization of the posterior dermal skull roof, in which the enlarged extrascapular bones cover the parietal bones. They are closely related to the bony fish groups Zeiformes, Beryciformes (*sensu stricto*), and Percomorpha, sharing with them the presence of pelvic fin spines (lost in certain stephanoberyciforms), as well as specialized features of their pelvic fin anatomy.

The fossil history of the Stephanoberyciformes is almost negligible in contrast to the more extensive fossil history of the Beryciformes. No fossil stephanoberyciform taxon has been erected to date, even though fossil otoliths (paired ear stones of the membranous inner ear labyrinth present in many fishes that can aid in detecting motion) and scant skeletal remains have been mentioned. Otoliths are difficult to identify because they lack features that are diagnostic of taxa of fishes; nonetheless, two unnamed species referred to the living genus *Melamphaes* have been recorded from the Tertiary period (some 50 million years ago) of France.

Physical characteristics

Stephanoberyciforms are a morphologically diverse lineage, ranging from the large-headed whalefishes to the small-headed megalomycterids and mirapinnids. In stephanoberyciforms generally, the head is somewhat large, with numerous bony ridges in the stephanoberycoids (giving them a highly armored look), usually with small eyes, the dorsal and anal fins end in opposition posteriorly on the body, the pectoral fins are of moderate size, and the caudal fin is truncated and not very large. All families have a single dorsal fin, either with very few spines (stephanoberycoids) or lacking them completely. Pelvic fins can be well developed (as in melamphaids), reduced (as in stephanoberycids), absent (as in cetomimids), and even winglike and aberrant (as in the hairyfish, *Mirapinna esau*). The caudal fin may have procurrent spines (as in stephanoberycoids), and is uniquely subdivided in *Mirapinna*. The jaw bones are rather weak and easily bent.

Many stephanoberyciform species appear velvety due to small protrusions from the epidermis (as in *Mirapinna*), or even from numerous spines on the scales (as in stephanoberycids) that are usually deciduous. Many species, such as the cetomimids, lack scales, others, such as the melamphaids, have large cycloid scales (their scales are rarely seen as they are easily lost on capture). The teeth are very small; numerous pores are usually visible on the head and lateral line, and some species may have luminous tissues and highly distensible stomachs, and are soft and flabby (as in whalefishes). One family, the Megalomycteridae, has extremely well-developed olfactory organs and nostrils. Species of the mirapinnid genera *Eutaeniophorus* and *Parataeniophorus*, and those in the family Megalomycteridae are morphologically quite distinct from other stephanoberyciforms.

The fishes of this order are usually small, rarely surpassing 9.8 in (25 cm) in length, and usually below 4.7 in (12 cm); only a few specimens above 13.8 in (35 mm) are known. Coloration is drab, brown, grayish black, or reddish (as in whalefishes). Many stephanoberyciform species are known from very few specimens, and sometimes these represent only a single sex or a juvenile stage; no adult mirapinnid has been collected to date.

Distribution

The fishes of this order are widely distributed throughout all major oceans, but are yet to be recorded from the Mediterranean Sea or from Arctic waters.

Habitat

Stephanoberyciforms are deepsea fishes, generally inhabiting bathypelagic depths down to about 13,123 ft (4,000 m). A few species, especially juvenile forms, have been captured closer to the surface in waters as shallow as 164 ft (50 m), probably an indication of vertical migrations. The only specimen of the hairyfish was captured near the surface. Stephanoberyciforms, especially whalefishes and melamphaids, form an important and very large proportion of fishes in the bathypelagic realm.

Behavior

Stephanoberyciforms undertake vertical migrations from deeper waters into shallower regions.

Feeding ecology and diet

Very little is known concerning the food preferences of stephanoberyciforms, as the stomachs of relatively few individuals have been found to contain food. Copepods and other small crustaceans have been found in the stomachs of a few specimens of different species, but many, including, stephanoberycoids and whalefishes, are probably capable of ingesting larger prey items because of their relatively wide gape. Stephanoberyciforms are probably eaten by larger fishes.

Reproductive biology

The reproductive biology of stephanoberyciforms is largely unknown, but both eggs and larvae are pelagic. Eggs are unknown in many species. One highly modified larval form was originally described as a separate genus, *Kasidoron*, and even given familial status, but it is now well established that it represents the larval form of the gibberfish (*Gibberichthys pumilus*). This larval form is remarkable, presenting a very long pelvic appendage that superficially resembles algae (such as *Sargassum*) or siphonophores. The appendage is lost by about 1.2 in (3 cm) standard length. The larvae of *Eutaeniophorus* and *Parataeniophorus* are remarkable in presenting very elongated caudal "streamers," long tape-like projections that may reach several times body length (somewhat smaller in *Parataeniophorus*); the streamer is lost in adults. Larvae also have been described for *Melamphaes*, *Poromitra*, *Scopeloberyx*, and *Scopelogadus*. These begin to display their generic characteristics from between 0.2 and 0.8 in (0.5 and 2 cm) standard length. As a general rule, larvae and juveniles appear to occur in more shallow water compared to adults of the same species, which tend to be more bathypelagic.

Conservation status

No species in Stephanoberyciformes are listed by the IUCN.

Significance to humans

Stephanoberyciforms are not eaten and are therefore of little direct importance economically. Whalefishes are very numerous, and believed to form the most numerous fish species in terms of biomass in the bathypelagic zone 3,281–13,123 ft (1,000–4,000 m) in depth; therefore they are probably important food items of other commercial species, such as the orange roughy (*Hoplostethus atlanticus*).

1. Hairyfish (*Mirapinna esau*); 2. Red whalefish (*Barbourisia rufa*); 3. Longjaw bigscale (*Scopeloberyx robustus*); 4. Pricklefish (*Stephanoberyx monae*). (Illustration by Brian Cressman)

Species accounts

Red whalefish
Barbourisia rufa

FAMILY
Barbourisiidae

TAXONOMY
Barbourisia rufa Parr, 1945, Gulf of Mexico.

OTHER COMMON NAMES
English: Velvet whalefish; Japanese: Aka-kujira-uo-damashi.

PHYSICAL CHARACTERISTICS
Length about 15.8 in (40 cm). Unusual, with very large mouth (maxillae extend posteriorly well beyond level of eyes); teeth present on entire length of jaws; single dorsal fin located far posteriorly on back, close to caudal fin, with 20–23 rays; anal fin terminating at same level of dorsal fin, with 14–18 rays; pelvics (with 6 rays) and pectorals (with 12–14 rays) very small; skin covered in minute protuberances (velvety to the touch); lateral line very clearly demarcated by large-pored scales; coloration a uniform bright red.

DISTRIBUTION
Widespread but uncommon, occur in every major ocean usually at low latitudes, but reported to reach as far north as Greenland.

HABITAT
Captured usually near the bottom or in midwater over continental slopes and seamounts in depths of 394–6,562 ft (120–2,000 m).

BEHAVIOR
Benthopelagic, believed to be capable of vertical midwater migrations for feeding.

FEEDING ECOLOGY AND DIET
Nothing known.

REPRODUCTIVE BIOLOGY
Presumably lays pelagic eggs.

CONSERVATION STATUS
Not threatened.

SIGNIFICANCE TO HUMANS
None known. ◆

Longjaw bigscale
Scopeloberyx robustus

FAMILY
Melamphaidae

TAXONOMY
Melamphaes robustus Günther, 1887, Eastern Atlantic.

OTHER COMMON NAMES
English: Ridgehead; Japanese: Tate-kabuto-uo.

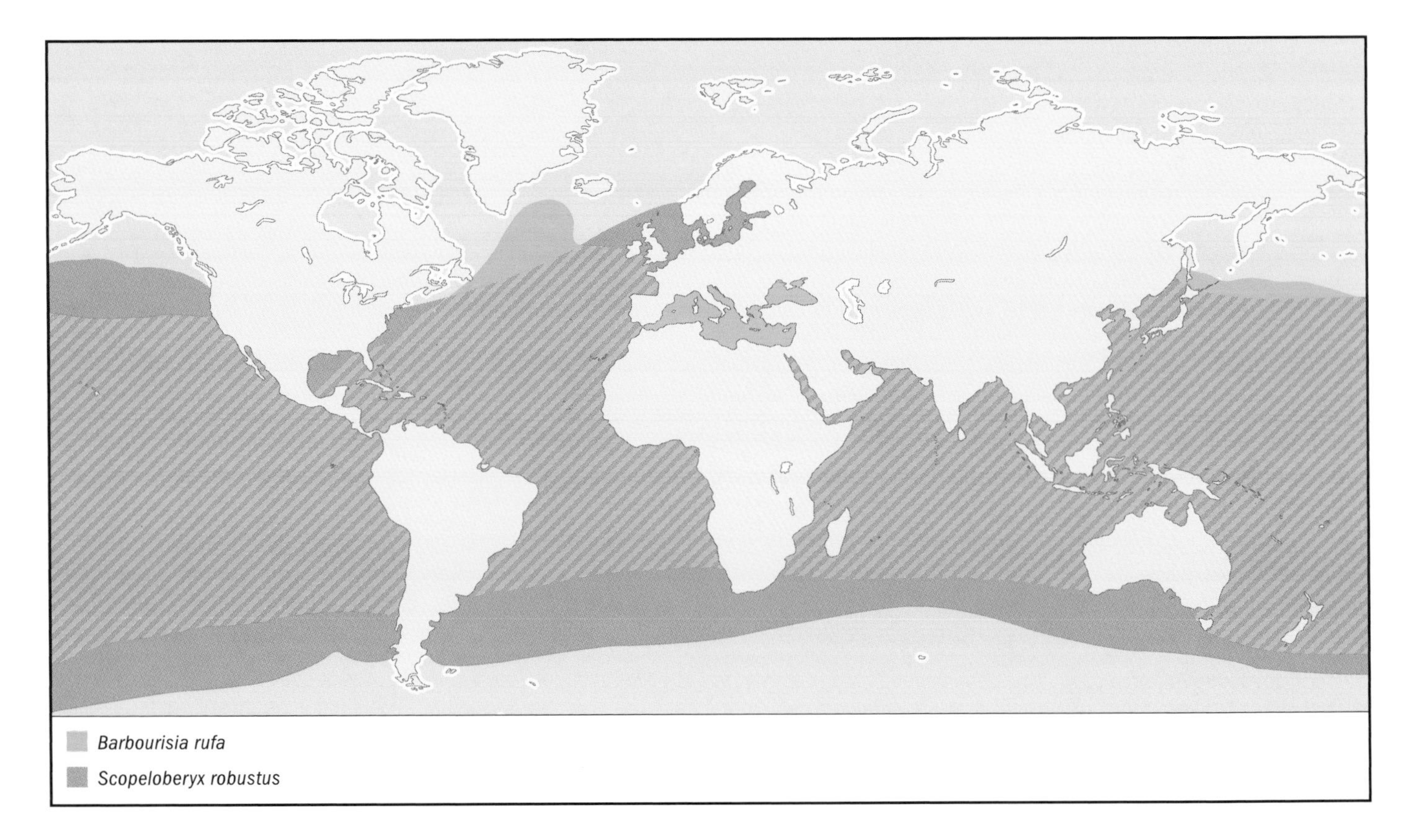

PHYSICAL CHARACTERISTICS
Length 2.8–3.9 in (7.3–10 cm). Head large (from 36 to 45% of standard length), with many bony ridges and deep sensory canals; elongate mouth reaching posteriorly well beyond relatively small eyes; large cycloid scales; 19–25 gill rakers; single dorsal fin originating at midlength, with 2–3 spines and 9–13 rays; anal fin posterior to dorsal, with a single spine and 7–9 rays; pelvic fins well posterior to head and with a single spine and 6–8 rays; pectoral fins elongate, reaching to level of mid-dorsal fin, with 11–14 rays; caudal fin with some 20 rays; small teeth in 2–4 rows in upper and lower jaws. Coloration dark brown (at least in preservative).

DISTRIBUTION
Cosmopolitan in all major oceans except the Mediterranean Sea and Arctic Ocean. Reported to be common around the Commander Islands.

HABITAT
Meso- and bathypelagic; adults usually at depths of 1,640–11,102 ft (500–3,384 m), juveniles in slightly shallower waters.

BEHAVIOR
Nothing known.

FEEDING ECOLOGY AND DIET
Unknown, but because of its relatively wide gape it may consume fishes.

REPRODUCTIVE BIOLOGY
Probably lays pelagic eggs.

CONSERVATION STATUS
Not threatened.

SIGNIFICANCE TO HUMANS
None known. ◆

Hairyfish
Mirapinna esau

FAMILY
Mirapinnidae

TAXONOMY
Mirapinna esau Bertelsen and Marshall, 1956, Eastern Atlantic.

OTHER COMMON NAMES
None known.

PHYSICAL CHARACTERISTICS
Size of only specimen 0.16 in (4 cm). Unusual appearance, head region proportionally small; mouth upturned; pelvic fins huge, projecting dorsally and located at throat, with 8 rays; small pectoral fins located more dorsally, with 13 rays; caudal fin divided into 2 distinct lobes that overlap; single dorsal fin posteriorly located, with 16 rays; anal fin opposite dorsal fin, with 14 rays; skin covered in hairlike protuberances; coloration dark brown.

DISTRIBUTION
Known from a single specimen, caught north of the Azores in the Eastern Atlantic.

HABITAT
Only known specimen was caught at the surface.

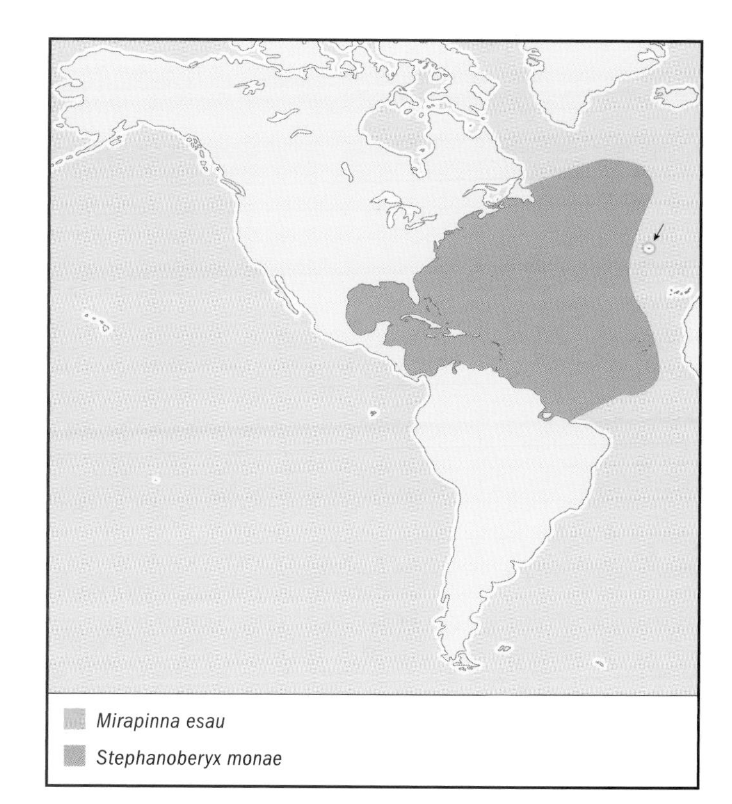

BEHAVIOR
Nothing known.

FEEDING ECOLOGY AND DIET
Copepods were the only food item in the stomach.

REPRODUCTIVE BIOLOGY
Unknown; only known specimen is a juvenile female.

CONSERVATION STATUS
Not listed by IUCN.

SIGNIFICANCE TO HUMANS
None known. ◆

Pricklefish
Stephanoberyx monae

FAMILY
Stephanoberycidae

TAXONOMY
Stephanoberyx monae Gill, 1883, Western North Atlantic.

OTHER COMMON NAMES
None known.

PHYSICAL CHARACTERISTICS
Length 3.9 in (10 cm). Head region proportionally large with many ridges; round eyes; mouth large extending posteriorly beyond level of eyes; single dorsal fin located posteriorly, with 1–3 poorly developed spines and 11–13 rays; anal fin terminating at level of dorsal fin, with 1–3 spines and 11–12 rays; pectoral fins with 12–13 rays; pelvic fins rudimentary and abdominal, with 5

rays; caudal fin with supports over and under caudal peduncle, with 8–11 spines on upper and lower aspects; body covered in scales that have small spines; coloration dark brown.

DISTRIBUTION
Western North Atlantic Ocean, the Caribbean Sea, and the Gulf of Mexico.

HABITAT
Presumably bathypelagic or demersal in relatively deep waters 1,115–15,673 ft (340–4,777 m).

BEHAVIOR
Nothing known.

FEEDING ECOLOGY AND DIET
Unknown, but its large gape suggests it may feed on fishes and invertebrates.

REPRODUCTIVE BIOLOGY
Essentially unknown; two specimens measuring 3 in (8 cm) were sexually mature.

CONSERVATION STATUS
Not threatened.

SIGNIFICANCE TO HUMANS
None known. ◆

Resources

Books

Bertelsen, E. "Families Mirapinnidae, Eutaeniophoridae." In *Fishes of the North-Eastern Atlantic and the Mediterranean.* Vol. II, edited by P. J. P. Whitehead, M.-L. Bauchot, J.-C. Hureau, J. Nielsen, and E. Tortonese. Paris: UNESCO, 1986.

Bertelsen, E., and N. B. Marshall. "Mirapinnatoidei: Development and Relationships." In *Ontogeny and Systematics of Fishes,* edited by H. G. Moser, W. J. Richards, D. M. Cohen, M. P. Fahay, A. W. Kendall, Jr., and S. L. Richardson. Special Publication No. 1. Lawrence, KS: American Society of Ichthyologists and Herpetologists, 1984.

Ebeling, A. W. "Family Melamphaidae." In *Smiths' Sea Fishes,* edited by M. M. Smith and P. C. Heemstra. Grahamstown, South Africa: J. L. B. Smith Inst. of Ichthyology, 1986.

Ebeling, A. W., and W. H. Weed. "Order Xenoberyces (Stephanoberyciformes)." In *Fishes of the Western North Atlantic,* edited by D. M. Cohen. Sears Foundation for Marine Research. Mem. No. 1, Part 6. New Haven: Yale University, 1973.

Gon, O. "Melamphaidae." In *Fishes of the Southern Ocean,* edited by O. Gon and P. C. Heemstra. Grahamstown, South Africa: J. L. B. Smith Inst. of Ichthyology, 1990.

Heemstra, P. C. "Family Stephanoberycidae." In *Smiths' Sea Fishes,* edited by M. M. Smith and P. C. Heemstra. Grahamstown, South Africa: J. L. B. Smith Inst. of Ichthyology, 1986.

Keene, M. J., and K. A. Tighe. *Beryciformes: Development and Relationships.* In *Ontogeny and Systematics of Fishes,* edited by H. G. Moser, W. J. Richards, D. M. Cohen, M. P. Fahay, A. W. Kendall, Jr., and S. L. Richardson. Special Publication no. 1. Lawrence, KS: American Society of Ichthyologists and Herpetologists, 1984.

Masuda, H., K. Amaoka, C. Araga, T. Uyeno, and T. Yoshino. *The Fishes of the Japanese Archipelago,* Vol. 1. Tokyo: Tokai University Press, Japan, 1984.

Maul, G. E. "Families Melamphaidae, Stephanoberycidae." In *Fishes of the North-Eastern Atlantic and the Mediterranean.* Vol. II, edited by P. J. P. Whitehead, M.-L. Bauchot, J.-C. Hureau, J. Nielsen, and E. Tortonese. Paris: UNESCO, 1986.

Mecklenberg, C. W., T. A. Mecklenberg, and L. K. Thorsteinson. *Fishes of Alaska.* Bethesda, MD: American Fisheries Society, 2002.

Moore, J., and J. R. Paxton. "Melamphaidae. Bigscales, Ridgeheads." In *FAO Species Identification Guide for Fishery Purposes: The Living Marine Resources of the WCP.* Vol. 4, *Bony Fishes. Part 2 (Mugilidae to Carangidae),* edited by K. E. Carpenter and V. H. Niem. Rome: FAO, 1999.

Patterson, C. "Osteichthyes: Teleostei." In *The Fossil Record 2,* edited by M. J. Benton. London: Chapman & Hall, 1993.

Paxton, J. R. "Families Cetomimidae, Rondeletiidae." In *Fishes of the North-Eastern Atlantic and the Mediterranean.* Vol. II, edited by P. J. P. Whitehead, M. -L. Bauchot, J. -C. Hureau, J. Nielsen, and E. Tortonese. Paris: UNESCO, Paris, 1986.

———. "Gibberichthyidae (Gibberfishes), Megalomycteridae (Bignose Fishes)." In *FAO Species Identification Guide for Fishery Purposes: The Living Marine Resources of the WCP.* Vol. 4, *Bony fishes. Part 2 (Mugilidae to Carangidae),* edited by K. E. Carpenter and V. H. Niem. Rome: FAO, 1999.

Paxton, J. R., and D. J. Bray. "Order Cetomimiformes." In *Smiths' Sea Fishes,* edited by M. M. Smith and P. C. Heemstra. Grahamstown, South Africa: J. L. B. Smith Inst. of Ichthyology, 1986.

Paxton, J. R., and O. Gon. "Cetomimidae." In *Fishes of the Southern Ocean,* edited by O. Gon and P. C. Heemstra. Grahamstown, South Africa: J. L. B. Smith Inst. of Ichthyology, 1990.

Periodicals

Bertelsen, E., and N. B. Marshall. "Mirapinnati, a New Order of Teleost Fishes." *Dana Report* 42 (1956): 1–34.

De Sylva, D. P., and W. N. Eschmeyer. "Systematics and Biology of the Deep Sea Fish Family Gibberichthyidae, a Senior Synonym of the Family Kasidoridae." *Proceedings of the California Academy of Sciences.* Series 4, 41 (1977): 215–231.

Ebeling, A. W. "Melamphaidae I. Systematics and Zoogeography of the Species in the Bathypelagic Fish Genus *Melamphaes* Günther." *Dana Report* 58 (1962): 1–164.

Ebeling, A. W., and W. H. Weed. "Melamphaidae III. Systematics and Distribution of the Species in the Bathypelagic Fish Genus *Scopelagadus* Vaillant." *Dana Report* 60 (1963): 1–58.

Kotlyar, A. N., and D. P. Andrianov. "Systematics and Biology of *Acanthochaenus luetkenii* (Stephanoberycidae)." *Journal of Ichthyology* 33, no. 6 (1993): 85–95.

Paxton, J. R. "Synopsis of the Whalefishes (Family Cetomimidae) with Descriptions of Four New Genera." *Records of the Australian Museum* 41 (1989): 135–206.

Paxton, J. R., G. D. Johnson, and T. Trnski. "Larvae and Juveniles of the Deepsea Whalefishes, *Barbourisia* and *Rondeletia* (Stephanoberyciformes: Barbourisiidae, Rondeletiidae) with Comments on Family Relationships." *Records of the Australian Museum* 53 (2001): 407–425.

Yang, Y.-R., B.-G. Zeng, and J. R. Paxton. "Additional Specimens of the Deepsea Fish *Hispidoberyx ambagiosus* (Hispidoberycidae, Berciformes [sic]) from the South China Sea, with Comments on the Family Relationships." *Jap. Soc. Ichthyol.* 38 (1988): 3–8.

Marcelo Carvalho, PhD

Beryciformes

(Roughies, flashlightfishes, and squirrelfishes)

Class Actinopterygii
Order Beryciformes
Number of families 7

Photo: Four splitfin flashlightfishes (*Anomalops katoptron*), one with its lights off. (Photo by Fred McConnaughey/Photo Researchers, Inc. Reproduced by permission.)

Evolution and systematics

The order Beryciformes encompasses 7 marine families, 29 genera, and about 140 species. The seven families are:

- Anomalopidae, the flashlightfishes or lanterneye fishes
- Anoplogasteridae (also spelled Anoplogastridae), the fangtooth fishes
- Berycidae, the alfoncinos and redfishes
- Diretmidae, the spinyfins
- Holocentridae, the squirrelfishes and soldierfishes
- Monocentridae, the pineapplefishes or pineconefishes
- Trachichthyidae, the roughies or slimeheads

The order Beryciformes falls at the base of a large grouping of fishes collectively known as percomorphs. These advanced fishes include the order Perciformes, which comprises a great diversity of fishes including cichlids, perches, blennies, and barracudas. Like all other percomorphs, the beryciforms have a characteristic linked arrangement of the pelvic and pectoral girdles. They differ from other percomorphs in the number of rays in the tail fin. Caudal rays in most percomorphs number 17, but fishes in the Beryciformes order have 18 or 19. Systematists believe that as the fishes advanced evolutionarily, the number of tail-fin rays decreased. The fact that the Beryciformes have a greater number of caudal rays places them at the base of the percomorph lineage.

At one time, the order Beryciformes was larger and included the beardfishes, whalefishes, gibberfishes, and pricklefishes. Systematists once classified beardfishes as primitive beryciforms, but have now placed them in their own order, the Polymixiiformes, which precedes the evolution of the Beryciformes. The whalefishes, gibberfishes, and pricklefishes are in the order Stephanoberyciformes, and accompany the Beryciformes at the base of the percomorph lineage. The genera of beryciform fishes have also undergone some changes in classification. For example, the squirrelfishes have at one time or another been classified under the genera *Adioryx*, *Flammeo*, *Holocentrus*, *Sargocentron*, and *Neoniphon*, but the latter three are currently used.

The fossil record indicates that beryciform fishes occurred at least as far back as the late Cretaceous period, and were abundant. The group has persisted and is still quite common.

Physical characteristics

The Beryciformes are small- to medium-sized, spiny-rayed fishes, 3–24 in (8–61 cm) long. They have big eyes, some have colorful scales, and some have light organs beneath the eyes. The order is also defined partly by the number of softer, flexible rays in the ventral fin.

Schooling in whitetip soldierfish. (Illustration by Wendy Baker)

The Holocentridae is the largest family within the Beryciformes. Encompassing both the squirrelfishes and soldierfishes, fishes in this family are typified by a reddish color from head to tail and a noticeably forked tail. The trailing edges of their scales often have spines (called spinoid scales); some have spines on their gill covers.

Fishes in the family Anomalopidae, the flashlight or lanterneye fishes, are distinguished primarily by the obvious light organ under each eye. These fishes and the bioluminescent bacteria that generate the light have developed a symbiotic relationship that offers the bacteria a place to live, while giving the fishes illumination, perhaps for attracting zooplankton during night feeding and for intraspecific communication. Light organs are also seen in other beryciform fishes, including those in the family Monocentridae.

Atlantic flashlightfish (*Kryptophanaron alfredi*) have light organs below their eyes. (Photo by Tom McHugh/Photo Researchers, Inc. Reproduced by permission.)

The monocentrids' whimsical common names of the pineapple and pineconefishes come from their beautiful, large scales. Usually yellow, each scale has its own dark outline, further accentuating an armorlike appearance. Like the holocentrids, species of the Monocentridae have spines poking backward from each scale.

Trachichthyids are known mostly from one species, the orange roughy (*Hoplostethus atlanticus*). This somewhat primitive-looking fish, as well as other members of this family, are distinguished by mucous cavities just beneath the skin of the head. This trait accounts for their less-than-flattering alternate common name of slimeheads.

Distribution

As a group, the beryciform fishes are found in tropical, subtropical, and temperate waters nearly around the globe. The holocentrids occur worldwide in tropical waters, although some venture further north or south. The alfoncinos and redfishes, spinyfins, and fangtooths prefer tropical to temperate waters worldwide. The pineapplefishes share this affinity, but remain in the Indian and Pacific Oceans. The flashlightfishes occur in the tropical seas from the western Atlantic to the Pacific; the roughy family favors the waters around Australia.

Habitat

Beryciform fishes cover a range of habitats. Some species, such as many of the squirrelfishes, gravitate toward shallow, tropical reefs. Others, such as the roughies, spend their lives in deep, dark, ocean waters. Most roughies inhabit continental-shelf and slope waters almost 5,280 ft (1,609 m) deep. Certain spinyfins and fangtooth species share this preference for deep waters, and live along the sea bottom 6,600 ft (2,012 m) down.

Beryciforms that live in shallow waters shun the light, usually tucking themselves under a coral overhang, backing into a cave, or hiding below another structure during the day. During daytime excursions, divers frequently encounter squirrelfishes poking out from some type of dark sanctuary. Shallow-dwelling species sometimes maintain a daily routine of descending into deep waters and remaining mostly inactive during the day, then rising into the shallows at night to feed. A few, such as the flashlightfishes, further avoid the light by limiting their shallow-water forays to nights of a new moon, or to the periods before the moon rises and after it sets.

A flashlightfish, with its light organ glowing under its eye in an otherwise dark background. Flashlightfish use the light organ at night to look for prey and to communicate with each other. (Illustration by Wendy Baker)

Behavior

Perhaps the most notable characteristic of the beryciform fishes is their ability to produce light, and in some cases, readily control it. The light is the result of bioluminescent bacteria that take up residence in pockets just below the skin of various species, including the flashlight, pineapple, and pineconefishes. Other beryciform fishes also have light organs, including members in the genera *Sorosichthys* and *Paratrachichthys.* As well as using the light to find and/or to attract prey during their nocturnal feeding forays, in some cases these fishes apparently employ the illumination as a means of communication between members of the their own species and as a method of confusing potential predators. This assumption is based on observations of alterations in the blinking pattern of the light when conspecifics approach one another or when a predator swims nearby. The eyelight fish (*Photoblepharon palpebratus*) can control its light production by lifting or dropping a flap of skin over the light organ. Other species have other ways of controlling the light.

Hawaiian squirrelfish (*Sargocentron xantherythrum*) in a shipwreck near Hawaii. (Photo by Andrew G. Wood/Photo Researchers, Inc. Reproduced by permission.)

Several beryciform species also make noises, either when interacting with members of their own species or with other fishes. The squirrelfishes are noted for their grunting and clicking sounds, which they produce with the swim bladder.

Feeding ecology and diet

Beryciformes feed on small fishes and various invertebrates. The shallow-dwellers are primarily nocturnal feeders, although some will feed on invertebrates passing through their diurnal retreats. For example, squirrelfishes primarily dine on the small fishes, various crabs, shrimps, and other crustaceans and zooplankton they find in the reef at night, but will take an invertebrate during the day if one happens to wander nearby.

A pineapplefish (*Cleidopus gloriamaris*) hides during the day in a rocky opening. (Illustration by Wendy Baker)

Blotcheye soldierfish (*Myripristus berndti*) school in Hawaii. (Photo by Andrew G. Wood/Photo Researchers, Inc. Reproduced by permission.)

Fishes in this order are known by their large eyes, which allow them to see in low-light conditions. Some have the added advantage of light organs, which assist in finding and perhaps attracting prey. Species in the Monocentridae and Anomalopidae use their light organs to fill their diet of crustaceans. The Australian pineapplefish (*Cleidopus gloriamaris*) has light organs near the mouth and uses them like bluish spotlights at night when it ventures out from its cave hideout to find food. The similar-appearing pineconefish (*Monocentris japonica*) is believed to use its light organs to lure light-responsive prey after it has seen them rather than to find them in the first place.

Predators for these fishes may include sea birds for shallow-dwelling species, as well as many of the larger, piscivorous fishes of their habitat.

Reproductive biology

Little is known about the reproductive biology of the Beryciformes. No observations or studies have shown hermaphroditism or sex reversal among the species; they are born male or female and remain that way throughout their lives. Scientists believe that all beryciform fishes have external fertilization.

More is known about the reproduction in squirrelfishes, because these fishes are common in reefs where they are frequently observed by divers. During mating in the Hawaiian squirrelfish (*Sargocentron xantherythrum*), a male and female grunt and click, align themselves side by side, and place their tails together while fanning out their heads to the left and right.

The mating of the red soldierfish (*Myripristis murdjan*) in the Holocentridae, has also been observed. The male and female courtship ritual involves an inward-spiraling swimming pattern between the two, followed by a quick, adjacent rise through the water when both eggs and sperm are ejected for fertilization.

Conservation status

No species of Beryciformes is listed by the IUCN.

Significance to humans

Various members of the Beryciformes are important in the pet trade. The colors of the pineapplefishes and squirrelfishes and the glowing organ of the flashlightfishes all draw interest from aquarium keepers. Divers also appreciate the reds of the shallow-dwelling squirrelfishes, even if they are mostly viewed in dark crevices and other hiding spots during the day. Several beryciforms, including the orange roughy, are commercially harvested as food.

1. Orange roughy (*Hoplostethus atlanticus*); 2. Common fangtooth (*Anoplogaster cornuta*); 3. Blackbar soldierfish (*Myripristis jacobus*); 4. Splitfin flashlightfish (*Anomalops katoptron*); 5. Pineconefish (*Monocentris japonica*); 6. Squirrelfish (*Holocentrus ascensionis*). (Illustration by Wendy Baker)

Species accounts

Splitfin flashlightfish
Anomalops katoptron

FAMILY
Anomalopidae

TAXONOMY
Anomalops katoptron Bleeker, 1856, Manado, Sulawesi [Celebes], Indonesia.

OTHER COMMON NAMES
English: Flashlightfish, great flashlightfish, Indian flashlightfish, lanterneye fish; twofin flashlightfish; German: Lanternenfisch.

PHYSICAL CHARACTERISTICS
Reaches length of nearly 12 in (about 30 cm). The smaller splitfins average about 4 in (10.2 cm) and live in shallower areas. Color brownish black. Has the typical large eye of the beryciforms. They have two dorsal fins; the hindmost fin is triangular and much larger than the front dorsal fin. Two light organs are noticeable just beneath each eye.

DISTRIBUTION
Western South Pacific, from Malaysia east to the Tuamotu Archipelago, and from the Great Barrier Reef up to southern Japan.

HABITAT
Prefers deeper reef areas of 650–1,300 ft (200–400 m), but is also seen in depths as shallow as 65 ft (20 m). During the day, it remains hidden from sunlight, either in deep water or in dark caves. In winter months, the species aggregates in the warmer, shallower waters of the Philippines.

BEHAVIOR
Fishes in this species have a light-producing organ, and regulate it using a muscular attachment that rotates the gland, either to allow the bioluminescent bacteria to shine forth or to hide the glow from view. The fishes can control the light, which they use to communicate with conspecifics. Splitfin flashlightfishes often travel in schools of 24 to 48 fish.

FEEDING ECOLOGY AND DIET
Uses its large light organ during feeding, which is primarily a nocturnal activity. These fishes shun even dim external light, opting to search for food before or after the moon has risen and set, or on nights of a new moon. Their diet is mainly zooplankton.

REPRODUCTIVE BIOLOGY
Little is known about the reproductive biology of the splitfin flashlightfish, but they probably do not guard eggs.

CONSERVATION STATUS
Not listed by the IUCN.

SIGNIFICANCE TO HUMANS
Part of the aquarium trade; sometimes used as bait fish. ◆

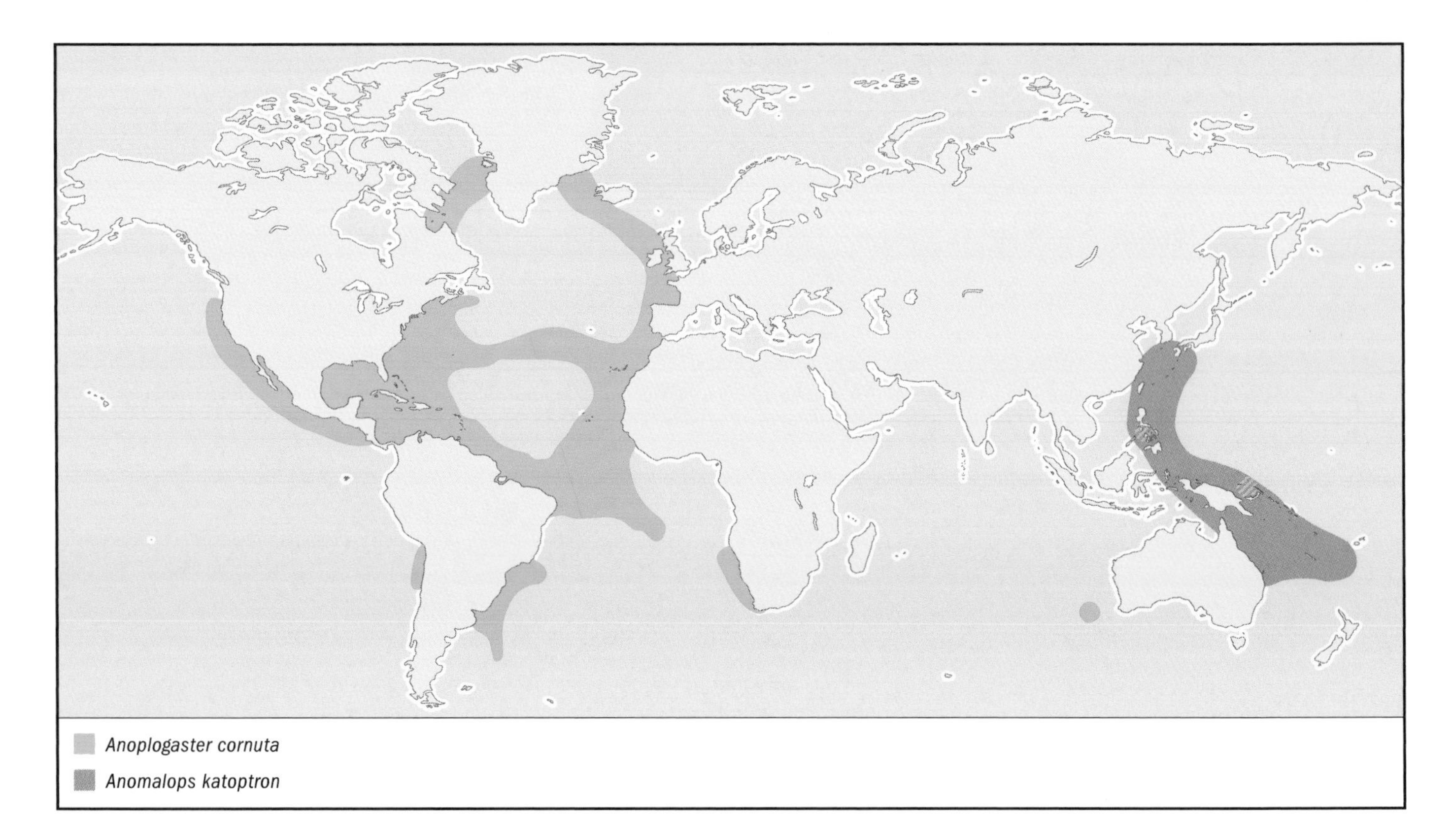

Common fangtooth

Anoplogaster cornuta

FAMILY
Anoplogasteridae

TAXONOMY
Anoplogaster cornuta Valenciennes, 1833, S. Atlantic.

OTHER COMMON NAMES
English: Common sabertooth, ogrefish; Spanish: Pez con colmillos.

PHYSICAL CHARACTERISTICS
Can reach up to 6 in (15.2 cm) in length. Their huge head, large mouth opening, and long, sharp teeth combine to give these fishes a rather frightening appearance. With long spines on their heads, the light-gray juveniles look quite different from adults, and were identified as a separate species for many years.

DISTRIBUTION
Temperate to tropical waters around the world.

HABITAT
Prefers deeper waters of 1,650–6,600 ft (500–2,000 m), but some occur as far down as 16,100 ft (4,900 m). Juveniles will venture almost to the surface.

BEHAVIOR
Fangtooths live singly or in small groups.

FEEDING ECOLOGY AND DIET
Adult fangtooths are mainly piscivores, feeding by opening their large mouth to draw in prey. The young primarily subsist on crustaceans.

REPRODUCTIVE BIOLOGY
Engages in external fertilization, and provides no parental care to eggs or young. The larvae are planktonic.

CONSERVATION STATUS
Not listed by the IUCN.

SIGNIFICANCE TO HUMANS
None known. ◆

Squirrelfish

Holocentrus ascensionis

FAMILY
Holocentridae

TAXONOMY
Holocentrus ascensionis Osbeck, 1765, Acension Island.

OTHER COMMON NAMES
English: Common squirrelfish, longjaw squirrelfish, French: Marignan coq; Spanish: Candil gallito.

PHYSICAL CHARACTERISTICS
A reddish, sometimes blotched, fish with a large dark eye. Typically grows to about 12 in (30.5 cm), although reports exist of specimens twice that size. Has a double dorsal fin, with the front fin comprising sharp spines and taking a yellowish hue. The back fin is much taller, with flexible rays.

DISTRIBUTION
Gulf of Mexico and the western Atlantic Ocean from New York to Brazil, and east to Bermuda. Eastern Atlantic in

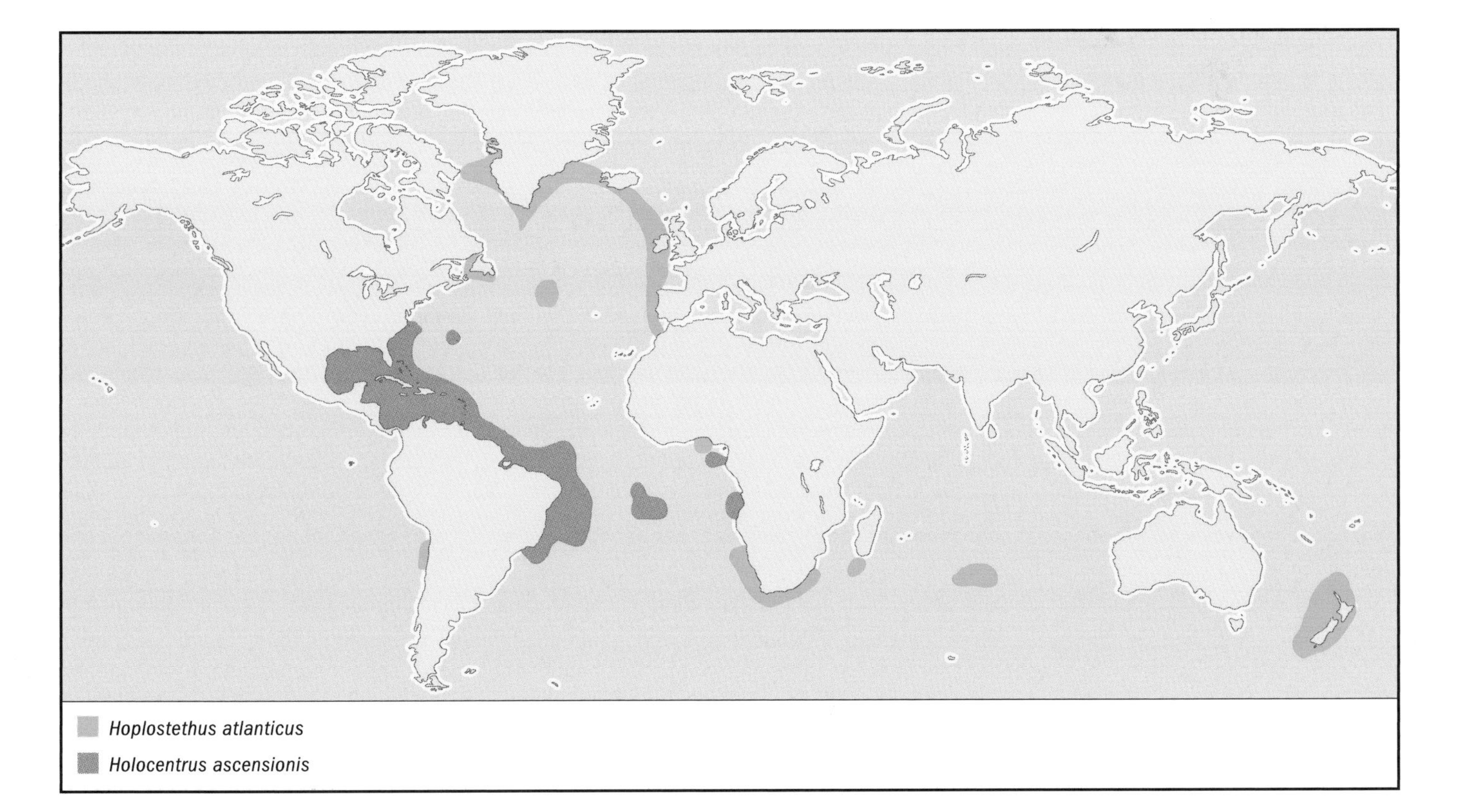

Gabon, Angola, St. Paul's Rocks, St. Helena, and Ascensión Island.

HABITAT
Subtropical reefs with suitable structures for hiding during the day. Commonly seen by divers in very shallow waters, but also found at depths down to 590 ft (180 m).

BEHAVIOR
Remains hidden from view during the day, either in deeper waters or in shallow-water crevices. May occur alone or in aggregations of up to a few dozen fish. This species makes grunting and clicking noises.

FEEDING ECOLOGY AND DIET
Mainly nocturnal feeders, consuming crustaceans and other invertebrates. Predators include sea birds and other fishes, such as the common dolphinfish (*Corphaena hippurus*), mutton snapper (*Lutjanus analis*), and yellowfin tuna (*Thunnus albacares*).

REPRODUCTIVE BIOLOGY
Mating behavior includes a pairing of male and female, during which they bring their tails together while their heads face away from one another. The eggs are pelagic. The larvae often venture far out to sea, and return to the reefs as adults.

CONSERVATION STATUS
Not listed by the IUCN.

SIGNIFICANCE TO HUMANS
Enjoyed by divers in their natural habitat, squirrelfishes are also part of the aquarium trade. ◆

Blackbar soldierfish
Myripristis jacobus

FAMILY
Holocentridae

TAXONOMY
Myripristis jacobus Cuvier, 1829, Martinique Island, West Indies; Brazil; Havana, Cuba.

OTHER COMMON NAMES
English: Bastard soldierfish, roundhead conga; French: Marignon mombin; Spanish: Candil de piedra.

PHYSICAL CHARACTERISTICS
Grow up to 10 in (25 cm) in length. Somewhat similar in appearance to the squirrelfish, they are red with a large eye, double dorsal fin, and forked tail, but also sport a brownish black, vertical bar behind the gillcover that extends to the pectoral fin.

DISTRIBUTION
Gulf of Mexico and the eastern Atlantic, north to North Carolina and south to Brazil.

HABITAT
Reefs and other structures, including piers. Commonly seen by divers in very shallow waters, but also found at depths to 330 ft (100 m).

BEHAVIOR
Usually a solitary species, but small groups of up to 36 individuals sometimes school. Under stress, blackbar soldierfishes will make clicking and grunting noises with the swim bladder.

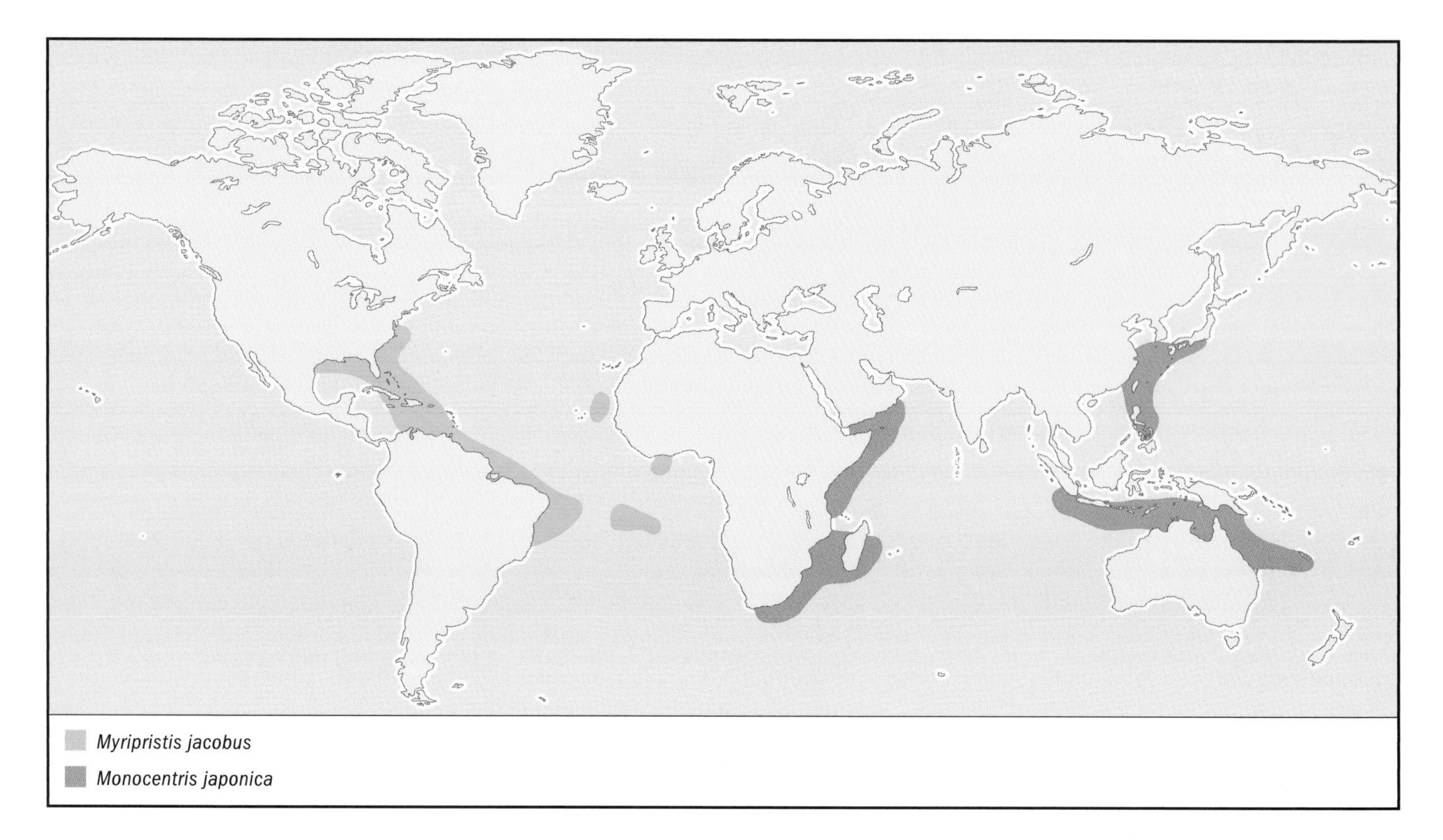

Divers have seen an occasional blackbar soldierfish swimming upside down, and they commonly swim upside down in caves.

FEEDING ECOLOGY AND DIET
Nocturnal feeder. Eats shrimp and zooplankton. Predators include such fishes as the horse-eye jack (*Caranx latus*), West Atlantic trumpetfish (*Aulostomus maculatus*), and Nassau grouper (*Epinephelus striatus*).

REPRODUCTIVE BIOLOGY
Engage in external fertilization on days following a full moon. While the adults prefer shallower reefs, the larvae may travel well out to sea.

CONSERVATION STATUS
Not listed by the IUCN.

SIGNIFICANCE TO HUMANS
Major part of the pet trade and a minor commercial food fish. ◆

Pineconefish
Monocentris japonica

FAMILY
Monocentridae

TAXONOMY
Monocentris japonica Houttuyn, 1782, Nagasaki, Japan.

OTHER COMMON NAMES
English: Dick bridegroom fish, Japanese pineapplefish, knight fish, pine sculpin; French: Poisson ananas; German: Japanischer Reuterfisch/Tannenzapfenfisch.

PHYSICAL CHARACTERISTICS
Can reach 6.5 in (16.5 cm) in length. Has large scales bordered in black. The scales are armed with spines that point toward the tail. The light organ is located near the mouth.

DISTRIBUTION
Southern Japan, spreading down to southern Australia, and west to the Red Sea.

HABITAT
Prefers rocky reefs. The young are found in shallow waters of 10–20 ft (3 to 6.1 m); the adults inhabit depths of 60–700 ft (18–213 m).

BEHAVIOR
Usually solitary or in pairs, but individuals sometimes congregate in schools of 50–100.

FEEDING ECOLOGY AND DIET
Nocturnal feeder. Searches sandy sea bottoms for prey, which includes small fishes, shrimps, and other invertebrates.

REPRODUCTIVE BIOLOGY
Mostly unknown, but they probably do not guard the eggs or young.

CONSERVATION STATUS
Not listed by the IUCN.

SIGNIFICANCE TO HUMANS
Pineconefishes are a minor part of the aquarium trade. ◆

Orange roughy
Hoplostethus atlanticus

FAMILY
Trachichthyidae

TAXONOMY
Hoplostethus atlanticus Collett, 1889, Florés, Azores, sta. 203, 1557 m.

OTHER COMMON NAMES
English: Deep sea perch, red roughy, slimehead; French: Hoplostète orange; German: Degenfisch, Granatbarsch, Kaiserbarsch; Spanish: Raloj anaranjado

PHYSICAL CHARACTERISTICS
Among the largest of the beryciform fishes, and possibly the largest, with some specimens up to 29.5 in (75 cm) in length. Has vivid orange-red coloration and a slightly jutting lower jaw.

DISTRIBUTION
Disjunct populations in tropical, subtropical, and temperate waters worldwide.

HABITAT
Prefers deep waters 590–5,900 ft (180–1,800 m) deep over steep or rough structures, particularly seamounts.

BEHAVIOR
Apparently sedentary rather than migratory. Remains in deep waters most of its life. Little is known about the young, but they are believed to primarily inhabit deep waters, too.

FEEDING ECOLOGY AND DIET
Diet is made up of small fishes, shrimps, squids, and other invertebrates. Predators include other fishes of their habitat, such as basketwork eels (*Diastobranchus* spp.).

REPRODUCTIVE BIOLOGY
Large groups of orange roughies congregate in annual mating schools. The schools may last one to two weeks, during which mature adults (30 years old and more) spawn and then leave the area. Females produce 10,000–90,000 eggs, which are fairly large at 0.08 in (2 mm) and spherical. The eggs drift toward the surface, where they hatch about two weeks later.

CONSERVATION STATUS
Although the orange roughy is not listed by the IUCN, catches have been restricted to help ensure a continued commercial harvest.

SIGNIFICANCE TO HUMANS
Best known as a commercial food fish, providing millions of pounds of harvest each year. Commercial harvests are now restricted, but past annual harvests have topped 100 million pounds (45 million kg) in the waters off Australia and New Zealand. ◆

Resources

Books

Deloach, Ned. *Reef Fish Behavior.* Jacksonville, FL: New World Publications, 1999.

Helfman, Gene S., Bruce B. Collette, and Douglas E. Facey. *The Diversity of Fishes.* Malden, MA.: Blackwell Science, 1997.

Michael, Scott W. "Family Anomalopidae: Flashlightfishes; Family Monocentridae: Pineapple Fishes; Subfamily Holocentrinae: Squirrelfishes; and Subfamily Myripristinae: Soldierfishes." In *Reef Fishes: A Guide to Their Identification, Behavior and Captive Care,*Vol. 1. Shelburne, VT: Microcosm Ltd., 1998.

Nelson, Joseph S. *Fishes of the World.* New York: John Wiley Sons, Inc., 1994.

Paxton, John R. "Squirrelfishes and Their Allies." In *Encyclopedia of Fishes.* 2nd edition, edited by John R. Paxton and William N. Eschmeyer. San Diego: Academic Press, 1998.

Snyderman, Marty, and Clay Wiseman. *Guide to Marine Life: Caribbean, Bahamas, Florida.* New York: Aqua Quest Publications, 1996.

Periodicals

Gauldie, R. W., and J. B. Jones. "Stocks, or Geographically Separated Populations of the New Zealand Orange Roughy, *Hoplostethus atlanticus*, in Relation to Parasite Infestation, Growth Rate, and Otolith Shape." *Bulletin of Marine Science* 67, no. 3 (2000): 949–972.

Leslie Ann Mertz, PhD

Zeiformes

(Dories)

Class Actinopterygii
Order Zeiformes
Number of families 6

Photo: A Japanese boarfish (*Pentaceros japonicus*) from costal Japan. (Photo by Mark Smith/Photo Researchers, Inc. Reproduced by permission.)

Evolution and systematics

Fossils identified as zeiform fishes are confined to marine habitats and are relatively young, ranging in age from Oligocene to Holocene deposits. They have been found in Europe, the West Indies, South Africa, and Indonesia. The phylogenetic relationships of the Zeiformes were explicated by Johnson and Patterson in their 1993 publication on the phylogenetic relationships of the percomorph fishes. They recognized Zeiformes as a monophyletic taxon characterized by the following shared derived characters. There is a distinctive configuration of upright columnar processes on the dorsal elements of the gill arch skeleton. Baudelot's ligament originates immediately under the vagus foramen of the exoccipitals. The distal part of the proximal middle radials of the dorsal fin pterygiophores is expanded laterally. The palatines have a specialized, mobile articulation with the ectopterygoid, which is truncated dorsally. The metapterygoid is extremely reduced, and a continuous median cartilage extends below the frontals and between the ethmoid cartilage and pterosphenoids.

Based on what they term "admittedly tenuous evidence," Johnson and Patterson considered the Zeiformes to be the sister taxon of a group making up the order Beryciformes and a huge conglomeration of spiny-rayed fishes known as the "percomorpha," including the Perciformes, Pleuronectiformes (flounders, soles, etc.), Tetraodontiformes (triggerfishes and pufferfishes, among others), Scorpaeniformes (scorpionfishes, gurnards, flatheads, and so on), Dactylopteriformes (helmet gurnards), Synbranchiformes (swamp eels, spiny eels, and others), Elassomatidae (pygmy sunfish), Gasterosteriformes (pipefish, trumpetfish, etc.), Mugiloidei (mullets), and Atherinomorpha (Atheriniformes, Beloniformes, and Cyprinodontiformes).

The order Zeiformes comprises six families: Zeidae, Parazenidae, Zeniontidae, Oreosomatidae, Grammicolepidae, and Caproidae. In their 1966 seminal paper on the phylogenetic relationships of teleost fishes, P. H. Greenwood and collaborators included the boarfishes (family Caproidae) in the Zeiformes, but they presented no evidence to support this assignment. Heemstra excluded the Caproidae from the Zeiformes in his 1980 taxonomic revision of the zeid fishes of South Africa, and in the book *Smiths' Sea Fishes*, Heemstra included the Caproidae in the order Perciformes. For the purposes of this publication, the Caproidae are included in the Zeiformes.

Much work remains to be done on the systematics of the Zeiformes, and the classification adopted here is tentative. The composition, definitions, and distinction of the families Zeidae, Zeniontidae, and Parazenidae are unsettled. The genus *Cyttomimus* (Gilbert, 1905) appears to be related closely to *Capromimus* (Gill, 1893). Although these two genera seem to be placed correctly in the Zeiformes, their affinity with any of the families recognized here is unclear. The genus *Macrurocyttus* (Fowler, 1933) may belong in the Zeniontidae, but the head of *Macrurocyttus acanthopodus* (Fowler, 1933) looks very different from those of the two species of *Zenion* that are known.

Physical characteristics

The body is deep, compressed, and oblong to disk-shaped. The upper jaw is more or less protrusible, and there are minute, slender, conical teeth in the jaws and vomer. Adults range in size from the dwarf dory (*Zenion hololepis*) at 4 in (10 cm) to the 3-ft (90 cm), 12-lb (5.3 kg) South African Cape dory (*Zeus capensis*). Most species are silvery, bronzy, brown, or reddish. The John Dory is silvery or bronzy, with indistinct longitudinal dark stripes from head to tail and a conspicuous white or yellow-edged black ocellus in the middle of the body. The juvenile buckler dory (*Zenopsis conchifer*) is silvery and covered with scattered, vaguely defined black spots. Dories can change from silvery to dark brown or gray in seconds. Males and females are colored similarly.

Zeiform fishes can be recognized by the following combination of characters: five to ten dorsal fin spines; zero to four anal fin spines; pelvic fins with one spine and five to seven soft rays or no spine and six to 10 soft rays; a caudal fin with 11, 13, or 15 principal (segmented) rays, of which 9, 11, or 13 rays are branched; and unbranched dorsal, anal, and pectoral fin rays. The orbitosphenoid bone is absent, and there is no subocular shelf or supramaxilla. There are seven or eight branchiostegal rays, 3.5 gills (no slit behind the last hemibranch), 25–46 vertebrae, and a gas bladder.

Distribution

The order is represented in the western Atlantic from Canada to Argentina, including the Gulf of Mexico and the Caribbean Sea. In the eastern Atlantic, zeiforms are known from the North Sea to South Africa, including the Mediterranean, Black Sea, Azores, Madeira, Canary Islands, Cape Verde Islands, and Saint Helena. In the Indian Ocean, zeiforms occur along the east coast of Africa to the Gulf of Aden, Oman, the Persian Gulf, Pakistan, India, Sri Lanka, Maldives, and Madagascar and eastward to the Andaman Sea, Burma, Thailand, Cambodia, Indonesia, Vietnam, Australia, New Zealand, New Guinea, the Philippines, China, Taiwan, Japan, and Korea. In the eastern Pacific, zeiforms range from Canada to Chile.

Habitat

Most zeiform fishes are demersal, living near the bottom of the continental shelf or upper continental slope region. They range in depth from 110 to 5,084 ft (35–1,550 m). The adult John dory (*Zeus faber*), and probably adults of some other zeiform species, frequently occurs in midwater or near the surface. Some species have a pelagic prejuvenile stage that lives near the surface in the open ocean. Adults are found over soft (sandy or muddy) or hard (rocky) substrata.

Behavior

Little is known of the behavior of zeiforms, as they live at depths where they are difficult to observe. Adults of the John dory (*Z. faber*) are mainly solitary. The buckler dory (*Z. conchifer*) often occurs in small aggregations.

Feeding ecology and diet

Zeiform fishes are carnivores; they feed mainly on a variety of fishes but also consume cephalopods and crustaceans. Juveniles of the larger species and adults of the dwarf dories (family Zeniontidae) and tinselfishes (family Grammicolepidae) feed on zooplankton (e.g., copepods, pteropods, fish and crustacean larvae). Adults of the larger zeiform fishes (John dory and the buckler dory) are eaten only by large piscivorous predators (e.g., some sharks, goosefishes [*Lophus* species], and lancetfishes [*Alepisaurus* species]). Juveniles and adults of small zeiforms (e.g., *Zenion* species) are subject to predation by a variety of piscivores.

Reproductive biology

The sexes are separate. Females grow larger than males. Spawning has not been observed, but zeiforms apparently are "broadcast spawners," with the eggs and sperm released into the water column and fertilization taking place in the sea. The eggs and larvae are pelagic and float near the surface; the eggs are spherical, 0.04–0.1 in (1–2.8 mm) in diameter, with a single oil globule. There are no reports of nests or egg guarding or parental care in zeiform fishes.

Conservation status

None of the zeiform fishes are listed by the IUCN. Some populations probably are overexploited in areas where intensive trawling takes place.

Significance to humans

The larger species of zeiforms are of some commercial importance as food fishes. Most zeiforms are caught by trawlers, but anglers also catch a few of the larger species.

1. Buckler dory (*Zenopsis conchifer*); 2. John dory (*Zeus faber*); 3. Red boarfish (*Antigonia rubescens*); 4. Thorny tinselfish (*Grammicolepis brachiusculus*); 5. Tinselfish (*Xenolepidichthys dalgleishi*). (Illustration by John Megahan)

Species accounts

Red boarfish

Antigonia rubescens

FAMILY
Caproidae

TAXONOMY
Antigonia rubescens Günther, 1860, Japan.

OTHER COMMON NAMES
English: Pink boarfish.

PHYSICAL CHARACTERISTICS
Attains a total length (including the tail fin) of about 6 in (15 cm). The head and body are deep, greatly compressed, and shaped like a disk or diamond; the body depth is more than twice the head length. The dorsal head profile and the predorsal region are concave, and there is a bulge on the nape. The body, cheeks, and operculum are covered with small ctenoid scales. The mouth is small, and the upper jaw is shorter than the eye diameter. The dorsal fin has nine spines and 27–30 soft rays; the anal fin has three spines and 24–28 rays. The tail fin has 10 branched rays. The body is pale reddish silvery, with a dark red bar from the dorsal fin origin to the origin of the anal fin; another red bar lies above and below the eye, and there is a red band on the body at the rear of dorsal and anal fin bases. The abdomen and lower rear part of the head are silvery white.

DISTRIBUTION
Japan, Midway Island northwest of Hawaii, Taiwan, Philippines, China, and Australia.

HABITAT
Adults are usually caught with trawls near the bottom at depths of 333–3,000 ft (100–900 m).

BEHAVIOR
The behavior of the red boarfish is poorly known, because it lives too deep to be observed easily. They occur in large aggregations, as many individuals may be caught in a single trawl haul.

FEEDING ECOLOGY AND DIET
No information has been published on the diet of the red boarfish. Probably feeds on plankton and small benthic invertebrates. Subject to predation by various piscivores that inhabit the outer shelf and slope region (e.g., sharks, lancetfishes, and gempylids).

REPRODUCTIVE BIOLOGY
Poorly known. Probably a broadcast spawner, with pelagic eggs and larvae.

CONSERVATION STATUS
Not listed by IUCN.

SIGNIFICANCE TO HUMANS
None known. ◆

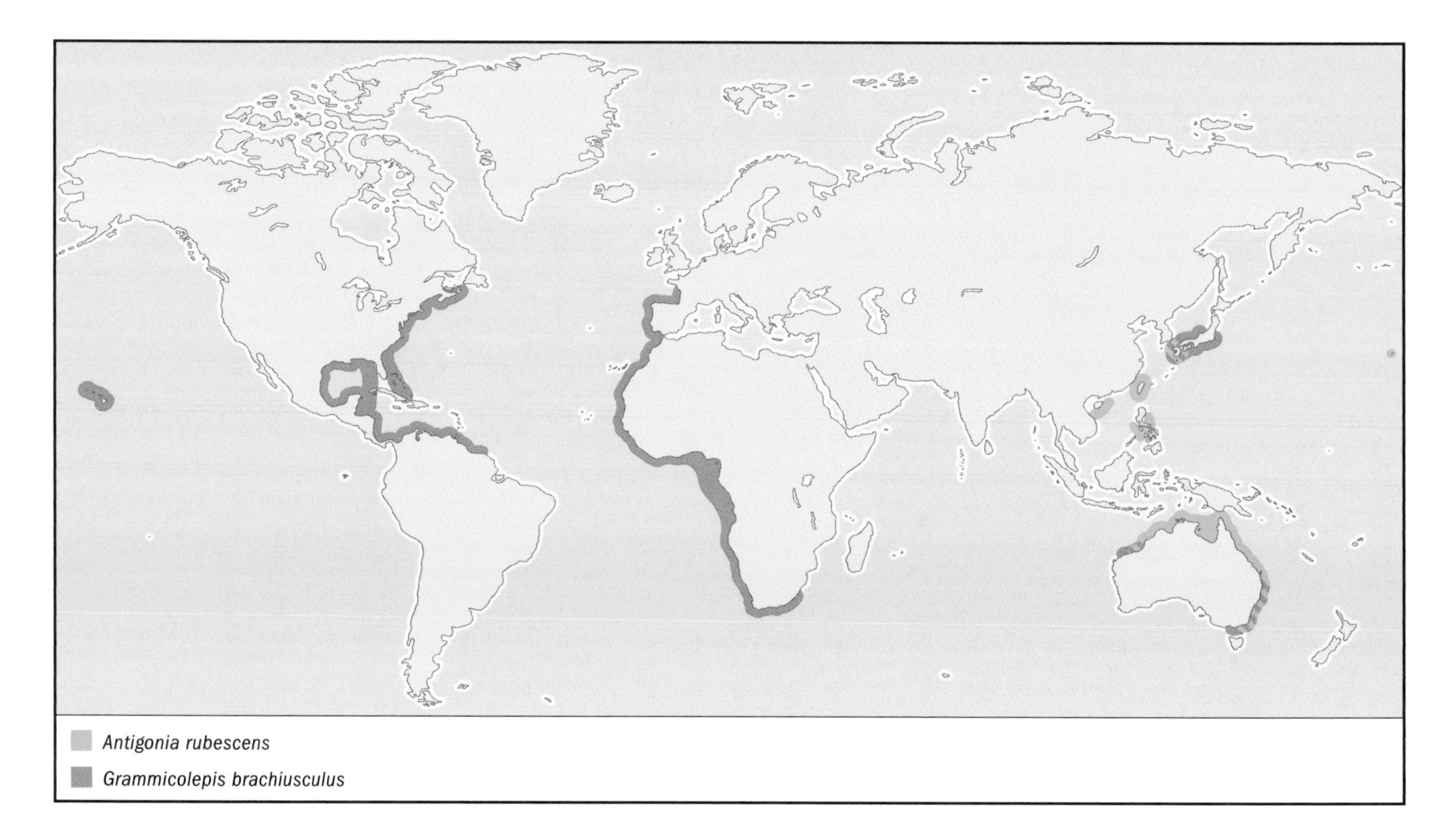

Thorny tinselfish
Grammicolepis brachiusculus

FAMILY
Grammicolepidae

TAXONOMY
Grammicolepis brachiusculus Poey, 1873, Cuba.

OTHER COMMON NAMES
English: Diamond dory, deepscale dory; Spanish: Palometa oropel.

PHYSICAL CHARACTERISTICS
Attains a total length (including the tail fin) of about 2 ft, 5 in (72 cm) and a weight of 9 lb (4 kg). The head and body are deep, very compressed, and shaped like an oblong disc, or dinner platter; juveniles are diamond-shaped. The body, cheeks, and operculum are covered with vertically elongated scales. The mouth is small, and the maxilla has two or three ridges, bound to the ascending processes of the premaxillae and loosely connected to the palatines. The jaws have one or two rows of minute, slender teeth, and the vomer may or may not have three or four minute, slender teeth. There are two dorsal fins, the first with six to seven spines and the second with 31–34 unbranched soft rays. The anal fin has two spines and 33–35 unbranched rays. The tail fin has 13 branched rays. The pelvic fins have one spine and six branched rays; there is a row of 34–36 small spines along each side of the dorsal and anal fin bases.

Juveniles have a greatly elongated first anal fin spine and second dorsal fin spine. The pelagic juvenile stage (4–8 in, or 10–20 cm in standard length) looks quite different from the adult. The body is more compressed and angular, with 10–13 prominent, flattened, bladelike, spiny scutes projecting laterally from the surface of each side of the body. Each scute is an outgrowth from a scale whose basal part is divided, overlapping both sides of the one behind. On the base of the larger scutes are retrorse spinules. The scutes become smaller as the fish grows, and they eventually shrink to nothing as the fish transforms to an adult at about 10–12 in (25–30 cm). Adults are silvery in color, with bronze reflections. Juveniles are silvery, with irregular black blotches on the body, black spots on the tail fin, and 5 black bars on anal fin.

DISTRIBUTION
Tropical and temperate waters of the eastern Atlantic from France to South Africa; also known from Japan, Hawaii, Australia, and the western Atlantic from Canada to Suriname.

HABITAT
Adults usually are caught with trawls near the bottom, at depths of 1,333–3,000 ft (400–900 m).

BEHAVIOR
The behavior of the thorny tinselfish is poorly known, as this species has yet to be observed in shallow water.

FEEDING ECOLOGY AND DIET
No information has been published on the diet of the thorny tinselfish. It probably feeds on plankton and small benthic invertebrates. Juveniles are subject to predation by a variety of piscivores. Adults are eaten by some large sharks and lancetfishes.

REPRODUCTIVE BIOLOGY
Poorly known. Probably a broadcast spawner with pelagic eggs and larvae.

CONSERVATION STATUS
Not listed by the IUCN. Its apparent rarity may be due to the difficulty in sampling fishes from depths of 1,333–3,000 ft (400–900 m).

SIGNIFICANCE TO HUMANS
None known. ◆

Tinselfish
Xenolepidichthys dalgleishi

FAMILY
Grammicolepidae

TAXONOMY
Xenolepidichthys dalgleishi Gilchrist, 1922, Natal, South Africa.

OTHER COMMON NAMES
English: Diamond dory; spotted tinselfish.

PHYSICAL CHARACTERISTICS
Attains a total length (including tail fin) of about 6 in (15 cm). The head and body are greatly compressed and vertically elongated, shaped like a flattened diamond. The body, cheeks, and operculum are covered with vertically elongated scales. The mouth is small and the maxilla, which has two or three ridges, is bound to the ascending processes of the premaxillae and loosely connected to the palatines. The jaws have one or two rows of minute, slender teeth; the vomer has a few minute, slender teeth. There are two dorsal fins, the first with five spines and the second with 27–30 unbranched soft rays. The anal fin has two spines and 27–29 unbranched rays. The tail fin has 13 branched rays, and the pelvic fins have one spine and six branched rays. There is a row of 29 small spines along each side of the dorsal fin and 26–27 small spines along the anal fin base. Juveniles have a greatly elongated (two or three times the length of the fish) first anal fin spine and second dorsal fin spine. The body is silvery with scattered, round black spots; the rear margin of tail fin is dusky.

DISTRIBUTION
Western Atlantic from Canada to southern Brazil and eastern Atlantic from Senegal to South Africa; also in Japan, Taiwan, Philippines, Australia, New Zealand, New Caledonia, Fiji, and Tonga.

HABITAT
Usually taken near the bottom in depths of 666–1,333 ft (200–400 m) but also taken in midwater and at the surface of the open ocean.

BEHAVIOR
The behavior of the tinselfish is poorly known, as this species is rarely observed in shallow water.

FEEDING ECOLOGY AND DIET
There is no information. Probably feeds on zooplankton and small benthic invertebrates. The tinselfish is prey for some sharks, lancetfishes, scombrids, carangids, and gempylids.

REPRODUCTIVE BIOLOGY
There is no information. Probably a broadcast spawner.

CONSERVATION STATUS
Not listed by IUCN.

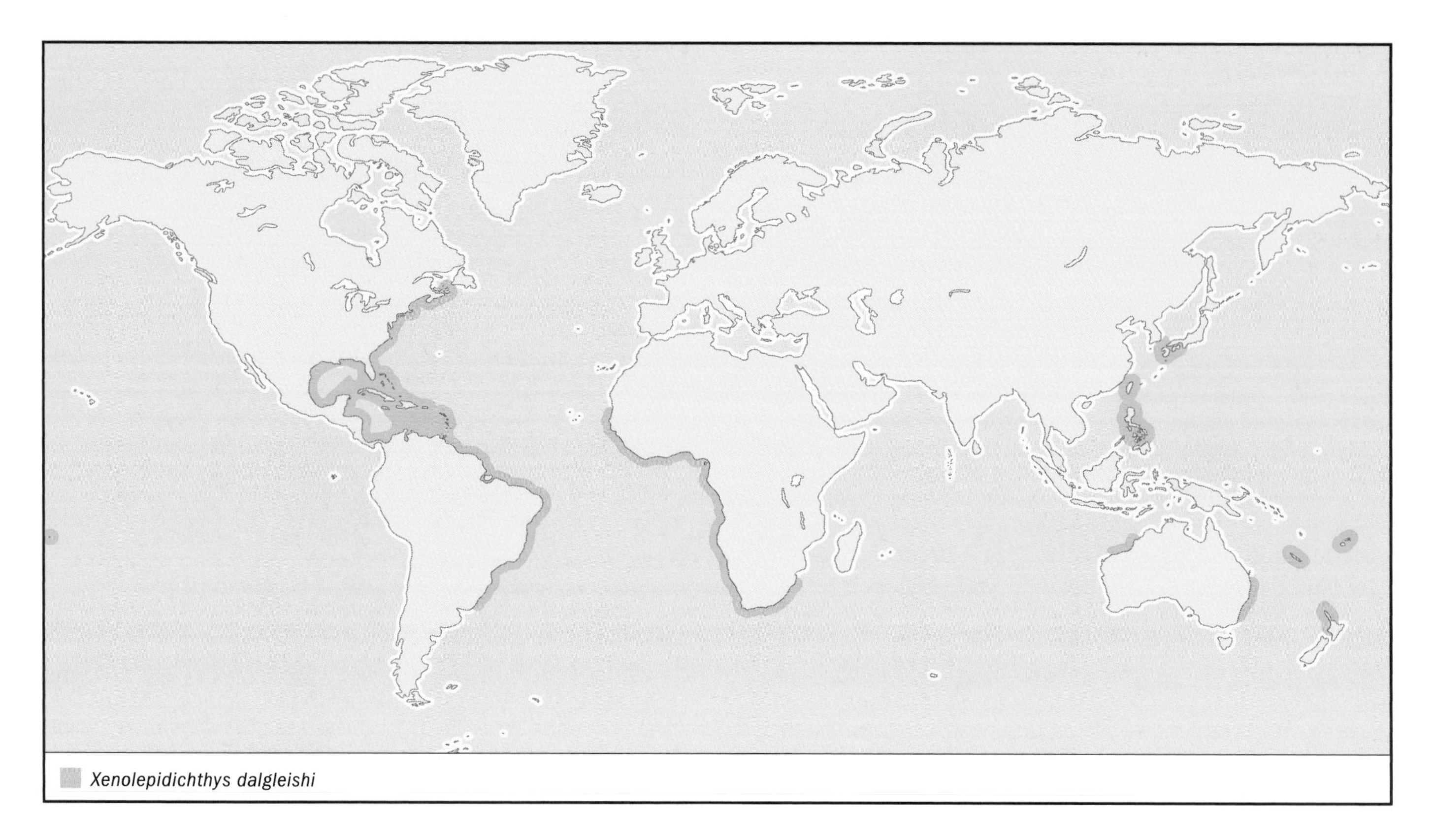

SIGNIFICANCE TO HUMANS
None known. ◆

Buckler dory
Zenopsis conchifer

FAMILY
Zeidae

TAXONOMY
Zeus conchifer Lowe, 1852, Madeira.

OTHER COMMON NAMES
English: Mirror dory (Australia), sailfin dory, silvery John Dory; Portuguese: Peixe galo, falo branco.

PHYSICAL CHARACTERISTICS
Attains a total length (including the tail fin) of 2 ft, 8 in (80 cm) and a weight of 9 lb (4 kg). The head and body are very compressed and shaped like an oblong disc, or dinner platter. The dorsal head profile is slightly concave. Dorsal fin spines of the adult are elongated and filamentous. The body has no scales. There is a row of five to eight bony bucklers, each with a strong spine, along each side of the bases of the anal and both dorsal fins. There are two or three bucklers on the isthmus and ventral midline of the chest in front of the pelvic fins and seven or eight pairs of bucklers (ridged bony scutes) along the ventral midline of the abdomen, from the base of the pelvic fins to the first anal fin spine. The thoracic region is compressed. The pelvic fins are large and close together, inserted on the chest below the eyes and well in front of the pectoral fins. Each has six or seven soft rays. The first pelvic ray could be considered a spine, because it is unbranched and not segmented, but (unlike the pelvic fin spine of *Zeus*) it is a biserial (double) ray. There are five to eight bony bucklers (enlarged, platelike scales) along the base of the soft dorsal and anal fins and seven to 10 pairs of spiny scutes along the belly, from the base of the pelvic fins to the anus. The dorsal fin has nine to 10 spines and 24–27 rays; the anal fin has three spines (the first two movable and the third fixed) and 24–26 rays. The caudal fin has 11 branched rays; the pectoral fins are much smaller than the pelvic fins, with 12 or 13 rays. Dorsal, anal, and pectoral fin rays are unbranched. Adults are silvery with a faint, dusky mid-lateral spot above the pectoral fin and below the lateral line. Small juveniles, 1–4 in (2–10 cm) long, are covered with scattered small black spots.

DISTRIBUTION
Mainly continental. Western Atlantic from Canada to Argentina, including the Gulf of Mexico and Caribbean Sea, and eastern Atlantic from France and British Isles to South Africa. Also in the Indian Ocean from South Africa to Kenya and India.

HABITAT
Adults are found over soft (sandy or muddy) or hard (rocky) substrata. They are demersal, usually caught near the bottom of the continental shelf or upper continental slope region at depths of 33–1,188 ft (10–360 m). Occasionally found in midwater well above the bottom.

BEHAVIOR
The behavior of the buckler dory is poorly known, as this species is rarely observed in shallow water. Adults usually are found in aggregations near the bottom. A slow, stalking, ambush-predator mode of hunting is assumed, and the greatly protrusible upper jaw compensates for the feeble swimming musculature.

FEEDING ECOLOGY AND DIET
Feeds on demersal fishes, crustaceans, and cephalopods and often makes excursions above the bottom to feed on midwater fishes. Prey species selection is influenced by availability and accessibility. The dominant species in the diet is likely to be the most abundant prey in the habitat. Adult buckler dory have few predators other than great white sharks and goosefish (*Lophius* spp.). Juveniles are likely prey for most piscivorous predators.

REPRODUCTIVE BIOLOGY
Reproduction is similar to that of the John dory (see following account).

CONSERVATION STATUS
Not listed by IUCN. The species is taken as bycatch in various trawl fisheries in the North Atlantic, off southern Brazil, Namibia, and South Africa. There are no fishing regulations or catch data that apply specifically to the buckler dory.

SIGNIFICANCE TO HUMANS
None known. ◆

John dory
Zeus faber

FAMILY
Zeidae

TAXONOMY
Zeus faber Linnaeus, 1758, habitat in Pelago.

OTHER COMMON NAMES
French: Saint-Pierre; German: Heringskönig; Spanish: Pez de San Pedro, gallo de San Pedro, barbero, gallo barbero; Portuguese: Peixe galo.

PHYSICAL CHARACTERISTICS
Attains a total length (including the tail fin) of 3 ft (90 cm) and a weight of 18 lb (8.2 kg). The head and body are very compressed and shaped like an oblong disc, or dinner platter. The dorsal head profile is straight. The dorsal fin spines of the adult are greatly elongated and filamentous. Scales are minute, cycloid, and embedded; there are five to eight bony bucklers (enlarged, platelike scales) along the base of the soft dorsal and anal fins and seven to 10 pairs of spiny scutes along the belly from the base of the pelvic fins to the anus. The dorsal fin has nine to 11 spines and 22–24 rays, and the anal fin has four spines and 20–23 rays. Dorsal, anal, and pectoral fin rays are unbranched, and the caudal fin has 11 branched rays. The pelvic fins are enlarged, situated below and slightly in front of the pectoral fins, with one spine and six or seven branched rays. The pectoral fins are much smaller than the pelvic fins.

The head and body are silvery to olive-brown, with a conspicuous yellow or white-edged, blue-black ocellus on the middle of the body below the lateral line. The body often has indistinct dark, wavy stripes. The dorsal fin spine filaments are white. Juveniles are small, about 1 in (2–3 cm) long. Color is brownish, with small black spots along the base of the median fins, curved white and black stripes on the head and body, and the usual round black spot on the middle of the body below the lateral line.

DISTRIBUTION
Eastern Atlantic from the North Sea to South Africa, including the Mediterranean Sea, Black Sea, Azores, Madeira, and Ca-

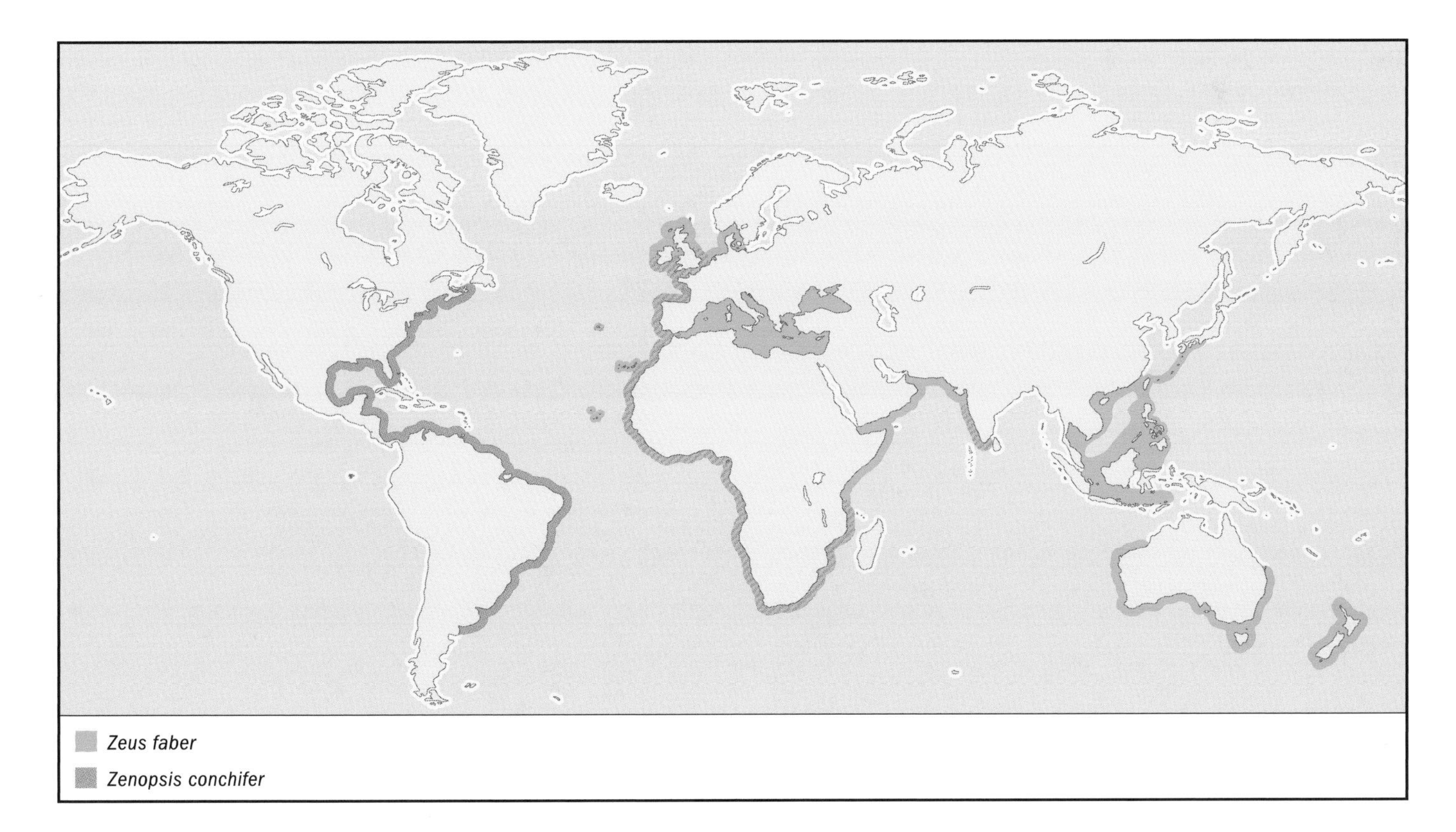

nary and Cape Verde Islands and along the west coast of Africa to Namibia. In the western Indian Ocean from South Africa to Somalia and India. In the western Pacific in Indonesia, the Philippines, China, Taiwan, Japan, Australia, and New Zealand. Not known from the eastern Pacific or western Atlantic.

HABITAT
Adults are found over soft (sandy or muddy) or hard (rocky) substrata; they are demersal, usually caught near the bottom of the continental shelf or upper continental slope region at depths of 33–1,188 ft (10–360 m). Adults also frequently occur in midwater or near the surface.

BEHAVIOR
Adults usually are solitary, although they probably congregate for spawning. They typically occur near the bottom, but they often make excursions above the bottom to feed on midwater fishes. With its greatly compressed head and body, large mouth, and extremely protrusible upper jaw, the John Dory is a successful ambush predator. It slowly approaches an unsuspecting small fish by means of undulating waves of the transparent soft dorsal and anal fins. In one quick motion it drops the "trapdoor" lower jaw, shoots out the upper jaw, and expands the gill cavity, sucking in the hapless prey along with a considerable volume of water.

FEEDING ECOLOGY AND DIET
The John dory is a carnivore, feeding primarily on fishes, crustaceans, and cephalopods. Prey species selection is influenced by the availability and accessibility of prey. The dominant species in the diet is likely to be the most abundant prey in the habitat. In the North Atlantic, juveniles, 3–10 in (8–25 cm), feed mainly on benthic fishes, for example, dragonets (family Callionymidae) and Norway pout (family Gadidae). Adults, 10–22 in (25–56 cm), switch to a diet of larger schooling fishes, primarily blue whiting, *Micromesistius poutassou* (Risso, 1826), and snipefish, *Macroramphosus scolopax* (Linnaeus, 1758), of the family Macroramphosidae. Adults have few predators, except for great white sharks and goosefishes (*Lophius* spp.). Juveniles are probably eaten by piscivorous predators.

REPRODUCTIVE BIOLOGY
The John dory is a broadcast spawner. The eggs and sperm are shed simultaneously into the water, where fertilization takes place. In the northeastern Atlantic, spawning occurs in April or May. The eggs and larvae are pelagic; the eggs are small (0.1 in or 2 mm diameter) and spherical, with a large oil globule. No nest building or parental care has been reported for the species. In the North Atlantic, John dory males mature in their second or third year at a length of about 10 in (26 cm), and they may live to an age of 13 years. Females are three or four years old when they become mature at a length of 14 in (35 cm). They grow larger than males do, and they may live 15 years. In New Zealand waters males attain maturity at 11.8 in (30 cm) in two years; their growth then slows considerably, with a maximum size of about 15.7 in (40 cm) and an age of nine years. Females grow faster and attain a maximum size of about 18.1 in (46 cm) and an age of nine years.

CONSERVATION STATUS
Not listed by IUCN. As of 1998 there were no regulations regarding John dory fishery in the English Channel or in Australian waters. The species is taken as bycatch in the trawl fisheries of the eastern Atlantic, South Africa, Australia, New Zealand, Indonesia, Philippines, Taiwan, and Japan.

SIGNIFICANCE TO HUMANS
The John dory is prized as a food fish and is of commercial importance because of its high price in fish markets. ◆

Resources

Books

Karrer, C., and P. C. Heemstra. "Family No. 140: Grammicolepididae." In *Smiths' Sea Fishes*, edited by M. M. Smith and P. C. Heemstra. Johannesburg: Macmillan South Africa, 1986.

Nelson, J. S. *Fishes of the World.* 3rd edition. New York: John Wiley & Sons. 1994.

Smith, M. M., and P. C. Heemstra, eds. *Smiths' Sea Fishes.* Johannesburg: Macmillan South Africa. 1986.

Periodicals

Barnard, H. K. "A Monograph of the Marine Fishes of South Africa. Part 1." *Annals of the South African Museum* 21 (1925): 1–105.

Bigelow, H. B., and W. C. Schroeder. "Fishes of the Gulf of Maine." *U.S. Fish and Wildlife Service Fishery Bulletin* 53 (1953): 1–577.

Dunn, M. R. "The Biology and Exploitation of John Dory, *Zeus faber* (Linnaeus, 1758), in Waters of England and Wales." *ICES Journal of Marine Science* 58, no. 1 (2001): 96–105.

Fowler, H. W. "The Buckler Dory and Descriptions of Three New Fishes from off New Jersey and Florida." *Proceedings of the Academy of Natural Sciences of Philadelphia* 86 (1934): 353–361.

Greenwood, P. H., D. E. Rosen, S. H. Weitzman, and G. S. Myers. "Phyletic Studies of Teleostean Fishes, with a Provisional Classification of Living Forms." *Bulletin of the American Museum of Natural History* 131, no. 4 (1966): 339–455.

Heemstra, P. C. "A Revision of the Zeid Fishes (Zeiformes: Zeidae) of South Africa." *Ichthyology Bulletin of the J. L. B. Smith Institute of Ichthyology* 41 (1980): 1–18.

Parin, N., and O. D. Borodulina. "Preliminary Review of the Bathypelagic Fish Genus *Antigonia* Lowe (Zeiformes, Caproidae)." *Transactions of the P. P. Shirsov Institute of Oceanology* 121 (1986): 1–105, 141–172.

Yoneda, M., S. Yamasaki, K. Yamamoto, H. Horikawa, and M. Matsuyama. "Age and Growth of John Dory, *Zeus faber* (Linnaeus, 1758), in the East China Sea." *ICES Journal of Marine Science* 59, no. 4 (2002): 749–756.

Phillip C. Heemstra, PhD

Gasterosteiformes

(Sticklebacks, seahorses, and relatives)

Class Actinopterygii
Order Gasterosteiformes
Number of families 11

Photo: The ornate, or harlequin, ghost pipefish (*Solenostomus paradoxus*) is a well-disguised seahorse relative that is rarely seen due to its highly effective camouflage. Here it is swimming by a sea fan. (Photo by Robert Yin/Corbis. Reproduced by permission.)

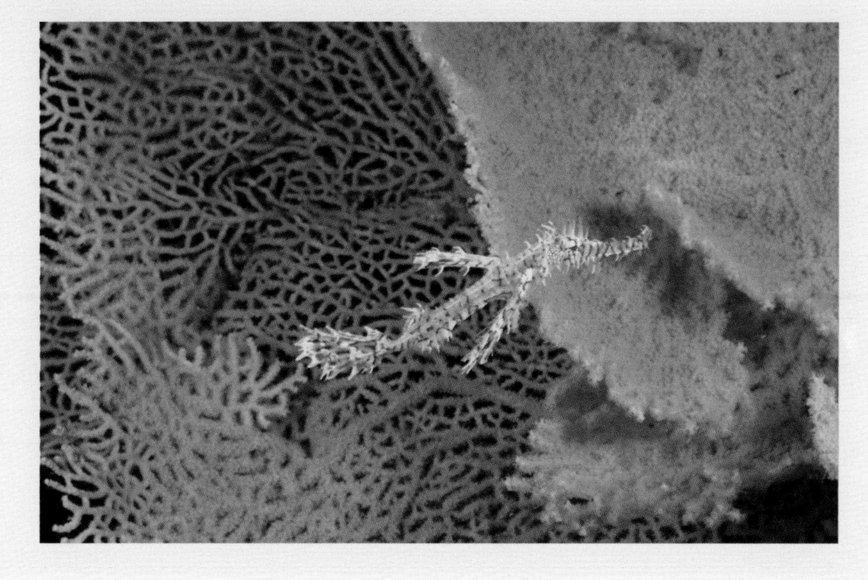

Evolution and systematics

The evolutionary affinities and composition of the Gasterosteiformes have been the subject of much debate; presently they are considered to be closely related to the teleost fish orders Synbranchiformes (swamp eels), Elassomatiformes (pygmy sunfish, *Elassoma*), Mugiliformes (mullets), and the collective Atherinomorpha (silversides, needlefishes, killifishes, and allies) and are placed with them in the larger group Smegmamorpha. Smegmamorphs are characterized by a unique configuration of the first vertebra and its associated intermuscular bone. Within this group, gasterosteiforms may be more closely related to the Synbranchiformes, but relationships within smegmamorphs are still the subject of controversy. The smegmamorphs, in turn, are closely related to the largest of all teleost groups, the Percomorpha. This evolutionary arrangement is the most recent and best-supported hypothesis based on morphological evidence, but alternative schemes of relationship have been proposed (including molecular studies), and a true consensus has not yet emerged.

The Gasterosteiformes, as described in this chapter, comprises two suborders, the Gasterosteoidei and the Syngnathoidei. Other authors sometimes have recognized these as separate orders, but evidence exists that they are each other's closest relative and therefore warrant being classified together. Gasterosteiformes share various specializations, such as pelvic bones without anterior processes, absence of Baudelot's ligament (a stout ligament connecting the shoulder girdle to either the posterior cranial base or an anterior vertebra—also absent in synbranchiforms), certain features of their branchial and caudal skeletons, and a particular configuration of their scales, which are represented by enlarged scutes, or plates. (Scales are absent in *Hypoptychus*, *Aulostomus* has small ctenoid scales, and *Fistularia* has embedded spines). The extent to which these features (and others) are truly indicative of a common ancestry for Gasterosteiformes is not completely understood.

The Gasterosteoidei includes the following families: Hypoptychidae (for the sand eel, *Hypoptychus dybowskii*), Aulorhynchidae (tubesnouts, two monotypic genera), Indostomidae (*Indostomus*, with three species), and Gasterosteidae (some five genera and, conservatively, seven spp.). The Syngnathoidei comprises the families Syngnathidae (seahorses and pipefishes, with about 52 genera and 220 spp.), Aulostomidae (trumpetfishes, *Aulostomus*, with some three spp.), Fistulariidae (cornetfishes, *Fistularia*, with four spp.), Macroramphosidae (snipefishes, three genera with 12 spp.), Centriscidae (shrimpfishes, two genera with four spp.), Solenostomidae (ghost pipefishes, *Solenostomus*, with some four spp.), and Pegasidae (seamoths, two genera with five spp.). Altogether, the order Gasterosteiformes is represented by 11 families, 70 genera, and at least 265 species, but undescribed species have been discovered (including some 20 species of pipefishes and seahorses), and numerous nominal species presently in synonymy, in fact, may be valid (such as for sticklebacks). Much work remains to be done concerning their taxonomy, and the phylogenetic position of *Indostomus* is still debated.

The fossil record of the Gasterosteiformes is extensive and dates back at least some 75 million years to the Calcare di Mellissano deposits near Nardò, in southeastern Italy (Apulia). This early fossil, *Gasterorhamphosus zuppichinii*, is known from skeletal remains and is similar to modern snipefishes. Additional gasterosteiform fossils, known from more or less complete skeletons, have been described from the extensive Monte Bolca beds of northeastern Italy (dating back some 52 million years). These include representatives of the Syngnathidae (at least five genera), Solenostomidae (some three

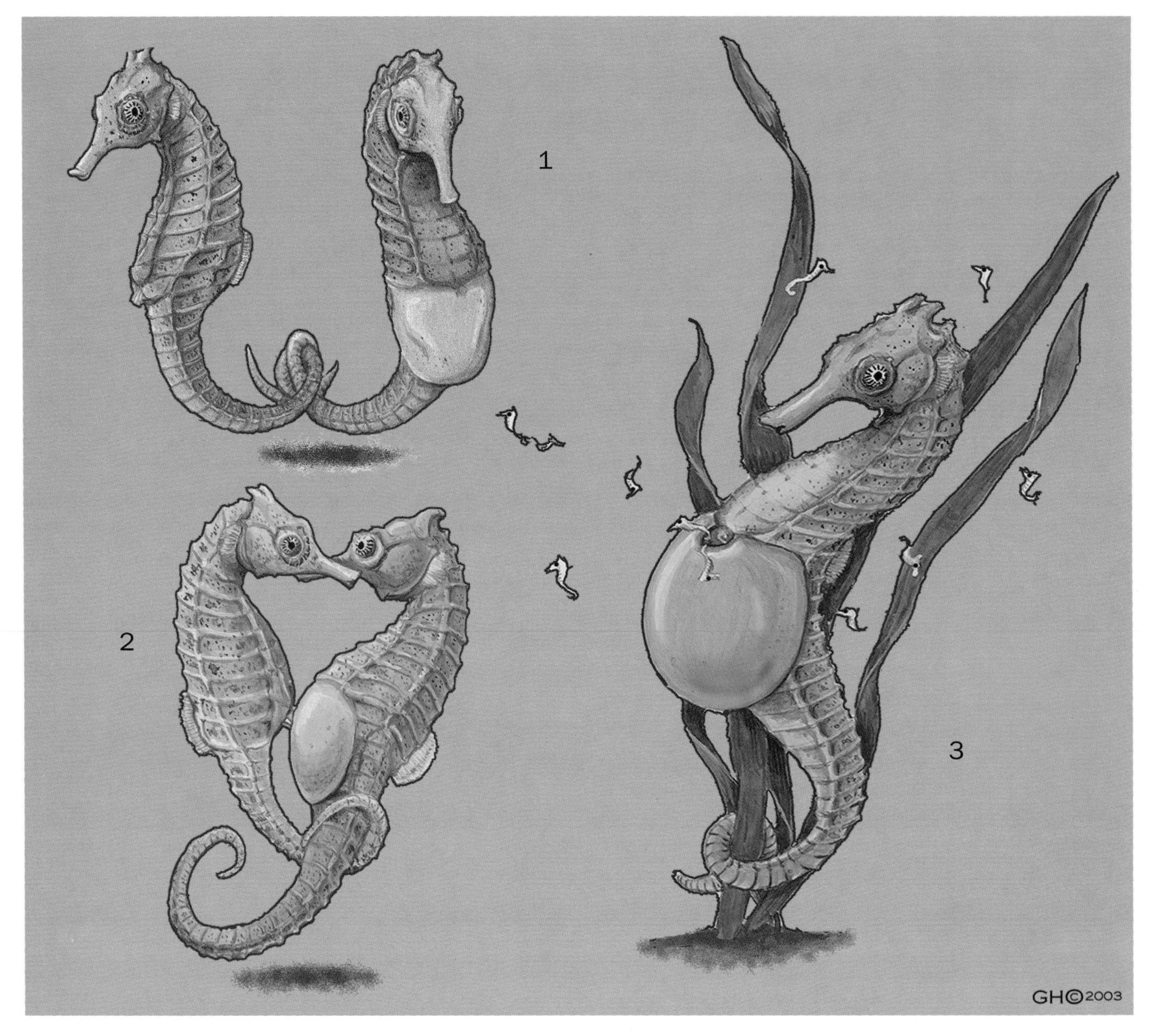

Breeding habits of the yellow seahorse (*Hippocampus kuda*): 1. Prenuptial courtship—male on right; 2. Nuptial embrace—female ejects eggs into male's pouch; 3. After gestation period, male shoots or pumps babies out of his pouch. (Illustration by Gillian Harris)

genera), Centriscidae (three genera), Aulostomidae (four genera), and Fistulariidae (some three genera and four spp.). The actual number of species from Monte Bolca is difficult to estimate with precision, but this formation represents the greatest extinct diversity of the order. The gasterosteiform species present in both the Nardò and Monte Bolca deposits were inhabitants of the former Tethys Ocean, which separated the extinct continents Laurasia and Gondwana during much of the Cretaceous and Tertiary periods. Other fossils, mostly represented by fragmentary material and allied to the Aulostomidae and Fistulariidae, are known from Turkmenia, in deposits almost contemporaneous with Monte Bolca. Fossil sticklebacks have been found in the Tertiary of California (Monterey Formation) and Siberia. Tertiary pipefishes have been described from the Modelo and Puente Formations of southern California as well as from the Caucasus and Carpathian Mountains of eastern Europe.

Physical characteristics

The morphological characteristics of various gasterosteiform families are highly modified and specialized; some of the members of this order are among the most morphologically interesting of all fishes. Some forms are more pelagic and streamlined (e.g., aulorhynchids), whereas others

A threespine stickleback (*Gasterosteus aculeatus*) male deposits his sperm in the nest after a female has spawned. (Photo by Animals Animals © D. Thompson, OSF. Reproduced by permission.)

are benthic and highly cryptic (e.g., pegasids). Many forms are elaborately camouflaged and blend in perfectly with their surroundings (some of the most notable examples are *Syngnathus typhle*, *Solenostomus cyanopterus*, and *Histiogamphelus cristatus*); some syngnathid species have been discovered only after their particular substrate was collected and examined. Additionally, some species carry algal growths and complex dermal projections that further aid in their concealment (the most striking example may be the leafy seadragon *Phycodurus eques*), and many species have the ability to change color at will (e.g., *Aulostomus chinensis*). Most species have the head and body on the same plane, as in "typical" fishes, but seahorses have the head at more or less a right angle to the body, a unique condition among fishes.

There is much morphological variation among the families and subfamilies, and many are recognized easily by their unique designs. Gasterosteids (sticklebacks) resemble more "typical fishes," with an unmodified head and clearly demarcated fins, but most gasterosteiforms have elongated snouts, with small upturned mouths, and may lack some fins or even all of them (*Bulbonaricus*). The snout may be absent (*Bulbonaricus*); truncated (*Histiogamphelus*); mildly elongated (*Indostomus*); or very long and tubular, reaching up to one-third of their total length (e.g., *Aulostomus* and *Fistularia*). The mouth usually is very small, located at the end of the snout (except in pegasids, where the mouth is underneath the snout), and most species lack teeth. The eyes typically are round and vary from large to small. Most species have very slender and elongated bodies, but gasterosteoids (except *Indostomus*) and some syngnathoids (e.g., Macroramphosidae, Centriscidae) are laterally compressed.

The fins typically are highly modified. They have either one or two dorsal fins; the single dorsal fin may be preceded by numerous spines (gasterosteoids, *Aulostomus*), but in some forms the spines represent the first dorsal fin (*Solenostomus*, centriscids, and macroramphosids). In most species, however, there is a single dorsal fin with soft rays only. Pelvic fins are absent in *Hypoptychus* and syngnathids, and the caudal fin is absent in most syngnathids. Many syngnathids have prehensile tails that enable them to cling to soft corals and algae. Most gasterosteiforms are covered in bony plates, scutes, or rings and have numerous elaborate dermal projections on the head, body, and tail. There is also great morphological variation internally (e.g., lack of true stomachs in seahorses and reduction of the number of cephalic bones in many species). Details of the morphological features of the different families are elaborated in the species accounts.

Gasterosteiforms are among the most colorful fishes, displaying a wide spectrum of colors and patterns. There is also much diversity in coloration within certain species, and some individuals may change color while breeding or to blend into different backgrounds. Every conceivable color pattern seems to be present within the order, as some species combine strong colors (e.g., red, orange, yellow, green, white, black, and blue) with spots, blotches, ocelli, mottlings, and stripes of various kinds. Gasterosteiformes also vary considerably in size, from some 0.8 in (2 cm) in length among pygmy seahorses to more than 4.9 ft (1.5 m) for cornetfishes (*Fistularia*).

A big-bellied seahorse (*Hippocampus abdominalis*) female deposits eggs into a male's pouch. (Photo by Animals Animals ©R. Kuiter, OSF. Reproduced by permission.)

A leafy seadragon (*Phycodurus eques*) near Kangaroo Island, Australia. Unlike seahorses, seadragons do not have a pouch for rearing the young. Instead, the male carries the eggs fixed to the underside of his tail until they hatch. (Photo by Animals Animals ©James Watt. Reproduced by permission.)

Distribution

Worldwide in tropical and temperate marine waters as well as in temperate freshwaters of the Northern Hemisphere. Species are more abundant in the tropical Indo-West Pacific region, where several undescribed pipefish species are known. Some species are widespread (e.g., the red cornetfish, *Fistularia petimba*), whereas others are very restricted in distribution (e.g., *Festucalex cinctus* off central New South Wales and southern Queensland, Australia). A few species with wide distributions, such as *Hippocampus erectus* and *Doryrhamphus dactyliophorus*, may require subdivision.

Habitat

Most species are coastal residents, present in shallow continental shelf areas in a variety of habitats, such as coral reefs, atolls, offshore reefs, sea grass meadows, kelp forests, tide pools, estuaries, bays, and lagoons and over sandy or muddy bottoms. Many species live cryptically, hiding among rocks and crevices in reefs or blending in with gorgonian corals or sea grasses. Many species have pelagic young, which eventually settle closer to the bottom. Some 20 gasterosteiform species are freshwater (at least one stickleback is exclusively so, along with *Indostomus* and pipefishes of the genera *Microphis*, *Hippichthys*, and *Dorichthys*) and are present in a variety of habitats, including lakes, coastal rivers, creeks, marshes, and protected coastal inlets. Some 40 species are euryhaline, found in brackish environments. As far as is known, no species occurs in deep-water environments.

Behavior

Gasterosteiforms are diurnal as far as is known. Most species are solitary or live in pairs or small groups or sometimes in larger groups (e.g., aulorhynchids). Many species have pelagic young, but most adults are benthic. A few species, such as macroramphosids and aulorhynchids, form schools, sometimes containing thousands of individuals. Most species remain in association with soft corals, hard corals (*Hippocampus barbouri*), algae, sea grasses, or other substrates, where they

A short-nosed seahorse (*Hippocampus breviceps*) male giving birth. (Photo by Animals Animals ©R. Kuiter, OSF. Reproduced by permission.)

are well camouflaged; in many cases they grasp the reef with their prehensile tails. Many species appear sluggish owing to their somewhat sedentary lifestyle and lack of ability to move very swiftly. Syngnathids swim by a combination of movements of their pectoral and dorsal fins, and some species swim by moving their tails from side to side. Many syngnathids appear to hover in one location, controlling their position by coordinated movements of their pectoral and dorsal fins. Some pipefishes (e.g., *Heraldia*) are able to swim in an upside-down position in caves and crevices. Centriscids remain in a vertical position, with their mouths directed toward the bottom, sometimes in association with urchins. Species of pipefishes, *Solenostomus*, *Aulostomus*, and macroramphosids also maintain a vertical position at times. Many species can change their coloration according to their background, using this ability to sneak up on prey or to hide from predators. Some species, especially pipefishes (e.g., *Doryrhamphus* spp.), have been documented to clean other fishes (e.g., moray eels and damselfishes), removing their parasites while in reefs.

Feeding ecology and diet

Most syngnathoids feed on a wide variety of small crustaceans (e.g., copepods and mysids), sometimes almost exclusively, as well as on other small invertebrates and the larvae of other fishes. The larger species, such as cornetfishes and trumpetfishes, also feed on larger fishes. Most gasterosteiform species ingest prey whole by quickly opening their mouths to produce a strong inward current, a suction mechanism called pipette feeding. In this manner, large prey items cannot be ingested, owing to the small terminal mouths, lack of teeth, and tubular snouts of many syngnathoid species. Most prey items are ingested from the substrate or when just hovering above it. Many gasterosteiforms rely on their highly developed camouflage to surprise prey items. This is the case in numerous species of pipefishes that slowly cruise over sandy bottoms with sea grasses, feeding on small crustaceans that fail to perceive them as a result of their cryptic appearance. Many species feed on small mysids and other crustaceans that are more free-swimming as well as fish larvae; they remain in strategic positions along the fringes of reefs, where they are exposed to currents that may contain these prey items. Some species, however, eat primarily in the water column (e.g., macroramphosids and aulorhynchids). Gasterosteiforms are preyed upon by larger carnivorous fishes, such as flatheads (Platycephalidae) and snappers (Lutjanidae). Because they move slowly, gasterosteiforms are ingested easily once they are discovered.

Reproductive biology

A great variety of reproductive strategies occur in this order, including some of the most elaborate known for any group of fishes. In a few families (Centriscidae, Macroramphosidae, Pegasidae, and perhaps Aulorhynchidae), the eggs and larvae are pelagic, and spawning probably is accomplished (at least in some of these families) in a style similar to broadcast spawning, in which eggs and sperm are released directly into the water column. In pegasids, spawning involves a courtship ritual whereby a female and a male swim vent to vent, releasing eggs and sperm simultaneously. In other families, reproduction is a far more complex process in which the male carries the eggs, sometimes in a special pouch, within which the eggs may be fertilized. In *Solenostomus*, the females carry the fertilized eggs in pouches made of their greatly extended pelvic fins, repeatedly opening and closing them to fan the eggs. As a preliminary behavior to spawning, many species employ complex courtship rituals, in which males compete for the female by dancing, inflating their pouches, or performing in some other way. A male stickleback lures the female into his nest and fertilizes the eggs there, sometimes attracting several females in succession. Some species are reproductively active throughout the year, whereas others are seasonal. Many species of syngnathids form monogamous pairs.

A ringed pipefish (*Doryrhamphus dactyliophorus*) over lava sand in the Lembeh Strait, Sulawesi, Indonesia. (Photo by Fred McConnaughey/Photo Researchers, Inc. Reproduced by permission.)

Reproduction has been studied in detail for many syngnathids. In seahorses, the females insert their eggs into the pouch of the males, using an abdominal projection known as the ovipositor (an everted egg duct); the eggs are fertilized by the sperm located in the pouches. During this process the male and female face each other. Inside the pouch the eggs are enveloped in tissue that supplies oxygen to the eggs (through diffusion from capillaries) as well as hormones by bathing the small portion of the egg that protrudes from the surrounding tissue. The pouch remains sealed. The male carries the eggs to term (incubation may last from 10 days to six weeks, depending on the species and surrounding temperature), during which time he appears very pregnant. The male actively expels the young (by forcefully moving back and forth) over a period of a few hours through an opening at the top of the pouch. Usually, about 100 young are born in this manner, but some species produce up to 400 young, whereas others produce between 10 and 50 (measuring close to 0.4 in, or 1 cm); pairs may have several broods in one year. The young do not receive any further parental care and are on their own immediately after birth. Young may congregate, and their sexes can be distinguished after a few months, when the pouches of the males become apparent (i.e., when they become sexually mature). Young resemble adults shortly after birth. As many as 1,000 young are produced per year by each couple in this manner, although the actual number varies between species, as reproduction may occur continuously.

Many pipefish males carry the eggs in the anterior tail region (just underneath the dorsal fin) without enclosing them in a pouch, fertilizing the eggs when they are deposited. The eggs are clearly visible in these cases. Courtship has been observed in detail in the species *Corythoichthys isigakius* from Japan. In this species, after a couple has consented to mate, the male and female repeatedly circle each other with their heads raised, exposing their undersides while remaining in an almost vertical position. When the male is ready to incubate the brood and the female is full of eggs, they practice egg transfer for several days until the process is actually enacted. The male's tail is flattened laterally as an indication that he is able to receive the eggs. The female places her small, greenish eggs on his tail, pushing them into place so that they remain attached. The eggs hatch after a few weeks. Hatching may be helped by the male, who shakes vigorously until all the eggs have separated. The young then are ready to begin their short pelagic life.

Conservation status

A relatively large number of gasterosteiform species are listed by the IUCN. Most of the threats that these species face are related to the pervasive seahorse trade, especially in southeastern and eastern Asia, and to widespread habitat degradation. Gasterosteiforms generally are not directly consumed in quantities that would otherwise place them in danger of overexploitation. Overexploitation of seahorses is a result of their use as ingredients for prepackaged medicines, where demand far exceeds supply, and for the curio and aquarium trade. Many species live in coastal or estuarine habitats, which typically are more affected by development and pollution. Furthermore, the vulnerability of seahorses is enhanced by their low fecundity, parental care, and complex social structures. The IUCN presently includes in their compilation of threatened taxa about 51 species of gasterosteiformes. The majority of species are listed either as Data Deficient or Vulnerable (all species of seahorses and many species of pipefishes). The Cape seahorse (*Hippocampus capensis*) is listed as Endangered (mostly due to commercial development of its restricted habitat), and the stickleback (*Pungitius hellenicus*) and the river pipefish (*Syngnathus watermeyeri*) are listed as Critically Endangered. Conservation efforts, mostly geared toward seahorses, are presently being undertaken by Project Seahorse, a praiseworthy initiative that aims to promote the sustainable exploitation of seahorses and their relatives.

Significance to humans

Numerous species of this order are popular fishes sought out by recreational divers. Many are important aquarium fishes and, fortunately, are now reared in captivity. Their popularity stems from their complex and highly modified morphological features and ornate coloration as well as the particular breeding habits of various species, in which there are elaborate courtship rituals and males fertilize the eggs and carry the young (in most species). Gasterosteiforms typically are not consumed as food fishes, but some syngnathids are commercialized heavily as curios. In particular, seahorses and seamoths are made into souvenirs (even as Christmas ornaments) and used as ingredients in traditional Chinese and related East Asian medicines. Some 47 countries worldwide participate in the seahorse market, and the total global consumption of seahorses in 1995 was at least 20 million specimens, roughly more than 62 tons (56 metric tonnes). The seahorse trade is not sustainable as it is currently implemented, leading to fears that many populations of seahorses have been exploited past the possibility of recovery.

1. Ringed pipefish (*Doryrhamphus dactyliophorus*); 2. Leafy seadragon (*Phycodurus eques*); 3. Weedy seadragon (*Phyllopteryx teaniolatus*); 4. Female lined seahorse (*Hippocampus erectus*); 5. Ornate ghost pipefish (*Solenostomus paradoxus*). (Illustration by Joseph E. Trumpey)

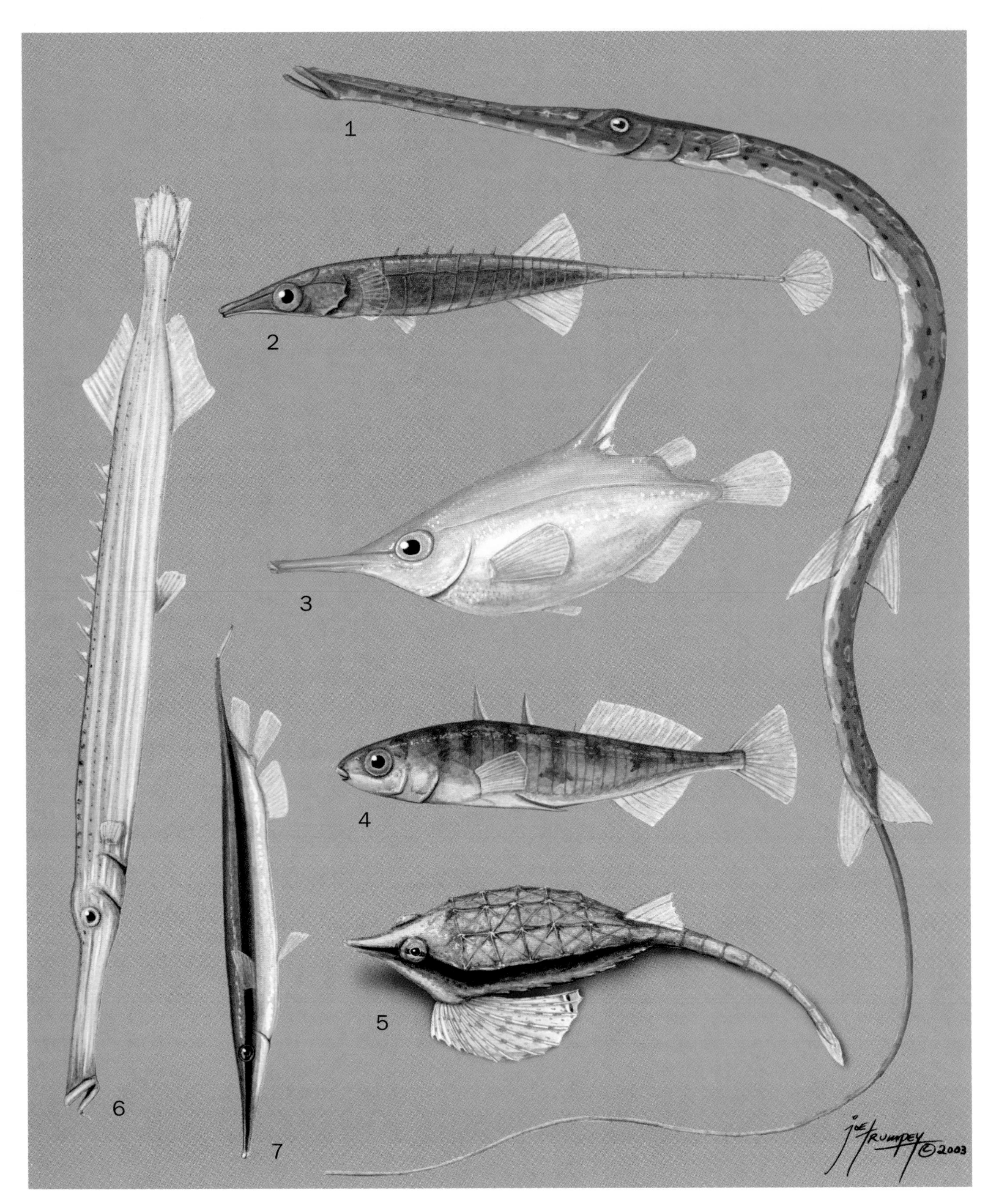

1. Blue-spotted cornetfish (*Fistularia tabacaria*); 2. Armored stickleback (*Indostomus paradoxus*); 3. Longspine snipefish (*Macroramphosus scolopax*); 4. Threespine stickleback (*Gasterosteus aculeatus*); 5. Sculptured seamoth (*Pegasus lancifer*); 6. West Atlantic trumpetfish (*Aulostomus maculatus*); 7. Common shrimpfish (*Aeoliscus strigatus*). (Illustration by Joseph E. Trumpey)

Species accounts

West Atlantic trumpetfish

Aulostomus maculatus

FAMILY
Aulostomidae

TAXONOMY
Aulostoma maculatum Valenciennes, 1837, West Atlantic.

OTHER COMMON NAMES
None known.

PHYSICAL CHARACTERISTICS
The trumpetfish is slender and elongate. The head is about one-third of the total length; the snout is tubular and the mouth large, terminal, and very upturned. The chin has a prominent barbel, and the eyes are relatively small and round. Single dorsal fin is preceded by eight to 13 short, well-spaced spines. Dorsal fin with 21–25 rays; anal fin opposite to dorsal fin and also with 21–25 rays; pelvic fins very posterior. Scales small and somewhat abrasive. Reaches at least 35.4 in (90 cm) in length. Coloration varies. May have dusky brown or reddish background with lighter stripes and darker spots, but some individuals are yellow or green with a blue snout.

DISTRIBUTION
Western Atlantic from Florida to southeastern Brazil; also in the Caribbean, Gulf of Mexico, Bermuda, and Antilles. Found in the eastern Atlantic at Saint Paul's Rocks.

HABITAT
Adults usually are found in coral reefs and associated habitats, at depths ranging from 6.6 to 82 ft (2–25 m); juveniles live in deeper water but also among sargassum.

BEHAVIOR
Trumpetfishes may hover in near vertical positions, with their heads pointing downward, sometimes almost motionless. They align their bodies with other linear objects, ranging from corals to other fishes to crevices, blending in with their surroundings. They also are capable of quick movements. Reported to change color according to their habitat. They also use larger non-piscivorous fishes such as parrotfish as mobile blinds, hiding behind them to approach within striking distance of prey.

FEEDING ECOLOGY AND DIET
Eats mostly smaller fishes and small invertebrates by employing stealthy movements and suction feeding (pipette feeding). May feed on larger fishes as well.

REPRODUCTIVE BIOLOGY
Mostly unknown; juveniles are known to inhabit deeper waters, but eggs and larvae probably are pelagic.

CONSERVATION STATUS
Not listed by the IUCN.

SIGNIFICANCE TO HUMANS
Sometimes kept in public aquaria. A harmless fish, sometimes marketed locally but of minor commercial importance. ◆

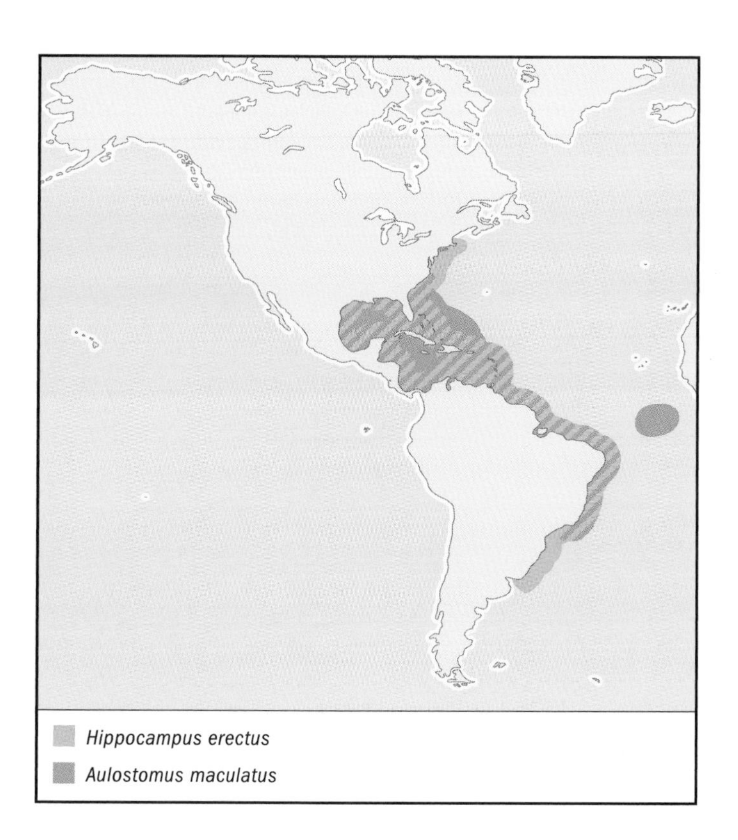

Common shrimpfish

Aeoliscus strigatus

FAMILY
Centriscidae

TAXONOMY
Amphisile strigata Günther, 1861, Java, Indonesia.

OTHER COMMON NAMES
English: Razorfish.

PHYSICAL CHARACTERISTICS
A very unusual species, with an elongate, highly compressed body. Snout long, almost filamentous, with a terminal mouth. Dorsal fin terminal, situated far posterior, with a long, posteriorly directed and hinged spine followed by two smaller spines and nine to 10 rays. Caudal fin underneath dorsal fin. Anal fin with 11–12 rays; pectoral fin with 11–12 rays. Body covered in plates. Coloration yellowish dorsally and laterally above a dark, horizontal stripe that runs through the eyes; white underneath stripe and ventrally. Reaches 5.5 in (14 cm) in length.

DISTRIBUTION
Widespread in the tropical western Pacific Ocean and northern Indian Ocean, reaching as far west as the Seychelles.

HABITAT
Usually found in shallow bays, coral reefs, and sea grass beds down to a depth of about 66 ft (20 m).

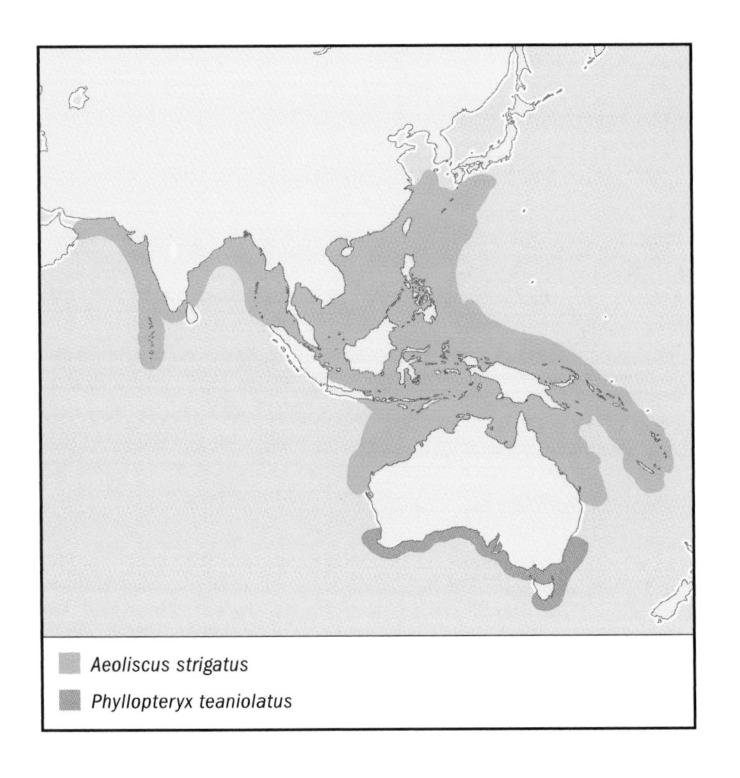

BEHAVIOR
Commonly encountered in large schools, typically hovering with its mouth pointed toward the bottom, maintaining a vertical position. Sometimes found in association with long-spined sea urchins of the genus *Diadema*.

FEEDING ECOLOGY AND DIET
Feeds on small invertebrates (e.g., polychaetes), including larvae of mollusks and crustaceans.

REPRODUCTIVE BIOLOGY
Mostly unknown. Pelagic larvae have been described and begin to resemble adults by about 0.7 in (17 mm) in length.

CONSERVATION STATUS
Not listed by the IUCN.

SIGNIFICANCE TO HUMANS
Not consumed but sometimes kept in aquaria. ◆

Blue-spotted cornetfish
Fistularia tabacaria

FAMILY
Fistulariidae

TAXONOMY
Fistularia tabacaria Linnaeus, 1758, North Atlantic.

OTHER COMMON NAMES
None known.

PHYSICAL CHARACTERISTICS
An unmistakable species, with an extremely elongated body and snout (about one-fourth of the body length). Oblique terminal mouth with many minute teeth. Eyes elliptical and large. Small, single, and triangular dorsal fin located far posterior on the back, opposite to the anal fin and with 13–18 rays. Anal fin also triangular, with 13–17 rays. Pectoral fins have 15 or 16

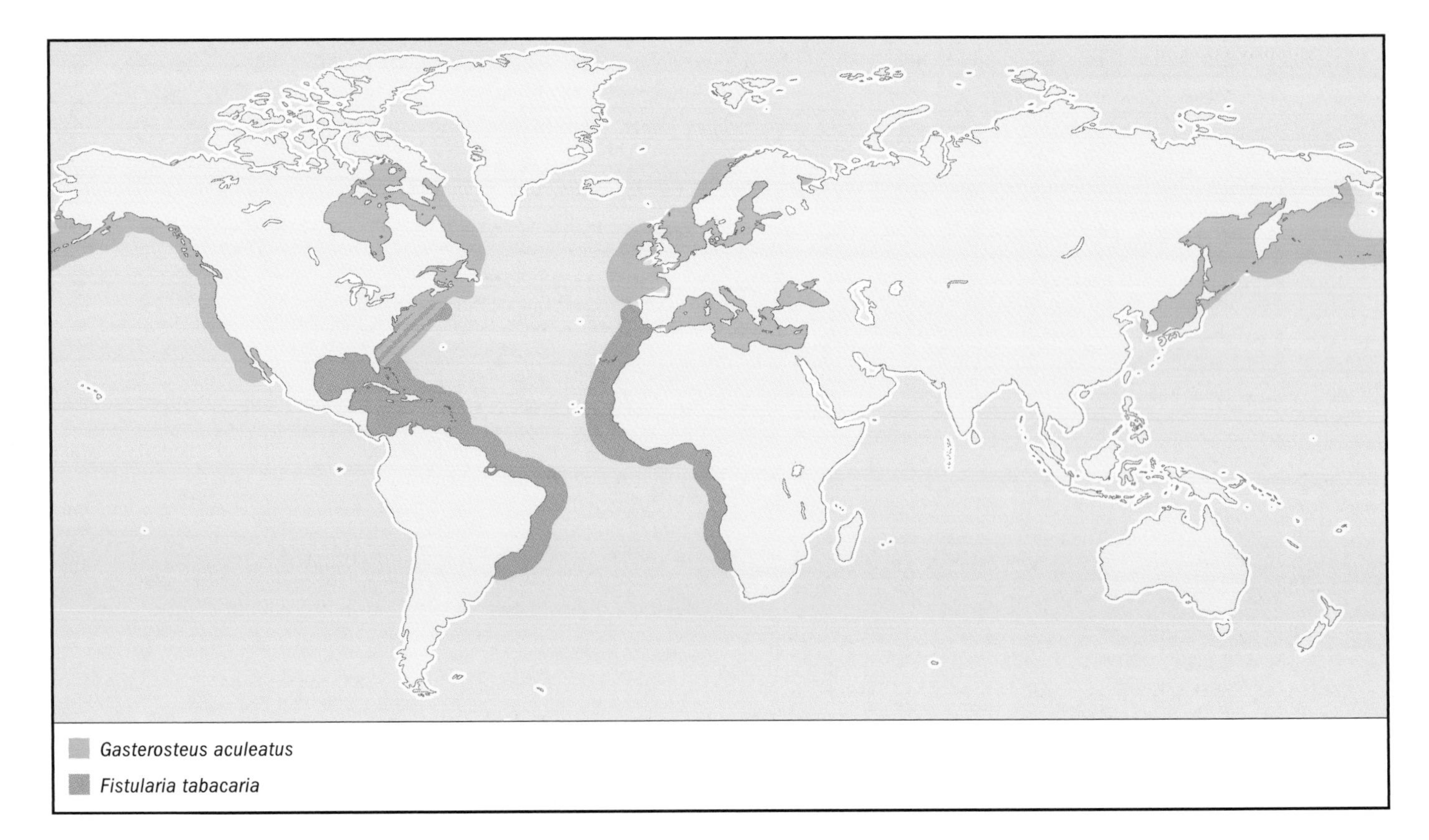

rays. Very small pelvic fins. Caudal fin has two clear forks, with upper and lower lobes divided by a long caudal filament (formed from caudal rays, about as long as the snout). Coloration is greenish brown, with numerous blue spots and some 10 darker cross bars. Caudal filament is blue (coloration can change quickly to mach the background). Reaches up to 5.9 ft (1.8 m) without the caudal filament.

DISTRIBUTION
Occurs in both sides of the tropical and warm temperate Atlantic Ocean. Young are reported to venture into colder temperate waters as far north as Nova Scotia in the Western Atlantic.

HABITAT
A mostly inshore species, occurring down to 656 ft (200 m); adults are found in coral reefs and sea grass meadows, over sandy and rocky bottoms, and even in estuaries.

BEHAVIOR
Mostly unknown. This species does not seek camouflage by aligning itself with objects, as does the trumpetfish. Usually solitary.

FEEDING ECOLOGY AND DIET
Eats mostly small shrimps and other fishes, which it can swallow whole, head first (by sucking them in a mode known as pipette feeding). Preyed upon by larger bony fishes (e.g., groupers) and sharks.

REPRODUCTIVE BIOLOGY
Little is known of its reproductive biology, but other species of *Fistularia* lay pelagic eggs in open waters; larval development probably is pelagic as well.

CONSERVATION STATUS
Not listed by the IUCN.

SIGNIFICANCE TO HUMANS
Not frequently consumed but a popular aquarium species. ◆

Threespine stickleback
Gasterosteus aculeatus

FAMILY
Gasterosteidae

TAXONOMY
Gasterosteus aculeatus Linnaeus, 1758, Europe.

OTHER COMMON NAMES
None known.

PHYSICAL CHARACTERISTICS
Head relatively small, with small upturned mouth anterior to the eyes; eyes round and somewhat large. Body about four times as long as it is deep, with slender caudal peduncle. Three strong spines precede the first dorsal fin (with 10–14 rays), the last of which is smaller and attached to the dorsal fin; the first two spines are very tall. Dorsal spines are widely spaced. Anal fin opposite soft dorsal fin, with one spine and eight to 10 rays. Pectoral fins ending at mid-body length, with 10 rays. Pelvics have strong spine and single ray. All spines can be locked into place or depressed. Caudal fin brush-shaped. Trunk covered by slender plates, with much variation in arrangement. Coloration silvery on sides and bluish black dorsally, with orange pelvic membranes. Males become more reddish when courtship commences and drab when it terminates. Reaches 3.5 in (90 mm) in length.

DISTRIBUTION
Widely distributed in temperate marine waters of the Northern Hemisphere but also in coastal rivers and lakes.

HABITAT
The threespine stickleback occurs in a wide variety of coastal habitats in both brackish and shallow marine waters. It can be captured occasionally in the open ocean as well. Typical habitats include tidal pools, coastal rivers and creeks, lakes, salt marshes, and protected coastal inlets. Adults typically are found in association with vegetation (e.g., eelgrass). Individuals may venture far out into the open oceans, as far as 621 mi (1,000 km) in the Pacific.

BEHAVIOR
Many populations remain most of the time in the open sea, venturing into coastal habitats to spawn and die in their second or third years. They swim by "rowing" with their pectoral fins and are capable of strong swimming motions, such as those required to ascend rivers. Moderately social outside of periods of reproductive activity, when males become strongly territorial.

FEEDING ECOLOGY AND DIET
Eats a wide variety of invertebrates and larvae, including copepods, gammarids, rotifers, branchiurans, oligochaetes, insects and their larvae, and even eggs of other sticklebacks. Sticklebacks are preyed upon by numerous other fishes as well as water birds, such as herons, mergansers, gulls, and loons.

REPRODUCTIVE BIOLOGY
The reproductive biology of this species has been studied in detail. Females from 2 to 2.8 in (5–7 cm) in standard length have between 116 and 838 eggs, measuring about 0.06 in (1.5 mm) in diameter; egg number increases with body weight. Before spawning, males establish a territory, building a nest on the substrate (which is accomplished by "gluing" together bits of vegetation with mucus and kidney secretions). The nest is complete when the male carves out a tunnel, at which time courtship begins. A dance is performed for a gravid female that has entered the territory, which consists of the male jumping toward and away from the female in a zigzag fashion, with spines erect and mouth open. Once a female is impressed, the male exhibits gluing behavior and fans the nest with his pectoral fins. He then zigzags back to the female, leads her to the nest, and points to it with his open mouth. The female enters the nest with her caudal peduncle protruding, allowing the male to begin quivering movements against her flank. After she has deposited her eggs, the male moves through the nest, fertilizing the eggs and expelling the female. He then pushes the eggs deeper into the nest, flattening the egg mass and repairing the nest at the same time, before fanning the eggs. The male attempts to induce other females to spawn in the same nest; the number of female partners may vary, but studies conducted in Quebec indicated that males having two or three female partners is the norm in that region. Incubation lasts roughly 14–20 days. After the eggs hatch, the male destroys the nest and guards the young (collecting any that may have fallen away). After his progeny become free-swimming, the male starts another courtship cycle.

CONSERVATION STATUS
Not listed by the IUCN.

SIGNIFICANCE TO HUMANS
Not a commercially important species. The threespine stickleback is studied intensely by fish ethologists. Often kept in aquaria, where it reproduces easily. ◆

Armored stickleback

Indostomus paradoxus

FAMILY
Indostomidae

TAXONOMY
Indostomus paradoxus Prashad and Mukerji, 1929, Lake Indawgyi, Myanmar.

OTHER COMMON NAMES
None known.

PHYSICAL CHARACTERISTICS
Body elongate and slender, with a spatulate snout and terminal mouth. Eyes large and round. The single dorsal fin contains six rays, is located at mid-body length, and is preceded by five short and evenly spaced spines. The anal fin is opposite the dorsal fin, also with six rays. The caudal fin contains 11 rays and is rounded posteriorly. Pelvic fins are very small and thoracic, with only four rays. The precaudal tail region is very long and slender. Large pectoral fins have 23 rays. Body is covered by a complex armor composed of numerous plates. Coloration is a light brown, with numerous small, darker blotches and irregular stripes. Reaches 1.2 in (3 cm) in length.

DISTRIBUTION
Northern Myanmar, possibly to Cambodia.

HABITAT
Rivers and lakes of northern Myanmar, including the Ayeyarwaddy River and Lake Indawgyi. Occurs close to the bottom in canals, swamps, ditches, and stagnant waters.

BEHAVIOR
Mostly unknown. This species apparently is sedentary in river or lake beds and moves about slowly.

FEEDING ECOLOGY AND DIET
Feeds mostly on worms and other small, slow-moving benthic invertebrates. Predators are presumably larger fishes.

REPRODUCTIVE BIOLOGY
The reproduction of this species has been observed under captive conditions. Males defended spawning sites of a few centimeters inside a tube and had a lighter reddish brown coloration (instead of the typical darker brown color of nonbreeding males). There is sexual dimorphism in the pelvic fins: males have longer and wider pelvics with inward curving rays. Males displayed by erecting their fins and shaking their tail regions. Females are lighter brown during spawning, with bulging abdomens due to the presence of eggs. Females approached the nesting sites of the male when they were about to spawn (indicated by their protruding genital papillae) and were encouraged to enter the tube by the male. The upside-down female deposited eggs, usually on the roof of the tube; the male fertilized the eggs while also upside-down. Spawning may take several hours, and the male guards the nest. From five to

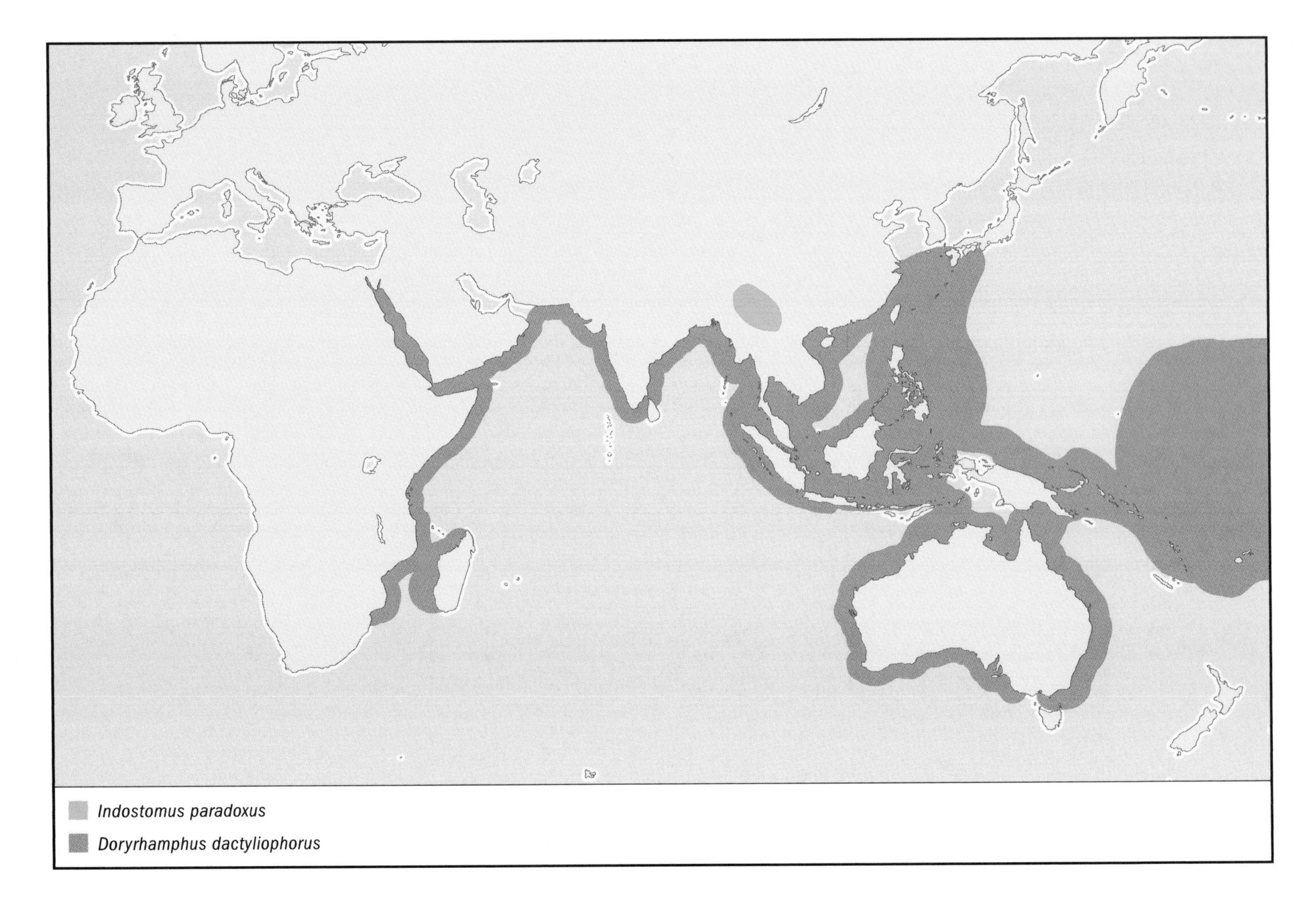

40 elliptical eggs were laid at a time, measuring about 0.08 in by 0.04 in (2 by 1 mm). Eggs hatched three days after spawning, and larvae were free-living four days after hatching, at about 0.14 in (3.5 mm). Larvae have an attachment organ on the tip of the yolk sac. Larval coloration is unique, composed of three black vertical bands, and adult coloration is attained after more than four weeks and at about 0.3 in (8 mm) in length.

CONSERVATION STATUS
Not listed by the IUCN.

SIGNIFICANCE TO HUMANS
Occasionally imported as an aquarium species. ◆

Longspine snipefish
Macroramphosus scolopax

FAMILY
Macroramphosidae

TAXONOMY
Balistes scolopax Linnaeus, 1758, Mediterranean Sea.

OTHER COMMON NAMES
None known.

PHYSICAL CHARACTERISTICS
Body is compressed laterally, with an elongated, tubular snout that terminates in a small upturned mouth lacking teeth. Eyes are round and large, with a single nostril on each side. Two dorsal fins, originating far posteriorly. First dorsal has extremely elongated and serrated second spine (four to eight spines, with each spine about as long as the snout); second dorsal fin has soft rays (10–14). Anal fin has 19–21 rays, pectoral fin has 18–21 rays, and pelvic fins have a single spine and four rays. Scales are very small. Pinkish or reddish dorsally; color fades to silvery on the sides. Reaches 9.1 in (23 cm) in total length.

DISTRIBUTION
Distributed worldwide, mostly in temperate latitudes.

HABITAT
Present mostly on the continental shelf in depths ranging from 164–902 ft (50–275 m). Juveniles are more pelagic, whereas adults are more demersal.

BEHAVIOR
Forms large schools, mostly while juvenile. Individuals may hover with their mouths pointing down toward the substrate, remaining in a stationary position.

FEEDING ECOLOGY AND DIET
Feeds on invertebrates, including copepods and other crustaceans, mollusk larvae, foraminifera, polychaete eggs, and mysids. Preyed upon by larger fishes, including tunas and blue sharks.

REPRODUCTIVE BIOLOGY
Eggs and larvae are pelagic, as are juveniles until they reach about 2 in (5 cm) in length. Adults are more demersal. In the eastern Atlantic, snipefish spawn on the continental shelf, over seamounts, and near islands between October and March. Larvae and juveniles remain in the surface layers during the day but migrate vertically to deeper water at night. At about 2 in (5 cm), some three months old, individuals move closer to the bottom.

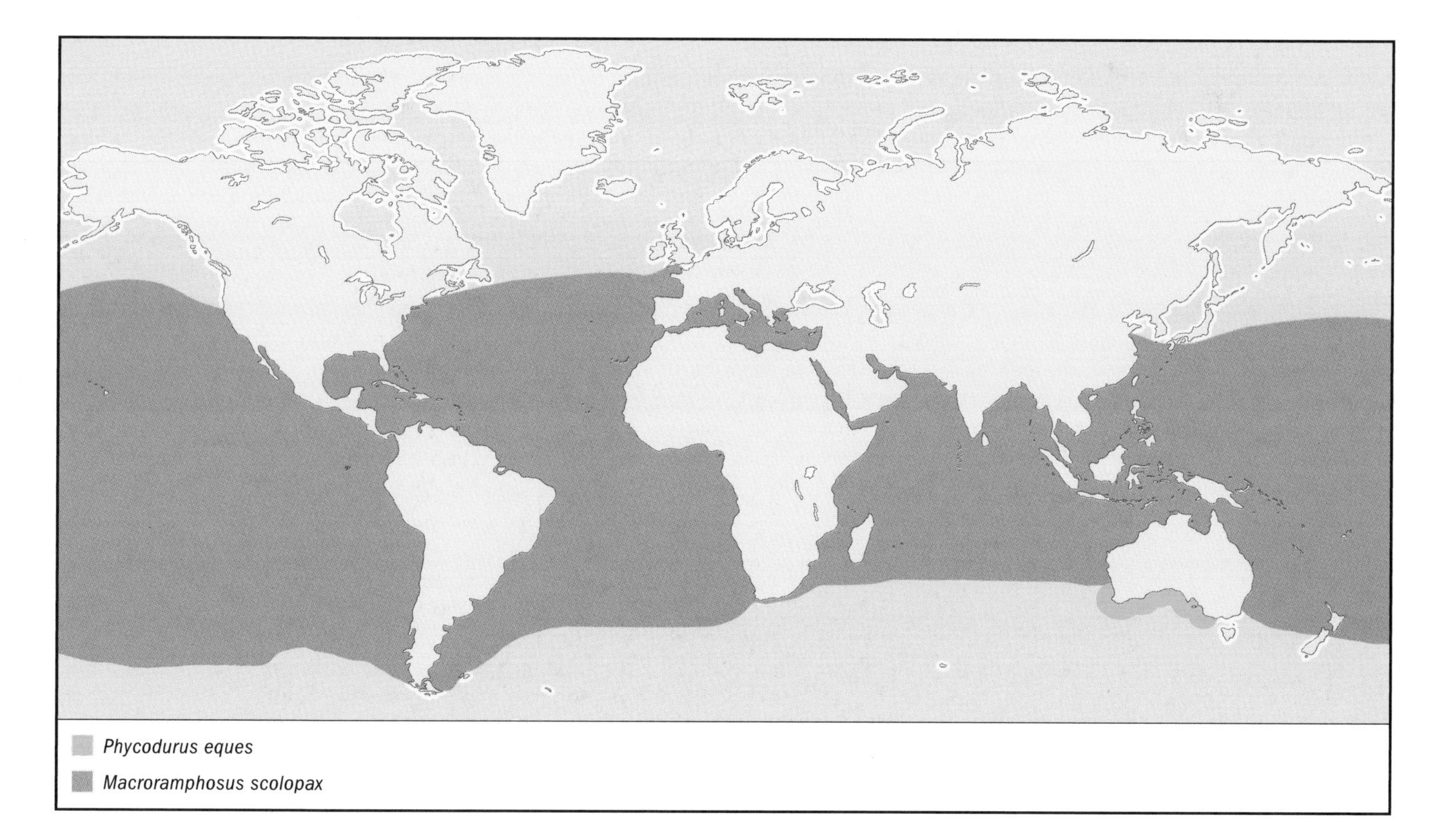

CONSERVATION STATUS
Not listed by the IUCN.

SIGNIFICANCE TO HUMANS
Not very important commercially, although it is consumed in the eastern Atlantic. This species has been kept in public aquaria. ◆

Sculptured seamoth
Pegasus lancifer

FAMILY
Pegasidae

TAXONOMY
Pegasus lancifer Kaup, 1861, Australia.

OTHER COMMON NAMES
None known.

PHYSICAL CHARACTERISTICS
Snout very slender but not very elongated (longer in males), ending in a spatulate tip; snout clearly demarcated from head and bearing four spiny ridges. Eyes large and round. Mouth small and protrusible, located ventral to the snout base. Head triangular anteriorly. Head and trunk regions are depressed and broad (broader in females), encased in fused bony plates that bear prominent ridges forming a star pattern. Single dorsal fin posterior to the trunk, with five rays; opposite anal fin also has five rays. Tail long and slender, with bony ridges, ending in a truncate caudal fin with eight to nine rays. Pectoral fins very wide and fanlike when expanded, with 18 rays. Pelvics resemble hooks, with one spine and three rays. Coloration sandy-brown or grayish above, with darker spots on pectorals, a dark longitudinal stripe on the trunk, and a dark stripe at base of the tail; pale underneath. Reaches about 4.7 in (12 cm) in length.

DISTRIBUTION
Restricted to southern Australian waters from South Australia to Tasmania.

HABITAT
A bottom-dwelling species occurring in many different coastal habitats, including estuaries, sea grass beds, and sandy bottoms, down to about 180 ft (55 m).

BEHAVIOR
This species can change its color to match that of its surroundings. It also can burrow into the substrate to escape predators. It often "walks" or "crawls" over the bottom in search of small crustaceans or other food items. Many individuals may congregate in estuaries.

FEEDING ECOLOGY AND DIET
Eats a variety of small invertebrates, such as polychaetes, mollusks, and crustaceans. Their mouths are somewhat protrusible,

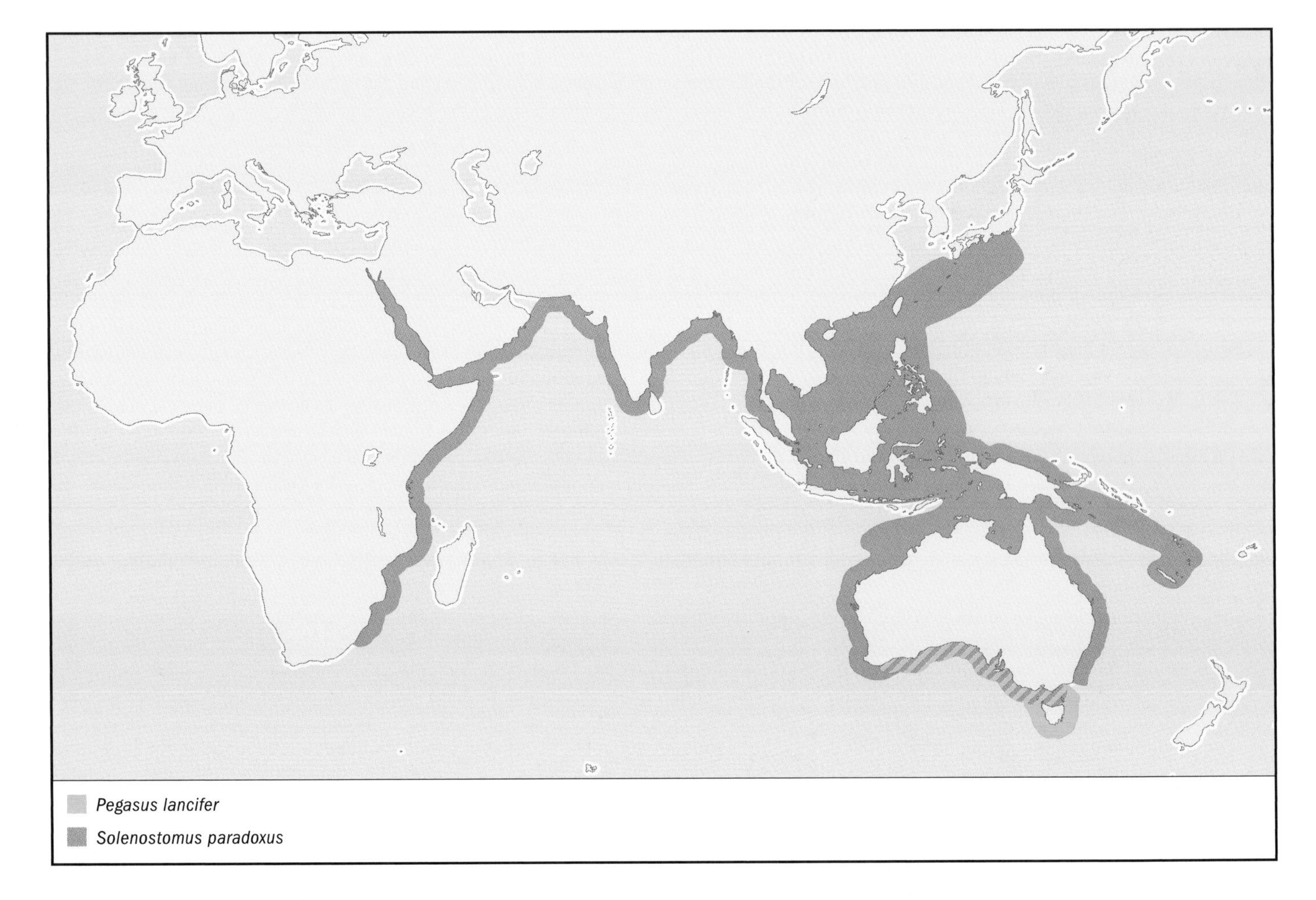

enabling them to snatch prey from the bottom with ease. Presumably eaten by larger fishes.

REPRODUCTIVE BIOLOGY
Enters sandy bays to breed in the spring, when courtship takes place. During courtship, a female and male remain on the bottom, side by side, until they rise together to spawn, vent to vent, about 3.3 ft (1 m) off the bottom. The posterior tips of the pectoral fins of males become yellow during the reproductive period. The eggs are pelagic, and larvae up to 0.1 in (2.5 mm) are enclosed in a dermal sac, probably an adaptation to a pelagic existence.

CONSERVATION STATUS
Listed as Data Deficient by the IUCN.

SIGNIFICANCE TO HUMANS
Not a commercial species. ◆

Ornate ghost pipefish
Solenostomus paradoxus

FAMILY
Solenostomidae

TAXONOMY
Fistularia paradoxa Pallas, 1770, Indonesia.

OTHER COMMON NAMES
English: Harlequin ghost pipefish.

PHYSICAL CHARACTERISTICS
A morphologically remarkable fish. Body is very slender and straight, and head is very elongate, about one-third the total length. Mouth terminal, small, and upturned. Eyes round. First dorsal fin situated over pelvic fins, with five very long spines; second dorsal fin more posterior, with 17–23 rays. Anal fin opposite second dorsal, with 17–22 rays; pectoral fins inconspicuous, with 24 rays; pelvic fins located just anterior to the middle of the body, with seven rays; caudal fin large and spur-shaped when rays are extended. There are numerous dermal projections on snout, trunk, and tail. Body encased in 31–35 segmented bony plates. Coloration spectacular and ornate. Background color varies from white or semitransparent to black, with elaborate pattern of red, orange, white, or yellow spots, blotches, and stripes. Reaches some 4.7 in (12 cm).

DISTRIBUTION
Present in much of the Indo-West Pacific Ocean from the Red Sea and East Africa to the Marshall Islands and New Caledonia; occurs as far north as southern Japan and as far south as Australia.

HABITAT
Usually found in coral reef habitats (down to about 115 ft, or 35 m, in depth) among gorgonian corals, weeds, and algae or with crinoids over sand, along reef edges, or in sheltered coastal waters and estuaries.

BEHAVIOR
Typically a solitary species, but small groups of up to six individuals have been observed. This species is highly camouflaged, blending in very well with many different backgrounds. Young are mostly pelagic, but adults are more benthic; post-pelagic specimens are more transparent. Individuals may hover in a near vertical position when they are among gorgonians or other substrates.

FEEDING ECOLOGY AND DIET
Feeds on mysids, small shrimps, and other benthic crustaceans. Predators unknown but presumably larger reef-dwelling fishes.

REPRODUCTIVE BIOLOGY
In contrast to other syngnathoids, females carry the brood in a pouch that is formed by the enlarged pelvic fins (which are missing from other syngnathoids). Eggs are numerous, small, spherical, and somewhat transparent. Females fan the eggs by opening and closing their pelvic fins. Young are expelled from the pouch at birth. Other details concerning their reproduction are unknown.

CONSERVATION STATUS
Not listed by the IUCN.

SIGNIFICANCE TO HUMANS
Sometimes kept as an aquarium fish; otherwise this species has no commercial value. ◆

Ringed pipefish
Doryrhamphus dactyliophorus

FAMILY
Syngnathidae

TAXONOMY
Syngnathus dactyliophorus Bleeker, 1853, Indonesia.

OTHER COMMON NAMES
English: Banded pipefish.

PHYSICAL CHARACTERISTICS
A very slender species, with both head and body on the same plane. Snout very slender, about one-eighth of the total length, with small upturned terminal mouth. Eyes round and relatively long. Single posterior dorsal fin, with 20–26 rays; small anal fin with four rays. Body has 15–17 bony rings, and tail has 18–22 bony rings. Caudal fin is round when expanded. Coloration composed of alternating black to reddish and yellow bars; caudal fin is white with a red circle. Reaches some 7.1 in (18 cm) in length.

DISTRIBUTION
This species is present throughout much of the Indo-West Pacific, in the Red Sea and East Africa to Samoa, and throughout Micronesia.

HABITAT
Inhabits a wide variety of coastal habitats, such as coral reefs, tide pools, lagoons, estuaries, sea grass meadows, and outer reef slopes. Frequently found in crevices and sheltered areas.

BEHAVIOR
A cryptic species, preferring to remain in caves or other shelters. Commonly seen in pairs, but occasionally large groups are formed. Young are pelagic and transparent (up to about 1.2 in, or 30 mm) and attain adult coloration only later, when they settle into a more benthic lifestyle.

FEEDING ECOLOGY AND DIET
Eats small crustaceans and may even clean larger fishes, such as morays, of small parasites. Consumed by larger fishes.

REPRODUCTIVE BIOLOGY
Males carry the eggs along their trunks ventrally. The eggs are small and spherical and reddish when fresh. The eggs are fertilized upon deposition on the male and hatch after a few weeks. Juveniles often are found in rock pools but typically are pelagic.

CONSERVATION STATUS
Listed as Data Deficient by the IUCN.

SIGNIFICANCE TO HUMANS
A popular aquarium species, which has been reared in captivity. ◆

Lined seahorse
Hippocampus erectus

FAMILY
Syngnathidae

TAXONOMY
Hippocampus erectus Perry, 1810, West Indies.

OTHER COMMON NAMES
English: Seahorse.

PHYSICAL CHARACTERISTICS
Body erect and somewhat sinuous, with head at a right angle in relation to the trunk and tail. Snout moderately long, with small upturned mouth devoid of teeth. Eyes round. Two pairs of spines present behind the eyes on the head. Small pectoral fins with 14–17 rays; single posterior dorsal fin with 16–20 rays; small anal fin with three to four rays; no pelvic fins. Prehensile tail tapers into a slender stalk without a caudal fin. Trunk encased in 10–12 bony rings, each with four spines; tail has 32–38 rings. Coloration varies widely—background light brown, black, gray, or yellow (sometimes red) with various small blotches, stripes, and spots. Area around the eyes has small white stripes radiating from the eyes. Reaches 7.9 in (20 cm) in length.

DISTRIBUTION
Western Atlantic from Cape Cod (sometimes Nova Scotia as strays) to Uruguay. The extensive range suggests that the name *H. erectus* may be applied to a complex of closely related species.

HABITAT
Lined seahorses are found in habitats with heavy vegetation, such as seaweeds and sargassum, in shallow waters and waters as deep as 240 ft (73 m). Present also in bays, piers, beaches, salt marshes, oyster beds, and other environments in which vegetation and shelter are present. They are capable of tolerating great variations in temperature and salinity.

BEHAVIOR
Seahorses swim slowly, in a vertical position, by undulating the dorsal and pectoral fins and tend to cling with their prehensile tails to vegetation, gorgonian corals, and so forth. They produce sounds to communicate with each other. Younger individuals tend to be pelagic, sometimes swimming in groups.

FEEDING ECOLOGY AND DIET
Food consists mostly of small crustaceans, such as copepods, amphipods, and larvae. They employ pipette feeding, after a swift sucking action following a sudden upswing of the head. They are eaten by many species of bony fishes, including cod, bluefish, remoras, and spiny and smooth dogfishes.

REPRODUCTIVE BIOLOGY
Between 250 and 400 eggs are deposited in the male's brood pouch during courtship (the larger the female, the greater the number of eggs). Males develop brood pouches by about 3 in (7.5 cm) in length, and males with eggs are recorded at 3.5 in (9 cm). In courtship the male and female closely follow each other, and the male presents his pouch to the female's genital area. As eggs are being transferred from the female into the male's pouch, they both rise in the water and may change color. Breeding takes place in the summer or year-round in tropical climates. Females deposit a few eggs at a time repeatedly. The eggs develop in the brood pouch and may derive nourishment from secretions within the pouch. Eggs are pear-shaped and light orange in color and may contain one or more oil droplets. Eggs are incubated for 12–14 days in the pouch; there is no true larval period; miniature seahorses are expelled, measuring about 0.24 in (6 mm) in length. Their tails become prehensile after one day, and they become mature after three months.

CONSERVATION STATUS
Listed as Vulnerable by the IUCN.

SIGNIFICANCE TO HUMANS
A common aquarium species whose commercialization requires monitoring. ◆

Leafy seadragon
Phycodurus eques

FAMILY
Syngnathidae

TAXONOMY
Phyllopteryx eques Günther, 1865, South Australia.

OTHER COMMON NAMES
None known.

PHYSICAL CHARACTERISTICS
One of the most distinctive of all fishes, the leafy seadragon has numerous, complex leaflike dermal projections from the extremities of its head, trunk, and tail spines. Body is elongate, slender but somewhat sinuous, and encased in ringlike bony plates that extend as spines. The head is long, directed at almost a right angle in relation to the body, with a frontal bony projection and an extremely elongated snout. The mouth is upturned and terminal. Eyes round. Dorsal fin, which is poseterior, has 34–38 rays; pectoral fins have 19–21 rays; anal fin has four rays. Tail slender and prehensile, with leaflike dorsal projections. Pelvic and caudal fins, lateral line, and scales are absent. Coloration is greenish brown or yellow, with vertical stripes along the trunk. Head has slight mask and varying dark blotches on leaflike projections. Specimens in deeper waters (about 98 ft, or 30 m) tend to be more reddish. Reaches some 13.8 in (35 cm) in length.

DISTRIBUTION
Coastal regions of southern Australia from Western Australia (below Perth) to southern Victoria.

HABITAT
A marine species inhabiting shallow (down to about 98 ft, or 30 m), temperate waters, usually sheltered among algae and reefs but also found over sandy areas.

BEHAVIOR
A sluggish species, appearing to float aimlessly in kelp beds, protected by its elaborate camouflage. It may move rhythmically back and forth in a manner similar to algae being swept by currents. Adults may congregate in shallow bays in late winter to pair and mate. Divers can approach this species slowly. Cleaner shrimp and clingfish have been observed to "clean" parasites off leafy seadragons.

FEEDING ECOLOGY AND DIET
Feeds mostly on mysids and other crustaceans (e.g., shrimps and squad-lobsters). Predators unknown but presumably larger fishes.

REPRODUCTIVE BIOLOGY
Reproduction has been observed in the wild in this species. (Aquarium-bred specimens take some two years to reach adult size.) Females have greatly swollen trunks before spawning. Males carry a brood containing an estimated 250–300 elliptical eggs underneath their tails, posterior to the anal fins. The eggs, which measure 0.3 by 0.2 in (7 by 4 mm), are maintained in honeycomb-like egg pockets in the abdominal skin of the males. Pregnant males usually are seen in November or December. The tail of the male becomes swollen and turns bright yellow to indicate readiness to mate, at which time sperm are released into the abdominal region. The courtship ritual is similar to that of some pipefishes (in which the abdominal area is displayed to the mate, while the head is maintained in an upright position). The female deposits her eggs onto the abdominal region of the male, pushing them into place. At that point, egg pockets form on the male to fasten the eggs securely in place. Incubation takes about eight weeks. Males deposit hatchlings in a wide area, as hatching takes about one week. Hatchlings are advanced and settle on the bottom (at about 1.4 in, or 35 mm, in length), sometimes remaining in small groups and venturing into shallower water.

CONSERVATION STATUS
Listed as Data Deficient by the IUCN.

SIGNIFICANCE TO HUMANS
An aquarium species that also attracts numerous recreational divers who want to see it up close (e.g., in Victoria Harbor, South Australia). This species cannot be collected off South Australia legally. It is somewhat difficult to maintain in aquaria, owing to its specialized feeding requirements, but aquarium-raised specimens may live for more than 10 years in captivity. Many aquarium specimens are bred in captivity. ◆

Weedy seadragon
Phyllopteryx teaniolatus

FAMILY
Syngnathidae

TAXONOMY
Syngnathus taeniolatus Lacepède, 1804, Bass Strait, Australia.

OTHER COMMON NAMES
English: Common seadragon.

PHYSICAL CHARACTERISTICS
Another remarkably distinct species. Body slender and elongate, highly arched between head and tail. Head at a slight angle to the body, with round eyes, very long snout, and small, upturned mouth. Encased in bony ringlike plates, many containing longer spines that have dermal leaflike flaps at the ends. Single dorsal fin situated posteriorly, with 27–34 rays; anal fin with four or five rays; pectorals with 20–23 rays. Tail prehensile, long, and slender, without a caudal fin; pelvic fins, lateral line, and scales absent. Coloration very ornate and somewhat varying—background usually reddish, with numerous closely packed yellow spots on head and body, bluish bars on sides of trunk and tail base, and darker dermal flaps. Reaches 18 in (46 cm) in length.

DISTRIBUTION
Southern Australian coast from central New South Wales to Rottnest Island (off Perth); also in Tasmania.

HABITAT
Typically found among algae, in kelp beds, and on rocky reefs, down to about 164 ft (50 m).

BEHAVIOR
A mostly solitary species, hovering among algae and sheltered rocky reefs, but also seen over sand. Individuals may move into deeper waters when food becomes less abundant.

FEEDING ECOLOGY AND DIET
As with the leafy seadragon, food principally consists of small crustaceans. Predators unknown but presumably larger fishes.

REPRODUCTIVE BIOLOGY
Many aspects of its reproduction are similar to that of the leafy seadragon. Males carry roughly equivalent numbers of eggs, which also are embedded in their skin. They usually have one brood per season, and mating begins from October to November. Young hatch after about two months, settling on the substrate. Hatchlings do not have the elongated snout of the adults, but it grows quickly; their elaborate dermal flaps are much smaller. Juveniles resemble adults by about 4.7 in (12 cm) in length.

CONSERVATION STATUS
Listed as Data Deficient by the IUCN.

SIGNIFICANCE TO HUMANS
A popular aquarium species. Most specimens displayed are bred in captivity. They can live for more than 10 years in aquarium conditions. ◆

Resources

Books

Allen, G. R. *Marine Fishes of Tropical Australia and South-east Asia.* Perth: Western Australian Museum, 1997.

Berra, T. M. *Freshwater Fish Distribution.* San Diego: Academic Press, 2001.

Browne, P. S. "Systematics and Morphology of the Gasterosteiformes." In *The Evolutionary Biology of the Threespine Stickleback,* edited by Michael A. Bell and Susan A. Foster. New York: Oxford University Press, 1996.

Dawson, C. E. *Indo-Pacific Pipefishes (Red Sea to the Americas).* Ocean Springs, MS: Gulf Coast Research Laboratory, 1985.

Fritzsche, Ronald A. "Gasterosteiformes: Development and Relationships." In *Ontogeny and Systematics of Fishes,* edited by H. G. Moser, W. J. Richards, D. M. Cohen, M. P. Fahay, A. W. Kendall, Jr., and S. L. Richardson. Special Publication no. 1. Lawrence, KS: American Society of Ichthyologists and Herpetologists, 1984.

Gomon, M. F., J. C. M. Glover, and R. H. Kuiter, eds. *The Fishes of Australia's South Coast.* Adelaide: State Print, 1994.

Kuiter, Rudie H. *Guide to Sea Fishes of Australia.* London: New Holland, 1996.

———. *Seahorses, Pipefishes and Their Relatives: A Comprehensive Guide to Syngnathiformes.* Chorleywood, U.K.: TMC Publishing, 2000.

Leis, J. M., and D. S. Renis. "Centriscidae, Fistulariidae." In *The Larvae of Indo-Pacific Coastal Fishes: An Identification Guide to Marine Fish Larvae,* edited by Jeffrey M. Leis and Brooke M. Carson-Ewart. Leiden: Brill, 2000.

Lieske, Ewald, and Robert Myers. *Coral Reef Fishes: Caribbean, Indian Ocean and Pacific Ocean: Including the Red Sea.* Princeton, NJ: Princeton University Press, 1996.

Lourie, S. A., A. C. J. Vincent, and H. J. Hall. *Seahorses: An Identification Guide to the World's Species and Their Conservation.* London: Project Seahorse, 1999.

Masuda, H., K. Amaoka, C. Araga, T. Uyeno, and T. Yoshino. *Nihon-san Gyorui Daizukan* (The Fishes of the Japanese Archipelago). 2 vols. Tokyo: Tokai University Press, 1984.

Nelson, J. S. *Fishes of the World.* 3rd edition. New York: John Wiley and Sons, 1994.

Orr, J. W., and T. W. Pietsch. "Pipefishes and Their Allies." In *Encyclopedia of Fishes,* edited by John R. Paxton and William N. Eschmeyer. San Diego: Academic Press, 1994.

Patterson, C. "Osteichthyes: Teleostei." In *The Fossil Record 2,* edited by M. J. Benton. London: Chapman and Hall, 1993.

Randall, John E., Gerald R. Allen, and Roger C. Steene. *Fishes of the Great Barrier Reef and Coral Sea.* Honolulu: University of Hawaii Press, 1997.

Reader, S. E., J. M. Leis, and D. S. Rennis. "Pegasidae." In *The Larvae of Indo-Pacific Coastal Fishes: An Identification Guide to Marine Fish Larvae,* edited by Jeffrey M. Leis, and Brooke M. Carson-Ewart. Leiden: Brill, 2000.

Smith, C. Lavett. *National Audubon Society Field Guide to Tropical Marine Fishes of the Caribbean, Gulf of Mexico, Florida, the Bahamas, and Bermuda.* New York: Knopf, 1997.

Trnski, T., and J. M. Leis. "Solenostomidae." In *The Larvae of Indo-Pacific Coastal Fishes: An Identification Guide to Marine Fish Larvae,* edited by Jeffrey M. Leis, and Brooke M. Carson-Ewart. Leiden: Brill, 2000.

Walker, H. J. "Aulostomidae." In *The Larvae of Indo-Pacific Coastal Fishes: An Identification Guide to Marine Fish Larvae,* edited by Jeffrey M. Leis, and Brooke M. Carson-Ewart. Leiden: Brill, 2000.

Periodicals

Banister, K. E. "The Anatomy and Relationships of *Indostomus paradoxus.*" *Bulletin of the British Museum of Natural History* 19 (1970): 179–209.

Britz, Ralf. "Aspects of the Reproduction and Development of *Indostomus paradoxus* (Teleostei: Indostomidae)." *Ichthyological Exploration of Freshwaters* 11, no. 1 (2000): 305–314.

Britz, Ralf, and G. David Johnson. "'Paradox Lost': Skeletal Ontogeny of *Indostomus paradoxus* and Its Significance for the Phylogenetic Relationships of Indostomidae (Teleostei, Gasterosteiformes)." *American Museum Novitates* 3383 (Dec. 2002): 43 pp.

Britz, Ralf, and Maurice Kottelat. "Two New Species of Gasterosteiform Fishes of the Genus *Indostomus* (Teleostei: Indostomidae)." *Ichthyological Exploration of Freshwaters* 10, no. 1 (1999): 327–336.

Dawson, C. E. "The Pipefishes (Subfamilies Doryrhamphinae and Syngnathinae)." *Memoirs of the Sears Foundation for Marine Research* 1, no. 8 (1982): 4–172.

Fritzsche, Ronald A. "A Review of the Cornetfishes, Genus *Fistularia,* (Fistulariidae), with a Discussion on Intrageneric Relationships and Zoogeography." *Bulletin of Marine Science* 26, no. 2 (1976): 196–204.

———. "A Revisionary Study of the Eastern Pacific Syngnathidae (Pisces: Syngnathiformes), Including Both Recent and Fossil Forms." *Proceedings of the California Academy of Sciences* 42, no. 6 (1980): 181–227.

Gosline, W. A. "Notes on the Osteology and Systematic Position of *Hypoptychys dybowskii* Steindachner and Other Elongate Perciform Fishes." *Pacific Science* 17 (1963): 90–101.

Herold, D., and E. Clark. "Monogamy, Spawning and Skin-Shedding of the Sea Moth, *Eurypegasus draconis* (Pisces: Pegasidae)." *Environmental Biology of Fishes* 37 (1993): 219–236.

Johnson, G. D., and C. Patterson. "Percomorph Phylogeny: A Survey of Acanthomorphs and a New Proposal." *Bulletin of Marine Science* 52, no. 1 (1993): 554–626.

Kuiter, Rudie. "The Remarkable Sea Moth." *Scuba Diver* 3 (1985): 16–18.

Lourie, S. A., J. C. Pritchard, S. P. Casey, S.-K. Truong, and A. C. J. Vincent. "The Taxonomy of Vietnam's Exploited Seahorses (Family Syngnathidae)." *Biological Journal of the Linnean Society* 66 (1999): 231–256.

Resources

Masonjones, H. D., and S. M. Lewis. "Courtship Behaviour in the Dwarf Seahorse, *Hippocampus zosterae*." *Copeia* 1996, no. 3 (1996): 634–640.

Orr, James W., and Ronald A. Fritzsche. "Revision of the Ghost Pipefishes, Family Solenostomidae (Teleostei: Syngnathoidei)." *Copeia* 1993, no. 1 (1993): 168–182.

Palsson, W. A., and T. W. Pietsch. "Revision of the Acanthopterygian Fish Family Pegasidae (Order Gasterosteiformes)." *Indo-Pacific Fishes* 18 (1989): 1–38.

Pietsch, T. W. "Evolutionary Relationships of the Sea Moths (Teleostei: Pegasidae) with a Classification of Gasterosteiform Families." *Copeia* 1978, no. 3 (1978): 517–529.

Sorbini, L. "The Cretaceous Fishes of Nardò. 1. Order Gasterosteiformes (Pisces)." *Bollettino del Museo Civico di Storia Naturale Verona* 8 (1981): 1–27.

Vari, R. P. "Seahorses (Subfamily: Hippcampinae)." *Fishes of the Western North Atlantic. Memoirs of the Sears Foundation for Marine Research* 1, no. 8 (1982): 173–189.

Vincent, A. C. J. "The Improbable Seahorse." *National Geographic* 186 (Oct. 1994): 126–140.

———. "Trade in Pegasids Fishes (Sea Moths), Primarily for Traditional Chinese Medicine." *Oryx* 31, no. 3 (1997): 199–208.

Vincent, A. C. J., and L. M. Sadler. "Faithful Pair Bonds in Wild Seahorses, *Hippocampus whitei*." *Animal Behaviour* 50 (1995): 1557–1569.

Other

"Project Seahorse." (17 Feb. 2003). <www.seahorse.mcgill.ca/intro.htm>

Marcelo Carvalho, PhD

Synbranchiformes
(*Swamp and spiny eels*)

Class Actinopterygii
Order Synbranchiformes
Number of families 3

Photo: An African eel (*Mastacembelus* sp.) from Cameroon. (Photo by Mark Smith/Photo Researchers, Inc. Reproduced by permission.)

Evolution and systematics

No synbrachiform fossil is known. The Mastacembeloidei were removed from the Perciformes and added to the Synbranchiformes after a phylogenetic analysis by Johnson and Patterson. These authors consider the Synbranchiformes to be monophyletic and related to Mugiliformes, Atheriniformes, Gasterosteiformes, and Smegmamorformes.

There are two suborders: Synbranchoidei and Mastacembeloidei, or Opisthomi. The Synbranchoidei has one family, the Synbranchidae; four genera; and 15 species. The Mastacembeloidei has two families: Chaudhuriidae, with four genera and five species, and Mastacembelidae. The latter family is divided into two subfamilies, Mastacembelinae, with two genera and 25 species, and Afromastacembelinae, with two genera and 42 species). There are a total of 87 species.

Physical characteristics

These eel-like fishes range in size from 8 to 48 in (20–150 cm). Although they are eel-like, they are not related to true eels (Anguilliformes). The premaxillae are present as distinct bones. The gills are poorly developed, and their openings are usually single, small, and confluent across the breast. Oxygen is absorbed through the membranes of the throat or intestine. The dorsal and anal fins are low and continuous around the tail tip. Pelvic fins, if present, are small and located on the throat. Scales are either absent or very small. They lack a swim bladder.

Distribution

These fishes are distributed in tropical America, tropical Africa, southeastern and eastern Asia, East Indies, and Australia. The three families each have a somewhat different distribution: The Synbranchidae are found in Mexico, Central and South America, West Africa (Liberia), Asia, and the Indo-Australian Archipelago. The Mastacembelidae are found in Africa and through Syria to the Malay Archipelago and China. The Chaudhuriidae are found in northeastern India through Thailand to Korea (including parts of Malaysia and Borneo).

Habitat

They usually are found in swamps, caves, and sluggish fresh and brackish waters. When found in pools, they typically are associated with leaf litter and mats of fine tree roots along the banks. Swamp eels are capable of overland excursions, and some can live out of water for extended periods of time. Some species are burrowers. Four species are found exclusively in caves: *Monopterus eapeni* and *M. roseni* from India, *Ophisternon candidum* from Australia, and *O. infernale* from Mexico. One species, *O. bengalense*, commonly occurs in coastal areas of southeastern Asia.

Behavior

Some species are considered air-breathing fishes because of their ability to breathe by highly vascularized buccopharyngeal pouches (pharynx modified for breathing air). They usually are active only at night.

Feeding ecology and diet

They feed on benthic invertebrates, especially larvae, and fishes.

The zig-zag eel (*Mastacembelus armatus*) is found in the weedy stream beds of Southeast Asia, Sri Lanka, southern China, and Sumatra. (Photo by Hans Reinhard. Bruce Coleman, Inc. Reproduced by permission.)

Reproductive biology

At least some of the species of the family Synbranchidae, that is, *O. infernale*, are sexually dimorphic. Adult males grow a head hump, and males are larger than females. These fishes lay about 40 spherical eggs per clutch. The eggs measure between 0.05 and 0.06 in (1.2–1.5 mm) in diameter and have a pair of long filaments for adhesion to the substrate. Reproduction takes place during the wet season, which lasts for several months, during which females probably spawn more than once. Data acquired from studying juvenile growth and the length of representative individuals within a population suggests that they are a short-lived species that matures during the first year, with few individuals surviving to the second breeding season.

Conservation status

As of 2002, five species were listed by the IUCN as species of special concern: *Macrognathus aral* (the one-stripe spiny eel), *Monopterus boueti* (Liberian swamp eel), *Monopterus indicus* (Bombay swamp eel), and *Ophisternon candidum* (the blind cave eel) have been classified as Data Deficient, meaning that they require more study to determine their conservation status; *O. infernale* (blind swamp cave eel) is classified as Endangered.

An undescribed spiny eel (*Mastacembelus* sp.) from Southeast Asia. (Photo by Animals Animals ©Dani/Jeske. Reproduced by permission.)

Significance to humans

In some parts of Asia, swamp eels and one species of spiny eel, *Mastacembelus erythrotaenia*, are valued as food and sometimes are kept in ponds or rice fields. Except for a few mastacembelids, they are rarely seen in home aquaria.

1. Fire eel (*Mastacembelus erythrotaenia*); 2. Swamp eel (*Monopterus albus*); 3. Marbled swamp eel (*Synbranchus marmoratus*); 4. Blind cave eel (*Ophisternon candidum*). (Illustration by John Megahan)

Species accounts

Fire eel
Mastacembelus erythrotaenia

FAMILY
Mastacembelidae

TAXONOMY
Mastacembelus argus Bleeker, 1850, Moluccan Archipelago.

OTHER COMMON NAMES
German: Feueraal; Vietnamese: Cá chachlua.

PHYSICAL CHARACTERISTICS
Grows to 39.4 in (100 cm). Soft-rayed portions of the median fins and pectoral fin have a sharply defined white distal margin. The basal portion of the dorsal, anal, and caudal fins is dark and that of the pectoral fin is dark or has broad vertical bars. Head and anterior part of the body have longitudinal red and black bands; the rest of the body has red spots or elongated marks on a black background.

DISTRIBUTION
In Asia, from Thailand and Cambodia to Indonesia.

HABITAT
A large lowland floodplain species occurring in slow-moving rivers and inundated plains. Also found in streams and lakes.

BEHAVIOR
Under aquarium conditions individuals tend to spend daylight hours in a preferred shelter spot.

FEEDING ECOLOGY AND DIET
Feeds on benthic insect larvae, worms, and some plant material. Under aquarium conditions, specimens larger than 12 in (30 cm) in total length become predatory, hunting and eating smaller fishes. Vulnerable to larger fish, water snakes, crocodilians, and fish-eating birds as well as fishermen.

REPRODUCTIVE BIOLOGY
Nothing is know of the reproductive biology of this species. Other mastacembelids are egg scatterers, depositing a few eggs at a time in fine-leafed aquatic plants.

CONSERVATION STATUS
Not listed by the IUCN. This species has become rare in recent years due to human consumption and overfishing.

SIGNIFICANCE TO HUMANS
Often seen in the aquarium trade. Bred in fish farms in Bangkok. ◆

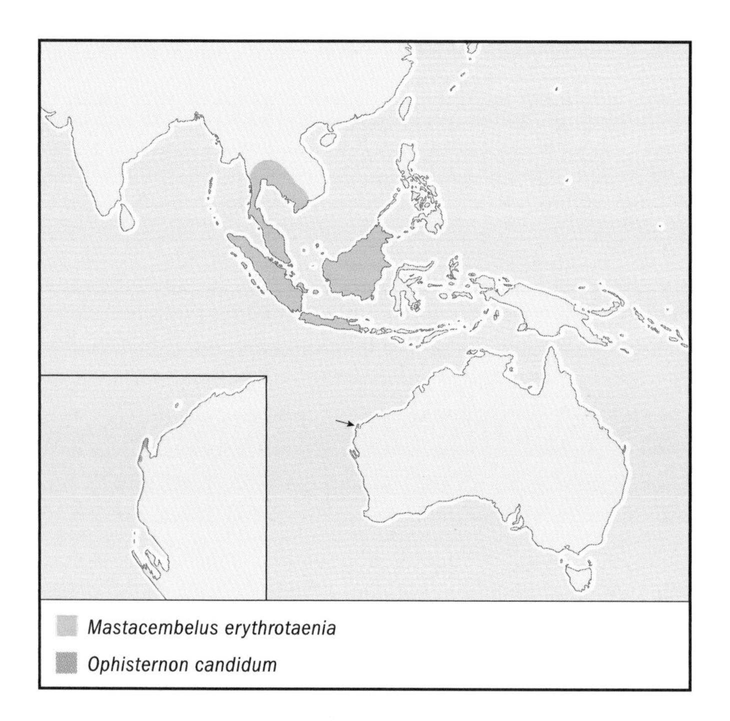

Swamp eel
Monopterus albus

FAMILY
Synbranchidae

TAXONOMY
Muraena alba Zouiev, 1793, type locality not specified.

OTHER COMMON NAMES
Cantonese: Wong sin; English: Rice (paddy field) eel; German: Ostasiatischer Kiemenschlitzaal; Japanese: Ta-unagi; Javanese: Welut; Khmer: Antong; Laotian: Pa lai; Malay: Belut; Thai: Pla lai; Vietnamese: Con lu'o'n, luon.

PHYSICAL CHARACTERISTICS
Grows to 39.4 in (100 cm). Eel-like body. It lacks scales and pectoral and pelvic fins. The dorsal, caudal, and anal fins are confluent and reduced to a skin fold. The gill openings merge into single slit underneath the head.

DISTRIBUTION
In India, China, Japan, Malaysia, and Indonesia. Probably also occurs in Bangladesh, Myanmar, and Thailand. Introduced populations in Florida, Georgia, and Hawaii in the United States.

HABITAT
It is a generalist that can be found in medium to large rivers, flooded fields, muddy ponds, swamps, canals and rice paddies; burrow in moist earth in dry season surviving for long periods without water.

BEHAVIOR
It burrows in moist earth at the beginning of the dry season, where it remains for long periods of time.

FEEDING ECOLOGY AND DIET
Feeds on detritus, plants, and small animals. Vulnerable to crocodilians, otters, and fish-eating birds.

REPRODUCTIVE BIOLOGY
External fertilization. Builds a bubble nest at the surface of the water near the shoreline. It is not known whether care is afforded to the eggs and fry. Spawning takes place in shallow

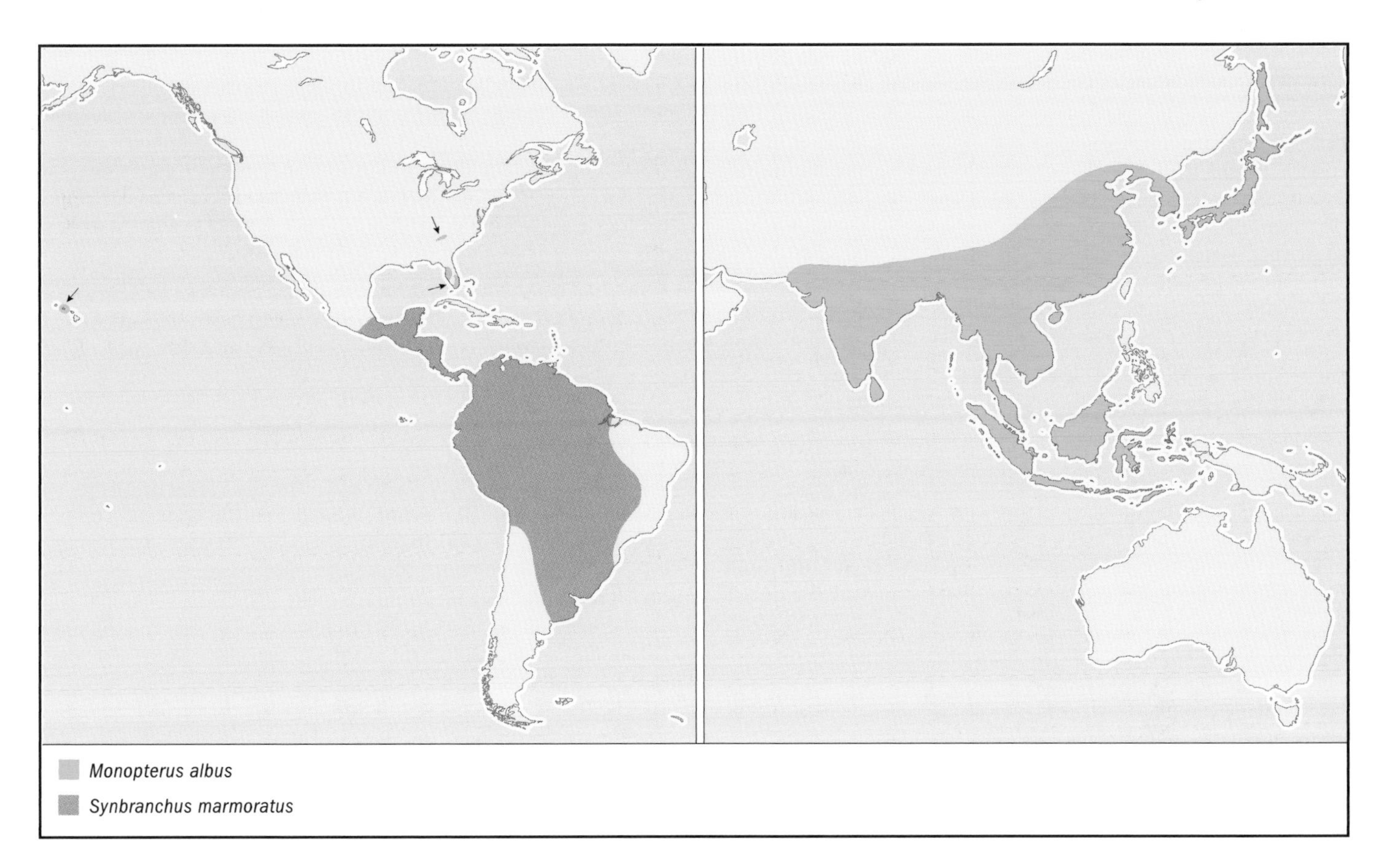

water. After spending part of their lives as females, some individuals undergo sex reversal, and change into males. Sex reversal is completed within eight to 30 weeks. All larger individuals are males.

CONSERVATION STATUS
Not listed by the IUCN.

SIGNIFICANCE TO HUMANS
Marketed fresh because of the good quality of its flesh. Stays alive for long periods of time as long as the skin is kept moist. Occasionally sold as an aquarium fish. ◆

Blind cave eel
Ophisternon candidum

FAMILY
Synbranchidae

TAXONOMY
Anommatophasma candidum Mees, 1962, Yardie Creek Station, North West Cape, Western Australia.

OTHER COMMON NAMES
Spanish: Anguila ciega.

PHYSICAL CHARACTERISTICS
Grows to 15.8 in (40 cm). These fishes have a very elongated, eel-like, and roundish body with no fins, except for a thin, rayless fin fold near and around the tip of the tail. The lateral line system is distinct and continues to near the tip of the tail. It is whitish in coloration, lacks externally visible eyes, and is scaleless.

DISTRIBUTION
This species used to be found in 11 locations (two now destroyed) in the western and northeastern coastal plain of the Cape Range Peninsula of Australia.

HABITAT
Inhabits wells, sinkholes, and caves and possibly also occurs in aquifers.

BEHAVIOR
No information is available.

FEEDING ECOLOGY AND DIET
This species feeds on invertebrates, both hypogean and epigean, that accidentally fall into its habitat. Due to its habitat, it is not subject to predation.

REPRODUCTIVE BIOLOGY
External fertilization. No additional information is available.

CONSERVATION STATUS
Listed as Data Deficient by the IUCN.

SIGNIFICANCE TO HUMANS
Of ecological and scientific interest only. ◆

Marbled swamp eel
Synbranchus marmoratus

FAMILY
Synbranchidae

TAXONOMY
Synbranchus marmoratus Bloch, 1795, Rio Negro, Brazil.

OTHER COMMON NAMES
English: Swamp eel; French: Anguille; German: Amerikanischer; Spanish: Anguila.

PHYSICAL CHARACTERISTICS
Grows to 50.1 in (150 cm). It has a long and cylindrical body, lacking pectoral and ventral fins and with vestigial dorsal and anal fins. The species has small eyes.

DISTRIBUTION
From Mexico to Central Argentina.

HABITAT
This species inhabits fresh and brackish waters in a variety of habitats, including streams, canals, drains, ponds, and rice fields. It can be seen in both clear and murky waters.

BEHAVIOR
This is a nocturnal fish usually found at the edge of the water. It can travel over land for considerable distances. It also burrows, especially during the dry season. During that time its metabolism is reduced considerably, but it still may flee if disturbed. After the first rains, it returns to larger bodies of water.

FEEDING ECOLOGY AND DIET
They feed on fish and invertebrates and are considered aggressive predators. They are vulnerable to crocodilians and fish-eating birds.

REPRODUCTIVE BIOLOGY
This is a species characterized by two unusual reproductive methods. In the first, many individuals undergo sequential hermaphroditism, where some fish function first as females and then as males, called terminal males. This condition is called protogy, and species that have such individuals are called protogynous. Those individuals that are males from the beginning are called primary males. Some individuals remain juvenile females, a condition termed diandric. They have external fertilization and show some level of genetic variability, which is consistent with the fact that this is a generalist species with a broad distribution in the New World. No specific seasonal reproductive data or parental care information has been published.

CONSERVATION STATUS
Because of its broad distribution, it is not considered threatened. Owing to its burrowing behavior, however, it may be missed in many faunal surveys.

SIGNIFICANCE TO HUMANS
Is not infrequent in public aquaria although it does not make a good exhibit because of its burrowing behavior. Because of its size it is difficult to keep in home aquaria. ◆

Resources

Books

Baensch, Hans A., and Rüdiger Riehl. *Aquarien Atlas.* Vol. 2. Melle, West Germany: Verlag für Natur- und Heimtierkunde, 1985.

Chan, S. T. H., F. Tang, and B. Lofts. "The Role of Sex Steroids on Natural Sex Reversal in *Monopterus albus.*" In *Proceedings of the International Congress of Endocrinology,* edited by Robert O. Scow. New York: American Elsevier Publishing Co., 1973.

Rainboth, Walter J. "Fishes of the Cambodian Mekong." *FAO Species Identification Field Guide for Fishery Purposes.* Rome: FAO, 1996.

Romero, Aldemaro, ed. *The Biology of Hypogean Fishes.* Dordrecht: Kluwer, 2001.

Periodicals

Humphreys, W. F. "The Distribution of Australian Cave Fishes." *Records of the Western Australian Museum* 19 (1999): 469–472.

Humphreys, W. F., and M. N. Feinberg. "Food of the Blind Cave Fishes of Northwestern Australia." *Records of the Western Australian Museum* 17 (1995): 29–33.

Johnson, G. D., and C. Patterson. "Percomorph Phylogeny: A Survey of Acanthomorphs and a New Proposal." *Bulletin of Marine Science* 52, no. 1 (1993): 554–626.

Kerle, R., R. Britz, P. K. L. Ng. "Habitat Preference, Reproduction and Diet of the Earthworm Eel, *Chendol keelini* (Teleostei: Chaudhuriidae), a Rare Freshwater Fish from Sundaic Southeast Asia." *Environmental Biology of Fishes* 57, no. 4 (2000): 413–422.

LoNostro F. L., and G. A. Guerrero. "Presence of Primary and Secondary Males in a Population of the Protogynous *Synbranchus marmoratus* Bloch, 1795, a Protogynous Fish (Teleost, Synbranchiformes)." *Journal of Fish Biology* 49 (1996): 788–800.

Roberts, T. R. "Systematic Review of the Mastacembelidae or Spiny Eels of Burma and Thailand, with Description of Two New Species of *Macrognathus.*" *Japanese Journal of Ichthyology* 33 (1986): 95–109.

Romero, A., and K. M. Paulson. "It's a Wonderful Hypogean Life: A Guide to the Troglomorphic Fishes of the World." *Environmental Biology of Fishes* 62 (2001): 13–41.

Romero, A., and P. B. S. Vanselow. "Threatened Fishes of the World: *Ophisternon candidum* (Mees, 1962) (Synbranchidae)." *Environmental Biology of Fishes* 58, no. 2 (2000): 214.

Sadovy, Y., and D. Y. Shapiro. "Criteria for the Diagnosis of Hermaphroditism in Fishes." *Copeia* 1987, no. 1 (1987): 136–156.

Sanchez, S., and A. Fenocchio. "Karyotypic Analysis in Three Populations of the South-American Eel Like Fish *Synbranchus marmoratus.*" *Caryologia* 49, no. 1 (1996): 65–71.

Other

"*Ophisternon infernale* (Hubbs, 1938)." 4 Dec. 2002 (31 Jan. 2003). <http://www.tamug.tamu.edu/cavebiology/fauna/bonyfish/O_infernale.html>

Romero, Aldemaro. "Guide to Hypogean Fishes." (31 Jan. 2003). <http://www.macalester.edu/envirost/ARLab/HypogeanFishes/synbranchidae.htm>

Aldemaro Romero, PhD

Scorpaeniformes I

(Gurnards and flatheads)

Class Actinopterygii

Order Scorpaeniformes

Number of families 4

Photo: An Oriental helmet gurnard (*Dactyloptena orientalis*) sleeping on the lava sand of the Lembeh Straits of Sulawesi Island, Indonesia. (Photo by Fred McConnaughey/Photo Researchers, Inc. Reproduced by permission.)

Evolution and systematics

The two suborders that comprise this chapter, the flatheads (Platycephaloidei) and flying gurnards (Dactylopteroidei) are mail-cheeked fishes (order Scorpaeniformes). This old group (first recognized in 1829), composed of approximately 1,400 species, is united by the presence of the suborbital stay. The suborbital stay is a bony strut that connects the bones under the eye with the front of the gill cover. Some authors have suggested that the suborbital stay in the flying gurnards evolved independently from the remainder of the scorpaeniform fishes. These authors place the flying gurnards in their own order (Dactylopteriformes).

In 1758 the father of binomial nomenclature, Carl Linnaeus, was the first naturalist to describe species of both the flying gurnards and the flatheads. Since the time of Linnaeus, six additional flying gurnards and 86 additional platycephaloids have been described. The seven flying gurnards are now classified into two genera (*Dactylopterus* in the Atlantic and *Dactyloptena* in the Indo-Pacific). The 88 platycephaloids have been classified variously into three to seven families, depending on the researcher. For the purposes of this review, the Platycephaloidei are split into three families: the flatheads (Platycephalidae, 64 species), the ghost flatheads (Hoplichthyidae, 11 species), and the deepwater flatheads (Bembridae, 11 species). At this time, there is considerable debate about the placement and classification of both of these suborders, so the classification follows Joseph Nelson's 1994 book, *Fishes of the World.*

Traditionally, the flying gurnards have been allied with the sea robins (Triglidae and Peristediidae), because both groups share enlarged pectoral fins and free pectoral rays, which these fishes use like legs to "walk" on the seafloor. Recently, researchers have suggested that the suborbital stay in flying gurnards evolved independently of the strut found in other mail-cheeked fishes. These ichthyologists have suggested that flying gurnards are related to the seahorses and their relatives (Sygnathoidei) or to the tilefishes (Malacanthidae). Because of these different views, the interrelationships of the flying gurnards remain unclear, and they are treated here as scorpaeniforms, pending resolution of their placement.

The placement of the flatheads and their relatives is not clear either. Historically, the three flathead families have been united because of their elongate bodies and depressed or "flattened" heads; hence their common name. Recent work by Hisashi Imamura has suggested that the sea robins and their relatives (Triglidae and Peristediidae) may have evolved from a flathead relative, suggesting that the sea robins should be placed in the Platycephaloidei. Clearly, further work using both morphological and DNA sequence data will shed light on the interrelationships and intrarelationships of the flatheads and their allies.

At more than 50 million years old, the enigmatic *Pterygocephalus paradoxus* from the Monte Bolca formation in Italy may be the oldest known flying gurnard, but the fossil *Prevolitans faedoensis* from Eocene deposits in northern Italy is the oldest clearly identifiable flying gurnard. Flatheads first appear in the fossil record with an Eocene otolith (ear stone) record identified as *Platycephalus janeti* from France. Whole fossilized skeletal specimens of *Platycephalus* date to the early Miocene in Tasmania.

Physical characteristics

Flying gurnards are one of the most recognizable of all spiny-rayed fishes because of their large heads, greatly developed winglike pectoral fins, and free pectoral rays. Their bodies are covered with ctenoid scales. They have a very characteristic "helmet-like" skull with a strong preopercular spine that gives them their other common name (helmet gurnards). Dactylopterids have a short snout with a subterminal mouth. Their jaws are filled with small, conical teeth. Flying gurnards have two dorsal fins separated by a deep notch, which typically have a couple of free dorsal spines detached from and preceding the first dorsal fin. As was found in the flatheads, sculpins (Cottoidei), and some sea robins, the dactylopterids lack true anal spines.

Typically, flying gurnards lie on the seafloor with their pectoral wings folded against the body. When dactylopterids have their wings folded, they are well camouflaged on the seafloor because of their dusky red and drab white, brown, and black markings. When startled, however, flying gurnards quickly spread their brightly colored pectoral fins (covered with bluish and whitish spots, for example) to distract would-be predators and make their quick escape. Flying gurnards can reach lengths of up to 15.7 in (40 cm).

Flatheads are recognized easily by their elongate bodies with depressed, broad heads. They typically are covered with ctenoid scales, except for their lower flanks and ventral surfaces, which often are covered by cycloid scales. The ghost flatheads lack scales, but their dorsal surface is overlaid with spiny, bony plates. All flatheads and their allies have large eyes placed close together high on their heads. The eyes of platycephalids often are covered with highly ornamented eye flaps. All platycephaloid heads are overlaid with bony or spiny ridges. These ridges are particularly well developed in the ghost flatheads. Platycephaloid jaws are filled with numerous bands of small, conical teeth, but *Ratabulus megacephalus* has strong canine teeth. Flatheads have two dorsal fins, the first with true spines. Bembrids and hoplichthyids usually have a single anal spine, but true anal spines are lacking in platycephalids. Most flatheads are dark dorsally and pale ventrally. The dark colors are various shades of brown, black, or gray, but the more colorful taxa can be red, purple, and green. The deepwater flatheads tend to be red, orange, and light brown, and the ghost flatheads are typically yellow, pink, or brown. Ghost flatheads and bembrids usually are smaller (11.8–15.7 in, or 30–40 cm) than the larger platycephalids (up to 27.6 in, or 70 cm).

Distribution

The flying gurnards are distributed in tropical and temperate waters in the Indo-Pacific and Atlantic Oceans. They usually are found in nearshore environments, whereas larval and juvenile dactylopterids are found in open ocean (pelagic) environments, where they can be collected at night using attracting lights and handheld dipnets.

The flatheads and relatives are distributed in the tropical and temperate waters of the Indo-Pacific, Mediterranean, and eastern Atlantic. Platycephalids are found on the continental shelf at depths to 984 ft (300 m), but most are found at less than 330 ft (100 m). Bembrids also live on the continental shelf, but they typically are taken at deeper depths (hence their common name, deepwater flatheads), down to 1,900 ft (581 m). Although there are significantly fewer ghost flatheads, they have a much wider depth distribution, ranging from 200 to 4,900 ft (60–1,500 m).

Habitat

Both the flying gurnards and flatheads typically are found on the continental shelf in mud or sandy bottoms. There is also a second group of flathead species commonly found in rocky shore habitats and coral reefs.

Behavior

Although the common name of the dactylopterids suggests that these fishes can fly, they cannot; only the true flying fishes (Exocoetidae) are capable of flight. It is for this reason that many authors have suggested that the common name be changed to helmet gurnards. It is likely that the giant pectoral fins are spread quickly, to scare "would-be" predators, to communicate with conspecifics, or for controlled gliding over the seafloor for prey. Dactylopterids can produce sound by stridulating the hyomandibular bone. Flatheads and their allies are typical "lie and wait" predators that spend most of their time buried completely or partially in the sand or mud.

Feeding ecology and diet

"Walking" quickly over the seafloor by moving their pelvic fins and their short pectoral fin rays alternately, flying gurnards sift through the sandy bottom in search of the next crustacean or fish, which they stir up and capture.

Lying on the seafloor buried partially beneath the sand, flatheads wait to ambush the next crab, fish, or shrimp that swims by. Like chameleons and flounders, flatheads have developed the ability to mimic the color pattern of the substrate on which they are lying. Platycephaloids are modified further for this "lie and wait" feeding strategy by specializations of the respiratory and pelvic structures that allow them to breathe while they are buried and to accommodate larger prey items. These numerous modifications make flatheads one of the premier ambush predators in the Indo-West Pacific.

Flatheads and gurnards are preyed upon by mammals (e.g. humans) and numerous species of larger, predatory fishes.

Reproductive biology

Very little is known about the reproductive biology of the flatheads and flying gurnards. All families produce pelagic, nonadhesive eggs. Japanese researchers have shown that some flatheads (e.g., *Suggrundus meerdervoortii* and *Cociella crocodila*) begin life as males and undergo sex reversal to become females as they grow older.

Conservation status

At the present time, no platycephaloid or dactylopterid species are included on the IUCN's Red List, although many species are quite rare.

Significance to humans

Of the four families, only the platycephalids are a commercially important food source for humans. Flying gurnards occasionally are fished for human consumption, but the fisheries are primarily for personal use. Bembrids and hoplichthyids do not have commercial fisheries, although the flesh of the ghost flatheads is supposed to be of good quality.

Although none of these fishes is as common in the aquarium trade as their lionfish or scorpionfish relatives, both dactylopterids (e.g., *Dactyloptena orientalis* and *Dactylopterus volitans*) and the crocodilefish (*Cymbacephalus beauforti*) occasionally are available in the aquarium trade.

1. Indian flathead (*Platycephalus indicus*); 2. Oriental helmet gurnard (*Dactyloptena orientalis*). (Illustration by Barbara Duperron)

Species accounts

Oriental helmet gurnard

Dactyloptena orientalis

FAMILY

Dactylopteridae

TAXONOMY

Dactyloptena orientalis Cuvier, 1829, Red Sea.

OTHER COMMON NAMES

French: Grondin volant oriental; Spanish: Alón oriental; Japanese: Semihôbô.

PHYSICAL CHARACTERISTICS

Grows to 15.7 in (40 cm) maximum length. Body moderately elongate with large, heavily armored head and greatly expanded winglike pectoral fins. Head broad and blunt, with large eyes and a small subterminal mouth. A unique fish that is gray to light brown, with dark brown and black spots on its back and upper sides. The elongate first dorsal spine is well separated from the second dorsal spine and the remainder of the first dorsal fin. The enlarged pectoral fins also are spotted, with striking blue wavy lines near the margins of the fins.

DISTRIBUTION

This widely distributed species ranges from the western Indian Ocean and Red Sea east to the Polynesian and Hawaiian Islands.

HABITAT

A benthic species that spends most of its time on sandy bottoms.

BEHAVIOR

A bottom-dwelling fish that quickly expands its pectoral fins as a defensive behavior.

FEEDING ECOLOGY AND DIET

Feeds primarily on benthic crustaceans, clams, and fishes that it stirs up as it "walks" along the seafloor.

REPRODUCTIVE BIOLOGY

Little is known about the reproductive biology of the Oriental flying gurnards in the wild, although one aquarium wholesaler has started offering aquacultured flying gurnards to aquarists.

CONSERVATION STATUS

Not listed by the IUCN.

SIGNIFICANCE TO HUMANS

A specialty item in the aquarium trade and incidentally fished in the Indo-Pacific, but not commercially collected for consumption. ◆

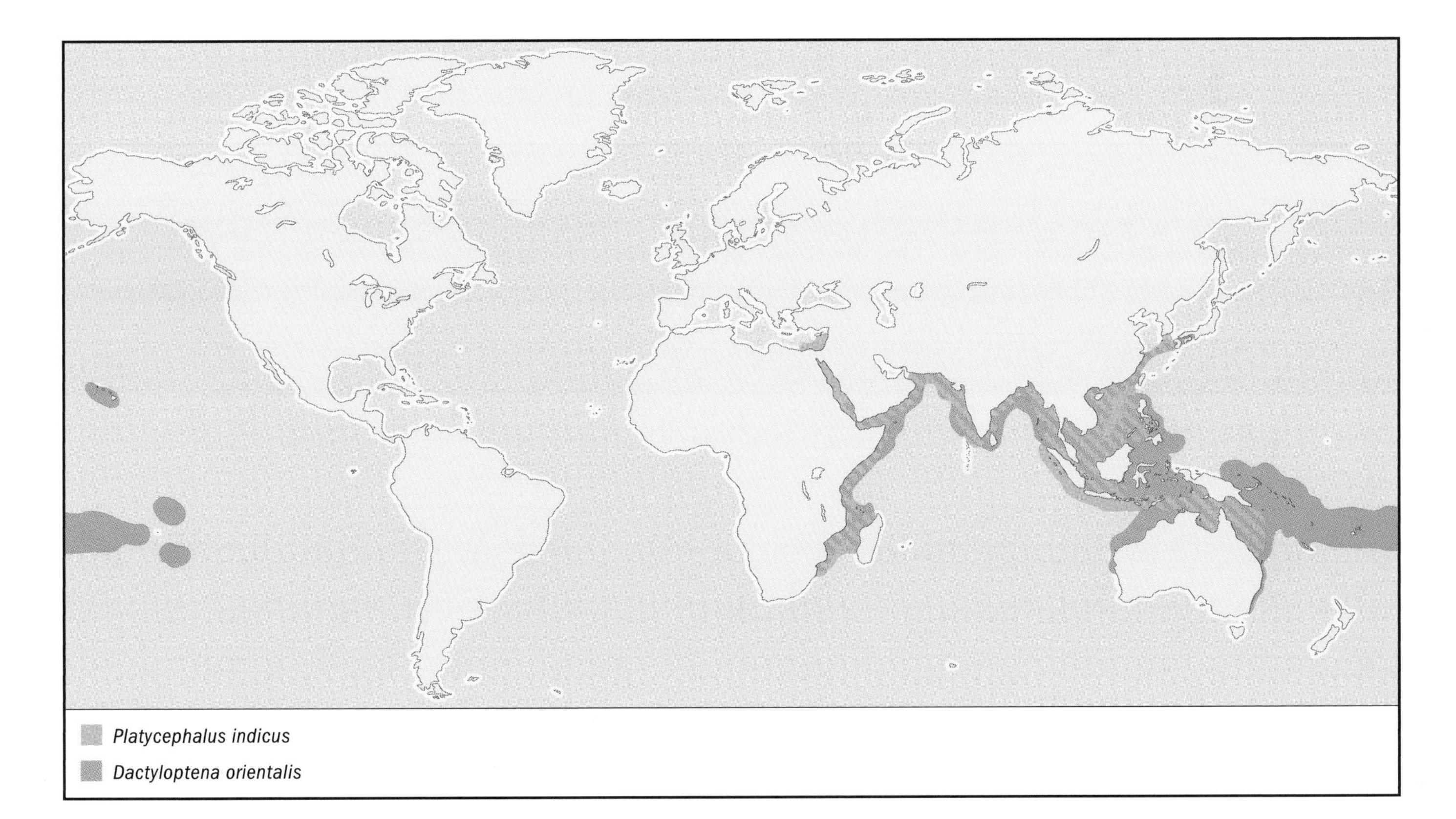

Indian flathead

Platycephalus indicus

FAMILY
Platycephalidae

TAXONOMY
Platycephalus indicus Linneaus, 1758, Asia.

OTHER COMMON NAMES
English: Bartail flathead; French: Platycéphale des Indes; Spanish: Chato índico; Japanese: Kochi.

PHYSICAL CHARACTERISTICS
Grows to 39.4 in (100 cm) in maximum length. Characteristic flattened head with various spinous ridges. Vomerine and palatine teeth are canine-like, with broad bands of villiform teeth in jaws. The two dorsal fins are well separated, with an isolated short spine before the first dorsal fin. Body covered with small ctenoid scales. Brownish coloration with eight or nine obscure dusky bands over the back. All fins, except the caudal fin, are covered with dusky spots on the rays. Caudal fin has a median longitudinal black band below and two oblique bands above.

DISTRIBUTION
Known from Israel and Egypt (entered the Mediterranean through the Suez Canal), eastern Africa, India, Indonesia, China, southern Japan, the Philippines, and Western Australia.

HABITAT
Typically found on rocky and soft bottoms at depths from 33 to 656 ft (10 to 200 m).

BEHAVIOR
A bottom-dwelling fish that spends most of its time buried beneath the sand or among rocks waiting to ambush its next prey.

FEEDING ECOLOGY AND DIET
Feeds primarily on benthic crustaceans and small fishes.

REPRODUCTIVE BIOLOGY
Little is known about the reproductive biology of the Indian flathead.

CONSERVATION STATUS
Not listed by the IUCN.

SIGNIFICANCE TO HUMANS
Occasionally harvested, but not a fish of great commercial importance. ◆

Resources

Books

Benton, M. J., ed. *The Fossil Record 2.* London: Chapman and Hall, 1993.

Breder, C. M., Jr., and D. E. Rosen. *Modes of Reproduction in Fishes.* Garden City, NY: Natural History Press, 1966.

Carpenter, K. E., and V. H. Niem, eds. *FAO Species Identification Guide for Fishery Purposes. The Living Marine Resources of the Western Central Pacific.* Vol. 4: *Bony Fishes.* Rome: FAO, 1999.

Nelson, J. S. *Fishes of the World.* 3rd edition. New York: John Wiley & Sons, 1994.

Paxton, J. R., and W. N. Eschmeyer, eds. *Encyclopedia of Fishes.* 2nd edition. San Diego: Academic Press, 1998.

Quéro, J. C., J. C. Hureau, C. Karrer, A. Post, and L. Saldanha, eds. *Check-list of the Fishes of the Eastern Tropical Atlantic.* Paris: UNESCO, 1990.

Randall, J. E., G. R. Allen, and R. C. Steene. *Fishes of the Great Barrier Reef and Coral Sea.* Honolulu, HI: University of Hawaii Press, 1996.

Periodicals

Corbett, K. D. "An Early Miocene Flathead (Pisces: Platycephalidae) from Wynyard, Tasmania." *Papers and Proceedings of the Royal Society of Tasmania* 114 (1980): 165–175.

Douglas, W. A., and W. J. R. Lanzing. "Color Change and Visual Cues in the Sand Flathead, *Platycephalus arenarius* (Ramsay and Ogilby)." *Journal of Fish Biology* 18 (1981): 619–628.

———. "The Respiratory Mechanisms of the Dusky Flathead, *Platycephalus fuscus* (Platycephalidae, Scorpaeniformes)." *Journal of Fish Biology* 18 (1981): 545–552.

Imamura, Hisashi. "Phylogeny of the Family Platycephalidae and Related Taxa (Pisces: Scorpaeniformes)." *Species Diversity* 1, no. 2 (1996): 123–233.

———. "An Alternate Hypothesis on the Phylogenetic Position of the Family Dactylopteridae (Pisces: Teleostei), with a Proposed New Classification." *Ichthyological Research* 47, no. 3 (2000): 203–222.

Imamura, Hisashi, and Leslie W. Knapp. "Review of the Genus *Bembras* Cuvier, 1829 (Scorpaeniformes: Bembridae) with Description of Three New Species Collected from Australia and Indonesia." *Ichthyological Research* 45, no. 2 (1998): 165–178.

William Leo Smith, MS

Scorpaeniformes II

(*Scorpionfishes and relatives*)

Class Actinopterygii

Order Scorpaeniformes

Number of families 14

Photo: A red lionfish (*Pterois volitans*) in the Coral Sea. The lionfish can give vemonous stings from its pelvic, dorsal, and anal spines. (Photo by JLM Visuals. Reproduced by permission.)

Evolution and systematics

The approximately 1,400 species of mail-cheeked fishes (order Scorpaeniformes), first grouped together by the naturalist Georges Cuvier, are united because they share a single remarkable feature. This feature, the suborbital stay, is a bony strut that connects the bones under the eye with the front of the gill cover. This character is found in all scorpaeniforms except the Australian prowfishes (Pataecidae). The gapers (Champsodontidae) are sometimes included in the Scorpaeniformes, but here are included in the Trachinoidei chapter.

This chapter covers the suborder Scorpaenoidei. Because scorpaenoids are the dominant group of venomous fishes, they have one of the oldest and best-documented natural histories (especially in terms of venomology), dating back almost 2,400 years. Even with this detailed historical record, the classification and taxonomy of the Scorpaenoidei remain some of the most difficult of all fish orders. The composition of the group presented here follows Joseph Nelson's 1994 book, *Fishes of the World*. The classification of the scorpionfishes and sea robins presented here is updated from Nelson's book and follows more recent studies.

The Scorpaenoidei is divided into two groups: the sea robins (composed of either one or two families) and the scorpionfishes and their relatives (composed of seven to thirteen families). Following Hisashi Imamura's 1996 analysis, there are two families of sea robins: the sea robins (Triglidae, about 110 species) and the armored sea robins (Peristediidae, about 40 species). Following Minoru Ishida's 1994 analysis and recent work by Randall Mooi and David Johnson, there are 12 families of scorpionfishes: longfinned waspfishes (Apistidae, three species), velvetfishes (Aploactinidae, about 40 species), orbicular velvetfishes (Caracanthidae, four species), pigfishes (Congiopodidae, nine species), red velvetfish (Gnathanacanthidae, one species), gurnard perches (Neosebastidae, 12 species), prowfishes (Pataecidae, three species), scorpionfishes (Scorpaenidae, some 200 species), rockfishes (Sebastidae, roughly 125 species), midwater scorpionfishes (Setarchidae, five species), stonefishes (Synanceiidae, about 35 species), and waspfishes (Tetrarogidae, about 40 species).

Traditionally, the scorpionfishes and their relatives have been grouped with the sea robins. Recent work suggests that the sea robins might be related more closely to the flatheads (Platycephaloidei); this hypothesis needs further testing. Generally, it is agreed upon that the Platycephaloidei and the Scorpaenoidei make up the "scorpaenoid lineage." All species in this lineage are united by the presence of an opercular spine (the largest bone that makes up the gill cover) that projects beyond the subopercle (the posterior margin of the gill cover) and a derived gas bladder muscle configuration. The Scorpaenoidei usually is treated as the most primitive group of scorpaeniforms, with the rockfishes (Sebastidae) representing the basic body form. Some researchers have argued that the rockfishes have numerous derived features, including live birth and modified gas bladder muscles, making them a poor choice for a primitive form. Because of these disagreements and the countless classifications that exist, a worldwide revision of the group is needed. In particular, an analysis of morphological and DNA sequence data is needed to resolve the remaining questions in scorpaenoid phylogeny and classification.

Regardless of the classification used, there are no known characters that unite all members of the Scorpaenoidei with or without the sea robins and relatives (Triglidae and Peristediidae). The scorpionfish and relatives clade can be broken into two groups. The first group is composed of the taxa that have the first and/or second dorsal spines and supports

A northern sea robin (*Prionotus carolinus*) "walking" on its fins in the Gulf of Maine. (Photo by Andrew J. Martinez/Photo Researchers, Inc. Reproduced by permission.)

articulating with the skull; this includes the families Aploactinidae, Congiopodidae, Gnathanacanthidae, Pataecidae, Synanceiidae, and Tetrarogidae. The remaining seven families have the traditional arrangement where the dorsal fin begins beyond the posterior margin of the skull. Recent work has suggested that the Australian prowfishes (Pataecidae) might not be related to the other scorpionfishes.

The scorpionfishes and their relatives first appear in the fossil record with an Eocene otolith (ear stone) record identified as *Scorpaenoideorum prominens* from the London Clay Formation in southern England. Another scorpaenoid fossil from the Miocene is the earliest scorpaenoid known from skeletal remains. This specimen is a fossil stonefish, *Eosynanceja brabantica*, from Belgium, which is known from a handful of bones, including portions of the jaw, cheek, and vertebral column.

Physical characteristics

Most sea robins and armored sea robins are medium-sized fishes, up to 15.7 in (40 cm). Their most conspicuous characters are their greatly expanded pectoral fins and a head that is completely encased in bony plates. The armored sea robins, as their name suggests, take this protection one step further. Their entire body is covered with spine-bearing plates. Like

A spotted scorpionfish (*Scorpaena plumieri*) is well camouflaged in the coral of the Caribbean. (Photo by Animals Animals ©Franklin J. Viola. Reproduced by permission.)

A Pacific spotted scorpionfish (*Scorpaena mystes*) mimicking the colorful sponges near Darwin Island, Galápagos Islands. (Photo by Fred McConnaughey/Photo Researchers, Inc. Reproduced by permission.)

the flying gurnards (Dactylopteridae) and some stonefishes (Synanceiidae), both families have detached, fingerlike pectoral rays that they use as tactile and chemoreceptive organs. These elongate, free rays also are used for "walking" along the seafloor in search of prey.

The scorpionfishes and their relatives have a wide range of sizes, from the smallest velvetfishes at 0.8–1.2 in (2–3 cm) to the largest rockfishes at about 39.4 in (about 100 cm). Their diversity of sizes is matched by their diversity in body forms, with only numerous sharp spines being common to all species, although most species could be described as "bass-like." Scorpaenoids fall into one of two categories. The first group is composed of the brightly colored, highly venomous species (e.g., the lionfishes) that hover around coral reefs displaying their warning coloration. The second category is made up of the cryptic species. These species tend to be dominated by colors that mimic their surroundings. Additionally, these species are covered with numerous cirri, fleshy appendages, spines, and ridges; these appendages provide additional camouflage.

Distribution

The Scorpaenoidei has a worldwide distribution in tropical and temperate marine waters. Most scorpaenoids (e.g., Scorpaenidae, Synanceiidae, Aploactinidae, and Tetrarogidae) are found primarily in the tropical Indo-Pacific. One tetrarogid, the bullrout (*Notesthes robusta*) from eastern Australia, is the only freshwater scorpaenoid. The other center of diversity for the scorpaenoids is the northern Pacific Ocean, which is dominated by the more than 80 species of rockfishes (Sebastidae).

Habitat

Almost all scorpaenoids are benthic predators that are found in rocky, sea grass, or coral habitats. Basically, scorpionfishes can be found in any topographically complex environment, where they can use their cryptic coloration and appendages to hide. Typically, sea robins, which are not as cryptic, are more common on sandy, rocky, or muddy habitats. Finally, the deep-water or midwater setarchids and armored gurnards are benthic species that are found at depths up to 6,562 ft (2,000 m), although most species are found at less than 2,297 ft (700 m).

Behavior

There are few generalizations that can be made about the behavior of scorpaenoid fishes. Most scorpaenoids are territorial and lead solitary lives, except for the formation of mating aggregations. Scorpaenoids are masters of disguise. Many species have cryptic coloration; numerous leaflike appendages, or cirri; epibiotic growth; and bony ridges that give the appearance of rocks. Other scorpaenoid species have specialized pectoral fins that allow them to bury themselves in the sand. All of these features help scorpaenoids blend into their environment as they lie waiting for prey.

In addition to their camouflage and mimicry, scorpaenoids are protected in their environment by their pungent, venomous dorsal spines. Almost all scorpaenoids, except such groups as prowfishes (Pataecidae) and sea robins and their allies (Triglidae and Peristediidae), have venom glands associated with the fin spines. The venom from these glands has both neurotoxic (affecting the nervous system) and hemotoxic (affecting the blood vessels) action, which has led to numerous human fatalities and given these species their infamous reputation. The venom of the colder-water species (e.g., the rockfishes, Sebastidae) generally has less severe effects, and that of the stonefishes (Synanceiidae) and lionfishes (*Pterois*) is most deadly. The stonefishes, in particular, have devised a particularly dastardly venom apparatus that has a hollowed-out dorsal spine with muscular control, which basically gives them a hypodermic needle to inject their deadly venom. Fortunately, we have learned that the effects of the venom are minimized if the affected area is soaked in very hot (not boiling) water to help denature the proteins; additionally, topical treatment with stonefish antivenoms limits the damage.

A tub gurnard (*Chelidonichthys lucerna*) on top of a European plaice (*Pleuronectes platessa*). (Photo by Tom McHugh/Photo Researchers, Inc. Reproduced by permission.)

Feeding ecology and diet

Almost all scorpaenoids are benthic predators that feed primarily on crustaceans and smaller fishes. Most species are typical lie-and-wait predators that use their highly camouflaged bodies and burst speed to capture prey. The sea robins (Triglidae) have excised pectoral rays that are chemoreceptive. These fishes run these fingerlike rays through the sand or mud to locate prey as they walk along the seafloor. Adult scorpaenoid fishes are rarely preyed upon because of their powerful venom, but venomous juveniles and less venomous species (e.g., rockfishes) are preyed upon by larger fishes, humans, and pinnipeds (e.g., California sea lion).

Reproductive biology

There are numerous reproductive modes in the Scorpaenoidei. All species are iteroparous (having more than one spawning event per lifetime) and have many sexual partners (at least the males). One of the most interesting scorpaenoid reproductive strategies is live birth, which is found in some

A red lionfish (*Pterois volitans*) eating a zebra turkeyfish (*Dendrochirus zebra*). (Photo by Tom McHugh/Steinhart Aquarium/Photo Researchers, Inc. Reproduced by permission.)

The face of a tub gurnard (*Chelidonichthys lucerna*) near Brittany, France, in the Atlantic Ocean. (Photo by Jeff Rtman/Photo Researchers, Inc. Reproduced by permission.)

rockfishes (Sebastidae). The live-born *Sebastes* embryos are interesting, because they derive nutrients both from the yolk sac and directly from the mother. A possibly related reproductive strategy is used by many scorpaenoids (e.g., *Pterois* and *Sebastolobus*). In this strategy, the male inseminates the female, and then the female extrudes the fertilized eggs in a gelatinous mass that floats at the surface. There are many scorpaenoid species that are typical broadcast spawners with planktonic larvae, for example, sea robins (Triglidae) and pigfishes (Congiopodidae). Last, there are two scorpaenoid families whose reproductive biology is unknown—orbicular velvetfishes (Caracanthidae) and velvetfishes (Aploactinidae).

Conservation status

At the present time, there are four scorpaenoids listed by the IUCN. First is the deepwater jack, *Pontinus nigropunctatus*, from Saint Helena, which is considered Vulnerable. The redfish, *Sebastes fasciatus*, from the northwest Atlantic and *Sebastolobus alascanus* from the northeast Pacific, are considered Endangered. Finally, the bocaccio, *Sebastes paucispinis*, from the northeast Pacific is considered Critically Endangered. These species, except for the deepwater jack, have been reduced to critical levels because of the pressures of overfishing. A recent report suggests that another eleven species of rockfishes (Sebastidae) also should be given protected status in North America because of continually declining numbers due to overfishing.

Significance to humans

Despite their venomous nature, many scorpaenoids support important commercial and recreational fisheries worldwide. The rockfish fishery in the northern Pacific and Atlantic is one of the best known. Unfortunately, many *Sebastes* species have been overfished. This exploitation is due to numerous factors, including life history traits, oceanography, and the difficulty in identifying larval species, which interferes with accurate population management. Another interesting scorpaenoid fishery is the commercial harvesting of the highly venomous stonefish, *Synanceia verrucosa*, for live fish markets in Hong Kong. This fishery has been so successful that there is serious discussion about aquaculturing the highly venomous stonefish species.

The lionfishes (*Pterois* and *Dendrochirus*) make up one of the dominant groups of fishes in the marine aquarium trade. These fishes are not bred in captivity; all are collected from the wild. In addition to the large number of lionfishes that are collected annually, many other scorpaenoids are collected occasionally for the aquarium trade (e.g., the weedy scorpionfish (*Rhinopias*), the sea robin (*Prionotus*), and the bearded ghoul (*Inimicus*).

1. Bearded ghoul (*Inimicus didactylus*); 2. Ocellated waspfish (*Apistus carinatus*); 3. Crested scorpionfish (*Ptarmus jubatus*); 4. Striped sea robin (*Prionotus evolans*); 5. Deepwater scorpionfish (*Setarches guentheri*); 6. South American pigfish (*Congiopodus peruvianus*); 7. Reef stonefish (*Synanceia verrucosa*); 8. Cockatoo waspfish (*Ablabys taenianotus*). (Illustration by Jonathan Higgins)

1. Belalang (*Gargariscus prionocephalus*); 2. Red lionfish (*Pterois volitans*); 3. California scorpionfish (*Scorpaena guttata*); 4. Red indianfish (*Pataecus fronto*); 5. Merlet's scorpionfish (*Rhinopias aphanes*); 6. Bocaccio (*Sebastes paucispinis*); 7. Red gurnard (*Chelidonichthys spinosus*); 8. Spotted coral croucher (*Caracanthus maculatus*). (Illustration by Jonathan Higgins)

Species accounts

Ocellated waspfish
Apistus carinatus

FAMILY
Apistidae

TAXONOMY
Apistus carinatus Bloch and Schneider, 1801, Tranquebar, India.

OTHER COMMON NAMES
English: Bearded waspfish; French: Rascasse ocellée; Spanish: Rascacio ocelado; Japanese: Hachi.

PHYSICAL CHARACTERISTICS
Reaches 6.7 in (17 cm) maximum length. An orange, bronze, or gray scorpionfish, with greatly expanded pectoral fins and free pectoral rays, similar to those seen in sea robins and stingfishes. Becomes increasingly white ventrally, with a characteristic ocellated black spot on the posterior half of the spinous dorsal fin.

DISTRIBUTION
Widely distributed species found throughout the Indian Ocean and the western Pacific. They range from South Africa north to the Red Sea and Persian Gulf in the western Indian Ocean and from China and Japan south to Australia in the Pacific.

HABITAT
Typically collected on the continental shelf on muddy or sandy bottoms at depths ranging from 66–197 ft (20–60 m).

BEHAVIOR
The ocellated waspfish, like sea robins and stingfishes, moves slowly over the seafloor, searching through the soft bottom with its free pectoral rays. It is highly venomous.

FEEDING ECOLOGY AND DIET
Like most scorpionfishes, the ocellated waspfish feeds primarily on crustaceans and fishes; it is preyed upon by humans.

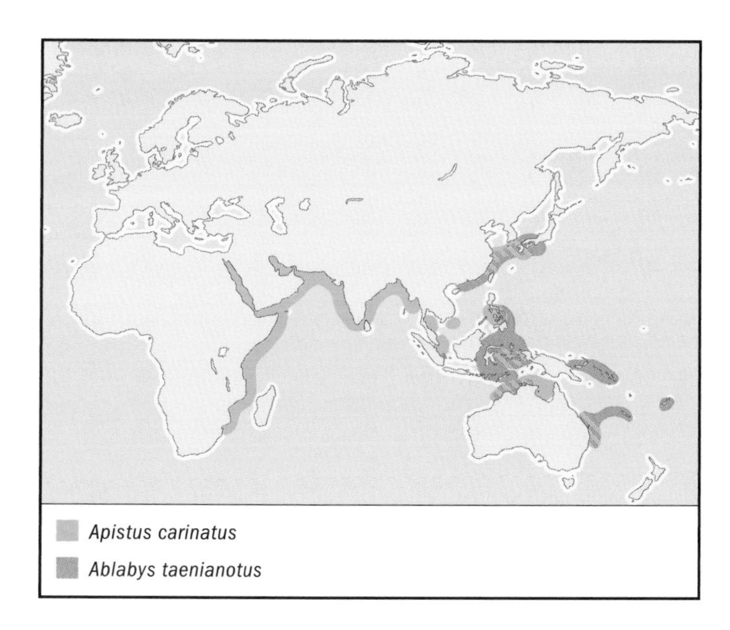

REPRODUCTIVE BIOLOGY
Little is known about the reproductive biology of the ocellated waspfish.

CONSERVATION STATUS
Not listed by the IUCN.

SIGNIFICANCE TO HUMANS
The ocellated waspfish is a commercially important food fish that is collected by trawls and seines in small quantities. This fish is sold in markets either fresh or dried and salted. ◆

Crested scorpionfish
Ptarmus jubatus

FAMILY
Aploactinidae

TAXONOMY
Ptarmus jubatus Smith, 1935, Natal Coast, South Africa.

OTHER COMMON NAMES
None known.

PHYSICAL CHARACTERISTICS
Grows to 3.9 in (10 cm) maximum length. As is implied by the common name of their family, velvetfishes such as the crested scorpionfish have a derived "knobby" scale morphological feature that makes them feel velvety to the touch. The single, continuous dorsal fin originates over the eye. This species typically varies in coloring from light brown to black. It can be covered with white spots and has a black stripe through its eye.

DISTRIBUTION
Can be found from southern Mozambique down to the Natal coast of South Africa.

HABITAT
Usually found in relatively shallow water, particularly in weedy areas.

BEHAVIOR
Nothing is known.

FEEDING ECOLOGY AND DIET
The diet of these fishes is unknown, but it is likely that they feed primarily on crustaceans. May be eaten by larger fishes.

REPRODUCTIVE BIOLOGY
Nothing is known.

CONSERVATION STATUS
Not listed by the IUCN.

SIGNIFICANCE TO HUMANS
Not collected commercially and reported to be nonvenomous. ◆

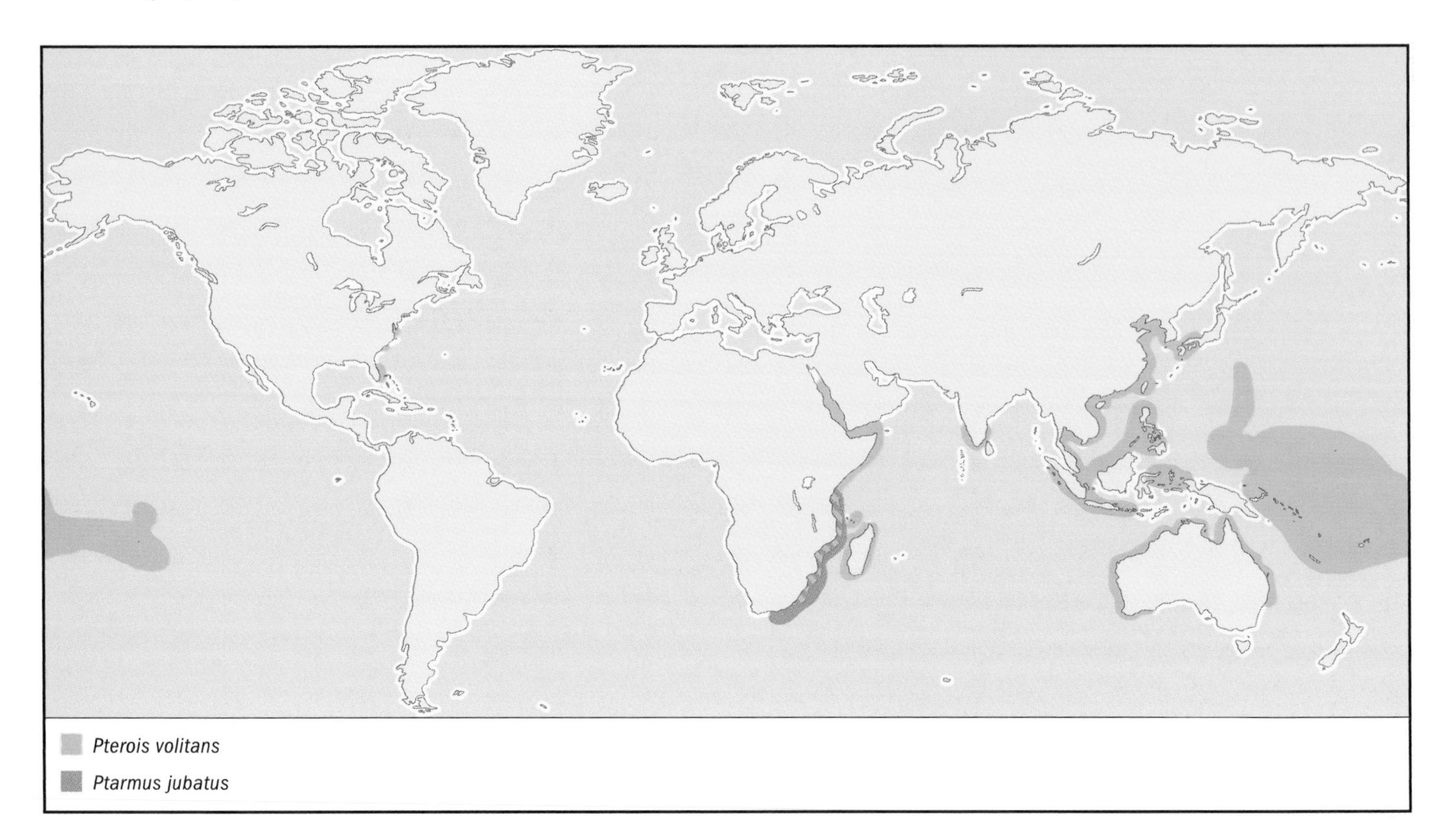

Spotted coral croucher
Caracanthus maculatus

FAMILY
Caracanthidae

TAXONOMY
Caracanthus maculatus Gray, 1831), Tuamotu Island, Polynesia.

OTHER COMMON NAMES
Japanese: Dango-okoze.

PHYSICAL CHARACTERISTICS
Reaches 1.6 in (4 cm) in maximum length. Small, rounded, laterally compressed fishes, with small pectoral fins and pelvic fins with one spine and three rays. Their bodies are covered with papillae, giving them a furry or velvet-like appearance. Typically gray laterally, with numerous red spots on the dorsal and lateral surfaces.

DISTRIBUTION
Can be found from southern Japan to southeastern Polynesia, Australia, and the East Indies.

HABITAT
These inconspicuous fishes can be found among the branches of *Acropora*, *Poecillopora*, and *Stylophora* corals.

BEHAVIOR
Coral crouchers spend most of their time among the branches of corals, rarely venturing away from the coral head. It is not known whether they are venomous.

FEEDING ECOLOGY AND DIET
Little is known about the feeding ecology and diet of these enigmatic scorpionfishes.

REPRODUCTIVE BIOLOGY
Little is known about the reproduction of the spotted coral croucher.

CONSERVATION STATUS
Not listed by the IUCN.

SIGNIFICANCE TO HUMANS
These fishes are too small to be commercially fished, but they are occasionally collected for the aquarium trade, where they are marketed as "gumdrops." ◆

South American pigfish
Congiopodus peruvianus

FAMILY
Congiopodidae

TAXONOMY
Congiopodus peruvianus Cuvier, 1829, San Lorenzo Island, Peru.

OTHER COMMON NAMES
Spanish: Pez chancho, chanchito; Japanese: Apachhi.

PHYSICAL CHARACTERISTICS
Grows to 11 in (28 cm) maximum length. Laterally compressed, with a single, sail-like dorsal fin. Snout has a single nostril on each side. The body is light brown with irregular dark blotches and spots, particularly on the dorsal fin. Additionally, there is a black band on the caudal peduncle and caudal fin.

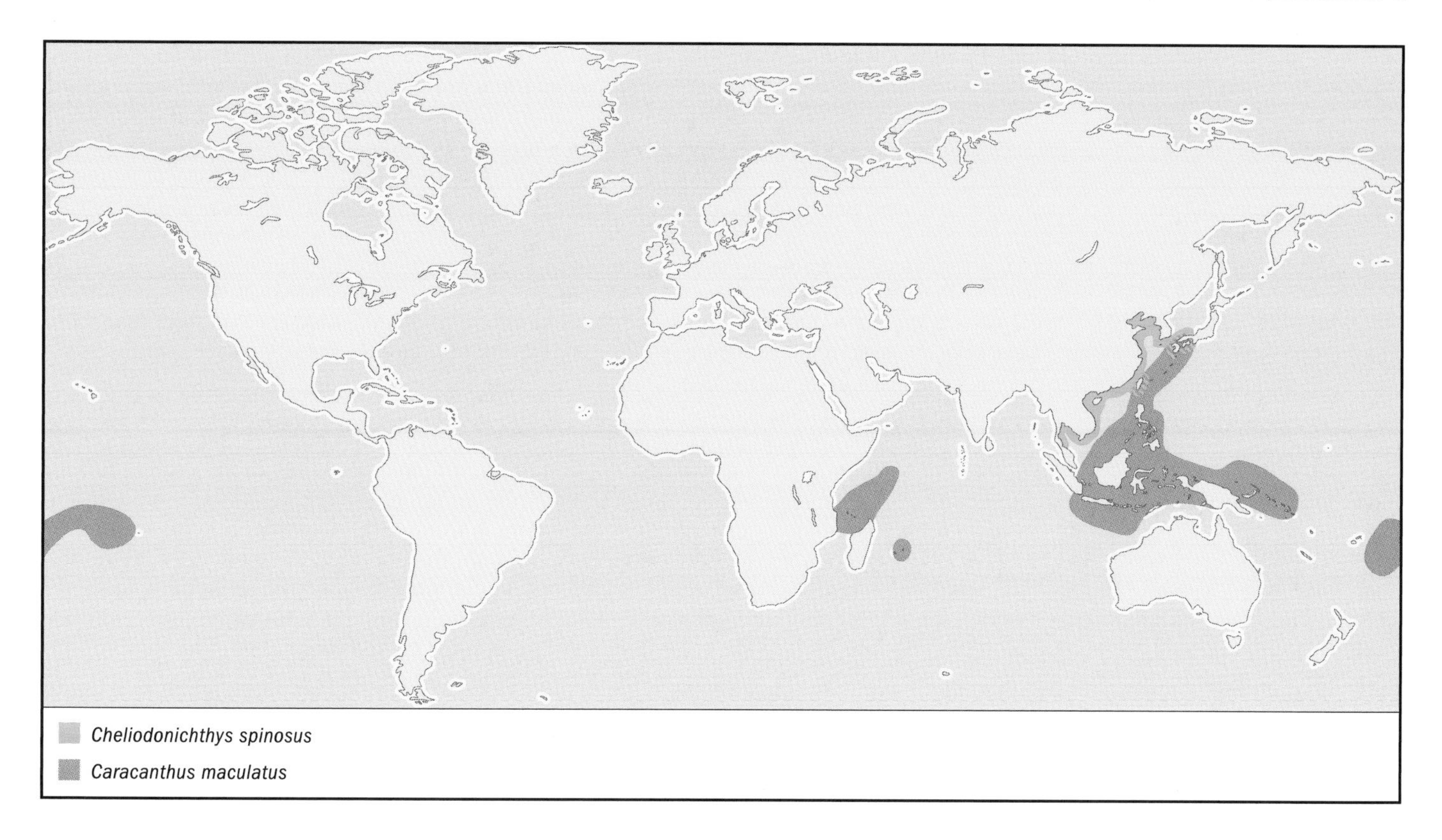

DISTRIBUTION
This pigfish can be collected off the coast of South America in both the southeastern Pacific and southwestern Atlantic.

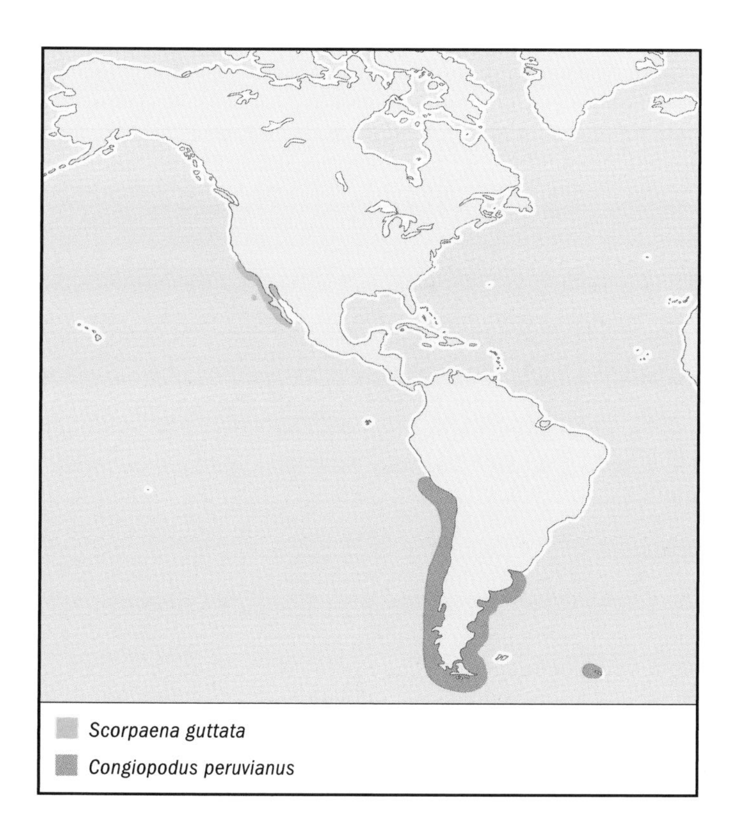

HABITAT
This is a demersal species that usually is taken at depths from 66–492 ft (20–150 m).

BEHAVIOR
As is seen in some other scorpionfishes, the molting or shedding of skin has been reported in congiopodids. These scorpionfishes are not venomous.

FEEDING ECOLOGY AND DIET
Very little is known about the diet of pigfishes. May be eaten by larger predatory fishes.

REPRODUCTIVE BIOLOGY
Nothing is known.

CONSERVATION STATUS
Not listed by the IUCN.

SIGNIFICANCE TO HUMANS
Although the flesh of this species is edible, it is rarely eaten. ◆

Red indianfish
Pataecus fronto

FAMILY
Pataecidae

TAXONOMY
Pataecus fronto Richardson, 1844, Southern Australia.

OTHER COMMON NAMES
None known.

PHYSICAL CHARACTERISTICS
Grows to 11.8 in (30 cm) maximum length. An unusual-looking fish that is highly compressed, with a high dorsal fin con-

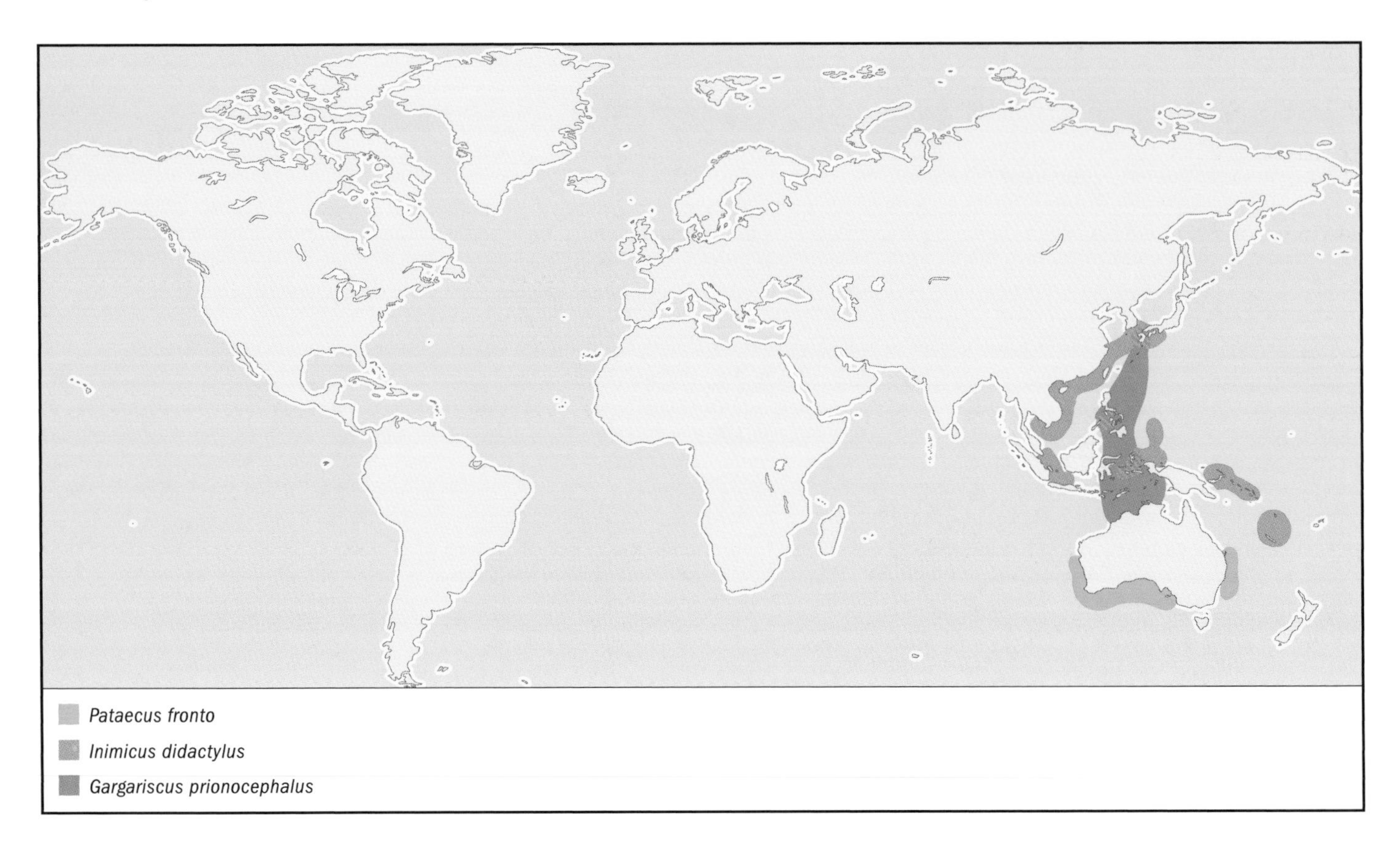

fluent with the caudal fin. The color varies but usually ranges from brownish orange to bright red. Their bodies often are covered by dark blotches dorsally.

DISTRIBUTION
Found in southern Australia from southern Queensland to eastern Victoria.

HABITAT
It is difficult to observe red indianfishes because of their excellent camouflage. When they are seen, they are found in rocky reefs and estuaries, often in similarly colored sponge beds.

BEHAVIOR
Red indianfishes often shed the outer layer of skin to help get rid of epibiotic growth (algae or bryozoans), which acts as camouflage. Additionally, these fishes have an unusual "swimming" style that mimics a dead leaf floating in the water; they basically twist and spin as the fall back to the sea floor. The red indianfish is not venomous.

FEEDING ECOLOGY AND DIET
Little is known about the diet of red indianfishes, but their diet probably consists primarily of shrimps and other crustaceans. May be eatern by larger predatory fishes.

REPRODUCTIVE BIOLOGY
Little is known about the reproductive biology of these fishes.

CONSERVATION STATUS
Not listed by the IUCN.

SIGNIFICANCE TO HUMANS
This species is not commercially fished, but they are occasionally captured in commercial trawl nets. ◆

Belalang
Gargariscus prionocephalus

FAMILY
Peristediidae

TAXONOMY
Gargariscus prionocephalus Duméril, 1869, Sulawesi, Indonesia.

OTHER COMMON NAMES
Japanese: Onikihôbô.

PHYSICAL CHARACTERISTICS
Reaches 11 in (28 cm) maximum length. Coloring is orange to red, with black bands on the dorsal and pectoral margins. Its body is entirely encased in spine-bearing plates. There are seven large barbels on the lower jaw and a pair of sculptured rostral projections on the snout.

DISTRIBUTION
Distributed from Japan and the Philippines south to northern Australia.

HABITAT
A deepwater species that typically is taken on the continental shelf. It is collected at depths greater than 657 ft (200 m).

BEHAVIOR
Little is known. They do not appear to be venomous.

FEEDING ECOLOGY AND DIET
The diet consists primarily of crustaceans. Preyed upon by larger predatory fishes.

REPRODUCTIVE BIOLOGY
Nothing is known about their reproductive biology, but other armored gurnards produce small pelagic eggs.

CONSERVATION STATUS
Not listed by the IUCN.

SIGNIFICANCE TO HUMANS
This species is rare, so it is not of commercial importance. ◆

Red lionfish
Pterois volitans

FAMILY
Scorpaenidae

TAXONOMY
Pterois volitans Linnaeus, 1758, Moluccas Island, Indonesia.

OTHER COMMON NAMES
English: Volitan lionfish, butterfly cod, red firefish, turkeyfish; French: Poisson volant; German: Rotfeuerfisch; Japanese: Hana-minokasago, ominokasago.

PHYSICAL CHARACTERISTICS
Reaches 13.8 in (35 cm) maximum length. One of the most easily recognized of all marine fishes. The most striking features of the red lionfish are its banded (reddish to black) head and body and its unique pectoral configuration. The long, flowing pectoral rays have varying degrees of connecting membranes, giving them the distinctive appearance of separate threadlike projections. The dorsal and anal fins are covered with dark rows of spots on a clear to yellowish background. All of the non-pelvic spines—these include the pectoral, anal, and dorsal spines—are venomous.

DISTRIBUTION
A wide-ranging species found throughout the Indo-Pacific region. Recently, populations of the red lionfish have become established on the Atlantic coast of the United States, presumably released by aquarium wholesalers in Florida following Hurricane Andrew in 1992. This introduced species has been collected from the Florida Keys north to North Carolina and the southern coast of Long Island, New York.

HABITAT
Usually found in lagoon and seaward reefs. Typically, it is a fairly shallow-water species, but it can be found as deep as 164 ft (50 m). They tend to hide among rocks or in caves during the day.

BEHAVIOR
The audacity of the colors of this species is a clear warning sign for its exceedingly venomous spines. This species flexes its pectoral fins quickly to charge aggressors with its extended dorsal spines.

FEEDING ECOLOGY AND DIET
This species, like most species of *Pterois*, is a voracious predator that feeds primarily on small fishes, shrimps, and crabs. It has been reported that it can eat as much as 8.2 times its body weight per year. Feeds primarily at night, when it uses its widespread pectoral fins to trap prey into a corner. Preyed upon by humans.

REPRODUCTIVE BIOLOGY
Primarily solitary, but a single male forms aggregations with females for mating. Courtship begins at twilight and is initiated by the male. Females generally produce two tubes composed of mucus and between 2,000 and 15,000 eggs. Shortly after the females release the eggs, the tubes swell with seawater and are penetrated by the male's sperm. Fertilization proceeds, and the larvae hatch after 36–48 hours.

CONSERVATION STATUS
Not listed by the IUCN.

SIGNIFICANCE TO HUMANS
Collected in large numbers for the aquarium trade. Despite its venomous nature, the red lionfish also is harvested commercially for food. ◆

Merlet's scorpionfish
Rhinopias aphanes

FAMILY
Scorpaenidae

TAXONOMY
Rhinopias aphanes Eschmeyer, 1973, New Caledonia.

OTHER COMMON NAMES
English: Weedy scorpionfish.

PHYSICAL CHARACTERISTICS
Grows to 10.2 in (26 cm) maximum length. A compressed, large-headed scorpaenid, with an upturned mouth. Typically yellow and black, forming a paisley or "mazelike" appearance. Predominately brown, green, and black specimens also have been recorded. The body and head of this species are covered with cirri and other fleshy appendages.

DISTRIBUTION
Has been collected in northeastern Australia, New Caledonia, New Guinea, and southern Japan but probably is more widespread in the western Pacific.

HABITAT
Little is known about this cryptic species, but it often is found sitting on corals and appears to be most common on coral slopes.

BEHAVIOR
Like many other scorpaenoid fishes, this cryptic fish appears to shed its skin periodically to prevent the buildup of too much epibiotic growth. The presence or absence of venom in this species has not been reported.

FEEDING ECOLOGY AND DIET
The diet is unknown, but most other species of *Rhinopias* feed primarily on crustaceans and small fishes.

REPRODUCTIVE BIOLOGY
Nothing is known.

CONSERVATION STATUS
Not listed by the IUCN.

SIGNIFICANCE TO HUMANS
Not commercially fished, but numerous other *Rhinopias* species are prized aquarium specimens. It is likely that Merlet's scorpionfish may already be imported for the aquarium trade. ◆

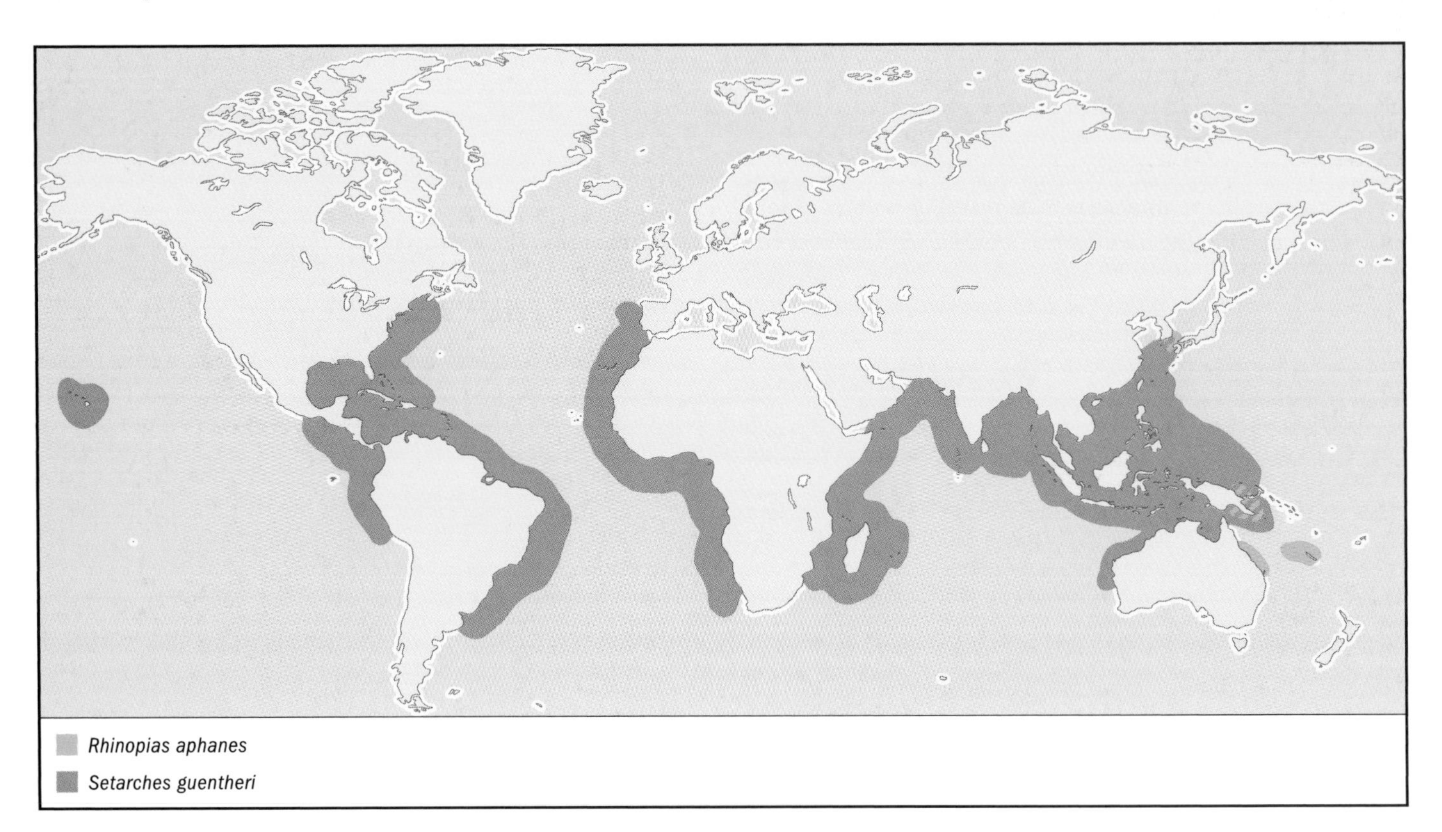

California scorpionfish
Scorpaena guttata

FAMILY
Scorpaenidae

TAXONOMY
Scorpaena guttata Girard, 1854, Monterey, California.

OTHER COMMON NAMES
English: Sculpin; French: Rascasse californienne; Spanish: Rascacio californiano, escorpión californiano.

PHYSICAL CHARACTERISTICS
Reaches 16.5 in (42 cm) maximum length. Like most scorpaenids, it is a well-camouflaged, spiny, massive fish. The coloring varies widely, from a deep red through light brown. Black, dark brown, and maroon spots cover the body and fins.

DISTRIBUTION
Found from Santa Cruz, California, south to southern Baja California. This species also is found in the Gulf of California.

HABITAT
These scorpionfishes are most abundant on hard bottoms, particularly rocky reefs, caves, and man-made structures, such as wrecked ships. Additionally, they can be found on muddy and sandy bottoms.

BEHAVIOR
Although most scorpionfishes and their allies are venomous, the California scorpionfish is the most venomous fish regularly collected off the California coast.

FEEDING ECOLOGY AND DIET
The diet of these predators primarily consists of crabs, but small fishes, octopi, and other crustaceans also are eaten.

REPRODUCTIVE BIOLOGY
Unlike most marine fishes, females produce eggs that are imbedded within the gelatinous walls of pear-shaped structures that float near the surface. After five days, the eggs hatch within these structures. The larval fishes that emerge have an integument that has an inflated appearance and is vesiculate. The larvae remain in plankton until they reach the length of 0.6–0.8 in (15–20 mm), upon which they settle in their adult habitat.

CONSERVATION STATUS
Although the species represents a fairly important fishery, they currently are not threatened as are some of their rockfish relatives.

SIGNIFICANCE TO HUMANS
Supports an important commercial and sport fishery in southern California and Ensenada, Baja California. Most fishes are taken in the spring and summer months, when commercial fishermen target spawning aggregations using hook and line, gill nets, and otter trawls. ◆

Bocaccio
Sebastes paucispinis

FAMILY
Sebastidae

TAXONOMY
Sebastes paucispinis Ayres, 1854, California.

OTHER COMMON NAMES
English: Rock salmon; Spanish: Rocote bocaccio.

PHYSICAL CHARACTERISTICS
Grows to 37.4 in (95 cm) maximum length. Bocaccios are one of the most elongate rockfishes in California and one of the least spiny rockfishes. They tend to be reddish brown on the dorsal surface, pink or brown on the flanks, and silver ventrally. Juveniles and small adults are reddish brown with dark spots.

DISTRIBUTION
Widespread from Alaska to Baja California. They are most abundant from British Columbia to Washington.

HABITAT
Juveniles typically are collected in shallow waters under drifting kelp mats that have broken free. Adults form benthic aggregations over hard and rocky bottoms at depths ranging from 164 to 984 ft (50–300 m).

BEHAVIOR
Bocaccios are a mobile rockfish. Tagged juveniles often are recaptured 60–80 mi (97–129 km) away from their point of origin. As with many other scorpaenoids, the bocaccio is venomous, but the venom is comparatively weak (although local fishermen suggest that they are the most venomous of the rockfishes).

FEEDING ECOLOGY AND DIET
Juveniles feed on small fishes, particularly other rockfishes. Adults feed on rockfishes, sablefishes (Anoplopomatidae), anchovies (Engraulidae), and squids. Eaten by larger fishes and pinnipeds.

REPRODUCTIVE BIOLOGY
As with all sebastids, the bocaccio is viviparous (live bearing). Large females can produce more than two million eggs per season, which are released as larvae in two or more batches. Rockfish larvae remain in the upper 263 ft (80 m) of the water column for several months. This stage is followed by a pelagic juvenile stage that lasts one to several months, after which the larvae settle.

CONSERVATION STATUS
The bocaccio is the only Critically Endangered scorpaenoid. This listing suggests that the population size has decreased by more than 80% in about the last ten years of the twentieth century, owing to the pressure of overfishing and the low minimum population doubling time, which is longer than 14 years.

SIGNIFICANCE TO HUMANS
As their population decline suggests, bocaccios traditionally have been a very important commercial and recreational food fish in the eastern Pacific. When they were more abundant, they represented more than 14% of the total marine recreational catch of California. ◆

Deepwater scorpionfish
Setarches guentheri

FAMILY
Setarchidae

TAXONOMY
Setarches guentheri Johnson, 1862, Madeira.

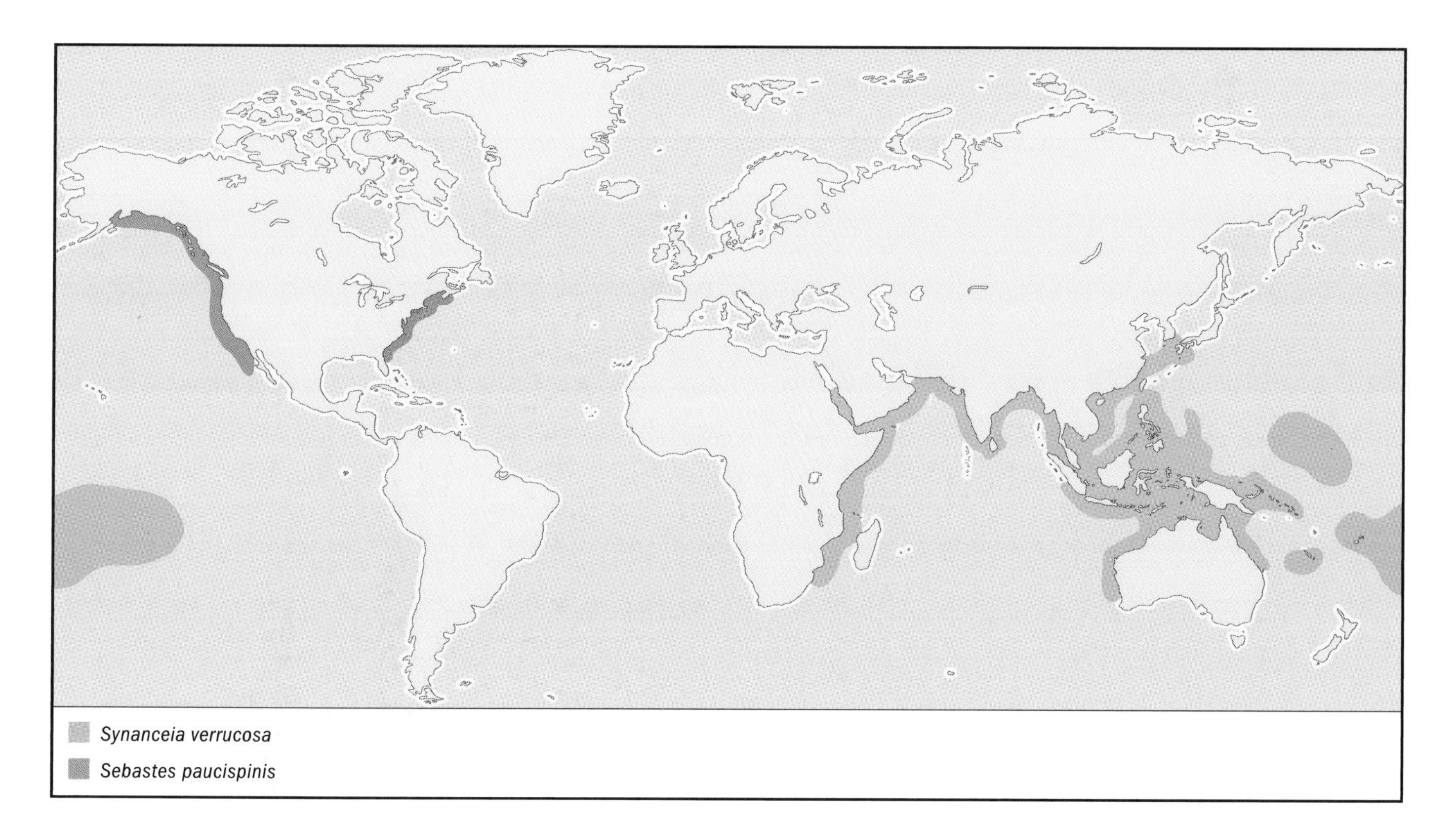

OTHER COMMON NAMES
English: Channeled rockfish; French: Rascasse serran; Spanish: Rascacio serrano; Japanese: Shirokasago.

PHYSICAL CHARACTERISTICS
Grows to 9.4 in (24 cm) maximum length. Typically gray or a shade of orange or pink. As is seen with many deepwater fishes, the skeleton is poorly ossified, and the head is cavernous.

DISTRIBUTION
Described as the most widely distributed scorpaenoid, because it has been collected worldwide in all tropical and temperate deep waters.

HABITAT
A benthic species that lives on or near the bottom at depths between 591–2,297 ft (180–700 m).

BEHAVIOR
Nothing is known about the behavior of this species.

FEEDING ECOLOGY AND DIET
Preliminary investigations into the diet of this species suggest that it eats deepwater crustaceans, including oplophorids and amphipods.

REPRODUCTIVE BIOLOGY
Nothing is known.

CONSERVATION STATUS
Not listed by the IUCN.

SIGNIFICANCE TO HUMANS
No commercial fishery exists for this deepwater species, although it can be found occasionally in eastern Atlantic fish markets. ◆

Bearded ghoul
Inimicus didactylus

FAMILY
Synanceiidae

TAXONOMY
Inimicus didactylus Pallas, 1769, Indian Ocean.

OTHER COMMON NAMES
English: Demon stinger, popeyed sea goblin, spiny devilfish; Japanese: Hime-oniokoze.

PHYSICAL CHARACTERISTICS
Grows to 7.9 in (20 cm) maximum length. The bearded ghoul is best recognized by the shape of its head and its elongate pectoral fins, which it uses for crawling along the bottom. This species can be distinguished by the pattern on the medial surface of its pectoral fin, which has a broad, dark, transverse bar that may be streaked.

DISTRIBUTION
Distributed from southern Japan and Indonesia to Australia, the Solomon Islands, and New Caledonia.

HABITAT
A benthic species found on open sandy or silty bottoms, particularly in estuaries, lagoons, and seaward reefs.

BEHAVIOR
A particularly venomous species that is capable of killing a human. Like most stonefishes, this species is a lie-and-wait predator that relies on its cryptic coloration and body form to surprise prey.

FEEDING ECOLOGY AND DIET
Feeds primarily on crustaceans, but small fishes also are preyed upon.

REPRODUCTIVE BIOLOGY
Little is known about the reproductive biology of this species.

CONSERVATION STATUS
Not listed by the IUCN.

SIGNIFICANCE TO HUMANS
Because of its small size, this species is not fished commercially; however, despite the fact that these fishes are highly venomous, they are collected occasionally for the aquarium trade, where they are sold as "popeyed sea goblins." ◆

Reef stonefish
Synanceia verrucosa

FAMILY
Synanceiidae

TAXONOMY
Synanceia verrucosa Bloch and Schneider, 1801, Indian Ocean.

OTHER COMMON NAMES
English: Stonefish, dornorn; French: Poisson pierre commun; Japanese: Oni-darumaokoze.

PHYSICAL CHARACTERISTICS
Reaches 15.7 in (40 cm) maximum length. The reef stonefish is among the best-camouflaged species in the world. Most of these fishes look like encrusted rocks or pieces of coral. Most specimens are dark brown or gray, but they usually have patches of yellow, red, or orange.

DISTRIBUTION
This is the most widely distributed stonefish. It can be found throughout the Indo-Pacific from Africa to the Tuomoto Archipelago.

HABITAT
Usually found living among rubble, on coral bottoms, or under rocks or ledges. They also are found on sandy or muddy bottoms, where they often bury themselves.

BEHAVIOR
This is the world's most venomous fish. Unlike most scorpaenoids, the stonefishes have grooves in their dorsal spines that act like syringes to deliver venom.

FEEDING ECOLOGY AND DIET
This species feeds primarily on small fishes and crustaceans that swim by.

REPRODUCTIVE BIOLOGY
Stonefishes are typically solitary creatures, but they do form larger aggregations for mating. The reef stonefish has external fertilization, and produces comparatively large eggs.

CONSERVATION STATUS
Not listed by the IUCN.

SIGNIFICANCE TO HUMANS
Despite the extreme danger associated with collecting this species, there are commercial fisheries for both dead and live fish markets. In addition to their commercial fishery, stonefishes make their way into the aquarium trade. The sting of the stone fish is extremely painful and is followed by rapid swelling around the wound. The severity of the response is related to the depth of the penetration by the spines. The treatment of the wound consists of bathing the stung area in very hot water until the victim can be hospitalized. For more serious stings, it is often advisable for stonefish antivenin to be given intramuscularly. The stonefish antivenin is composed of purified antibodies against stonefish venoms and venom components. These antibodies are harvested from laboratory animals, which are given small doses of the venom. Eventually, these animals build up a tolerance to the venom, which allows them to produce the large quantity of venom antibodies required for the antivenin. ◆

Cockatoo waspfish
Ablabys taenianotus

FAMILY
Tetrarogidae

TAXONOMY
Ablabys taenianotus Cuvier, 1829, Mauritius.

OTHER COMMON NAMES
English: Rogue fish.

PHYSICAL CHARACTERISTICS
Grows to 5.9 in (15 cm) maximum length. A strongly compressed waspfish with a sail-like dorsal fin that originates above the eye. Cockatoo waspfishes are reddish brown with black lines and black and white blotches sprinkled over the dorsal and lateral surfaces.

DISTRIBUTION
Widely distributed in the western Pacific as far north as Japan and south to Australia and Fiji. Can be found as far west as India and as far east as the Society Islands.

HABITAT
A cryptic species that typically is collected among seaweed in littoral or rocky intertidal habitats in shallow depths to 263 ft (80 m).

BEHAVIOR
Cockatoo waspfishes tend to be more active at dusk or night. As is seen in some other scorpaenoids, this species often is found rocking back and forth in response to the surge, to mimic the seaweed that surrounds it. This species is quite venomous.

FEEDING ECOLOGY AND DIET
Like many waspfishes, this species feeds primarily on small crustaceans, particularly shrimps, and smaller fishes.

REPRODUCTIVE BIOLOGY
Little is known.

CONSERVATION STATUS
Not listed by the IUCN.

SIGNIFICANCE TO HUMANS
Not commercially fished but collected for the aquarium trade. ◆

Red gurnard
Chelidonichthys spinosus

FAMILY
Triglidae

TAXONOMY
Chelidonichthys spinosus McClelland, 1844, China.

OTHER COMMON NAMES
Japanese: Hôbô.

PHYSICAL CHARACTERISTICS
Reaches 15.7 in (40 cm) maximum length. Head encased in bony armor with short rostral projections. Body coloration varies from brown to light orange or red when under stress. The dorsal surface of this species has brown patches, and the inner surface of the pectoral fins is olive to dark green, with scattered blue or white spots.

DISTRIBUTION
Found from southern Hokkaido (Japan) south to the South China Sea.

HABITAT
Found on sandy bottoms.

BEHAVIOR
The red gurnard, like all sea robins, spends much of its time "walking" on the seafloor, using its free pectoral rays to probe the sediment for food. Capable of making sounds using drumming muscles that are beaten against the gas bladder. Venom has not been found.

FEEDING ECOLOGY AND DIET
The diet consists mainly of various crustaceans and mollusks. Preyed upon by larger predatory fishes.

REPRODUCTIVE BIOLOGY
The red gurnard produces large pelagic eggs. There have been numerous reports of increased male grunting during the breeding season, suggesting that auditory signals are important in mate selection.

CONSERVATION STATUS
Not listed by the IUCN.

SIGNIFICANCE TO HUMANS
An excellent food fish. Taken by trawlers throughout its range. ◆

Striped sea robin
Prionotus evolans

FAMILY
Triglidae

TAXONOMY
Prionotus evolans Linnaeus, 1766, North or South Carolina, United States.

OTHER COMMON NAMES
French: Grondin volant; Spanish: Rubio volador.

PHYSICAL CHARACTERISTICS
Reaches 17.7 in (45 cm) maximum length. Characterized by a large bony head with many ridges and spines and a bifurcate lateral line on the tail. The striped sea robin is white ventrally, with various shades of golden, brown, and orange on the sides and dorsal surface. It often has dark saddles underneath the dorsal fins and is best distinguished from other species by the two thin, black stripes running along the side of the body. The dorsal stripe runs the entire length of the fish along the lateral line, and the smaller, incomplete stripe is situated below. The tail has two black bars with a light area between them.

DISTRIBUTION
Found from southern Nova Scotia down the Atlantic Coast of North America to northeastern Florida. Typically collected at depths of less than 200 ft (61 m) but have been found at depths as great as 550 ft (168 m).

HABITAT
Usually found on sandy bottoms. It often can be taken in inshore estuaries or over reefs, particularly in summer months.

BEHAVIOR
The striped sea robin uses its free pectoral rays to "walk" along the seafloor in search of prey. The sea robins along the Atlantic Coast of North America are famous for their ability to produce sounds by "beating" the swim bladder muscles against the gas-filled swim bladder, making a characteristic grunting noise. Typically, the striped sea robin is found offshore, but the species makes migrations into the deeper, more saline estuaries for breeding. This species is not venomous.

FEEDING ECOLOGY AND DIET
The diet consists mainly of crustaceans, mollusks, and fishes. Preyed upon by larger predatory fishes.

REPRODUCTIVE BIOLOGY
Produces pelagic eggs. This species appears to breed in deeper estuarine environments, typically in the summer months from May to October. It seems clear that sound plays a role in choice of mate.

CONSERVATION STATUS
Not listed by the IUCN.

SIGNIFICANCE TO HUMANS
The striped sea robin is a commercially important fish. It is collected and used for human consumption, fishmeal, bait, pet food, and fertilizer. Smaller specimens often are collected for the aquarium trade, though they grow too large for most home aquaria. ◆

Resources

Books

Benton, M. J., ed. *The Fossil Record 2.* London: Chapman and Hall, 1993.

Breder, C. M., Jr., and D. E. Rosen. *Modes of Reproduction in Fishes.* Garden City, New York: Natural History Press, 1966.

Carpenter, K. E., and V. H. Niem, eds. *FAO Species Identification Guide for Fishery Purposes.* Vol. 4, *The Living Marine Resources of the Western Central Pacific.* Rome: FAO, 1999.

Halstead, Bruce W. *Poisonous and Venomous Marine Animals of the World.* Washington, DC: Government Printing Office, 1965–1970.

Love, Milton S. *Probably More Than You Want to Know About the Fishes of the Pacific Coast.* 2nd edition. Perm: Izd-vo Permskogo Universiteta, 2001.

Love, Milton S., Mary Yoklavich, and Lyman Thorsteinson. *The Rockfishes of the Northeast Pacific.* Berkeley: University of California Press, 2002.

Mandritsa, S. A. *Lateral Line System and Classification of Scorpaenoid Fishes.* Perm, Russia: Izd-vo Permskogo Universiteta, 1994.

Masuda, H., K. Amaoka, C. Araga, T. Uyeno, and T. Yoshino, eds. *The Fishes of the Japanese Archipelago.* Tokyo: Tokai University Press, 1984.

Nelson, J. S. *Fishes of the World.* 3rd edition. New York: John Wiley & Sons, 1994.

Paxton, John R., and William N. Eschmeyer, eds. *Encyclopedia of Fishes.* 2nd edition. San Diego: Academic Press, 1998.

Randall, J. E., G. R. Allen, and R. C. Steene. *Fishes of the Great Barrier Reef and Coral Sea.* Honolulu, HI: University of Hawaii Press, 1997.

Whitehead, P. J. P., M. L. Bauchot, J. C. Hureau, J. Nielsen, and E. Tortonese. *Fishes of the North-Eastern Atlantic and the Mediterranean.* Paris: UNESCO, 1986.

Periodicals

Fishelson, L. "Ethology and Reproduction of the Pteroid Fishes Found in the Gulf of Aqaba (Red Sea), Especially *Dendrochirus brachypterus* (Cuvier) Pteroidae (Teleostei)." *Publ. Stat. Zool. Napoli* 39 (1975): 635–656.

Imamura, H. "Phylogeny of the Family Platycephalidae and Related Taxa (Pisces: Scorpaeniformes)." *Species Diversity* 1 (1996): 123–233.

Ishida, M. "Phylogeny of the Suborder Scorpaenoidei (Pisces: Scorpaeniformes)." *Nansei National Fisheries Research Institute* 27 (1994): 1–112.

Mooi, R. D., and G. D. Johnson. "Dismantling the Trachinoidei: Evidence of a Scorpaenoid Relationship for the Champsodontidae." *Ichthyological Research* 44 (1997): 143–176.

Other

U.S. Geological Survey. "Nonindigenous Aquatic Species." (30 Dec. 2002). <http://nas.er.usgs.gov>.

William Leo Smith, MS

Scorpaeniformes III

(Greenlings, sculpins, and relatives)

Class Actinopterygii
Order Scorpaeniformes
Number of families 10

Photo: The red Irish lord (*Hemilepidotus hemilepidotus*) is one of the most beautiful sculpins. It can be found in a number of different colors. (Photo by David Hall/Photo Researchers, Inc. Reproduced by permission.)

Evolution and systematics

The Scorpaeniformes, the mail-cheeked fishes, are united by the presence of a bony ridge called a "suborbital stay" on the cheek, running horizontally below the eye and providing an armored look to the head of most species. The suborbital stay is a posterior extension of the second infraorbital (eye socket) bone. In the sculpins (Cottoidei), this suborbital stay tends to be very prominent, while in the greenlings (Hexagrammoidei) it is not easily seen.

The ancestral scorpaeniform stock is considered to have derived from a generalized percoid (perch) fish. That is to say, cottoid (sculpin) and hexagrammoid (greenling) fishes are among the most recently evolved, advanced fishes. The Scorpaeniformes is the fourth largest order of fishes, including about 29 families, about 260 genera, and 1,400 species. Most of these fishes are bottom-dwelling or live near the seabed. Nine families are covered in this chapter, with the remainder covered in Scorpaeniformes I and II.

Among the scorpaeniform fishes, the suborder Cottoidei includes between seven and 13 families in modern classifications. Historically, various additional cottoid families have been distinguished, often containing only one species. The most prominent families include the sculpins (Cottidae, with 305 species in 70 genera), fathead sculpins (Psychrolutidae, with 11 species, five genera), poachers (Agonidae, with 49 species, 20 genera), lumpfishes (Cyclopteridae, with 27 species, eight genera), and snailfishes (Liparididae, with 195 species, 13 genera). Also within the Cottoidei are two families of Lake Baikal fishes, the Cottocomephoridae (Baikal sculpins, with 24 species, eight genera), and Comephoridae (Baikal oilfishes, with two species, one genus). The Baikal oilfishes are distinguished by being the only viviparous (live-bearing) cottoid fishes. A monotypic (one species) family, the Normanichthyidae, consists of one marine species off Chile that is often included within the Cottoidei. That species, *Normanichthys crockery*, would be the only cottoid fish having a swim bladder, but the anatomical description by Yabe and Uyeno (1996) concludes that the family is of uncertain systematic position within the Scorpaeniformes, and not correctly within the Cottoidei. Other families common in various classifications include the Cottunculidae (here included within the Psychrolutidae), the Icelidae (sometimes distinguished from Ereuniidae, both within the Cottidae here), and the Hemitripteridae (here included within the Cottidae). The species in the Psychrolutidae are usually included within the Cottidae in older works, but Jackson and Nelson (1998) have described the unique features of sensory canals and associated bones on the head which, together with other characters, distinguish the fathead sculpins from the other sculpin species. Some classifications place the liparidid fishes within the Cyclopteridae. Cottoid fishes probably first appeared in the North Pacific and only invaded the Arctic and North Atlantic Oceans 3.5 million years ago.

The zoogeography of cottoid fishes centers on the Pacific coast of North America. The other scorpaeniform suborder considered within this chapter is the Hexagrammoidei, the greenlings, which includes the largest family of fishes endemic to the North Pacific, the Hexagrammidae (11 species, five genera). Some classifications separate the combfishes into a separate family, the Zaniolepidae. Another family usually included within the Hexagrammoidei is the Anoplopomatidae, which includes the sablefish (*Anoplopoma fimbria*) and the skilfish (*Erilepis zonifer*). Sometimes the skilfish is separated into the monotypic family Erilepidae, also with these two families separated from other hexagrammoids into the Anoplopomatoidei.

A tidepool sculpin (*Oligocottus maculosus*) in the Pacific Ocean, near the United States. Each sculpin changes color to match its tidepool, or buries itself in sand. If it is washed out of its tidepool, it uses its sense of smell and returns to its own pool. (Photo by Nancy Sefton/Photo Researchers, Inc. Reproduced by permission.)

Physical characteristics

The hexagrammoid fishes are perhaps the most generalized scorpaeniform fishes, resembling many perciform fishes. Hexagrammoid fishes have no head spines, and they include some of the only pelagic species, fish which have a fusiform body shape adapted to swimming in open water. Fishes of the genus *Hexagrammos* have multiple lateral lines, although Wonsettler and Webb (1997) have demonstrated that only the central trunk canal is innervated, as in other teleost fishes, so the biological significance of the other four, non-functional lateral lines, has become a mystery. As with the more diverse cottoids, most hexagrammids are adapted to dwelling on the sea bottom. They have no swim bladder and the pectoral fins are enlarged, with the lower edge attached further forward than the top edge, so that the fish can rest on these fins like a pair of elbows.

In the Cottoidei, extreme divergence of structure and function have yielded some of the most unusual-looking of all fishes. All cottoids lack the swim bladder and most have a flattened head, as well as enlarged pectoral fins like the greenlings. There are many skin adaptations, including loss of scales or modification of scales into bony plates. The Psychrolutidae (fathead sculpins) have flaccid bodies (puffy skin) with reduced skeletal structure. The Agonidae (poachers) have armor plates over their body, have all fin rays unbranched, and have lost the suborbital stay. The Cyclopteridae (lumpfishes or lumpsuckers) have a globular body covered with tubercles and with the pelvic fins modified into a sucking disc. The Liparididae (snailfishes) also have a pelvic sucking disc, but the body is without scales and is elongate. The Cottidae (sculpins) include the greatest divergence in body form and size, which is perhaps why various families historically have been separated from this taxonomic grouping. All cottid fishes have a single pelvic fin spine (except one species lacking pelvic fins) and no anal fin spine; they usually have naked skin, sometimes with cirri, sometimes with scales, and sometimes with prickles. Cottids lack swim bladders and tend to have large, flattened heads, often with pronounced spines and often with large eyes.

Color and fin shape lend toward a camouflaged appearance in most of these fishes. Cottoid and hexagrammoid fishes often display sexual dimorphism in color, body size, fin shape, or other features. Many cottid species have prominent male copulatory organs.

Distribution

The Cottoidei and Hexagrammoidei center their distribution in the eastern North Pacific Ocean, but antitropical (both hemispheres) distributions occur for cottoids in some instances. The Psychrolutidae include North Pacific and North Atlantic species, plus several species off New Zealand, Australia, and South Africa. Cottids similarly occur in the Northern Hemisphere (but in both marine and fresh waters) and off New Zealand, eastern Australia, and Argentina. Liparidids occur in both warm and cold marine waters of all the world's oceans, although rarely in the Indian Ocean. The Cyclopteridae occur only in cold, marine waters of the Northern Hemisphere. Agonids are found in the North Pacific, North Atlantic, and off southern South America.

The unique families of Lake Baikal cottoid fishes have been mentioned. Molecular genetic studies of Baikal cottoids (Kirilchik and Slobodyanyuk, 1997) reveal close relatedness with freshwater cottids such as *Cottus bairdii* from Lake Michigan and *Cottus cognatus* from Lake Michigan and Siberia. The genus *Cottus* is very widely distributed across North America and Eurasia, and the different species are all a very generalized cottid body type. The Baikal cottoids, although closely related to other freshwater cottids, have radiated into 22 endemic Baikal species (another four cottocomephorid species also occurring in other drainages of Siberia).

Habitat

The microdistribution of any species relates to habitat preference. The Northeast Pacific distribution of most cottoid species is reflected in an analysis of fish community structure of rocky shorelines of the North Pacific, which is dominated by cottids. The rocky intertidal of the Pacific coast of North America is dominated by various species of *Artedius*, *Clinocottus*, and *Oligocottus*. Other cottid genera have very narrow depth preferences in shallow subtidal marine waters. For example, the longfin sculpin, *Jordania zonope*, was considered extremely rare prior to the advent of scuba diving, because it inhabits vertical rock surfaces that are not amenable to sampling with nets. Similarly, the manacled sculpin, *Synchirus gilli*, was considered very rare until divers discovered it on the feather boa kelp, *Egregia menziesi*, on outer Pacific coast shores. The feather boa kelp does not occur, however, in protected inland seas like the Strait of Georgia (British Columbia), where the manacled sculpin spawns in the holdfasts of another kelp species, *Alaria marginata*. For the majority of cottoid species, very little is known of precise habits and habitat preferences.

The diverse snailfishes (Liparididae) occur broadly in the world ocean over a very wide depth range from the intertidal zone to greater than 23,000 ft (7,010 m) depth. Again, very little precise habitat information is available for the vast majority of species. Since most snailfishes possess a pelvic sucking disc, however, it is presumed that smooth surfaces of plants, animals, or rocks provide substrate for many of these species. As with other cottoid fishes, different species may prefer shores exposed to waves and tidal currents versus protected shores.

The Lake Baikal sculpins have been investigated since the eighteenth century. The diversification of these species seems in great part related to different depth preferences. Lake Baikal, at 5,315 ft (1,620 m) depth, is the deepest lake in the world.

The reason why the less diverse freshwater cottoid fishes have been studied more intensely than the vast diversity of marine species is that freshwater habitats are more accessible. Perhaps the most accessible marine habitat is the tidepool, and intertidal cottoids have received disproportionate study, except in comparison with fish species of commercial importance. A problem with the study and interpretation of tidepool fishes, however, has to do with the human perspective. Since people can most readily work around tidepools during low tides and calm weather, our interpretation of tidepools tends to focus on ecological advantages of the tidepool during a low tide (in calm weather) rather than during high tides and storms. Thus, tidepools are considered to offer refuge from subtidal predators during low tides, rather than to offer refuge from turbulence during high tides. Habitat can have widely divergent values and characteristics under different conditions, so the interpretation of habitat preferences needs to be tempered by perspective.

Discussions of fish habitat implicitly consider only adults in most studies. Tidepool studies reveal, however, that early juvenile stages of cottid species can have different habitat requirements than adults. Larval stages occupy the planktonic realm, a completely different habitat than that of most adult cottoids. Larvae of rocky intertidal cottids of various species have been shown to avoid drifting with currents away from the shoreline or along the shoreline, so behavior during early life stages may be directed toward providing access to required habitat in later stages. The precise substrate preference of a cottoid larva at settlement from the plankton may not be the same substrate preference as an adult, but it will be a substrate that is a component of the adult habitat. Thus, growth stages shift only their relative position on the seabed within a general habitat.

Niche partitioning between species in terms of habitat preference (as with food preference) is examined for mechanisms that allow ecological separation that could have led to speciation. Again, interpretation often focuses only on adults. In addition, intertidal studies of cottids have led to different interpretations of competition than subtidal studies. The flathead sculpin, *Artedius lateralis*, and the padded sculpin, *Artedius fenestralis*, occupy very similar habitats in the rocky intertidal, yet their larvae occupy different depth ranges along the shoreline. The flathead sculpin settles from the plankton in surface waters, directly onto substrates in the intertidal, whereas larvae of the padded sculpin occupy slightly deeper water, so that they settle subtidally. Thus, habitat preferences in terms of depth provide ecological segregation of these two cottid species during the larval and early juvenile stages. Adults of these species do not necessarily have to display any behavioral differentiation in order for their speciation to be explained.

Behavior

As mentioned, developmental stages of a species may be expected to shift behavior, even though most investigation concerns only the adult stage. Since cottoid and hexagrammoid fishes lack a swim bladder, the planktonic larval and pelagic juvenile stage (if occurring) need to display behavioral solutions to negative buoyancy. In some species those behaviors reflect morphological specializations for the particular life stage, but other species may simply have to swim very energetically, which will affect food requirements and feeding behavior. Schooling only occurs in adults of some of the few pelagic species in these taxa, but larval stages appear capable of schooling in many of the species observed in aquarium settings.

The cryptic appearance of most cottoids, together with bottom dwelling habits, leads to predation threats that require adaptive behaviors. Chance observations have led to the discovery that the tadpole sculpin, *Psychrolutes paradoxus*, apparently has an emetic flavor that causes predators to cough them out upon ingestion. This species relies first on its cryptic appearance to prevent predation, then secondarily upon its noxious taste to cause rejection by predators.

Even though cottoids lack a swim bladder, they have big extrinsic swim bladder muscles. The buffalo sculpin, *Enophrys bison*, vibrates when grasped. Another defense behavior in this and other species is the flaring of the gill cover (with the bony suborbital stay) to expose spines that could deter ingestion.

Homing behavior is well documented in cottids that inhabit tidepools. Tagging studies have not been directed as much at subtidal species, but the demonstration of topographic familiarity in tidepool species may extend to other species inhabiting subtidal reefs with significant landmarks.

A painted greenling (*Oxylebias pictus*) in southern California. During mating season, the males often turn so dark in color that they appear black. (Photo by Gregory Ochocki/Photo Researchers, Inc. Reproduced by permission.)

A male lumpfish (*Cyclopterus lumpus*) guarding its eggs in the Gulf of Maine. (Photo by Andrew J. Martinez/Photo Researchers, Inc. Reproduced by permission.)

The existence of homing behavior proves that learned familiarity with surrounding habitat is of significant survival value and that these fish are capable of exploiting that sort of advantage. Seasonal migrations encompass developmental shifts in habitat preference of young stages as well as reproductive behavior of adults.

Feeding ecology and diet

Because the many cryptic species of cottoid and hexagrammoid fishes tend to hide on the bottom of the sea, their feeding habits reflect the closeness of their prey species, which frequently include crustaceans such as amphipods, crabs, or shrimp. The functional morphology of the diverse mouth types in these taxa have been investigated in detail. Most studies of community feeding ecology in these fishes have been conducted with tidepool species.

In terms of mouth types, these fishes tend to feed either by pouncing on and engulfing their prey (ramming) or by drawing the prey into their mouth together with a stream of water by means of very rapid expansion of the gill covers (sucking). The third major type of prey capture in fishes (biting off a piece of a prey) is less well documented in cottoid and hexagrammoid fishes as a strict mode of feeding, although worms as prey may require biting behavior. Species that feed by suction tend to have small mouth openings, whereas species that engulf their prey have large, broad mouths. Different species can combine aspects of these different types of feeding behavior, and relative effectiveness of capture methods influences dietary preferences.

Beyond ingestion of prey, manipulation may be required to enable digestion. In *Asemichthys taylori* the vomerine teeth of the upper jaw are modified to enable punching holes in the shells of snail or clam prey so that soft tissues can be digested.

Two cottid species have been documented as switching to a substantially herbivorous diet as adults. These two species, *Enophrys bison* (buffalo sculpin) and *Clinocottus globiceps* (mosshead sculpin), have only been documented in terms of their gut contents. Gut morphology and digestive mechanisms such as gut acidity have not been studied in detail for these species. Ingestion of some seaweed may accompany predation on crustaceans inhabiting those seaweeds, but gut contents in these two sculpins, as in other taxa, can show the seaweed itself to be the preferred item. In the mosshead sculpin, the biting behavior that enables it to consume algae also enables it to remove pieces of tentacles from intertidal sea anemones (which have algae cells in their tissues).

The large eyes of most cottoid and hexagrammoid fishes would aid in detection of prey in very dim light. The sense of distant touch (lateral line and head canal system), however, enable detection of movement by potential prey in the vicinity of the fish. Psychrolutid fishes have particularly well-developed head canals with large pores. In the Agonidae, various species have the lower rays of their pectoral fins elongated beyond the webbing of the fin, so that they can rake through sand and gravel with these rays much as if they were fingers. Whether movement of the prey they disrupt and ingest is detected by these fin rays or by distant touch is not known. Many morphological features of these fishes probably relate to feeding adaptations.

In terms of predators, little is known about predation of the less common species such as the skilfish, but the larger of these bottom-dwelling species tend to be prey to larger fishes (like lingcod) as well as to seals and sea lions. Sablefishes are quite cannibalistic, as are lingcod.

Reproductive biology

Most greenlings, sculpins, and related species lay adhesive masses of eggs that either adhere to rocky substrate, cluster around plant or animal stalks and tubes, or are wedged into crevices. The eggs always adhere to each other, but not always to the spawning substrate. Among the sculpin relatives, only the comephorid sculpins of Lake Baikal give live birth to hatched larvae. Perhaps the most remarkable reproductive specialization among cottoid fishes is the phenomenon of internal gametic association, which has been demonstrated for sculpins (Munehara et al., 1997) and poachers (Munehara, 1997). With internal gametic association, the fishes copulate, but the sperm does not fertilize the egg until introduced to the calcium ions in seawater. This enables a female to repeatedly deposit small egg masses in specific ways over a long period of time, up to dozens or hundreds of depositions over weeks of time, based on one mating. Many highly specialized spawning substrates have evolved in various cottoid species, perhaps on the basis of internal gametic association in more cases than have yet been demonstrated. Proof of internal gametic association merely requires dissection of eggs from ovaries and placement in seawater, then incubation and observation for embryonic development.

Another typical characteristic of sculpin reproduction is the guarding of a cluster of different egg masses by a single territorial male. Male greenlings and lingcod also guard one or more egg masses. In some sculpins, however, the male exhibits haremic behavior in which both the nest site and a group of females are guarded together. In haremic species there is no evidence yet of internal gametic association. With the scalyhead sculpin, *Artedius harringtoni*, this author, while diving, observed several rotund little fish, presumed females, dart inside the empty shell of a giant barnacle, immediately followed by a much larger individual, presumed male, which curled its body and spread its fins to close off entry to the barnacle shell. This species lays its eggs inside giant barnacle shells, and there are typically several different colors of egg mass. Females of various sculpin species lay only one of various characteristic colors of egg. In haremic species of the genus *Artedius*, a cluster of egg masses may number over a dozen masses, but only with a few different colors of egg mass, corresponding to the number of females in the harem. In other genera (e.g. *Scorpaenichthys*, *Enophrys*, *Hemilepidotus*) where the male guards clusters of egg masses of differing colors, it is not known whether haremic behavior is involved, because groups of females have not been observed remaining near the guarding male.

Whereas haremic species tend to lay egg masses that adhere to a rock surface or to previously laid egg masses in a cluster at the guarded nest site, species with internal gametic association tend to lay smaller egg masses of less characteristic size. That is, the female can extrude whatever number of eggs is required to fill an interstitial space or to form a ring around a stalk, then move on to search for another deposition site, the sites being dispersed according to availability. In agonid species there appears to be a tendency to spawn inside sponges, whereas liparidid species tend to spawn in seaweeds or inside shells.

As much as the spawning characteristics vary among sculpins and relatives, their larvae also demonstrate diverse adaptations. Lacking a swimbladder at any stage, these species have evolved diverse mechanisms to enable early growth while inhabiting the water column. Enlarged pectoral fins exist in larvae of many species, whereas larvae of other species have a flaccid, globular body with relatively large volumes of low-density, buoyant body fluid.

Conservation status

Of the families covered in this chapter, the IUCN 2002 Red List includes nine species of the genus *Cottus*: one is categorized as Extinct (*C. echinatus*); two as Critically Endangered; four as Vulnerable; and two as Data Deficient. Sculpins and greenlings tend to inhabit rocky, marine shorelines that are less subject to alteration by human activities than estuarine habitats. For a species like the staghorn sculpin (*Leptocottus armatus*), which uses estuaries as nursery habitat for juveniles, habitat loss can affect populations locally, but the species as a whole is widespread and abundant. For many species that are rarely encountered by people, too little is known of habitat or true abundance to enable determination of conservation status. Where human developments eliminate all natural shoreline, as in municipal harbor areas, intertidal fish species generally lose their natural abundance. It is not known whether greenlings and sculpins are any more sensitive to pollution than other marine fish species. In freshwater lakes and streams, sculpins tend to be depleted both by habitat destruction and by introduction of alien species, as well as by pollution. The widespread distribution of various species of *Cottus* has prevented extinction from occurring at the species level, but geographically significant populations do become threatened.

Only a limited number of these fishes are directly sought in fisheries as food for humans. In the cases of the cabezon (*Scorpaenichthys marmoratus*), the sablefish (*Anoplopoma fimbria*), and the lingcod (*Ophiodon elongatus*), commercial fisheries have led to localized depletion that has necessitated fishing restrictions. In the cases of the cabezon and the lingcod, extended periods of restricted fishing have not led to population recoveries to original levels of abundance. Overexploitation has tended to occur first in more densely populated areas along inland seas and more southerly waters of the Pacific coast of North America.

A recent trend in North Pacific fisheries is to land live fishes for Asian markets. The lack of a swimbladder and bottom-dwelling habits render species of greenling and sculpin hardy in this trade. In addition, the head and skeleton are recovered after filleting and used in making soup stock, so that sculpins with large heads are marketable in the live trade. Greenlings and larger sculpins that are of little value in traditional fisheries are becoming increasingly exploited for live seafood trade, owing to the higher prices paid for live fish.

Significance to humans

The most obvious significance to humans of greenlings and sculpins is the importance as food of the few commercially sought species. Before the availability of ice or refrigeration systems, the lingcod (*Ophiodon elongatus*) was the target of fishing with hook and line by boats with flooded holds for keeping the fish live (and therefore fresh) until delivery to processing plants. The lack of a swimbladder enabled keeping the lingcod alive during extended fishing trips.

Diversification of fish farming that currently involves salmon species may soon include various relatives of greenlings. The sablefish or Alaska blackcod, *Anoplopoma fimbria*, is the subject of aquaculture research in British Columbia, and at the turn of the millennium the first commercially produced, cultured fingerlings were being grown to market size by salmon farmers.

The unusual body forms of various sculpins and related fishes lends aesthetic value to them. They are popular species for display in public aquariums. Seacoast tourism depends to some extent on the attraction people feel toward exploring tidepools along rocky shores, and intertidal sculpins are among the most readily observed species. Marine biological research has long encompassed study of intertidal sculpins because of their accessibility, and to a similar extent, because of their robust ability to withstand manipulation like tagging or maintenance in aquarium tanks.

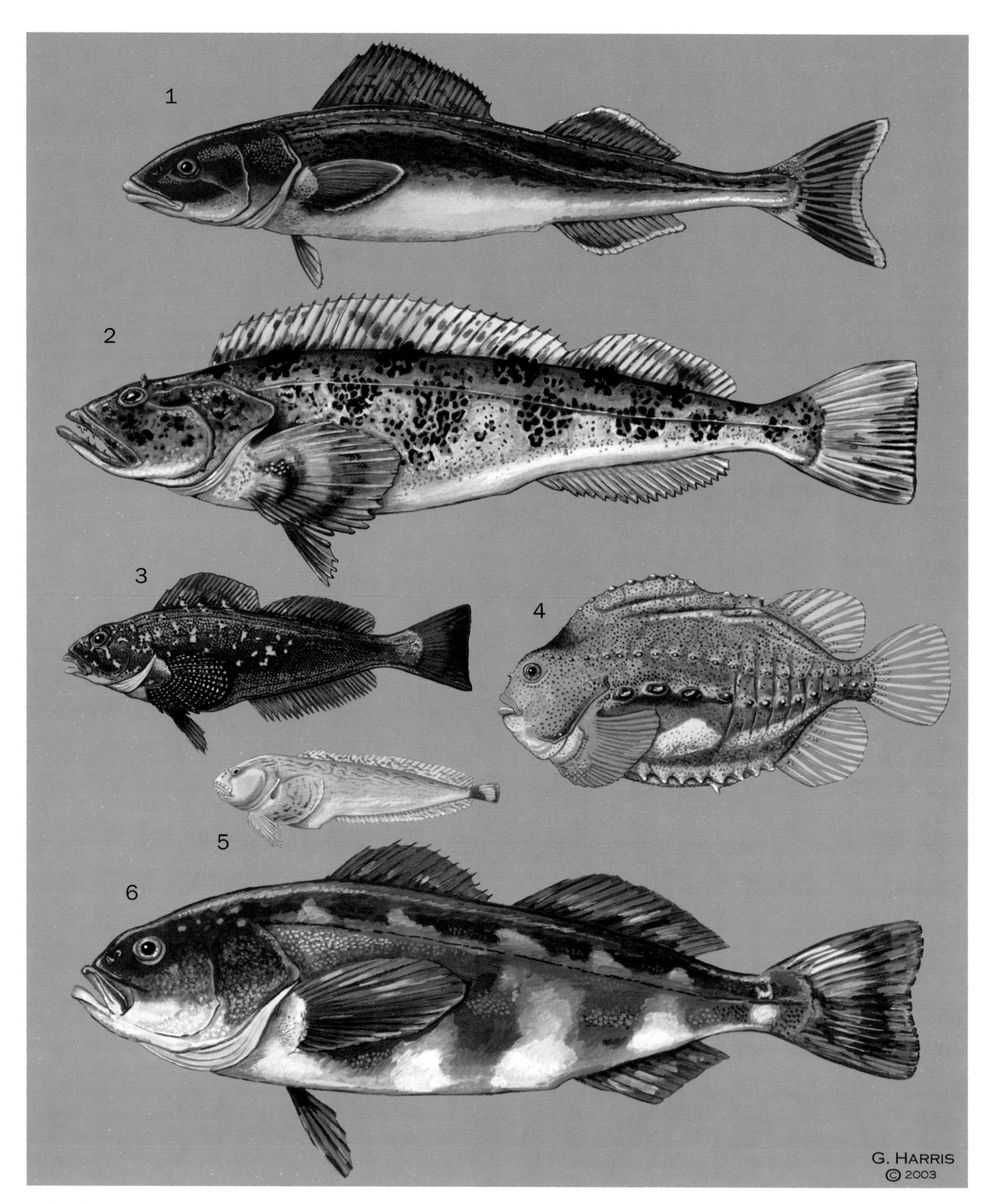

1. Sablefish (*Anoplopoma fimbria*); 2. Lingcod (*Ophiodon elongatus*); 3. Breeding male kelp greenling (*Hexagrammos decagrammus*); 4. Breeding male lumpfish (*Cyclopterus lumpus*); 5. Slipskin snailfish (*Liparis fucensis*); 6. Skilfish (*Erilepis zonifer*). (Illustration by Gillian Harris)

1. Tidepool sculpin (*Oligocottus maculosus*); 2. Soft sculpin (*Psychrolutes sigalutes*); 3. Rockhead (*Bothragonus swani*); 4. Cabezon (*Scorpaenichthys marmoratus*); 5. Grunt sculpin (*Rhamphocottus richardsoni*); 6. Sailfin sculpin (*Nautichthys oculofasciatus*). (Illustration by Gillian Harris)

Species accounts

Rockhead
Bothragonus swani

FAMILY
Agonidae

TAXONOMY
Bothragonus swani Steindachner, 1877, previously called *Bothragonus swanii*, Port Townsend, Puget Sound, Washington, United States.

OTHER COMMON NAMES
English: Pithead poacher, pit-headed poacher, pitted poacher, deep-pitted poacher, deep-pitted sea-poacher.

PHYSICAL CHARACTERISTICS
Short, stout body of 3.5 in (9 cm), covered with armor plates and with small fins other than the pectorals. Like other poachers (Agonidae), all fin rays are unbranched, and tail bones are fused. The rockhead, or pithead poacher, is most obvious for its wide, flattened head with a deep indentation in the top. Eyes and mouth are small. As a larva, the pectoral fins become enlarged like butterfly wings, and the body is covered with fine spines. The pelagic juvenile develops heavier, recurved body spines that hook toward the tail. These sharp hooks flatten into armor plates after settlement to the bottom.

DISTRIBUTION
Northern California to Kodiak Island, Alaska.

HABITAT
Rocky shorelines exposed to waves, down to 66 ft (20 m) depth. Found under rocks, in crevices, or among kelp holdfasts.

BEHAVIOR
Very little is known of this elusive fish. When first settled, the hooked body plates serve to prevent the fish from being washed backward by wave surge. It is not known whether the adult braces itself against overhead rock protrusions with the pit in its head. This poacher relies on close hiding quarters and cryptic appearance rather than flight, for predator evasion. If picked up, it vibrates.

FEEDING ECOLOGY AND DIET
Larval rockheads tend to feed on copepods or fish larvae. Adult rockheads feed on small crustaceans.

REPRODUCTIVE BIOLOGY
Orange eggs are deposited within a kelp holdfast. It is not known whether copulation and internal gametic association are required to permit the female to extrude eggs into such interstitial spaces.

CONSERVATION STATUS
Not listed by the IUCN. This poacher is too rarely encountered to permit evaluation, but the exposed, rocky shorelines it inhabits tend not to face human development.

SIGNIFICANCE TO HUMANS
The rockhead is not sought for any purpose and is only rarely displayed in aquariums, since it tends to hide from view. It is poorly studied. ◆

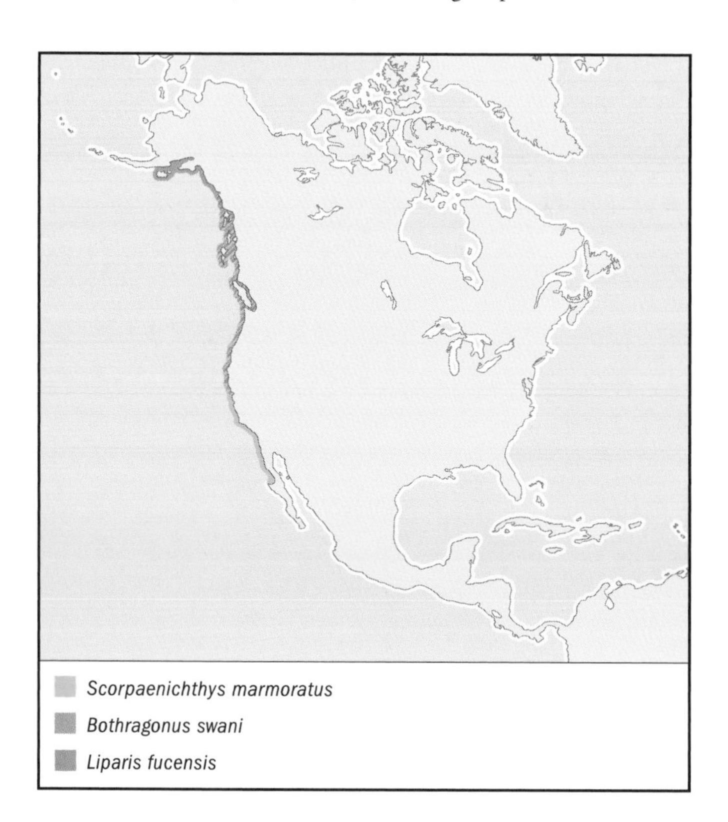

Sablefish
Anoplopoma fimbria

FAMILY
Anoplopomatidae

TAXONOMY
Anoplopoma fimbria Pallas, 1811, San Francisco, United States. Family Anoplopomatidae placed either within suborder Hexagrammoidei or suborder Anoplopomatoidei.

OTHER COMMON NAMES
English: Blackcod, Alaska blackcod, coalfish.

PHYSICAL CHARACTERISTICS
Sablefishes have separate dorsal fins of equal size and a forked tail fin on a streamlined body of gray to black color. Sablefishes grow up to 42 in (107 cm) in length and over 125 lb (57 kg) in weight.

DISTRIBUTION
Deep, offshore waters of the North Pacific, from Baja California (Mexico), to the Bering Sea, and across to southern Japan. Sablefishes range thousands of miles during their lives and occur at depths of over a thousand feet, abundant down to 3,000 ft (914 m). Their young occur in more inshore waters.

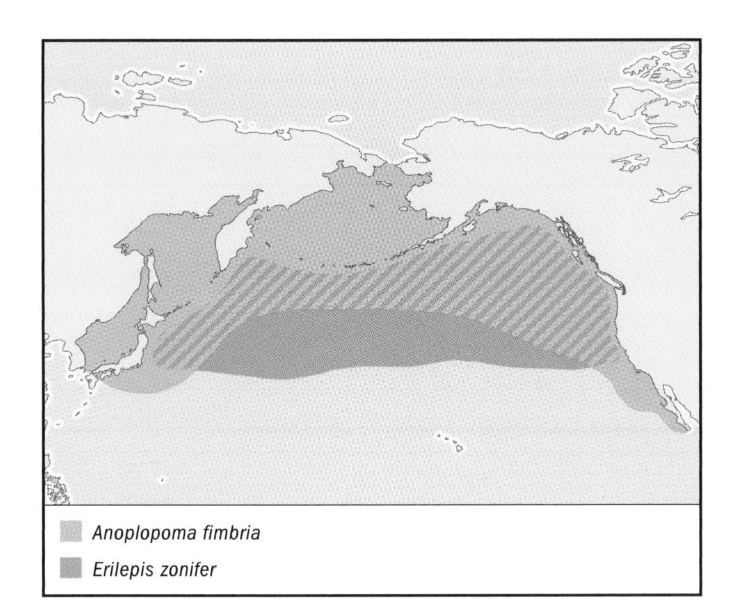

HABITAT
Open, deep ocean water of the North Pacific. They often feed near the bottom in association with deep-sea fishes like rattails and grenadiers.

BEHAVIOR
Sablefishes swim at relaxed speeds and approach and school with fishes of other species, sometimes as a prelude to a sideways lunge for a head-first swallowing of the unsuspecting prey. They cannibalize their young when they are abundant in inshore waters.

FEEDING ECOLOGY AND DIET
Sablefishes eat crustaceans, worms, small fishes, and any larger fishes they can capture, which can include salmon almost as long as the sablefish.

REPRODUCTIVE BIOLOGY
Spawning occurs during winter, and the pelagic eggs float in deep water of about 3,000 ft (914 m) where seawater remains constantly cold. Larvae hatch before functional eyes, jaws, or gut have formed and remain deep until absorbing their large yolk. Larvae with large pectoral fins grow in surface waters during spring.

CONSERVATION STATUS
Not listed by the IUCN. Heavily fished through the last century by American, Canadian, Russian, and Japanese longliners and trawlers, sablefishes are now recognized as a depleted species. Catch quotas are now a small fraction of the peak landings that occurred during the middle of the last century. Russian landings from the Bering Sea were reported to have been 38 million pounds (17,000 metric tons) during 1967. Canadian landings at that time were lowest, around one million pounds (454 metric tons).

SIGNIFICANCE TO HUMANS
Sablefishes have been valued greatly as a smoked fish (smoked Alaska blackcod). Their flesh is quite oily. Native North Americans sundried the sablefish. Because they adapt well to living in tanks and net pens, blackcod are being developed as a high-value species for diversification of salmon farms. ◆

Skilfish
Erilepis zonifer

FAMILY
Anoplopomatidae

TAXONOMY
Erilepis zonifer Lockington, 1880, Monterey, California, United States. This species is sometimes placed alone in the family Erilepidae in the suborder Anoplopomatoidei, but the more accepted classification puts them together with sablefishes in the Anoplopomatidae and within the Hexagrammoidei.

OTHER COMMON NAMES
None known.

PHYSICAL CHARACTERISTICS
The largest of the greenlings and sculpins, skilfishes resemble a heavier version of a sablefish with mottled blue of dark and light shades. Older fishes reach 70 in (178 cm) length and 200 lb (91 kg) weight.

DISTRIBUTION
Deep water of the North Pacific from Monterey Bay, California, to central Honshu Island, Japan and north to the Gulf of Alaska and Kamchatka.

HABITAT
Young are sometimes caught in offshore, surface waters. Adults are typically caught in deeper water.

BEHAVIOR
Not known.

FEEDING ECOLOGY AND DIET
Not known.

REPRODUCTIVE BIOLOGY
Not known.

CONSERVATION STATUS
Not listed by the IUCN. Little is known of the behavior, food habits, or reproduction of this species. It is not the subject of directed fisheries, so there is little basis for determining whether incidental bycatch in other high seas fisheries might have detrimental impact upon the abundance of the skilfish.

SIGNIFICANCE TO HUMANS
None known. ◆

Sailfin sculpin
Nautichthys oculofasciatus

FAMILY
Cottidae

TAXONOMY
Nautichthys oculofasciatus Girard, 1857, Fort Steilacoom, Puget Sound, Washington, United States. Has been listed as *Nautichthys oculo-fasciatus* and has been segregated from the Cottidae into the family Hemitripteridae on the basis of scales modified into embedded spines.

OTHER COMMON NAMES
English: Sailor fish.

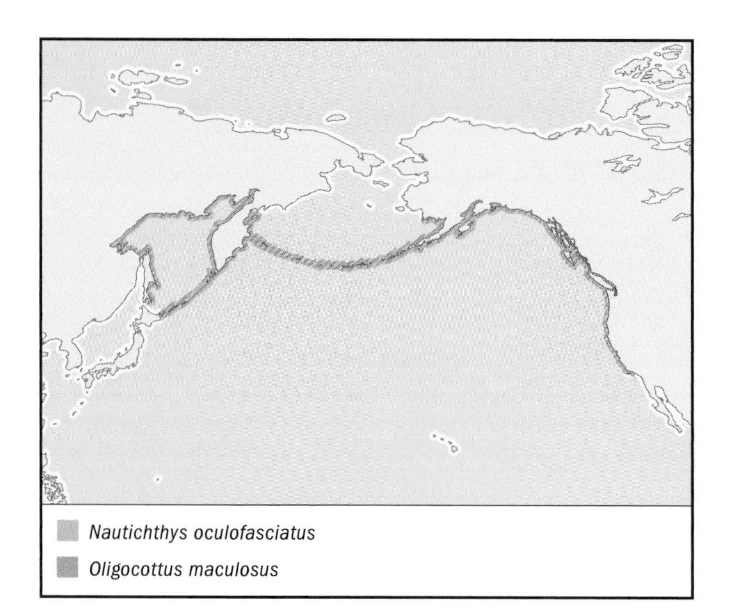

PHYSICAL CHARACTERISTICS
The elongated rays of the first dorsal fin and the long second dorsal and anal fins, together with a dark bar through the eye joining a dark flap of skin over each eye, cause the cream or brown sailfin sculpin to appear very cryptic, especially among seaweed. Sailfins reach 8 in (20 cm).

DISTRIBUTION
From southern California to the Sea of Okhotsk.

HABITAT
Sailfins live on rocky reefs and outcroppings, and on adjacent sand bottom, from shallow water down to over 360 ft (110 m) depth.

BEHAVIOR
In shallow water, the sailfin sculpin waves its first dorsal fin back and forth in synchrony with the motion of adjacent seaweeds in the surge. Between waves of the dorsal fin, the sailfin hops forward by rippling the second dorsal fin and sculling with the pectoral fins. In captivity, sailfins that have never experienced any surge perform the same combination of hopping forward between sweeps of the first dorsal fin. This disruptive mimicry of seaweed obscures the movement of the fish and may enable close approach to prey. This species is nocturnally active.

FEEDING ECOLOGY AND DIET
Feeds on small shrimps and other crustaceans. Individuals slowly approach their prey prior to hopping forward and engulfing them.

REPRODUCTIVE BIOLOGY
Sailfin sculpins copulate, and internal gametic association enables the female to repeatedly extrude the bright orange eggs into interstitial spaces among mussels in the intertidal over a period of weeks during winter. The female must migrate into shallow water during high tides in order to reach the mussel beds, where the eggs are periodically exposed to air but are kept cool and damp by the mussels. Larval sailfin sculpins develop extremely elongate pectoral fins that are spread like butterfly wings and used to glide down through the water column while the larva forages on zooplankton. This behavior enables sailfin sculpins to reach relatively large sizes before permanently settling during late spring.

CONSERVATION STATUS
Not listed by the IUCN. Elimination of mussel beds by harvesting or pollution will limit the reproduction of the sailfin sculpin, but the small fish is not directly taken for any purpose other than use in aquarium displays. As with the grunt sculpin, larvae of the sailfin sculpin are easily cultured.

SIGNIFICANCE TO HUMANS
Sailfin sculpins are popular with night divers and as display species in public aquariums. ◆

Tidepool sculpin
Oligocottus maculosus

FAMILY
Cottidae

TAXONOMY
Oligocottus maculosus Girard, 1856, Neah Bay, Washington, United States.

OTHER COMMON NAMES
English: Tidepool johnny.

PHYSICAL CHARACTERISTICS
Typical sculpin body form, with a small, elongate body, to 2 in (5.1 cm), and a relatively large head. Tidepool sculpins have varying numbers and sizes of cirri on their body, singly or in pairs, especially in the head region. They have one forked cheek spine. Color is banded light and dark gray, sometimes with red or green shades. Newly settled young have red fin rays on the tail.

DISTRIBUTION
Coastal waters from the Los Angeles Bight to the Bering Sea and the Sea of Okhotsk.

HABITAT
Many scientific studies of tidepool sculpins living in tidepools have led to the opinion that this species is an obligate dweller of tidepools. Ironically, spawning is much more dense in protected areas without tidepools than on exposed headlands where pools are formed. Furthermore, tidepool sculpins are abundant in inlets where no tidepools occur, and they strand under rocks during low tides in areas without tidepools. Thus, the tidepool sculpin is a facultative inhabitant of tidepools where they occur, but does not require them to make a living.

BEHAVIOR
When it occurs in tidepools, the tidepool sculpin exhibits homing behavior when displaced from a home pool. The sense of smell appears to assist in home site recognition.

FEEDING ECOLOGY AND DIET
Small crustaceans like amphipods or harpacticoid copepods.

REPRODUCTIVE BIOLOGY
Although it is not known whether this species has internal gametic association, the female mates and then deposits egg clusters in spaces between barnacles or mussels. The eggs are either emerald green, dark green, or maroon. Maroon eggs are laid on shores exposed to wave action, but the same females, removed from exposed shores, in captivity lay green eggs the next season. On small stretches of shore that either gradate or abruptly shift from wave exposure to protection from waves,

the proportion of maroon to emerald eggs similarly gradates or shifts abruptly. The egg pigment may reflect some physiological response of the female to the gas saturation of the seawater, but the subject remains a mystery, as does the polymorphism for egg color in many other sculpin species.

CONSERVATION STATUS
Not threatened. This is the most abundant and commonly occurring shoreline fish in many parts of the Pacific Northwest. It would probably be one of the last species to disappear in the face of environmental degradation.

SIGNIFICANCE TO HUMANS
The tidepool sculpin is the most easily observed fish in many tidepools, and it has been of great interest to students of intertidal biology. ◆

Grunt sculpin
Rhamphocottus richardsoni

FAMILY
Cottidae

TAXONOMY
Rhamphocottus richardsoni Günther, 1874, Fort Rupert, western North America (Prince Rupert, British Columbia, Canada). Sometimes classified alone in the family Rhamphocottidae.

OTHER COMMON NAMES
English: Grunt-fish, pigfish.

PHYSICAL CHARACTERISTICS
With a short, stout body and a large head with elongate snout, the resemblance of a grunt sculpin to a pig is only enhanced by the way it uses its large pectoral fins to scoot around on the seabed. The lower rays of the pectoral fin are elongate and have almost no webbing. The body is covered with small, bristly spines. All of the fin rays are unbranched and the tail bones are fused, as in poachers (Agonidae). These fish grow to 3.3 in (8.3 cm).

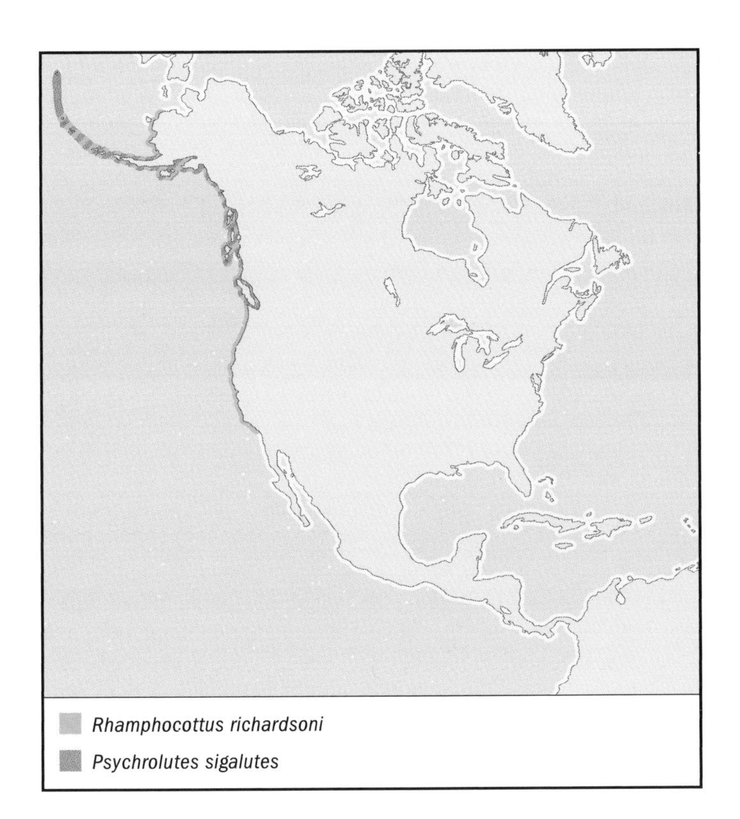

DISTRIBUTION
From Santa Barbara, California, to the Bering Sea coast of Alaska.

HABITAT
Occurs in shallow coastal waters down to 660 ft (200 m) depth, on rocky or shell/sand bottoms, often associated with encrusting invertebrates like giant barnacles.

BEHAVIOR
The elongate pectoral rays can function like fingers for crawling over coarse surfaces. Grunt sculpins are more frequently observed walking over the seabed than swimming. They hide in empty barnacle shells and may appear to be mimicking live barnacles.

FEEDING ECOLOGY AND DIET
Video analysis of predation by grunt sculpins reveals a highly adapted method of capturing relatively large shrimp by approaching obliquely, then snapping the head sideways while extending the snout and sucking water in together with the prey.

REPRODUCTIVE BIOLOGY
Nesting often occurs inside barnacle shells, and controversy exists over whether the female or the male tends the eggs, or whether a mated pair trade off guarding duties. Parental care includes assisting hatching, the parent sucking out hatchlings and spitting them upward into the water. A young female spawns a single mass of eggs in late winter or spring, but with increasing age the females start reproducing with increasing frequency until, at a decade of age, captive females reproduce year-round.

CONSERVATION STATUS
Not listed by the IUCN. The only conservation concern relates to possible temporary depletion at popular diving reefs by past collections for display aquariums. Grunt sculpins are easily cultured, however, and most specimens for public display are now propagated in laboratories at major aquariums.

SIGNIFICANCE TO HUMANS
The grunt sculpin is a popular quarry of scuba divers on rocky reefs along the Pacific Northwest. Similarly, they are popular for display in aquariums because of their unusual appearance and interesting habits. ◆

Cabezon
Scorpaenichthys marmoratus

FAMILY
Cottidae

TAXONOMY
Scorpaenichthys marmoratus Ayres, 1854, San Francisco, California, United States. Has at times been separated into its own family, Scorpaenichthyidae.

OTHER COMMON NAMES
English: Giant marbled sculpin, giant sculpin.

PHYSICAL CHARACTERISTICS
This relatively huge fish (to 30 in [76 cm] and 30 lb [13.6 kg]) has a marbled color pattern that incorporates light and dark gray, brown and beige, or olive green shades. The young often have shades of red. Scales are covered by skin. This is the only sculpin with an unpaired flap of skin on the tip of the snout. The pelagic juvenile is metallic blue, shaped like a blunt-nosed, short salmon smolt.

DISTRIBUTION
Central Baja California, Mexico, to Sitka, southeast Alaska.

HABITAT
Inhabits rocky reefs from very shallow depths down to the limits of kelp growth, sometimes deeper. They tend to hide in kelp beds, sometimes literally hanging in the seaweed. Juveniles occur in tidepools.

BEHAVIOR
An ambush predator, the cabezon tends to be sedentary, relying on the camouflage of its coloring. If stranded in kelp during low tide, a cabezon flares its gill covers and holds still. Pelagic juveniles are attracted to turbulence in a laboratory situation, and will strike at prey in the most rapid flows they can find, which may lead them to settle on exposed shorelines.

FEEDING ECOLOGY AND DIET
Kelp crabs, other crabs, shrimp, snails, clams, worms, or small fish.

REPRODUCTIVE BIOLOGY
Males guard clusters of egg masses during late winter and spring. The egg masses vary in color from burgundy to purple to dark green. The eggs are toxic to birds and mammals that might otherwise predate on them during low tide exposure. Small sculpins crowd in to predate on cabezon larvae hatching at the edges of the nest site, without the male paying any attention to their activity.

CONSERVATION STATUS
Not listed by the IUCN. Cabezons show signs of being overfished through their southern range. Depletion is evident in southern British Columbia, where they are not the target of either sport or commercial fishing, although bycatch landings occur in both fisheries. Commercial setline fishing has been directed at cabezons in California for over half a century, and they are considered a top sport angling species there as well. Cabezon spearfishing is popular because of the ease of spearing these big fish.

SIGNIFICANCE TO HUMANS
The firm flesh of the cabezon is favored by many. It is a popular sport fish with bait anglers, especially in California. ◆

Lumpfish
Cyclopterus lumpus

FAMILY
Cyclopteridae

TAXONOMY
Cyclopterus lumpus Linnaeus, 1758, Baltic Sea and North Sea.

OTHER COMMON NAMES
English: Henfish, lumpsucker; French: Grosse poule de mer.

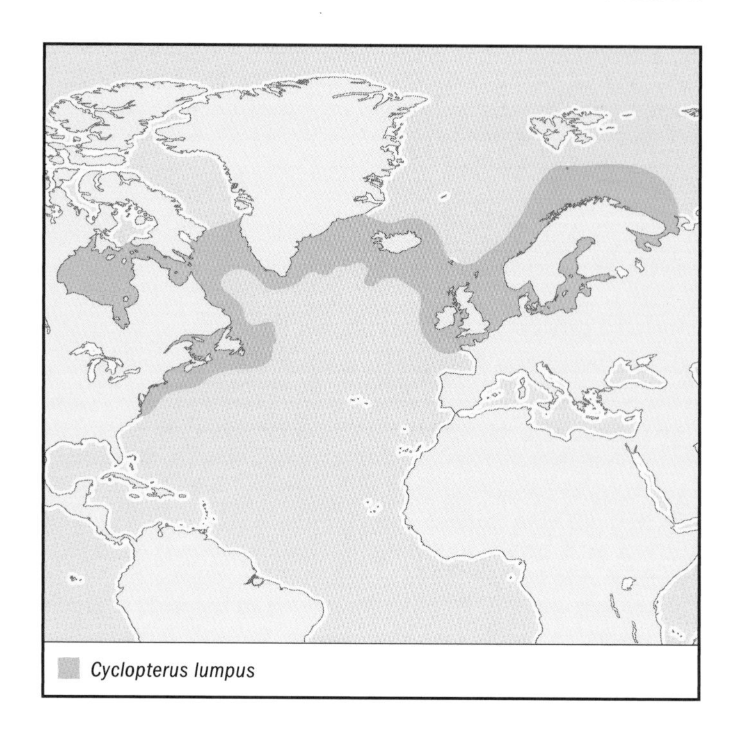
Cyclopterus lumpus

PHYSICAL CHARACTERISTICS
Thick body with rows of hard, conical tubercles, and with a soft, cartilaginous hump on the back, embedding the first dorsal fin. The tubercles are in a single row along the back and in three lateral rows on each side. Pelvic fins are modified into a suction disc. The fish is green, gray, blue, or brown, with red highlights on breeding males. Breeding females become distended with eggs. Lumpfishes have reached 2 ft (61 cm) and 21 lb (9.5 kg).

DISTRIBUTION
Across the North Atlantic Ocean from Chesapeake Bay to Hudson Bay, Greenland, Iceland, the White Sea, and south through the British Isles to France.

HABITAT
This is a bottom fish that inhabits cold waters on rock bottom. Early in life the hatchlings, fully formed as juveniles, are in surface waters, then older juveniles are semipelagic, living under seaweed.

BEHAVIOR
From the time of hatching, lumpfishes are capable of using their pelvic sucking disc to adhere to smooth surfaces, which tend first to be seaweed blades, then rocks.

FEEDING ECOLOGY AND DIET
Crustacean plankton like krill, amphipods, and copepods, as well as jellyfishes and small fishes.

REPRODUCTIVE BIOLOGY
Early in spring the lumpfish female deposits an adhesive egg mass of over 100,000 eggs on rocky bottom among seaweed. Lumpfishes spawn more than one batch of eggs, and hatching occurs from May to July, after six to ten weeks of incubation. Egg color is polymorphic between different females, including colors like brown, red, pink, orange, yellow, green, and purple.

CONSERVATION STATUS
Not listed by the IUCN. Early accounts listed the lumpfish as being of no economic value, but they were always a bycatch in

the inshore cod trap and gillnet fishery, and their ripe eggs were marketed as caviar starting in the late 1960s. In North America the flesh is sold as dog food, but it is eaten by people in Europe. The eggs are stripped from ripe females and packed in brine for preparation as lumpfish caviar.

Overfishing has occurred. In Newfoundland the catch rate was 229 lb (104 kg) per net per year in 1979 (total landings 85 tonnes) at 16 nets per boat, whereas by 1996 the catch rate was down to 21 lb (9.5 kg) per net per year (total landings still 82 tonnes) at 40 nets per boat, and nearly ten times as many boats in the same region.

SIGNIFICANCE TO HUMANS
With the decline of beluga sturgeon and other traditional sources of caviar, lumpfish caviar has become highly valued. ◆

Kelp greenling
Hexagrammos decagrammus

FAMILY
Hexagrammidae

TAXONOMY
Hexagrammos decagrammus Pallas, 1810, formerly *Chiropsis decagrammus* Pallas, Cape St. Elias, Alaska, United States.

OTHER COMMON NAMES
English: Tommy cod, speckled sea trout, greenling sea trout.

PHYSICAL CHARACTERISTICS
Reaches 24 in (61 cm), 4.6 lb (2.1 kg), males smaller. Five lateral lines occur along each side of body, and the male is blue-brown with prominent light blue spots on top of head and anterior body; the female is yellow-brown with small red or orange spots.

DISTRIBUTION
Santa Monica Bay, southern California, to Amchitka Island (Aleutians), Alaska.

HABITAT
Inhabits shallow, subtidal seabed on rocky shores, often around kelp beds.

BEHAVIOR
Fairly active; males in particular cruise around a home area during daylight. More aggressive than the whitespotted greenling, *Hexagrammos stelleri*, which sometimes cohabits with kelp greenlings, the kelp greenling even feeds on unguarded lingcod eggs.

FEEDING ECOLOGY AND DIET
Small shrimps as well as worms and small fishes.

REPRODUCTIVE BIOLOGY
Courtship occurs in late fall, with spawning from October through December, eggs hatching in January through February. The male guards the nest site, at which successive females will be courted for spawn deposition. Eggs are blue to purple with a pale center when first laid, becoming golden brown when embryos are eyed. Larval and pelagic juvenile stages last about two months, with these early stages a gun barrel blue, with snub snout shape. Juveniles school over rock faces prior to settlement and metamorphosis to a benthic juvenile resembling the adult. Sexually dimorphic color is evident during the first summer.

CONSERVATION STATUS
Not listed by the IUCN. Kelp greenlings are caught as bycatch in hook and line fisheries for live rockfishes, and have achieved their own localized market demand in Asian communities in Vancouver, Seattle, San Francisco, and Los Angeles. Little potential exists for overharvest by sport anglers unless using small tackle, owing to small mouth size. Serious depletion of larger groundfish species could prelude the depletion of greenlings in a localized area, since hook, live market, and line fishery includes landings of greenlings as well as larger groundfishes.

SIGNIFICANCE TO HUMANS
Fillets of kelp greenling are high quality. Kelp greenlings are desired in the retail marketing of live fish (primarily rockfishes). Kelp greenlings are too small to provide enjoyable sport fishing unless light tackle is used, but greenlings are easy to catch and therefore provide good sport for small children. ◆

Lingcod
Ophiodon elongatus

FAMILY
Hexagrammidae

TAXONOMY
Ophiodon elongatus Girard, 1854, San Francisco, California, United States. Sometimes placed in family Ophiodontidae.

OTHER COMMON NAMES
English: Cultus cod, ling.

PHYSICAL CHARACTERISTICS
Lingcod are large, up to 5 ft (1.5 m) in length and 100 lb (45 kg) weight (males smaller), and have a large mouth extending

behind the eyes. The mouth has prominent teeth. The spiny and soft dorsal fins are joined to form one long, moderately notched dorsal fin running the entire length of the body behind the head, and the tail fin is squared, not forked. The color is brown, rarely blue-green, with a staggered array of black blotches along the body midline and top.

DISTRIBUTION
From Ensenada, Mexico (Baja California), to the Alaska Peninsula (Shumigan Islands).

HABITAT
Lingcod spawn on rocky reefs along the shoreline, usually at depths of about 33–99 ft (10–30 m), but spawning has been observed in the intertidal and by submarine at much greater depths. The females migrate onto sand and mud bottoms at greater depths up to 330 ft or more (100 m), except when they return inshore for spawning, whereas males tend to remain all year on the spawning reefs. Lingcod will hide in crevices. Young lingcod tend to be more generally distributed near the shoreline, avoiding areas occupied by adults. Recently settled lingcod have been collected in eelgrass beds and have been seen from a submarine on flat bottom at the base of a cliff over 360 ft (110 m) deep.

BEHAVIOR
Aside from male territoriality, female seasonal migrations, and rapacious predatory behavior, lingcod tend to be sedentary ambush predators. They rest near rocks and wait for prey to swim near. During salmon migrations, however, lingcod have been observed predating at the surface over great depths, so relative abundance and position of prey affect behavior.

The life history of lingcod relates closely to that of Pacific herring. Larval lingcod settle from the plankton at the time during spring when herring larvae are becoming silver juveniles. Young lingcod that have not settled permanently from a swimming habit search in school formation during the twilight hours of dawn and dusk, and young herring are their favorite prey.

FEEDING ECOLOGY AND DIET
Although lingcod will eat invertebrates such as crabs, shrimps, and octopi, they mainly feed on other fishes, including younger lingcod. A lingcod engulfs another fish headfirst. The throat rapidly opens while the mouth engulfs the prey, so that a fish about two-thirds the length of the lingcod will be swallowed immediately into the entire length of the stomach, with only the tail protruding from the mouth. During years of abundant prey, growth is rapid. After two years lingcod of both sexes tend to reach about 1.5 ft (46 cm) in length, after which males grow more slowly than females, perhaps owing to the seasonal feeding migration that only females undertake.

REPRODUCTIVE BIOLOGY
Males become jet black and fight over territory during winter, prior to arrival of ripening females. Males will successively spawn with different females, guarding up to three egg masses at a time. Males are capable of spawning at two years of age, and females at three, but most females do not lay eggs until they are four. If larger females are not abundant, then females tend to mature and spawn a very small egg mass at three years of age. In British Columbia, peak abundance of guarded egg masses is during February, although spawning can occur from December through April. Spawning occurs later in more northerly latitudes. Older females of 10–15 years of age can spawn a half million eggs, and they spawn earlier and deeper than the younger fish. Larvae spawned by the largest females tend to be slightly larger than larvae of small females, which could confer advantage under certain feeding conditions in the plankton. Thus, a population with a full demographic spread from young to old fish will have greater chances of survival of young under a variety of environmental conditions.

CONSERVATION STATUS
Not listed by the IUCN. Lingcod have been extremely depleted since the 1980s in Puget Sound, and since the 1990s in the Strait of Georgia. Outer coast populations have become overfished in more recent years. It has been demonstrated mathematically that even the earliest hand-line fisheries prior to World War II led to significant reduction in lingcod biomass in inland seas around Vancouver and Seattle. More efficient otter trawls in the 1940s greatly increased levels of landings, which in British Columbia exceeded eight million pounds per year (over 3,700 metric tons). Landings in the Strait of Georgia were negligible when the commercial fishery closed in 1990, but since then it has become evident that sport fishing alone can prevent population recovery near metropolitan areas. Lingcod are of interest for management strategies that include protection within sanctuaries (marine protected areas).

SIGNIFICANCE TO HUMANS
The common name "cultus" is a Coast Salish term meaning "cheap," which indicates that original levels of abundance ensured that lingcod could be caught for use as food when preferred species like halibut became less available. As mentioned, lingcod has always been valued as a fresh fish. Aquaculture is possible but not yet economical. Appreciation of the value of lingcod as a sport species tends to increase as the availability of this and other groundfish species declines in a given area. ◆

Slipskin snailfish
Liparis fucensis

FAMILY
Liparididae

TAXONOMY
Liparis fucensis Gilbert, 1895, Strait of Juan de Fuca, 109 fathoms. Is classified sometimes within the Cyclopteridae, and the family name Liparididae was formerly Liparidae.

OTHER COMMON NAMES
English: Juan de Fuca liparid.

PHYSICAL CHARACTERISTICS
This snailfish grows to 7 in (18 cm) and has a lobe at the front of its dorsal fin and the anal fin barely extending onto the tail fin. Color is from brown to olive. Like other snailfish, it has a tadpole shape to the body, large pectoral fins with extended lower rays, and a small pelvic suction disc. Larvae become spherical in a globular bubble of body fluid beneath the skin. Immediately upon settlement there is a metamorphic change (shrinkage) to a snailfish shape like a tadpole.

DISTRIBUTION
Northern California to southeast Alaska.

HABITAT
Has been collected over a wide range of depths from the shore down to 1,270 ft (388 m).

BEHAVIOR
The larval stage grows for an extended period in the planktonic realm by means of neutral buoyancy conferred by the globby, bubble shape of the body.

FEEDING ECOLOGY AND DIET
Small crustaceans including shrimp.

REPRODUCTIVE BIOLOGY
The male slipskin snailfish guards a cluster of egg masses inside an empty mussel shell during spring. Egg color varies between egg masses, from tan to pink or orange.

CONSERVATION STATUS
Not listed by the IUCN. This snailfish is relatively common in occurrence.

SIGNIFICANCE TO HUMANS
None known. ◆

Soft sculpin
Psychrolutes sigalutes

FAMILY
Psychrolutidae

TAXONOMY
Psychrolutes sigalutes Jordan and Starks, 1895, Puget Sound near Port Orchard, Washington, United States. Species is listed as *Gilbertidia sigalutes* in older species compendiums, and is listed as a member of the family Cottidae.

OTHER COMMON NAMES
None known.

PHYSICAL CHARACTERISTICS
The soft sculpin has flaccid skin in which the single dorsal fin seems embedded. The lateral line and head canal pores of adults are large and obvious, especially around the jaws and cheeks. Adults are dark brown or translucent beige. Adult males are much larger than females and often have scars over their large heads. Males can approach 3.5 in (9 cm) in length. The pelagic young have relatively larger eyes, no obvious pores, and are purple with orange pectoral fins.

DISTRIBUTION
Coastal waters from Puget Sound, Washington, to the Aleutian Islands, Alaska.

HABITAT
Larvae of the soft sculpin migrate to feed at the surface at dawn and dusk. Then as pelagic juveniles, they alternate between settling to soft bottom and migrating to the surface on dark nights when surface plankton is abundant. Females grow to 75% of mature body size while exploiting the plankton. Soft sculpins permanently settle during late spring and occupy deep crevices or recesses in multilayered rock rubble, usually on shorelines protected from wave surge.

BEHAVIOR
The male soft sculpin continues growing after settlement, whereas the female starts developing ripe ovaries. The male is highly territorial, rotating his body in a circle and stuttering his head to keep other males away from his territory. Head biting occurs among males.

FEEDING ECOLOGY AND DIET
Feeding is visual during early larval stages, then the pelagic juveniles switch to use of distant touch at night as the eyes become relatively smaller and the head canal pores enlarge. Adults in crevices use distant touch to detect crustacean prey like amphipods. Adults cannibalize young soft sculpins that enter their territory.

REPRODUCTIVE BIOLOGY
The male soft sculpin attracts a harem of several females and courts them in synchrony prior to copulating with each female in sequence. Females simultaneously lay a group, monolayer egg mass on the underside of a rock surface. The females tend the eggs communally, the fanning behavior of one female stimulating similar behavior in adjacent females. The male remains to one side and guards the nest site against other fishes or invertebrates. At hatching, the females suck larvae from the egg shells, swim away from the nest crevice, and spit the larvae toward the surface.

CONSERVATION STATUS
Not listed by the IUCN. Detection of soft sculpins tends to occur rarely. It is doubtful whether human activities directly affect this species.

SIGNIFICANCE TO HUMANS
None known. ◆

Resources

Books

Horn, Michael H., Karen L. M. Martin, and Michael A. Chotkowski, eds. *Intertidal Fishes: Life in Two Worlds.* San Diego: Academic Press, 1999.

Periodicals

Jackson, K. L., and J. S. Nelson. "*Ambopthalmos*, a New Genus for '*Neophrynichthys*' *angustus* and '*Neophrynichthys*' *magnicirrus*, and the Systematic Interrelationships of the Fathead Sculpins (Cottoidei, Psychrolutidae)." *Canadian Journal of Zoology* 76 (1998): 1344–1357.

Kirilchik, S. V., and S. Ya. Slobodyanyuk. "Evolution of the Cytochrome *b* Gene Fragment from Mitochondrial DNA in Some Baikalian and Non-Baikalian Cottoidei fishes." *Molecular Biology (Molekulyarnaya Biologiya)* 31 (1997): 141–148.

Munehara, H. "The Reproductive Biology and Early Life Stages of *Podothecus sachi* (Pisces: Agonidae)." *Fishery Bulletin* 95 (1997): 612–619.

Munehara, H., Y. Koya, Y. Hayakawa, and K. Takano. "Extracellular Environments for the Initiation of External Fertilization and Micropylar Plug Formation in a Cottid Species, *Hemitripterus villosus* (Pallas) (Scorpaeniformes) with Internal Insemination." *Journal of Experimental Marine Biology and Ecology* 211 (1997): 279–289.

Wonsettler, A. L., and J. F. Webb. "Morphology and Development of the Multiple Lateral Line Canals on the Trunk in Two Species of *Hexagrammos* (Scorpaeniformes, Hexagrammidae)." *Journal of Morphology* 233 (1997): 195–214.

Yabe, M., and T. Uyeno. "Anatomical Description of *Normanichthys crockeri* (Scorpaeniformes, *incertae sedis*: Family Normanichthyidae)." *Bulletin of Marine Science* 58 (1996): 494–510.

Jeffrey Burton Marliave, PhD

Percoidei I

(Perches and darters, North American basses and sunfishes, pygmy sunfishes, and temperate basses)

Class Actinopterygii
Order Perciformes
Suborder Percoidei
Number of families 4

Photo: The pumpkinseed sunfish (*Lepomis gibbosus*) is found in large numbers in the shallow sheltered areas of ponds and lakes. (Photo by Animals Animals ©Stouffer Prod. Reproduced by permission.)

Evolution and systematics

The suborder Percoidei contains more than 70 families and 2,800 species. This chapter and the four Percoidei chapters that follow highlight a representative sample of Percoidei taxa. This chapter focuses on four families:

- Percidae, the perches, with 10 genera and 162 species
- Centrarchidae, the sunfishes, with eight genera and 29 species
- Elassomatidae, the pygmy sunfishes, with one genus and six species
- Moronidae, often grouped under the umbrella descriptor "temperate basses," with two genera and six species

All four families fall within the massive order Perciformes, although the placement of Elassomatidae in this order has been disputed. The current classification positions Percidae, Centrarchidae, and Moronidae within the 71-family suborder Percoidei, and the pygmy sunfishes in the single-family suborder, Elassomatoidei. Some taxonomists have suggested moving the elassomatids out of the percomorphs altogether and placing them into the atherinomorphs, a broad grouping that includes the killifishes (cyprinodonts), but at least one genetic analysis (Jones and Quattro, 1999) discounts this change. Past arrangements had placed the pygmy sunfishes together as a subfamily within Centrarchidae, but their similarities in appearance have since been attributed to evolutionary convergence.

Changes have also occurred in other families. The moronids, for example, now include species formerly within the family Percichthyidae. Some biologists consider them to be most closely related to the snooks and giant perches in the family Centropomidae, because both share two additional lateral lines on the tail, above and below the main lateral line. Little doubt exists that all four families—Percidae, Centrarchidae, Moronidae, and Elassomatidae—will undergo further taxonomic alterations in the future, as scientists learn more about the phylogenetic relationships between species, genera, and higher classifications.

Physical characteristics

The perch family is a large one, making generalizations about physical appearance rather difficult. These species do, however, have some unifying traits, such as an elongate body. All percids except the genus *Zingel* have either two fully separate dorsal fins or two dorsal fins that are only minimally connected. The opercle has one sharp spine. Most fishes in this family, including the North American darters, are on the small side. One exception is the walleye (*Stizostedion vitreum*), a popular game fish that can grow to 42 in (107 cm).

Sunfishes, many of which are described by the angler simply as "panfish," are typically broad animals with bodies one-third to one-half as tall as they are long. The single dorsal fin usually has 10–12 spines, although the range runs from 5–13. Some, like the longear (*Lepomis megalotis*) and pumpkinseed sunfishes (*L. gibbosus*), are colorful in hues of orange, green, and blue, while others are quite drab. Maximum adult sizes range from less than 4 in (10 cm) in two *Enneacanthus* species, to 39 in (nearly 1 m) in the largemouth bass (*Micropterus salmoides*).

Moronid species all have two dorsal fins, including a spiny front fin and a mostly soft-rayed hind fin. Other distinguishing

Male *Elassoma okefenokee* in full breeding color. (Illustration by Michelle Meneghini)

features include a two-spined opercle and a long lateral line with two additional lateral lines on the tail. Maximum adult sizes range from nearly 1.5 ft (46 cm) in the white bass (*Morone chrysops*), to 6.6 ft (2 m) in the striped sea bass (*M. saxatilis*).

Members of the pygmy sunfish family have a rounded tail fin and a three- to five-spined dorsal fin with eight to 13 soft rays. They lack a lateral line on the body. Pygmy sunfishes are less broad than the typical centrarchid and considerably smaller. Maximum adult sizes range from 1.3 in (3.3 cm) in the Carolina pygmy sunfish (*Elassoma boehlkei*), to 1.9 in (4.7 cm) in the banded pygmy sunfish (*E. zonatum*).

Sunfish nests. (Illustration by Emily Damstra)

Distribution

Most species in this group occur in North America. Some percid species extend throughout the Northern Hemisphere, and the family Moronidae contains members that naturally occur in Europe and northern Africa. In addition to the native populations of the four families, numerous species, particularly among the centrarchids and percids, have been purposely introduced as game and/or food fishes in new geographic areas nearly around the world. As a result, various centrarchid species are now found in lakes and ponds throughout temperate Europe, Japan, and southern Africa and Australia.

Habitat

Fishes in the families Centrarchidae and Elassomatidae live in strictly freshwater habitats. Centrarchids prefer a temperate climate, and are especially common to lakes, streams, and other inland waterways in the northern United States and Canada. Many seek out weedy areas or other protective cover, such as swamps with fallen trees. Elassomatids similarly inhabit swampy areas with heavy vegetation, but favor the warmer environs of the southeastern United States.

The large family Percidae is primarily a freshwater group, with species in just about every type of waters, from lakes and swamps to both fast- and slow-moving streams and other bodies of water. Some live in the brackish waters of estuaries and in salt lakes. Moronids are the most versatile of the four families in regard to habitat, with members existing in fresh, brackish, and marine waters along coasts.

Various species within these families may coexist in a single body of water. For example, a typical inland lake in the Great Lakes region may be home to numerous species, including largemouth bass, bluegill (*Lepomis macrochirus*), and

Largemouth bass (*Micropterus salmoides*) cleaning the eggs in its nest. (Illustration by Emily Damstra)

Two males and one female (large, center fish) walleye (*Stizostedion vitreum*) spawning. (Illustration by Emily Damstra)

others in the family Centrarchidae, as well as the yellow perch (*Perca flavescens*), walleye, and others in the Percidae.

Behavior

Because of the breadth of this four-family group, behavior varies widely. Some species school; some exist alone. Some are broadcast spawners; others build nests and mate one-on-one. Numerous species provide no parental care; others build nests and guard eggs and/or young.

The centrarchids are particularly known for their nest-building and parental care. Nest-building typically involves the male setting up a small territory, fanning its tail to create a depression in the substrate, then enticing a receptive female to his nest to lay her eggs, which he fertilizes. In many centrarchids, as well as darter perches, more than one female may be attracted to a male's territory. This often results in the male mating with different females in a single breeding season. Male centrarchids defend the nest at least until the eggs hatch, and the males of a few species continue to watch over the young for several weeks.

One unusual behavior has gained some attention. Largemouth bass in the family Centrarchidae will frequently eat whirligig beetles (*Dineutes hornii*), which are flat, round insects that move along calm water surfaces in looping, swirling patterns. These beetles secrete a noxious slime that repels most predators. The bass, however, have developed a way to get around the goo. According to researchers, they rinse off the beetle by repeatedly taking the insect into the mouth, gargling it, and spitting it out until the unpleasant taste is gone, or until the bass gives up. If the bass is persistent enough to rinse off the slime (which can take more than a minute) it eats the beetle. If not, the beetle escapes.

The darters comprise a huge group within the perch family, and the 146 species live in waters of North America. Members are characterized by their wary nature: Whenever the slightest threat arises, they quickly "dart" for cover. They also exhibit other interesting behaviors. The johnny darter (*Etheostoma nigrum*), for instance, has atypical spawning habits. The male prepares a spot under some cover, perhaps a rock overhang. When a mate arrives, the male and female both swim to the site and turn upside down to spawn. The female's eggs stick to the overhead cover. The male's job is to tend and guard the eggs for the next several weeks until they hatch. Other fishes in the Percidae may mate in pairs like the johnny darter, spawn in small assemblages of one female and several males, or engage in mass spawning.

The breeding season for moronids begins with male courtship displays, usually involving one female and several males. The routine typically includes the males closely following a female and nudging her vent area, and in some species, stereotypically swimming in circles. When appropriately stimulated, the female rises to the surface with the males in tow, and spawning occurs. Occasionally, some *Morone* species will hybridize with each other.

Perhaps the most interesting behavior among the elassomatids occurs during territorial and courtship displays. The territorial display takes place when a male approaches another's territory. The territorial male displays toward the intruder by expanding and/or rapidly flitting his fins, turning sideways as to present his largest view, bending his head and tail toward the intruder, and as a last resort, striking with his head. During courtship, a male elassomatid displays toward a female by bobbing up and down and waving his fins, while swimming toward the spawning site. When a female approaches, the male develops a bright spawning coloration and begins to tremble, while bumping and nipping at the female. Courtship continues for several minutes before the release of eggs and milt.

Feeding ecology and diet

Many centrarchids and percids are near or at the top of the food chain in their habitats. Largemouth bass and walleye, for example, are typically top predators, feeding on herbivorous as well as smaller carnivorous fishes. Even the much smaller yellow perch and bluegill are primarily carnivorous and will take invertebrates and minnows. Centrarchids also include a number of specialized mollusk-feeding forms, such as the longear and the red-ear (*Lepomis microlophus*) sunfishes. Many centrarchids and percid species are known as crepuscular feeders, but anglers often take considerable numbers of

School of yellow perch (*Perca flavescens*). (Illustration by Emily Damstra)

The rainbow darter (*Etheostoma caeruleum*) has been found in almost every state of the United States east of the Mississippi River and west of the Appalachian Mountains. (Photo by Animals Animals ©Raymond A. Mendez. Reproduced by permission.)

these species at midday (although fishing does peak around dawn and dusk).

Some centrarchids are also known to alter their feeding ecology depending on the species composition of their neighbors. Bluegills living in waters with predatory largemouth basses, for instance, are more likely to forage in weeds rather than in open water. This action may protect them from the jaws of a bass, but it also limits their ability to find food. Additional studies of bluegill and pumpkinseed sunfishes report that they will shift their diet to zooplankton or bottom-dwelling prey in deference to the apparently dominant green sunfish (*L. cyanellus*) when all three occur together. Elassomatids and moronids are also carnivorous. Both eat invertebrates, especially crustaceans, and adult moronids are also piscivorous.

Predation on the species within these four families is primarily by larger piscivorous fishes. For example, northern pike (*Esox lucius*) will eat even quite large yellow perch, which will, in turn, feed on smaller fishes. Other predators of shallow water fishes include piscivorous birds, such as herons and ospreys.

Reproductive biology

Percid reproduction varies. Some scatter their eggs and milt into the water or over vegetation, while others spawn over gravel nests. Some species deposit individual eggs; others, such as the yellow perch and the European perch (*Perca fluviatilis*), lay their eggs in masses. Some species, such as the yellow perch and sauger (*Stizostedion canadense*), leave the area almost immediately after spawning; others, such as the johnny darter and tessellated darter (*Etheostoma olmstedi*), remain with their eggs until hatching. Among centrarchids, the bluegill and longear sunfishes prefer to spawn in still waters, others, including many darters, opt for the running water of a stream.

During mating season, males generally develop brighter coloration, which apparently assists in attracting females and announcing their territories to other males. Sunfishes spawn in one of three ways. A male may make a nest and attract females for mating there. Females may spawn with smaller males known as satellite and sneaker males. During breeding season, satellite males take on coloration of a female and trick the nesting male into allowing them to approach, giving them access to females coming to the nest. Sneaker males hide among vegetation near a nesting male's site, watch the females arrive, then dash through the nest site ejecting milt as they reach the females. Non-nesting males mature earlier than nesting males, commonly going through the sneaker stage when they reach the age of two years, then becoming satellite males in later years. Nesting males are usually older individuals that top the seven-year mark. Some controversy exists over whether elassomatids are nest builders. Scientists have reported spawning with and without nests in some species, leading to the hypothesis that substrate conditions determine whether nesting occurs.

Of the moronids, perhaps most is known about the behavior of the striped sea bass, probably because of its popularity as a sport fish. These fishes inhabit coastal waters along both the Atlantic (its native distribution) and Pacific coasts. The Pacific introduction can be traced to 1879 and 1882, when the fishes were translocated to the San Francisco shore. Striped sea basses are migratory, moving to cooler northern waters in summer and warmer southern waters in the fall. As breeding season approaches, they return to their home stream to spawn, apparently via olfactory cues. Feeding ceases prior to the annual breeding season, during which the fishes engage in broadcast spawning. One study of the European sea bass (*Dicentrarchus labrax*) indicated that its gender can be influenced by temperature. Experiments showed that temperature differences during the development of eggs could affect the ultimate sex ratio.

Bluegills (*Lepomis macrochirus*) occur in all three drainage basins in Wisconsin (Lake Michigan, Mississippi River, and Lake Superior). (Photo by Animals Animals ©E. R. Degginger. Rerpoduced by permission.

Conservation status

Careful monitoring of species fished for sport maintains these populations in adequate numbers. Nongame species, and those with very small distributions, fare less well. For example, numerous darter species, which are neither sport fishes nor broadly distributed, are listed as threatened or endangered. Some species in these families have been widely introduced outside of their geographic ranges. The successful naturalization of percids and centrarchids in areas well outside of their aboriginal ranges has often had deleterious effects upon faunas of indigenous fishes.

The IUCN Red List categorizes 1 species from this group as Extinct; 2 species as Critically Endangered; 4 species as Endangered; 30 species as Vulnerable; 12 species as Lower Risk/Near Threatened; and 7 as Data Deficient.

Significance to humans

A great number of species within these four families are prized sport fishes. Examples are walleye and perch in the family Percidae, as well as both large- and smallmouth bass in the family Centrarchidae. A few, such as the banded pygmy sunfish (*Elassoma zonatum*) are considered indicator species, and provide scientists with information about the environmental health of a particular waterway. Several members of these families, particularly the yellow perch and walleye (in the family Percidae), are commercial food fishes.

1. Logperch (*Percina caprodes*); 2. Breeding male rainbow darter (*Etheostoma caeruleum*); 3. Ruffe (*Gymnocephalus cernuus*); 4. Blue-spotted sunfish (*Enneacanthus gloriosus*); 5. Rock bass (*Ambloplites rupestris*); 6. Female black crappie (*Pomoxis nigromaculatus*). (Illustration by Emily Damstra)

1. Banded pygmy sunfish (*Elassoma zonatum*); 2. Male yellow perch (*Perca flavescens*); 3. Walleye (*Stizostedion vitreum*); 4. Largemouth bass (*Micropterus salmoides*); 5. Male longear sunfish (*Lepomis megalotis*); 6. Bluegill (*Lepomis macrochirus*); 7. Striped sea bass (*Morone saxatilis*). (Illustration by Emily Damstra)

Species accounts

Rock bass
Ambloplites rupestris

FAMILY
Centrarchidae

TAXONOMY
Ambloplites rupestris Rafinesque, 1817, Lakes of New York, Vermont, United States. No subspecies are recognized.

OTHER COMMON NAMES
English: Goggle eye, northern rock bass, redeye; French: Crapet de roche; German: Gemeiner felsenbarsch, gemeiner sonnenbarsch.

PHYSICAL CHARACTERISTICS
Maximum total length 17 in (43.2 cm), typically little more than one-half that. Large-mouthed, red-eyed fish with rows of small, chocolate-colored squares along the sides of its greenish to brownish body.

DISTRIBUTION
North America from the Mississippi Valley almost to the Atlantic coast, and Lake Winnipeg to Missouri and the northern boundaries of Georgia and Alabama. Also widely introduced worldwide.

HABITAT
Heavily vegetated areas of freshwater lakes and ponds, as well as clear, rocky streams.

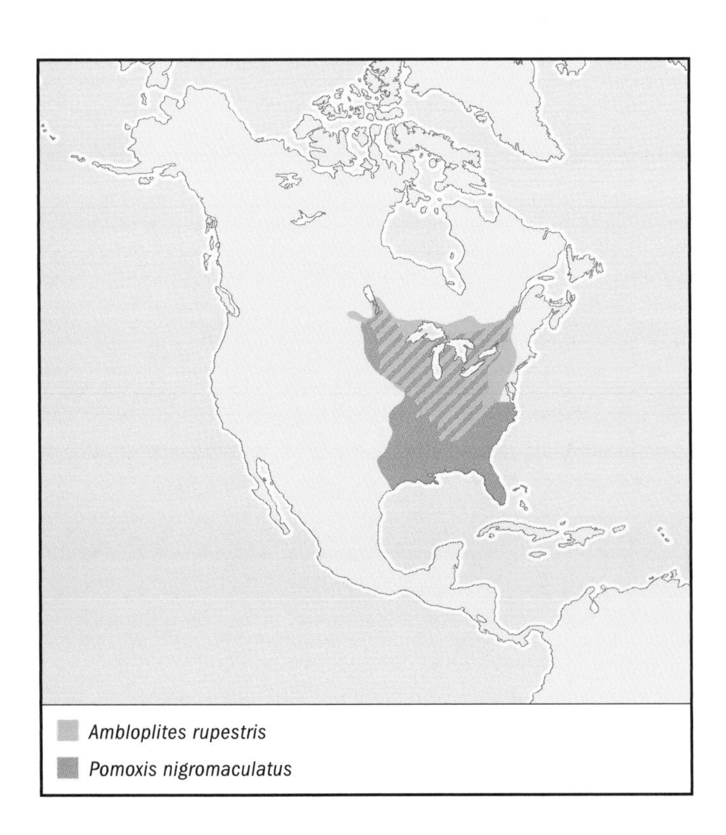

BEHAVIOR
Schools in winter, becomes solitary when breeding season commences in the spring.

FEEDING ECOLOGY AND DIET
Feeds on aquatic plants, invertebrates, and fishes, occasionally conspecifics.

REPRODUCTIVE BIOLOGY
Spawns in late spring and early summer over male-constructed and guarded nests. The demersal, adhesive eggs hatch in three to four days. The male continues to protect the young as long as they remain in the nest area. Once they scatter, usually within a few days, protection ceases. They reach sexual maturity at three to five years.

CONSERVATION STATUS
Not threatened.

SIGNIFICANCE TO HUMANS
Minor sport and commercial fishes. ◆

Blue-spotted sunfish
Enneacanthus gloriosus

FAMILY
Centrarchidae

TAXONOMY
Enneacanthus gloriosus Holbrook, 1855, South Carolina, Georgia, and Cooper Rivers, South Carolina, United States. No subspecies are recognized.

OTHER COMMON NAMES
German: Kiemenfleck-diamantbarsch.

PHYSICAL CHARACTERISTICS
Maximum total length 3.7 in (9.5 cm). Similar in general appearance to a bluegill, but with proportionally longer fins; dark "teardrop" band beneath eye; vertical, dark banding on body; and numerous metallic blue spots in the head, body, and vertical fins.

DISTRIBUTION
Eastern United States, from southern New York to western Florida.

HABITAT
Prefers vegetated freshwater lakes, ponds, pools, and stream backwaters.

BEHAVIOR
Solitary; seldom strays far from cover. Easily dominated by other centrarchids; typically found in substantial numbers only in habitats *Lepomis* species do not find congenial, such as the highly acidic black waters of the New Jersey pine barrens.

FEEDING ECOLOGY AND DIET
Normally feeds on snails and other invertebrates near cover of vegetation.

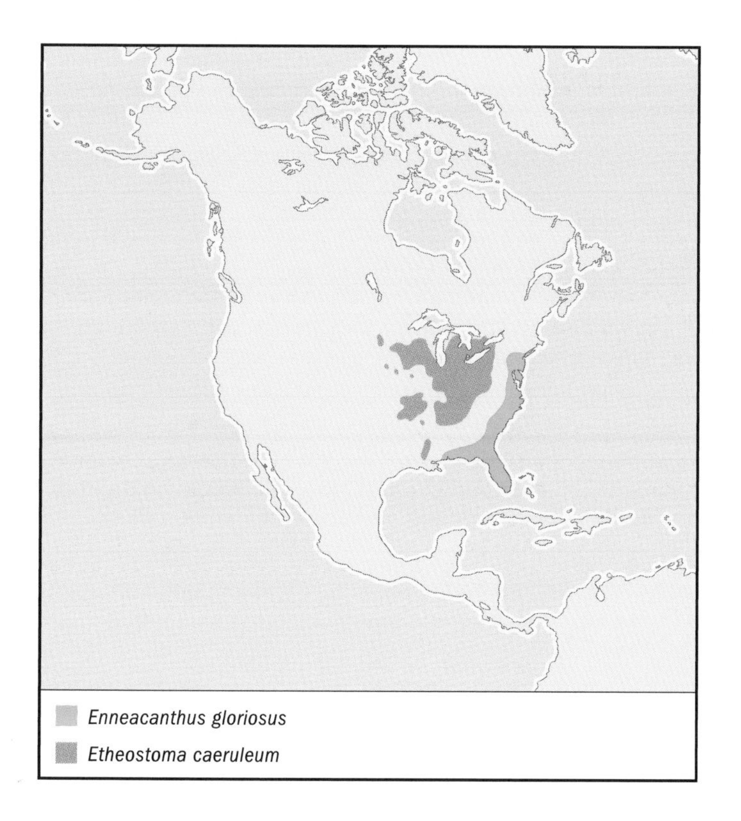

REPRODUCTIVE BIOLOGY
Males build nests 4–12 in (10.2–30.5 cm) in diameter in algae or soft substrate. Spawning occurs in spring, perhaps extending from late winter to early summer. Little is known of its reproductive biology in the wild.

CONSERVATION STATUS
Not listed by the IUCN.

SIGNIFICANCE TO HUMANS
Sometimes kept as an aquarium fish. ◆

Bluegill
Lepomis macrochirus

FAMILY
Centrarchidae

TAXONOMY
Lepomis macrochirus Rafinesque, 1819, Ohio River, United States. Two or three subspecies are recognized.

OTHER COMMON NAMES
English: Bluegill sunfish, sunfish; German: Blauer Sonnenbarsch; Spanish: Pez sol.

PHYSICAL CHARACTERISTICS
Maximum total length 16 in (41 cm). Broad, rather flat fish with a small mouth. Distinguished from its relatives by an all-black opercular flap, general gray-blue coloration, a dark spot at the rear edge of the soft dorsal fin, and dark banding on the sides of the body.

DISTRIBUTION
Common in North America from southern Canada to northern Mexico; widely introduced throughout the world.

HABITAT
Freshwater, inland waters from large lakes to small ponds, also slow-moving streams. Prefers some type of cover, such as rocky or vegetated areas.

BEHAVIOR
Schooling fishes. Schools of several dozen smaller fishes ranging up to 4 in (10.2 cm) long are commonly seen along lake shores in 1–2 ft (0.3–0.6 m) of water, darting from beneath docks and boats. Larger fishes generally remain further from shore in deeper water. Cleaning behavior has been recorded for Florida populations.

FEEDING ECOLOGY AND DIET
Mainly diurnal feeders. Diet comprises invertebrates and small fishes.

REPRODUCTIVE BIOLOGY
Breeds in the late spring and early summer, when groups of males enter shallow water to begin building nests, which are depressions in the substrate. The male guards its nest. Females may also spawn with smaller males known as satellite and sneaker males, which take on the female coloration and fool nesting males into allowing them to approach and mate with females coming to the nest. Sneaker males may also lie in ambush in vegetation near a nesting male's site, wait for females to arrive, then quickly swim through the nest site, ejecting milt. The spherical, demersal eggs, which are laid singly or in small clusters, typically hatch in two to three days. Bluegills in the field typically attain sexual maturity at two to three years and about 4–5 in (10.2–12.7 cm) in length. Bluegills hybridize with pumpkinseeds and many other sunfish species.

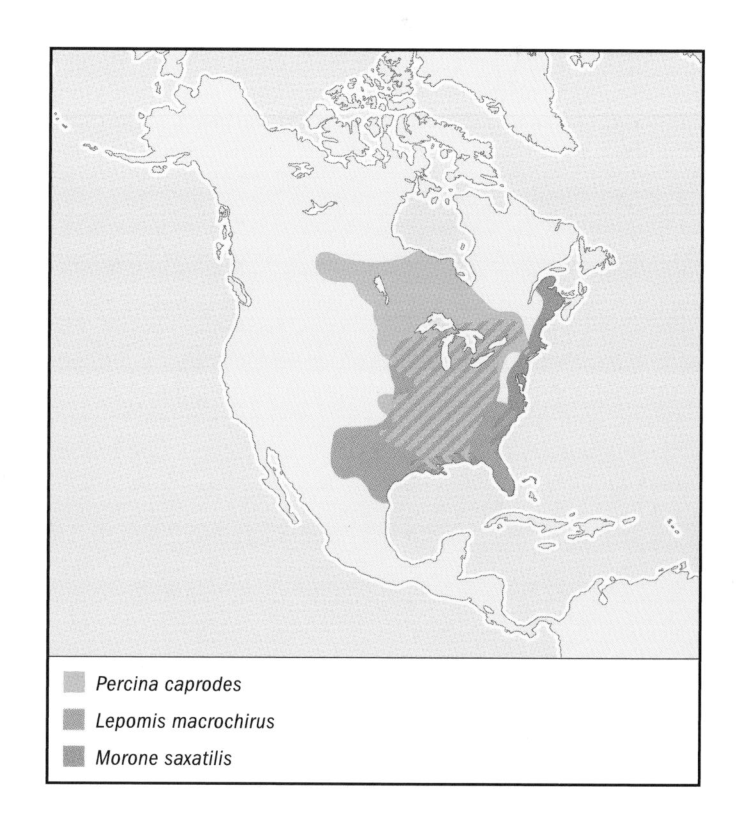

CONSERVATION STATUS
Not threatened.

SIGNIFICANCE TO HUMANS
Sport and minor commercial food and aquarium fishes. ◆

Longear sunfish
Lepomis megalotis

FAMILY
Centrarchidae

TAXONOMY
Lepomis megalotis Rafinesque, 1820, Licking and Sandy Rivers, Kentucky, United States. Five or six subspecies are recognized.

OTHER COMMON NAMES
English: Creek perch, Great Lakes longear, northern longear.

PHYSICAL CHARACTERISTICS
Maximum total length 9.5 in (24 cm). Similar in body shape to bluegill, but with a notably larger opercular flap, especially in the adult male. Young olive with yellow specks; adults red above and orange below, and decorated on the body with small, blue spots and on the opercle with blue, curving lines.

DISTRIBUTION
Northeast North America, from northeastern Mexico and north to the Great Lakes.

HABITAT
Prefers shallow, weedy waters of lakes and ponds; also found in quiet streams.

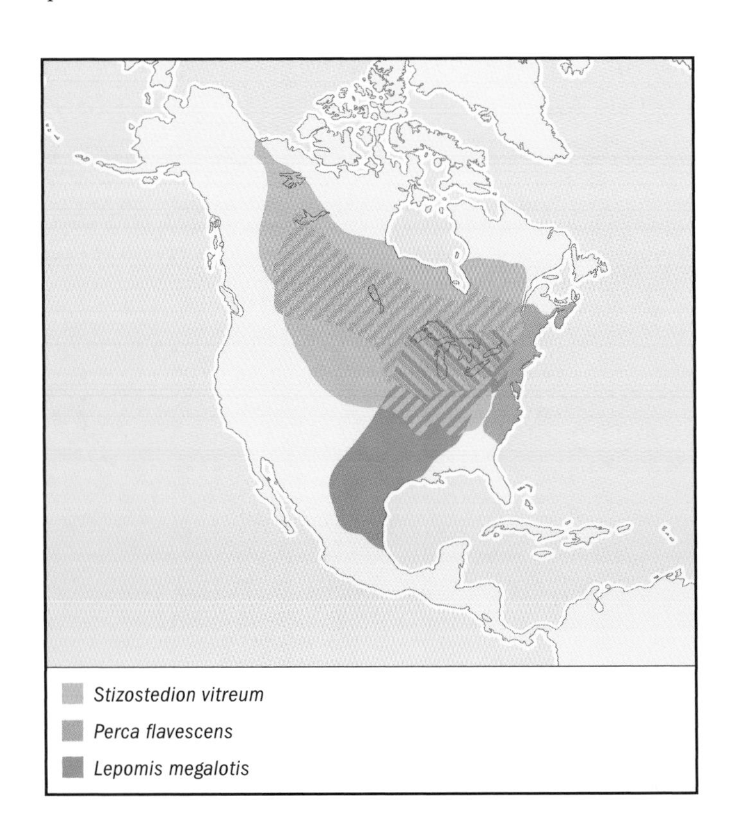

BEHAVIOR
During the spring-to-summer breeding season, males prepare nests in shallow waters along the shoreline. More than one females may approach one male's nest. Spawning occurs over the nests, and the male offers parental care to the young. In addition, some males engage in the sneaker-spawning behavior typical of many centrarchids.

FEEDING ECOLOGY AND DIET
Feeds in shallow waters on small mollusks, insects, and small fishes. Predators include other piscivorous fishes.

REPRODUCTIVE BIOLOGY
Females lay demersal, adhesive eggs, which hatch in three to seven days. Sexual maturity occurs at about two to three years.

CONSERVATION STATUS
American and Canadian populations are not considered to be at risk. Mexican populations are threatened by degradation of their habitat due to poor water-management practices.

SIGNIFICANCE TO HUMANS
The basis for a recreational fishery; sometimes kept as an aquarium fish. ◆

Largemouth bass
Micropterus salmoides

FAMILY
Centrarchidae

TAXONOMY
Micropterus salmoides Lacepède, 1802, Carolinas, United States. Two subspecies are recognized.

OTHER COMMON NAMES
English: Green bass, largemouth black bass, northern largemouth bass; French: Achiganà grande bouche; German: Forellenbarsch; Spanish: Huro, lobina negra.

PHYSICAL CHARACTERISTICS
Maximum total length 38 in (97 cm). About one-third as wide as long, distinguished in part by a deeply cut dorsal fin. Differs from its relative the smallmouth bass (*M. dolomieui*) by its lack of horizontal striping on the head, the presence of a dark horizontal stripe on each side of the body instead of vertical banding, and a maxillary that reaches just past the eye.

DISTRIBUTION
North America from the Great Lakes east to the Atlantic coast, and from Lake Winnipeg south to northern Mexico. Also widely introduced throughout the United States and around the world, including Europe, South America, and Africa.

HABITAT
Freshwater fish, prefers lakes, ponds, swamps, and river/stream backwaters with considerable hiding places, including thick vegetation or rocky structures.

BEHAVIOR
Juveniles school, but adults are solitary animals that remain near cover, such as logs or heavy vegetation and seldom venture into waters deeper than 20 ft (6 m).

FEEDING ECOLOGY AND DIET
Diurnal feeder on crustaceans and other invertebrates, also fishes.

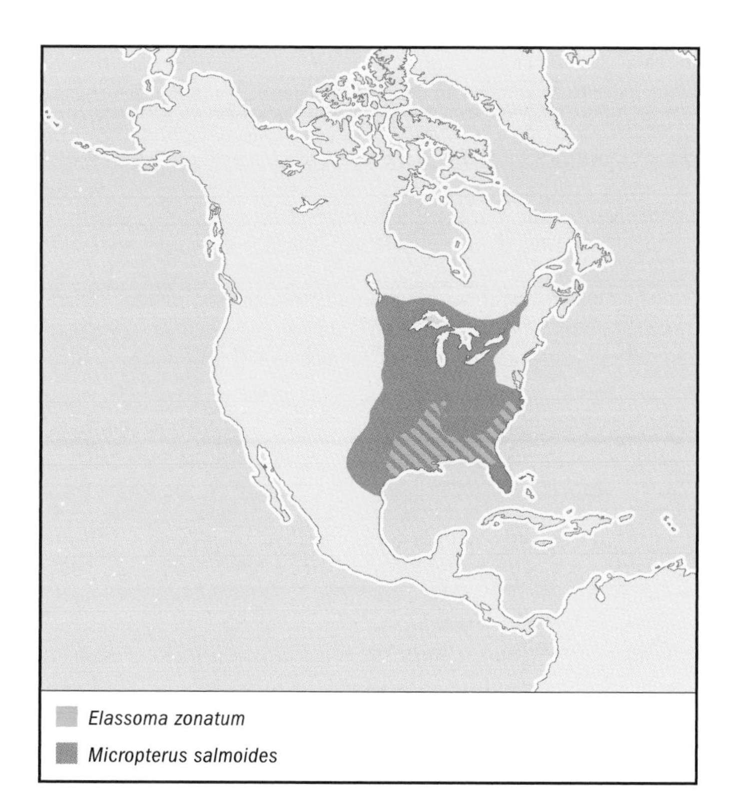

REPRODUCTIVE BIOLOGY
Spawns in the late spring and early summer. The male becomes territorial and makes depressions in the substrate of weedy areas to serve as nests. A single female may lay eggs over several nests. Both males and females provide parental care, and have been known to guard the eggs and young for up to a month after hatching. Parental care continues as long as the young fishes remain schooled.

CONSERVATION STATUS
Not threatened.

SIGNIFICANCE TO HUMANS
Part of a large, popular fishing industry in the United States and Canada. ◆

Black crappie
Pomoxis nigromaculatus

FAMILY
Centrarchidae

TAXONOMY
Pomoxis nigromaculatus Lesueur, 1829, Wabash River, Indiana, United States. No subspecies are recognized.

OTHER COMMON NAMES
English: Calico bass, grass bass, moonfish, oswego bass, speckled bass, strawberry bass; French: Marigane noire.

PHYSICAL CHARACTERISTICS
Maximum total length 19.3 in (49 cm). Rather flat, broad, silvery fishes with sloping foreheads and a black, mottled pattern on the sides. Fins are also noticeably mottled. The similar white crappie (*P. annularis*) is less mottled and has vertical banding on its sides.

DISTRIBUTION
Central and eastern North America, south to Florida and Texas, United States, and north to Quebec and Manitoba, Canada. Widely introduced throughout the United States and other countries.

HABITAT
Freshwater species. Prefers clear, weedy lakes, ponds, and slow-stream backwaters.

BEHAVIOR
Schools during the day in deep water around structures. Crepuscular feeder; moves to shallower water to feed. Exciting sport fish that puts up a good fight when hooked by anglers.

FEEDING ECOLOGY AND DIET
Feeds both amidst vegetation and in open waters on small fishes and invertebrates. Primarily feeds at dawn and dusk.

REPRODUCTIVE BIOLOGY
Spawning occurs in late spring and early summer. Males make nests, sometimes near other males, in the substrate of weedy or rocky areas. Females may mate with several males. Females lay spherical, demersal, adhesive eggs singly or perhaps in small clumps. Males guard the eggs and young, which hatch in two to three days. Reach sexual maturity by two to four years old.

CONSERVATION STATUS
Not threatened.

SIGNIFICANCE TO HUMANS
Fished for sport in the United States and Canada. ◆

Banded pygmy sunfish
Elassoma zonatum

FAMILY
Elassomatidae

TAXONOMY
Elassoma zonatum Jordan, 1877, Little Red River, Arkansas; R. Brazos, Texas; United States. No subspecies are recognized.

OTHER COMMON NAMES
German: Gebänderter Zwergbarsch.

PHYSICAL CHARACTERISTICS
Maximum total length 1.9 in (4.8 cm). White and grayish black, speckled fish with two stripes on the anterior sides of the body. One stripe extends from the mouth through the eye, and the other begins above and behind the eye. Tail and fins are also speckled.

DISTRIBUTION
South-central United States south of a line from southern Illinois to North Carolina.

HABITAT
Muddy-bottomed freshwaters, including swamps and other weedy aquatic areas.

BEHAVIOR
Normally solitary; males in limited space will defend their territories against encroachment by other males. Territorial displays include a more intense body coloration, a rapid beating

of the tail and pectoral fins, and if necessary, a quick strike at the intruder.

FEEDING ECOLOGY AND DIET
Carnivore; feeds on crustaceans, worms, and other invertebrates.

REPRODUCTIVE BIOLOGY
During breeding season, the male takes on brighter coloration and begins enticing a female to his spawning site by wiggling all his fins in a distinctive pattern. Once a female appears receptive, the male trembles and "points" to the site with his snout, and then gently prods the female. Nest building may or may not occur, perhaps depending on the suitability of the substrate. Relatively little is known of its reproductive biology in the wild.

CONSERVATION STATUS
Not listed by the IUCN.

SIGNIFICANCE TO HUMANS
Of minor importance to the pet trade, also important as indicator species of environmental quality. ◆

Striped sea bass
Morone saxatilis

FAMILY
Moronidae

TAXONOMY
Morone saxatilis Walbaum, 1792, New York, United States. No subspecies are recognized.

OTHER COMMON NAMES
English: Linesider, roccus, rock, rockfish, striped bass, striper bass; French: Bar d'Amérique, bar rayé; Spanish: Lubina estriada.

PHYSICAL CHARACTERISTICS
Maximum total length 6.6 ft (2 m). Body silver, with more or less interrupted, black, parallel stripes along the sides of the body.

DISTRIBUTION
Tributaries along the U.S. Atlantic coast, also Gulf of Mexico west to Louisiana; widely introduced throughout the United States and other countries.

HABITAT
Fresh and brackish bays and tributaries, also inland coastal waterways. Found in rivers during spawning.

BEHAVIOR
Social; typically lives in large size-graded schools. When handled or threatened, responds with grunts and clicks as it attempts an escape.

FEEDING ECOLOGY AND DIET
Carnivores; larvae feed on zooplankton, juveniles eat various invertebrates, and adults take invertebrates and fishes.

REPRODUCTIVE BIOLOGY
Broadcast spawners, produce fairly buoyant eggs during an annual reproductive season. Eggs are laid in moving water, which keeps the eggs afloat until they hatch in two to seven days. No parental care for eggs or young. Males reach maturity at about two to three years and 12–15 in (30.5–38 cm); females at about three to four years and 20–24 in (51–61 cm) long.

CONSERVATION STATUS
Not threatened.

SIGNIFICANCE TO HUMANS
Important sport and minor commercial food fishes. ◆

Rainbow darter
Etheostoma caeruleum

FAMILY
Percidae

TAXONOMY
Etheostoma caeruleum Storer, 1845, Fox River, Illinois, United States. No subspecies are recognized.

OTHER COMMON NAMES
German: Regenbogen-Springbarsch.

PHYSICAL CHARACTERISTICS
Maximum total length 3 in (7.7 cm). Brownish orange fish with about a dozen vertical, green bands on the sides of its body and green markings on its face. It has two dorsal fins.

DISTRIBUTION
Eastern North America from southern Ontario, Canada to Mississippi and Louisiana, and from Minnesota to West Virginia in the United States.

HABITAT
Bottom-dwelling, freshwater fishes that inhabit the swift currents of creeks, as well as average-to-small rivers.

BEHAVIOR
Solitary, stations in the lee of rocks as it moves over the bottom in search of food. Individuals are very wary and seek shelter among rocks at the slightest disturbance.

FEEDING ECOLOGY AND DIET
Feeds primarily on crustaceans and invertebrates, but also eats fish eggs and larvae. Feeding can involve suction, in which they draw water and prey into their mouths, or diving head-first toward sandy and gravely substrate after prey.

REPRODUCTIVE BIOLOGY
Males become more brightly colored during breeding season, which occurs in the spring. Females wait in slower waters of a stream, half-buried in the substrate, while males enter stream riffles. As a female approaches the spawning site, a male swims up, following her closely and defending her from any other interested males. He continues to maintain this "moving territory" until mating occurs. During mating, males align themselves next to the female and spawn while both vibrate. Females typically lay three to seven eggs at a time, but spawn multiple times, sometimes laying up to 1,000 eggs per season. Females typically bury their eggs, which hatch in 10–12 days. They reach sexual maturity at one year.

CONSERVATION STATUS
Not listed by the IUCN.

SIGNIFICANCE TO HUMANS
None known. ◆

Ruffe

Gymnocephalus cernuus

FAMILY
Percidae

TAXONOMY
Gymnocephalus cernuus Linnaeus, 1758, European lakes. No subspecies are recognized.

OTHER COMMON NAMES
English: Pope; French: Frash, grémille; German: Kaulbarsch, Pfaffenlaus; Spanish: Acerina.

PHYSICAL CHARACTERISTICS
Maximum total length 9.8 in (25 cm). Small, dark-spotted, brownish fish with 15 to 19 long dorsal spines. Dorsal fins are fused.

DISTRIBUTION
France to eastern Siberia. Accidentally introduced to Lakes Superior and Michigan in the United States.

HABITAT
Prefers deep waters of lakes and ponds, but also found in streams.

BEHAVIOR
Releases alarm pheromones that alert conspecifics to danger. Adults have particularly well-developed neuromasts, sensory organs that can detect even slight vibrations in the water and thus allow the fishes to hunt for food in the cover of darkness.

FEEDING ECOLOGY AND DIET
Nocturnal feeders of shallow-water zooplankton.

REPRODUCTIVE BIOLOGY
Spawns in shallow water, where strands of sticky, whitish yellow, demersal eggs adhere to weeds and/or rocks along the bottom. No parental care for eggs or young, which hatch in 5–12 days. They attain sexual maturity in two to three years at 4.3–4.7 in (11–12 cm) in length. In warmer areas, sometimes become sexually mature in just one year.

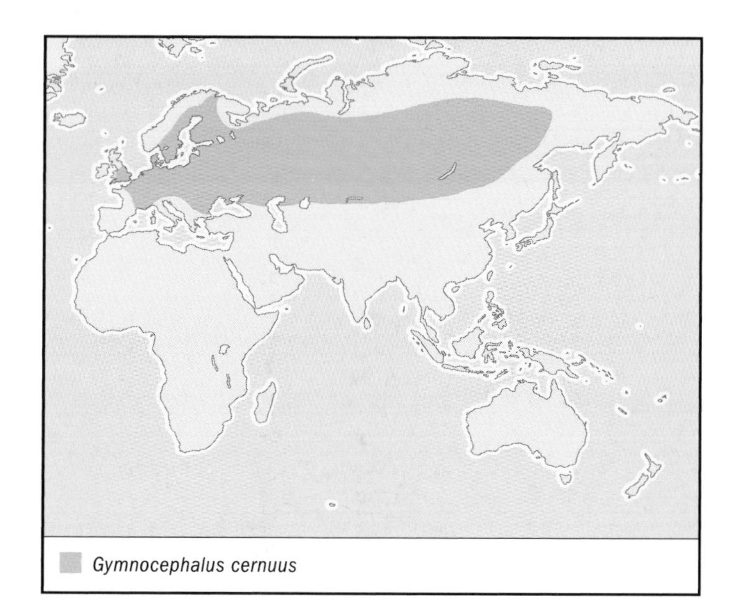

CONSERVATION STATUS
Not threatened.

SIGNIFICANCE TO HUMANS
Sport and minor food fishes. Considered an undesirable, invasive species in the Great Lakes, where it likely arrived via ballast water from Eurasian ships. ◆

Yellow perch

Perca flavescens

FAMILY
Percidae

TAXONOMY
Perca flavescens Mitchill, 1814, New York, United States. No subspecies are recognized.

OTHER COMMON NAMES
English: American perch, lake perch, perch; French: Perchaude, perche canadienne, perche jaune; German: Amerikanischer Flußbarsch.

PHYSICAL CHARACTERISTICS
Maximum total length 19.7 in (50 cm), but anglers rarely see one longer than 12 in (30.5 cm). Greenish yellow, full-bodied, fusiform fishes with six to nine vertical dark bars along the sides.

DISTRIBUTION
North America from northwestern Canada to the far northwestern United States, through central and southern Canada, the north-central United States, and the Great Lakes region, and into the southeastern Canadian provinces, as well as the eastern United States south to South Carolina. Also introduced to various locations around the world. In North America, for instance, it has been introduced to the western and southern United States and has spread into British Columbia in western Canada.

HABITAT
Freshwater streams, rivers, ponds, and lakes. Prefers sites with submerged vegetation, also inhabits brackish water and salt lakes.

BEHAVIOR
Shoals of young fishes are commonly seen in very shallow waters near shore, often darting between docks in water less than 3 ft (0.9 m) deep. Shoals of larger fishes prefer somewhat deeper waters.

FEEDING ECOLOGY AND DIET
Feeds on small to large insects and other invertebrates, fish eggs, and fishes, sometimes taking bait almost as large as themselves. Predators include larger fishes, walleye, smallmouth bass, northern pike (*Esox lucius*), and numerous salmon species.

REPRODUCTIVE BIOLOGY
Spawns from late winter to midsummer. They frequently lay their long, spiral egg masses in vegetation. There is no parental care.

CONSERVATION STATUS
Not threatened, although stocks in many areas have been severely overexploited, and these fisheries are increasingly regulated by fishery managers.

SIGNIFICANCE TO HUMANS
Commercial food and sport fishes. ◆

Logperch
Percina caprodes

FAMILY
Percidae

TAXONOMY
Percina caprodes Rafinesque, 1818, Ohio River, United States. Three subspecies are recognized.

OTHER COMMON NAMES
English: Manitou darter, zebrafish; French: Dard-perche, fouille-roche.

PHYSICAL CHARACTERISTICS
Maximum total length 7 in (18 cm). Long, fusiform fishes with conical nose and banding on the sides of the body that becomes more irregular posteriorly. Body color is greenish brown.

DISTRIBUTION
North America from Hudson Bay east to the Atlantic coast, west to Saskatchewan, and south through the Great Lakes region to the Gulf of Mexico.

HABITAT
Spread through many habitats, but primarily gravel- or sandy-bottomed, medium-sized rivers. Also bottom dweller in weedy lakes, usually some distance from the shoreline.

BEHAVIOR
Perhaps their most notable behavior occurs during feeding, when they frequently use their noses to root prey out of the substrate and from beneath stones.

FEEDING ECOLOGY AND DIET
Feeds on zooplankton, particularly copepods.

REPRODUCTIVE BIOLOGY
Moves into shallow waters or rivers to spawn in the spring. The male becomes more vividly colored and develops tubercles on its ventral side. A gravid female approaches a school of males, which follows her to the substrate. Following the release of eggs and milt, the female and often the males engage in rapid wriggling, which serves to churn up the bottom and bury the eggs. No parental care for eggs or young.

CONSERVATION STATUS
Not listed by the IUCN.

SIGNIFICANCE TO HUMANS
None known. ◆

Walleye
Stizostedion vitreum

FAMILY
Percidae

TAXONOMY
Stizostedion vitreum Mitchill, 1818, Cayuga Lake, Ithaca, New York, United States. Two subspecies are recognized.

OTHER COMMON NAMES
English: Walleyed pike, blue pike, gray pike, green pike, yellow pike, pickerel, dory, glass eye, marble eye, pikeperch; French: Doré doré jaune; German: Amerikanischer Zander.

PHYSICAL CHARACTERISTICS
Maximum total length 42 in (107 cm), but fishes below (often well below) 24 in (61 cm) are the norm. Long, fusiform body with large, translucent eyes. Body is typically brownish.

DISTRIBUTION
Most of the United States and Canada, except for far northern and eastern Canada, and far southern and western United States.

HABITAT
Freshwater (rarely brackish); inhabits lakes, ponds, and larger rivers, prefers sites with shallow, somewhat murky waters.

BEHAVIOR
Beginning in late winter, small groups of up to a half-dozen males will begin chasing one or two gravid females. The females are noticeably larger than the pursuing males. Eventually, the females will lead the groups to very shallow water, sometimes only a few inches below the surface, for a single night of spawning.

FEEDING ECOLOGY AND DIET
Feeding sessions begin at sunset and carry on into the night. An opportunistic predator, the diet comprises fishes, especially yellow perch and invertebrates, but it is also known to take frogs and mudpuppies.

REPRODUCTIVE BIOLOGY
A broadcast spawner, reproduces over a variety of substrates, including sandy, gravelly, or rocky lake and river bottoms, as well as stream vegetation, usually in an area where the water is moving either via a slow current or shoreline waves. The female scatters her eggs. Breeding occurs in spring to early summer, sometimes even in late winter. Egg production in the females begins much earlier, and anglers find females well laden with eggs in early winter. Eggs hatch in one to two weeks. No parental care of eggs or young.

CONSERVATION STATUS
Not threatened.

SIGNIFICANCE TO HUMANS
Major sport fishes and minor commercial food fishes. ◆

Resources

Books

Helfman, Gene S., Bruce B. Collette, and Douglas E. Facey. *The Diversity of Fishes.* Malden, MA: Blackwell Science, 1997.

Johnson, G. D., and A. C. Gill. "Perches and Their Allies." In *Encyclopedia of Fishes,* edited by John R. Paxton and William N. Eschmeyer. 2nd edition. San Diego: Academic Press, 1998.

Nelson, Joseph S. *Fishes of the World.* New York: John Wiley & Sons, Inc., 1994.

Periodicals

Jones, W. J., and J. M. Quattro. "Phylogenetic Affinities of Pygmy Sunfishes (*Elassoma*) Inferred from Mitochondrial DNA Sequences." *Copeia* 199, no. 2 (1999): 470–474.

Koumoundouros, G., M. Pavlidis, L. Anezaki, C. Kokkari, A. Sterioti, P. Divanach, and M. Kentouri. "Temperature Sex Determination in the European Sea Bass, *Dicentrarchus labrax* (L., 1758) (Teleostei, Perciformes, Moronidae): Critical Sensitive Ontogenetic Phase." *Journal of Experimental Zoology* 292, no. 6 (2002): 573–579.

Maniak, P. J., R. D. Lossing, and P. W. Sorensen. "Injured Eurasian Ruffe, *Gymnocephalus cernuus,* Release an Alarm Pheromone that Could Be Used to Control Their Dispersal." *Journal of Great Lakes Research* 26, no. 2 (2000): 183–195.

Organizations

American Fisheries Society. 5410 Grosvenor Lane, Suite 110, Bethesda, MD 20814-2199 USA. Phone: (301) 897-8616. Fax: (301) 897-8096. E-mail: main@fisheries.org Web site: <http://www.fisheries.org/>

North American Native Fishes Association. 1107 Argonne Dr., Baltimore, MD 21218 USA. Phone: (410) 243-9050. E-mail: nanfa@att.net Web site: <http://www.nanfa.org>

Other

"Animal Diversity Web." University of Michigan Museum of Zoology [cited January 20, 2003]. <http://animaldiversity.ummz.umich.edu>

Finley, L. "An Introduction to *Enneacanthus obesus* (Girard), the Banded Sunfish (with Special Reference to Rhode Island Distribution)." The North American Native Fishes Association [cited January 20, 2003]. <http://www.nanfa.org/articles/acobesus.htm>

"TNHC: The North America Freshwater Fishes Index" [cited January 20, 2003]. <http://www.tmm.utexas.edu/tnhc/fish/na/naindex.html>

Leslie Ann Mertz, PhD

Percoidei II

(Bluefishes, dolphinfishes, roosterfishes, and remoras)

Class Actinopterygii
Order Perciformes
Suborder Percoidei
Number of families 4

Photo: A leaping dolphinfish (*Coryphaena hippurus*) near Baja California. Dolphinfish are also known as mahi-mahi and dorado. (Photo by Lawrence Naylor/Photo Researchers, Inc. Reproduced by permission.)

Evolution and systematics

The suborder Percoidei contains more than 70 families and 2,800 species. This chapter focuses on the families Pomatomidae, Coryphaenidae, Nematistiidae, and Echeneidae. The bluefish (Pomatomidae) and the dolphinfishes (Coryphaenidae) are two small families of streamlined pelagic fishes. The bluefish is monotypic and dates from the Miocene epoch. The dolphinfishes consist of just one genus, *Coryphaena*, and two species. These fishes likely date from the Eocene and sometimes are grouped together with the Nematistiidae (roosterfish), Echeneidae (remoras), Rachycentridae (cobia), and Carangidae (jacks and trevallys). The roosterfish, *Nematistius pectoralis*, is a monotypic species within its family. Remoras consist of four genera and eight species. The genera are *Echeneis* (two species), *Phtheirichthys* (one species), *Remora* (four species), and *Remorina* (one species).

Physical characteristics

Percoids have elongated, fusiform bodies with forked or slightly forked caudal fins. Bluefishes have two dorsal fins, a long anal fin, and a mouth well armed with sharp, compressed teeth arrayed in a single series. They grow to about 51 in (130 cm) in length. Adult male dolphinfishes have pronounced bony crests on the heads; these fishes also have long, continuous dorsal and anal fins and brilliant yellow, green, and blue coloration. They grow to about 83 in (210 cm) in length. The roosterfish has an elongated, fusiform body with a raised head profile dorsally and a dorsal fin containing seven remarkably long spines. The remoras are streamlined, elongated, or club-shaped, with a flattened head that has a sucking disc modified from a dorsal spine. The disc has between 10 and 28 movable lamina that allow it to grasp the body surface of a host. The dorsal fin is positioned just ahead of the caudal peduncle and has 18–40 soft rays. The anal fin also has 18–40 soft rays. The scales are cycloid and small, and the swim bladder is absent. Remoras have a highly modified dorsal surface on the head that acts as a suction disc. This disc allows remoras to attach themselves to the body surfaces of various hosts. Body coloration ranges from black to combinations of blue, black and white. Body sizes range from 20 to over 44 in (50 to over 110 cm). Remoras are remarkably adapted for a commensal relationship with numerous host organisms and clean their hosts in exchange for "hitching rides" and feeding on leftover food items. The bluefish, the dolphinfishes, and the roosterfish have evolved to become efficient predators in open waters or inshore habitats.

Distribution

The bluefish ranges widely across three oceans and is found in marine and brackish waters of the eastern and western Atlantic Ocean, the Mediterranean Sea, the Black Sea, the Indian Ocean, and the southwest Pacific Ocean; it is absent from the eastern and northwestern Pacific and almost all of the central Pacific. The dolphinfishes occur in tropical and warm temperate marine waters of the Atlantic, Indian, and Pacific Oceans. The roosterfish is distributed in the eastern Pacific from southern California south to Baja California, Mexico, the Galapagos Islands, and Peru. The remoras occur in tropical and warm temperate waters of the Atlantic, Indian, and Pacific Oceans.

Habitat

The bluefish is pelagic but generally is found inshore off headlands and beaches and enters brackish water habitats, such as estuaries, to forage. Dolphinfishes are pelagic in surface

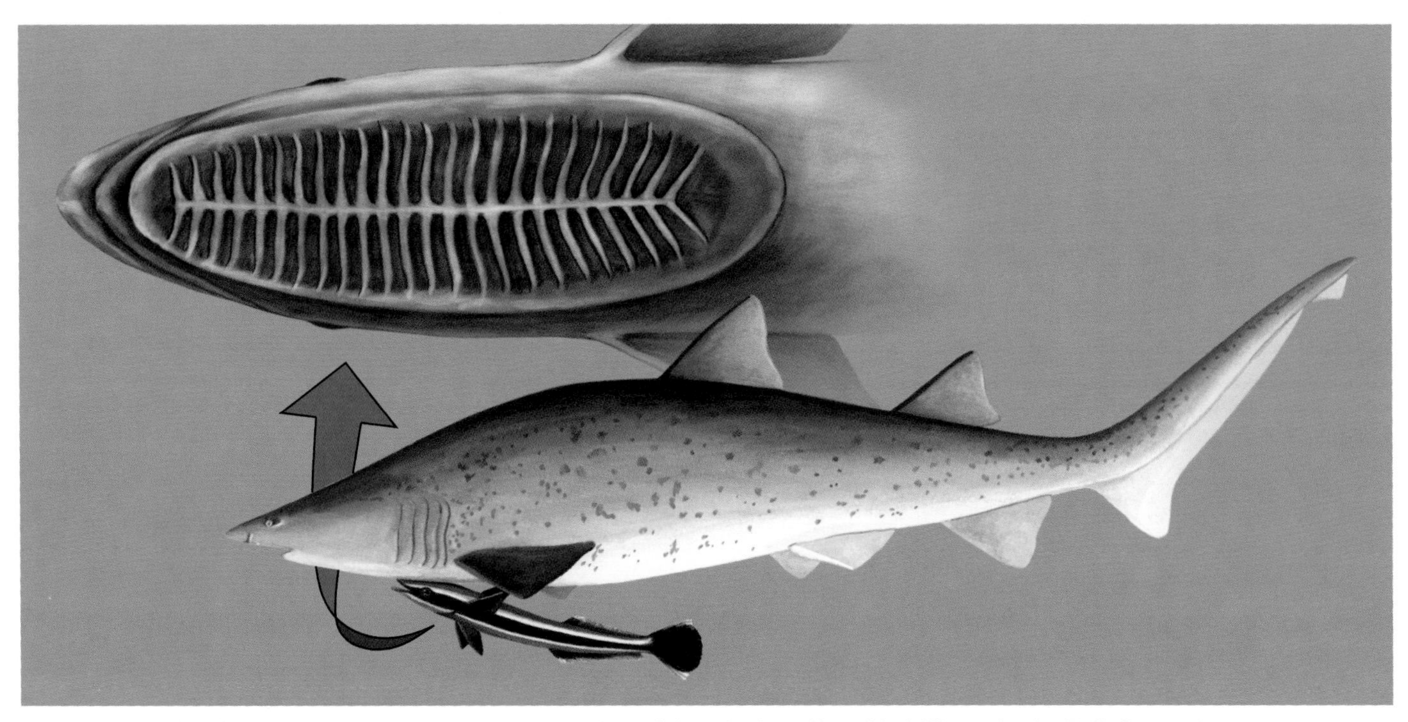

Modified dorsal fin of the remora, which allows it to stick to large fish and take a "free ride." (Illustration by Emily Damstra)

waters but also move inshore to forage, usually at insular localities where deep water is close to shore. The roosterfish prefers inshore areas, particularly sandy beaches; this species also frequents the water column away from beaches. Juveniles appear in tide pools. The remoras utilize their hosts as a microhabitat, whether swimming in surface waters of the pelagic realm or in inshore habitats.

Behavior

Bluefishes form schools or loose aggregations that are effective in hunting prey. These schools migrate seasonally, moving predictably to warm water during the cool or winter season and to cooler water during the warm or summer season. Dolphinfishes form schools that migrate over large distances on a seasonal basis. Small groups hover under floating objects, such as tree trunks or fish aggregation devices, and schools also follow boats and ships. The roosterfish may be solitary and usually swims inshore along sandy beaches as it forages for food. Remoras utilize their specialized sucking discs to attach themselves to sharks, skates, rays, billfishes, and other large fishes as well as sea turtles, dolphins, and whales. Occasionally, they are seen swimming freely. The attachment behavior allows them to function as transportation parasites, in that their host species do most of the swimming while the remora goes along for the ride, moving only to change position along the host's body, when feeding or moving between hosts, or when reproducing.

Feeding ecology and diet

Bluefishes, dolphinfishes, and roosterfishes are predators. Bluefishes are voracious predators of fishes, squids, and other shoaling animals and continue to attack prey even after satiation. Dolphinfishes feed upon small fishes, squids, and other pelagic prey. The roosterfish feeds on smaller fishes and other pelagic prey inshore or just offshore. In contrast, remoras feed on external parasites, mainly copepods, of host organisms but also take leftover items not consumed by their predatory hosts.

Reproductive biology

Bluefishes court and spawn in schools or groups, spawn seasonally, and produce pelagic eggs and larvae. Dolphinfishes court and spawn in schools or groups and produce pelagic eggs and larvae. One species has been spawned in captivity. The reproductive biology of roosterfishes apparently is not well known. They are likely pelagic spawners, perhaps in groups, and they produce pelagic eggs and larvae. Little is known about the reproductive biology of remoras. They presumably pair-spawn. Eggs and larvae probably are pelagic.

Conservation status

None of these fishes is listed by the IUCN, but they are subject to overfishing, either commercial or sport, or are caught as by catch.

Significance to humans

The bluefish and the dolphinfishes are very important commercial, recreational, and subsistence fishes. Both the bluefish and the common dolphinfish have been raised by aquaculture. The roosterfish is an important game fish that also is taken in subsistence and artisanal fisheries. Remoras may be taken incidentally for display in public aquaria.

1. Bluefish (*Pomatomus saltatrix*); 2. Common dolphinfish (*Coryphaena hippurus*); 3. Pompano dolphinfish (*Corphaena equiselis*); 4. Roosterfish (*Nematistius pectoralis*); 5. Common remora (*Remora remora*). (Illustration by Gillian Harris)

Species accounts

Pompano dolphinfish
Coryphaena equiselis

FAMILY
Coryphaenidae

TAXONOMY
Coryphaena equiselis Linnaeus, 1758, type locality not specified.

OTHER COMMON NAMES
English: Pompano dolphin; Japanese: Ebisu-shiira.

PHYSICAL CHARACTERISTICS
Body is fusiform and elongate, with a vertical head profile and a bony crest in adults, especially males; it is much less pronounced than in the common dolphinfish. The single dorsal fin has 52–59 soft rays and extends from just behind the gills down to the caudal peduncle. The anal fin is long, with 24–28 soft rays, and extends from the anus to the caudal peduncle. The caudal fin is forked. Body color is metallic blue and green on the back and silver with gold spots and a golden sheen on the flank, with a dark dorsal fin. Upon death, the dorsal color fades to gray. The caudal fin margin of juveniles is white. Grows to 50 in (127 cm) in length but lives only 4 years.

DISTRIBUTION
Tropical and some subtropical seas worldwide.

HABITAT
Pelagic but ventures inshore to forage off reefs or emergent rocks.

BEHAVIOR
A schooling species that swims in pelagic surface waters but also ventures inshore. Follows boats or hovers under floating structures, such as tree trunks or palm leaves. Attracted to fish aggregation devices.

FEEDING ECOLOGY AND DIET
Preys on smaller fishes and squids in addition to miscellaneous pelagic prey, which it hunts in schools or groups.

REPRODUCTIVE BIOLOGY
Reproduction is not well known, but apparently it courts and spawns in groups or aggregations and produces pelagic eggs and larvae.

CONSERVATION STATUS
Not listed by the IUCN, but may be vulnerable to overfishing. Recognized under Annex I of the 1982 Convention on the Law of the Sea because of its migratory habits.

SIGNIFICANCE TO HUMANS
A minor commercial species sought after by consumers for its excellent food quality. It is highly prized in sport fisheries, especially off South America. Also taken in subsistence fisheries. ◆

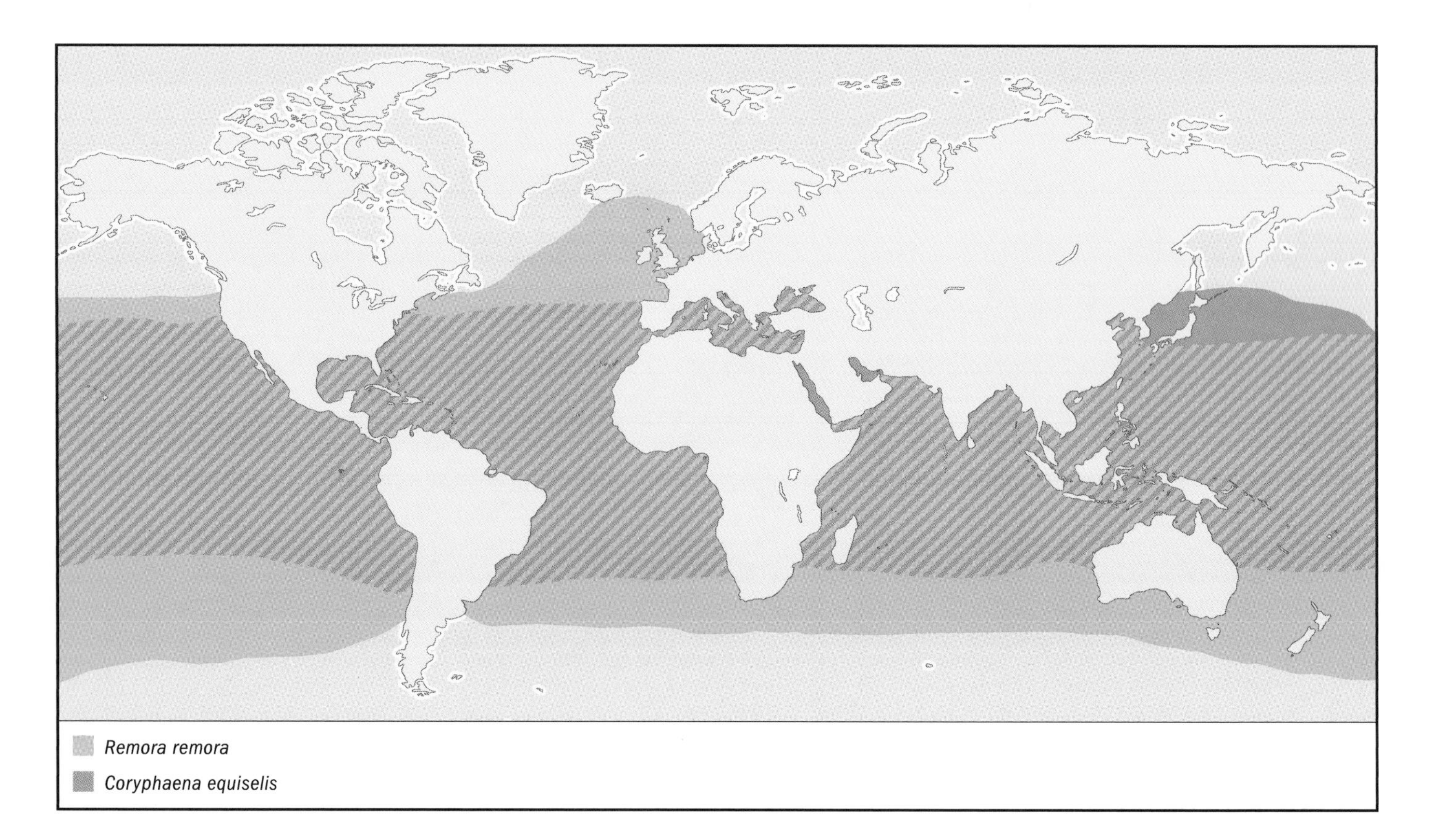

Common dolphinfish
Coryphaena hippurus

FAMILY
Coryphaenidae

TAXONOMY
Coryphaena hippurus Linnaeus, 1758, type locality not specified.

OTHER COMMON NAMES
English: Dolphin, dorado, mahi-mahi; Spanish: Dorado; Hawaiian: Mahi-mahi; Japanese: shiira.

PHYSICAL CHARACTERISTICS
Elongate and fusiform body. A long dorsal fin, with 58–66 soft rays, arises above the eye and extends to the caudal peduncle. The anal fin is concave, has 25–31 soft rays, and extends from the anus to the caudal peduncle. The pectoral fin is more than half the length of the head, and the caudal fin is forked. Adult males have a pronounced bony crest on the front of the head. Brilliantly colored, with metallic greens and blues on the back and flanks, gold on the flanks, and yellow and white shades on the venter. Juveniles and young adults have vertical bars on the flanks. Grows to longer than 83 in (210 cm) but lives for only about 5 years.

DISTRIBUTION
Tropical, subtropical, and warm temperate waters worldwide.

HABITAT
Inhabits pelagic surface waters but ventures inshore to forage. Reported to enter brackish waters, such as estuaries.

BEHAVIOR
Forms highly migratory schools, aggregations, or groups. Attracted to boats, ships, and floating objects, such as palm fronds or tree trunks. Also attracted to fish aggregation devices.

FEEDING ECOLOGY AND DIET
A predator upon smaller fishes and squids but also takes pelagic crustaceans and even macroplankton.

REPRODUCTIVE BIOLOGY
This species matures sexually in the wild at only 4–5 months. Females mature at about 13.5 in (34.3 cm) and males at 16.5 in (42 cm) fork length (the distance between the snout and the midpoint in the "fork" of the caudal fin). Eggs are pelagic, buoyant, spherical, about 0.05 in (1.4 mm) in diameter, and clear in color; the oil globule is yellow. Eggs hatch after 1.5 days and larvae measure just under 0.16 in (4 mm) at hatching. Melanophores develop on the head, trunk, and tail soon afterwards. Larvae begin to feed upon planktonic copepods after the yolk sac is absorbed, and growth is rapid.

CONSERVATION STATUS
Not listed by the IUCN, but may be vulnerable to overfishing. Recognized under Annex I of the 1982 Convention on the Law of the Sea because of its migratory habits.

SIGNIFICANCE TO HUMANS
A highly prized commercial and game fish that also is taken in subsistence fisheries. Reported to be ciguatoxic in some areas. Ciguatera poisoning is caused by the cumulative deposition of a class of polyether toxins within the tissues of fishes. The toxins are produced by certain microscopic dinoflagellate organisms of the genus *Gambierdiscus*, and are transmitted by the mechanism of the food chain, increasing in intensity by a factor

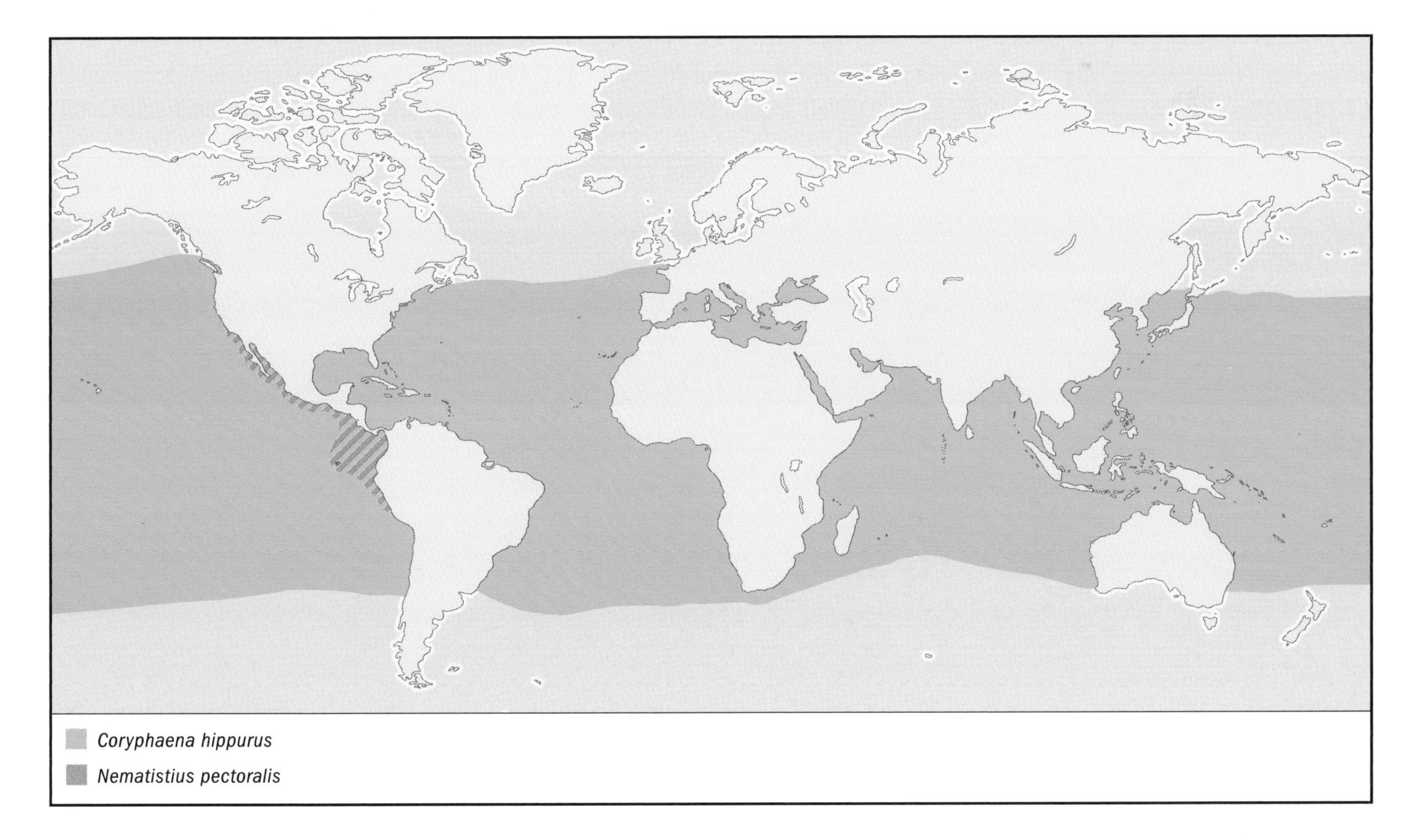

of ten in each successive level within the chain. The common dolphinfish acquires these toxins from prey fishes, which in turn acquire them from smaller fishes or zooplankton that have ingested dinoflagellate during feeding. If this fish is consumed by man, the concentrated poison contained within its tissues causes neurological damage that may be fatal. ◆

Common remora

Remora remora

FAMILY
Echeneidae

TAXONOMY
Remora remora Linnaeus, 1758, Indian Ocean.

OTHER COMMON NAMES
Japanese: Nagakoban.

PHYSICAL CHARACTERISTICS
Body is elongate, robust, and clublike, with a large, flat head and a large sucker disc. Dorsal and anal fins are positioned just forward of the caudal peduncle. There are 22–26 soft rays in the dorsal fin and 22–24 soft rays in the anal fin. The caudal fin is large and somewhat truncated. The coloring is a dark, brownish gray. Reaches nearly 35 in (90 cm) in length and just over 2.2 lb (1 kg) in weight.

DISTRIBUTION
Circumglobal in tropical, subtropical, and warm temperate waters of the Pacific, Atlantic, and Indian Oceans. Also found in the Mediterranean. Reported from Iceland and Scandinavia, presumably during warmer months.

HABITAT
Attaches to host organisms, such as sharks, rays, seaturtles, billfishes, other large fishes, and even ships. Pelagic but also appears inshore.

BEHAVIOR
Attaches itself to host organisms and hitches rides. Swims freely, on occasion. Otherwise, behavior is not well known.

FEEDING ECOLOGY AND DIET
Feeds on parasitic copepods attached to host organisms. Also takes leftover scraps from the host's feeding events.

REPRODUCTIVE BIOLOGY
Not well known. Likely pair-spawns pelagic eggs and has pelagic larvae. Spawning may be seasonal, especially at higher latitudes.

CONSERVATION STATUS
Not listed by IUCN.

SIGNIFICANCE TO HUMANS
May be taken incidentally for public aquaria or as bycatch from shark fisheries. ◆

Roosterfish

Nematistius pectoralis

FAMILY
Nematistiidae

TAXONOMY
Nematistius pectoralis Gill, 1862, Cape Saint Lucas, Baja California, Mexico.

OTHER COMMON NAMES
None known.

PHYSICAL CHARACTERISTICS
Body fusiform and elongate, with a raised profile of the head that extends to the dorsal fin. The first dorsal fin has seven long spines that retract into a groove; it may have a signal function, or it may be used in maneuvering when attacking prey. A second dorsal fin has 25–28 soft rays. There are three spines and 15–17 soft rays in the anal fin. The caudal fin is forked. Body color is blue dorsally and silvery white ventrally. There are black stripes positioned obliquely on the dorsal surface; one runs up onto the upper lobe of the caudal fin. Grows to nearly 65 in (165 cm) in length, with a weight of nearly 115 lb (52 kg).

DISTRIBUTION
Eastern Pacific from southern California, where it is rare, south through Baja California, Mexico, and the Gulf of California, where it is more common, and down to Peru and west to the Galápagos Islands.

HABITAT
Mainly inshore or near-coastal habitats, especially off sandy beaches.

BEHAVIOR
Not well known. Usually solitary or travels in small groups.

FEEDING ECOLOGY AND DIET
A predator upon fishes and other smaller pelagic organisms.

REPRODUCTIVE BIOLOGY
Reproduction not well known. Likely a pelagic spawner, with pelagic eggs and larvae. May spawn in groups. Spawning probably is seasonal, especially at higher latitudes.

CONSERVATION STATUS
Not listed by the IUCN, but vulnerable to overfishing.

SIGNIFICANCE TO HUMANS
An important game fish that also is taken as a commercial and subsistence food fish. Some collected incidentally for public aquaria. ◆

Bluefish

Pomatomus saltatrix

FAMILY
Pomatomidae

TAXONOMY
Pomatomus saltatrix Linnaeus, 1766, Carolina, United States.

OTHER COMMON NAMES
English (Australia): Tailor.

PHYSICAL CHARACTERISTICS
Body fusiform and elongate, with two dorsal fins, an elongate anal fin, and a slightly forked caudal fin. There are seven to eight spines in the first dorsal fin and one spine and 13–28 soft rays in the second dorsal fin. The anal fin has two to three spines and 12–27 soft rays. Both the dorsal and anal fin soft rays have a scaly appearance. There is a black blotch at the base of each of the pectoral fins. The mouth has sharp and

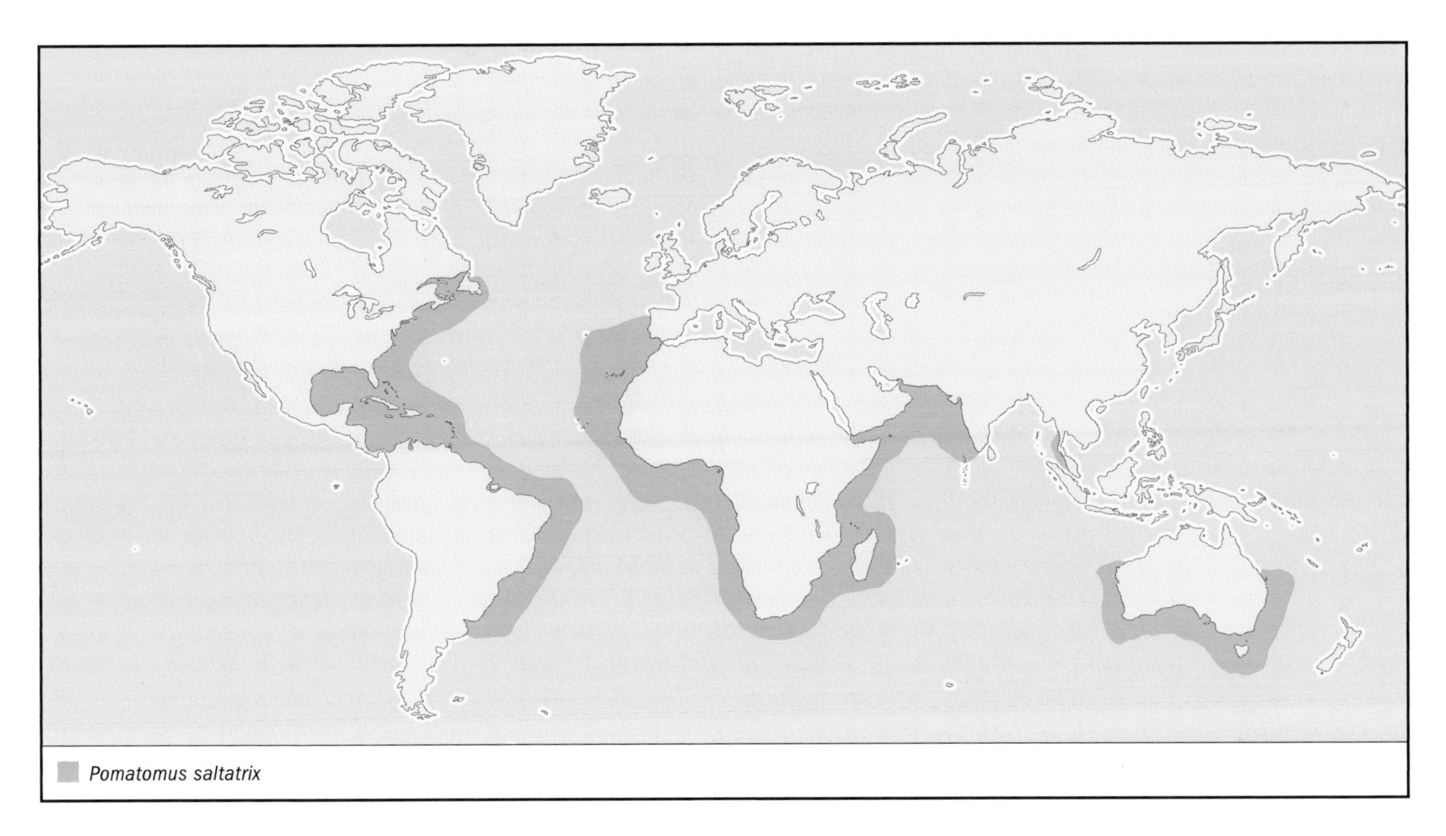

Pomatomus saltatrix

compressed jaw teeth arranged in a single prominent series. The preopercle has a membranous flap that extends over the subopercle. Color is silvery blue or greenish blue on the back, with silvery flanks and belly. Grows to 51 in (130 cm) length and lives as long as nine years.

DISTRIBUTION
Favors subtropical waters but enters tropical and temperate waters seasonally. In the eastern Atlantic it occurs from Portugal south to South Africa, including Madeira and the Canary Islands; it also may be found in the Mediterranean and Black Seas. In the western Atlantic, it is distributed from Canada to Florida and Bermuda and south as far as Argentina. In the Pacific, it is found throughout most of Australia, except for the Northern Territory but is absent from the northwest Pacific, virtually all of the central Pacific, and the eastern Pacific. In the Indian Ocean, this species ranges from East Africa and Madagascar north to southern Oman and east to southwest India, the Malay Peninsula, and Western Australia. Records from Hawaii and Taiwan require verification, and those from New Guinea, Indonesia, and the Northern Territory of Australia probably are in error.

HABITAT
Pelagic but prefers inshore waters and often is found off rocky or sandy headlands, beaches, and breakwaters. Enters estuaries. Moves with the tide.

BEHAVIOR
Adults form schools or loose aggregations. Individuals are reported to move in association with sharks and billfishes. Juveniles form schools. Schools migrate seasonally.

FEEDING ECOLOGY AND DIET
Highly predatory. Schools or aggregations forage inshore and in open water, attacking schools of prey, mainly smaller fishes and squids, and continues to attack after having fed to satiation. Dangerous if handled, because it bites out of water and reportedly has severed fingers or caused serious wounds. Is reported to attack splashing swimmers, too.

REPRODUCTIVE BIOLOGY
Spawning is seasonal, usually in spring and summer months (May–August in the northwest Atlantic; March–May and September–November in the southwest Atlantic, north of the equator; September–December in the Southern Hemisphere), and occurs serially. Females are remarkably fecund, with egg counts ranging from 400,000–2,000,000 depending upon body size of the female. Eggs are pelagic and buoyant, about 0.04 in (1.09 mm) in diameter, and hatch after 1.5–2 days. Larvae are pelagic, possess large heads and mouths, and acquire teeth on both jaws by the time they reach 0.13 in (3.3 mm) in length. The yolk sac disappears soon afterwards and larvae become predacious. Dorsal, anal, and caudal fins differentiate at about 0.25 in (6.35 mm), and fin rays appear at about 0.33 in (8.4 mm). Dorsal spines develop more fully at 1.1 in (27.9 mm). Pigmentation is apparent on the head dorsally and on the gut at 0.33 in (8.4 mm), with very small dots appearing on the entire body at 1.1 in (27.9 mm).

CONSERVATION STATUS
Not listed by the IUCN, but vulnerable to overfishing on both commercial and recreational scales. Has been caught for sport but is wasted because the flesh degrades quickly without proper handling. Population sizes are cyclical.

SIGNIFICANCE TO HUMANS
Very important commercial species that also is a significant and prized game fish. ◆

Resources

Books

Eschmeyer, W. N., ed. *Catalog of Fishes.* 3 vols. San Francisco: California Academy of Sciences, 1998.

Helfman, Gene S., Bruce B. Collette, and Doug E. Facey. *The Diversity of Fishes.* Malden, MA: Blackwell Science, 1997.

Kuiter, R. H. *Coastal Fishes of South-eastern Australia.* Honolulu: University of Hawaii Press, 1993.

Masuda, H., K. Amaoka, C. Araga, T. Uyeno, and T. Yoshino, eds. *The Fishes of the Japanese Archipelago.* Tokyo: Tokai University Press, 1984.

Myers, R. F. *Micronesian Reef Fishes: A Field Guide for Divers and Aquarists.* 3rd edition. Barrigada, Guam: Coral Graphics, 1999.

Neira, F. J., A. G. Miskiewicz, and T. Trnski, eds. *The Larvae of Temperate Australian Fishes: A Laboratory Guide for Larval Fish Identification.* Perth: University of Western Australia Press, 1998.

Nelson, J. S. *Fishes of the World.* 3rd edition. New York: John Wiley and Sons, 1994.

Smith, M. M., and P. C. Heemstra, eds. *Smiths' Sea Fishes.* Berlin: Springer-Verlag, 1986.

Organizations

IUCN/SSC Coral Reef Fishes Specialist Group. International Marinelife Alliance-University of Guam Marine Laboratory, UOG Station, Mangilao, Guam 96913 USA. Phone: (671) 735-2187. Fax: (671) 734-6767.
E-mail: donaldsn@uog9.uog.edu
Web site: <http://www.iucn.org/themes/ssc/sgs/sgs.htm>

Terry J. Donaldson, PhD

○

Percoidei III

(Grunters, temperate basses and perches, snooks and giant perches, and relatives)

Class Actinopterygii

Order Perciformes

Suborder Percoidei

Number of families 7

Photo: A common snook (*centropomus undecimalis*) swimming in the Florida Keys, USA. (Photo by Larry Lipsky. Bruce Coleman, Inc. Reproduced by permission.)

Evolution and systematics

The suborder Percoidei contains more than 70 families and 2,800 species. This chapter focuses on the following seven families: the grunters or tigerfishes (Terapontidae: 15 genera; 48 subspecies), the temperate basses and austral perches (Percichthyidae: 11 genera; 31 subspecies), the blackfishes (Gadopsidae: 1 genus; 2 subspecies), the pygmy perches (Nannopercidae: 3 genera; 6 subspecies), the Chilean perches (Percilidae: 1 genus; 2 subspecies), the kuhlias or flagtails (Kuhliidae: 2 genera; 10 subspecies), and the snooks and giant perches (Centropomidae: 4 genera; 23 subspecies). The Percichthyidae is a loosely organized family in need of revision. Previous authors have placed the Gadopsidae, Nannopercidae, and Percilidae within this family. These fishes are recognized in this chapter as separate from the percichthyids because of disparate fin counts or their unique patterns of endemism. The blackfishes and pygmy perches are endemic to Australia, while the Chilean perches are endemic to Chile in South America. This separation has been recognized in Allen et al. (2002), and in part in Froese and Pauly (2002). The families Percichthyidae, Gadopsidae, Nannopercidae, and Percilidae, all likely date from the Eocene period. The Centropomidae date from the Cretaceous period. Freshwater grunters date back to the late Cretaceous or early Tertiary periods. Allen et al. (2002) reported that a fossil terapontid unearthed in Queensland, Australia, was dated from the Oligocene, thus indicating that grunters had been present in Australia 30 million years ago. The history of the Kuhliidae is uncertain, but the family likely dates from the Eocene.

Physical characteristics

The fishes grouped in this chapter are typically bass or perchlike in their morphology and appearance. Grunters have bodies that are oblong and compressed slightly, a sloping head with an operculum bearing two spines, and conical or flattened teeth on the jaws. Some species have enlarged "blubber lips" as adults. The dorsal fin is notched with 11–13 spines and 9–11 soft rays. There are 3 spines on the anal fin and 7–10 soft rays. The pelvic fins have 1 spine and 5 soft rays, and are positioned just behind the base of the pectoral fins. The caudal fin may be emarginate, truncate, or rounded. Grunters derive their common name from their ability to contract muscles adjacent to the swim bladder (which acts as an amplifier) to produce gruntlike sounds when alarmed, stressed, or removed from the water. They may reach at least 31.5 in (80 cm) in total length. The temperate basses and austral perches include some of the largest of freshwater fishes, and certainly among the largest perciform fishes. Many have elongate bodies and large mouths, although some have a pronounced hump or steeply sloping head. The dorsal fin is single and notched, and the anterior pelvic fin rays are elongate. Their lateral line is continuous and complete, and their scales are small, primarily ctenoid, but cycloid to a lesser extent.

Percichthyids may reach up to 71 in (180 cm) total length. Blackfishes have slender, elongate bodies; a somewhat blunt snout; a long single dorsal fin; and a somewhat long anal fin. The pelvic fins are jugular xand consist of just a single branched ray. The caudal fin is rounded. The lateral line is reasonably well developed and the scales are minute. Blackfishes may grow up to 23.6 in (60 cm) total length. Pygmy perches have small, slender bodies; small mouths; interrupted lateral lines that are poorly developed or even absent; and a notched dorsal fin. These fishes are all less than 3.9 in (10 cm) total length, and usually much shorter. Chilean perches are also small, usually less than 3.9 in (10 cm) total length, and have slender, perchlike bodies with truncate or slightly emarginate caudal fins.

Kuhlias are usually oval shaped and compressed, have large eyes relative to the size of their heads, and have a silver or grayish color. The caudal fin is forked or nearly so and often distinctly marked. The dorsal fin has 10 spines, 9–12 soft rays, and is deeply notched. The anal fin has 3 spines and 9–12 soft rays. The opercle has 2 spines, and the pelvic fins lack an axillary process. The lateral line is well developed.

Snooks and giant perches have large, elongate, perchlike bodies. The snout may be concave. Most species are silvery in color. The lateral line is continuous and extends from just behind the gill well onto the caudal fin. The caudal fin may be forked, truncate, or rounded. The dorsal fin is either deeply notched or has a pronounced gap between the first part (7–8 spines) and second part (1 spine and 8–11 soft rays). The anal fin has 3 spines and 6–9 soft rays, and the pelvic fin has 1 spine and 5 soft rays. Males and females are sexually dimorphic for body size, with females being larger. Adults range in size from 7.9 in to over 78.7 in (20–200 cm) total length. The Nile perch is the largest and grows to at least 441 lb (200 kg).

Distribution

The Terapontidae occurs in freshwater, brackish, and coastal marine waters of the Indo-Pacific. A number of them, including *Amniataba* (one species), *Bidyanus* (two subspecies), *Hannia* (one species), *Hephaestus* (13 subspecies), *Pelates* (three subspecies), *Pelsartia* (one species), *Pingalla* (one species), *Scortum* (three subspecies), *Syncomistes* (four subspecies), and *Varichthys* (two subspecies), have limited or endemic freshwater and brackish-water (some also marine) distributions in Australia and New Guinea. Others, such as *Pelates* and *Terapon* (three subspecies), are widely distributed from East Africa east to Southeast Asia, Japan, Australia, New Caledonia, and Lord Howe Island, or more narrowly in the western Pacific (for example, the monotypic *Rhynchopelates oxyrhynchus*). The genus *Mesopristes* has one species endemic to Madagascar and another to Fiji, in addition to two that occur more widely in the western Pacific. The genus *Leiopotherapon* has one species endemic to Luzon Island, Philippines, and another two that are endemic to Australia, whereas *Lagusia micracanthus* is endemic to Sulawesi, Indonesia. *Pelates quadrilineatus*, a wide-ranging Indo-West Pacific species, has also colonized the eastern Mediterranean from the Red Sea by way of the Suez Canal.

The Percichthyidae occurs in tropical, subtropical, and temperate marine and fresh waters of the Atlantic, Indian, and Pacific Oceans. A number have limited or endemic distributions in the fresh or brackish waters of Australia and include *Bostockia* (one species), *Guyu* (one species), *Maccullochella* (three subspecies, but one with two subspecies), and *Macquaria* (four subspecies). *Percichthys* (five subspecies) is endemic to freshwaters in Argentina and Chile. *Coreoperca*, (three subspecies), *Coreosiniperca* (one species) and *Sinoperca* (five subspecies) are endemic to freshwaters of China, Korea, and Japan. The genus *Lateolabrax* has two species limited to coastal, brackish, or fresh waters of Japan. *Bathysphyraenops simplex* is widely distributed in coastal waters of the Indo-Pacific and Atlantic regions. The four species of the genus *Howella* are bathypelagic in the open ocean. The closely related Gadopsidae (genus *Gadopsis*, two subspecies) and Nannopercidae (genus *Edelia*, one species; genus *Nannatherina*, one species; and genus *Nannoperca*, four subspecies) are endemic to Australia. The Percilidae (genus *Percilia*; two subspecies) are endemic to Chile in South America.

The Kuhliidae are distinguished by wide-ranging Indo-Pacific species (including *Kuhlia mugil*, *K. marginata*, and *K. rupestris*), but there are also a number of limited-distribution or endemic species. *Kuhlia caudavittata* is limited to fresh and brackish waters of Reunion, Mauritius, Rodriguez, and Madagascar in the Indian Ocean. The Pacific Ocean counterpart is *K. boninensis*, which occurs in the Ogasawara and Ryukyu Islands of Japan, but is also found at larger islands in Micronesia and also at Tahiti. Three species endemic to their localities in the Pacific region include *K. munda* from fresh and brackish waters in New Caledonia, *K. nutabunda*, a marine species from Easter Island, and *K. sandivicensis* from marine, brackish, and fresh waters of the Hawaiian Islands. A monotypic genus, *Parakuhlia*, has been reported from the Atlantic.

The family Centropomidae includes 12 species of snooks (*Centropomus* spp.) found in warm marine and brackish waters of the western Atlantic and eastern Pacific, from the southern United States south to Brazil in the Atlantic, and Baja California and the Gulf of California south to Peru in the Pacific. *Hypoterus macropterus* is endemic to Western Australia in coastal waters of the southeast Indian Ocean. Six species of *Lates* are endemic to the Rift Lakes of East Africa (Lakes Albert, Rudolf, Tanganyika, and Turkana) and a seventh, the notorious Nile perch (*L. niloticus*) is present in some Rift Lakes, but also many of the major river systems of Africa and brackish water lakes of Egypt. This species was introduced into Lake Victoria with highly disastrous effects upon the endemic and greatly diverse fish fauna. Two other species of *Lates* are distributed in marine, brackish, and fresh waters of the Indo-West Pacific. The Japanese giant perch (*L. japonica*) is found in coastal waters and streams of Japan. The highly prized barramundi (*L. calcarifer*) occurs in Australia and New Guinea, but can also be found from the Arabian Gulf east to China and southern Japan, and south through Indonesia. Another Indo-West Pacific species is *Psammoperca waigiensis*, which ranges in marine and brackish waters from the Bay of Bengal in the Indian Ocean east to Southeast Asia, China, Japan, the Philippines, and Indonesia, and south to northern Australia.

Habitat

Members of this group are found in marine, brackish, and fresh waters. Among the grunters, *Amniataba* occur in ponds, lakes, and reservoirs, as well as fast-flowing streams or pools in those streams; one species also ventures into mangroves in estuaries. *Bidyanus* and *Syncomistes* also occur in rivers, lakes, or reservoirs. Inhabitants of fast or occasionally slow-flowing rivers or rocky creeks include *Hannia*, *Hephaestus*, *Leiopotherapon*, and *Pingalla*. One species of *Leiopotherapon* has adapted to high-salinity desert waters. *Scortum* subspecies prefer clear streams or rivers, but are also found in lakes. Inhabitants of slow-flowing, turbid rivers and streams, or swamps include *Pingalla*. Lake-dwelling grunters include *Hephaestus*. *Lagusia*, *Mesopristes*, *Pelates*, *Pelsartia*, *Rhynchopelates*, and *Terapon* can be found in fresh and brackish water reaches of rivers and streams, or in bays.

Percichthyids are found mainly in freshwater streams, rivers, or lakes. A few species occur in reservoirs. Some species, such as *Lateolabrax*, also enter estuaries and are found in coastal waters. However, *Howella* spp. are an exception. They are pelagic in mid- or deepwater depths and migrate towards the surface at night. Some species may be benthic as adults. The blackfishes are found mainly in clear rivers or streams, sometimes at higher elevations. Pygmy perches prefer either streams and rivers, ponds, or wetlands. The Chilean perches are also found in streams.

Kuhlias inhabit coastal marine and brackish waters, usually in the water column. Some species have adapted to freshwater habitats, mainly rivers and streams, and one, *K. rupestris*, is well adapted to upper freshwater reaches. Juveniles of marine species frequent tide pools. The snooks and giant perches are either resident in marine or estuarine waters, usually in association with mangroves or lower freshwater reaches of rivers (i.e., *Centropomus*, *Hypopterus*, *Lates*, and *Psammoperca*) or freshwater lakes, reservoirs, and rivers (such as *Lates*). Adults of some species may frequent deeper water than juveniles.

Behavior

The grunters are quite variable in their general behavior. Some species are solitary and frequently associate with structures such as rocks, logs, flooded trees, or emergent vegetation. Others form large schools, often up in the water column and over deep water, or hugging the bottom in shallower water. The temperate basses and austral perches are often solitary or occur in small groups, usually in association with structure. At least one species is segregated by sex; female Australian basses (*Macquaria novemaculeata*) move to the upper reaches of streams, whereas males move downstream. Some species, such as the nightfishes, are nocturnal. *Howella* move up and down in the water column in relation to night and day. The blackfishes hug the bottom, patrol home ranges, and are active at night. Pygmy perches are often solitary and swim in midwater where they forage for prey or lurk near shelter, such as emergent vegetation. Little is known about the behavior of Chilean perches, but it is presumed that they associate with structure or the bottom. Kuhlias usually aggregate in schools and move in and out of surge zones. Species in freshwater streams often associate with structure or aggregate, as juveniles and young adults, in pools. Snooks and giant perches are often solitary, but may also occur in small groups. Many undertake seasonal migrations into estuaries, usually in relation to patterns of freshwater runoff, but some marine species also migrate into freshwater.

Barramundis (*Lates calcarifer*) are important commercial fish. (Photo by Tom McHugh/Photo Researchers, Inc. Reproduced by permission.)

Feeding ecology and diet

Grunters prey upon fishes, insects, and benthic invertebrates, including crustaceans, worms, and mollusks, but some are also omnivorous or herbivorous and will feed upon benthic algae, plant roots, palm berries, or other plant life. Members of the genus *Syncomistes* have mouths modified for scraping algae off rock surfaces. Adult percichthyids are largely carnivores, although juveniles may feed upon zooplankton. Freshwater species feed upon insects, mollusks, crustaceans, fishes, fish eggs and fry, or amphibians. Some larger species will also take reptiles, birds, and even aquatic mammals. Marine and estuarine species also feed upon fishes and crustaceans; deep-dwelling species feed upon macroplankton, crustaceans, and fishes in the water column. Blackfishes feed upon a wide range of items, including worms, insects, crustaceans, small fishes, and even fish eggs. Pygmy perches feed on terrestrial and aquatic insects (including their larvae) and microcrustaceans. Chilean perches likely also feed upon insects, crustaceans, and small fishes. Kuhlias feed upon small crustaceans and fishes, but those living in freshwater may also feed upon insects and even fruits, such as figs, that fall into the water. Snooks and giant perches are predatory fishes that feed mainly upon other fishes and crustaceans; juveniles in freshwater will also feed upon insects. They are skillful at ambushing prey. Members of the families featured in this chapter may be preyed upon by larger predator fishes and birds, and in some cases by marine and aquatic reptiles and mammals.

Reproductive biology

Mating strategies, courtship, spawning, and postspawning investment by parents is variable within this group. Freshwater grunters migrate upstream or into the shallows of lakes to spawn, usually during summer months (wet season), although occasionally between autumn and late winter (dry season). Migrations are usually triggered by changes in water temperature, rising water levels in response to seasonal rains, or both. Spawning takes place during daylight or night, may be in groups, and results in the release of eggs that fall onto the substrate. Eggs hatch within 36 hours, and the larvae develop relatively rapidly and disperse soon after. Marine species spawn in the sea and the juveniles migrate into fresh or brackish water.

The diversity of this group dictates that the reproductive biology of these fishes is variable. Age of maturity will vary both between species and within species. An example of variation within species is the nightfish *Bostockia porosa.* Males of this species mature after their first year, but females are not mature until their second year. Generally, freshwater percichthyids spawn during the spring and summer months, and fishes often migrate upstream or downstream to spawning sites. Many of these traditional sites have been obstructed by dams or other barriers to migration, and this has resulted in corresponding declines in population sizes of various species. Reproductive effort may be considerable in these fishes. Members of the genus *Coreoperca* have male parental care of eggs and larvae. The trout cod (*Maccullochella macquariensis*) spawns demersal eggs that are large and adhesive onto hard surfaces such as rocks. Parental care is practiced by males of this and other species in this genus. In rivers, care extends until flooding occurs; flooding makes food items, such as insect larvae, available to the postlarval fishes. In the golden perch (*Macquaria ambigua*), eggs are floating and no care is practiced. Others of this genus scatter eggs that are either pelagic or that sink into interstices within the substrate, and are not cared for. The reproductive biology of other freshwater species is not that well known. Among marine species, members of the genus *Howella* likely breed in aggregations and scatter pelagic eggs into the water column. *Lateolabrax* spp. spawn during winter and scatter eggs on deeper rocky reefs of coastal waters. Juveniles may migrate upriver after recruiting from the ocean. The blackfishes mature in two to three years and practice pair spawning during summer months. Females enter a male's nest and lay eggs, usually one batch that varies in number from 20–500, depending upon the species and her size, and these and the subsequent larvae are guarded by the male until they disperse. Pygmy perches are mature in about one year. Depending upon the species, spawning begins late in the austral winter (dry season) and continues through early summer (wet season). Males are territorial, assume bright colors, defend nest sites, and attract females to spawn. Females spawn multiple small batches of eggs every few days during a period that can last for several weeks. The eggs are spawned upon aquatic algae or the substrate within a male's territory. The eggs hatch within two to four days. Most adults of at least one species, *Nannatherina balstoni*, die after spawning. The reproductive biology of the Percilidae is not well known. Presumably, spawning also occurs sometime between the austral spring and summer. Eggs are likely demersal and may be guarded by the male.

Kuhlias in freshwater are catadromous, which means that they migrate downstream to spawn in estuaries or the open ocean. Marine species also spawn in the ocean. Tropical species may spawn all year long but seasonal peaks may occur. Spawning behavior is not well known. The spawning mode is pelagic, with spawning in groups. The eggs and larvae are pelagic; larvae of freshwater species migrate into freshwater streams and rivers.

Marine and brackish water centropomids form spawning aggregations in estuaries or reef passes. *Lates* spp. are catadromous and spawn in groups or pairs (not always within groups) in estuaries and nearshore waters. Females are generally larger than males, and two or more males have been observed courting much larger females. Eggs and larvae are pelagic. The reproductive behavior of African lake species of this genus is not well known, but courtship and spawning in groups or pairs, usually after migration to a specific site, with the release of pelagic eggs, is likely. Both *Hypoterus macropterus* and *Psammoperca waigiensis* probably migrate to estuaries or inshore waters to spawn pelagic eggs.

Conservation status

A good number of species of the families Percichthyidae, Gadopsidae, Nannopercidae, and Terapontidae are endemic to specific localities or have relatively small distributions, thus rendering them vulnerable to extirpation or extinction because of habitat destruction, overfishing, or the introduction of exotic species that either compete with or prey upon some life stage of the native species. For example, among the Percichthyidae, the Mary River cod (*Maccullochella peelii mariensis*) is listed as Critically Endangered by the IUCN. The trout cod (*M. macquariensis*) and the Eastern freshwater cod (*M. ikei*) are listed as Endangered. The mountain perch (*Macquaria australasica*) is classified as Data Deficient. In the Nannopercidae, the Oxleyan pygmy perch (*Nannoperca oxleyana*) is listed as Endangered, the Yarra pygmy perch (*N. obscura*) and variegated pygmy perch (*N. variegata*) are listed as Vulnerable. Among the Terapontidae, the silver perch (*Bidyanus bidyanus*), Adamson's grunter (*Hephaestus adamsoni*), and Yamur Lake grunter (*Varia jamoerensis*), are all listed as Vulnerable.

Significance to humans

A number of species are important in commercial, recreational, and subsistence fisheries, while others may be taken incidentally as aquarium fishes. Some species of grunters, temperate basses, and snooks and giant perches are also cultured for food or for release as game fishes. Certain grunters and percichthyids are also utilized in Chinese medicine.

1. Mountain perch (*Macquaria australasica*); 2. Target perch (*Terapon jarbua*); 3. Sooty grunter (*Hephaestus fuliginosus*); 4. Murray cod (*Maccullochella peelii*); 5. Golden perch (*Macquaria ambigua*); 6. Japanese perch (*Lateolabrax japonicus*); 7. River blackfish (*Gadopsis marmoratus*). (Illustration by Marguette Dongvillo)

1. Nile perch (*Lates niloticus*); 2. Barramundi (*Lates calcarifer*); 3. Western pygmy perch (*Edelia vittata*); 4. Southern pygmy perch (*Nannoperca australis*); 5. Flagtail kuhlia (*Kuhlia mugil*); 6. Jungle perch (*Kuhlia rupestris*); 7. Common snook (*Centropomus undecimalis*). (Illustration by Marguette Dongvillo)

Species accounts

Common snook
Centropomus undecimalis

FAMILY
Centropomidae

TAXONOMY
Centropomus undecimalis Bloch, 1792, Jamaica, West Indies.

OTHER COMMON NAMES
English: Robalo blanco; French: Crossie blanc; Spanish: Robalo blanco.

PHYSICAL CHARACTERISTICS
Total length 55 in (140 cm); maximum weight 53.6 lb (24.3 kg). Body relatively large, elongate, and robust. Pronounced sloping forehead and snout. Color silvery with faint olive or greenish hues dorsally, lateral line is black and highly visible. Caudal fin is large and somewhat forked. There are 8–9 spines and 10 soft rays in the dorsal fin, and 3 spines and 6 soft rays in the anal fin.

DISTRIBUTION
Western Atlantic region, in the United States from North Carolina south to Florida, and west along the coast of the Gulf of Mexico coast as far west as Texas; south through several islands in the Caribbean to the coasts of Central and South America, and further south as far as Rio de Janeiro in Brazil.

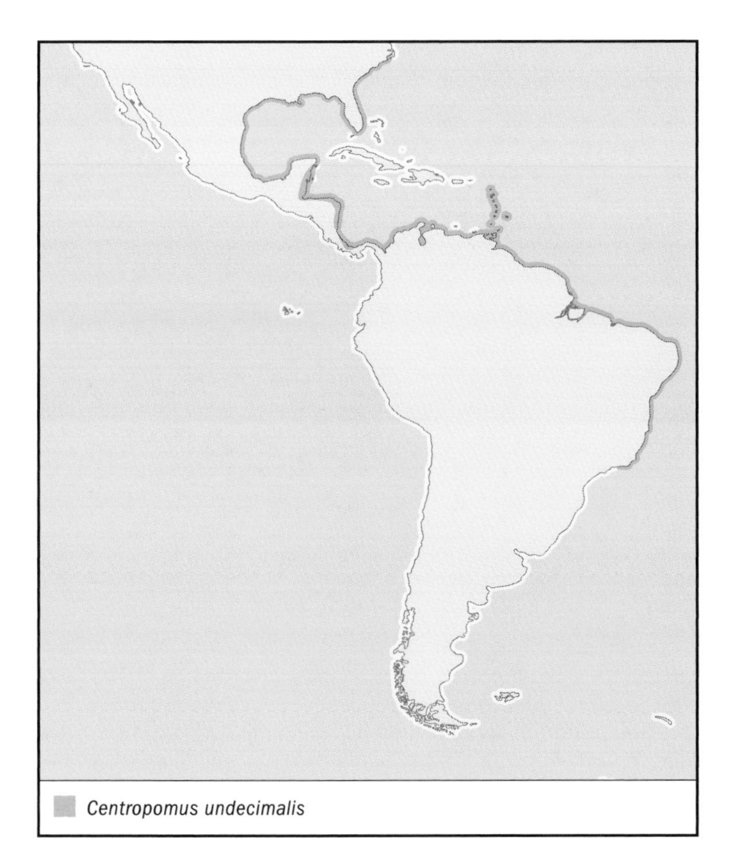
Centropomus undecimalis

HABITAT
Shallow coastal marine, brackish, and fresh waters, mainly in estuaries and lagoons and usually in association with mangroves.

BEHAVIOR
Solitary, usually in association with structure but also swims in the water column. Migrates for spawning but will also migrate, for an as yet unexplained reason, into fresh water seasonally.

FEEDING ECOLOGY AND DIET
Prey on smaller benthic and inshore pelagic fishes, as well as crustaceans such as shrimps and crabs.

REPRODUCTIVE BIOLOGY
Maturity occurs in three to five years, with smaller males maturing sooner than larger females. Lives to at least seven years. Fishes migrate seasonally, usually between May to September (but also August to July in Cuba), to estuaries or mouths of passes to court and spawn. Eggs are scattered over the bottom, and larvae are pelagic.

CONSERVATION STATUS
Not listed by the IUCN, but vulnerable to overfishing and habitat destruction.

SIGNIFICANCE TO HUMANS
A very important commercial, subsistence, and game fish. Also raised by aquaculture. ◆

Barramundi
Lates calcarifer

FAMILY
Centropomidae

TAXONOMY
Lates calcarifer Bloch, 1790, Japan.

OTHER COMMON NAMES
English: Asian seabass, Barramundi perch; French: Brochet de mer.

PHYSICAL CHARACTERISTICS
Total length 79 in (200 cm). Body large, elongate, and stout, with pronounced concave dorsal profile in head and a prominent snout. Color silvery but may be greenish or bluish gray on dorsal surfaces. Fins blackish or dusky brown. Juveniles have mottled pattern of brown with three white stripes on head and nape, and white blotches irregularly placed on back. There are 8–9 spines and 10–11 soft rays in the dorsal fin, 3 spines and 7–8 soft rays in the anal fin, and 7–8 soft rays in the pectoral fin. Caudal fin is truncate.

DISTRIBUTION
Indo-West Pacific from East Africa through tropical and warm temperate Asia, including southern Japan, south through Indonesia to northern Australia (from Shark Bay north in Western Australia, and the Mary River in southern Queensland in the east). Insular localities must have sufficient stream development.

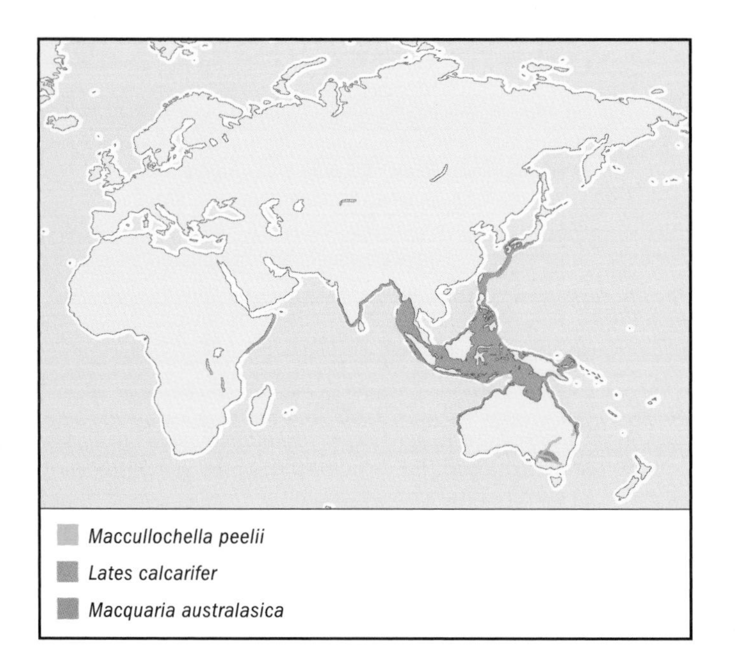

HABITAT
Rivers and larger streams, billabongs, submerged floodplains, estuaries, and coastal waters. Juveniles migrate from the ocean upstream to the upper reaches of rivers and creeks where they shelter in vegetation, undercut banks, and other forms of structure. Adults tend to be found in lower reaches of turbid rivers and utilize holes or structure, such as submerged timber, rocks, and mangroves.

BEHAVIOR
Adults and juveniles tend to be solitary, patrol home ranges near structure, and may be territorial. Migration is seasonal.

FEEDING ECOLOGY AND DIET
Voracious predator of fishes and crustaceans. Juveniles also feed on insects. Skilled at stalking or ambushing prey.

REPRODUCTIVE BIOLOGY
Catadromous. Migrates downstream to shallow mudflats in estuaries during the wet season. Spawning occurs between September and March in Australia, with peaks in November to December and again in February to March. Females are larger than males, are highly fecund, and may be courted by one or more males at the same time. Eggs are pelagic, hatch within 24 hours, and the larvae grow quickly as they move into mangrove areas, mudflats, and floodplain lagoons. Juveniles move into coastal waters after one year, then migrate upstream where adults reside for three to four years. Populations landlocked by dams migrate to the dam face, but do not spawn. Raised extensively by aquaculture as food or for game fish-stocking programs.

CONSERVATION STATUS
Not listed by the IUCN, but has been threatened by habitat destruction and overfishing.

SIGNIFICANCE TO HUMANS
Important as a commercial and subsistence food fish but also as a game fish. The most important commercial fish of Australia, and the most sought after game fish, generates millions of dollars per year in revenue for the sportfishing and tourist industries. Cultured fish are replacing wild-caught fishes as food in parts of Australia where netting is banned. Also important in the cultures of aboriginal Australians. ◆

Nile perch
Lates niloticus

FAMILY
Centropomidae

TAXONOMY
Lates niloticus Linnaeus, 1758, Egypt.

OTHER COMMON NAMES
English: African snook; French: Capitaine; German: Nilbarsch; Spanish: Perca del Nilo.

PHYSICAL CHARACTERISTICS
Total length 76 in (193 cm); weight 441 lb (200 kg). Body large, elongate, and robust. Color silvery, with grayish blue along back and grayish silver along flank and belly. Caudal fin and portion of pectoral fin black to dark gray. There are 7–8 spines and 10–14 soft rays in the dorsal fin; the caudal fin is rounded. Free edge of operculum bears a large spine.

DISTRIBUTION
Africa, in fresh and occasionally brackish waters. Present in the Rift Lakes of East Africa, including Lakes Albert, Rudolph, and Tana. Introduced into Lake Victoria and others within and outside of the region, usually with catastrophic results for native fish faunas. Also occurs in major river systems including the Chad (and Lake Chad), Congo, Nile, Senegal, and Volta. Present in Lake Mariout, a brackish water body outside Alexandria, Egypt, near the Nile River.

HABITAT
Large lakes and major rivers and their larger tributaries; also occurs in channels and irrigation canals, brackish-water lakes, and estuaries. Larger adults prefer deeper water, smaller fishes are found in the shallows.

BEHAVIOR
Swims in the water column, but may also associate with structure. Generally solitary.

FEEDING ECOLOGY AND DIET
Highly voracious predator of fishes, especially freshwater herrings in its native range, but also of cichlid fishes. Smaller fishes feed upon crustaceans and insects in shallow water.

REPRODUCTIVE BIOLOGY
Maturity comes in two to three years; larger females reach maturity later than smaller males. The spawning season varies with latitude, temperature, and type of water body, and ranges from February to November, although shorter seasons have been reported. Presumably migrates to a specific site to court and spawn. Eggs are scattered over the bottom or in the water column, and are buoyant. Larvae are pelagic, there is no parental care.

CONSERVATION STATUS
Not listed by the IUCN, but could be vulnerable to commercial overfishing in its native range.

SIGNIFICANCE TO HUMANS
An important commercial and subsistence food fish and game fish, raised by aquaculture for food and stocking. Stocking out-

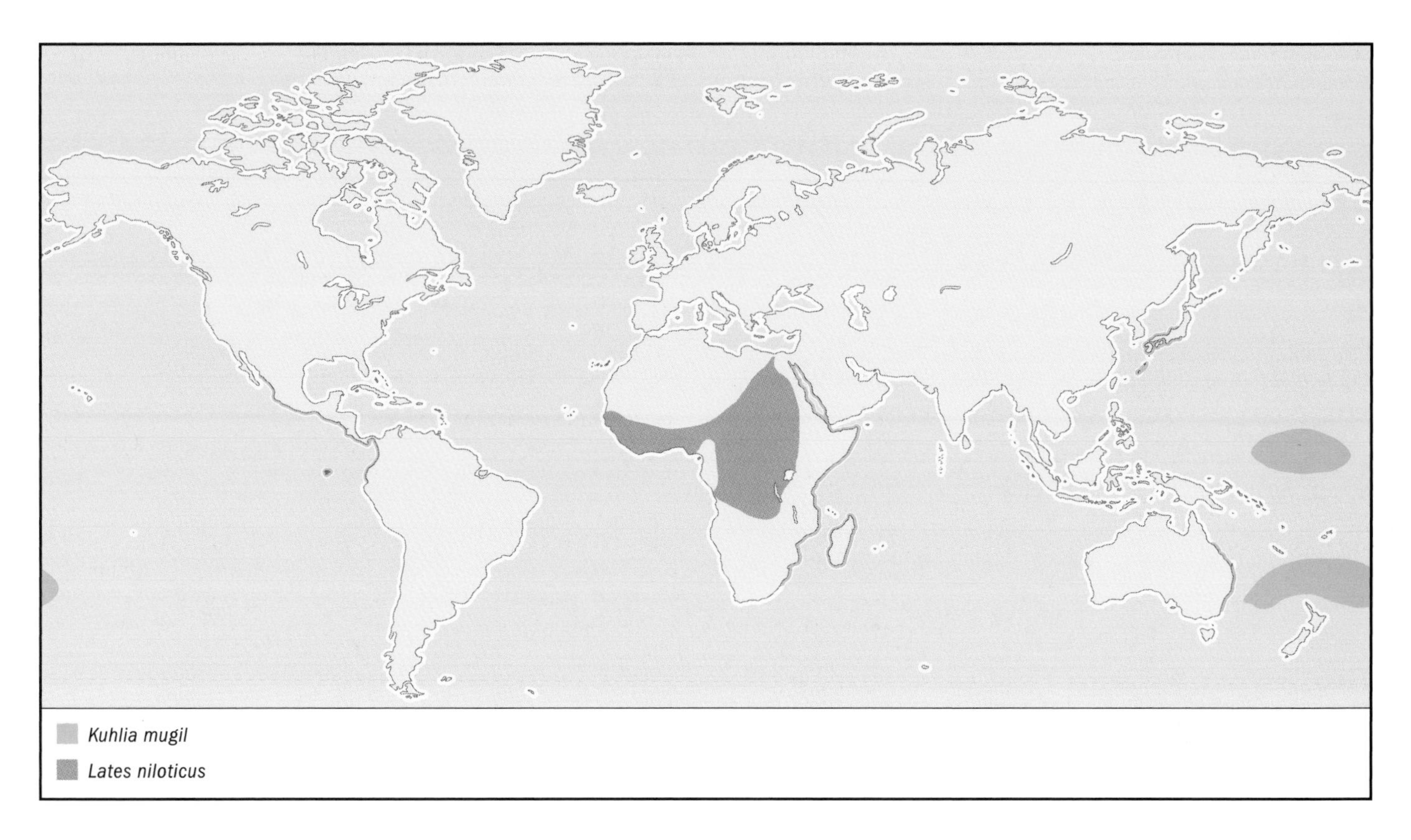

side its natural range is greatly discouraged. Introduction of this species into Lake Victoria resulted in the decimation of most of a highly diverse and evolved endemic species flock of cichlid fishes, in addition to the loss of other species endemic to that great lake. Similar effects reported elsewhere. ◆

River blackfish
Gadopsis marmoratus

FAMILY
Gadopsidae

TAXONOMY
Gadopsis marmoratus Richardson, 1848, Murray River, and other rivers in South Australia.

OTHER COMMON NAMES
English: Marbled river cod; German: Aalrute, Süßwasserdorsch.

PHYSICAL CHARACTERISTICS
Total length 23.6 in (60 cm) but most, especially in smaller streams, are smaller. Body elongate, slender, and compressed laterally. Snout and caudal fin both rounded. Color is dark gray to pale brown with darker mottling on the flanks and basal portions of fins, with milky or clear color on outer portions of the fins. There are 6–8 spines and 25–28 soft rays in the dorsal fin, 3 spines and 17–19 soft rays in the anal fin, and 15 soft rays in the pectoral fins.

DISTRIBUTION
Murray-Darling River system of Australia, from Victoria into parts of New South Wales, Queensland and South Australia;

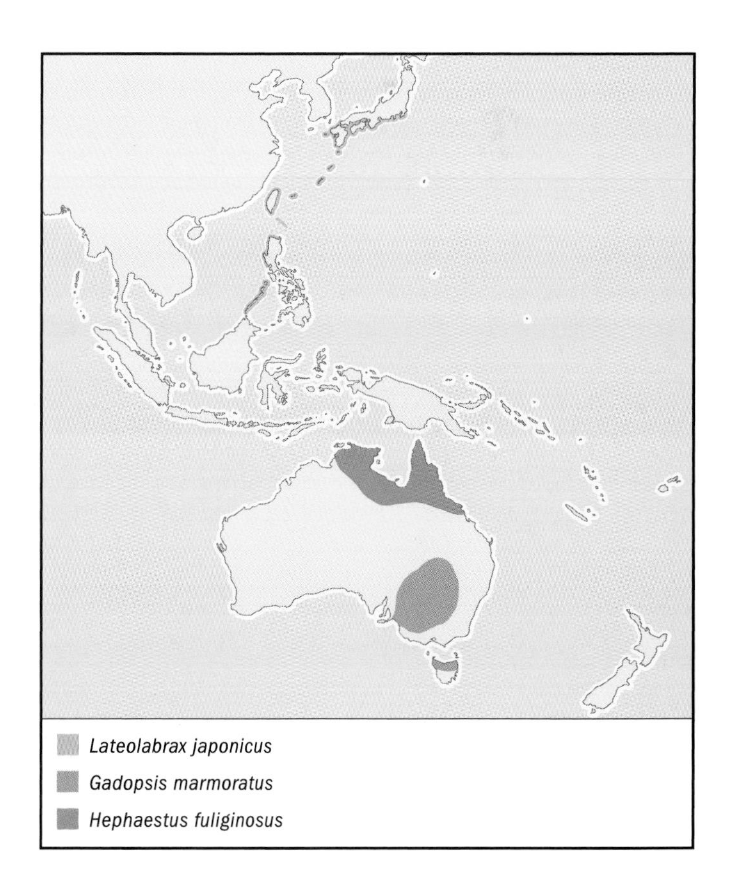

also northern Tasmania. Introduced into some rivers of southern Tasmania.

HABITAT
Submerged timber or cover in gently flowing, clear-water streams. Tolerates slightly brackish water in upper tidal reaches of streams. Tolerates extreme temperature ranges from 41–77°F (5–25°C).

BEHAVIOR
Patrols home ranges and is strongly site attached.

FEEDING ECOLOGY AND DIET
Forages for insects, crustaceans, mollusks, and smaller fishes.

REPRODUCTIVE BIOLOGY
Matures in two to three years, females take longer than males. Spawns during austral spring and summer. Female spawns a single, small batch of eggs (20–500 per batch depending upon her size) in a protected place, usually a rock crevice or hollow log within a male's home range. Hatching occurs in about 16 days and larvae remain attached to the egg case for up to 21 more days. The male defends the nest and larvae until they disperse.

CONSERVATION STATUS
Not listed by the IUCN, but vulnerable to habitat destruction and overfishing.

SIGNIFICANCE TO HUMANS
A game fish in the southern portion of its range, but the flesh is oily and not esteemed. May be collected incidentally for aquaria. ◆

Flagtail kuhlia
Kuhlia mugil

FAMILY
Kuhliidae

TAXONOMY
Kuhlia mugil Forster, 1801, Tahiti, Society Islands, Polynesia.

OTHER COMMON NAMES
English: Barred flagtail, flagtail aholehole; French: Croco drapeau; Spanish: Dara bandera.

PHYSICAL CHARACTERISTICS
Standard length 15.7 in (40 cm). Body oval shaped and compressed, the color silver, with black and white markings on the caudal fin. There are 10 spines and 9–11 soft rays in the dorsal fin, and 3 spines and 8–10 soft rays in the anal fin. The caudal fin is forked.

DISTRIBUTION
Widespread in the Indo-Pacific, ranging from the Red Sea and East Africa, including Madagascar, east to the Galápagos Islands, Mexico and Ecuador; also north to southern Japan, and south to Australia and Lord Howe Island. Occurs throughout Micronesia and most of Polynesia as far as Rapa.

HABITAT
Inshore waters along rocky and coral reefs, cliff faces, and rocks. Also enters estuaries and freshwater. Juveniles enter tide pools.

BEHAVIOR
Forms tightly packed schools that swim close to structure, generally just below or in the surge zone.

FEEDING ECOLOGY AND DIET
Feeds upon small swimming crustaceans in the water column.

REPRODUCTIVE BIOLOGY
Not well known. Probably spawns in estuaries or inshore, perhaps in schools, year round in the tropics but seasonally at higher latitudes. Eggs and larvae are pelagic.

CONSERVATION STATUS
Not listed by the IUCN.

SIGNIFICANCE TO HUMANS
Of minor importance as a commercial and subsistence food fish. Collected for bait and for aquaria. ◆

Jungle perch
Kuhlia rupestris

FAMILY
Kuhliidae

TAXONOMY
Kuhlia rupestris Lacepède, 1802, Gol Ravine, Reunion Island, Indian Ocean.

OTHER COMMON NAMES
English: Freshwater aholehole, freshwater kuhlia, rock flagtail; French: Crocro sauvage; German: Felsen-Flaggenbarsch.

PHYSICAL CHARACTERISTICS
Total length 17.7 in (45 cm), but fishes in upper reaches of freshwater streams are much smaller. Body elongate and robust compared to others in this family. Mouth is relatively large, caudal fin slightly emarginate. Color silvery or light gray with black scale margins, becomes very silvery in marine waters. Caudal fin is either black or clear,with black blotches on each caudal lobe. There are 10 spines and 11 soft rays in the dorsal fin, 3 spines and 9–10 soft rays in the anal fin, and 13–14 soft rays in the pectoral fin.

DISTRIBUTION
Indo-West Pacific, from East Africa east to the Ryukyu, Ogasawara, and Mariana Islands, south and east through Micronesia, Melanesia, and Polynesia on islands with freshwater streams; also to Australia as far south as Fraser Island.

HABITAT
Freshwater streams in coastal drainages, usually in clear waters. Also in estuaries and coastal waters seasonally.

BEHAVIOR
Solitary or in small groups in fresh water. Somewhat cryptic in areas with overhanging brush, undercut banks, or other forms of structure.

FEEDING ECOLOGY AND DIET
Omnivorous, feeds upon small fishes, insects, crustaceans, and fruit that falls on the water.

REPRODUCTIVE BIOLOGY
Catadromous, migrates to estuaries on nearshore waters to breed. Behavior not well known. Courtship and spawning

might be in schools or aggregations. Eggs and larvae are pelagic. Post-larvae migrate upstream.

CONSERVATION STATUS
Not listed by the IUCN, but insular populations vulnerable to habitat destruction and overfishing.

SIGNIFICANCE TO HUMANS
Subsistence food fish, may also be taken in minor commercial fisheries. An important freshwater game fish in Australia and esteemed as table fare. ◆

Southern pygmy perch
Nannoperca australis

FAMILY
Nannopercidae

TAXONOMY
Nannoperca australis Günther, 1861, Murray River, Australia.

OTHER COMMON NAMES
German: Südaustralischer Zwergbarsch.

PHYSICAL CHARACTERISTICS
Total length nearly 3.5 in (9 cm). Body small and elongate with relatively large, rounded caudal fin. Color silvery or pale gold (occasionally light brown), mottled with a greenish brown hue, the fins are clear. During breeding season, males have hints of orange at base of the dorsal and caudal fins, posterior portion of belly, and above eyes. Pelvic and anal fins are marked with black. There are 7–9 spines and 7–10 soft rays in the dorsal fin, 3 spines and 7–8 soft rays in the anal fin, and 11–14 soft rays in the pectoral fin.

DISTRIBUTION
Australia, generally coastal drainages of Victoria west to mouth of the Murray River in South Australia. Also east to inland reaches of the Murrumbidgee and Murray Rivers in Victoria and New South Wales, and south across Bass Strait to Flinders Island, King Island, and northern Tasmania. The range has contracted because of human interference with natural river flows.

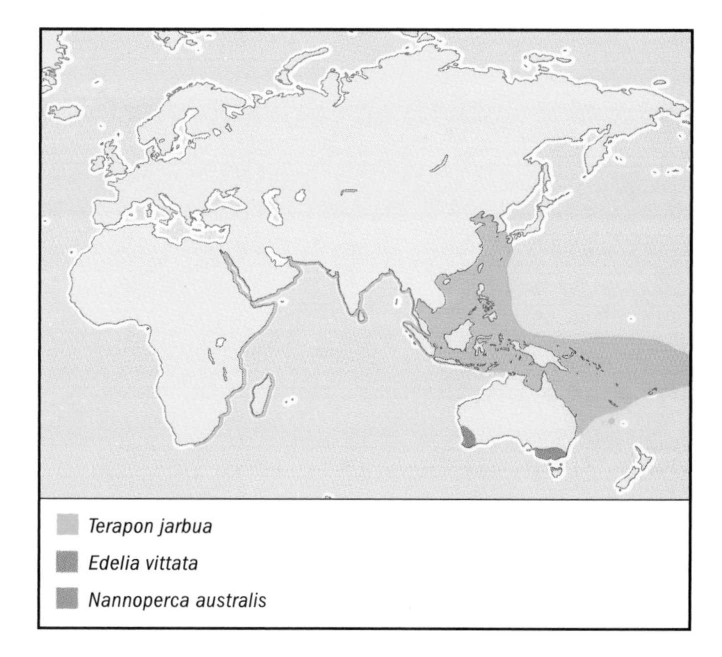

HABITAT
Frequents submerged and emerging shoreline vegetation in streams, drains, billabongs of larger rivers, swamps, and reservoirs. Prefers still or slightly flowing water.

BEHAVIOR
Forms small groups.

FEEDING ECOLOGY AND DIET
Prefers small crustaceans, and larval and adult insects.

REPRODUCTIVE BIOLOGY
Probably matures after the first year and lives as long as five years. Likely moves to preferred spawning sites to spawn demersal eggs in small batches over the course of the season (September to January).

CONSERVATION STATUS
Not listed by the IUCN, but vulnerable to habitat destruction.

SIGNIFICANCE TO HUMANS
May be collected for aquaria or for use as bait. An important prey of introduced species of game fishes. ◆

Western pygmy perch
Edelia vittata

FAMILY
Percichthyidae

TAXONOMY
Edelia vittata Castelnau, 1873, Western Australia.

OTHER COMMON NAMES
German: Westaustralischer Zwergbarsch.

PHYSICAL CHARACTERISTICS
Total length less than 2.75 in (7 cm). Body small and somewhat ovate to elongate. Mouth small, caudal fin relatively large and rounded. Color is mottled, variably olive, brown, or gold with two orange stripes on the flank and a whitish belly. Males assume darker colors dorsally, fins become blackish, flanks mottled gold, and orange stripes deepen during breeding season. Females also change color somewhat during breeding season, assuming a bluish tinge over the bodies. Lateral line interrupted and interspersed with tubed scales. There are 7–9 spines and 8–9 soft rays in the dorsal fin, 3 spines and 6–8 soft rays in the anal fin, and 11–13 soft rays in the pectoral fin.

DISTRIBUTION
Endemic to the southwestern part of Western Australia; common in a number of coastal drainages, from the Philipps River near the city of Albany, north to the Arrowsmith River, about 186 mi (300 km) north of Perth.

HABITAT
Occurs in a variety of permanent water bodies and is somewhat tolerant of slightly brackish or tannin-stained, as well as clear, waters. Shelters in aquatic vegetation along the shorelines of lakes, ponds, wetlands, rivers, and creeks.

BEHAVIOR
Generally solitary in close association with benthic algae.

FEEDING ECOLOGY AND DIET
Carnivore; feeds on benthic microcrustaceans, and adult and larval insects.

REPRODUCTIVE BIOLOGY
Becomes mature after the first year and lives for about five years. Migrates into smaller tributary creeks or submerged floodplains to spawn between July and November. Females lay demersal eggs in small batches for several weeks during the season.

CONSERVATION STATUS
Not listed by the IUCN, but its relatively limited geographic distribution could make it vulnerable to habitat destruction.

SIGNIFICANCE TO HUMANS
May be collected as an aquarium fish and also used to control aquatic insects such as mosquitoes. ◆

Japanese perch
Lateolabrax japonicus

FAMILY
Percichthyidae

TAXONOMY
Lateolabrax japonicus Cuvier, 1828, seas of Japan.

OTHER COMMON NAMES
English: Japanese seaperch, Japan sea bass; Japanese: Suzuki.

PHYSICAL CHARACTERISTICS
Total length 40 in (102 cm). Body elongate, compressed, and perchlike. Mouth is large, lower jaw projects beyond upper jaw. Two dorsal fins. Color silvery but with fine black spots or large black spots on adults from some localities. Juveniles have small black spots on dorsal fin and back. There are 12–15 spines and 12–14 soft rays in the dorsal fin, 3 spines and 7–9 soft rays in the anal fin. Caudal fin is truncate.

DISTRIBUTION
Marine, brackish, and fresh waters of Japan south and west to the South China Sea.

HABITAT
Inshore rocky reefs, usually in moving water. Juveniles enter brackish and fresh waters.

BEHAVIOR
Solitary or in small groups. Will migrate.

FEEDING ECOLOGY AND DIET
Adults and older juveniles feed upon crustaceans, usually prawns and shrimps, and smaller fishes. Younger juveniles feed upon zooplankton.

REPRODUCTIVE BIOLOGY
During winter months, Japanese perch migrate downstream to the sea, join conspecifics migrating from other streams, and spawn on deeper rocky reefs.

CONSERVATION STATUS
Not listed by the IUCN, but may be vulnerable to habitat destruction, particularly in rivers and estuaries.

SIGNIFICANCE TO HUMANS
Commercial fishes and game fishes, also raised by aquaculture and collected for aquaria. Used in Chinese medicine. ◆

Murray cod and Mary River cod
Maccullochella peelii peelii and *M. peelii mariensis*

FAMILY
Percichthyidae

TAXONOMY
Maccullochella peelii peelii Mitchell, 1838, Peel River, New South Wales, Australia. *M. peelii mariensis* Rowland, 1993, Mary River, Queensland, Australia.

OTHER COMMON NAMES
English: Goodo, ponde.

PHYSICAL CHARACTERISTICS
Total length (both subspecies) 80 in (180 cm); weight 250 lb (113 kg). Body large, elongate, and heavy. Body color ranges from olive to cream, occasionally yellowish, with reticulated mottling of dark gray or green along body and onto fins. Fins brownish with white edges, pelvic fin is white. Dorsal fin positioned posteriorly along back. Caudal peduncle of Mary River cod is shorter and the pelvic fins are longer than in Murray cod. Caudal fin is rounded in both subspecies. There are 10–12 spines and 13–16 soft rays in dorsal fin, 3 spines and 11–13 soft rays in anal fin, and 19–20 soft rays in pectoral fin.

DISTRIBUTION
Murray cod: Murray-Darling River system of Australia; absent from upper reaches of these rivers and from the smaller tributaries; its range has shrunk because of habitat destruction. Has been stocked elsewhere. Mary River cod: River Mary in Queensland, Australia, although attempts have been made to establish populations elsewhere, such as the Brisbane River.

HABITAT
Both subspecies favor structure, mainly submerged logs, rocks, or undercut banks, as well as deeper holes in turbid, slow flowing, or clear, rocky reaches of rivers.

BEHAVIOR
Territorial and will defend a piece of habitat, usually a hole or some form of structure. Those in lakes will defend a larger area. Juveniles are migratory.

FEEDING ECOLOGY AND DIET
Carnivorous as adults and older juveniles, feeds upon macroinvertebrates, fishes, amphibians, reptiles, birds, and small terrestrial mammals that enter water. Young juveniles feed upon zooplankton.

REPRODUCTIVE BIOLOGY
Both subspecies mature in about five to six years and may live to 60 years. Courtship and spawning occurs during the austral spring and early summer. Eggs are demersal and guarded by the male. Hatching occurs in one to two weeks to coincide with flooding that brings food for the post-larvae.

CONSERVATION STATUS
The Murray cod is not listed, but is vulnerable to habitat destruction and overfishing. The Mary River cod is classified as Critically Endangered by the IUCN.

SIGNIFICANCE TO HUMANS
Culturally important to aboriginal peoples of Australia. Formerly important as commercial species, with overfishing and habitat destruction leading to its decline. The Murray cod is an important but highly regulated game fish. ◆

Golden perch
Macquaria ambigua

FAMILY
Percichthyidae

TAXONOMY
Macquaria ambigua Richardson, 1845, Western Australia (probably an error, although Australia is the correct country).

OTHER COMMON NAMES
English: Yellowbelly, Murray bream; German: Australischer Goldbarsch.

PHYSICAL CHARACTERISTICS
Standard length 29.5 in (75 cm); weight 53 lb (24 kg); most are smaller. Base color is olive, bronze, or brownish, with distinct yellow ventral surface. Body compressed and elongate, but with concave dorsal profile of the head just above eyes, protruding lower jaw, and strongly arched nape. Caudal fin is truncate, pelvic fins have elongate filaments. There are 8–11 spines and 11–13 soft rays in the dorsal fin, 3 spines and 7–10 soft rays in the anal fin, and 15–18 pectoral fin soft rays.

DISTRIBUTION
Lower reaches of the Murray-Darling River system, Dawson-Fitzroy River system in southeastern Queensland, and Bulloo River and Lake Eyre drainages in New South Wales and western Victoria, Australia (the population from the latter locality may be a separate subspecies). Stocked extensively elsewhere, especially in reservoirs.

HABITAT
Turbid, slow-flowing rivers, billabongs, and backwaters, or in clear, fast-flowing rivers. Also in reservoirs. Favors fallen or submerged timber, overhanging banks, and rocky ledges.

BEHAVIOR
Solitary; often associated with structure.

FEEDING ECOLOGY AND DIET
Adults and older juveniles favor macrocrustaceans, mollusks, and smaller fishes; juveniles forage on zooplankton over submerged floodplains.

REPRODUCTIVE BIOLOGY
Females mature in four or more years, males in two to three years. Migrates upstream to spawn and will travel over 1,243 mi (2,000 km) to reach its spawning grounds. Courtship and spawning take place in flooded backwaters. Eggs float after spawning, hatch within 24–36 hours, and larvae disperse downstream. Barriers to migration and corresponding habitat destruction have caused serious declines in recruitment. Cultured artificially for stocking programs.

CONSERVATION STATUS
Not listed by the IUCN, but threatened by habitat destruction.

SIGNIFICANCE TO HUMANS
Mainly a game fish, but a limited commercial fishery exists in the Murray-Darling River system. ◆

Mountain perch
Macquaria australasica

FAMILY
Percichthyidae

TAXONOMY
Macquaria australasica Cuvier, 1830, Macquarie River at Bathurst, New South Wales, Australia.

OTHER COMMON NAMES
English: Black bream, Macquarie perch; German: Macquaries Barsch, Silberauge.

PHYSICAL CHARACTERISTICS
Standard length 17 in (43 cm); maximum weight 7.7 lb (3.5 kg). Base color is dark gray, silvery gray, or olive-brown, and may be mottled; the belly is often pale or light gray, and the scale margins are dark gray. Body is compressed and elongate, the dorsal profile of the head concave, the nape arched, the body deep, the eyes and jaws both large; large mucous cavities are found around the eyes and also on the preoperculum. The dorsal fin has 8–10 spines and 11–14 soft rays; the anal fin has three spines and 8–11 soft rays; there are 14–17 pectoral fin soft rays.

DISTRIBUTION
Middle and upper reaches of Murray River system in New South Wales and Victoria, Australia. Also known from the Yarra River system in Victoria. Introduced elsewhere. A genetically distinct population occurs in the Shoalhaven and Hawkesbury Rivers of New South Wales that may comprise a separate species.

HABITAT
Frequents deep holes on the bottom, but will move up into midwater in streams or reservoirs.

BEHAVIOR
Adults are solitary but form aggregations inshore and when migrating.

FEEDING ECOLOGY AND DIET
Forages for aquatic insects, crustaceans, and mollusks. Post-larvae feed on zooplankton.

REPRODUCTIVE BIOLOGY
Males mature in two years, females in three. Females larger than males. Forms small aggregations that migrate upstream to spawn. Spawning takes place between October and December. Courtship occurs over rocky or gravel bottoms above riffle stretches and demersal eggs are released to scatter onto the bottom; eggs usually slip between interstices of substrate. Eggs hatch in 13–18 days. No parental care.

CONSERVATION STATUS
Classified as Data Deficient by the IUCN. Likely vulnerable to habitat destruction and losses from introduced parasites.

SIGNIFICANCE TO HUMANS
Primarily a game fish, but also a minor component of commercial and aquarium fisheries. ◆

Sooty grunter
Hephaestus fuliginosus

FAMILY
Terapontidae

TAXONOMY
Hephaestus fuliginosus Macleay, 1883, Upper Burdekin River, north Queensland, Australia.

OTHER COMMON NAMES
English: Black bream; German: Rußiger Grunzbarsch.

PHYSICAL CHARACTERISTICS
Reaches 17.7 in (45 cm) total length, but usually smaller in smaller streams. The body is typical of freshwater grunters. Silvery olive in color. There are 11–12 spines and 12–14 soft rays in the dorsal fin, 3 spines and 8–10 soft rays in the anal fin, and 15–17 pectoral fin rays. Lower jaw is distinguished by a discontinuous lip fold.

DISTRIBUTION
Australia, from north coastal Queensland (Cape Hillsborough) to the Gulf of Carpentaria and west to the Daly River in the Northern Territory. Replaced by *H. jenkinsi* in northwestern Australia, including the Kimberly Region of Western Australia.

HABITAT
Sandy or rocky bottoms in the upper reaches of large, flowing streams; also in reservoirs. Tolerates a considerable range of temperatures and moderately acidic conditions.

BEHAVIOR
Solitary or in groups, usually in association with submerged plants or other forms of structure.

FEEDING ECOLOGY AND DIET
Omnivorous, feeds upon invertebrates, amphibians, algae, roots of emergent plants, and fruits that fall upon the water.

REPRODUCTIVE BIOLOGY
Migrates to spawn in groups during summer months; spawning migration is triggered by rains. Eggs are demersal.

CONSERVATION STATUS
Not listed by the IUCN.

SIGNIFICANCE TO HUMANS
Important as a game fish in Australia. ◆

Target perch
Terapon jarbua

FAMILY
Terapontidae

TAXONOMY
Terapon jarbua Forsskål, 1775, Jidda, Saudi Arabia, Red Sea.

OTHER COMMON NAMES
English: Crescent-banded tiger-fish, tiger grunter, tiger perch; French: Peau d'âne, relégué, violon jarbua; Spanish: Baraonga jarbúa.

PHYSICAL CHARACTERISTICS
Total length 14.2 in (36 cm). Body typically perchlike but more elongate. Silvery to light tan in color, with cream on belly and series of black stripes along flank and onto caudal fin. Caudal fin has black tips. There are 11–12 spines and 9–11 soft rays in the dorsal fin, 3 spines and 7–10 soft rays in the anal fin. Lower-most spine on the operculum extends beyond the opercular flap.

DISTRIBUTION
Indo-Pacific, from the Red Sea and East Africa east to Samoa, north to southern Japan, and south to Australia and Lord Howe Island.

HABITAT
Marine and brackish waters, usually in bays, mangroves, and estuaries; will enter rivers. Both adults and juveniles frequent sand flats inshore, but juveniles will enter intertidal zone.

BEHAVIOR
Solitary or in small groups that move about the bottom or around structure.

FEEDING ECOLOGY AND DIET
Omnivorous, feeds upon fishes, insects, benthic invertebrates, and algae.

REPRODUCTIVE BIOLOGY
Catadromous in rivers. Probably spawns in groups in the saltwater. Eggs and larvae are pelagic. Juveniles migrate into lower reaches of rivers.

CONSERVATION STATUS
Not listed by the IUCN.

SIGNIFICANCE TO HUMANS
Minor commercial species also taken in subsistence fisheries; incidental game fishes. Juveniles are collected for the aquarium trade. Also raised in aquaculture. ◆

Resources

Books

Allen, G. R., S. H. Midgley, and M. Allen. *Field Guide to the Freshwater Fishes of Australia.* Perth: Western Australian Museum, 2002.

Eschmeyer, W. N., ed. *Catalog of Fishes.* 3 vols. San Francisco: California Academy of Sciences, 1998.

Helfman, G. S., B. B. Collette, and D. E. Facey. *The Diversity of Fishes.* Oxford: Blackwell Science, 1997.

Leis, J. M., and B. M. Carson-Ewart, eds. *The Larvae of Indo-Pacific Coastal Fishes.* Boston: Brill, 2000.

Masuda, H., K. Amaoka, C. Araga, T. Uyeno, and T. Yoshino, eds. *The Fishes of the Japanese Archipelago.* Tokyo: Tokai University Press, 1984.

Myers, R. F. *Micronesian Reef Fishes.* 3rd edition. Barrigada, Guam: Coral Graphics, 1999.

Nelson, J. S. *Fishes of the World.* 2nd edition. New York: John Wiley & Sons, 1994.

Thresher, R. E. *Reproduction in Reef Fishes.* Neptune City, NJ: T.F.H. Publications, 1984.

Organizations

IUCN/SSC Coral Reef Fishes Specialist Group. International Marinelife Alliance-University of Guam Marine Laboratory, UOG Station, Mangilao, Guam 96913 USA. Phone: (671) 735-2187. Fax: (671) 734-6767. E-mail: donaldsn@uog9.uog.edu Web site: <http://www.iucn.org/themes/ssc/sgs/sgs.htm>

Terry J. Donaldson, PhD

Percoidei IV

(Goatfishes, butterflyfishes, angelfishes, chubs, and relatives)

Class Actinopterygii
Order Perciformes
Suborder Percoidei
Number of families 18

Photo: A queen angelfish (*Holacanthus ciliaris*) among the coral near Cozumel, Mexico. (Photo by Animals Animals ©Mickey Gibson. Reproduced by permission.)

Evolution and systematics

The suborder Percoidei contains more than 70 families and 2,800 species. This section on the Percoidei deals with 18 families of fishes. The families are the Mullidae (goatfishes), Toxotidae (archerfishes), Dichistiidae (galjoens), Kyphosidae (sea chubs or rudderfishes), Paracorpididae (jutjaws), Drepaneidae (sicklefishes), Monodactylidae (monos or moonfishes), Chaetodontidae (butterflyfishes), Pomacanthidae (angelfishes), Enoplosidae (oldwives), Pentacerotidae (armorheads and boarfishes), Nandidae (leaffishes), Oplegnathidae (knifejaws), Cirrhitidae (hawkfishes), Chironemidae (kelpfishes), Aplodactylidae (seacarps or marblefishes), Cheilodactylidae (morwongs), and Latridae (trumpeters).

Fossil records indicate that perciform fishes date back as far as the Cenozoic era (about 65 million years ago), with percoids evolving rapidly during the Eocene epoch. This rapid evolution is evident in such families as the Chaetodontidae, Pomacanthidae, Mullidae, Kyphosidae, and Cirrhitidae, which contain the most species in the group. A number of families are placed in groups with variable degrees of formalization. For example, the Squammipinnes is an unranked group, denoted by having rows of scales covering the base of the dorsal and anal fins, and includes the Monodactylidae, Toxotidae, Chaetodontidae, and Pomacanthidae. The Kyphosidae consists of four subfamilies, three of which were recognized previously as separate families. (A new molecular study of this polyphyletic group suggests that the subfamilies, which appear to be closely related to the Kuhlidae, might be recognized again as separate families.) The Nandidae also consists of three subfamilies recognized on the basis of their geographical distribution, which parallels the breakup of Gondwana (the supercontinent, made up of South America, Africa, Antarctica, India, and Australia), thus making the nandids an ancient group. Leaffishes have been linked falsely with labyrinth fishes of the suborders Anabantoidei (gouramies and allies) and Channoidei (snakeheads). The superfamily Cirrhitoidea is denoted by the presence of pectoral fins with five-to-eight unbranched and elongated lower fin rays on the pectoral fin. Members include the Cirrhitidae, Chironemidae, Aplodactylidae, Cheilodactylidae, and the Latridae.

Physical characteristics

As with other Percoidei fishes, these fishes are characterized by the presence of spines in the dorsal, pelvic, and anal fins; the presence of two dorsal fins but no adipose fin; the abdominal placement of pelvic fins; pectoral fins that are placed laterally and oriented vertically; the presence of 17 or less principal rays in the caudal fin; and, usually, ctenoid scales (some species possess cycloid scales). These fishes are further characterized by a maxilla that is excluded from the gape of the jaw, a gas bladder that is physoclistous (lacking a connection between the gas bladder and the gut), and bone that is acellular. They also lack four specific bone types: the orbitosphenoid (within the orbital region of the skull), the mesocoracoid (within the pectoral girdle; this bone helps to position the pectoral fin obliquely rather than vertically, as in the percoids), the epileural (riblike intermuscular bones extending below the vertebral column), and the epicentral (primitively ligamentous, riblike intermuscular bones). Body shapes are quite divergent, and range from deep-bodied and highly compressed (Kyphosidae, Drepaneidae, Monodactylidae, Pentacerotidae, Chaetodontidae, Pomacanthidae, and Cheilodactylidae) to elongated (Mullidae) forms. Similarly, there is considerable

Yellowfin goatfishes (*Mulloidichthys vanicolensis*) moving in tight formation over the Dolphin Reef in the Red Sea near Egypt. (Photo by Fred McConnoughey/Photo Researchers, Inc. Reproduced by permission.)

variation in the ecomorphology of the head, jaws, or other structures, both between and within families. Some families, such as the Chaetodontidae, Pomacanthidae, and Cirrhitidae, include many species with brightly colored or highly conspicuous color patterns. Other families (Nandidae) are noted for the cryptic coloration of many species.

Goatfishes are characterized by pairs of long, chemosensory barbels under the lower jaw, a terminal mouth, and the presence of large ctenoid scales. Goatfish larvae are laterally compressed and moderate in depth or elongate. The barbels begin to form at about a length of 0.3–0.35 in (0.8–0.9 cm). Archerfishes have moderately compressed bodies, a pointed head, a dorsal that originates well back towards the caudal fin, and a grooved palate that can direct a stream of water into the air. Their larvae are moderate in depth initially, but become deeper, compressed, and heavily pigmented with age. Galjoens have deep bodies, small ctenoid scales (even on the fins), and a scaleless snout. Sea chubs have a small head with a terminal mouth, a deep and moderately compressed body, a dorsal fin that is continuous, and a caudal fin that is either emarginate (notched) or forked. Their larvae are initially elongate, but become moderate in depth with age, with moderate-to-heavy pigmentation. There is some variation between subfamilies. The jutjaw has an oval-shaped, compressed body, small ctenoid scales, and a large mouth with a projecting lower jaw. Sicklefishes have strongly compressed, oval-shaped bodies covered with cycloid scales and protusile mouths. Monos have deep, highly compressed bodies that are covered with small deciduous scales. Pelvic fins are present in juveniles, but become rudimentary or absent in adults. Their larvae are specialized, having head spination that is moderately developed and elongate pelvic fins that form rather early.

The butterflyfishes have deep, highly compressed bodies; small protractile mouths; somewhat small ctenoid scales; a single dorsal fin that is continuous; and a caudal fin that is rounded or slightly emarginate. The post-larvae have a distinctive *tholichthys* stage, denoted by the presence of bony plates and protruding spines on the head. Angelfishes also have deep, compressed bodies, along with small ctenoid scales, small mouths, and a single dorsal fin that is unnotched. Characters that differentiate angelfishes from the butterflyfishes are a single prominent spine located at the corner of the preopercle; smaller spines on the preopercle, interopercle, and preorbital; and the absence of a *tholichthys* postlarval stage. The oldwife, as an adult, has remarkably large dorsal, pelvic, and anal fins, as well as small ctenoid scales, and a deep, compressed body. Sharp spines are present on the head, anal fin, and dorsal fin; the dorsal spines contain poison and thus the oldwife must be handled with care. Larvae are elongate to moderately long, and are heavily pigmented.

Boarfishes and armorheads resemble the adult oldwife. They have deep, moderately to strongly compressed bodies, exposed head bones that form rough, striated plates, and may have elongated mouths and fin rays. Juveniles may have tall dorsal fins and elongated anal and pelvic fins, but these are reduced as adults. Some species have poisonous spines and should be handled with care. Their larvae are denoted by elongate pelvic fins and extensive, well-developed spination on the head. Leaffishes possess large heads and mouths (the mouth is protactile), continuous dorsal fins, and rounded caudal fins. Knifejaws (Oplegnathidae) have oblong, moderately compressed bodies, tiny ctenoid scales, a single dorsal fin, and a parrotlike beak reminiscent of that on the distantly related parrotfishes (Scaridae). Knifejaw larvae lack subopercular or interopercular spines and are lightly pigmented. Hawkfishes resemble small groupers and their allies (Serranidae) and are characterized by the presence of filaments arranged in small tufts on the tips of the dorsal fin spines and by lower pectoral fin rays that are thickened. Members of the genus *Paracirrhites*, especially *P. arcatus*, *P. hemistictus*, and *P. forsteri*, are polychromatic. Two fixed color morphs occur in the former two species, while the latter exhibits two fixed color patterns in juveniles and a wide range of patterns that have no apparent relationship to sex, adult body size, or geographic locality. However, geographical color variation has been reported for the wide-ranging *Cirrhitus pinnulatus*. Hawkfish larvae have a pigmented chip barbel that disappears with age, cirri on the tips of the dorsal fin spine membrane, and serrations on the preopercle.

Kelpfishes are grouper or hawkfish-like in appearance, with cycloid scales that are moderately large, large pectoral fins, filaments on the tips of dorsal fins, and tufts of filaments on the dorsal spines of some species. Larvae are elongate to moderate in length, lack head spines, have heavy pigmentation in the postflexion stage, and are unspecialized. Sea carps have robust, elongate bodies, small cycloid scales, and pectoral fins with thickened lower rays positioned well forward under the gill covers. Their larvae are long to very elongate, lacking in head spines, heavily pigmented in the postflexion stage, and rather unspecialized. Many morwongs have elongated pectoral rays, forked caudal fins, an elevated dorsal fin, and large rubbery lips. A humplike head and back is present in a number of species. The postlarval stage of morwongs is silvery in color, deep-bodied, and very thin. This stage is retained in most species until they reach 17.7–23.6 in (45–60 cm) in total length. Although somewhat similar in shape to morwongs, the trumpeters have bodies that are elongate and compressed, with small pectoral fins, dorsal fins that are deeply notched, and a protruding mouth. Their postlarval stage resembles that of the morwongs.

Mated pair of blackback butterflyfish (*Chaetodon melanotus*) feeding. (Illustration by Joseph E. Trumpey)

Distribution

These fishes occur in coastal and estuarine waters of the Atlantic, Pacific, and Indian Oceans. Some species also occur in freshwater habitats in Africa, Australia, South America, and Southeast Asia. The Mullidae (six genera and 55 species) is found in tropical and warm-temperate coastal waters of the Atlantic, Pacific, and Indian Oceans. The Toxotidae (one genus and six species) is distributed in coastal marine, brackish, and fresh waters of India, east to the Philippines and Vanuatu, and south to northern Australia. The Dichistiidae (one genus and three species) has a distribution limited to coastal and brackish waters in South Africa and Madagascar. The Kyphosidae (15 genera and 42 species) is mostly tropical or subtropical in distribution, but two species have adapted to temperate waters off California. The Paracorpididae (one genus and at least one species) is known from Mozambique and South Africa. The Drepaneidae (one genus and three species) ranges from the Indo-West Pacific to West Africa. The Monodactylidae (two genera and five species) includes popular aquarium fishes found in tropical and warm temperate waters. One wide-ranging species, *Monodactylus argenteus*, occurs in brackish, marine, and freshwater habitats from the Red Sea and East Africa, east to Samoa, north to the Yaeyama Islands of southern Japan, and south to Australia and New Caledonia. A second species, *M. falciformis*, is limited to the western Indian Ocean, from the Red Sea, down to South Africa, and another, *M. kottelati*, is found only at Sri Lanka. *Monodactylus sebae* is found in the eastern Atlantic, from the Canary Islands and Senegal south to Angola. The two remaining species, both in the genus *Schuettea*, are found in temperate brackish and marine waters of southern Australia; one species off southeastern Australia, and the other off Western Australia. The Chaetodontidae (10 genera and as many as 125 species) and the Pomacanthidae (nine genera and at least 73 species) are distributed in tropical, subtropical, and warm temperate coastal waters. Among the chaetodontids, at least 13 species occur in the Atlantic, four species in the eastern Pacific, and the remainder in the Indo-West Pacific region. A similar pattern exists for the pomacanthids. The Enoplosidae (one genus and species) is found in coastal temperate waters of southern Australia. The Pentacerotidae (eight genera and 13 species) occurs primarily in temperate waters of the Indo-Pacific and southwestern Atlantic. The Nandidae (seven genera and 10 species) contains primarily freshwater fishes, but may also occur in brackish-water estuaries. Two species occur in riverine freshwaters of tropical western Africa. *Polycentrus schomburgkii* is found in fresh and brackish water from Trinidad and the Guyanas south to the Amazon Basin, while *Monocirrhus polyacanthus* is found from Guyana south to Brazil and west to the Peruvian Amazon. The remaining species occur from southern Asia east to Southeast Asia, including Indonesia. The Oplegnathidae (one genus and six species) is found in the coastal warm-temperate waters of the Indo-Pacific region, mainly in Japan, southern Australia (including Tasmania), the Galápagos Islands, Peru, and South Africa. One species has also been reported from the volcanic far-northern Mariana Islands. Among cirrhitoid fishes, the Cirrhitidae (12 genera and 33 species) occurs mainly in the Indo-West Pacific region, but also in the tropical and subtropical Atlantic (three species), and eastern Pacific (three species). Two species have distributions that range from the Red Sea east to Mexico and south to Colombia. Three other species range from the Red Sea east to Hawaii and eastern French Polynesia. On the other hand, a number of hawkfishes have limited distributions. For example, the giant hawkfish (*Cirrhitus rivulatus*) is restricted to the Eastern Pacific, including the Galápagos. Similarly, *Itycirrhitus wilhelmi* is found only at Easter and Pitcairn

A pair of redtail butterflyfish (*Chaetodon collare*) near the Similan Islands (Photo by Fred McConnaughey/Photo Researchers, Inc. Reproduced by permission.)

Islands at the eastern margin of the central Pacific. The splendid hawkfish (*Notocirrhitus splendens*) is endemic to Lord Howe Island, but strays west to Sydney and the coast of New South Wales in Australia, but also occurs at the Kermadec Islands of New Zealand. The red-barred hawkfish (*Cirrhitops fasciatus*) has a disjunct distribution and is found only in Hawaii, southern Japan, Mauritius, and Madagascar. Another species with a disjunct distribution, *Paracirrhites hemistictus*, occurs sporadically throughout the central and western Pacific, but is also found at Christmas and Cocos-Keeling Islands in the eastern Indian Ocean. The Chironemidae (two genera and four species) is distributed in temperate coastal waters of Australia and New Zealand, with one species endemic to Chile, and another limited to the warm-temperate Lord Howe, Norfolk, and the Kermadec Islands. The Aplodactylidae (three genera and five species) is another Southern Hemisphere family found in the temperate marine waters of Australia, New Zealand, Peru, and Chile. The Cheilodactylidae (four genera and 20 species; *Goniistius* of some authors, treated here as a subgenus of *Cheilodactylus*) has an antitropical distribution. Most species are found in Southern Hemisphere waters (Australia, New Zealand; the Kermadec, Lord Howe, Norfolk, Easter, Rapa, St. Paul's, and Amsterdam Islands; Ilots de Bass, southern Africa, Chile and Peru in the Indo-Pacific; Argentina, southern Africa, Tristan da Cunta Island and Vima Mount in the Atlantic) but members of one genus, *Cheilodactylus*, also occur off Japan, China, and the Hawaiian Islands. Finally, another Southern Hemisphere family, the Latridae (three genera and nine species), is distributed in coastal southern Australia, New Zealand, Chile, and the southern Atlantic.

Habitat

The habitats of these fishes are highly variable, and range from coastal and deepslope or seamount marine habitats, to estuaries, rivers, lakes, ponds, swamps, and even water-filled ditches. Three terms describe the use of these habitats. Benthic fishes are those found in close association with the bottom (including structure); benthopelagic fishes swim on or just above the bottom; and pelagic fishes swim in the water column well above the bottom. Goatfishes are benthopelagic fishes that frequent coral reefs, rocky reefs, and sand, rubble, and mud flats. Archerfishes are pelagic and swim frequently just beneath the surface of shallow marine and brackish waters, and in freshwater lakes and streams. They prefer coastal and estuarine mangroves or weed beds, and flooded trees in freshwaters. Three species are restricted to freshwater habitats. Galjoens are pelagic shoaling fishes found in turbid water along rocky coastlines. Sea chubs are pelagic or benthopelagic, although some pelagic species may be benthopelagic at night. They occur on coral and rocky reefs and frequent high-energy or surge zones. Jutjaws are pelagic fishes of coastal waters. Sicklefishes are benthopelagic in coastal and estuarine waters, and frequent mudflats. Monos are pelagic or benthopelagic, and may frequent mangroves or other structures when not swimming in the water column of bays, back bays, estuaries, or the lower reaches of rivers. Butterflyfishes are either benthopelagic or pelagic over coral reefs (most species), rocky reefs, walls and deep slopes (pelagic species, such as *Hemitaurichthys polycanthus*, that feed on zooplankton), and flats. Juveniles and adults of some species have been reported from mangroves. Similarly, the angelfishes are either benthopelagic or semipelagic, the latter (i.e., *Genicanthus* spp.) moving up into the water column to feed upon zooplankton. A number of benthopelagic angelfishes, such as some members of the genus *Centropyge*, are often found among coral rubble, dead corals, or rocks, where filamentous algae, a component of their diet, is found. Adults of the single species of oldwife (*Enoplosus armatus*) are found on inshore and offshore rocky reefs, and on sea grass beds and in estuaries, but juveniles occur mainly in estuaries. The boarfishes and armorheads are pelagic and occur on deep slopes of inshore reefs, often in or near caves, and seamounts offshore. The leaffishes occur largely in freshwater. All are benthopelagic. Three riverine species are also found in brackish water in estuaries. The most famous leaffish, the *Badis badis*, lives solitarily in rivers, ponds, ditches, and swamps. All leaffishes make use of structure, usually plant materials, for shelter and ambush sites. The knifejaws are found swimming among rocks, underwater cliffs, and walls on rocky reefs. The hawkfishes are benthic, although one species, *Cyprinocirrhites polyactis*, hovers in the

Sea chubs schooling. (Illustration by Brian Cressman)

water column to feed on zooplankton, but darts into holes on the bottom to avoid danger or to shelter for the night. Two species of hawkfishes, *Neocirrhites armatus* and *Oxycirrhites typus*, are obligate coral-dwelling fishes. The former species occurs in shrublike corals of the genus *Pocillopora*, while the latter occurs in gorgonians (sea fans) and black corals found on deep slopes and walls of coral and rocky reefs. Species of the genus *Paracirrhites* are faculative coral-dwelling fishes that shelter in or perch on larger shrub-like corals. If such corals are absent, they will also perch upon coral boulders, ledges and rocks, or hide in holes. Kelpfishes are bottom dwelling, usually in inshore kelp forests, algal beds, and surge zones, or in intertidal reaches of rocky reefs. Seacarps are bottom dwelling and poor swimmers. They are found in small or large loosely maintained aggregations, in surge zones of rocky reefs, and in tidal passes in estuaries. The morwongs occur in a variety of habitats and depth ranges; most species are found on coral or rocky reefs inshore to about 197 ft (60 m), but *Nemadactylus* species have been reported from deep-slope habitats down to 1,312 ft (400 m). The trumpeters occur on rocky reefs and bottoms, although members of the genus *Mendosoma* forage for zooplankton in the water column.

Behavior

There is considerable variety in social organization, locomotion, and other behavior among members of this group. General patterns characteristic of each family are presented here with some greater detail given to certain families. Goatfishes move purposefully along the bottom while foraging and may accompany other bottom-feeding or swimming species for opportunistic feeding, and other fishes, such as wrasses (Labridae), may join them for the same reason. Many goatfishes aggregate in small groups, although members of certain genera, such as *Mulloidichthys* form large schools that seem to hover in the water column when not foraging on the bottom. Others move about singly, however, and some of these, such as *Parupeneus cyclostomus* (a species with two color morphs, golden yellow and pale blue), are remarkably active swimmers. Many species also rest upon the bottom for long periods of time, while some are more active at night than during daytime. The archerfishes use stealth when hunting prey in the manner discussed, but they may be territorial as they move back and forth encountering their neighbors. Galjoens move about in small groups, as do the sea chubs, although the latter can form exceptionally large groups that move about in the water column. The behavior of the jutjaws is poorly known, but they likely form aggregations or schools that swim in the water column. Sicklefishes do the same closer to the bottom. Monos form large schools in the mouths of rivers or in bays, but some occur singly or in small groups under the cover of mangroves in brackish water or brush in freshwater, where some species may be highly territorial.

The behavior of butterflyfishes, perhaps because of their bright coloration and high level of conspicuousness, has been the subject of considerable research. Largely diurnal, butterflyfishes may occur singly and be highly territorial, or live in aggregations with a general home range and relatively little territorial or aggressive behavior. Most, however, may be found in pairs, usually monogamous, that patrol home ranges or protect shared territories. Territorial defense, in general, may be directed towards conspecifics, other butterflyfishes with similar feeding requirements or preferences, or other species of fishes that also share their diet. Species that live singly or in pairs may form aggregations during certain times of the day or season, and these may be related to reproduction. At least one species, *Chaetodon lunula*, which may form aggregations during daytime, is active at night.

Pyramid butterflyfish (*Chaetodon polylepis*) cluster over coral rocks near New Georgia Islands. (Photo by Fred McConoughey/Photo Researchers, Inc. Reproduced by permission.)

Archerfish (*Toxotes jaculatrix*) can spit water up to five feet (1.5 m) to knock insects into the water for a meal. (Photo by Animals Animals ©Stephen Dalton. Reproduced by permission.)

The angelfishes mostly form social and mating groups consisting of a single male and multiple females. They change sex from female to male (protogynous hermaphroditism), usually under social control. Some species may be monogamous or facultatively monogamous. For example, the emperor angelfish (*Pomacanthus imperator*) may often be encountered in pairs that consist of a male and female, yet additional females may reside in the pair's home range and join them prior to courtship at dusk. The regal angelfish (*Pygoplites diacanthus*) is often seen alone, but will form pairs or mating groups during sunset courtship periods. Some intra- and interspecific territorial behavior may occur, usually during encounters while patrolling a home range.

The behavior of the oldwife is not well known. Juveniles occur singly, but tend to form schools as they age. Adults are encountered either singly or in pairs, but large schools may also form. Shallow-dwelling boarfishes occur in pairs or small-to-large aggregations, depending upon the species, but the rarely seen juveniles occur singly. The behavior of their deep-water cousins, the armorheads, is poorly known, but they likely aggregate or form schools. The leaffishes are well known for the way they mimic leaves drifting in the water column. This allows them to avoid predation and to ambush potential prey. Little is known about the behavior of knifejaws except that they swim about rocks and will enter caves.

The hawkfishes are territorial and organized socially into mating groups consisting of a single male and a variable number of females (usually two to five) that maintain territories or home areas inside the male's territory. Two species, however, are facultatively monogamous as a consequence of habitat use. For example, obligate coral-dwelling species, such as *Oxycirrhites typus* and *Neocirrhites armatus*, are facultatively monogamous if their coral (black coral or gorgonian sea fan, in the case of *O. typus*) is too small to support more than two fish, and if neighboring corals are too far away to allow a male to safely move to it to court a resident female. However, if the coral is large enough to support more than two fishes, or if suitable corals are close by, then a multiple-female group is possible. Juvenile obligate coral-dwelling hawkfishes often recruit to less-favorable corals nearby and then attempt, as do facultative and noncoral-dwelling species, to gain entry into the mating group. Usually, resident females will attack the new arrival and try to expel it from the coral or the group. This behavior is also directed towards adults, usually emigrants from fragmented groups elsewhere, that attempt to join the group. If the emigrant is recognized as another male, the resident male will attempt to expel it; if a female, it may be attacked by other females. As with other fishes that have a similar social organization, all hawkfishes likely change sex from female to male (protogynous hermaphroditism). Sex change may proceed through social control or maturation. The effect of sex change upon social organization can be profound, because, in addition to allowing for succession by the dominant female with the loss of the male, it can also allow the dominant female to fragment the group should it become too large for the male to control effectively. Similarly, if the dominant male cannot control his group, then neighboring males and "rogue" males (those that have lost a group or were recent females that could not fragment their own group after changing sex) can attempt to sequester females. Another interesting behavior is practiced by the plankton-feeding *Cyprinocirrhites polyactis*, which mimics fairy basslets (Serranidae: Anthiinae) that aggregate, often in considerable numbers, in the water column.

Kelpfishes may establish large aggregations that form loosely over reefs, particularly surge zones, but individuals and small groups are more likely in the intertidal zone. The seacarps also aggregate, either in small or large, loosely organized groups that rest upon the bottom. They swim with the dorsal fin raised when seeking food or changing location. Most morwongs move about the bottom singly or in small, loosely formed aggregations, but members of the genus *Nemadactylus* form large aggregations. Trumpeters are mostly schooling fishes, which move about in small groups or in aggregations of hundreds of individuals.

Feeding ecology and diet

These fishes have considerable variation in both feeding methods and diet. Although most direct their attention towards bottom feeding or planktivory, in which they feed upon phytoplankton or zooplankton in the water column, there are some remarkable exceptions. Goatfishes utilize their barbels, which bear chemosensory organs, to stir up the bottom and

Schooling bannerfishes (*Heniochus diphreutes*) feed on plankton and are widespread in the tropical Indo-Pacific. (Photo by David Hall/Photo Researchers, Inc. Reproduced by permission.)

detect prey. They also feed upon smaller invertebrates in sand, rubble, algal beds, or on coral pavement, and some species may take small fishes that they disturb. Archerfishes swim near the surface of the water and can detect visually, with a correction for light refraction, terrestrial prey on overhanging branches of trees and shrubs or blades of grass at the water's edge. Upon detecting the prey, usually an insect, archerfishes direct a stream of water drops toward that prey, knocking it to the water where it can be captured with the mouth and consumed. The drops are formed by compression of water by the gill covers and shot by forcing them along a groove formed by the tongue and palate. Archerfishes also consume floating fruits and flowers. Although galjoens have small mouths, they are capable of preying upon ascidians, mussels, barnacles, and other crustaceans. Galjoens also consume seaweeds. The sea chubs have a mixed feeding strategy and diet. Members of the subfamilies Kyphosinae (except the genus *Gras*) and Girellinae graze or pluck algae. Those in the subfamilies Scorpidinae and Microcanthinae are carnivores that feed upon benthic invertebrates. The jutjaws strain zooplankton from the water column. The sicklefishes feed upon small invertebrates they extract from soft sediments, usually mud. Monos have small, obliquely positioned mouths with a combination of brushlike teeth on their jaws, villiform teeth on the vomer, and palatines within the mouth. They use this combination to capture and feed upon small fishes, invertebrates, or plankton. The butterflyfishes have been collectively described as microconsumers. Some species feed exclusively upon coral polyps, others upon small invertebrates on corals, rocks, coral pavement, or crevices, and still others, such as *Hemitaurichthys polylepis*, forage for zooplankton in the water column. There is some plasticity in diet for a number of species. While *Chaetodon trifascialis* feeds only upon acroporid corals, *C. punctatofasciatus* will feed upon a variety of benthic invertebrates, corals, and even filamentous algae. The forcepsfish (*Forcipiger flavissimus*) uses its tubular snout to feed upon hydroids, fish eggs, and small crustaceans, and to excise pieces of the tentacles of polychaete worms, the tube feet of starfishes, and the pedicilaria of sea urchins. On the other hand, the longnose butterflyfish (*F. longirostris*), with its longer snout, feeds mainly upon small crustaceans that are taken whole. Angelfishes also

A pair of yellow goatfish (*Mulloidichthys martinicus*) foraging in bottom sediments for food. The barbels situated under their chins have chemosensory organs that allow goatfishes to detect potential prey. (Illustration by Joseph E. Trumpey)

A juvenile emperor angelfish (*Pomacanthus imperator*) near Gavutu Island, in the Solomon Islands. (Photo by Fred McConnaughey/Photo Researchers, Inc. Reproduced by permission.

have considerable plasticity in their diet, although perhaps less so compared to butterflyfishes. Pygmy angelfishes (*Centropyge* spp.) feed mostly upon filamentous algae, while *Genicanthus* angelfishes forage mainly for zooplankton in the water column and supplement their diet with benthic invertebrates or algae. Others in the family feed upon sponges, small soft-bodied invertebrates, fish eggs, and algae.

Little is known about the feeding habits and diets of the oldwife, boarfishes, and armorheads, but they appear to feed upon benthic invertebrates and fishes. Leaffishes, which can conceal themselves even in open water thanks to their ability to mimic leaves downed in water, are remarkably good predators upon fishes and invertebrates (mostly crustaceans, insects, and worms). However, one larger species from Southeast Asia, the catopra (*Pristolepis fasciata*), is primarily a herbivore that feeds upon filamentous algae, submerged land plants, seeds, and fruits, but will also take crustaceans and aquatic insects. Knifejaws use their parrotlike beaks to break open and feed upon barnacles and mollusks. The hawkfishes feed upon small crustaceans and fishes they ambush from a perched position on a coral or rock, or from a resting position on the bottom. One species, *Cyprinocirrhites polyactis*, plucks zooplankton while hovering in the water column, and a second species, *Cirrhitichthys aprinus*, has been observed taking zooplankton as it darts out from soft corals into the water column along current-swept reef walls. Kelpfishes exhibit feeding behavior similar to hawkfishes and feed upon small invertebrates. The seacarps feed upon algae and other seaweeds they grab from the bottom with their mouths. The small mouths and thick lips of morwongs are used to feed upon small benthic or pelagic invertebrates, such as polychaete worms, crustaceans, mollusks, and echinoderms, or in some species, algae. At least one species, *Chirodactylus grandis*, feeds upon squid and small fishes. Trumpeters feed upon benthic invertebrates, although the members of the genus *Mendosoma* feed upon zooplankton.

Reproductive biology

Although studies on reproductive biology of these fishes are far from complete, there is considerable information, which characterizes some families in detail. Overall, they are either gonochorists or sequential hermaphrodites. Gonochoristic fishes begin life either as females or males and remain that way their entire lives. Examples include goatfishes, archerfishes, galjoens, sea chubs, jutjaws, sicklefishes, monos, butterflyfishes, the oldwife, boarfishes and armorheads, leaffishes, knifejaws, and morwongs. Sequential hermaphrodites change sex. Most Percoidei with this life strategy appear to be protogynous hermaphrodites, in that they begin as females but may change sex and become males. The control of sex change may be social or as a consequence of maturation, and despite detailed research on some groups (i.e., angelfishes and hawkfishes), much remains to be learned about the circumstances that trigger sex change. The status of kelpfishes, seacarps, and trumpeters is uncertain.

Spawning may occur in monogamous pairs, as sequential pairs within a haremic mating group or spawning aggregation, or promiscuously. Most of these fishes appear to spawn pelagically, the eggs are fertilized in the water column during pair spawning, and there is no parental care. Sneaking (partial fertilization of a pair's eggs by a second male that sneaks up on a rising pair) and group spawning are also possible. Pelagic spawning has been verified for the goatfishes (paired and group spawning), archerfishes, sea chubs, monos (in part), butterflyfishes, angelfishes, the oldwife, hawkfishes, morwongs, and trumpeters. Pelagic spawning is presumed in the galjoens, jutjaws, sicklefishes, boarfishes and armorheads, knifejaws, kelpfishes, and seacarps. Demersal spawning, in which eggs are scattered on the bottom, placed in nests or bubble nests (a specialized behavior of some leaffishes), brooded in holes or caves, or attached to plants, shells, rocks, or some form of structure, may involve parental care to some degree or not at all. Parental care is pronounced, especially in freshwater fishes such as the leaffishes, but is largely absent or minimal in marine fishes. Leaffishes spawn demersally, or in bubble nests, and some monos also have demersal eggs.

There is considerable variability in the temporal patterns of spawning of these fishes. They may spawn only seasonally in relation to latitude and water temperature but not exclu-

An Amazon leaffish (*Monocirrhus polyacanthus*) eating a smaller fish. (Photo by Animals Animals ©M. Gibbs, OSF. Reproduced by permission.)

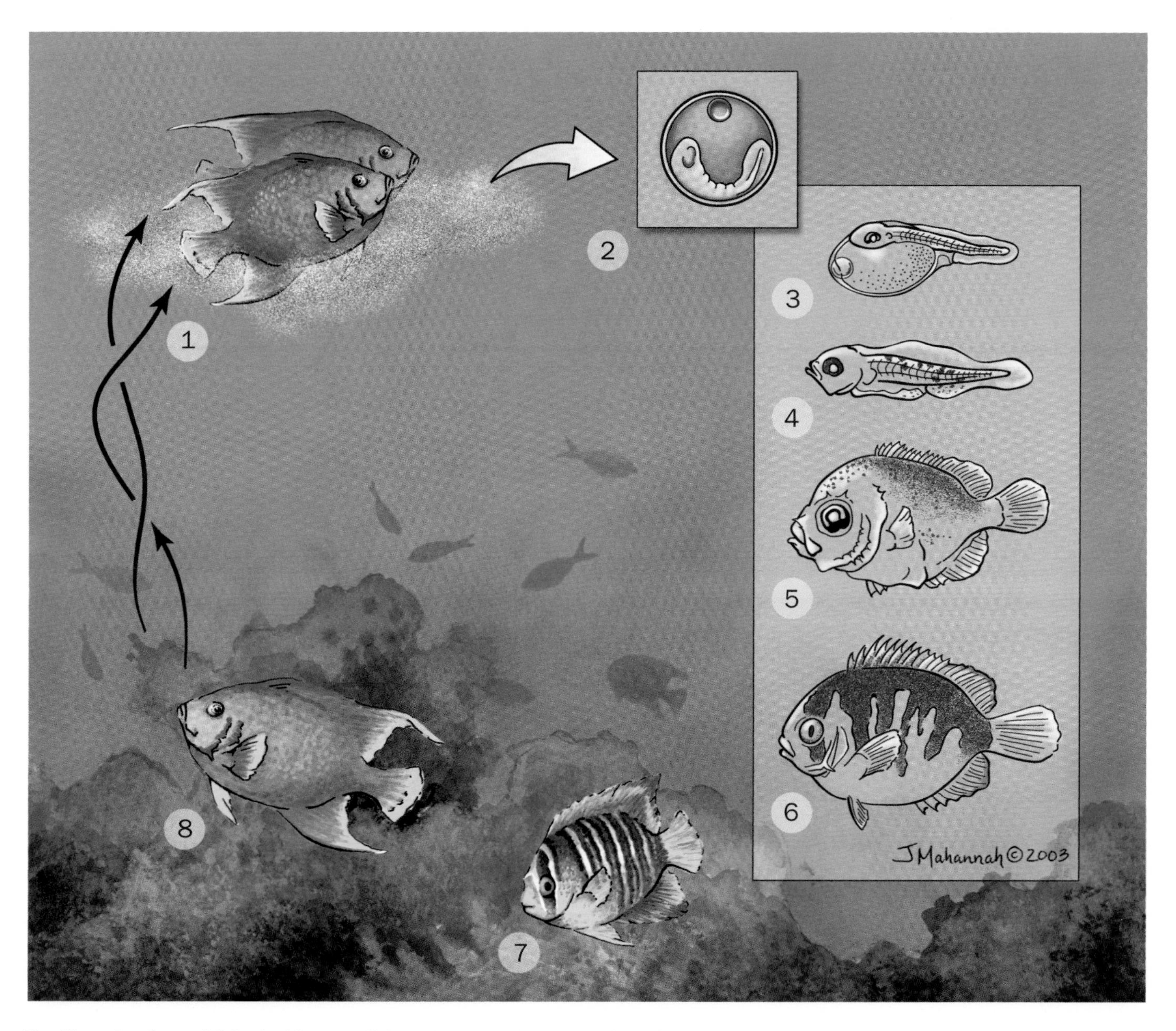

The life cycle of a reef fish, the blue angelfish (*Holacanthus bermudensis*). 1. A pair spirals toward the surface, where eggs are released; 2. Pelagic fertilized egg with single oil droplet; 3. Newly hatched larva, 0.06 in (1.58 mm); 4. 0.10 in (2.5 mm) larva; 5. 0.17 in (4.2 mm) larva; 6. 0.5 in (13 mm) post-larva; 7. Juvenile; 8. Adult. (Illustration by Jacqueline Mahannah)

sively. For example, fishes of some families might spawn seasonally over a short period of time once or twice a year regardless of latitude. Others, such as the hawkfishes, may spawn daily at lower latitudes, but seasonally at higher latitudes. Those fishes that produce pelagic eggs and larvae have the potential for relatively long-distance dispersal in marine systems. Demersally spawned eggs usually also hatch to become pelagic larvae, but these may not disperse as far. Of course, there are plenty of exceptions to the rule in both cases.

Conservation status

One species of angelfish (*Centropyge resplendens* [Pomacanthidae]), and five species of butterflyfish (*Chaetodon flavocoronatus*, *C. litus*, *C. marleyi*, *C. obliquus*, and *C. robustus* [Chaetodontidae]) are listed as Vulnerable on the IUCN Red List. Other species may be vulnerable or threatened, but the data simply do not exist to verify the conservation status of many of these fishes. Certainly, species that live in close association with corals (such as butterflyfishes, angelfishes, and hawkfishes) might be at risk because of the effects of coral bleaching, the harvest of corals, and the general destruction of coral habitats. The same might be said of species that live in mangroves (such as archerfishes) threatened with destruction. The leaffishes may be at risk from the effects of pollution, hydroelectric development, so-called stream improvements, the draining of swamps, destructive agricultural practices, mining, logging, and other forms of development that contribute to the

A pair of red goatfish (*Mullus surmuletus*). (Photo by Tom McHugh/ Photo Researchers, Inc. Reproduced by permission.)

sedimentation of rivers and streams. Galjoens require considerable protection from overfishing in South Africa. Similarly, armorheads that dwell on seamounts or on deep-slope reefs and grounds are likely candidates for overfishing and should be of considerable concern. Overfishing of popular aquarium species (such as leaffishes, butterflyfishes, angelfishes, and hawkfishes), especially those that are rare, have limited distributions locally or geographically, or have mating systems or other life history characteristics that make them especially vulnerable to unregulated harvests, is also of great concern.

Significance to humans

Fishes of the following families are taken in major or minor commercial food fisheries: Mullidae, Parascorpididae, Enoplosidae, Pentacerotidae, Oplegnathidae, Cheilodactylidae, Chironemidae, and Latridae. Fishes that may be found in subsistence food fisheries include the Mullidae, Toxotidae, Kyphosidae (prized at some localities but avoided at others), Monodactylidae, Chaetodontidae, Pomacanthidae, Nandidae, larger Cirrhitidae, and Cheilodactylidae. Game fishes include members of the Dichistiidae (their numbers heavily depleted in South Africa), Nandidae, Oplegnathidae, Cheilodactylidae, and the Latridae. Some members of the Chironemidae and Aplodactylidae are not taken generally for food because they apparently taste bad. A number of families are taken for the aquarium trade, either as juveniles or adults. Important species include members of the Toxotidae, Nandidae, Monodactylidae, Chaetodontidae, Pomacanthidae, and Cirrhitidae. Of minor importance are members of the Mullidae, Kyphosidae, Enoplosidae, and Cheilodactylidae.

1. Pebbled butterflyfish (*Chaetodon multicinctus*); 2. Adult emperor angelfish (*Pomacanthus imperator*); 3. Juvenile emperor angelfish; 4. Hawaiian morwong (*Cheilodactylus vittatus*); 5. Longnose hawkfish (*Oxycirrhites typus*); 6. Striped boarfish (*Evistias acutirostris*). (Illustration by Joseph E. Trumpey)

1. Cortez chub (*Kyphosus elegans*); 2. Banded archerfish (*Toxotes jaculatrix*); 3. Amazon leaffish (*Monocirrhus polyacanthus*); 4. Mono (*Monodactylus argenteus*); 5. Gangetic leaffish (*Nandus nandus*); 6. Yellow goatfish (*Mulloidichthys martinicus*). (Illustration by Joesph E. Trumpey)

Species accounts

Pebbled butterflyfish
Chaetodon multicinctus

FAMILY
Chaetodontidae

TAXONOMY
Chaetodon multicinctus Garrett, 1863, Hawaiian Islands.

OTHER COMMON NAMES
Hawaiian: Kikakapu.

PHYSICAL CHARACTERISTICS
Deep, compressed body, whitish in color, with four to six narrow brown or gold-brown bars and many faint olive spots on both the body and the fins; a gold-brown eye bar on the head, a black and gold bar on the caudal peduncle, and a solid black bar on the caudal fin. To 4.5 in (12 cm) total length.

DISTRIBUTION
Limited to the Hawaiian Islands and Johnston Atoll in the east-central Pacific.

HABITAT
Seaward or lagoon coral reefs, usually with considerable stands of *Porites* and *Pocillopora* corals, between 16 and 98 ft (5–30 m) in depth.

BEHAVIOR
Forms heterosexual pairs but occasionally occurs in small aggregations.

FEEDING ECOLOGY AND DIET
Omnivorous, its diet consisting of coral polyps, polychaete worms, small crustaceans, and algae.

REPRODUCTIVE BIOLOGY
Gonochoristic. Paired courtship just before or after sunset into early evening, with pelagic spawning in the water column. Spawning pairs are occasionally joined by one or more intruding males who attempt to spawn with the female. The intruders may be either rogue males or paired males who have temporarily abandoned their mates. Courtship occurs between December and July, but courtship and spawning is most pronounced between March and July. Eggs and larvae are pelagic. The eggs are spherical and small (0.023–0.029 in [0.6–0.75 mm] in diameter). Larvae are around 0.059 in (1.5 mm) in length at hatching and have a large yolk sac. The mouth is unformed and the eyes are unpigmented. With growth, the larva develops fused head plates that extend over the trunk of the body. This is known as the thoichthys stage of development; it adapts the larva to a long pelagic phase. Other adaptations include spine formation on the head, dorsal, and pelvic fins.

CONSERVATION STATUS
Not listed by the IUCN. Coral habitats may be threatened by coral bleaching, pollution, sedimentation, and other forms of degradation that may negatively impact populations of this limited-distribution species.

SIGNIFICANCE TO HUMANS
May be collected for the aquarium trade, but this and many other butterflyfishes generally do poorly in most aquaria. ◆

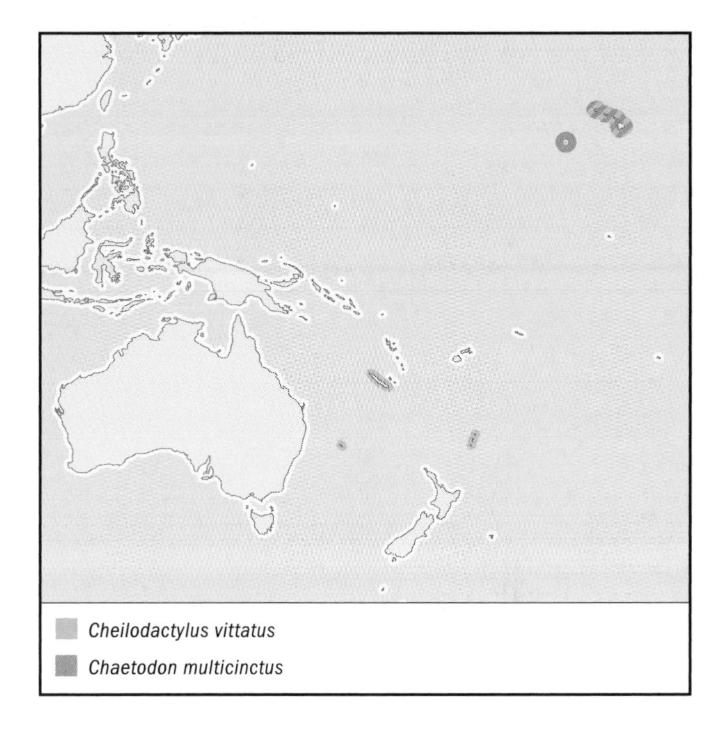

Hawaiian morwong
Cheilodactylus vittatus

FAMILY
Cheilodactylidae

TAXONOMY
Cheilodactylus vittatus Garrett, 1864, Hawaiian Islands.

OTHER COMMON NAMES
Hawaiian: Kikakapu.

PHYSICAL CHARACTERISTICS
Laterally compressed with a high forehead and long, sloping back. There are prominent bony knobs on the head. The caudal fin is forked. There is an oblique pale black bar on the head, and there are four oblique dark-black bars, all against a white background color, along the head and flank. The dorsal fin spines are black, white, and orange-red; the dorsal rays are white.

DISTRIBUTION
Hawaiian Islands in the Northern Hemisphere, and New Caledonia, Lord Howe Island, and the Kermedec Islands in the Southern Hemisphere.

HABITAT
Coral and volcanic rocky reefs, in areas of coral, rocks, pavement and rubble. Depth usually below 65 ft (20 m).

BEHAVIOR
Occur singly or in small groups or aggregations.

FEEDING ECOLOGY AND DIET
Benthic carnivores that forage on the bottom for invertebrates, mainly crabs and shrimps, amphipods, polychaete worms, and gastropod mollusks.

REPRODUCTIVE BIOLOGY
Gonochoristic, with females tending to be larger than males. Courtship and spawning at night, with the release of pelagic eggs and larvae. Eggs are spherical and range in size from about 0.035 to 0.043 in (0.9–1.1 mm) in diameter. Hatching larvae are approximately 0.1–0.13 in (2.5–3.3 mm) in length, have unpigmented eyes, an unformed mouth, and a large yolk sac. Pigmentation develops with absorption of the yolk. The body is elongate but becomes deeper and compressed with growth, and silvery in color (paper-fish phase). The ventral keel also becomes prominent with growth. The gas bladder is small to moderate in size. The mouth becomes smaller and oblique with development. Larvae are adapted for a relatively long pelagic life.

CONSERVATION STATUS
Not listed by the IUCN. This species is rare in the Hawaiian Islands and may yet be proven distinct from those of the Southern Hemisphere populations.

SIGNIFICANCE TO HUMANS
Taken as a food fish, but smaller individuals may also be collected for the aquarium trade. ◆

Longnose hawkfish
Oxycirrhites typus

FAMILY
Cirrhitidae

TAXONOMY
Oxycirrhites typus Bleeker, 1857, Ambon Island, Moluccas, Indonesia.

OTHER COMMON NAMES
German: Gestreifter Schützenfish; Japanese: Kudagonbe.

PHYSICAL CHARACTERISTICS
Body slender, with an elongated snout; body whitish in color with a tartanlike pattern in red along body, snout, and dorsal fin; males larger than females within mating groups. To 5.1 in (13 cm).

DISTRIBUTION
Red Sea east to Panama; southern Japan south to New Caledonia.

HABITAT
Usually deep slope or wall habitats in seaward coral or rocky reefs, passes, and lagoons; shallow rocky reefs in southern Baja California Sur, Baja Peninsula, Mexico, and occasionally found on shallow coral reefs elsewhere, particularly in areas of upwelling. Otherwise, more common in deeper reef habitats. Usually found in gorgonians (sea fans) and black corals.

BEHAVIOR
Occurs singly or in facultatively monogamous pairs on gorgonians and black corals. Will have a haremic mating system if the gorgonian or black coral is large or if others are close by.

FEEDING ECOLOGY AND DIET
Plucks small benthic and planktonic crustaceans with its long snout.

REPRODUCTIVE BIOLOGY
Protogynous hermaphrodites. Paired courtship and pelagic spawning just prior to or after sunset. If a male has a haremic mating group, he will move back and forth between females to court them until they spawn. Eggs and larvae are pelagic. Eggs are spherical, about 0.027 in (0.69 mm) in diameter, and hatch after 15 hours incubation at 80.6°F (27°C). The yolk sac is

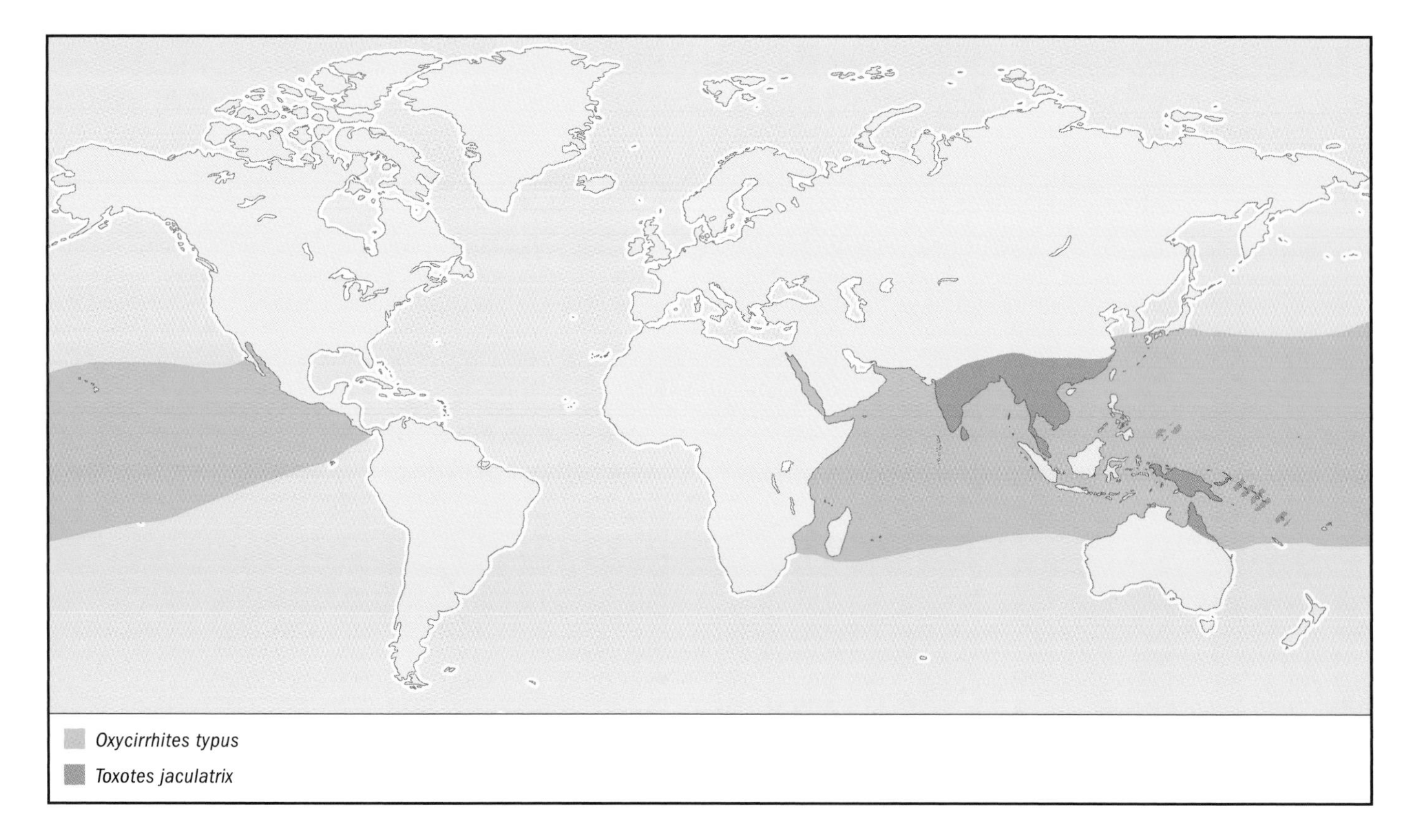

large, and the jaws and eyes undeveloped. Pigmentation develops with absorption of the yolk. Larvae are elongate and compressed laterally. Possess a gas bladder that is lost in adults. Specializations for pelagic life include serrations on the preopercle, a chin barbel (which disappears with growth), and the presence of cirri on the spinous dorsal fin membrane and also on the anterior nostril.

CONSERVATION STATUS
Not listed by the IUCN. Not commonly seen, despite its wide distribution, and population levels likely to be limited by available microhabitat. The collection of black corals for the jewelry trade and gorgonians for the ornamental trade poses a threat, as do unregulated harvests by the aquarium trade.

SIGNIFICANCE TO HUMANS
A highly prized aquarium species. ◆

Cortez chub
Kyphosus elegans

FAMILY
Kyphosidae

TAXONOMY
Kyphosus elegans Peters, 1869, Mazatlan, Sinaloa, Mexico.

OTHER COMMON NAMES
Spanish: Chopas.

PHYSICAL CHARACTERISTICS
Deep, compressed body with silver-gray color marked by faint brown stripes between scale rows on the flanks. Can quickly produce a pattern of white spots about the diameter of the eye. The jaw teeth are fixed and not freely movable. To 15 in (38 cm) total length.

DISTRIBUTION
Gulf of California south to Panama and west to the Galápagos Islands.

HABITAT
Rocky and coral reefs, usually in high-energy areas below the surf zone.

BEHAVIOR
Often found in large schools swimming in the water column.

FEEDING ECOLOGY AND DIET
Omnivorous but feeds mainly upon benthic algae.

REPRODUCTIVE BIOLOGY
The eggs are pelagic and about 0.039–0.043 in (1–1.1 mm) in diameter. Larvae hatch at 0.094–0.114 in (2.4–2.9 mm), are elongate, have a large yolk sac, unformed jaws, and unpigmented eyes. With growth, the head and mouth become large. Possesses a gas bladder. Head spines develop as a specialization for pelagic life.

CONSERVATION STATUS
Not listed by the IUCN.

SIGNIFICANCE TO HUMANS
May be taken in subsistence fisheries or local commercial fisheries as a food fish. ◆

Mono
Monodactylus argenteus

FAMILY
Monodactylidae

TAXONOMY
Monodactylus argenteus Linnaeus, 1758, East Indies.

OTHER COMMON NAMES
English: Diamondfish, fingerfish, moony, silvery moony, Natal moony; Japanese: Himetsubameuo.

PHYSICAL CHARACTERISTICS
Deeply compressed body with silvery color marked by a slender dark-colored bar on the head. The anterior edges of the dorsal and anal fins are also dark colored, the fins are pale yellow. The pelvic fin is absent in adults. To 8.5 in (22 cm) total length.

DISTRIBUTION
Red Sea east to Samoa; Yaeyama Islands (southern Japan), south to Australia and New Caledonia.

HABITAT
Brackish waters, estuaries, and silty reefs.

BEHAVIOR
Found singly or in small groups among mangroves or other structures, but in large schools in estuaries and silty reefs where they swim in the water column.

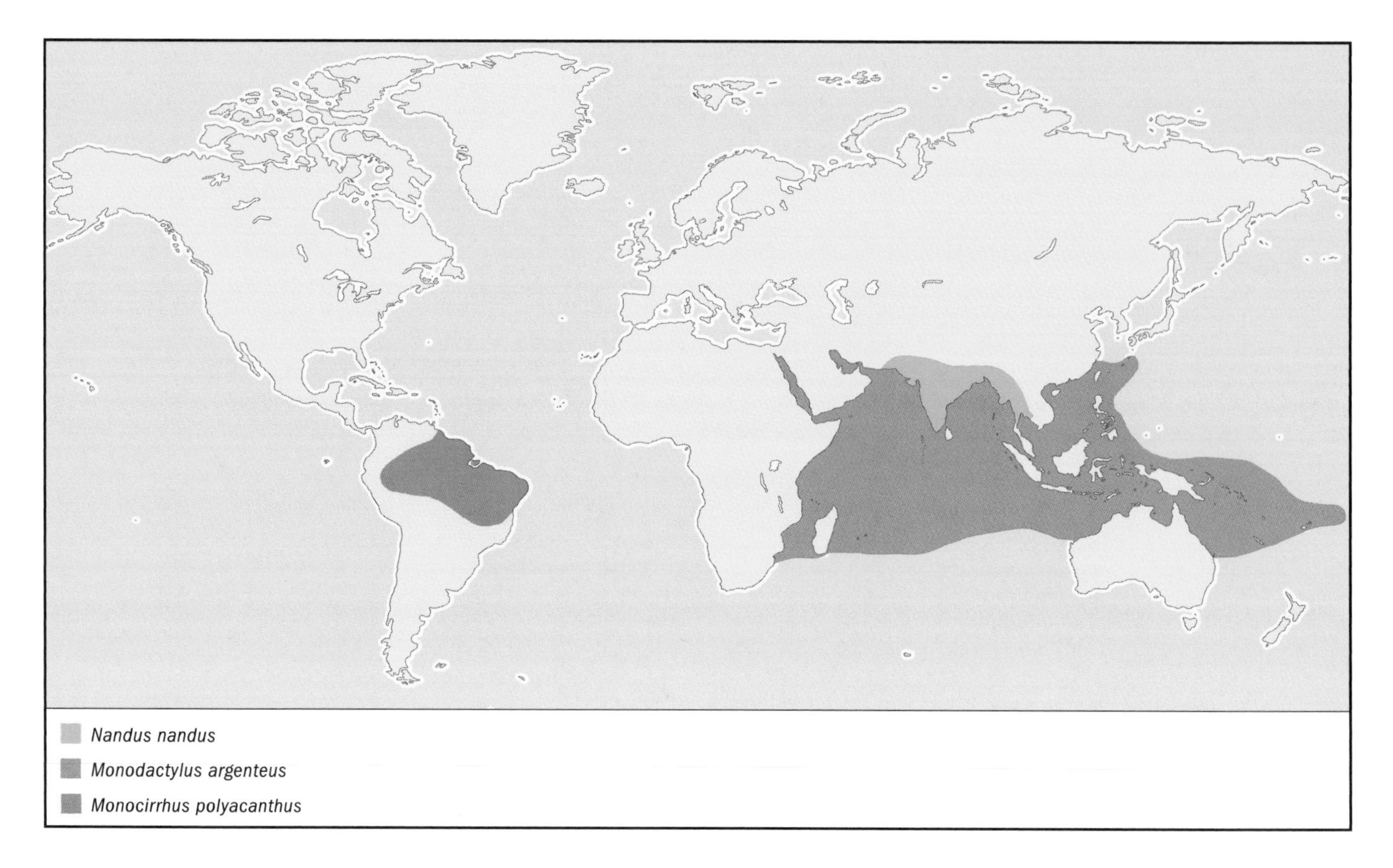

FEEDING ECOLOGY AND DIET
Feed upon plankton in the water column and detritus on the bottom.

REPRODUCTIVE BIOLOGY
Spawns adhesive, demersal eggs in freshwater. Others in the genus spawn pelagic eggs in seawater. Eggs are spherical and approximately 0.039 in (1 mm) in diameter. The larvae hatch around 0.071 in (1.8 mm) in length, are compressed and moderate in depth, have a large yolk sac, an unformed mouth, and unpigmented eyes. With growth, the head becomes large with a convex dorsal profile that becomes steep over time. Possesses a small gas bladder. Larvae have pelvic fins that are reduced in adults. This character, plus the development of spines on the head, are viewed as specializations for pelagic life.

CONSERVATION STATUS
Not listed by the IUCN.

SIGNIFICANCE TO HUMANS
Popular as an aquarium fish. Maybe taken in subsistence fisheries as a food fish. ◆

Yellow goatfish
Mulloidichthys martinicus

FAMILY
Mullidae

TAXONOMY
Mulloidichthys martinicus Cuvier and Valenciennes, 1829, Martinique Island, West Indies.

OTHER COMMON NAMES
French: Poisson chèvre jaune.

PHYSICAL CHARACTERISTICS
Twin barbels on the lower jaw, a relatively short snout, the body elongate with two well-separated dorsal fins, two spines in the anal fin, and a forked caudal fin. There are 36–37 lateral line scales from the upper gill opening to the base of the caudal fin. These fishes are bronze and pale yellow in color with a yellow median stripe that extends along the length of the body, and grow to 12 in (30 cm) total length.

DISTRIBUTION
Western Atlantic, from Bermuda southwest to Florida and the Gulf of Mexico, and south from the Bahamas to Brazil.

HABITAT
Coral reefs and flats of sand or rubble.

BEHAVIOR
Swims lazily or hovers in schools above the bottom during the day and forages after dark.

FEEDING ECOLOGY AND DIET
Feeds by probing the bottom with its barbels in search of prey, usually at night or during low-light periods. The diet includes annelid worms, crabs, ophiurans, and to a lesser extent, small fishes.

REPRODUCTIVE BIOLOGY
Spawns in aggregations and produces pelagic eggs and larvae. Eggs are spherical and small, between 0.024–0.036 in (0.63–0.93 mm) in diameter. Larvae hatch between 0.063–0.134 in (1.6–3.4 mm) in length, are elongate, and have a large yolk sac, unformed jaws, and unpigmented eyes. Body pigmentation develops with absorption of the yolk sac. Possesses a gas bladder and a short

gut. The barbels form at about 0.67–0.71 in (17–18 mm) but remain recessed under the lower jaw. Post-larvae appear capable of a relatively long pelagic phase prior to settlement.

CONSERVATION STATUS
Not listed by the IUCN.

SIGNIFICANCE TO HUMANS
Highly esteemed as a food fish and taken in subsistence and local commercial fisheries. ◆

Amazon leaffish
Monocirrhus polyacanthus

FAMILY
Nandidae

TAXONOMY
Monocirrhus polyacanthus Heckel, 1840, Marabitanos, Brazil.

OTHER COMMON NAMES
English: Barbeled leaf fish, South American leaf-fish; German: Blattfisch; Spanish: Pez hoja.

PHYSICAL CHARACTERISTICS
Total length 3.9 in (10 cm). Body somewhat elongated, compressed, and basslike. Profile of forehead is steep and rises to the first dorsal spine at a point directly over the posterior edge of pelvic fin. A narrow, lurelike organ extends forward from the lower part of a protrusible mouth. Sharp dorsal, anal, and pelvic spines. Body color yellowish bronze to light or dark brown. Outer edges and tips of the anal, pelvic, and caudal fins are dark brown. Narrow dark brown stripe runs obliquely from a point midway up the forehead down through the eye, a second similar stripe runs obliquely from the lower operculum to the eye. A third, fainter band is found mid flank below the dorsal fin.

DISTRIBUTION
South America, from Guyana south to Brazil and west into the Peruvian Amazon.

HABITAT
Benthopelagic, primarily streams and rivers; often in association with or near submerged vegetation.

BEHAVIOR
Cryptic. Often found floating motionless, like a leaf, or in the cover of aquatic vegetation.

FEEDING ECOLOGY AND DIET
Carnivorous. An ambush predator that preys upon smaller fishes, and aquatic insects and their larvae. Engulfs prey rapidly by extending its large, protrusible mouth.

REPRODUCTIVE BIOLOGY
Gonochoristic. Demersal eggs are spawned and fertilized in a nest prepared by the male. Male parental care of eggs. Eggs hatch after three to four days and are free swimming by five days.

CONSERVATION STATUS
Not listed by the IUCN.

SIGNIFICANCE TO HUMANS
Collected commercially for the aquarium trade. ◆

Gangetic leaffish
Nandus nandus

FAMILY
Nandidae

TAXONOMY
Nandus nandus Hamilton, 1822, Gangetic Provinces of India.

OTHER COMMON NAMES
None known.

PHYSICAL CHARACTERISTICS
Typically basslike; pale olive-bronze color with loosely arranged vertical brownish olive bands along the flank and extending into dorsal fin. Noted for its sharp dorsal, pelvic, and anal spines.

DISTRIBUTION
From Pakistan east to Thailand.

HABITAT
Benthopelagic in freshwater ponds, ditches, flooded fields, lakes, reservoirs, canals, rivers, and estuaries.

BEHAVIOR
Often cryptic and holding to cover from where it can ambush prey.

FEEDING ECOLOGY AND DIET
Gangetic leaffishes are predators that feed upon smaller fishes and aquatic insects.

REPRODUCTIVE BIOLOGY
Gonochoristic. Males clean a spawning site, usually a rock or leaf, and then court the female, who spawns around 200–400 demersal eggs that are fertilized by the male. Male parental care has been observed in aquaria.

CONSERVATION STATUS
Not listed by the IUCN.

SIGNIFICANCE TO HUMANS
A prized food fish. Juveniles may also be taken for the aquarium trade. ◆

Striped boarfish
Evistias acutirostris

FAMILY
Pentacerotidae

TAXONOMY
Evistias acutirostris Temminck and Schlegel, 1844, Nagasaki, Japan.

OTHER COMMON NAMES
English: Whiskered armorhead; Japanese: Tenguidai.

PHYSICAL CHARACTERISTICS
Body deep with a sail-like dorsal fin; the lower jaw has many small barbels. Color pattern denoted by a dark brown face and alternating dark brown and white vertical bands along the

body; all fins are yellow except for the pelvic fins, which are dark brown. To 25.6 in (65 cm) total length.

DISTRIBUTION
Antitropical distribution, with populations in central and southern Japan, including the Izu and Ogasawara Islands, and Hawaii in the Northern Hemisphere and Lord Howe Island (Australia), and the islands of northern New Zealand in the Southern Hemisphere.

HABITAT
Generally a deep-slope fish, from 65 to almost 1,000 ft (20 to almost 330 m) deep, depending upon locality. Found among rocky and coral reefs, usually in or near caves.

BEHAVIOR
May be found in pairs or small groups.

FEEDING ECOLOGY AND DIET
Uses its snout and barbels to probe the bottom for various invertebrates.

REPRODUCTIVE BIOLOGY
Gonochoristic. Presumed to spawn pelagically, but little is known of its reproductive behavior and ecology. Eggs and larvae are pelagic.

CONSERVATION STATUS
Not listed by the IUCN. Might be vulnerable because of its status as a commercial food fish.

SIGNIFICANCE TO HUMANS
Esteemed as a food fish, but juveniles and young adults are also taken, although infrequently, for the aquarium trade. ◆

Emperor angelfish
Pomacanthus imperator

FAMILY
Pomacanthidae

TAXONOMY
Pomacanthus imperator Bloch, 1787, Japan.

OTHER COMMON NAMES
Afrikaans: Keiser-engelvis; French (Polynesia): Poisson empereur; paraharaha; Japanese: Tatejima-kinchakudai.

PHYSICAL CHARACTERISTICS
Deeply compressed, with longish pelvic fins and a long preopercule spine. Adults and juveniles have distinct color phases. Adults are yellow with horizontal blue stripes along the flank, a deep purplish mask over the eye and on the flank above the pectoral fin, a white snout and mouth, and yellow fins with purplish blue stripes on the anal fin. Juveniles are deep navy blue to almost black, with concentric white circles on the flank and white bands on the head. One of the largest angelfishes, to over 15.7 in (40 cm) total length.

DISTRIBUTION
Red Sea east to the Line Islands and the Tuamotu Archipelago; southern Japan south to northern Australia, and southeast to Lord Howe Island and the Austral Islands. Reported from Hawaii, but either very rare as a stray or introduced from aquarium stock.

HABITAT
Coral reefs. Juveniles on patch reefs in lagoons, protected areas in lagoon passes, and among corals, in holes or under ledges

on outer reefs or reef flats; subadults move to surge channels and holes on reef fronts; and adults are found usually near caves and ledges of seaward reefs, passes, or lagoons.

BEHAVIOR
Usually observed paired or singly as they patrol a large home range; may display some intra- or interspecific territorial behavior (usually towards other similarly sized angelfishes). Have an haremic mating group, with subordinate females residing within the male's home range.

FEEDING ECOLOGY AND DIET
Feed almost exclusively on sponges and tunicates.

REPRODUCTIVE BIOLOGY
Protogynous hermaphrodites. Sequential courtship and spawning among members of a mating group occur at dusk. Eggs and larvae are pelagic. Eggs are round, approximately 0.023–0.039 in (0.6–1 mm) in diameter. Larvae hatch around 0.05–0.1 in (1.3–2.6 mm) in length, have a large yolk sac, an unformed mouth, and unpigmented eyes. With growth, they assume a deep, laterally compressed shape, and the profile of the head becomes steep. Possesses a gas bladder. The development of head spination and spinules (spine-like structures on the skin) are specializations for pelagic life.

CONSERVATION STATUS
Not listed by the IUCN. Because of apparent low population densities at many localities, this species is subject to overfishing by the aquarium trade.

SIGNIFICANCE TO HUMANS
Highly prized, especially the juveniles, as an aquarium fish. Adults are also taken occasionally in subsistence food fisheries. ◆

Banded archerfish

Toxotes jaculatrix

FAMILY
Toxotidae

TAXONOMY
Toxotes jaculatrix Pallas, 1767, Indian Ocean.

OTHER COMMON NAMES
None known.

PHYSICAL CHARACTERISTICS
Body is compressed and almost triangular, with a pointed head. Color is silvery white with five black bands across the upper sides of the body, the dorsal fin, and the caudal peduncle, and yellow on anal, caudal, and dorsal fins. These fishes grow to 8 in (20 cm) total length.

DISTRIBUTION
Widespread from India and Sri Lanka, east through Southeast Asia, the Ryukyu Islands, Palau and Yap in Micronesia, Melanesia, and parts of Queensland in northern Australia; also in Vanuatu.

HABITAT
Generally found in bays, estuaries, and the lower reaches of rivers and creeks, usually among mangrove branches and roots.

BEHAVIOR
Occurs singly or in small groups around shelter. Swims slowly as it hunts for prey both below and above the surface of the water. Will swim rapidly on occasion to challenge conspecifics. As with other archerfishes, the banded archerfish will hunt by shooting a stream of water drops at terrestrial prey above the surface.

REPRODUCTIVE BIOLOGY
Gonochoristic, with pair spawning and the release of demersal eggs near the bottom. Others in the genus release either pelagic or demersal eggs. Demersal eggs are larger compared to pelagic eggs in this species, about 0.023–0.031 in (0.6–0.8 mm) in diameter. Larvae hatch at 0.12 in (3 mm) in length, are pelagic, possess a large yolk sac, have an unformed mouth, and lack pigmented eyes. With growth, the body becomes moderate in depth and the mouth large. Possesses a gas bladder. The development of head spines is regarded as a specialization for pelagic life.

CONSERVATION STATUS
Not listed by the IUCN.

SIGNIFICANCE TO HUMANS
Taken in subsistence fisheries and as a light-tackle gamefish, but more importantly as an aquarium fish. ◆

Resources

Books

Allen, G. R., and D. R. Robertson. *Fishes of the Tropical Eastern Pacific.* Honolulu: University of Hawaii Press, 1994.

Allen, G. R., and R. Swainston. *The Marine Fishes of North-Western Australia.* Perth: Western Australia Museum, 1988.

Allen, G. R., S. H. Midgley, and M. Allen. *Field Guide to the Freshwater Fishes of Australia.* Perth: Western Australian Museum, 2002.

Bohlke, J. E., and C. C. G. Chaplin. *Fishes of the Bahamas and Adjacent Tropical Waters.* 2nd ed. Austin: University of Texas Press, 1993.

Donaldson, T. J. "Assessing Phylogeny, Historical Ecology, and the Mating Systems of Hawkfishes (Cirrhitidae)." In *Proceedings of the 5th Indo-Pacific Fish Conference, Noumea 1997*, edited by B. Seret and J. Y. Sire. Paris: Societe Francaise d'Ichtyologie & Institut de Recherche pour le Developement, 1999.

Eschmeyer, W. N., ed. *Catalog of Fishes.* 3 vols. San Francisco: California Academy of Sciences, 1998.

Helfman, G. S., B. B. Collette, and D. E. Facey. *The Diversity of Fishes.* Oxford: Blackwell Science, 1997.

Kuiter, R. H. *Coastal Fishes of South-Eastern Australia.* Honolulu: University of Hawaii Press, 1993.

Leis, J. M., and B. M. Carson-Ewart, eds. *The Larvae of Indo-Pacific Coastal Fishes.* Leiden: Brill, 2000.

Masuda, H., K. Amaoka, C. Araga, T. Uyeno, and T. Yoshino, eds. *The Fishes of the Japanese Archipelago.* Tokyo: Tokai University Press, 1984.

Resources

Myers, R. F. *Micronesian Reef Fishes.* 3rd ed. Barrigada, Guam: Coral Graphics, 1999.

Neira, F. J., A. G. Miskiewicz, and T. Trnski, eds. *Larvae of Temperate Australian Fishes. Laboratory Guide for Larval Fish Identification.* Perth: University of Western Australia Press, 1998.

Nelson, J. S. *Fishes of the World.* 3rd ed. New York: John Wiley & Sons, 1994.

Sadovy, Y., and A. S. Cornish. *Reef Fishes of Hong Kong.* Hong Kong: Hong Kong University Press, 2000.

Sadovy, Y., and A. C. J. Vincent. "Ecological Issues and the Trades in Live Reef Fishes." In *Coral Reef Fishes: Dynamics and Diversity in a Complex Ecosystem,* edited by P. F. Sale. San Diego: Academic Press, 2002.

Sakurai, A., Y. Sakamoto, and F. Mori. *Aquarium Fish of the World. The Comprehensive Guide to 650 Species.* San Francisco: Chronicle Books, 1993.

Schafer, F. *Aqualog: Reference Fish of the World. All Labyrinths: Bettas, Gouramis, Snakeheads, Nandids.* Morfelden-Walldorf, Germany: Verlag A. C. S. GmbH, 1997.

Smith, M. M., and P. C. Heemstra, eds. *Smiths' Sea Fishes.* Berlin: Springer-Verlag, 1986.

Thomson, D. A., L. T. Findley, and A. N. Kerstitch. *Reef Fishes of the Sea of Cortez.* 2nd ed. Tucson: University of Arizona Press, 1987.

Thresher, R. E. *Reproduction in Reef Fishes.* Neptune City, NJ: T. F. H. Publications, 1984.

Periodicals

Donaldson, T. J. "Facultative Monogamy in Obligate Coral-Dwelling Hawkfishes (Cirrhitidae)." *Environmental Biology of Fishes* 26 (1989): 295–302.

———. "Reproductive Behavior and Social Organization of Some Pacific Hawkfishes (Cirrhitidae)." *Japanese Journal of Ichthyology* 36 (1990): 439–458.

Donaldson, T. J., and P. L. Colin. "Pelagic Spawning of the Hawkfish *Oxycirrhites typus.*" *Environmental Biology of Fishes* 24 (1989): 295–300.

Donaldson, T. J., J. T. Moyer, R. F. Myers, and P. J. Schuup. "Zoogeography of the Fishes of the Mariana, Ogasawar, and Izu Islands: A Preliminary Assessment." *Natural History Research Special Issue* 1 (1994): 303–332.

Lobel, P. S. "Spawning Behavior of *Chaetodon multicinctus* (Chaetodontidae); Pairs and Intruders." *Environmental Biology of Fishes* 25 (1989): 12–130.

Patterson, C., and G. D. Johnson. "The Intermuscular Bones and Ligaments of Teleostean Fishes." *Smithsonian Contributions in Zoology* 559 (1995): 1–83.

Randall, J. E. "Review of the Hawkfishes (Family Cirrhitidae)." *Proceedings of the United States National Museum* 114 (1963): 389–451.

———. "A Review of the Fishes of the Subgenus *Goniistius,* Genus *Cheilodactylus,* with Description of a New Species from Easter Island and Rapa." *Occasional Papers of the Bernice P. Bishop Museum* 25, no. 7 (1983): 1–24.

———. "Revision of the Generic Classification of the Hawkfishes (Cirrhitidae), with Descriptions of Three New Genera." *Zootaxa* 12 (2001): 1–12.

Sadovy, Y., and T. J. Donaldson. "Sexual Pattern of *Neocirrhites armatus* (Cirrhitidae) with Notes on Other Hawkfish Species." *Environmental Biology of Fishes* 42 (1995): 143–150.

Yagishita, N., T. Kobayashi, and T. Nakabo. "Review of Monophyly of the Kyphosidae (Sensu Nelson 1994), Inferred from the Mitochondrial ND2 Gene." *Ichthyological Research* 49 (2002): 103–108.

Organizations

IUCN/SSC Coral Reef Fishes Specialist Group. International Marinelife Alliance-University of Guam Marine Laboratory, UOG Station, Mangilao, Guam 96913 USA. Phone: (671) 735-2187. Fax: (671) 734-6767. E-mail: donaldsn@uog9.uog.edu Web site: <http://www.iucn.org/themes/ssc/sgs/sgs.htm>

Terry J. Donaldson, PhD

○

Percoidei V

(Groupers, sea basses, trevallys, snappers, emperors, and relatives)

Class Actinopterygii
Order Perciformes
Suborder Percoidei
Number of families 29

Photo: Blue banded sea perches (*Lutjanus kasmira*) near Palau. (Photo by Animals Animals ©Joyce & Frank Burek. Reproduced by permission.)

Evolution and systematics

The suborder Percoidei contains more than 70 families and 2,800 species. This section on percoid fishes is widely divergent and includes the following 29 families: Ambassidae (glassfishes; Chandidae of some authors), Polyprionidae (wreckfishes), Serranidae (fairy basslets, groupers, hamlets, perchlets, sand perches, sea basses, soapfishes), Grammatidae (basslets), Callanthiidae (splendid perches), Pseudochromidae (dottybacks and eel blennies), Plesiopidae (longfins and roundheads), Glaucosomatidae (pearl perches), Opistognathidae (jawfishes), Priacanthidae (bigeyes), Apogonidae (cardinalfishes), Sillaginidae (sillagos or smelt whitings), Malacanthidae (tilefishes), Rachycentridae (cobias), Carangidae (jacks, pompanos, and trevallys), Menidae (moonfish), Leiognathidae (ponyfishes or slipmouths), Bramidae (pomfrets), Lutjanidae (snappers), Caesionidae (fusiliers), Lobotidae (tripletails), Gerreidae (mojarras), Haemulidae (grunts and sweetlips), Dinopercidae (cave basses or lampfishes), Sparidae (porgies), Lethrinidae (emperors), Nemipteridae (monocle or threadfin breams), Polynemidae (threadfins), and Sciaenidae (croakers and drums). As with other percoid fishes, these families date largely from the Lower Tertiary during the Eocene or Miocene. Fossil records from the Eocene have been discovered for the Serranidae, Pseudochromidae, Lutjanidae, Haemulidae, and Lethrinidae. The Priacanthidae dates from the Middle Eocene. The Apogonidae, Sillaginidae, Carangidae, Menidae, and Sparidae date from the Lower Eocene. The Malacanthidae and Sciaenidae date from the Miocene, and the Bramidae from the Upper Miocene. The Leiognathidae, however, dates from the Oligocene.

Diverse and speciose families include the Serranidae, with 62 genera and at least 449 species. This family is divided into three subfamilies, the Serraninae (sea basses), the Anthiinae (fairy basslets and perchlets), and the Epinephelinae, with the latter arranged into five tribes. These are the Niphonini (Japanese ara or grouper), Epinephelini (groupers and coral trouts), Liopropomini (Swissguard basslets), and the Diploprionini and Grammistini (both soapfishes). Other speciose families include the Apogonidae, with 22 genera and at least 207 species; the Carangidae, with 33 genera and 140 species; the Lutjanidae, with 17 genera and 103 species; the Haemulidae, with 17 genera and 150 species; the Sparidae, with 35 genera and 112 species; and the Sciaenidae, with 70 genera and at least 266 species.

Physical characteristics

Fishes of the family Ambassidae have perch-like bodies, a dorsal fin that is notched deeply, and a forked caudal fin. The bodies of many species are transparent or partially so. Body sizes range up to 10.2 in (26 cm) in total length, but most species are considerably smaller. The six species comprising the Polyprionidae are all large, robust grouper-like fishes with large heads and mouths, two rounded spines on the opercle, pelvic fins with one spine and 5 soft rays, and continuous lateral lines. Color patterns range from silvery to blue-gray to mottled. Body sizes range from 59 in (150 cm) to over 98 in (250 cm) in total length.

The Serranidae has considerable variation in morphology and body size. Most species have a single dorsal fin bearing spines and soft rays, small ctenoid scales, 2–3 flattened spines on the opercula, a continuous lateral line, a maxillary (upper jaw) that is exposed fully, and a lower jaw that extends beyond the maxillary. Members of the subfamily Anthiinae are rather small and quite colorful. Most water column-dwelling species have lunate caudal fins, while benthic-dwelling species tend towards caudal fins that are truncate. Males are often sexually dimorphic; that is, they are distinguished in appearance from females with regard to body size, fin elongation, and color pattern. Groupers and soapfishes (tribe Epinephelini) are usually robust, bass-like fishes, although some soapfishes are elongate and slender. Most species have large heads and mouths, well-developed spines in the dorsal fin, and caudal fins that range from truncate to emarginate, lunate, or rounded. Swissguard basslets (tribe Liopropriónini) tend to be small and slender with flattened heads, and are often colorful. Soapfishes have dermal glands that secrete a toxin that is an effective antipredator mechanism. Groupers, soapfishes, and Swissguard basslets tend to be sexually dimorphic for body size, with males being larger than females. Adult body sizes range from less than 3.9 in (10 cm) to over 118 in (300 cm) in total length.

Species of the Grammatidae tend to be small and colorful. They have an interrupted lateral line, arranged in two segments; alternately, the lateral line is absent. There are 11–13 spines in the dorsal fin and one spine and five soft rays in the pelvic fin. Body sizes are usually less than 3.9 in (10 cm) in total length. The Callanthiidae is distinguished by having an compressed, oblong-shaped body, a single dorsal fin with 11 spines and 9–10 soft rays, an opercle with 1 or 2 spines, a lateral line running along the base of the dorsal fin, and truncate, emarginate, or excessively lunate caudal fins. Most have color patterns of bright orange, yellow, and red. Adults range in size from about 3.1 in (8 cm) to over 23.6 in (60 cm). Members of the Pseudochromidae follow one of two body plans. Those in the subfamily Pseudochrominae have somewhat elongate bodies; a long, continuous dorsal fin; small scales (cycloid anteriorly and ctenoid posteriorly); and many have rather brilliant color patterns. Some species lack a lateral line while others have one that is interrupted. Most species are less than 4.3 in (11 cm) in total length. Those in the subfamily Congrogadinae are eel-like in appearance and rather drably colored. Both dorsal and anal fins are long and continuous. They reach up to 20 in (50 cm) in total length. Species in the family Plesiopidae tend to be similar in shape to those in the Pseudochromidae, except that their bodies may be deeper in some genera. The dorsal fins have considerably more spines (11 to 14), too. The dorsal, anal, and caudal fins are large and exaggerated in the genera *Calloplesiops* and *Paraplesiops*. Most species are drably colored in comparison to the pseudochromids, but some species are remarkably colorful. Body sizes are usually less than 10 in (25 cm) in total length, many much smaller.

The Glaucosomatidae resembles deep-bodied serranids or haemulids. They have 8 spines and 12–14 soft rays in the dorsal fin and 3 spines and 12 soft rays in the anal fin. The maxilla is scaled, and the lateral line is straight and reaches the caudal fin. The caudal fin is either lunate or truncate. Color patterns tend to be dull gray or silver. Some have horizontal stripes that are black or yellow in color; one species, *Glaucosoma hebraicum* as a juvenile, has a distinctive black chevron stripe on the head and through the eye. Adults range from 15.7 in (40 cm) to over 47 in (120 cm) in total length. The Opistognathidae is recognized by having elongate, slender, or tapering bodies (although some are larger and robust), an enlarged head, large eyes, a continuous dorsal fin with 9 to 12 spines, pelvic fins placed ahead of the pectoral fins, and a lateral line that runs just under the dorsal fin and terminates halfway along the body. The scales are cycloid and the head is scaleless. Body sizes range up to 20 in (50 cm) in total length, but many species are much smaller. Members of the family Priacanthidae have deep, compressed bodies with very large eyes, a large, obliquely positioned mouth, a lower jaw that projects outward, rough scales, and scales on the head. Color patterns are usually red or coppery red and can be changed behaviorally to silver or mottled red and silver. Adults range in size to over 20 in (50 cm) in total length, but most species are less than 12 in (30 cm) long. Fishes of the family Apogonidae are generally small and compressed laterally, although some are elongate in shape. There are two dorsal fins, the first with 6–8 spines and the second comprised of soft rays. The anal fin has 2 spines. The mouth is relatively large and placed obliquely. The eyes are also large and adapted for low light conditions. Many species are colorful, but some are nearly transparent with faint shades of red, yellow, silver, or bronze. Adults range in size up to 10 in (25 cm) in total length, but the majority of species are considerably smaller.

The Sillaginidae is distinguished by having elongate bodies with two dorsal fins, the first bearing 10–13 spines and the second a single spine and 16–27 soft rays. The anal fin is elongate, with two spines and 14–26 soft rays. Color patterns tend to be silvery, white, or tan to match the color of the sea bottom. Fishes of the family Malacanthidae follow two body plans. Those in the genera *Hoplolatilus* and *Malacanthus* have relatively slender, elongate bodies with slightly rounded or pointed heads, truncate or somewhat-forked caudal fins, continuous dorsal fins, small scales that are largely ctenoid, and one spine or the opercle. *Hoplolatilus* species are often quite colorful, while *Malacanthus* tend to be striking despite coloration limited to black, white, some blue, or pale green. Tilefishes in the genera *Branchiostegus*, *Caulolatilus*, and *Lopholatilus* tend to be stockier, with larger, blunter heads. Colors range from pale or drab shades of brown, bronze, and green to pink. Body sizes range up to 49 in (125 cm) in total length. The single species of Rachycentridae, *Rachycentron canadum*, has an elongate body and depressed head. There are 6–9 short free spines positioned ahead of the long dorsal fin. There are 1–3 spines and 26–33 soft rays in the dorsal fin and 2–3 spines and 22–28 soft rays in the long anal fin. The caudal fin is lunate, with the lower lobe shorter than the upper. The body has a pattern of three darkly colored lateral stripes along the flank and, in juveniles and young adults, a long whitish silver stripe runs dorsally from the snout to the caudal peduncle. This species grows to over 79 in (200 cm) in total length.

The Carangidae is diverse in body form. Most species have deep, compressed bodies, two dorsal fins, forked caudal fins, and slender caudal peduncles bearing bony scutes. Some species are elongate and fusiform. Colors are typically silver or gray with ornamentation in black, blue, green, olive, or yellow. The pilotfish, *Naucrates ductor*, is bright yellow with black vertical stripes, however. Most species are less than 39 in (100 cm) in total length, but some will grow to over 98 in 250 cm). The Menidae has a disc-like body, a sharply angled, deep breast, and a nearly horizontal dorsal surface. The dorsal and anal fins are spineless but bear 43–45 and 30–33 soft rays, respectively. The first pelvic fin ray of adults is elongate, and the caudal fin is deeply forked. Color is blue dorsally and

Bar jacks (*Caranx ruber*) preying on herring near Bonaire Island in the Netherlands Antilles. (Photo by Andrew J. Martinez/Photo Researchers, Inc. Reproduced by permission.)

white ventrally, with 2–3 rows of dark spots along the dorsal flank down to the lateral line. The Leiognathidae is distinguished by strongly compressed bodies; small cycloid scales; gill membranes that are unified at the isthmus; a continuous dorsal fin with 8–9 spines and 14–16 soft rays; an anal fin with 3 spines and 14 soft rays; and a forked caudal fin. These fishes also possess luminous organs on the throat and the ability to secrete mucus from their skin. Members of the Bramidae are deeply compressed with long, continuous dorsal and anal fins, long pectoral fins, and forked or widely forked caudal fins. Color patterns range from silver to bronze, dull red, or black. Members of the genus *Pteraclis*, the fan fishes, have excessively large dorsal and anal fins that give these fishes a fan-like shape. Adults range up to 39 in (100 cm) in total length, although most species are half of that size.

Species of the Lutjanidae have ovate or elongate bodies that are compressed moderately. The single dorsal fin is notched with 10 spines and 8–18 soft rays. The anal fin has 3 spines and 7–11 soft rays. The caudal fin is either truncate, emarginated, or forked deeply. Color patterns are highly variable. Most species are less than 39 in (100 cm) in total length, but some will grow to over 59 in (150 cm). The colorful Caesionidae is distinguished by slender and rather streamlined bodies, with small mouths and a protrusible upper jaw. There is a single dorsal fin with 9–15 spines and 9–21 soft rays. The anal fin has 3 spines and 9–13 soft rays. The caudal fin is forked deeply. Color patterns range from blue to silvery blue, with yellow, pink, or red accents. Most species are less than 23.6 in (60 cm) in total length. The single species of Lobotidae, *Lobotes surinamensis*, has an oval or oblong and compressed body, a single dorsal fin with 11–12 spines and 15–16 soft rays, an anal fin with 3 spines and 11–12 soft rays, and 17 soft rays in the pectoral fin. The dorsal, anal, and caudal fins are all rounded. The scales are ctenoid. Adults are dark brown or greenish yellow along the back, and silverfish gray along the flanks; juveniles tend to be brown and yellow, and are usually mottled. This species grows to about 39 in (100 cm) in total length.

Members of the Gerreidae are silvery in color, have moderately deep and compressed bodies, a head that is concave in profile ventrally, and protractile mouths. The scales are ctenoid and large, and the caudal fins are forked. Most species are less than 14 in (35 cm) in total length. The Haemulidae resembles the Lutjanidae in body shape, but a number of species tend to be more robust; have smaller mouths, thicker lips, and conical teeth; and lack canines. Their dorsal fins are continuous with 9–14 spines and 11–26 soft rays, and the anal fins are much shorter with 3 spines and 6–18 soft rays. The caudal fins are truncate to slightly emarginate. Color patterns are variable. Most species are less than 23.6 in (60 cm) but some grow to at least 39 in (100 cm) in total length. The Dinopercidae is distinguished by an oval, compressed body, a protruding lower jaw, a continuous dorsal fin with 9–11 spines and 18–20 soft rays, an anal fin with 3 spines and 12–14 soft rays, and a truncate caudal fin. Scales are ctenoid and cover the body, head, and fins. Color patterns range from a dull blackish brown with white specks to a barred pattern of alternating dark and whitish colors. Size ranges of adults are from 12 in (30 cm) to over 30 in (75 cm) in total length. The diverse Sparidae is snapper-like in appearance, with compressed oblong or ovate bodies, a dingle dorsal fin with 10–13 spines and 8–15 soft rays, an anal fin with 3 spines and 8–14 soft rays, a forked or emarginate caudal fin, weakly ctenoid scales, scaly cheeks and opercula, and conical, incisiform or molar teeth. Color patterns are variable but often have a metallic sheen. Adult body sizes range from about 12 in (30 cm) to over 79 in (200 cm) in total length. The Lethrinidae resembles both the Lutjanidae and Haemulidae in appearance. There are 10 spines and 9–10 soft rays in the dorsal fin and 3 spines and 8–10 soft rays in the anal fin. The caudal fin is emarginate or forked. The lips are thick, the mouth terminal, and the front of the jaws support canine teeth while conical or molariform teeth are positioned along the side of the jaws. Color patterns are largely drab in most species, ranging from gray to silvery gray, olive, yellow, or brown. Some are distinctively black or yellow and have red, blue, black, yellow, or white markings or stripes. Adult body sizes range from about 12 in (30 cm) to over 39 in 100 cm) in total length.

Fishes of the family Nemipteridae have slender or ovate bodies, 10 spines and 9–10 soft rays in the dorsal fin, 3 spines and 7–8 soft rays in the anal fin, and forked caudal fins. Their eyes are relatively large. Color patterns are variable and often bright and distinctive. Body sizes of most adults are less than 14 in (35 cm) in total length. The very distinctive Polynemidae is distinguished by a blunt, rounded snout, elongate body, two dorsal fins, and a deeply forked caudal fin. The unusual pectoral fin has a detached lower portion consisting of 3–7 free rays that may be used to detect prey in turbid water. Color patterns tend to be, for the most part, drab shades of olive and silver. Adults can grow up to 71 in (180 cm) in total length. The highly diverse Sciaenidae has long notched dorsal fins, with 6–13 spines in the anterior portion and 1 spine and 20–35 soft rays in the anterior portion. There are 1–2 weak spines and 6–13 soft rays in the anal fin. The caudal fins are mainly emarginate or rounded. Color patterns vary from silvery white or gray to light brown, yellow, pale pink, or pale blue. Some species, such as those in the genus *Equetus*, are striking, however. Adult sizes range from less than 3.9 in (10 cm) to over 79 in (200 cm) in total length.

Distribution

The Ambassidae occurs in tropical and subtropical marine, brackish, and fresh waters of the Indo-West Pacific; freshwater species are especially prominent in Australia, India, and Southeast Asia. The Polyprionidae has a scattered distribution in temperate and subtropical waters of the Atlantic, Indian and Pacific Oceans. The Serranidae is distributed in tropical, subtropical, and temperate waters of the Atlantic, Pacific, and Indian Oceans; some species occur in freshwater. The Grammatidae is limited to tropical and subtropical waters of the western Atlantic and western Pacific Oceans. The Callanthiidae occurs in temperate, subtropical, and tropical waters of the eastern Atlantic, including the Mediterranean, and the Indian and Pacific Oceans. The Pseudochromidae occurs in tropical marine, rarely brackish, waters of the Indo-Pacific region. The Plesiopidae is found in tropical and subtropical waters of the Indian and Pacific Oceans. The Glaucosomatidae appears to be a continental family that ranges from Western Australia east to Japan. Both the Opistognathidae and the Priacanthidae occur in tropical and subtropical waters of the Atlantic, Indian, and Pacific Oceans. Fishes of the family Apogonidae are mainly marine in the tropical and subtropical Atlantic, Indian, and Pacific Oceans, but some species occur in fresh or brackish waters in the western Pacific. The Sillaginidae is distributed in tropical and warm temperate coastal and estuarine waters of the Indo-West Pacific, from Africa east to Australia and New Caledonia, and north to Japan. Most species are continental, and at least one species has entered the eastern Mediterranean via the Suez Canal. The Malacanthidae occurs largely in tropical and warm temperate waters of the Atlantic, Indian, and Pacific Oceans; one species occurs in brackish and marine waters of New Guinea, however. The Rachycentridae is found in marine and, to a lesser extent, brackish waters of the tropical, subtropical, and temperate continental waters of the Atlantic, Indian, and Pacific Oceans, and are absent from most insular localities on the Pacific Plate. The Carangidae is widely distributed in tropical, subtropical, and warmer temperate waters of the Atlantic, Indian, and Pacific Oceans; some species will enter brackish and coastal rivers. The Menidae is distributed in the Indo-West Pacific, but limited from the east coast of Africa east to Indonesia and Southeast Asia. The Leiognathidae is distributed in coastal waters of the Indo-West Pacific; one species has entered the Mediterranean Sea via the Suez Canal, however. The Bramidae occurs in deeper oceanic waters of the tropical and temperate Atlantic, Indian, and Pacific Oceans. The Lutjanidae is found mainly in tropical and subtropical waters of the Atlantic, Indian, and Pacific Oceans, but some species also enter brackish and freshwater reaches; at least one species has been introduced successfully into freshwater lakes in Australia. The Caesionidae is distributed in tropical and subtropical waters of the Indo-West Pacific. The Lobotidae occurs in pelagic and coastal waters of the tropical Atlantic, Indian, and Pacific Oceans. The Gerreidae is found in coastal tropical and subtropical marine and, to a lesser extent, brackish waters of the Atlantic, Indian, and Pacific Oceans. Most species in the family Haemulidae occur in tropical and subtropical regions of these oceans as well, but some species also occur in brackish and coastal fresh waters. The Dinopercidae is restricted to warm temperate and subtropical localities of the western Indian Ocean and in the southeastern Atlantic off the coast of southern Africa from Angola to South Africa. The Sparidae is distributed largely in continental marine waters of the tropical and temperate Atlantic, Indian, and Pacific Oceans, but is also rare in fresh and brackish water. With one exception, the Lethrinidae is limited to tropical and subtropical waters of the Indian and Pacific Oceans. A single species occurs off West Africa in the Atlantic Ocean. The Nemipteridae is found in the tropical and subtropical Indo-West Pacific. The Polynemidae is found in tropical and warm temperate marine, brackish, and freshwaters of the Atlantic (including the Mediterranean Sea), Pacific, and Indian Oceans. The Sciaenidae occurs in tropical, subtropical, and temperate waters. Most species are found in marine or brackish waters of the Atlantic, Indian, and Pacific Oceans, but some occur in freshwater drainages with current or historical connections to the sea.

Habitat

The Ambassidae occurs in coastal marine, brackish, and fresh waters, mainly in protected areas with overhanging or emergent vegetation, such as mangroves; swamps, ponds, ditches, billabongs, creeks, and deep holes in rivers are among the freshwater habitats where these fishes may be found. The Polyprionidae frequents deep slope rocky reefs and pinnacles, rock bottoms, or sand flats and kelp beds at depths of 16–1,968 ft (5–600 m), or more, depending upon the species. These fishes also associate with shipwrecks, and at least one species has been found inhabiting structure around a deep water thermal vent system.

The Serranidae, owing to its great diversity, frequents a wide variety of habitats in tropical and temperate marine, brackish, and freshwaters. Many species are found on seaward or protected coral or rocky reefs, often hiding in caves, holes, and crevices, under corals and ledges. Some species hover above some form of structure or swim actively in the water column. Some frequent sand, mud, rubble, mangrove, sea grass, or algal flats in estuaries and rivers, as well as on reefs. Others, such as most fairy basslets (subfamily Anthiinae), hover over deep slopes or pinnacles on reefs. Soapfishes (tribe Grammistini) and Swissguard basslets (tribe Liopropomini) are often associated with caves, crevices, or holes. Depth ranges vary, depending upon the species, from one to over hundreds of meters.

The Grammatidae lives in close association with structure, mainly holes and corals on coral reefs at depths down to over 1,198 ft (365 m); most species are found below 98 ft (30 m). The Callanthiidae prefers coral and rocky reef habitats at depths usually greater than 66 ft (20 m). The Pseudochromidae is often associated with structure on coral reefs, usually in holes, tubes, or caves, under rocks and corals, or in crevices, but some species may also be found on rubble flats. One species of eel-blenny (subfamily Congrogadinae) lives among sea urchin spines. Depth ranges vary from one to over 180 ft (55 m). The Plesiopidae is also found among holes, under rocks, and in caves but, these fish emerge at night to move along the bottom. The Glaucosomatidae occurs on deeper offshore rocky reefs, hard flat bottoms in deeper water, and possibly coral reefs as well. Depth ranges are 33 ft (10 m) to well over 656 ft (200 m). The Opistognathidae excavates burrows in sand and gravel on coastal reefs and flats.

The Priacanthidae dwells among rocks or corals on seaward coral and rocky reefs, occasionally on deeper flats, at depths of a few to over 656 ft (200 m). The highly diverse Apogonidae utilizes an equally diverse array of habitats. A number of species associate with structure, usually branching or eroded corals and rocks, on coral and rocky reefs. Others frequent holes, caves, crevices, ledges, rubble, silty or sandy bottoms, algal beds, sea grasses, mangroves, sponges, and even sea urchin and crown-of-thorns starfish spines. Still others occur in estuaries, rivers, creeks, ponds, and lakes, usually in association with structure. Depending upon the species and habitat, depths range from 3 ft (1 m) to over 262 ft (80 m). The Sillaginidae is found in shallow coastal waters, usually over sand or mud flats, and often along beaches; some species enter estuaries. The Malacanthidae may be found on coral and rocky reefs, deep sand and rubble flats, and, at least for one species, in shallow brackish water habitats. Depths range from 33 ft (10 m) to 1,640 ft (500 m), but most species occur in less than 656 ft (200 m) of water. The Rachycentridae swims in the pelagic water column but is also associated with structure, such as oil or sulfur drilling platforms, offshore piers, and drifting logs. The Carangidae is associated with a variety of coral and rocky reef habitats that range from sand, rubble, mud, algal and sea grass flats, boulder fields and old lava flows, and the water column inshore and in the open ocean. Some species enter estuaries and rivers or brackish water ponds. Depth ranges vary from 3 ft (1 m) to over 656 ft (200 m). The Menidae is found on inshore sand and mud flats, off deeper coral reefs, or in estuaries. The Leiognathidae frequents inshore sand and mud flats; some species enter freshwater. The Bramidae is pelagic, dwelling on deep-slopes and shelves, and is usually found at depths of over 656 ft (200 m).

The Lutjanidae is found mainly on seaward or protected coral and rocky reefs, usually in association with coral formations and rocks, but also hovering in the water column. Other species frequent sea grass, algal, rubble, and sand flats. A few species occur in estuaries, and one species ranges from marine to freshwaters, where it shelters in mangroves or similar kinds of emergent vegetation. Deep water species are associated with pinnacles. Depth ranges are from 3 ft (1 m) to over 1,476 ft (450 m). The Caesionidae swims in the water column over coral reefs, especially along outer slopes and lagoon pinnacles. The Lobotidae occurs in two widely divergent habitats. Juveniles and smaller adults may be found in the open sea, often around floating vegetation, flotsam, and jetsam. Alternately, these fishes may also be found on inshore flats and estuaries. The Gerreidae frequents inshore sand and mud flats, and may enter brackish or freshwater on occasion. The Haemulidae may be found on coral and rocky reefs, where they shelter near or under ledges. Some species are also found in brackish and freshwater. Members of the Dinopercidae are found in association with rocky and coral reefs, usually in caves or under ledges, to a depth of about 164 ft (50 m); juveniles may be found around rocky shorelines. The Sparidae is found on coral and rocky reefs, rubble and sand flats, or, rarely, in brackish and freshwater habitats. The Lethrinidae and the Nemipteridae also frequent coral and rocky reefs, as well as rubble, sand, and sea grass flats. The Polynemidae can be found on mud and sand flats in marine and brackish water, although some species also occur in similar habitats in rivers. The Sciaenidae inhabits a variety of habitats including mud, sand, and rubble flats and beaches, shell reefs, coral and rocky reefs, and flooded salt marshes. Some species may also be found in freshwater rivers and lakes. Depth ranges vary from 3 ft (1 m) to over 328 ft (100 m).

Feeding ecology and diet

The Ambassidae preys upon benthic invertebrates; freshwater species will also feed upon aquatic and terrestrial insects, and algae. Feeding activity occurs during the night and, to a lesser extent, during the day. Predators include larger fishes, wading and diving birds, and, in freshwater habitats, reptiles. The Polyprionidae feeds upon benthic fishes, cephalopods, and large crustaceans. Predators are likely larger fishes that feed upon juveniles, but some toothed whales, including sperm whales, prey upon adults. Many members of the Serranidae are predators upon smaller fishes, crustaceans, and cephalopods. Others, particularly members of the subfamily Anthiinae, are planktivores that feed upon zooplankton in the water column. Larger fishes, including sharks and even other serranids, are predators upon these fishes. With the exception of the anthiines, most predation likely takes place upon juveniles, however. The Callanthiidae feeds upon zooplankton, mainly crustaceans. Predators of juveniles and adults include larger fishes that forage in the water column during daylight. The Pseudochromidae and Plesiopidae feed upon benthic invertebrates and small fishes. They likely fall prey to benthic ambush predators such as groupers and scorpionfishes. The Glaucosomatidae feeds upon smaller fishes, crustaceans, and cephalopods. Members of this family may be preyed upon by larger fishes; juveniles are likely to be more susceptible. The Opistognathidae feeds upon benthic invertebrates but also plucks zooplankton out of the water column. Predators of adults likely include ambush predators such as groupers, and possibly moray eels and sea snakes that investigate their burrows.

Fishes of the family Priacanthidae are nocturnal predators that feed upon both invertebrates and smaller fishes. In turn, they likely fall prey to larger predatory fishes. The Apogonidae feeds mainly upon zooplankton or benthic invertebrates, usually at night, but members of the genus *Cheilodipterus* are predatory upon smaller fishes as well. Predators of apogonids are usually larger ambush and foraging fishes such as groupers, scorpionfishes, and trevallys. Members of the Sillaginidae feed upon benthic invertebrates that they take from sand or other soft sediments. Predators include larger roving predatory fishes. The Malacanthidae feeds upon benthic invertebrates or zooplankton. Members of this family are likely preyed upon by larger fishes, especially when young. The Rachycentridae feeds upon smaller fishes, cephalopods, and crustaceans in the water column or around structure. Juveniles are probably more susceptible to predation from other pelagic fishes than adults, although sharks might prey upon the latter. Most members of the Carangidae are swift-moving predators of smaller fishes, crustaceans, and cephalopods. Some species feed exclusively in the pelagic realm, while most others feed on benthic or epibenthic prey. Members of the genus *Decapterus* strain zooplankton from the water column. Juvenile *Scomberoides* feed on the scales of inshore fishes such as mullets (Mugiloididae). The pilotfish, *Naucrates ductor*, accompanies sharks and feeds upon scraps leftover by these predators. This, and some other species in this family, will also swim alongside rays while foraging opportunistically for prey disturbed by the rays' movements. Carangids, especially juveniles and smaller species, are preyed upon by larger fishes and may also fall prey to some dolphins or other smaller toothed-whales.

Both the Menidae and Leiognathidae forage upon benthic invertebrates, although members of the latter family may also feed upon larger zooplankton in the water column at night. They may be preyed upon by larger fishes, such as sharks and mackerels. The Bramidae feeds upon small fishes, large planktonic crustaceans, and cephalopods in the water column. Larger pelagic fishes likely prey upon them in return. Most species of Lutjanidae are predatory upon smaller fishes, crustaceans, mollusks, or worms. A number are planktivores, however. Members of this family are susceptible to predation by larger fishes, especially when juveniles or young adults. The Caesionidae feeds in schools or aggregations upon zooplankton in the water column. Larger pelagic or epibenthic fishes are their predators. The Lobotidae feeds upon benthic crustaceans and small fishes inshore or near floating objects and *Sargassum* patches in the open sea. Members of this family often float sideways to mimic plant life, such as leaves or fronds, and then ambush their prey. Juveniles are probably more susceptible to predation than adults, although the latter may be preyed upon when drifting near the surface by pelagic predators. Members of the Gerreidae use their protrusible mouths to root out, sort, and feed upon benthic invertebrates from sand or other soft sediments. They are preyed upon by larger roving or ambush predatory fishes. Members of the family Haemulidae are accomplished at feeding upon hard-shelled benthic invertebrates such as mollusks and crustaceans, but some species also feed upon smaller fishes and benthic worms. Juveniles are likely to be more susceptible to predation than adults in the larger species; otherwise, larger fishes are their chief predators. The Dinopercidae also feeds upon benthic invertebrates and possibly smaller fishes. Their predators are doubtless larger fishes, and juveniles are more likely to be preyed upon than adults. The Sparidae usually feeds upon hard-shelled benthic invertebrates (mollusks and crustaceans). Their main predators are probably larger fishes.

The Lethrinidae usually feeds at night upon smaller fishes and benthic invertebrates that range from crustaceans and mollusks to polychaete worms, tunicates, and starfishes and their relatives. Larger fishes, including roving and ambush predators, are their likely predators, and juveniles are probably more susceptible to predation than adults. The Nemipteridae also has a variable diet, for these fishes feed upon crustaceans, polychaete worms, cephalopods, or small fishes; some species feed upon zooplankton. Members of this family are preyed upon by larger fishes. The Polynemidae sifts through soft sediments with elongated pelvic fin rays in search of benthic invertebrates and small fishes. Larger fishes such as sharks, and, depending upon the locality, large carnivorous reptiles such as estuarine crocodiles likely prey upon these fishes. Juveniles probably fall prey to ambush or roving predatory fishes such as flatheads (Platycephalidae) and trevallys (Carangidae). The Sciaenidae consists of benthic predators of small fishes, crustaceans, and other benthic invertebrates. Their predators range from larger fishes to wading birds.

Behavior

The Ambassidae gathers in aggregations, some times quite large, under shelter. At night, these fishes become active and disperse as they feed. Fishes of the family Polyprionidae are probably territorial and patrol rather large home ranges. Within the Serranidae, the subfamilies Serraninae and Epinephelinae are largely solitary and territorial. Species that form haremic mating systems have multiple territories within that of a single male. Most species make good use of shelter or the bottom, from where they can avoid predation and also ambush prey. Some species are active swimmers in the water column, however. Many fairy basslet species (Anthiinae) aggregate in the water column but seek shelter on the bottom or against the faces of steep reef slopes, while others move, often cryptically, along the bottom. Males are territorial. Soapfishes (Diploprionini and Grammistini) and Swissguard basslets (Liopropomini) hover or rest in caves and holes, although some species move freely through the water column just above the bottom. These fishes seem to be more active at night. The Callanthiidae hovers in the water column singly or in groups, but the Grammatidae, Pseudochromidae, and Plesiopidae all tend to hide in holes, under rocks or corals, or in some other form of shelter, where they wait to ambush prey. Many colorful pseudochromids hover outside of their shelters, however. The Plesiopidae forages outside of shelter at night as well. Fishes in these families tend to be territorial. The behavior of the Glaucosomatidae is not well known. Members of this family shelter in caves or holes when approached and are likely to patrol a territory or home range. Adults and juveniles tend to move to shallower waters seasonally during cooler months. Jawfishes (Opistognathidae) excavate burrows with their large mouths and use them for shelter and nesting sites. When not in a burrow, they may be seen hovering above it in the water column; they enter the burrow tail first.

The Priacanthidae and the Apogonidae generally associate with structure during daylight but move into the water column to forage at night. Their relatively large eyes are used to detect both prey and predators. The Sillaginidae forages singly or in aggregations on sand or mud bottoms. The Malacanthidae lives singly, in pairs, or in haremic social groups and excavates burrows in the sand, where these fishes live when they are not hovering in the water column. Some of these burrows are distinguished by rather large mounds of rubble. The Rachycentridae is active in open water but will associate with structure. The Carangidae moves singly, in pairs or small aggregations, or in large schools either along the bottom or up in the water column. The Menidae forages in schools. The Leiognathidae forms schools and forages over the bottom during daylight; at night, these fishes move into the water column and may communicate with one another (directly or indirectly) by light flashes generated by bioluminescent organs on their throats.

Little is known about the behavior of the Bramidae because of the depths in which they live. Presumably, they form aggregations or schools that move up and down in the water column during night and day, respectively, as they follow their prey. The Lutjanidae hides under shelter, hovers in the water column, or forms aggregations that move lazily over the bottom. Some species are territorial and others patrol home ranges. The Caesionidae forms aggregations or schools that swim actively in the water column. A number of species appear to be able to change their color patterns behaviorally. Lobotidae juveniles and young adults often swim on their sides and hover under floating vegetation or logs, and may mimic leaves as well. Adults tend to be solitary. The Gerreidae forms small or large aggregations and forages actively along the bottom. The Haemulidae, depending upon the species, occurs singly or forms small or large aggregations. Members of this family seek shelter under ledges or in large holes during daylight but forage after dark. Alternately, they may form aggregations that swim lazily along the reef. Some aggregating species make daily migrations at dawn and dusk, and knowledge of the paths of these migration routes is transmitted culturally within social groups. The Dinopercidae shelters during daylight but likely moves about after dark. Members of this family can make a drumming sound by contracting muscles; the sound of the contraction is amplified by the swim bladder. The Lethrinidae moves singly, in small aggregations, or in large schools along the bottom; some species swim or hover up in the water column. The Nemipteridae either swims singly or in aggregations well up in the water column, rather like the Caesionidae, or these fishes dart about or hover alone or in groups just above the bottom. The Polynemidae swims just above the bottom, and these fishes use their specialized pelvic fins to detect prey as they forage. Their behavior is not well known. The Sciaenidae occurs singly or in groups, sometimes large aggregations, and members of the family swim actively along the bottom as they search for prey. As with the Dinopercidae, these fishes can communicate by the production of drumming sounds. Their hearing is well developed, too, which is useful for detecting prey, predators, and conspecifics (other members of the same species).

Reproductive biology

The Ambassidae spawns demersal eggs that are scattered on vegetation in freshwater; in marine and brackish waters, these fishes appear to spawn pelagic eggs. The larvae are pelagic, although those of freshwater species hold close to shelter. Some species of the Polyprionidae reportedly aggregate to spawn during summer months. Their eggs and larvae are pelagic. Details about deep-dwelling species are largely unknown, but it is assumed that they have a similar life history pattern. The reproductive behavior and ecology of the Serranidae is complex owing to the diversity of taxa within this family. Serranine fishes are hermaphroditic, but unlike other serranids this hermaphroditism is simultaneous rather than sequential. Thus, mature fishes can produce both eggs and sperm simultaneously. Courtship in these fishes may involve considerable ritual, as in the hamlets (*Hypoplectrus*), or virtually none at all, as in the genus *Serranus*. In the latter case, it has been hypothesized that a rapid spawning ascent without much visible courtship is a mechanism that prevents rivals from parasitizing spawning events by sneaking or streaking during the spawning rush. Sometimes, however, triad or group spawning occurs. Eggs and larvae of these fishes are probably pelagic. Anthiine fishes include protogynous hermaphrodites, small single-male or larger multi-male haremic mating systems, pelagic spawning, and pelagic eggs and larvae. Courtship and spawning begin around sunset. Spawning is seasonal at higher latitudes but may spawn nightly at low latitudes. Epinepheline fishes include protogynous hermaphrodites (sex change from male to female) and secondary gonochorists (primary males within the same species). Mating systems may be haremic (e.g., in the genus *Cephalopholis* or in smaller *Epinephelus* species) or in pairs or groups within spawning aggregations (e.g., larger *Epinephelus* or *Plectropomus* species). At low population densities, some aggregating species appear to have haremic mating systems. Courtship usually commences prior to sunset with spawning after sunset and into darkness. Spawning, eggs, and larvae are pelagic. Little is known about the reproductive biology of soapfishes (tribes Diploprionini and Grammistini) and Swissguard basslets (tribe Liopropomini). They may be either protogynous hermaphrodites or secondary gonochorists (derived from hermaphroditic ancestors). Mating systems may be monogamous or haremic. Spawning is presumed to be pelagic, as are the eggs and larvae.

Little is known about the reproductive behavior of the Callanthiidae. Eggs and larvae are pelagic, however. Members of the Pseudochromidae are demersal spawning fishes. Females lay a ball of eggs that are fertilized and then guarded by males; alternately, some species are mouthbrooders. The Plesiopidae spawns demersal eggs on the undersides of rocks; alternately, they are mouthbrooders. Eggs are bound together into a small mass by chorionic filaments. The Glaucosomatidae do not change sex. Although the details of this family's reproductive biology are not well known, courtship and spawning are pelagic and eggs are broadcast over the bottom. Larvae are pelagic. In western Australia, one species spawns through the summer. The Opistognathidae practices mouthbrooding and cares for a ball or small mass of eggs that is tightly bound together by chorionic filaments. Although the details of courtship and spawning are largely unknown, the Priacanthidae may spawn in aggregations and produce pelagic eggs and larvae. Most species of Apogonidae are mouthbrooders. Eggs are spherical or spindle-shaped and range in size depending upon the species, and the larvae disperse (albeit poorly in a number of species) pelagically. The larvae of one species, the Banggai cardinalfish, *Pterapogon kauderni*, do not disperse; this life history trait explains this species' limited geographical distribution.

The Sillaginidae spawns pelagically, producing pelagic eggs and larvae; some species have been cultured artificially. The Malacanthidae has a mating system of either monogamy or haremic polygyny (one male and multiple females). Spawning is pelagic, as are the eggs and larvae. Little is known of the reproductive biology of the Rachycentridae. Spawning is presumed to be pelagic. Spawning aggregations have been reported for some species of Carangidae, and this trait may likely be true throughout the family. Eggs and larvae are pelagic. The spawning mode of the Menidae is unknown but is presumed to produce pelagic eggs and larvae. Details about spawning behavior of the Leiognathidae are also few, but pelagic eggs and larvae are produced. Little is known about reproduction in the Bramidae. Presumably, members of the family have pelagic eggs and larvae. Spawning aggregations have also been reported for the Lutjanidae and the Caesionidae. Courtship and spawning usually takes place near dusk. Some species reportedly do this in groups that split off from the main aggregation and spawn in the water column. Eggs and larvae are pelagic. The reproductive biology of the Lobotidae is not well known, but it is assumed that spawning, eggs, and larvae are all pelagic. The Gerreidae spawns either pelagically or by scattering eggs over the bottom. Some species form spawning aggregations over sand late in the afternoon during periods around the full moon. Their larvae are pelagic. Some Haemulidae have been observed forming spawning aggregations around the new moon in late spring. Spawning, eggs, and larvae are also pelagic. The Dinopercidae likely also spawn pelagically; little is known of their reproductive biology.

The reproductive biology of the Sparidae is complex. Many species are hermaphroditic, changing from one sex to the other with growth, while others are simultaneously so. Most species spawn pelagic eggs but some deposit demersal eggs in nests. The larvae are pelagic. The Lethrinidae include a number of species that are protogynous hermaphrodites. Spawning aggregations have been reported for some species, with courtship commencing after sunset. Spawning for many species occurs at peaks around the new moon and also seasonally. Spawning is pelagic, as are the eggs and larvae. Protogynous hermaphroditism has also been reported for the Nemipteridae. Spawning, eggs, and larvae are also pelagic. The reproductive biology of the Polynemidae is largely unknown. These fishes appear to be pelagic spawners; their eggs and larvae are pelagic. Although considerable effort has been devoted to the study of the eggs and larvae of the Sciaenidae (mainly for aquaculture purposes), surprisingly little is known about their reproduction in nature. A number of species are known to form spawning aggregations during summer and well into late autumn at lower latitudes. Males often produce drumming noises during courtship of females. Spawning is known to be pelagic for these species, as are their eggs and larvae, but other species have larvae, and perhaps eggs and spawning, that are demersal.

Conservation status

The 2002 IUCN Red List categorizes numerous species from these families as Critically Endangered (5 species), Endangered (3 species), or Vulnerable (12 species). In addition, fishes important to commercial (food and aquarium), subsistence, and recreational fisheries, either as target species or as bycatch, are vulnerable to overfishing.

Significance to humans

Members of the following families are important commercial, recreational, and subsistence fisheries species: Ambassidae, Polyprionidae, Serranidae, Callanthiidae, Glaucosomatidae, Priacanthidae, Apogonidae, Sillaginidae, Malacanthidae, Rachycentridae, Carangidae, Menidae, Leiognathidae, Bramidae,

Lutjanidae, Caesionidae (also used as bait in tuna fisheries), Lobotidae, Gerreidae, Haemulidae, Dinopercidae, Sparidae, Lethrinidae, Nemipteridae, Polynemidae, and Sciaenidae. Species important in the aquarium trade include members of the following families: Ambassidae, Serranidae, Grammatidae, Callanthiidae, Pseudochromidae, Plesiopidae, Opistognathidae, Apogonidae, Malacanthidae, Lutjanidae, Caesionidae, Haemulidae, Sparidae, Lethrinidae, Nemipteridae, and Sciaenidae. Some species within a few families are also cultured for food or for release in the wild to enhance recreational fisheries (e.g., Serrandiae, Sillaginidae, Lutjanidae, and Sparidae). Others, particularly larger individuals of certain species, have been implicated in cases of ciguetara poisoning in humans.

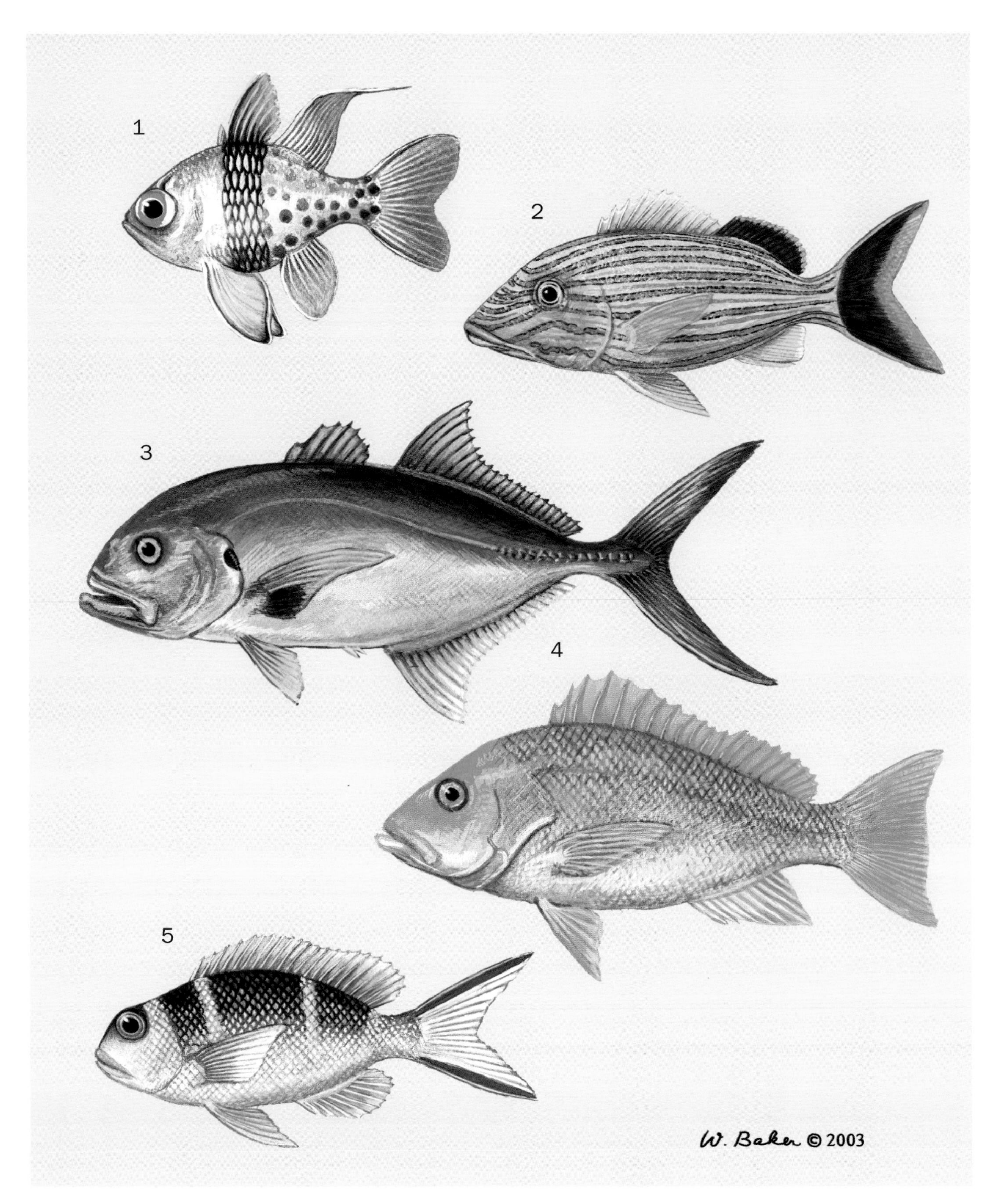

1. Pajama cardinalfish (*Sphaeramia nematoptera*); 2. Bluestriped grunt (*Haemulon sciurus*); 3. Crevalle jack (*Caranx hippos*); 4. Northern red snapper (*Lutjanus campechanus*); 5. Humpnose big-eye bream (*Monotaxis grandoculis*). (Illustration by Wendy Baker)

1. Blacksaddled coral grouper "tiger morph" (*Plectropomus laevis*); 2. Sixline soapfish (*Grammistes sexlineatus*); 3. Goggle eye (*Priacanthus hamrur*); 4. Red drum (*Sciaenops ocellatus*); 5. Nassau grouper (*Epinephelus striatus*). (Illustration by Emily Damstra)

Species accounts

Pajama cardinalfish
Sphaeramia nematoptera

FAMILY
Apogonidae

TAXONOMY
Sphaeramia nematoptera (Bleeker, 1856), Manado, Sulawesi (Celebes), Indonesia.

OTHER COMMON NAMES
English: Coral cardinalfish, polka-dot cardinalfish; Japanese: manjû-ishimochi; Malay: Capungon; Tagalog: Suga.

PHYSICAL CHARACTERISTICS
Body somewhat deep with large eyes; large, extended fins; and a slightly forked caudal fin. The skin has bioluminescent bands. There are two dorsal fins; the first has 7 spines and the second 1 spine and 9 soft rays. The anal fin has 2 spines and 9 soft rays. There are 12–14 soft rays on the pectoral fin. The head and gills are yellow that grades into a broad brown band that extends from the dorsal fin to the posterior portion of the pelvic fin. Posterior to the band, the body is a pale luminescent white with numerous brown spots. The second dorsal fin, anal fin, and caudal fin are clear but edged with luminescent white. The anterior portion of the pelvic fin is yellow and the eye is bright red. This species is sexually dimorphic for body size, with females slightly larger on average but males with deeper bodies and larger heads. Grows to 3.1 in (8 cm) in total length.

DISTRIBUTION
Western Pacific, from Java, Indonesia, east to New Guinea, Palau, and Pohnpei in the Caroline Islands, and north to the Mariana and Ryukyu islands.

HABITAT
Usually found in protected bays, lagoons, and backreefs among the branches of *Porites* corals.

FEEDING ECOLOGY AND DIET
Feeds upon plankton in the water column. Preyed upon by larger roving and ambush predatory fishes.

BEHAVIOR
Aggregates in coral branches during daylight and disperses along the bottom at night.

REPRODUCTIVE BIOLOGY
Not well known in nature. May possess a promiscuous mating system with multiple spawning events during a season. Males incubate the eggs orally. The incubation period varies with water temperature; eggs of a congener hatched after eight days at water temperatures of 80–86° F (27–30° C). The larvae are pelagic. This species has also been bred in captivity.

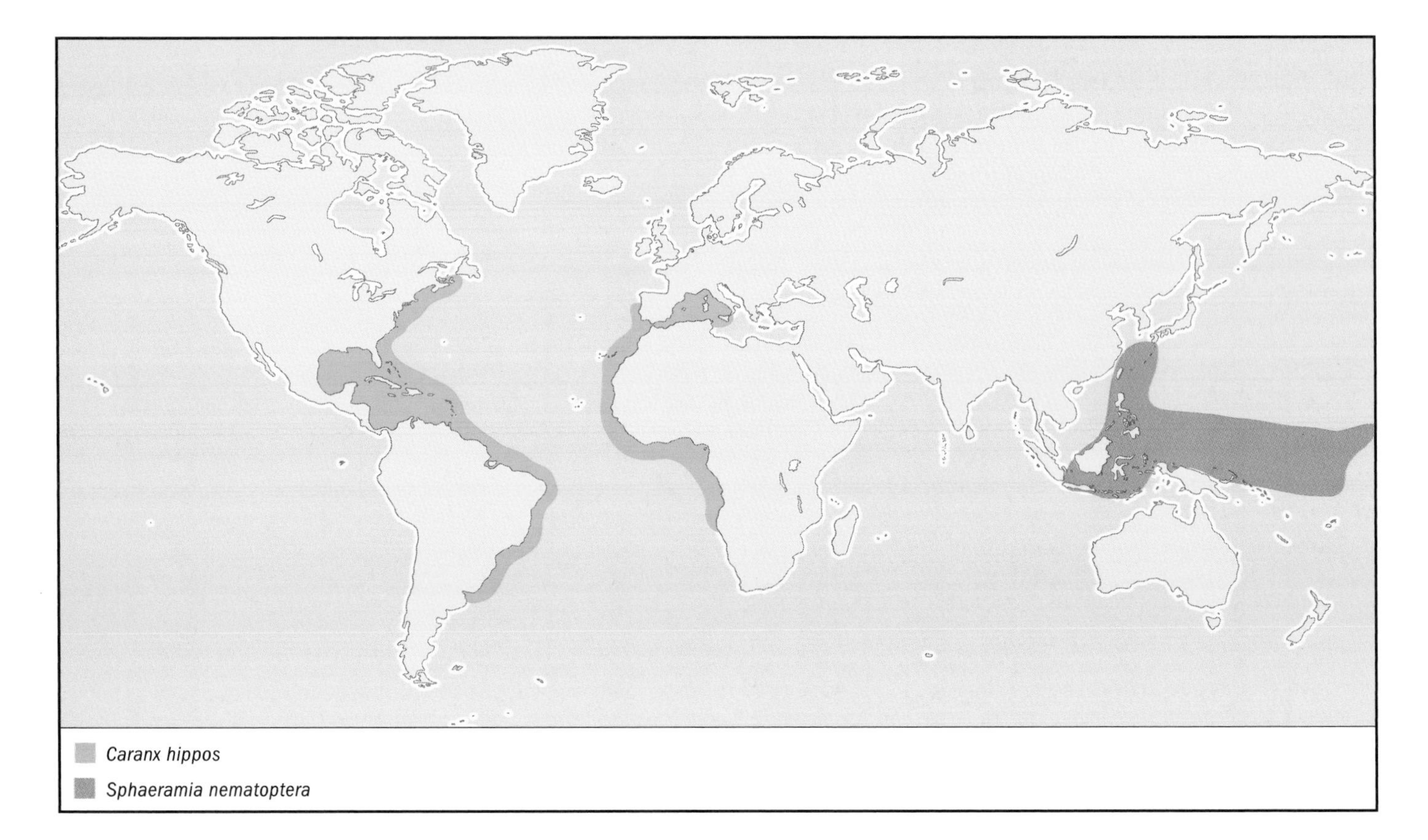

CONSERVATION STATUS
Not listed by the IUCN.

SIGNIFICANCE TO HUMANS
Popular in the aquarium trade. ◆

Crevalle jack
Caranx hippos

FAMILY
Carangidae

TAXONOMY
Caranx hippos (Linnaeus, 1766), Carolina, United States.

OTHER COMMON NAMES
English: Jack crevalle, common jack, couvalli jack; French: Carangue crevalle; Spanish: Cavalla; Portuguese: Coa.

PHYSICAL CHARACTERISTICS
Body deeply compressed with a steep forehead, two dorsal fins, a narrow caudal peduncle, and a slender forked caudal fin. The caudal peduncle is reinforced with a series of scutes (25–42) formed from modified bone. There are 9 spines in the first dorsal fin and 19–21 soft rays in the second dorsal fin, and 3 spines and 15–17 soft rays in the anal fin. The second dorsal fin and the anal fin are both elevated. The eye has an adipose eyelid. Scales are cycloid and small. Body color is silvery to brassy, the dorsal surface olive or bluish green, and the caudal fin yellowish. There is a black spot on the gill cover at equal height with the eye. Grows to about 49 in (124 cm) in total length.

DISTRIBUTION
In the eastern Atlantic, from Portugal to Angola and into the western Mediterranean. In the western Atlantic, from Nova Scotia south to Uruguay and including the Gulf of Mexico; absent from the eastern Lesser Antilles.

HABITAT
In the lower water column on coral and rocky reefs, over mud, sand, and rubble bottoms, and into brackish estuaries, canals, and rivers.

FEEDING ECOLOGY AND DIET
Highly predatory, feeding upon smaller fishes, shrimp, crabs, and other macroinvertebrates. Juveniles may be preyed upon by larger fishes, wading birds, and sea birds, while adults may be taken by sharks or other large predatory fishes.

BEHAVIOR
Forms aggregations, although larger individuals are often solitary or paired.

REPRODUCTIVE BIOLOGY
Forms spawning aggregations at predictable locations during peak times annually, usually April thru May. Eggs and larvae are pelagic.

CONSERVATION STATUS
Not listed by the IUCN.

SIGNIFICANCE TO HUMANS
Highly prized as a game fish but also harvested by commercial and subsistence fisheries; also collected for display in larger aquaria. May be ciguatoxic in some areas. ◆

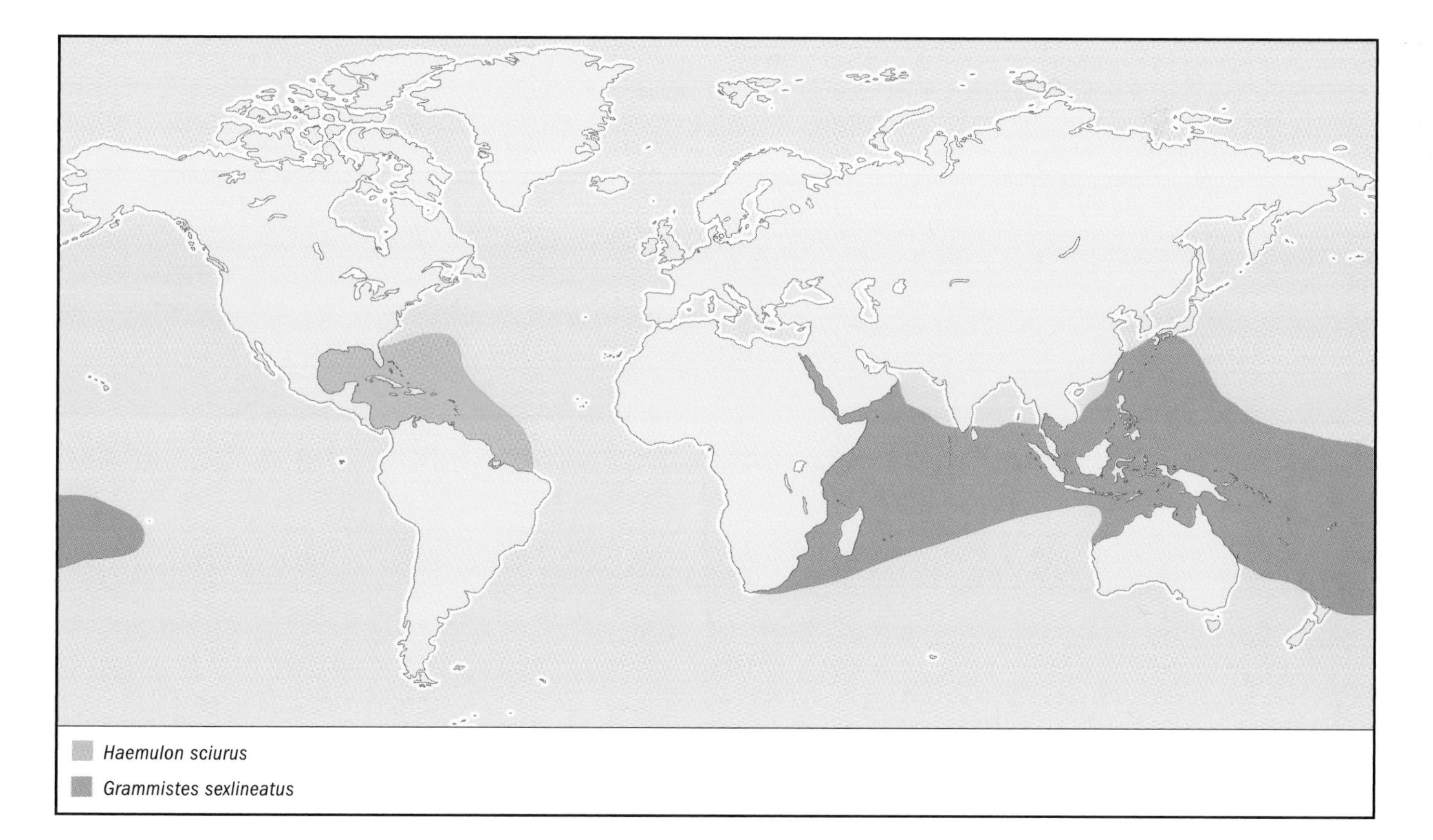

Bluestriped grunt
Haemulon sciurus

FAMILY
Haemulidae

TAXONOMY
Haemulon sciurus (Shaw, 1803), Antilles, Caribbean.

OTHER COMMON NAMES
English: Golden grunt, yellow grunt; French: Gorette catire; Spanish: Ronco catire; Dutch: Neertje; Portugese: Biquara.

PHYSICAL CHARACTERISTICS
Typically perch or bass-like; the body color is yellow with a series of blue stripes that run from the head to the caudal peduncle. Fins are yellowish, except for the posterior portion of the dorsal fin and also the caudal fin, which are black. The caudal fin margin may be yellow. There are 12 spines and 16–17 soft rays in the dorsal fin, and 3 spines and 9 soft rays in the anal fin. The caudal fin is emarginate. Grows to about 18 in (46 cm) in total length.

DISTRIBUTION
Western Atlantic, from Florida south to northern Brazil, west through the Gulf of Mexico and throughout the Caribbean.

HABITAT
Juveniles are found in *Thalassia* sea grass beds. Adults occur over coral and rocky reefs and near drop-offs. Depth range is 3–98 ft (1–30 m).

FEEDING ECOLOGY AND DIET
Mainly nocturnal predators of crustaceans, bivalves, and small fishes.

BEHAVIOR
Forms small groups that migrate twice daily along predictable routes at dawn and dusk. Knowledge of the locations of these routes has been demonstrated to be transmitted culturally by older fishes to younger ones in a related species.

REPRODUCTIVE BIOLOGY
Little is known. Likely forms spawning aggregations at predictable locations annually (autumn through spring, probably around the full moon); spawning is doubtless pelagic, as are the eggs and larvae.

CONSERVATION STATUS
Not listed by the IUCN.

SIGNIFICANCE TO HUMANS
A minor commercial and recreational species; also collected for the aquarium trade and for public aquaria. Ciguatoxic in some areas. ◆

Humpnose big-eye bream
Monotaxis grandoculis

FAMILY
Lethrinidae

TAXONOMY
Monotaxis grandoculis (Forsskål, 1775), Jidda, Saudi Arabia, Red Sea.

OTHER COMMON NAMES
English: Big-eye barenose, big-eye bream; French: Emperor bossu; Japanese: Yokushima-kurodai.

PHYSICAL CHARACTERISTICS
The body is oblong with a strongly convex profile of the head anterior to the eye; the snout is steeply sloped. The mouth is relatively large with pronounced canines and molars that are used for grasping and crushing prey, respectively. The eye is large; juveniles have a prominent black stripe through the eye. The dorsal fin has 10 slender spines and 10 soft rays, the anal fin has 3 spines and 9 soft rays, and the pectoral fin has 14 rays. The caudal fin is forked in adults and somewhat lunate in juveniles. Body color is light brown to bluish grey; ventral surfaces are white. Three prominent black or dark brown saddles cover the flank dorsally. Fins and the caudal peduncle range from yellow or reddish orange to clear or dusky. The lobes of the caudal fin may be pink in adults. Able to switch between dark and light color forms by behavioral control, usually in response to the color of the sea bottom. Grows to 23.6 in (60 cm) in total length.

DISTRIBUTION
Indo-West Pacific, from the Red Sea and East Africa east to the Hawaiian Islands, southeast to French Polynesia, south to northern Australia, and north to southern Japan.

HABITAT
Tropical coral and rocky reefs, over coral, sand, and rubble.

FEEDING ECOLOGY AND DIET
Feeds at night upon gastropods, echinoderms (mainly sea stars and brittle stars, but also sea cucumbers), crabs, polychaete worms, and tunicates.

BEHAVIOR
Often solitary in the water-column; juveniles closer to the bottom. Adults also form large aggregations that swim lazily over the reef or reef slope.

REPRODUCTIVE BIOLOGY
Little is known. Probably forms spawning aggregations and produces pelagic eggs and larvae.

CONSERVATION STATUS
Not listed by the IUCN.

SIGNIFICANCE TO HUMANS
Taken in commercial, subsistence, and recreational fisheries; juveniles are collected infrequently for the aquarium trade, and adults are collected for large public aquaria. May be ciguatoxic in some areas, such as the Marshall Islands. ◆

Northern red snapper
Lutjanus campechanus

FAMILY
Lutjanidae

TAXONOMY
Lutjanus campechanus (Poey, 1860), Campeche, Mexico.

OTHER COMMON NAMES
English: Red snapper; French: Vivaneau campèche; Spanish: Pargo de golfo.

Epinephelus striatus
Monotaxis grandoculis

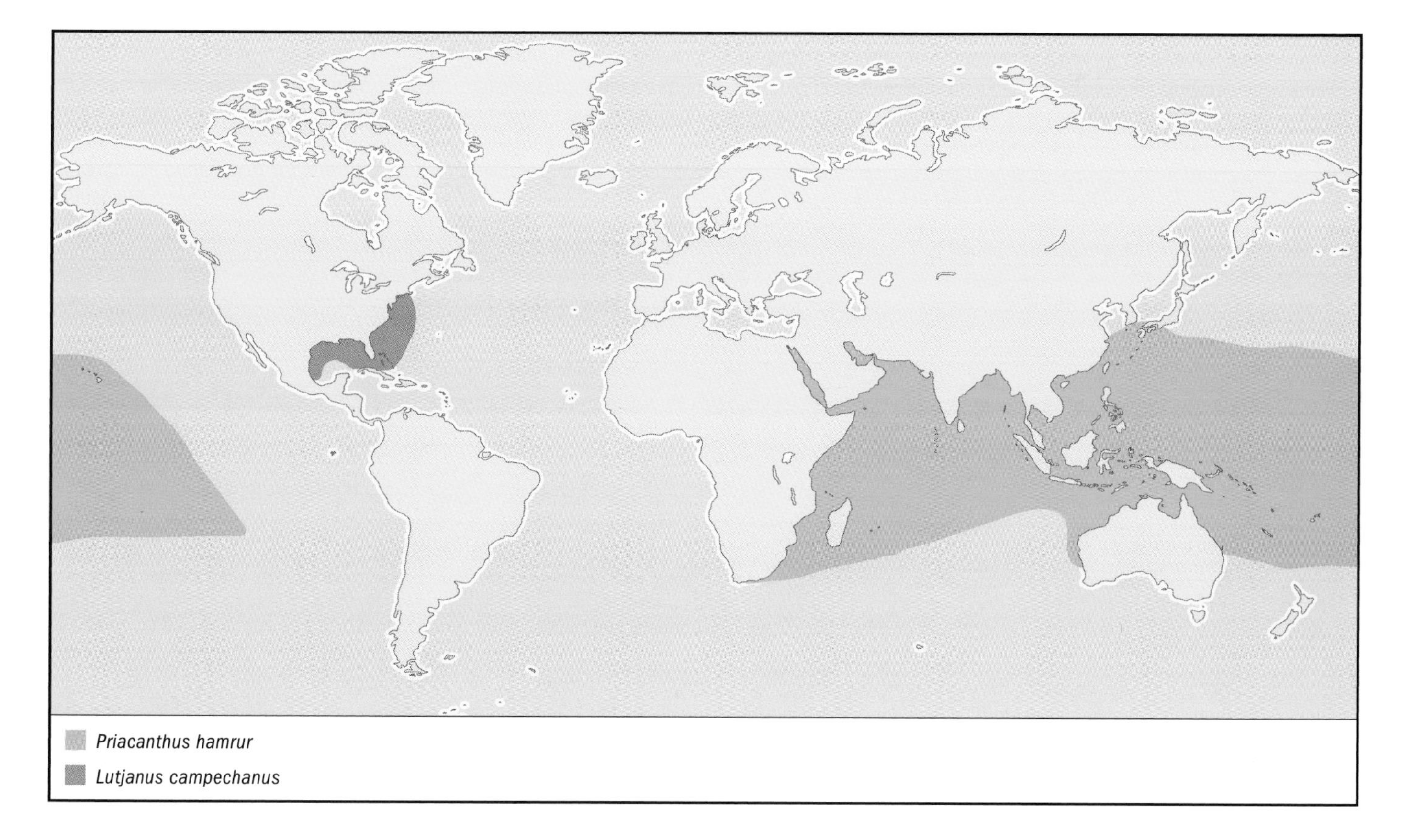
Priacanthus hamrur
Lutjanus campechanus

PHYSICAL CHARACTERISTICS
Typically bass-like in appearance, with a deep body, a single dorsal and anal fin, and an emarginate caudal fin. There are 10 spines and 14 soft rays in the dorsal fin, 3 spines and 8–9 soft rays in the anal fin, and the pectoral fin is elongate, almost reaching the anus, with 17 rays. The eyes are relatively small. Body color is red with orangish red fins. Grows to about 39 in (100 cm) in total length.

DISTRIBUTION
Western Atlantic, from Massachusetts (rarely) south through the Carolinas to Florida, west through the Gulf of Mexico to the Yucatan Peninsula and southeast to the northern edge of Cuba.

HABITAT
Juveniles frequent inshore waters, usually over sand or mud bottoms. Adults prefer rocky bottoms. Depth range of adults is 33–623 ft (10–190 m).

FEEDING ECOLOGY AND DIET
Accomplished predator that feeds upon smaller fishes, crabs, shrimps, cephalopods, polychaete worms, and gastropods and urochordates in the water column.

BEHAVIOR
Swims alone or in aggregations just above the bottom. May be idle during daylight and active at night.

REPRODUCTIVE BIOLOGY
Dioecious; there is no sex change. Males and females migrate to specific locations to form spawning aggregations between the months of April and December. Spawning is pelagic, as are the eggs and larvae. Eggs hatch in about a day.

CONSERVATION STATUS
Not listed by the IUCN but has been shown to be vulnerable to overfishing as a primary target species and, for juveniles, as bycatch in shrimp trawls. Fisheries are regulated in U. S. waters.

SIGNIFICANCE TO HUMANS
Important commercial and recreational species. ◆

Goggle eye
Priacanthus hamrur

FAMILY
Priacanthidae

TAXONOMY
Priacanthus hamrur (Forsskål, 1775), Jidda, Saudi Arabia, Red Sea.

OTHER COMMON NAMES
English: Lunar tail bigeye, moontail bullseye; French: Beauclaire miroir; Japanese: Hoseki-kintoki.

PHYSICAL CHARACTERISTICS
Body deep and compressed, with rough scales, large eyes, relatively large fins, and a caudal fin that is slightly emarginate. There are 10 spines and 14–15 soft rays in the dorsal fin, and 3 spines and 14–15 soft rays in the anal fin. Body color is a red or coppery red that fades to a mottled pattern of silver and red in darkness. Grows to 18 in (45 cm) in total length.

DISTRIBUTION
Indo-West Pacific, from the Red Sea and East Africa east to the Marquesas and Mangareva in French Polynesia, and Easter Island. Also found from southern Japan in the Northern Hemisphere to Australia and Lord Howe Island in the Southern Hemisphere.

HABITAT
Frequents ledges, crevices, caves, and the lower water column of outer reef slopes, passes, and deep lagoons; also found around pinnacles in lagoons and offshore patch reefs.

FEEDING ECOLOGY AND DIET
Feeds mainly at night upon smaller fishes, crustaceans, cephalopods, and larger zooplankton. Preyed upon by larger predatory fishes.

BEHAVIOR
Often solitary, hovering in or next to shelter or in the lower water column in daylight but more active at night. Changes color from red to silver or mottled-silver and red in darkness. The large eyes of this species are advantageous in low-light conditions, both for feeding and predator avoidance.

REPRODUCTIVE BIOLOGY
Little is known. May form spawning aggregations. The eggs and larvae are reportedly pelagic.

CONSERVATION STATUS
Not listed by the IUCN.

SIGNIFICANCE TO HUMANS
Taken in commercial and subsistence fisheries and incidentally in recreational fisheries. Sometimes collected for larger aquaria. ◆

Nassau grouper
Epinephelus striatus

FAMILY
Serranidae

TAXONOMY
Epinephelus striatus (Bloch, 1792), Martinique, West Indies.

OTHER COMMON NAMES
French: Mérou rayé; Spanish: Cherna criolla.

PHYSICAL CHARACTERISTICS
Robust body with sloping forehead, large fins, and a somewhat truncated caudal fin (rounded in juveniles). There are 11–12 spines, the third or fourth being the longest, and 16–18 soft rays in the notched dorsal fin, and 3 spines and 8 soft rays in the anal fin. Body color is tawny brown (shallow water) to pinkish brown or red (deeper water). There are alternating dark (brown or olive) and pale bands along the flanks and onto the dorsal fins, with similar dark bands extending along the head to the snout, a dark saddle on the upper caudal peduncle, and dark spots around the eye. Two color phases, pale and dark, are controlled behaviorally, and change between one and the other is rapid. Grows to about 47 in (120 cm) in total length.

DISTRIBUTION
Western Atlantic, from Bermuda and Florida south to the Bahamas and the Yucatan Peninsula, throughout the Caribbean,

and south to northern Brazil. Absent from most of the Gulf of Mexico.

HABITAT
Juveniles usually found in sea grass beds, while adults prefer coral and rocky reefs to a depth of 295 ft (90 m).

FEEDING ECOLOGY AND DIET
An ambush or hunting predator that feeds upon smaller fishes, crustaceans (mainly crabs), and large mollusks. Vulnerable to natural predation mainly as juveniles; adults vulnerable to larger predators such as sharks and large barracudas.

BEHAVIOR
Generally solitary as adults, although may form aggregations (especially for spawning). Site specific but probably with a large territory. Changes color pattern from one phase to the other depending upon circumstances. Not especially wary and may be friendly towards divers.

REPRODUCTIVE BIOLOGY
Matures at about 16–18 in (40–45 cm) in standard length, somewhere between 4–8 years of age. A protogynous hermaphrodite, but primary males have also been found. Spawns pelagically in large aggregations that form at specific locations annually depending upon lunar phase and water temperature. Courtship occurs within the aggregation just prior to or after sunset with spawning soon after sunset. The two color phases are used at this time to indicate submissive roles that reduce aggregation and promote courtship behavior. Females assume a dark color phase during courtship and lead spawning events. Events occur usually in subgroups of 3–25 fish, with a spawning ascent well into the water column. Eggs and larvae are pelagic.

CONSERVATION STATUS
Listed as Endangered on the IUCN Red List. Spawning aggregations of this species should be protected and fisheries harvests greatly restricted.

SIGNIFICANCE TO HUMANS
Important in commercial, subsistence, and recreational fisheries but now severely overfished throughout most of its range. Especially vulnerable when in spawning aggregations. Also collected for large aquaria. May be ciguatoxic in some areas. ◆

Sixline soapfish
Grammistes sexlineatus

FAMILY
Serranidae

TAXONOMY
Grammistes sexlineatus (Thunberg, 1792), type locality not specified.

OTHER COMMON NAMES
English: Black and white–striped soapfish, gold-striped soapfish, six-stripe soapfish; French: Poisson savon bagnard.

PHYSICAL CHARACTERISTICS
Body typically grouper or perch-like but somewhat stout. The head is relatively large. There are 7 spines and 13–14 soft rays in the dorsal fin and 2 spines and 9 soft rays in the anal fin. The caudal fin is truncate. The base color is dark brown to black with a series of yellow stripes running from the snout back to the caudal peduncle. With age, some stripes may break up into dashes. Juveniles have small spots. Fins are pinkish in color. Grows to about 12 in (30 cm) in total length.

DISTRIBUTION
Indo-West Pacific, from the Red Sea east to the Marquesas and Mangareva Islands, north to southern Japan and south to northern New Zealand.

HABITAT
This species occurs on coral and rocky reefs, usually in or near caves and under ledges to a depth of 425 ft (130 m).

FEEDING ECOLOGY AND DIET
Generally an ambush predator, and quite voracious as it feeds upon smaller fishes and crustaceans. May be preyed upon by larger predatory fishes but usually rejected immediately because of the secretion of grammistin, a toxin secreted from glands in the skin that is used as an antipredator mechanism.

BEHAVIOR
Usually solitary, preferring to hide during daylight while foraging at night.

REPRODUCTIVE BIOLOGY
Little is known. Likely a protogynous hermaphrodite with a haremic mating system, pair spawning, and pelagic eggs and larvae.

CONSERVATION STATUS
Not listed by the IUCN.

SIGNIFICANCE TO HUMANS
An interesting aquarium species, although because of its voracious appetite it must be kept with much larger fishes. Also taken as a minor commercial and subsistence species in some localities. May be ciguatoxic in some areas. ◆

Blacksaddled coral grouper
Plectropomus laevis

FAMILY
Serranidae

TAXONOMY
Plectropomus laevis (Lacepede, 1801), type locality not specified.

OTHER COMMON NAMES
English: Blacksaddled coral trout, giant coral trout, tiger coral trout; French: Mérou sellé.

PHYSICAL CHARACTERISTICS
Body elongate and robust, with the outer margin of the anal fin straight, and a large and slightly emarginate caudal fin. The mouth is relatively large with prominent canines. There are 8 spines and 11 soft rays in the dorsal fin, 3 spines and 8 soft rays in the anal fin, and 16–18 rays in the pectoral fin. This species has two color phases. The "tiger" or pale phase consists of a base color of white with four black bars or saddles, some incomplete, along the flank, yellow fins and mouth parts, and small blue spots with dark edges on the caudal penduncle and caudal fin. The dark phase is reddish brown with many small blue spots with dark edges scattered over the body and fins, and less prominent bars along the flanks. Grows to 39 in (100 cm) in total length.

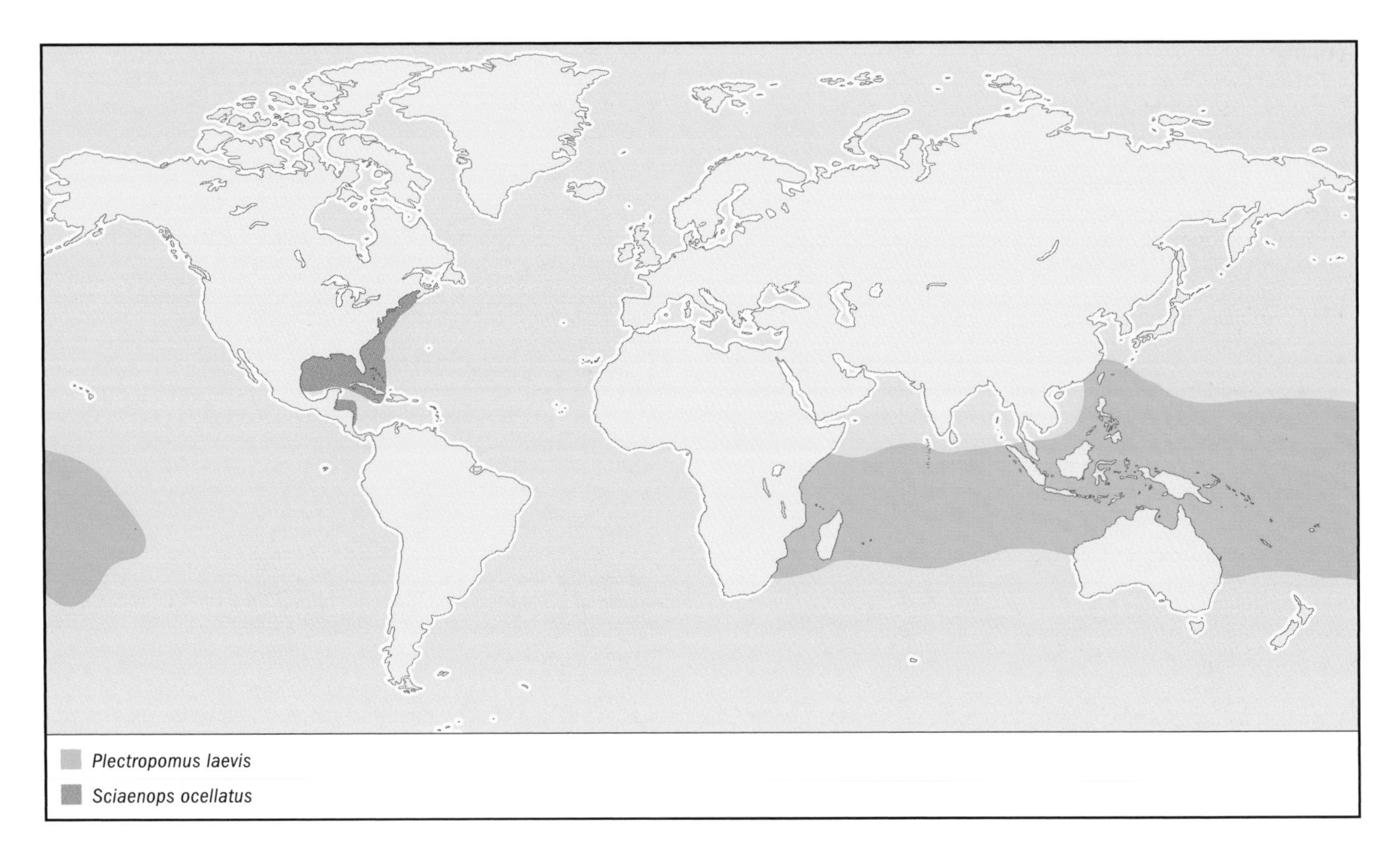

DISTRIBUTION
Indo-West Pacific, from Kenya and Mozambique east to the Tuamotu Archipelago in Polynesia, north to the Ryukyu Islands of Japan, and south to Queensland in Australia; absent from the Red Sea and Persian Gulf.

HABITAT
Juveniles are often found in turbid areas of deeper lagoons and back reefs, while adults prefer the clear water of seaward reefs, lagoons, and passes. Will utilize holes and crevices in the reef for shelter.

FEEDING ECOLOGY AND DIET
An efficient predator that feeds primarily upon smaller fishes and, to a lesser extent, large crustaceans. Larger individuals take larger prey, however, and often these are fishes (e.g., parrotfishes, large wrasses, surgeonfishes) that may be up to nearly half their body length in size.

BEHAVIOR
Juveniles and young adults may be somewhat gregarious and often hover above the bottom, but will retreat to shelter when threatened. Larger adults utilize habitat as shelter but may be found patrolling territories or home ranges.

REPRODUCTIVE BIOLOGY
Protogynous hermaphrodite that forms spawning aggregations prior to courtship and spawning. Spawning occurs between October and November on the northern Great Barrier Reef. Spawning is pelagic, as are the eggs and larvae.

CONSERVATION STATUS
Not listed by the IUCN but potentially vulnerable to overfishing at some localities.

SIGNIFICANCE TO HUMANS
A species of particular importance to commercial, recreational, and subsistence fisheries. Large individuals may be highly ciguatoxic. ◆

Red drum
Sciaenops ocellatus

FAMILY
Sciaenidae

TAXONOMY
Sciaenops ocellatus (Linnaeus, 1766), Carolina, (United States).

OTHER COMMON NAMES
English: Redfish, channel bass; French: Tambour rouge; Spanish: Corvinón ocelado.

PHYSICAL CHARACTERISTICS
Body elongate with a somewhat large head, subterminal mouth, two dorsal fins, and a truncate caudal fin. The anal fin has two spines and the lateral line is continuous. Body color is coppery-orange to light red; ventral surfaces are white. The upper caudal peduncle has a prominent black spot. Grows to about 61 in (155 cm) in total length, although less so inshore compared to fishes living around barrier islands.

DISTRIBUTION
Western Atlantic, from Massachusetts south to southern Florida and west to northern Mexico in the Gulf of Mexico.

HABITAT
Inshore coastal waters and estuaries, over sand, mud, or oyster-shell bottoms, among flooded marsh grasses, or in the surf zone.

FEEDING ECOLOGY AND DIET
Forages along the bottom in search of crustaceans, mollusks, and smaller fishes; will also form aggregations and attack schools of baitfish in shallow water.

BEHAVIOR
Occurs singly or in aggregations. Communicates by using muscle contractions to make a drumming noise that is amplified by the swim bladder.

REPRODUCTIVE BIOLOGY
Forms spawning aggregations, mainly from August through November. Males produce a drumming noise when courting females. Eggs are scattered and the larvae are pelagic.

CONSERVATION STATUS
Not listed by the IUCN. Fisheries are heavily regulated in most places, as this species was severely overfished in the Gulf of Mexico during the 1980s.

SIGNIFICANCE TO HUMANS
Once important in commercial fisheries but less so by the beginning of the twenty-first century. An important recreational species that is collected also for large public aquaria. Raised in aquaculture.

Resources

Books

Allen, G. R. *Snappers of the World.* FAO Species Catalog Vol. 6. Rome: Food and Agriculture Organization of the United Nations, 1985.

Allen, G. R., S. H. Midgley, and M. Allen. *Field Guide to the Freshwater Fishes of Australia.* Perth: Western Australian Museum, 2002.

Bohlke, J. E., and C. C. G. Chaplin. *Fishes of the Bahamas and Adjacent Tropical Waters.* 2nd ed. Austin: University of Texas Press, 1993.

Carpenter, K. E., and G. R. Allen. *Emperor Fishes and Large-Eye Breams of the World (Family Lethrinidae).* FAO Species Catalog Vol. 9. Rome: Food and Agriculture Organization of the United Nations, 1989.

Eschmeyer, W. N., ed. *Catalog of Fishes,* 3 vols. San Francisco: California Academy of Sciences, 1998.

Francis, M. *Coastal Fishes of New Zealand.* 3rd ed. Auckland: Reed, 2001.

Helfman, G. S., B. B. Collette, and D. E. Facey. *The Diversity of Fishes.* Oxford, UK: Blackwell Science, 1997.

Hutchins, B., and M. Thompson. *The Marine and Estuarine Fishes of South-Western Australia.* Perth: Western Australian Museum Press, 1995.

Kuiter, R. H. *Coastal Fishes of South-Eastern Australia.* Honolulu: University of Hawaii Press, 1993.

Lieske, E., and R. Myers. *Coral Reef Fishes: Indo-Pacific and Caribbean.* Rev. ed. London: HarperCollins Publishers, 2001.

Masuda, H., K. Amaoka, C. Araga, T. Uyeno, and T. Yoshino, eds. *The Fishes of the Japanese Archipelago.* Tokyo: Tokai University Press, 1984.

Myers, R. F. *Micronesian Reef Fishes.* 3rd ed. Barrigada, Guam: Coral Graphics, 1999.

Neira, F. J., A. G. Miskiewicz, and T. Trnski, eds. *Larvae of Temperate Australian Fishes: Laboratory Guide for Larval Fish Identification.* Perth: University of Western Australia Press, 1998.

Nelson, J. S. *Fishes of the World.* 3rd ed. New York: John Wiley and Sons, 1994.

Randall, J. E., G. R. Allen, and R. C. Steene. *Fishes of the Great Barrier Reef and Coral Sea.* Rev. ed. Honolulu: University of Hawaii Press, 1996.

Smith, M. M., and P. C. Heemstra, eds. *Smiths' Sea Fishes.* Berlin: Springer-Verlag, 1986.

Thresher, R. E. *Reef fish: Behavior and Ecology on the Reef and in the Aquarium.* St. Petersburg, FL: Palmetto Publishing Co., 1980.

———. *Reproduction in Reef Fishes.* Neptune City, NJ: T. F. H. Publications, 1984.

Periodicals

Colin, P. L. "Reproduction of the Nassau Grouper, *Epinephelus striatus* (Pisces: Serranidae) and Its Relationship to Environmental Conditions." *Environmental Biology of Fishes* 34 (1992): 357–377.

Donaldson, T. J. "Pair Spawning of *Cephalopholis boenack* (Serranidae)." *Japanese Journal of Ichthyology* 35 (1989): 497–500.

———. "Courtship and Spawning Behavior of the Pygmy Grouper, *Cephalopholis spiloparaea* (Serranidae: Epinephelinae), with Notes on *C. argus* and *C. urodeta.*" *Environmental Biology of Fishes* 43 (1995): 363–370.

Fischer, E. A., and C. W. Petersen. "The Evolution of Sexual Patterns in the Seabasses." *Bioscience* 37 (1987): 482–489.

Sadovy, Y. and A.-M. Eklund. "Synopsis of Biological Data on the Nassau Grouper, *Epinephelus striatus* (Bloch, 1792), and the Jewfish, *E. itajara* (Lichtenstein)." *NOAA Technical Report NMFS* 146 (1999).

Shapiro, D. Y. "Social Behavior, Group Structure, and the Control of Sex Reversal in Hermaphroditic Fish." *Advances in the Study of Behavior* 10 (1979): 43–102.

Organizations

IUCN/SSC Coral Reef Fishes Specialist Group. International Marinelife Alliance-University of Guam Marine Laboratory, UOG Station, Mangilao, Guam 96913 USA. Phone: (671) 735-2187. Fax: (671) 734-6767. E-mail: donaldsn@uog9.uog .edu Web site: <http://www.iucn.org/themes/ssc/sgs/sgs.htm>

Society for the Conservation of Reef Fish Aggregations. c/o Department of Ecology and Biodiversity, University of Hong Kong, Pok Fu Lam Road, Hong Kong, China. E-mail: scrfa@hkucc.hku.hk Web site: <http://www.scrfa.org/>

Other

Froese, R., and D. Pauly, eds. *Fishbase 2002.* ICLARM World Wide Web Electronic Publisher. 2002. [cited April 22, 2003]. <www.fishbase.org>

Terry J. Donaldson, PhD

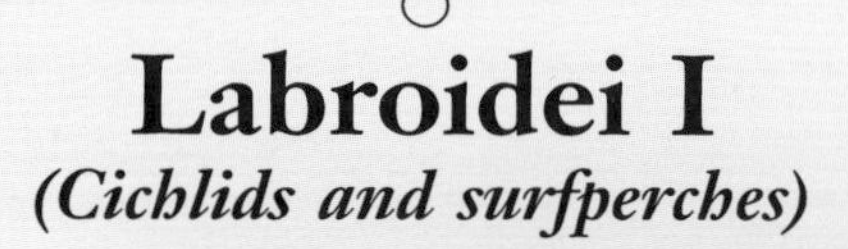

Labroidei I

(*Cichlids and surfperches*)

Class Actinopterygii
Order Perciformes
Suborder Labroidei
Number of families 2

Photo: Tampico cichlid (*Herichthys carpintis*) adult and juvenile. (Photo by Hans Reinhard/Okapia/Photo Researchers, Inc. Reproduced by permission.)

Evolution and systematics

The families Cichlidae (cichlids) and Embiotocidae (surfperches) have traditionally been grouped in the suborder Labroidei, along with Labridae (wrasses), Scaridae (parrotfishes), Pomacentridae (damselfishes), and Odacidae (western Pacific butterfishes). These groups were first suggested to share a common ancestry by Müller in 1843. All possess a pharyngeal jaw apparatus (a second set of jaws in the throat) that aids in the processing of food, and the morphological characters supporting their close relationship are all a part of this functional complex. Characters used to unite labroids include fused fifth ceratobranchials that bear teeth, a muscle sling suspending the lower pharyngeal jaw from the skull, an articulation between the upper pharyngeal jaws and the base of the skull without intervening muscle, and an undivided sheet of sphincter esophagi muscle. The reliability of these characters has been questioned, and the reality of the Labroidei has been challenged for several reasons. One is that none of the characters are unique to labroids (in fact, a whole suite of pharyngeal jaw features are found in the quite distantly related Beloniformes); another is that not all labroids present all the characters. In addition, characters supporting the Labroidei that are independent of the pharyngeal jaw apparatus, and perhaps less strongly influenced by selective forces related to food processing, have eluded detection. Given these challenges, it isn't surprising that recent molecular studies have failed to uphold the Labroidei as a natural group.

Nonetheless, this chapter considers two of the traditional labroid families, the cichlids and surfperches. Surfperches are temperate, almost entirely marine fishes found only in the northern Pacific, and are represented by 13 genera and 24 species. Cichlids are mostly tropical freshwater fishes, with a distribution on fragments of the supercontinent of Gondwana. There are about 105 genera of cichlids and 1,405 described species, although some estimates place the total number of cichlid species to be in excess of 2,500 if a full accounting is made of the exceptional species diversity in the three African Great Lakes. Different geologic processes and time scales have shaped these three lakes and their species flocks. Two of the lakes, Malawi and Tanganyika, formed in a rift created by plate movements that are slowly ripping apart the African continent. Lake Malawi is roughly 700,000 years old and contains 500–1,000 cichlid species; Lake Tanganyika is 9–12 million years old and contains about 200 cichlid species. Lake Victoria, formed by processes of mountain building that blocked the flow of rivers draining eastern Kenya, was probably entirely dry as recently as 12,400 years ago, leaving precious little time for the subsequent evolution of its 500 or so haplochromine cichlid species from a single ancestral species.

Evolutionary biologists have been intrigued by the question of how so many cichlid species evolved together in African lakes. Clearly, the ability of cichlids to finely partition resources provides some explanation, but the lakes themselves also play a role. Lake Victoria is relatively shallow and turbid, with few rocky microhabitats along its shoreline, but the two rift lakes are clear and deep, with abundant patches of rocky microhabitat scattered along their shores. Within Tanganyika, there are three distinct communities: the pelagic, containing six fish species; the benthic, with 80 fish species; and most diverse, the littoral, with 207 species of fishes. Most of the species diversity is restricted to inshore regions. Malawi's piscine endowment is similarly biased; so species poor are the pelagic waters in comparison with the littoral rocky communities that two zooplanktivorous species of the genus *Diplotaxodon* make up 71% of the open-water fish biomass. In Malawi and Tanganyika, populations in rocky littoral habitats separated by sandy bottom and deepwater barriers wider than 1.2 mi (2 km) exhibit little or no gene flow. Thus, a model of microallopatric speciation may be appropriate for the rift lakes, whereby the numerous and fragmented littoral rocky habitats, coupled with attributes of some of the cichlids, such as

Kribensis pair (*Pelvicachromis pulcher*) with young. (Photo by Mark Smith/Photo Researchers, Inc. Reproduced by permission.)

resource specialization and adherence to circumscribed territories, have fostered explosive speciation. Another speciation mechanism, sexual selection, has been considered most important in Victoria due to its paucity of isolated microhabitats, and may also be important in the rift lakes.

Age differences between the lakes are also manifest in the cichlid species flocks. The two younger lakes, Malawi and Victoria, contain relatively more homogenous assemblages of strictly mouth-brooding cichlids. In contrast, Lake Tanganyika contains a very diverse assemblage of both mouth brooders and substrate spawners descended from numerous colonization events. In a 1998 paper, Sturmbauer contrasted the relatively young lake Victoria species flock, which lacks extreme morphotypes and has many intermediate species, with the ancient Lake Tanganyika flock, characterized by few intermediates and high morphological distances between species, and proposed that the two lakes represent opposite ends of a species flock evolutionary continuum. He suggested that in older lakes with mature flocks (as in Tanganyika), very diverse communities sharing most of their species will predominate, because dispersal events (facilitated by lake level fluctuations) have, over long stretches of time, mixed together species that arose in isolation. Consequently, lesser competitors in shared niches will have alternately diverged or perished.

Despite the present diversity of cichlids, both cichlids and surfperches are poorly represented in the fossil record. Fossil surfperches are first known from the upper Miocene epoch (5–10 million years ago), from the Monterey Formation in California. Only a few species of fossil cichlids are known, the earliest of which are 45 million years old (Eocene epoch) from Tanzania, East Africa. Cichlid material from Maranhao, Brazil, may also date from the Eocene, and putatively cichlid material of similar age has been found in Vicenza, Italy. If, as many researchers suggest, the present geographic distribution of cichlids is an artifact of their being widely distributed on the supercontinent of Gondwana, then cichlids must be at least 130 million years old, leaving an inexplicable gap of 85 million years in the fossil record. This has led some to argue that cichlids are actually a younger group that has dispersed across marine barriers to achieve their present distribution, an argument supported by the ability of several cichlid species to survive in sea water and the fact that cichlid distribution is not strictly Gondwanan.

Physical characteristics

Surfperches are relatively deep bodied, laterally compressed fishes with an undivided series of lateral line scales. They achieve lengths of 17.5 in (45 cm). All surfperches have cycloid scales, a deeply forked caudal fin, and three anal fin spines. Most species have scales extending onto the dorsal fin. Although many species are uniformly silvery, others have blue, black, or reddish stripes and bars. Cichlids are more variable, presenting both cycloid and ctenoid scales, a divided lateral line scale series (except in the African riverine genera *Teleogramma* and *Gobiocichla*), numerous fin shapes and lengths (the smallest species is 1 in [2.5 cm], while the largest is 36 in [91.4 cm]), and anywhere from three to 15 anal fin spines. Cichlids are also unusual in that they have only a single nostril on each side of the head, unlike most teleosts, which have two pairs of nostrils.

Cichlids exhibit an impressive range of morphological diversity: there are dinner plate–shaped discus and cigar-shaped pike-cichlids, horse-faced *Tropheus* that scrape algae from rocks, and canine-endowed *Cichla* that voraciously prey on other fishes. A diversity of tooth morphologies in both the oral and pharyngeal jaws allow cichlids to utilize every conceivable source of food. Cichlids can be drab colored or come in brilliant hues, and many exhibit magnificent color patterns that are rivaled only in coral-reef fishes. Despite their variability, however, cichlids share numerous distinctive morphological characters supporting the family as a natural group derived from a common ancestor. For instance, all cichlids have a transversus dorsalis muscle partitioned into four parts, a functionally decoupled premaxilla and maxilla, microbranchiospines on the outer faces of the gill arches, and an expanded head of the fourth epibranchial.

Distribution

Surfperches are distributed in coastal areas of the North Pacific: three species are found around Japan and Korea, and 21 species are found in North America from Alaska to the Baja Peninsula. One species, the shiner perch (*Cymatogaster aggregata*), enters brackish and fresh waters. Another species, the tule perch (*Hysterocarpus traskii*) is entirely confined to freshwater lakes and rivers in California.

Cichlids are distributed in the tropical and subtropical regions of most Gondwanan fragments, including much of South America and Africa, Madagascar, India, and Sri Lanka. They also occur naturally in Syria, Iran, Israel, Central America, North America (Texas), and the West Indies. Human introductions have led to the establishment of numerous cichlid species in localities outside their natural range. Owing to their popularity in aquaculture, the tilapiines in particular have been spread around the globe, often with devastating consequences for endemic fishes. Various tilapiines now thrive in the fresh waters of Madagascar, Mexico, Florida, and much of Southeast Asia, as well as the coastal waters of southern California.

Habitat

Surfperches are mostly restricted to inshore marine environments. Many species are found in the surf near sandy beaches, but members of the family also frequent sea-grass beds, rocky outcroppings, and piers. Less frequently, surfperches are known to enter brackish and fresh waters, and one species lives its whole life in lakes and rivers.

Cichlids also exhibit a wide salinity tolerance, and some species are found in brackish and marine waters. Some cichlids are capable of withstanding rather extreme conditions. For example, some *Oreochromis* species live in high-temperature, high-alkalinity salt lakes in Kenya and Tanzania. Most cichlids are freshwater fishes, however, and members of the family can be found in most every conceivable freshwater habitat within their range, including open waters of lakes, oxygen-deprived depths, rocky inshore areas, aquatic plant beds, swamps, small streams, and large rivers, including rapids habitats.

Behavior

Surfperches are unusual among marine teleosts in that they do not go through a planktonic dispersal stage as eggs or larvae; they are somewhat large when they are born and stay in the general vicinity as they mature. Many cichlids, especially those that occupy specialized lacustrine habitats, behave similarly. Although most cichlids do not defend feeding territories, some cichlids in the African Great Lakes do defend separate feeding and breeding territories. Those that defend feeding territories are predominantly algae feeders, and some herbivorous species subvert such territoriality by forming large schools that barrage an algal mat with a few individuals at a time, diluting the aggression of the territory holder so that algae can be stolen. In captivity, the territoriality and aggression of many cichlids has given the family a particular notoriety among aquarists.

In some cichlids, older siblings assist their parents in guarding new clutches of fry, an investment that can be explained evolutionarily because it increases the share of both the parents' and helpers' genes in the next generation. Consider, however, the few examples of cichlids that help guard the offspring of another species of cichlids, a behavior that at first defies explanation. One tantalizing explanation for such behavior on the part of the moga (*Cichlasoma nicaraguense*), males of which sometimes help guard the fry of the guapote (*C. dovii*), is that larger numbers of piscivorous *C. dovii* help to increase the reproductive success of *C. nicaraguense* by preying upon and reducing the numbers of poor man's tropheus (*Neetroplus nematopus*), a nest-site competitor of the helper species. This interpretation, while intriguing, is still controversial.

Blue discus (*Symphysodon aequifasciata haraldi*) are popular aquarium fish. (Photo by Hans Reinhard/Okapia/Photo Researchers, Inc. Reproduced by permission.)

Feeding ecology and diet

Surfperches feed variously on shrimps, amphipods, crabs, and other crustaceans, as well as mollusks and worms. To feed on such organisms, many surfperches take indiscriminate bites out of the algae and debris found on the bottom, and then winnow out their invertebrate prey within the oropharyngeal cavity, spitting out the remainder. The small kelp perch (*Brachyistius frenatus*) feeds largely on ectoparasites picked off other fishes, and several other surfperches supplement their regular diets by engaging in such cleaning activities.

Cichlid feeding ecologies are extravagantly diverse, and may partly explain the species diversity within the family. Although most cichlids can opportunistically feed on a wide range of foods, they are also specialized to feed on certain types of food with particular efficiency. Of course, many cichlids specialize in hunting other fishes and eating them whole. *Tilapia* is the only genus of African cichlid to specialize on phytoplankton, by collecting the particles (with the assistance of sievelike gill rakers for filtering) in balls of mucous secreted in the mouth and then swallowing them. Deposit feeders take a very similar approach, but with sedimented phytoplankton and even disintegrated hippopotamus droppings. Cichlids who feed on epilithic algae, or aufwuchs, scrape the algae from rocks, typically employing multiple rows of fine teeth used like a file. Periphyton feeders scrape algae off living plants. For example, the giant haplochromis *Hemitilapia*

Breeding habits of the cichlid *Cyathopharynx furcifer.* 1. Male drags long pelvic fins along the bed of the nest, showing female where to lay eggs; 2. Female lays eggs while male waits to fertilize them; 3. Female takes fertilized eggs into her mouth; 4. Hatched fry are released from the female's mouth. (Illustration by Gillian Harris)

oxyrhynchus in Malawi feeds by placing the grasslike leaves of the plant *Vallisneria* between its jaws and nibbling off the algae without damaging the leaf. Singularly robust dentition characterizes leaf choppers, while mollusk feeders have large pharyngeal bones covered with flat, stocky teeth to crush shells. *Labidochromis vellicans* of Malawi, a benthic arthropod feeder, has very long, sharp, outward pointing teeth that resemble forceps, and uses them to pluck insects from the substrate. Alternatively, the greenface sandsifter (*Lethrinops furcifer*) of Malawi has a pointed, protractile mouth and sievelike gill rakers. It rams its head into the sand, filling the buccal cavity, then separates out and swallows insect larvae by expelling the sand across the gill rakers and through its opercular openings. Some species have developed protractile mouths specially suited for inhalant feeding on zooplankton. There exist fin-nipping feeders, and in Victoria and Malawi, species that feed on the eggs and larvae of other mouth brooders. Many species have acquired large molariform teeth on the pharyngeal jaws for crushing the shells of snails. Additionally, in all three African Great Lakes, cichlids with differing tooth morphologies have evolved to feed exclusively on the scales of other fishes; some even mimic the color patterns of their prey to avoid detection.

Interesting selective forces are at work within the scale-feeding Lake Tanganyika genus *Perissodus.* To a greater or lesser degree, all seven species exhibit laterally asymmetrical mouths that are angled either to the right or left, and this trait has a genetic basis. Left-handed individuals nip scales off the right flank of their prey, and vice versa for right-handed individuals. Incredibly, the differential alertness of prey species appears to exert selective pressure on the scale eaters that acts to maintain about equal numbers of left- and right-handed individuals within a population. A study on *Perissodus microlepis* found that whenever one or the other form becomes significantly less abundant, prey species become more vigilant about watching for scale eaters from the side favored by the more abundant form. In this way, the less-abundant form is conferred a selective advantage, and the genetic polymorphism responsible for the two forms is maintained.

Cross-section of a female *Oreochromis* mouth-brooder. (Illustration by Michelle Meneghini)

Reproductive biology

Embiotocids are viviparous (give birth to live young), and males have a thickened anterior portion of the anal fin that aids in internal fertilization. They display elaborate courtship behaviors, reminiscent of a small cross-section of the varied courtship behaviors found in cichlids. Some cichlids, for example, members of the genera *Crenicara*, *Sarotherodon*, and *Etroplus*, are capable of changing sex. When a group of females finds themselves without access to a male, the dominant female is able to generate testicular tissue sufficient to fertilize the eggs of the remaining females.

Primitively, cichlids likely formed monogamous pairs, in which both parents guarded eggs laid on the substrate. Cichlids exhibiting this behavior include the Indian etroplines, most Central American taxa, many South American taxa (such as *Pterophyllum* and *Symphysodon*), as well as several African groups (including *Hemichromis* and the lamprologines). Some cichlids have evolved polygamy and mouth brooding as embellishments of the original reproductive mode. Mouth brooders allow the eggs and young to develop inside the safety of the buccal cavity of one or both parents. As the young mature, they make feeding excursions outside, but quickly retreat into the adult's mouth at the first sign of danger. It is fascinating that following egg deposition, females of some mouth-brooding species collect the eggs into their mouths with such alacrity that males do not have time to fertilize them externally. On the anal fin of many such males, however, are conspicuously colored spots, called egg dummies, which bear a remarkable resemblance to the actual eggs. The female notices these spots and nips at them (in an apparent attempt to collect them into her mouth, some have argued), and the male releases sperm at the same time. Thus the eggs are fertilized inside the mouth of the female.

In one instance, a noncichlid has devised a means of capitalizing on the mouth-brooding habit of cichlids at their expense. *Synodontis multipunctatus*, a mochokid catfish, is a brood parasite on several species of mouth-brooding cichlids in Lake Tanganyika. Catfish eggs are concealed among the cichlid eggs during spawning, and are then taken into the mouth of the adult cichlid. As development proceeds, the young catfishes devour the cichlid fry from the safety of the parental buccal cavity, so that ultimately the cichlid parents find themselves caring for a small brood of catfish and none of their own progeny!

Many cichlids expend enormous energy in altering their surroundings prior to reproduction. Elaborate sand nests, in the form of pits or cone-shaped mounds, are constructed by the males of some cichlid species for spawning. Some nest mounds are enormous, and require many days work by the male, moving sand one mouthful at a time. In certain species, lekking occurs: males build their nests in aggregations of various sizes. In Lake Malawi, male *Copadichromis eucinostomus* come together in groups as large as 50,000 individuals, stretched over several miles of sand, each guarding a cone-shaped nest. In Lake Tanganyika, numerous species from the tribe Lamprologini take advantage of abundant snail shells on the lake bottom as places of refuge and spawning sites. *Neolamprologus callipterus*, a haremic species, is noteworthy in that the males are dramatically larger than the females (as much

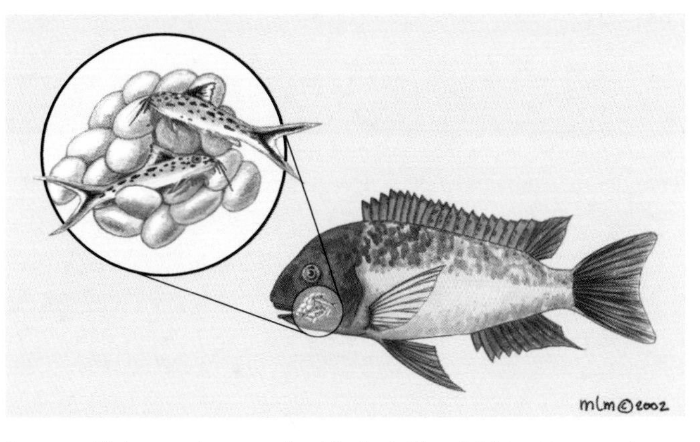

Some catfish eggs have evolved to look like cichlid eggs, so when a female cichlid (*Tropheus moorii*) collects her own eggs in her mouth, she inadvertantly collects some catfish eggs and raises them too. When the catfish eggs hatch, the young catfish eat the cichlid eggs. (Illustration by Michelle Meneghini)

Male rift lake cichlids have spots on their anal fins. When a female swims around collecting her eggs in her mouth after releasing them to be fertilized (1), she often mistakes the spots on the male's fin to be eggs. She bites the anal fin (2), and the eggs in her mouth are then fertilized. (Illustration by John Megahan)

30 times by weight), and only females occupy shells during spawning. Selection seems to maintain this size dimorphism because males need to be large enough to carry shells to their territories, while females need to be small enough to spawn inside the shells.

Conservation status

Although none of the surfperches are listed by the IUCN, hundreds of cichlid species are either extinct or in great peril. There are 133 cichlids on the IUCN Red List, 44 of which are listed as Extinct and five of which are listed as Extinct in the Wild. Of the 37 Critically Endangered cichlids, all are from Africa or Madagascar, and most are from either Lake Barombi Mbo in Cameroon or the East African lakes Victoria, Nabugabo, Kanyaboli, and Nawampasa. Humans are responsible for the plights of these fishes, which have declined due to the introduction of exotic species, pollution from industrial sources and sewage, sedimentation related to deforestation, and overfishing. The story of Lake Victoria is emblematic in this regard, as hundreds of species have been affected by the extensive degradation of the lake's ecosystem, and thus is considered here in some detail.

Gill nets were introduced to fishermen on Lake Victoria in 1905 to catch the endemic tilapiine cichlids *Oreochromis esculentus* and *Oreochromis variabilis*. Increasing fishing activity, aided by gill nets, quickly led to declines in catches of numerous species, including the tilapiines, catfishes, and lungfishes. By the 1950s, endemic tilapia had been all but wiped out, so the Nile perch (*Lates niloticus*) and four nonnative tilapiine species were introduced: *Tilapia zilii*, *T. melanopleura*, *Oreochromis niloticus*, and *O. leucostictus*. Gill-net fisheries had since shifted to endemic catfishes, characins, cyprinids, and mormyrids, and as those stocks declined, fishermen were forced by the 1970s to shift their focus toward the abundant haplochromine cichlids and the pelagic cyprinid *Rastrineobola argentea*. Beach seines, which had been introduced to the region during the 1960s, exacerbated the problem of overfishing by disrupting the littoral spawning nests of tilapiine and haplochromine cichlids. Soon after a haplochromine trawl fishery was established in Tanzania, and a factory was built to convert the catch into chicken feed, the haplochromine fishery began its precipitous collapse. This was in the late 1970s, and was coincident with a population explosion of the much-maligned Nile perch. Lake Victoria fisheries were on the verge of total collapse, but were revitalized by the now-abundant Nile perch, supplemented by *Rastrineobola* and *Oreochromis niloticus*.

Small, bony haplochromines were not valued highly by local fishers, so creative schemes were concocted for utilizing them. Early on, it was suggested that haplochromines could be trawled and used as manure; later efforts at canning haplochromines as fish meal for livestock were hampered by the fact that the can was worth more than its contents. In areas where haplochromine trawling was implemented, however, populations declined dramatically. Their biology helps to explain why. Haplochromines are mouth brooders, and females produce a relatively small number of eggs. As narrow specialists with low-standing stocks and a low reproductive potential, they are peculiarly susceptible to collapse when nonselective fishing methods like trawling are used.

Although numerous factors, including overharvesting, appear to have precipitated the collapse of the haplochromine fishery, strong evidence implicates the Nile perch. In Mwanza Gulf, which was continuously surveyed during the 1980s, the Nile perch boom documented at the beginning of the decade was immediately followed by the total collapse of open-water haplochromine stocks. Haplochromines were selectively eaten until they were exhausted. Once Nile perch had colonized a part of the lake, the larger and less common haplochromine species were first to disappear there, especially those species with the most habitat overlap with Nile perch.

Another factor that has played a significant role in haplochromine cichlid collapse is eutrophication. The causes of Lake Victoria's eutrophication are numerous: deforestation, urban and industrial development around the shore, a rise in lake level, and even positive feedback from post haplochromine collapse food-web alterations (many haplochromines were specialized phytoplankton feeders and kept blooms in check). Since 1960, the lake's primary productivity has more than doubled, and more than half the bottom area of the lake is now devoid of oxygen below 98 ft (30 m). Drinking water for the human population on the lake's shores is now threatened by toxic blue-green algae blooms. In addition, the fact that large *Lates* cannot be sun dried like haplochromines, but instead require smoking over a wood fire, has accelerated deforestation and sedimentation in the basin, as the Nile perch fishery has expanded.

Eutrophication further threatens Lake Victoria's haplochromines in a very unexpected way. In a 1997 paper, Seehausen and collaborators argued that the decreased water transparency brought about by eutrophication has caused rock-dwelling cichlid declines by eroding between-species mating barriers. They point out that among the African Great Lake basins, only those lakes with clear water have given rise to cichlid species flocks. The sympatric haplochromine cichlids of Lake Victoria lack postmating reproductive barriers among closely related species, and have historically avoided hybridization through mate choice; this barrier is now breaking down as turbidity prevents females from recognizing males of their own species. Males of similar species are always manifestly distinct in coloration, and in experiments with monochromatic light, females lose their ability to discern conspecifics. Within different regions of the lake, *Haplochromis* color morphs behave as distinct species where the water is clear, and blend into homogenous hybrid populations where the water is cloudy.

Significance to humans

Both cichlids and surfperches are sought after by humans as sources of food. Tilapiine cichlids are easy to culture artificially, and so have been introduced around the world in fish-farming ventures. They supply an important source of protein to many human populations. However, exotic tilapiines have also adversely affected native fish populations in regions where they have been introduced by humans. Cichlids are enormously popular in the aquarium trade, and help support local economies in parts of the Americas and Africa through revenues from wild-caught fish exports. Numerous strains of certain species are also cultivated in captivity.

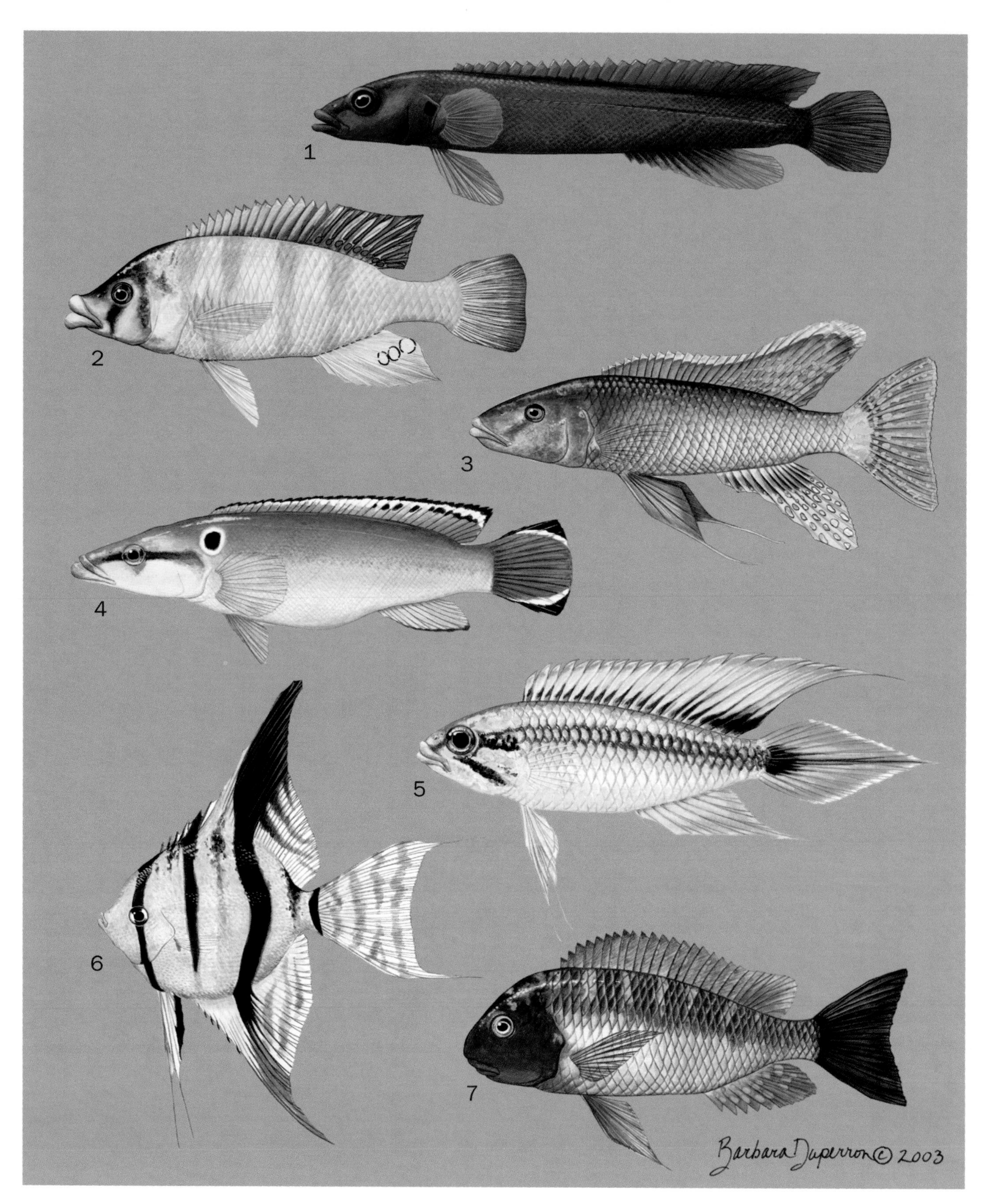

1. Worm cichlid (*Teleogramma gracile*); 2. *Paralabidochromis chilotes*; 3. Trout cichlid (*Champsochromis caeruleus*); 4. Millet (*Crenicichla alta*); 5. Agassiz's dwarf cichlid (*Apistogramma agassizii*); 6. Freshwater angelfish (*Pterophyllum scalare*); 7. Blunthead cichlid (*Tropheus moorii*). (Illustration by Barbara Duperron)

1. Rainbow seaperch (*Hypsurus caryi*); 2. Ngege (*Oreochromis esculentus*); 3. Trondo mainty (*Ptychochromoides betsileanus*); 4. Giant cichlid (*Boulengerochromis microlepis*); 5. Blue discus (*Symphysodon aequifasciata*); 6. Speckled pavon (*Cichla temensis*); 7. *Lepidiolamprologus kendalli*. (Illustration by Barbara Duperron)

Species accounts

Agassiz's dwarf cichlid
Apistogramma agassizii

FAMILY
Cichlidae

TAXONOMY
Geophagus (Mesops) agassizii Steindachner, 1875, Curupira, Codajas, Rio Poti de Sao Joao, Lago Maximo, and Lago Manacapuru, Brazil.

OTHER COMMON NAMES
German: Agassiz Zwergbuntbarsch.

PHYSICAL CHARACTERISTICS
Maximum length 2 in (5 cm). Dorsal and anal fins somewhat elongate, caudal fin spade shaped. Head somewhat rounded. Female coloration is lemon yellow; males brilliantly colored with yellows and blues anteriorly, a dark midlateral band, and red, black, and white bands on posterior fins.

DISTRIBUTION
Amazon basin along the Amazon River from Peru to the Capim River in Brazil.

HABITAT
Clear, black, and whitewater rivers.

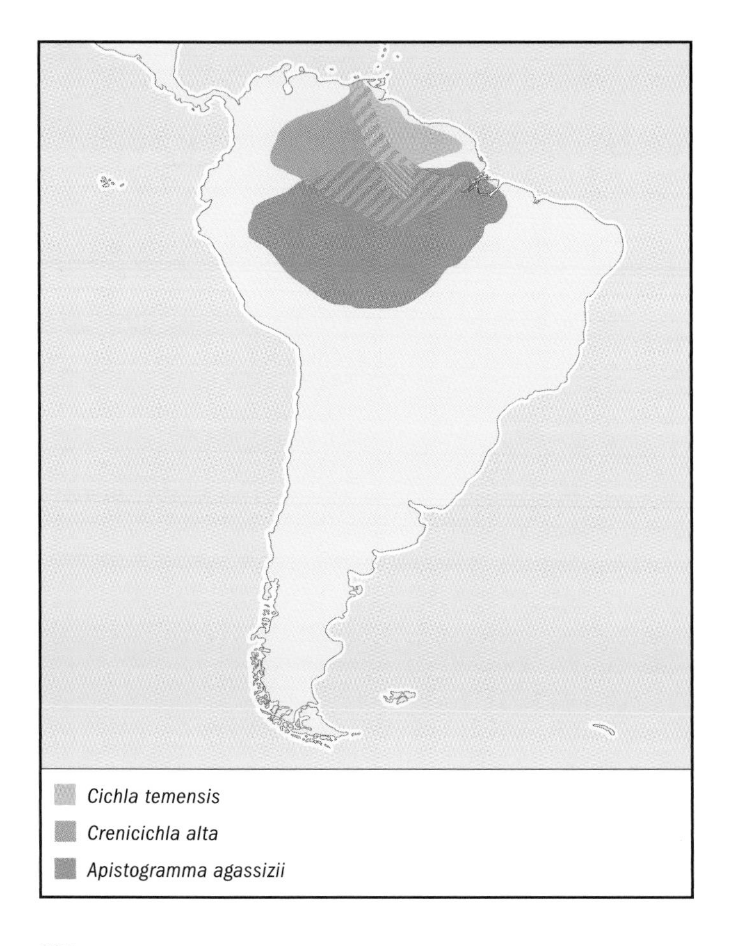

BEHAVIOR
Males exhibit ritualized behaviors in defense of their territories; females in the harem of a single male compete for his attention.

FEEDING ECOLOGY AND DIET
Feeds on aquatic macroinvertebrates.

REPRODUCTIVE BIOLOGY
Eggs laid in caves, or in flowerpots in aquaria; they hatch in about four days. Practices harem polygyny.

CONSERVATION STATUS
Not listed by the IUCN.

SIGNIFICANCE TO HUMANS
Owing to its brilliant colors and peaceable demeanor, this fish is popular in the aquarium trade. ◆

Giant cichlid
Boulengerochromis microlepis

FAMILY
Cichlidae

TAXONOMY
Tilapia microlepis Boulenger, 1899, Lake Tanganyika at Moliro, Democratic Republic of the Congo, Africa.

OTHER COMMON NAMES
English: Yellowbelly cichlid; German: Riesenbuntbarsch; Swahili: Kuhe.

PHYSICAL CHARACTERISTICS
Maximum length 25.6 in (65 cm); largest African cichlid. Background color yellowish, with numerous dark but faint vertical bars. Four dark blotches along middle of flanks. Caudal fin lunate.

DISTRIBUTION
Widely distributed in Lake Tanganyika, East Africa.

HABITAT
Deeper water over sand as well as open water, enters shallower waters to breed.

BEHAVIOR
Cruises open waters in pursuit of schools of small fishes.

FEEDING ECOLOGY AND DIET
Feeds on smaller fishes.

REPRODUCTIVE BIOLOGY
Substrate spawners, uses rocks or abandoned nests of other cichlids to deposit eggs. Both parents aggressively guard fry, have even been known to attack scuba divers who approached too close to a nest. Lays 5,000–12,000 eggs during spawning; eggs hatch in about three days. Some evidence suggests that pairs do not feed again after spawning, and spawn only once in a lifetime.

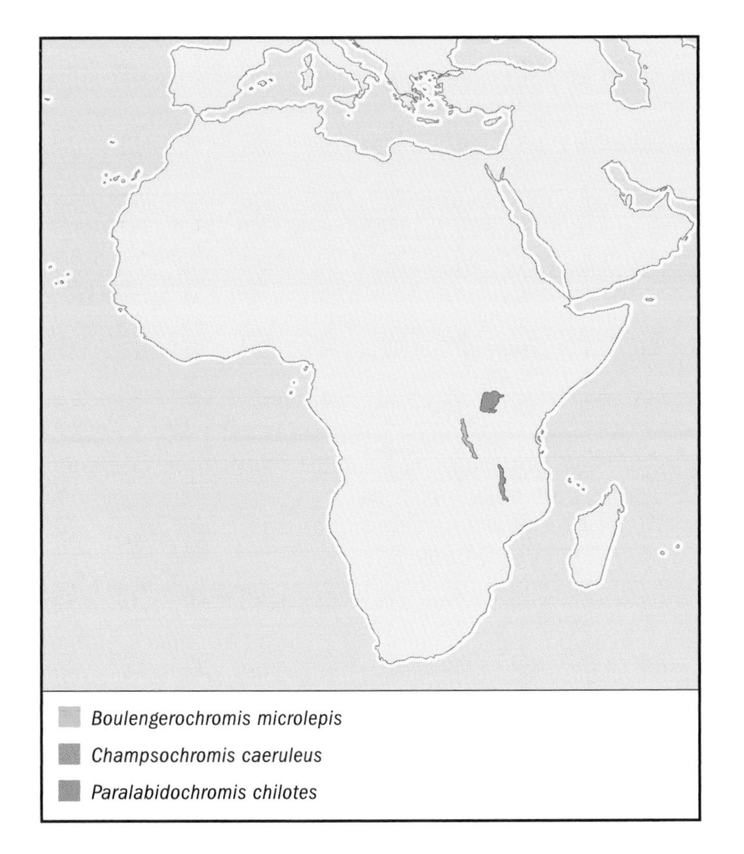

CONSERVATION STATUS
Not listed by the IUCN.

SIGNIFICANCE TO HUMANS
Popular sport fishes with excellent tasting flesh. ◆

Trout cichlid
Champsochromis caeruleus

FAMILY
Cichlidae

TAXONOMY
Paratilapia caerulea Boulenger, 1908, Lake Malawi.

OTHER COMMON NAMES
German: Forellen-Cichlide; Nyanja: Ndunduma.

PHYSICAL CHARACTERISTICS
Maximum length 12.6 in (32 cm). Large, predatory cichlid with slender body. Dorsal, anal, and pelvic fins somewhat elongate in adults. Immature males and females silver gray, with dark band running obliquely along uppermost part of body. Breeding males acquire brilliant greenish blue coloration, and edges of the dorsal and anal fins become orange.

DISTRIBUTION
Widely distributed in Lake Malawi, East Africa.

HABITAT
Inshore waters to depths of 180 ft (55 m).

BEHAVIOR
Juveniles school together, but adults usually cruise solitarily.

FEEDING ECOLOGY AND DIET
Feeds on smaller fishes.

REPRODUCTIVE BIOLOGY
Mouth brooder. Male forms a pit over sandy substrate. Female lays eggs in the pit and then collects them; eggs are fertilized in the female's mouth.

CONSERVATION STATUS
Not listed by the IUCN.

SIGNIFICANCE TO HUMANS
Locals occasionally catch these fish using baited hooks. ◆

Speckled pavon
Cichla temensis

FAMILY
Cichlidae

TAXONOMY
Cichla temensis Humboldt, 1821, Rio Temi, Venezuela.

OTHER COMMON NAMES
English: Peacock bass, tucunare; German: Humboldtcichlide; Spanish: Pavón.

PHYSICAL CHARACTERISTICS
Maximum length 39 in (99 cm); perhaps the world's largest cichlid. Elongate with large head, has superficial appearance of a bass. Spinous and soft-rayed dorsal fin subdivided. Background color yellow green to tan, with three dark vertical bars and a series of horizontal rows of cream-colored spots. Ocellus on caudal fin; red on pelvic, anal, and lower lobe of caudal fins. Males acquire nuchal hump during spawning.

DISTRIBUTION
Amazon Basin in Rio Negro and Rio Uatumã systems; Orinoco basin in Venezuela and Colombia; introduced in Florida and Texas.

HABITAT
Along banks of main river channels over sand and rocks, and in deeper inshore waters of lagoons.

BEHAVIOR
Voracious predators; in aquaria, they ignore any fishes that are too large to swallow.

FEEDING ECOLOGY AND DIET
Feeds on other fishes.

REPRODUCTIVE BIOLOGY
Monogamous substrate spawners. Both parents care for eggs and young. Eggs may number up to 4,000, and are laid on logs, rocks, or in excavated pits.

CONSERVATION STATUS
Not listed by the IUCN.

SIGNIFICANCE TO HUMANS
Popular sport and food fishes. Also found in the aquarium trade. ◆

Millet
Crenicichla alta

FAMILY
Cichlidae

TAXONOMY
Crenicichla alta Eigenmann, 1912, Gluck Island, Guyana.

OTHER COMMON NAMES
English: Spangled pike cichlid.

PHYSICAL CHARACTERISTICS
Maximum length 6.3 in (16 cm). Elongate, pikelike, with pointed snout. Lower jaw extends beyond upper. Color grayish on flanks, with dark stripe running from tip of snout to caudal fin. Abdomen rose colored, cheek yellow. Ocelli on caudal fin and shoulder; dorsal, anal, and caudal fins have black and white edges.

DISTRIBUTION
Rio Branco River drainage, Brazil, and Essequibo River drainage, Guyana.

HABITAT
Rivers and streams.

BEHAVIOR
Juveniles school together, adults are solitary.

FEEDING ECOLOGY AND DIET
Feeds on smaller fishes.

REPRODUCTIVE BIOLOGY
Spawns in caves, which males dig into the bank. Both parents care for eggs and fry.

CONSERVATION STATUS
Not listed by the IUCN.

SIGNIFICANCE TO HUMANS
Occasionally found in the aquarium trade. ◆

No common name
Paralabidochromis chilotes

FAMILY
Cichlidae

TAXONOMY
Paratilapia chilotes Boulenger, 1911, Jinja, Ripon Falls, Uganda.

OTHER COMMON NAMES
German: Viktoria-Wulstlippen-Maulbrüter.

PHYSICAL CHARACTERISTICS
Maximum length 5.8 in (14.8 cm). Elongate, unicuspid, forward-directed teeth used like forceps to capture insect larvae. Lips dramatically swollen and enlarged. Color pattern varies geographically; most populations are yellow-gray over most of body. Males with green and blue on flanks, orange on dorsal, anal, and caudal fins, and chest. Females sometimes exhibit piebald coloration.

DISTRIBUTION
Lake Victoria, East Africa.

HABITAT
Sheltered inshore waters with rocky bottom to 55.8 ft (17 m) depth

BEHAVIOR
In aquaria, observed digging under overhanging rocks for prey. Presses large lips against cracks in rocks when feeding. Not as territorial as many other haplochromines.

FEEDING ECOLOGY AND DIET
Feeds mainly on mayfly larvae, though also eats larvae of Trichoptera (caddisflies) and Diptera (true flies).

REPRODUCTIVE BIOLOGY
Mouth-brooder.

CONSERVATION STATUS
Listed as Vulnerable by the IUCN.

SIGNIFICANCE TO HUMANS
Found in the aquarium trade; one of the Lake Victoria haplochromines that has survived the Nile perch boom and other perturbations that have befallen the lake. ◆

No common name
Lepidiolamprologus kendalli

FAMILY
Cichlidae

TAXONOMY
Lamprologus kendalli Poll and Stewart, 1977, Lake Tanganyika, northwest of Mutondwe Island, Zambia.

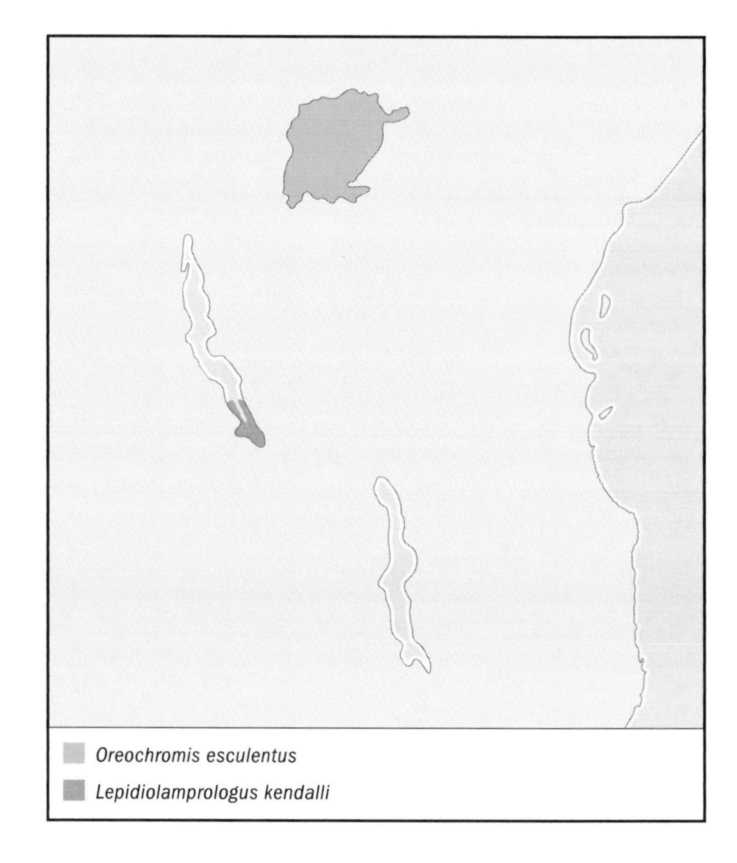

OTHER COMMON NAMES
German: Kendalls Tanganjikasee-Buntbarsch.

PHYSICAL CHARACTERISTICS
Maximum length 6.3 in (16 cm). Elongate and torpedo shaped. Dark background color, with white blotching on head and fins, and white blotches forming several almost continuous horizontal bands on flanks. Hints of blue on face, pectoral fins yellowish.

DISTRIBUTION
Southern part of Lake Tanganyika, East Africa.

HABITAT
Shallow rocky areas as well as deeper waters, up to about 148 ft (45 m).

BEHAVIOR
Adults are solitary, and cover large distances in search or prey.

FEEDING ECOLOGY AND DIET
Feeds on other fishes.

REPRODUCTIVE BIOLOGY
Substrate spawner. Females may release up to 500 eggs. In the wild, spawns at a depth of 148 ft (45 m) over a rocky bottom covered with thick sediment.

CONSERVATION STATUS
Not listed by the IUCN.

SIGNIFICANCE TO HUMANS
Exported for the aquarium trade. ◆

Ngege
Oreochromis esculentus

FAMILY
Cichlidae

TAXONOMY
Tilapia esculenta Graham, 1928, Kenya, Lake Victoria.

OTHER COMMON NAMES
None known.

PHYSICAL CHARACTERISTICS
Maximum length 19.7 in (50 cm). Relatively deep bodied, with small head. Color olive-brown to dull green, becoming whitish ventrally. Breeding males become red dorsally and black ventrally, except that dorsal fin is also black.

DISTRIBUTION
Lake Victoria, Lake Nabugabo, Lake Kioga, Lake Kwania, the Victoria Nile above Murchison falls, and the Malawa River. Also Lake Gangu, west of Lake Victoria. Introduced into Tanzanian and Ugandan reservoirs.

HABITAT
Inshore waters with aquatic vegetation when young, move into open water as adults, to depths of 164 ft (50 m). Found over muddy bottoms.

BEHAVIOR
Adults school together in open water, following plankton blooms.

FEEDING ECOLOGY AND DIET
Plankton feeder, follows diatom blooms in lakes. Secretes mucous in the mouth that helps trap plankton, forming a bolus for swallowing.

REPRODUCTIVE BIOLOGY
Mouth brooder; fertilization takes place in female's mouth. Spawns year round. Males aggregate around spawning grounds, where they occupy basin-shaped nests and defend a small territory. Once eggs are fertilized, females move to weed beds with the brood.

CONSERVATION STATUS
Listed as Vulnerable by the IUCN.

SIGNIFICANCE TO HUMANS
Historically an important food fishery in Lake Victoria, but was overfished to the point of commercial extinction in the lake, where introduced tilapiines now predominate. ◆

Freshwater angelfish
Pterophyllum scalare

FAMILY
Cichlidae

TAXONOMY
Zeus scalaris Lichtenstein, 1823, Brazil.

OTHER COMMON NAMES
English: Black angelfish, longfin angelfish, veil angelfish; French: Poisson ange, scalaire; German: Dumerils, Perlscalar, Segelflosser; Spanish: Pez angel.

PHYSICAL CHARACTERISTICS
Maximum length 3 in (7.5 cm). Very deep bodied and laterally compressed, with diamond-shaped body. Dorsal, anal, and

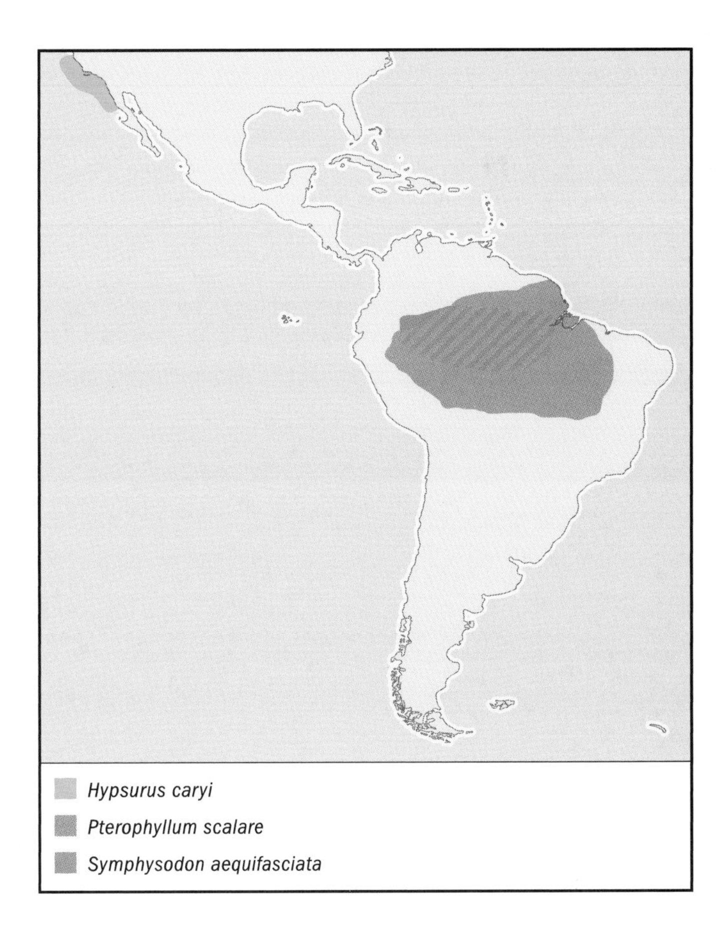

pelvic fins very elongate. Adults silvery in color with four dark vertical bars; juveniles have seven bars.

DISTRIBUTION
Widely distributed in the Amazon basin; found in Peru, Colombia, and Brazil in the Ucayali, Solimões and Amazonas Rivers. Also in the rivers of Amapá, Brazil, the Oyapock in French Guiana, and the Essequibo River in Guyana.

HABITAT
Lakes, swamps, and flooded forests with dense aquatic vegetation and minimal current.

BEHAVIOR
Peaceable, gregarious fishes that take refuge in aquatic vegetation.

FEEDING ECOLOGY AND DIET
Feeds mainly on benthic crustaceans, including shrimps, prawns, and *Artemia nauplii.*

REPRODUCTIVE BIOLOGY
Forms monogamous pairs. Female spawns on thick plant leaves, and both parents care for eggs and young. Male and female nibble at eggs to release wrigglers about a day and a half after spawning, and may move brood to a pit in the substrate.

CONSERVATION STATUS
Not listed by the IUCN.

SIGNIFICANCE TO HUMANS
Exceedingly popular in the aquarium trade, with millions of specimens sold every year; numerous color and fin varieties have been selectively bred. ◆

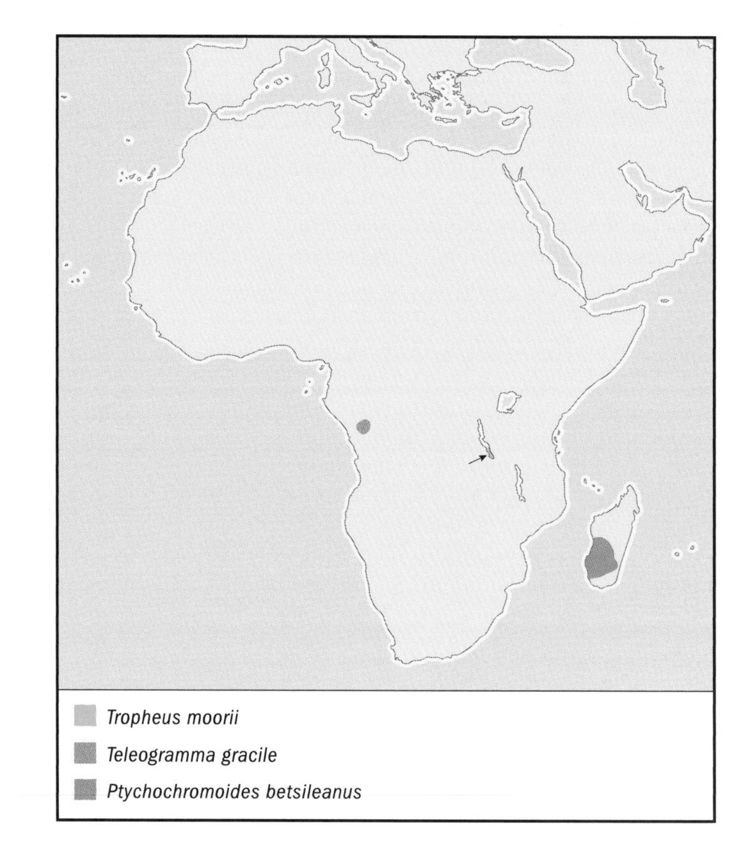

Trondo mainty
Ptychochromoides betsileanus

FAMILY
Cichlidae

TAXONOMY
Tilapia betsileana Boulenger, 1899, Betsileo, Madagascar.

OTHER COMMON NAMES
French: Marahrely à bosse.

PHYSICAL CHARACTERISTICS
Maximum length 9.4 in (24 cm). A fairly deep-bodied, robust fish, usually with a well-developed occipital hump. Color black or dark gray, without stripes or bars. Females appear to be speckled and lighter in color, with white blotches on the cheek.

DISTRIBUTION
Central highlands of Madagascar, from the Betsileo region to Lake Itasy in the Merina district.

HABITAT
Cool, clear, well-oxygenated waters. In rivers, often found near rapids and waterfalls; found in deep waters with rocky substrate in Lake Itasy.

BEHAVIOR
Very little is known, except that the species seeks out riffles with rocky substrate.

FEEDING ECOLOGY AND DIET
Omnivorous, feeds on shrimp, aquatic insect larvae, aquatic plants, and phytoplankton.

REPRODUCTIVE BIOLOGY
Spawns on substrate; both parents care for fry for an extended period. Reproduces from October to early November. Females lay many hundreds of eggs, preferably on large rocks.

CONSERVATION STATUS
Listed as Critically Endangered by the IUCN.

SIGNIFICANCE TO HUMANS
Human encroachment has caused the extinction of this species in the northern part of its range, and the remaining populations are decimated. ◆

Blue discus
Symphysodon aequifasciata

FAMILY
Cichlidae

TAXONOMY
Symphysodon discus aequifasciata Pellegrin, 1904, Teffé and Santarém, Brazil.

OTHER COMMON NAMES
English: Brown discus, green discus; German: Blauer Diskus, Grüner Diskus, Scheibenbarsch; Spanish: Disco azul, pez disco.

PHYSICAL CHARACTERISTICS
Maximum length 5.5 in (14 cm). Extremely laterally compressed, almost perfectly round, dinner-plate shaped. Background yellow-green to tan, with brilliant undulating streaks of turquoise blue and numerous dark vertical bands running in parallel.

DISTRIBUTION
Amazon basin from the Rio Putumayo in Colombia and Peru to the Rio Tocantins in Brazil.

HABITAT
Calm waters in and around rock crevices, roots, and aquatic vegetation.

BEHAVIOR
Rather shy, schooling fish, most comfortable in or near tangles of submerged roots and branches or rock crevices. Becomes somewhat territorial during spawning.

FEEDING ECOLOGY AND DIET
Feeds on insects and insect larvae, as well as zooplankton.

REPRODUCTIVE BIOLOGY
Forms monogamous pairs in which both parents care for the young. Females lay clutches of about 100 eggs, which hatch in roughly 60 hours, on branches or sturdy leaves. Within 4–5 days, fry become mobile and begin feeding on mucous secreted from flanks of the parents, showing preference for the male. In captivity, this mode of feeding can last as long as 8 weeks, although other foods supplement the habit.

CONSERVATION STATUS
Not listed by the IUCN.

SIGNIFICANCE TO HUMANS
Extremely popular in the aquarium trade; many captive-bred strains exist. ◆

Worm cichlid
Teleogramma gracile

FAMILY
Cichlidae

TAXONOMY
Teleogramma gracile Boulenger, 1899, Matadi, Democratic Republic of Congo, Africa.

OTHER COMMON NAMES
None known.

PHYSICAL CHARACTERISTICS
Maximum length 3.2 in (8 cm). Very elongate, cigar shaped; teeth small and unicuspid, with a few enlarged canines anteriorly. Uncharacteristically for cichlids, has single, continuous row of lateral line scales, instead of upper and lower divisions. Brownish in color, with thin red margin to dorsal fin and upper lobe of caudal fin. Females exhibit broad black margin above red patch in upper lobe of caudal fin.

DISTRIBUTION
Lower Congo River rapids.

HABITAT
Turbid, fast-flowing waters in lower Congo.

BEHAVIOR
Displays extreme territoriality toward conspecifics in aquaria. Hides under rocks and in caves.

FEEDING ECOLOGY AND DIET
Feeds on aquatic macroinvertebrates.

REPRODUCTIVE BIOLOGY
Spawns in caves, female tends to eggs. Polygynous.

CONSERVATION STATUS
Not listed by the IUCN.

SIGNIFICANCE TO HUMANS
Very rarely exported for the aquarium trade. ◆

Blunthead cichlid
Tropheus moorii

FAMILY
Cichlidae

TAXONOMY
Tropheus moorii Boulenger, 1898, Kinyamkolo, Lake Tanganyika, Africa.

OTHER COMMON NAMES
English: Brabant cichlid, moorii; German: Brabantbuntbarsch.

PHYSICAL CHARACTERISTICS
Maximum length 5.7 in (14.5 cm). Moderately deep bodied, with blunt head and snout and down-turned mouth. Bicuspid outer row of teeth, tightly spaced. Exhibits wide range of geographical color variation, with populations ranging from uniformly dark brown with vertical bars to brilliant yellow with red fins.

DISTRIBUTION
Patchily distributed along the southern shores of Lake Tanganyika, Africa.

HABITAT
Shallow inshore waters with rocky substrate.

BEHAVIOR
Sometimes forms large schools composed of hundreds of individuals, which patrol rocky habitats in search of algal mats. Males are territorial, and will defend feeding territories.

FEEDING ECOLOGY AND DIET
Tears strands of algae from rocks.

REPRODUCTIVE BIOLOGY
Maternal mouth brooder; after laying 5–17 eggs, female collects them in her mouth and positions herself near male's genital area to receive sperm. Fertilization takes place in female's mouth. Eggs hatch after about four weeks. Females continue to guard offspring for a few days after they are first released from the mouth.

CONSERVATION STATUS
Not listed by the IUCN.

SIGNIFICANCE TO HUMANS
Popular in the aquarium trade. ◆

Rainbow seaperch
Hypsurus caryi

FAMILY
Embiotocidae

TAXONOMY
Embiotoca caryi Agassiz, 1853, San Francisco Bay, California, United States.

OTHER COMMON NAMES
English: Rainbow surfperch; Spanish: Perca.

PHYSICAL CHARACTERISTICS
Maximum length 11.8 in (30 cm). Deep bodied and laterally compressed, lateral line scales uninterrupted. Red and blue stripes on body and about 10 reddish brown bars on upper part of flanks. Pelvic and anal fins reddish orange with blue edges.

DISTRIBUTION
Pacific coast from Cape Mendocino, California, United States, to Isla San Martin, Baja California.

HABITAT
Around rocky shores, reefs, piers, and kelp beds, to depths of 132 ft (40.2 m).

BEHAVIOR
In the fall, gathers in large aggregations prior to breeding; females move into shallow coastal waters to give birth the following summer. Sometimes function as cleaner for other fishes by picking off parasites.

FEEDING ECOLOGY AND DIET
Feeds during day; eats mostly amphipods and copepods. Described as an oral winnower, meaning it takes mouthfuls of sand or turf, selectively removes food items, and spits out the remainder.

REPRODUCTIVE BIOLOGY
Viviparous; nutrient and gas exchange occurs between mother and the developing embryos. Females gravid from April until mid-September, and carry 9–22 young. Juveniles born at lengths of about 2.7 in (6 cm). Males develop swelling in tissues surrounding first anal fin spine during breeding, creating an intromittent (copulatory) organ used in internal fertilization. After mating, sperm are stored for several months in female's ovarian cavity prior to fertilization.

CONSERVATION STATUS
Not listed by the IUCN.

SIGNIFICANCE TO HUMANS
Minor commercial importance as food fishes. ◆

Resources

Books

Barlow, G. W. "Mating Systems Among Cichlid Fishes." In *Cichlid Fishes: Behaviour, Ecology and Evolution*, edited by M. H. A. Keenleyside. London: Chapman & Hall, 1991.

———. *The Cichlid Fishes: Nature's Grand Experiment in Evolution.* Cambridge: Perseus, 2000.

Boschung, H. T., Jr., J. D. Williams, D. W. Gotshall, D. K. Caldwell, and M. C. Caldwell. *The Audubon Society Field Guide to North American Fishes, Whales, and Dolphins.* New York: Alfred A. Knopf, 1983.

Bugenyi, F. W. B., and K. M. Magumba. "The Present Physicochemical Ecology of Lake Victoria, Uganda." In *The Limnology, Climatology and Paleoclimatology of the East African Lakes*, edited by T. C. Johnson and E. O. Odada. Amsterdam: Gordon and Breach, 1996.

Eccles, D. H., and E. Trewavas. *Malawian Cichlid Fishes: The Classification of Some Haplochromine Genera.* Herten, West Germany: Lake Fish Movies, 1989.

Fryer, G., and T. D. Iles. *The Cichlid Fishes of the Great Lakes of Africa.* Edinburgh: Oliver and Boyd, 1972.

Harrison, I. J., and M. L. J. Stiassny. "The Quiet Crisis: A Preliminary Listing of the Freshwater Fishes of the World That Are Extinct or 'Missing in Action.'" In *Extinctions in Near Time*, edited by R. D. E. MacPhee. New York: Kluwer Academic/Plenum, 1999.

Katunzi, E. F. B. "A Review of Lake Victoria Fisheries with Recommendations for Management and Conservation." In *The Limnology, Climatology and Paleoclimatology of the East African Lakes*, edited by T. C. Johnson and E. O. Odada. Amsterdam: Gordon and Breach, 1996.

Kawanabe, H., M. Hori, and M. Nagoshi, eds. *Fish Communities in Lake Tanganyika.* Kyoto: Kyoto University Press, 1997.

Konings, A. *Malawi Cichlids in Their Natural Habitat.* El Paso: Cichlid Press, 1995.

———. *Tanganyika Cichlids in their Natural Habitat.* El Paso: Cichlid Press, 1998.

Kudhongania, A. W., D. L. Ocenodongo, and J. O. Okaronon. "Anthropogenic Perturbations on the Lake Victoria Ecosystem." In *The Limnology, Climatology and Paleoclimatology of the East African Lakes*, edited by T. C. Johnson and E. O. Odada. Amsterdam: Gordon and Breach, 1996.

Loiselle, P. V. *The Cichlid Aquarium.* Melle, Germany: Tetra-Press, 1985.

Nelson, J. S. *Fishes of the World,* 3rd edition. New York: John Wiley & Sons, 1994.

Patterson, C. "Osteichthyes: Teleostei." In *The Fossil Record 2*, edited by M. J. Benton. London: Chapman & Hall, 1993.

Paxton, J. R., and W. N. Eschmeyer, eds. *Encyclopedia of Fishes*, 2nd edition. San Diego: Academic Press, 1998.

Ribbink, A. J. "Distribution and Ecology of the Cichlids of the African Great Lakes." In *Cichlid Fishes: Behavior, Ecology, and Evolution*, edited by M. H. A. Keenleyside. London: Chapman & Hall, 1991.

Riehl, R., and H. A. Baensch. *Aquarium Atlas.* Melle, Germany: Baensch, 1986.

Rossiter, A. "The Cichlid Fish Assemblages of Lake Tanganyika: Ecology, Behaviour and Evolution of Its Species Flocks." In *Advances in Ecological Research.* Vol. 26,

Resources

edited by M. Begon and A. H. Fitter. London: Harcourt Brace and Company, 1995.

Stiassny, M. L. J. "Phylogenetic Intrarelationships of the Family Cichlidae: An Overview." In *Cichlid Fishes: Behavior, Ecology and Evolution*, edited by M. H. A. Keenleyside. London: Chapman & Hall, 1991.

Witte, F., P. C. Goudswaard, E. F. B. Katunzi, O. C. Mkumbo, O. Seehausen, and J. H. Wanink. "Lake Victoria's Ecological Changes and Their Relationships with the Riparian Societies." In *Ancient Lakes: Their Cultural and Biological Diversity*, edited by H. Kawanabe, G. W. Coulter, and A. C. Roosevelt. Ghent: Kenobi Productions, 1999.

Periodicals

Arnegard, M. E., J. A. Markert, P. D. Danley, J. R. Stauffer, Jr., A. J. Ambali, and T. D. Kocher. "Population Structure and Color Variation of the Cichlid Fish *Labeotropheus fuelleborni* Ahl Along a Recently Formed Archipelago of Rocky Habitat Patches in Southern Lake Malawi." *Proceedings of the Royal Society of London*, Series B, 266 (1999): 119–130.

Bernardi, G., and G. Bucciarelli. "Molecular Phylogeny and Speciation of the Surfperches (Embiotocidae, Perciformes)." *Molecular Phylogenetics and Evolution* 13 (1999): 77–81.

Casciotta, J., and G. Arratia. "Tertiary Cichlid Fishes from Argentina and Reassessment of the Phylogeny of New World Cichlids (Perciformes: Labroidei)." *Kaupia Darmstaedter Beitraege Zur Naturgeschichte* 2 (1993):195–240.

Cohen, A. S. "Extinction in Ancient Lakes: Biodiversity Crises and Conservation 40 years After J. L. Brooks." *Advances in Limnology* 44 (1994): 451–479.

Coulter, G. W., and R. Mubamba. "Conservation in Lake Tanganyika, with Special Reference to Underwater Parks." *Conservation Biology* 7, no. 3 (1993): 678–685.

Drucker, E. G., and J. S. Jensen. "Functional Analysis of a Specialized Prey Processing Behavior: Winnowing by Surfperches (Teleostei: Embiotocidae)." *Journal of Morphology* 210 (1991): 267–287.

Galis, F., and J. A. J. Metz. "Why Are There So Many Cichlid Species?" *Trends in Ecology and Evolution* 13, no. 1 (1998): 1–2.

Hori, M. "Frequency-Dependent Natural Selection in the Handedness of Scale-Eating Cichlid Fish." *Science* 260 (1993): 216–219.

Kaufman, L. S. and K. F. Liem. "Fishes of the Suborder Labroidei (Pisces: Perciformes): Phylogeny, Ecology, and Evolutionary significance." *Breviora* 472 (1982): 1–19.

Liem, K. F. "The Pharyngeal Jaw Apparatus of the Embiotocidae (Teleostei): A Functional and Evolutionary Perspective." *Copeia* 2 (1986): 311–323.

Murray, A. M. "The Fossil Record and Biogeography of the Cichlidae (Actinopterygii: Labroidei)." *Biological Journal of the Linnean Society* 74 (2001): 517–532.

———. "The Oldest Fossil Cichlids (Teleostei: Perciformes): Indication of a 45 Million-Year-Old Species Flock." *Proceedings of the Royal Society of London*, Series B, 268 (2001): 679–684.

Ogutu-Ohwayo, R. "The Decline of the Native Fishes of Lakes Victoria and Kyoga (East Africa) and the Impact of Introduced Species, Especially the Nile Perch, *Lates niloticus*, and the Nile Tilapia, *Oreochromis niloticus*." *Environmental Biology of Fishes* 27 (1990): 81–96.

Poll, M. "Classification des Cichlidae du Lac Tanganika: Tribus, Genres et Espèces." *Académie Royale de Belgique. Mémoires de la classe des sciences* 45 (1986): 1–163.

Reinthal, P. N., and M. L. J. Stiassny. "Revision of the Madagascan Genus *Ptychochromoides* (Teleostei: Cichlidae), with Description of a New Species." *Ichthyological Exploration of Freshwaters* 7 (1997): 353–368.

Schliewen, U. K., D. Tautz, and S. Pääbo. "Sympatric Speciation Suggested by Monophyly of Crater Lake Cichlids." *Nature* 368 (1994): 629–632.

Seehausen, O., J. J. M. van Alphen, and F. Witte. "Cichlid Fish Diversity Threatened by Eutrophication That Curbs Sexual Selection." *Science* 277 (1997): 1,808–1,811.

Stiassny, M. L. J. "The Phyletic Status of the Family Cichlidae (Pisces, Perciformes): A Comparative Anatomical investigation." *Netherlands Journal of Zoology* 31, no. 2 (1981): 275–314.

———. "Cichlid Familial Intrarelationships and the Placement of the Neotropical Genus *Cichla* (Perciformes, Labroidei)." *Journal of Natural History* 21 (1987): 1311–1331.

Stiassny, M. L. J., and J. S. Jensen. "Labroid Intrarelationships Revisited: Morphological Complexity, Key Innovations, and the Study of Comparative Diversity." *Bulletin of the Museum of Comparative Zoology* 151, no. 5 (1987): 269–319.

Streelman, J. T., and S. A. Karl. "Reconstructing Labroid Evolution with Single-Copy Nuclear DNA." *Proceedings of the Royal Society of London*, Series B, 264 (1997): 1,011–1,020.

Sturmbauer, C. "Explosive Speciation in Cichlid Fishes of the African Great Lakes: A Dynamic Model of Adaptive Radiation." *Journal of Fish Biology*, Supplement A, 53 (1998): 18–36.

Thompson, A. B., E. H. Allison, and B. P. Ngatunga. "Distribution and Breeding Biology of Offshore Cichlids in Lake Malawi/Niassa." *Environmental Biology of Fishes* 47 (1996): 235–254.

Turner, G. F., O. Seehausen, M. E. Knight, C. J. Allender, and R. L. Robinson. "How Many Species of Cichlid Fishes Are There in African Lakes?" *Molecular Ecology* 10 (2001): 793–806.

Robert Schelly, MA

Labroidei II

(Damselfishes, wrasses, parrotfishes, and rock whitings)

Class Actinopterygii
Order Perciformes
Suborder Labroidei
Number of families 4

Photo: Threespot dascyllus (*Dascyllus trimaculatus*) pair spawning over coral rocks near the Florida Islands, Solomon Islands. (Photo by Fred McConnaughey/Photo Researchers, Inc. Reproduced by permission.)

Evolution and systematics

The four families in this grouping are part of the suborder Labroidei. The families are: Labridae (the wrasses); Scaridae (the parrotfishes); Pomacentridae (the damselfishes); and Odacidae (the rock whitings, or butterfishes, of western Pacific waters; the butterfishes of North America represent a different family).

The family Labridae is a large one, with approximately 500 species in 60 genera. The next largest of the four families is the Pomacentridae, with more than 320 species in 28 genera, followed by the Scaridae, with 83 species in 9 genera, and finally the Odacidae, with 12 species in 4 genera.

The taxonomy of these four families is under dispute, and several studies are under way to iron out the relationships. In fact, some taxonomists feel the scarids and odacids, which are believed to have evolved from labrids, should actually be listed as subfamilies within Labridae. Other researchers have focused their work on the evolutionary relationships among genera within families. As the genetic makeup of individual species becomes a more prominent tool in defining evolutionary relationships, there is little doubt that some shaking looms ahead for these branches of the tree of life.

Physical characteristics

This grouping of fishes is varied, but vibrant color is a trademark of many species in the four families, and has made wrasses and damselfishes the highlights of reef-diving excursions. In addition, the scarids, labrids, and odacids all share unusual pharyngeal modifications that serve as additional chewing surfaces to grind kelp, hard-shelled invertebrates, and other dietary items.

Generalizations can also be made about each of the four families. Most labrids, or wrasses, have gapped and outward-projecting teeth on protrusible upper and lower jaws. The dorsal fin typically has less than 15 spines; some have as many as 21 spines and 6–21 soft rays. The anal fin has 4–6 spines and 7–18 soft rays. Other than these characteristics, little else is noticeably uniform among this highly diverse family. Maximum adult size in the family ranges from about 2 in (5 cm) in a variety of species, to more than 8 ft (2.5 m) in the humphead wrasse (*Cheilinus undulatus*).

A brindlebass (*Epinephelus lanceoletus*) having its teeth cleaned by a bluestreak cleaner wrasse (*Labroides dimidiatus*). (Photo by Mark Smith/Photo Researchers, Inc. Reproduced by permission.)

The pomacentrids, or damselfishes, are characterized by a deep, compressed body; a small mouth with short canine teeth; a usually two-spined anal fin; one dorsal fin with spines numbering 8–17, and 10–18 rays; and a lack of teeth on the palate. Maximum adult size ranges from 1.6 in (4 cm) in some *Chromis* and *Chrysiptera* species, to over 12 in (30 cm) in the garibaldi (*Hypsypops rubicundus*) and giant damselfish (*Microspathodon dorsalis*).

Scarids, or parrotfishes, have a rather distinctive "parrot's beak" resulting from a fusion of the teeth and often fleshy lips. The dorsal fin has 9 spines and 10 soft rays, the anal fin has 3 spines and 9 soft rays, and the pectoral fin has one spine and 5 soft rays. Large scales are evident even from some distance.

The small Odacidae family is still quite diverse, and is occasionally described as being intermediate between the scarids and labrids. The jaw teeth are fused in a manner similar to the parrotfishes, but they have a different dorsal fin pattern, with 12–23 spines, compared to the 9 spines in scarids. Odacids, the rock whitings or butterfishes, have the more elongate body seen in many labrids, but do not have the protrusible jaws.

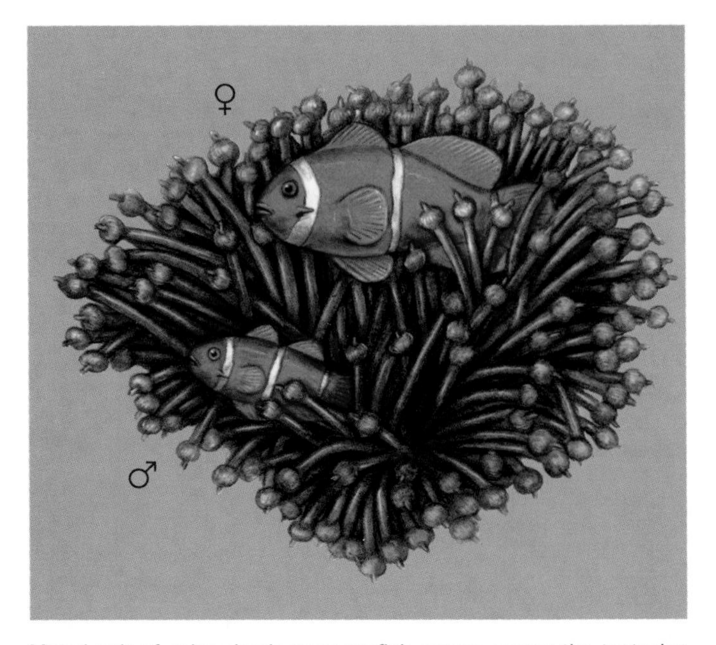

Mated pair of spine-cheek anemonefish moves among the tentacles of its host, the bubble tip anemone. (Illustration by Joseph E. Trumpey)

Distribution

Parrotfishes, wrasses, and damselfishes occupy mainly tropical, marine waters of the Atlantic, Pacific, and Indian Oceans, but some species, particularly wrasses, extend well into cooler waters, and a few damselfishes exist in estuaries and occasionally venture into fresh water. The rock whitings are found in the temperate, coastal waters of Australia and New Zealand.

Habitat

This group comprises primarily reef-associated fishes, although a few make their homes in the open sea. The reefs provide a source of refuge, as many of the species use the nooks and crannies in the coral as daytime hiding places from predators. A number of species also gain nourishment directly from the coral by nipping off polyps and grinding them with their pharyngeal jaws. Some species, particularly the juveniles, prefer the shallower, protected waters of lagoons and bays, and the adults of many reside in waters of steep, outer, reef slopes.

A number of parrotfishes live among beds of sea grass or in rocky reefs, and wrasses frequently require living coral reefs, but some do very well among dead coral. Damselfishes generally live in rocky and coral reefs, but individual species may prefer open water as deep as 300 ft (91 m) or more, or shallow habitats of as little as 12 ft (3.7 m) in either the open sea or in sheltered areas with calmer waters. Some species within these four families prefer very specific habitats. Anemonefishes (genera *Amphiprion* and *Premnis*), for example, may require the presence of a particular species of anemone.

A threespot damselfish (*Stegastes planifrons*) in its nest amidst fire coral near Bonaire, Netherlands Antilles. (Photo by Charles V. Angelo/ Photo Researchers, Inc. Reproduced by permission.)

Behavior

Two of the most well-known behaviors of this group of fishes involve that of certain damselfishes and their symbiotic relationship with large sea anemones, and that of parrotfishes, which produce a mucus "cocoon" that surrounds them while they sleep at night.

Known as anemonefishes, or clownfishes, the 30 damselfishes in the genera *Amphiprion* and *Premnas* move among and between the dangerous tentacles of the sea anemone, even sleeping within them. Stinging cells (nematocysts) on the surface of the tentacles normally sting and paralyze prey fish for easy consumption by the anemone. The anemonefishes, however, live peacefully among the anemones. Current consensus is that a protective mucus coat on the anemonefish shields it from the sting of the tentacle. The relationship is symbiotic because both species gain from it. The anemonefish receives shelter from predators, while helping to maintain the health of the anemone by picking off organisms and detritus from its tentacles. The relationship is more important to the anemonefishes than to the anemones, because the fishes are unable to escape predation without the sanctuary of the anemone, but anemones can and do survive in the absence of the fishes.

Parrotfishes, on the other hand, have their own presumably protective behavior. At night, these diurnal animals prepare for sleep by generating a tube of mostly clear mucus that surrounds the body. The tube forms in about 30–60 minutes. There is some debate over whether cocoon formation is a behavioral trait or simply the result of normal mucus secretions that accumulate because the fish is stationary. Either way, the mucus tube appears to provide the fishes with some protection from predators by masking them, and perhaps by surrounding them with an unpleasant-tasting barrier.

A parrotfish sleeps in a mucous cocoon in the reefs of Fiji. (Photo by Stuart Westmorland/CORBIS. Reproduced by permission.)

Feeding ecology and diet

Herbivores and carnivores exist within this group, with the diet of many species comprising both plants and animals. Feeding habits vary. The odacids are mainly herbivores, with a few dining on kelp, a coarse seaweed shunned by most other species. One study of the butterfish *Odax pullus* showed that this herbivorous odacid has an active symbiotic relationship with gut microbes that assists in the digestion of the plant matter. Several damselfishes hold feeding territories and even cultivate algal beds. Many scarids and odacids are herbivorous grazers, using their parrotlike beaks to scrape algae from rocks and coral, or to crush open a hard-shelled invertebrate, such as a mussel or sea urchin. A few even eat coral polyps. Some parrotfishes are so aggressive in their feeding habits that they are considered a cause of reef erosion, as they not only alter the reef, but also excrete a great deal of silty sediment that coats the reef structure. As a rule, species that dine on coral, mollusks, and other crunchy prey have more rounded teeth attuned to grinding. The teeth of species that favor softer dietary items are more pointed.

A slingjaw wrasse (*Epibulus insidiator*) sucks up prey, a humbug damsel (*Dascyllus aruanus*), through its long jaws. (Illustration by Marguette Dongvillo)

The wrasses are primarily carnivores, and generally choose from a variety of hard-shelled, invertebrate prey. Some, however, prefer that their food come to them, and take their diet of plankton from the water column. One of the most unusual feeding behaviors in this group is seen in the cleaner wrasses.

A cleaner wrasse cleans a fish (left inset). A cleaner mimic is a blenny that mimics the cleaner wrasse, but instead of cleaning the larger fish, it rips out a gill filament to eat (right inset). (Illustration by Brian Cressman)

This group of small fishes in the genus *Labroides* have specific sites where they provide cleaning services to other fishes. Fishes come to these so-called cleaning stations, and announce their desire to be cleaned by exhibiting stereotyped behaviors through movements of their mouths or bodies. The cleaner wrasses strike a deal by responding with their own behaviors, including brushing the "client" with their fins, and the cleaning begins. The wrasses pick over the body, fins, and head of the client fish, and may even enter the gill chamber and mouth to remove crustacean ectoparasites, mucus, dead skin, dislodged scales, and other detritus. Both client and wrasse benefit from the arrangement: the cleaner is fed, and the client is cleaned.

Other wrasses also provide cleaning services, particularly juveniles of the genus *Thalassoma.* A study of Noronha wrasse (*T. noronhanum*) and a client fish, the piscivorous coney (*Cephalopholis fulva*) indicated that client fishes sometimes take advantage of the situation and eat the cleaner. In this study, scientists observed predation in two instances, both of which occurred when the wrasse was tending the coney away from its normal cleaning station.

Fishes in these four families may fall prey to larger bony fishes. The primary predators include larger serranids, synodontids, aulostomids, and members of other families.

Reproductive biology

Perhaps most notable characteristics of this group of fishes are the three separate color phases and sex reversal associated with many species. The typical three-phase lifestyle begins with a juvenile (immature) phase, then an adult initial phase, and finally the terminal phase. Each has a distinctive appearance that differs from species to species. In the bluehead wrasse (*Thalassoma bifasciatum*), for example, juveniles are bright yellow, initial-phase adults are yellow with black stripes, and terminal-phase males are deep blue with a green rear and midbody, and bold, black and white bands behind the head. The initial-phase adults in many species are almost all females, and in most cases, the few initial-phase males look the same as their female counterparts. The initial and terminal phases are usually so different that inexperienced divers

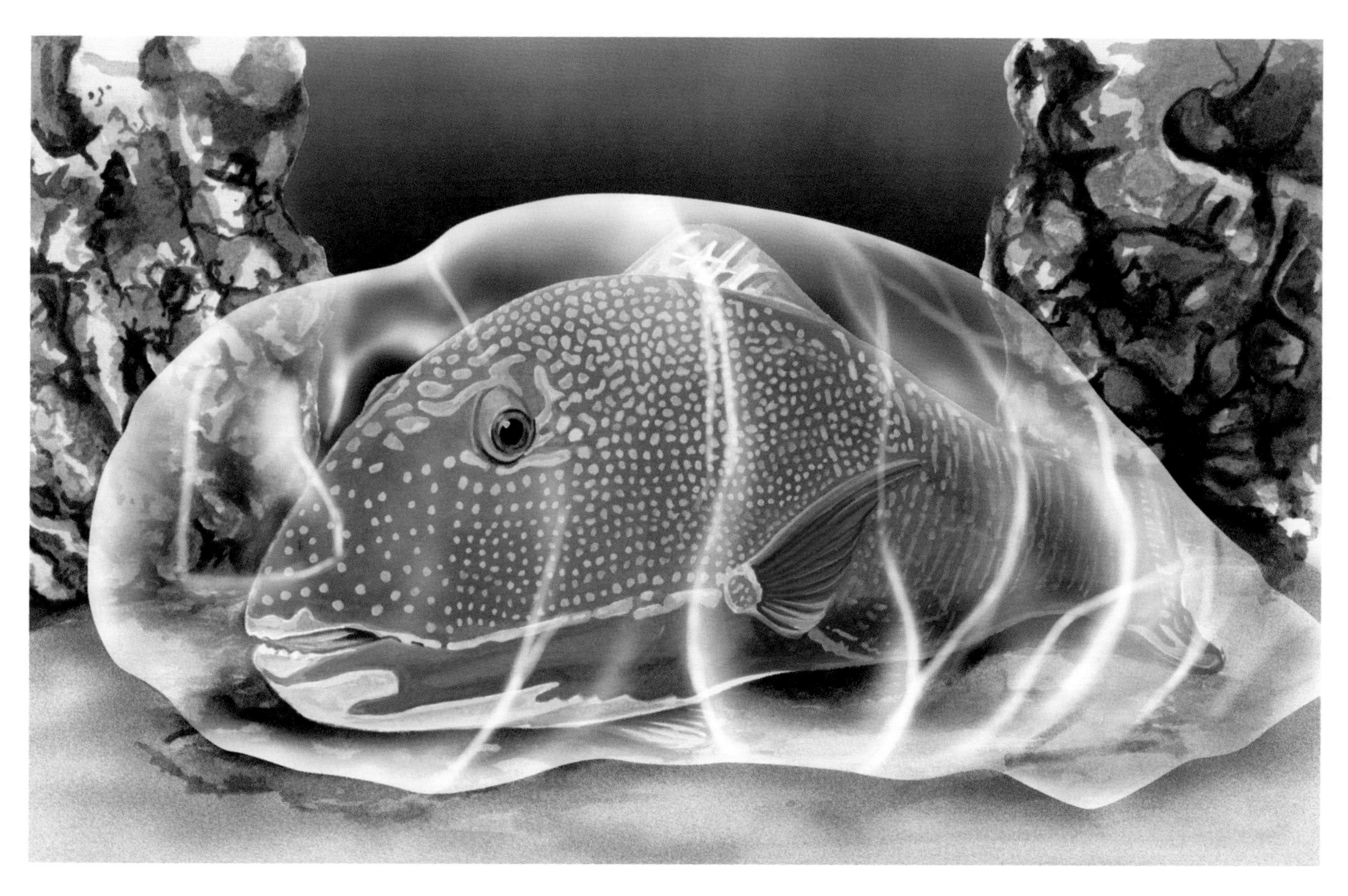

Detail of a parrotfish cocoon. (Illustration by Patricia Ferrer)

frequently assume they are two species. This three-phase lifestyle is characteristic of the wrasses and parrotfishes. Damselfishes typically shift gradually from a colorful juvenile pattern into a the more drab suit of an adult.

Along with the three color phases in the parrotfishes and wrasses, these two families engage in sex reversal. Here, the initial-phase adults usually are mostly female, often living in a small grouping, or harem. Each harem has one terminal-phase male, which mates with the adult females. The females form a hierarchy within the harem, with one dominant female followed by a second-ranking female, a third, and so on. If the male is removed from the group through predation or other means, the dominant female steps up to become, socially and physiologically, the terminal-phase male. The social change can occur in as little as a few hours. The physiological change, including the development of functional testes, may take a couple of weeks. The hierarchy among the females continues, with the second-ranking female moving into the position of dominant female, and the rest moving up a step in the hierarchy.

Initial-phase males are also capable of becoming the terminal-phase male, and this frequently occurs in nonharemic groups. In this reproductive arrangement, the terminal-phase males set up a territory where females may enter for one-on-one mating. When the terminal-phase male is removed, an initial-phase male or female may take his place. Another reproductive strategy is available for initial-phase males, which are sexually mature. Initial-phase males will form schools, and chase individual females as they make their way to a terminal male's territory, and "sneak-spawn" with them. Although sole paternity is out of the question for individual sneak-spawners, at least the initial-phase male can add his milt to the mix and perhaps fertilize a few eggs. Many of these fishes also engage in mass spawning, in which both terminal-phase and initial-phase males participate.

Damselfishes may not have the obvious phase differences, but they do have interesting reproductive behavior. As a group, these fishes are quite territorial of their mating sites and will mount vicious attacks on intruders, complete with rushes, grunts, and bites. The male typically clears a nest site, engages in ritualistic visual and tactile displays to attract a female, mates, and then cares for the eggs until they hatch, when the young are on their own. In an unusual twist, the marine damselfish (*Acanthochromis polyacanthus*) continues its care of the young for several months after hatching. Anemonefishes are noted because they reverse gender like the parrotfishes and wrasses, but in the opposite direction. In an anemone, this species exists in small groups with one large female, one large male, and several small, immature males. Should the large female be removed, the most-dominant immature male can develop to take her place.

Little is known about reproduction in odacids overall, but the butterfish *O. pullus* does change sex from female to male and exists in small harems. Females lay their eggs directly into a plankton column.

Conservation status

Eight species are included on the IUCN Red List, all of them categorized as Vulnerable. The eight species are: in the Pomacentridae, *Chromis sanctaehelenae*, *Stegastes sanctaehelenae*, and *S. sanctipauli*; in the Labridae, *Cheilinus undulatus*, *Lachnolaimus maximus*, *Thalassoma ascensionis*, and *Xyrichtys virens*; and in the Scaridae, *Scarus guacamaia*.

Significance to humans

Aesthetic, particularly to divers, and often popular in the aquarium trade. A few are minor commercial food fishes.

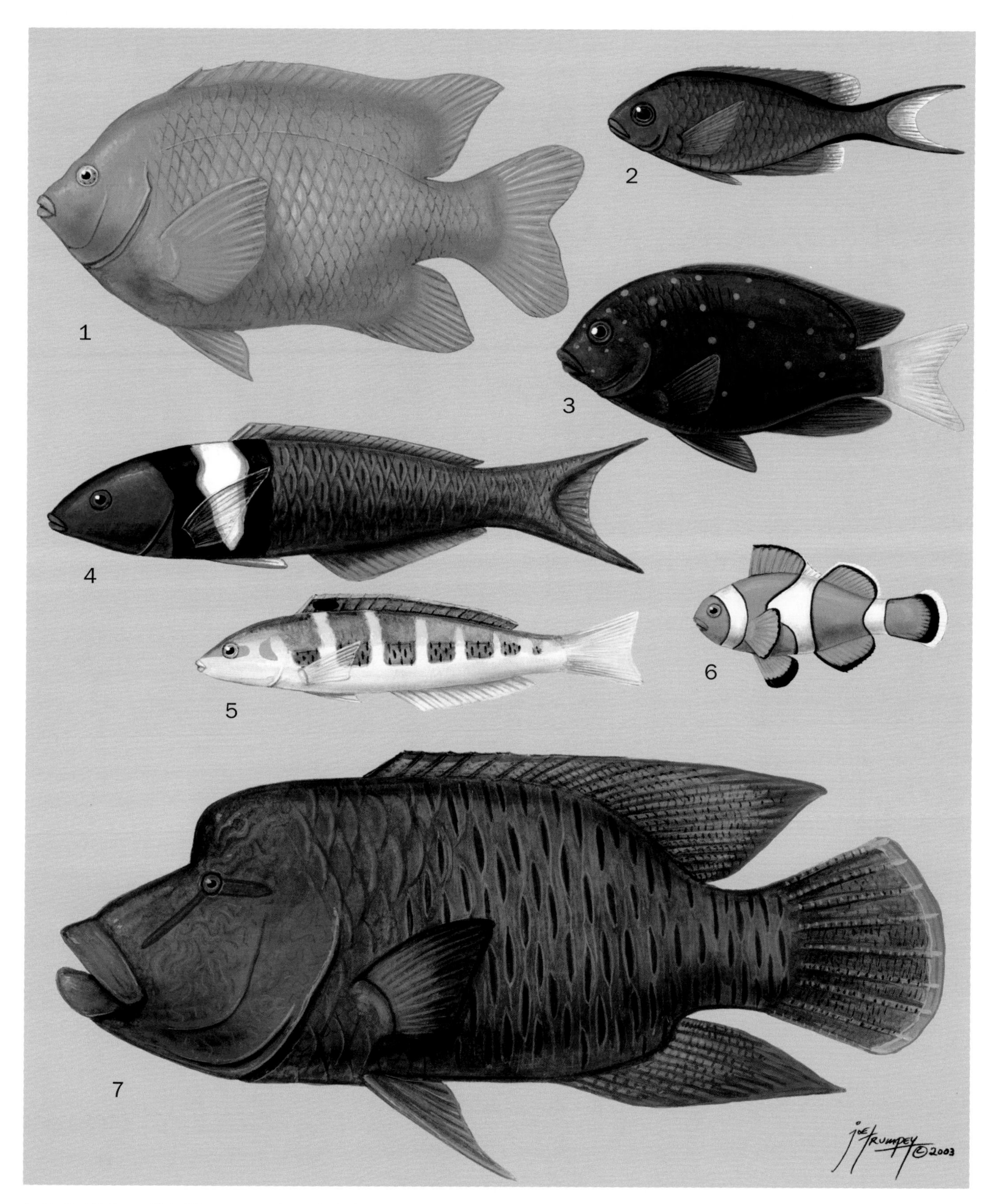

1. Garibaldi damselfish (*Hypsypops rubicundus*); 2. Blue chromis (*Chromis cyanea*); 3. Yellowtail damselfish (*Microspathodon chrysurus*) intermediate stage; 4. Bluehead adult (*Thalassoma bifasciatum*); 5. Bluehead (*Thalassoma bifasciatum*) intermediate stage; 6. Clown anemonefish (*Amphiprion ocellaris*); 7. Humphead wrasse (*Cheilinus undulatus*). (Illustration by Joseph E. Trumpey)

1. California sheephead (*Semicossyphus pulcher*); 2. Bluestreak cleaner wrasse (*Labroides dimidiatus*); 3. Hogfish (*Lachnolaimus maximus*); 4. Striped parrotfish (*Scarus iseri*); 5. Butterfish (*Odax pullus*); 6. Juvenile stoplight parrotfish (*Sparisoma viride*); 7. Initial phase of the stoplight parrotfish; 8. Terminal phase of the stoplight parrotfish. (Illustration by Patricia Ferrer)

Species accounts

Humphead wrasse
Cheilinus undulatus

FAMILY
Labridae

TAXONOMY
Cheilinus undulatus Rüppell, 1835, Jidda, Saudi Arabia, Red Sea.

OTHER COMMON NAMES
English: Maori wrasse, Napoleon wrasse; French: Napoléon; Spanish: Napoleón.

PHYSICAL CHARACTERISTICS
Length 5 ft (about 1.5 m), although a few have been recorded at more than 8.2 ft (2.5 m); one of the largest reef fishes. The large adult has distinctive hump on the top of the blue-fronted, small-eyed, thick-lipped head; rest of the body is yellowish green. Juveniles pale green, with horizontally placed spots or bar extending down the sides of the body.

DISTRIBUTION
Much of the Red Sea and Indian Ocean, east through Indonesia to Tuamotus, French Polynesia, and north to southern Japan.

HABITAT
Juveniles frequent sea grass beds and reef lagoons. Adults prefer deeper reef areas to 325 ft (100 m) deep.

BEHAVIOR
Shy, diurnal, remains among reef refuges at night. Usually solitary, but will sometimes live in pairs or in small groups.

FEEDING ECOLOGY AND DIET
Feeds during the day, primarily on mollusks, but also takes fishes, as well as other invertebrates such as brittle stars and sea urchins.

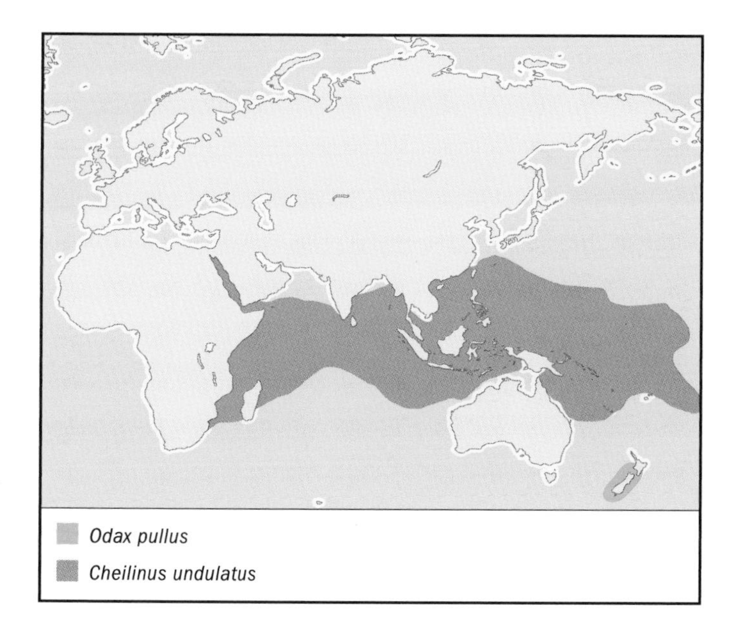

REPRODUCTIVE BIOLOGY
Both small and large spawning aggregations form, but fishes pair up for mating. No parental care. Sex reversal has been noted, in which females develop into mature males.

CONSERVATION STATUS
Listed as Vulnerable by the IUCN.

SIGNIFICANCE TO HUMANS
Commercial food fish whose intensive exploitation has led to concern for the survial of local populations of this species. Also popular aesthetically to divers. ◆

Bluestreak cleaner wrasse
Labroides dimidiatus

FAMILY
Labridae

TAXONOMY
Labroides dimidiatus Valenciennes, 1839, El Tûr, Sinai Coast, Egypt, Gulf of Suez; Mauritius.

OTHER COMMON NAMES
English: Bridled beauty, cleaner wrasse; French: Bande bleue, nettoyeur à poisson doctère, poisson nettoyeur commun.

PHYSICAL CHARACTERISTICS
Total length 4.5 in (11.5 cm). Adults mostly light blue, with long black stripe running along each side of body, widening as it approaches the tail. Juveniles black with blue dorsal stripe.

DISTRIBUTION
Throughout the Indo-Pacific region, west to the Red Sea and East Africa, north to the southern tip of Japan, and as far east as the Marquesas and Ducie Islands in the south central Pacific Ocean.

HABITAT
Prefers coral reef areas from surface waters to 130 ft (40 m) deep.

BEHAVIOR
Known for its cleaning habits, picking at and removing ectoparasites and assorted detritus on various species of fishes. Individuals or pairs set up stations, where they remove material from the bodies, gills, and even mouths of their "clients." Stereotyped signals between the two fishes help ensure that the cleaning goes smoothly, and the cleaner does not end up as the client's dinner.

FEEDING ECOLOGY AND DIET
Attains bulk of the diet from ectoparasites and detritus on other fishes.

REPRODUCTIVE BIOLOGY
Sometimes lives in pairs, but frequently in harems of one dominant male and 6–10 females. Sex reversal occurs when domi-

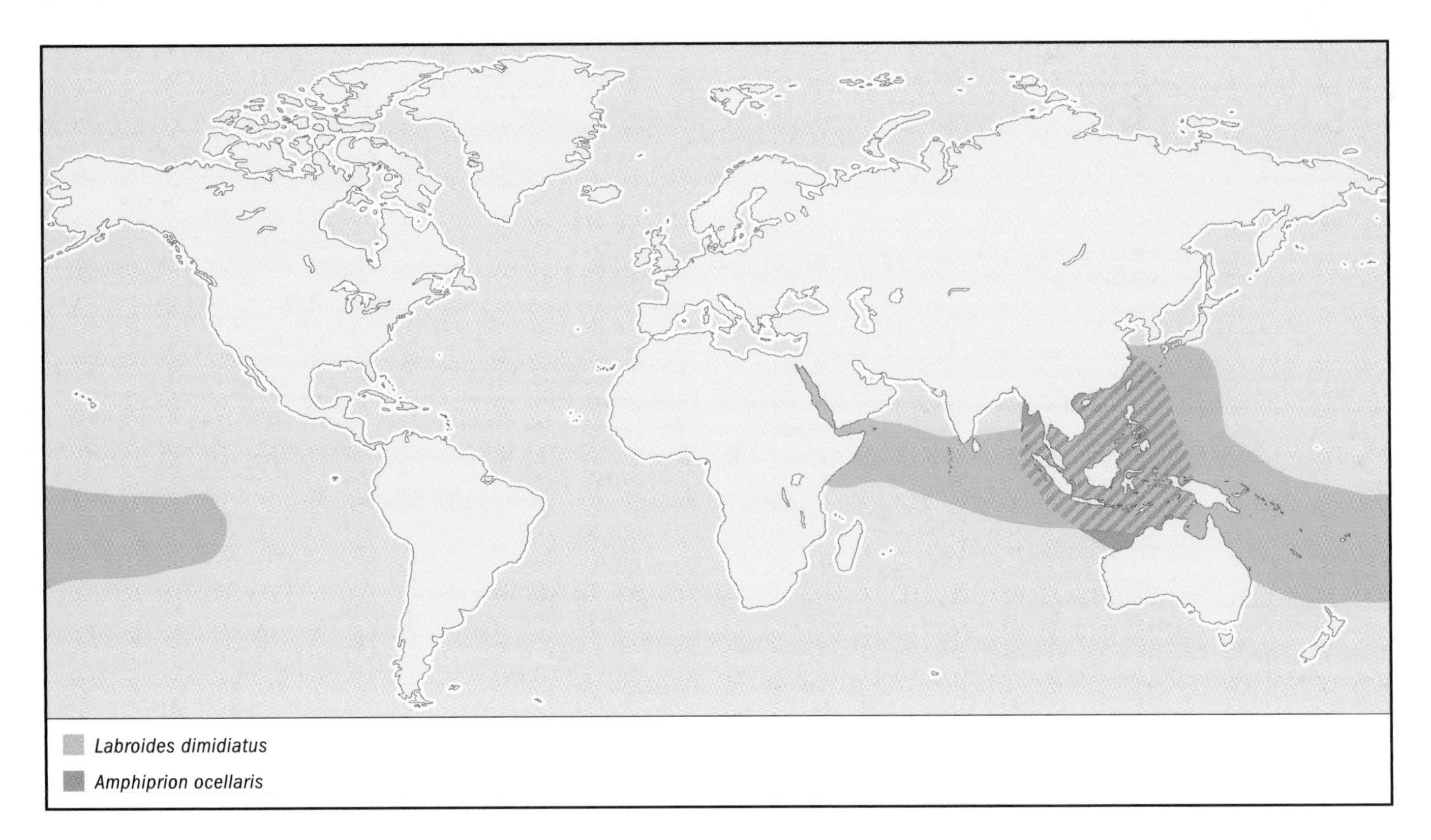

nant male is removed from a harem. The dominant female assumes his place, and becomes a functional male in about two weeks.

CONSERVATION STATUS
Not listed by the IUCN.

SIGNIFICANCE TO HUMANS
Common aquarium fish. ◆

Hogfish
Lachnolaimus maximus

FAMILY
Labridae

TAXONOMY
Lachnolaimus maximus Walbaum, 1792, type locality not specified (probably Bahamas or Carolinas).

OTHER COMMON NAMES
French: Labre capitaine; Spanish: Doncella de pluma.

PHYSICAL CHARACTERISTICS
Total length 36 in (91 cm). Often mottled, light orangey brown, but can adjust its color to match immediate surroundings. Adult dorsal fin sports three unusually long spines on anterior end. Dark stripe extends through eye to base of dorsal fin just behind head.

DISTRIBUTION
Bermuda, along the Atlantic coast of the Americas, from Nova Scotia to northern Venezuela, and into the southern and eastern Gulf of Mexico.

HABITAT
Sandy reef areas.

BEHAVIOR
Males are territorial, holding narrow areas about 100 yards (90 m) long. Around a dozen females share the territory with the male.

FEEDING ECOLOGY AND DIET
Prefers mollusks, crabs, and other hard-shelled invertebrates. Diurnal; during feeding ejects water streams into substrate to root out buried invertebrates.

REPRODUCTIVE BIOLOGY
Sex reversal is common. All fishes begin life as females, which can develop into males. No parental care is given to the pelagic eggs.

CONSERVATION STATUS
Listed as Vulnerable by the IUCN.

SIGNIFICANCE TO HUMANS
Commercial food fish. The large adult size and relatively sedentary character of this species make it particularly vulnerable to overharvesting.◆

California sheephead
Semicossyphus pulcher

FAMILY
Labridae

TAXONOMY
Semicossyphus pulcher Ayres, 1854, San Diego, California, United States.

OTHER COMMON NAMES
French: Labre californien; Spanish: Vieja de California.

PHYSICAL CHARACTERISTICS
Length 36 in (91 cm). Large teeth are a trademark. Initial-phase adults are orange with a white chin, terminal-phase

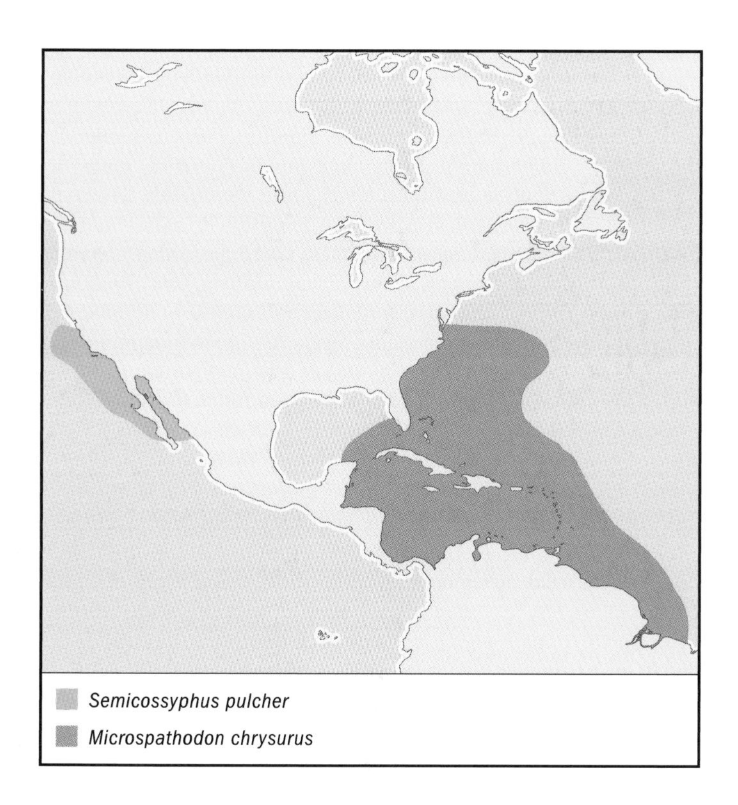

adults retain orange coloration at midbody at white chins, but have black heads and rear bodies.

DISTRIBUTION
Subtropical waters in the Gulf of California, and off the west coast of North America, from Monterey Bay in central California, United States, south about 600 mi (966 km) to Guadalupe Island, Mexico.

HABITAT
Lives among kelp beds, in shallow, rocky-bottomed waters.

BEHAVIOR
Sex reversal is common, with females developing into males as they grow older.

FEEDING ECOLOGY AND DIET
Feeds on mollusks, crabs, sea urchins, and other invertebrates.

REPRODUCTIVE BIOLOGY
Protygynous, mating takes place each summer. No parental care.

CONSERVATION STATUS
Not listed by the IUCN.

SIGNIFICANCE TO HUMANS
Commercial food fish. ◆

Bluehead
Thalassoma bifasciatum

FAMILY
Labridae

TAXONOMY
Thalassoma bifasciatum Bloch, 1791, East India (actually western Atlantic).

OTHER COMMON NAMES
English: Bluehead wrasse, tiki tiki; Spanish: Cara de cotorra.

PHYSICAL CHARACTERISTICS
Length 9.9 in (25 cm). Begins life as yellow, eventually develops black, horizontal stripes. Large terminal-phase males have vivid blue heads followed by white and black vertical bands, then green rear bodies and a bluish tail.

DISTRIBUTION
Western Atlantic Ocean near Bermuda, also from northern South America to the West Indies and southern Florida, United States, and into the Gulf of Mexico.

HABITAT
Coral reefs. Also seen among sea grass beds.

BEHAVIOR
Juveniles often provide cleaning services to other fishes. Among adults, spawning occurs differently depending on the size of the reef. On large reefs, group spawning is the rule. On small reefs, terminal-phase males utilize mating territories. Males from ensuing generations frequently use their ancestral mating site, apparently as a result of social convention. Dozens of females may select the same spawning site, and thus mate with the same male. Groups of smaller males often mate with egg-laden females by ambushing them on their way to the territorial mating sites.

FEEDING ECOLOGY AND DIET
Primarily feeds on drifting zooplankton, but also takes crabs, shrimps, sea urchins, and sea stars. Sometimes engage in cleaning of other species.

REPRODUCTIVE BIOLOGY
Engage in sex reversal. Over a period of several weeks, the black-striped, yellow males or females take on full coloration of large, terminal-phase males. Initial-phase females can take on the role of the terminal-phase male and begin producing sperm in as few as eight days.

CONSERVATION STATUS
Not listed by the IUCN.

SIGNIFICANCE TO HUMANS
Aquarium fish. ◆

Butterfish
Odax pullus

FAMILY
Odacidae

TAXONOMY
Odax pullus Forster, 1801, Queen Charlotte Sound, New Zealand.

OTHER COMMON NAMES
English: Greenbone; Maori: Mararii.

PHYSICAL CHARACTERISTICS
Length up to 27.5 in (70 cm). Long dorsal fin begins just behind head, becoming increasingly wide as it draws to its posterior end. Males bluer than females, and have longer dorsal and anal fin tips.

DISTRIBUTION
Throughout New Zealand waters, with fisheries in the area of Cook Strait and Stewart Island.

HABITAT
Mostly occurs near the surface to 33 ft (10 m) deep, in waters heavy with kelp. Some live in sheltered areas, others prefer tidal zones.

BEHAVIOR
Diurnal, lives in small groups of one male and several females.

FEEDING ECOLOGY AND DIET
Adults move into more open reefs at dawn to begin grazing on red, green, and brown algae, particularly kelp, which they clip with beak-like mouths, then grind with pharyngeal jaws. Juveniles eat red algae and crustaceans.

REPRODUCTIVE BIOLOGY
Begins life as a female, and reverses sex as it matures. Mating occurs several times each year between females and territorial males, from late winter to early spring. A pelagic spawner.

CONSERVATION STATUS
Not listed by the IUCN.

SIGNIFICANCE TO HUMANS
Sportfish and minor commercial food fish. ◆

Clown anemonefish
Amphiprion ocellaris

FAMILY
Pomacentridae

TAXONOMY
Amphiprion ocellaris Cuvier, 1830, Sumatra, Indonesia.

OTHER COMMON NAMES
English: Common clownfish, false clown anemonefish.

PHYSICAL CHARACTERISTICS
Total length 4.3 in (11 cm). Distinctive orange fish with three wide, vertical, white bands encircling the body just behind the eyes, at midback and in front of the tail. The fins and rounded tail are outlined in black, then edged in grayish white. Similar in general appearance to *A. percula.*

DISTRIBUTION
Coastal waters surrounding Indonesia; north and west to Burma, north and east past the Philippines to southern Japan, and as far south as northern Australia.

HABITAT
Prefers the sheltered shallow waters of lagoons, where it takes up residence among sea anemones.

BEHAVIOR
Most well known for its symbiotic relationship with the sea anemone. The clown anemonefish lives safely among the stinging cells, or nematocysts, of the anemone due at least in part to the specialized mucus layer that coats the fish. In this arrangement, the fish gains protection from predators, while the anemone receives a regular cleaning from the clown anemonefish. The fish has such a symbiotic relationship with four different species of sea anemones.

FEEDING ECOLOGY AND DIET
Eats mainly invertebrates that it often finds among the sea anemone's tentacles. Usually ventures only short distances from the shelter of the anemone, which has stinging tentacles. The sting deters most other species, making the anemone a safe haven.

REPRODUCTIVE BIOLOGY
Protandrous, lives in small groups in which all but two fishes are sexually immature males. The largest two in each group are sexually mature, with the larger being the only female in the group. If the female is removed, the sexually mature male develops into a sexually mature female, and the next largest has a growth spurt and becomes the sexually mature male. A dominance hierarchy controls the shift from male to female, and from immature to sexually mature.

CONSERVATION STATUS
Not listed by the IUCN.

SIGNIFICANCE TO HUMANS
Very popular marine aquarium fish. Many of the individuals sold are captive-bred. ◆

Blue chromis
Chromis cyanea

FAMILY
Pomacentridae

TAXONOMY
Chromis cyanea Poey, 1860, Cuba.

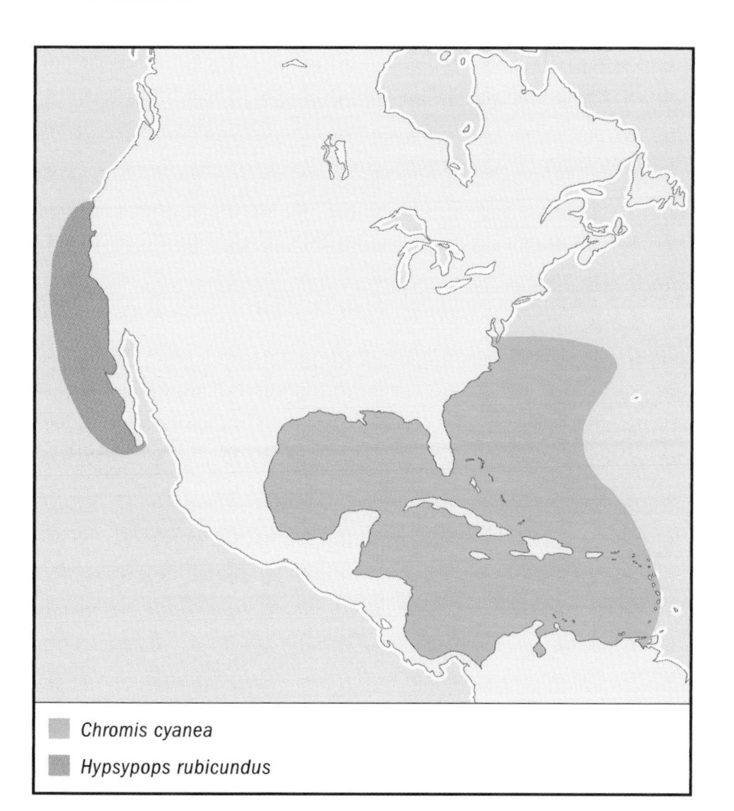

OTHER COMMON NAMES
Spanish: Cromis azul.

PHYSICAL CHARACTERISTICS
Total length 6 in (15 cm). Shimmering, blue with black dorsal shading, strongly forked tail, and dark eyes. Tail and belly also have some black coloration.

DISTRIBUTION
Western Atlantic Ocean near Bermuda, also from northern South America to the West Indies and into the Gulf of Mexico.

HABITAT
Reef fish, prefers deeper waters up to 200 ft (61 m) deep with coral overhangs and crevices where it can take shelter when threatened.

BEHAVIOR
Travels in sometimes multispecies schools, especially when feeding. Usually shy, will dash to the cover of reef when it feels threatened, and has been observed to dim its bright blue coloration to a duller gray.

FEEDING ECOLOGY AND DIET
Engages in group feeding of drifting zooplankton, but is territorial when feeding on algae and will defend an area against other algae-feeding species.

REPRODUCTIVE BIOLOGY
Male may spawn with several females and guards all eggs in his nest until hatching, which takes up to one week.

CONSERVATION STATUS
Not listed by the IUCN.

SIGNIFICANCE TO HUMANS
A popular commercial aquarium fish. ◆

Garibaldi damselfish
Hypsypops rubicundus

FAMILY
Pomacentridae

TAXONOMY
Hypsypops rubicundus Girard, 1854, Monterey, California, United States.

OTHER COMMON NAMES
English: Garibaldi; French: Chauffet Garibaldi; Spanish: Jaqueta vistosa.

PHYSICAL CHARACTERISTICS
Standard length 11.8 in (30 cm). Deep bodied, small mouthed, almost uniformly orange.

DISTRIBUTION
Occurs in subtropical waters off the west coast of North America from Monterey Bay in central California south about 600 mi (966 km) to Guadalupe Island, Mexico.

HABITAT
Lives in rocky-bottomed, reef-associated waters up to 100 ft (30.5 m) deep among and near caves, crevices, and other cover.

BEHAVIOR
Adults defend a home territory.

FEEDING ECOLOGY AND DIET
Feeds on vegetation and invertebrates, especially crustaceans, sponges, and worms.

REPRODUCTIVE BIOLOGY
Spawns in a bed of red algae, which the male prepares and tends. Courtship includes male visual and tactile displays. The female lays the eggs, the male guards them.

CONSERVATION STATUS
Not listed by the IUCN.

SIGNIFICANCE TO HUMANS
State marine fish of California. ◆

Yellowtail damselfish
Microspathodon chrysurus

FAMILY
Pomacentridae

TAXONOMY
Microspathodon chrysurus Cuvier, 1830, St. Thomas Island.

OTHER COMMON NAMES
English: Jewelfish (as juveniles); French: Chaffet queue jaune; Spanish: Jaqueta rabo amarillo.

PHYSICAL CHARACTERISTICS
Length 8.3 in (21 cm). Juveniles typically dark blue with sky blue spots on all but the yellow tail. Adults are golden brown with dark-outlined scales and a yellow tail.

DISTRIBUTION
Western Atlantic Ocean near Bermuda, also from northern South America to the West Indies and southern Florida, United States, and into the southern Gulf of Mexico.

HABITAT
Shallow waters of shelter-filled coral reefs, usually associated with yellow stinging coral.

BEHAVIOR
When food is abundant, shares feeding area with other wrasses, but will weakly defend a territory if food is limited. Juveniles sometimes engage in cleaner activities.

FEEDING ECOLOGY AND DIET
Prefers grazing on algae, but also takes invertebrates, especially coral polyps.

REPRODUCTIVE BIOLOGY
Spawns mostly during semiannual periods. As males prepare nests, their coloration lightens. Female coloration brightens as they arrive to lay their eggs. Males tend and guard the eggs.

CONSERVATION STATUS
Not listed by the IUCN.

SIGNIFICANCE TO HUMANS
Occasional food fish and aquarium fish. ◆

Striped parrotfish
Scarus iseri

FAMILY
Scaridae

TAXONOMY
Scarus iseri Bloch, 1789, St. Croix Island, Virgin Islands, West Indies.

OTHER COMMON NAMES
English: Gray chub, mottlefin parrotfish; French: Perroquet rayé; Spanish: Jabón, loro rayado.

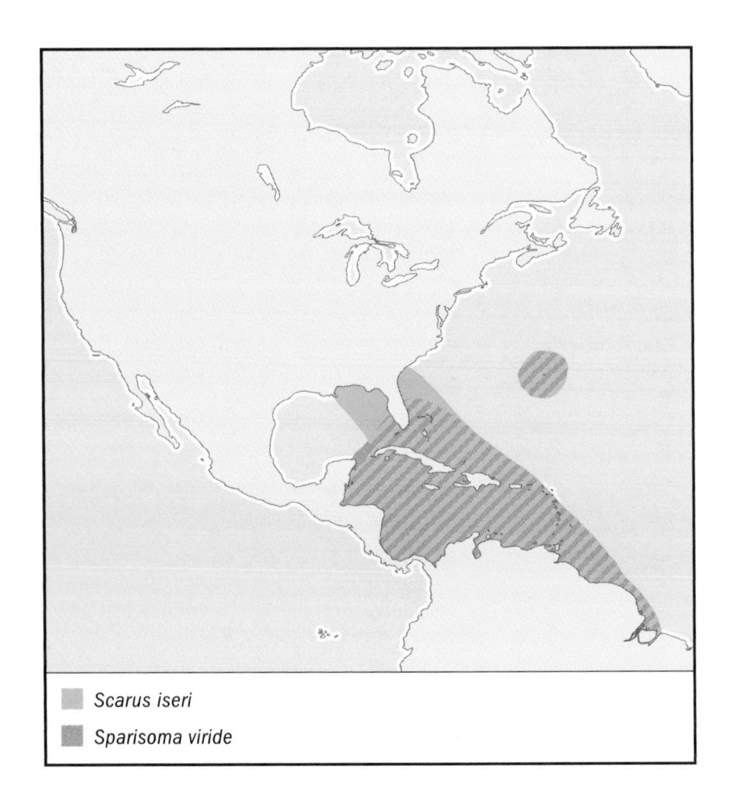

PHYSICAL CHARACTERISTICS
Total length 13.7 in (35 cm). Dark horizontal stripes are key feature of initial-phase individuals. Most prominent stripe is typically a center marking extending through the eye and nearly to the tail, where it narrows and fades. Juveniles are similarly patterned. Terminal-phase males are turquoise.

DISTRIBUTION
Lesser and Greater Antilles to southern Florida, United States, south along the South American coast to Brazil.

HABITAT
Prefers reef waters from the surface to 100 ft (30.5 m) deep.

BEHAVIOR
Typically schools, sometimes defends feeding territories if resources are limited or fish population numbers are high. Conspecific territorial displays include fanning of the ventral fins, opening of the mouth, and noncontact rushes toward one another.

FEEDING ECOLOGY AND DIET
Forms schools for feeding on algae, which it scrapes from rocks and other surfaces.

REPRODUCTIVE BIOLOGY
Initial-phase males and females may become terminal-phase males; female-to-male switch takes less than two weeks. Mating occurs year round, either in pairs or in groups. A broadcast spawner that gives no parental care to its pelagic eggs.

CONSERVATION STATUS
Not listed by the IUCN.

SIGNIFICANCE TO HUMANS
Minor commercial and aquarium fish. ◆

Stoplight parrotfish
Sparisoma viride

FAMILY
Scaridae

TAXONOMY
Sparisoma viride Bonnaterre, 1788, Bahamas.

OTHER COMMON NAMES
English: Moontail, parrot chub, redbelly; French: Perroquet feu; Spanish: Loro viejo.

PHYSICAL CHARACTERISTICS
Total length 25 in (64 cm). Females and younger adult males have mostly blue-gray body scales outlined in dark gray; reddish belly, tail, and fins. Colorful terminal-phase males mostly green, with blue and reddish horizontal stripes on head and lower ventral body, and yellow and blue markings on the tail, which is shaped like a crescent moon.

DISTRIBUTION
Western Atlantic from southern Florida, United States, to Brazil, also around Bermuda.

HABITAT
Juveniles prefer shallow sea grass beds and vegetated areas, adults inhabit coral reefs.

BEHAVIOR
Diurnal; lives alone or in small groups; moves to the sea bottom at night. Has been known to accept cleaning services from other species.

FEEDING ECOLOGY AND DIET
Feeds mainly on algae and other plants, but will eat corals. Feeding occurs only during the day. Will become territorial of feeding areas if food is limited.

REPRODUCTIVE BIOLOGY
Initial-phase individuals may be either male or female; terminal phase is male. Initial-phase females require three weeks to become terminal-phase males. Mating occurs year round.

CONSERVATION STATUS
Not listed by the IUCN.

SIGNIFICANCE TO HUMANS
Minor commercial and aquarium fish. ◆

Resources

Books

Allen, Gerald R. "Damselfishes." In *Encyclopedia of Fishes*, edited by John R. Paxton and William N. Eschmeyer. 2nd edition. San Diego: Academic Press, 1998.

Choat, J. Howard, and David R. Bellwood. "Wrasses and Parrotfishes." In *Encyclopedia of Fishes*, edited by John R. Paxton and William N. Eschmeyer. 2nd edition. San Diego: Academic Press, 1998.

Helfman, Gene S., Bruce B. Collette, and Douglas E. Facey. *The Diversity of Fishes.* Malden, MA: Blackwell Science, 1997.

Michael, Scott W. *Reef Fishes: A Guide to Their Identification, Behavior and Captive Care.* Volumes 1–3. Shelburne, VT: Microcosm Ltd., 1998.

Randall, John E., Gerald R. Allen, and Roger C. Steene. *Fishes of the Great Barrier Reef and Coral Sea*, 2nd edition. Honolulu: University of Hawaii Press, 1997.

Periodicals

Bshary, R., and D. Schäffer. "Choosy Reef Fish Select Cleaner Fish That Provide High-Quality Service." *Animal Behaviour* 63 (2002): 557–564.

Donaldson, T. J., and Y. Sadovy. "Threatened Fishes of the World: *Cheilinus undulatus* Rüppell, 1835 (Labridae)." *Environmental Biology of Fishes* 62 (2001): 428.

Francini-Filho, R. B., R. L. Moura, and I. Sazima. "Cleaning by the Wrasse *Thalassoma noronhanum*, with Two Records of Predation by its Grouper Client *Cephalopholis fulva*." *Journal of Fish Biology* 56, no. 4 (2000): 802–809.

Kavanagh, K. D. "Larval Brooding in the Marine Damselfish *Acanthochromis polyacanthus* (Pomacentridae) Is Correlated with Highly Divergent Morphology, Ontogeny and Life-History traits." *Bulletin of Marine Science* 66, no. 2 (2000): 321–337.

Organizations

IUCN/SSC Grouper and Wrasse Specialist Group. Department of Ecology and Biodiversity, The University of Hong Kong, Hong Kong, China. Phone: (852) 2859 8977. Fax: (852) 2517 6082. E-mail: yjsadovy@hkusua.hku.hk Web site: <http://www.hku.hk/ecology/GroupersWrasses/iucnsg/index.html>

Leslie Ann Mertz, PhD

○

Zoarcoidei

(Eelpouts and relatives)

Class Actinopterygii
Order Perciformes
Suborder Zoarcoidei
Number of families 9

Photo: An ocean pout (*Zoarces americanus*) in the Gulf of Maine. (Photo by Andrew J. Martinez/ Photo Researchers, Inc. Reproduced by permission.)

Evolution and systematics

The zoarcoids, except for a few cryptic tidepool species, are a relatively obscure group of fishes of unverified affinity. On the basis of overall shape, fin structures, and position, and their ecological preference for living on sea bottoms in cold water, the most primitive family of zoarcoids, the ronquils, (Bathymasteridae) are thought to be close to the Antarctic "cods" or icefishes (suborder Notothenioidei). The zoarcoids and notothenioids share several important anatomical features, such as having only a single pair of nostrils, females with one ovary, lacking a gas bladder, and in having many similar skull characteristics. But they differ in other skull features and in the head sensory canal system slightly.

The time of appearance of the zoarcoids is unknown. Only the ear stones (otoliths) of an Upper Pliocene (3.2 to 1.9 million years ago) eelpout from southern California are known for the entire suborder's fossil record, and that species, *Lycodes pacificus*, is extant. If the zoarcoids are indeed related to the notothenioids, then we might construct an evolutionary scenario like this: In their 1966 classification of teleost fishes, Humphry Greenwood and colleagues noted that by the Eocene (57 to 35 million years ago), or possibly earlier (the view now taken by more evidence), marine inshore fish faunas had become composed of the groups that make up today's faunas. Sea floor spreading along tectonic plate boundaries separated Antarctica from Australia at the close of the Eocene, and for about the next 15 million years Antarctica drifted to its present position. As the Drake Passage deepened a circumpolar current, the Antarctic Polar Front, developed lying between latitudes 50° and 60° south, isolating Antarctica climatically. This spurred the creation of its ice cap and frigid sea temperatures. The notothenioids subsequently diversified rapidly here. This implies that prior to the Eocene, an ancestral group common to both zoarcoids and notothenioids split into two subgroups that developed physiological adaptations to the cold. One of these radiated throughout the Northern Hemisphere (zoarcoids) and the other across the cold Southern Hemisphere (notothenioids and the more primitive sandperches). Later re-invasions also took place, such as by deep-sea eelpouts (Zoarcidae) into the Southern Hemisphere and sandperches (Pinguipedidae) into the tropics.

As of 2002, science recognizes about 325 species of zoarcoids in about 100 genera and nine families, although further taxonomic work is needed and new species are discovered almost every year. Three families, the quillfish (Ptilichthyidae), the prowfish (Zaproridae), and the graveldiver (Scytalinidae) have only one species. Four of the familes are very small, with 6–7 ronquils (Bathymasteridae), five wolffishes (Anarhichadidae), four wrymouths (Cryptacanthodidae) and 14 gunnels (Pholidae). Thus only two families are sizeable, with 20% of all zoarcoids in the family Stichaeidae (pricklebacks) and 70% in the Zoarcidae (eelpouts).

An analysis of the relationships within Zoarcoidei has never been published. The ronquils were seen to be the most primitive family in having no advanced features common among the other families. The gunnels and graveldiver share a common, elongate shape and have lost the pleural ribs. Pricklebacks, gunnels, and the graveldiver have characteristic elongated crania not shared by the other families except for a few advanced eelpouts. However, many of the primitive pricklebacks, particularly the subfamily Stichaeinae, have many similarities to ronquils. The prowfish and wolffishes have similar shapes, large body size, fin structures, and features of the skeleton except for the skull and may be just generally primitive zoarcoids that split off the main lineage early in the group's evolution at two different stages. Finally, the bizarre quillfish is so highly modified that its relationships are obscure. It is an extremely elongated fish (over 230 vertebrae) with many modifications of the head and fin structures and may represent another early offshoot of the main lineage that became highly modified rapidly, unlike the prowfish and wolffishes.

Physical characteristics

Two basic body shapes exist among zoarcoids. Primitively, a somewhat elongated, torpedo-like shape characterizes ronquils, some pricklebacks, wolffishes, and the prowfish; however, most pricklebacks and the wolffish *Anarrhichthys ocellatus* have attained a much elongated eel-like shape. All the eelpouts, gunnels, wrymouths, graveldiver, and the quillfish are eel-like in shape. Some eelpouts, especially the genus *Lycenchelys* and the quillfish are extremely elongated.

Most zoarcoids are small fishes reaching lengths of less than 15.8 in (40 cm). Some eelpouts attain 23.6 in (60 cm), and the largest of these, *Zoarces americanus*, reaches 46.5 in (118 cm). Other "giants" among the zoarcoids are the prowfish, reaching just over 3.3 ft (1 meter), and the wolffishes *Anarhichas* spp., reaching 47–57 in (120–145 cm), and *Anarrhichthys ocellatus*, the largest zoarcoid, which reaches 80 in (203 cm).

Deep-living, bottom-dwelling zoarcoids are rather somber in color, exhibiting more or less uniform shades of gray, brown, black, or purplish. Shallow-dwelling species can be more colorful, such as the prickleback genus *Chirolophis*, with shades of red, orange, and yellow, the ronquil genus *Rathbunella*, with its beautiful yellow and blue anal fin, and most of the gunnels, which show shades of green and brown to red reflecting the colors of the seaweeds among which they live. The dwarf wrymouth, *Cryptacanthodes aleutensis*, is uniformly red, and some North Pacific eelpouts, such as *Andriashevia aptera* and *Puzanovia rubra*, have also evolved this red to pinkish coloration. Sexual differences in coloration either do not exist or are usually subtle, with spawning males having dark anal fins, as in some gymneline eelpouts. However, the male prickleback, *Opisthocentrus ocellatus*, is normally drab grayish with dark markings but turns red with darker variegations in the spawning season.

Distribution

Ranges of the various zoarcoid groups can be confusing, especially in attempting to understand the pricklebacks and eelpouts; thus it is best to examine distributions on a family basis.

Bathymasteridae (ronquils): Central Sea of Japan, around the North Pacific rim to northern Baja California, Mexico.

Zoarcidae (eelpouts): In the Pacific from the northern Yellow Sea, east around the Pacific rim to the tip of South America and across the Scotia Arc to Antarctica. Four species are known from deep water in the tropical western Pacific. Distributed throughout the Arctic, especially the genera *Gymnelus* and *Lycodes* (*Lycodes frigidus* has been photographed in the Chukchi Sea at 78°28'N at a depth of 1.7 mi (2,653 m) and is probably widespread under the polar ice cap in suitable habitats). They live in the Atlantic from the Arctic south along the eastern United States to the Gulf of Mexico and Caribbean Sea; a few species are known from the Mid-Atlantic Ridge at hydrothermal vents. They also live in the eastern Atlantic from northern Europe, along the west African coast to South Africa. A few species of the deep-water genus *Pachycara* are known from the northern Indian Ocean abyss and may be widespread there. A few species of the mesopelagic genera *Lycodapus* and *Melanostigma* are circumpolar in subantarctic waters but do not reach coastal Antarctica, which has an endemic eelpout fauna.

Ptilichthyidae (quillfish): Hokkaido Island, Japan, around the North Pacific rim to central Oregon, United States.

Zaproridae (prowfish): Hokkaido Island, Japan, around the North Pacific rim to San Miguel Island, California.

Anarhichadidae (wolffishes): In the Pacific from Hokkaido, Japan, to the eastern Bering Sea (*Anarhichas lupus*) and the eastern Bering Sea to off southern California (*Anarrhichthys ocellatus*). In the Atlantic (three species of *Anarhichas*) from Massachusetts to southern Greenland and east to the Kara Sea. The American wolffish, *Anarhichas lupus*, reached as far south as New Jersey in the nineteenth century.

Cryptacanthodidae (wrymouths): In the Pacific throughout the Sea of Japan north to Sakhalin Island, Russia (*Cryptacanthodes bergi*); northern California to the Bering Sea (*C. aleutensis* and *C. giganteus*). In the Atlantic from New Jersey to Labrador (*C. maculatus*).

Stichaeidae (pricklebacks): In the Pacific from the northern Yellow Sea around the North Pacific rim to Baja California, Mexico. In the Atlantic from Massachusetts to Iceland, northern Norway, Britain, and throughout the Baltic Sea. A few species enter the Canadian Arctic; *Lumpenus medius* is circumpolar and even reaches the Sea of Japan.

Pholidae (gunnels): In the Pacific from the Yellow Sea around the North Pacific rim to Guadalupe Island, Baja California, Mexico. In the northwestern Atlantic from Delaware to Hudson Strait and the northeastern Atlantic from southern France to the Kara Sea including Iceland and Jan Mayen Island.

Scytalinidae (graveldiver): Central California (San Luis Obispo County) north to the Aleutian Islands, Alaska.

Habitat

The zoarcoids inhabit a wide variety of ecological niches and some have adapted to rather extreme habitats. At one end of this spectrum several gunnels, pricklebacks, and the graveldiver are found above the high tide line in rock pools with broad daily variation in temperature and salinity. Graveldivers also can be found burrowing in sand or gravel beaches above the surf wash. Most gunnels and pricklebacks are found in rocky areas with growths of seaweeds or invertebrate colonies, where they keep well hidden by day. In the opposite extreme many eelpouts are abyssal, or found on the bottom of the

A mosshead warbonnet (*Chirolophis nugator*) peering out from its hole. (Photo by Gregory Ochocki/Photo Researchers, Inc. Reproduced by permission.)

world's ocean basins, at depths of more than 4,375 yd (4,000 m). The deepest catch of a zoarcoid is that for a specimen of the eelpout *Lycenchelys antarctica*, trawled at 5,818 yd (5,320 m) in the Peru–Chile Trench by Russian researchers. Most eelpouts inhabit continental slopes of the North Pacific and North Atlantic, where bottom temperatures are below 44.6°F (7°C). Two eelpout genera, *Lycodapus* and *Melanostigma*, have adapted to a deep-water, free-swimming habit (in the mesopelagic zone), but have been photographed near the bottom. The ronquils are generally found in rocky reefs from inshore to about 301 yd (275 m). Nowhere very abundant, they are mostly seen by scuba divers in calm waters either in the rocks or over sandy patches. The wrymouths are mud burrowers and may be more active at night than in the day. The three Pacific wrymouths are found only in moderately deep water 33–383 yd (30–350 m), but the Atlantic species may be found in intertidal areas. The Atlantic wolffishes (genus *Anarhichas*) occur in rocky areas, never over muddy bottoms, generally on the outer shelf 109–328 yd (100–300 m), although the American species, *A. lupus*, has been seen in tidepools. The maximum depth is around 503 yd (460 m), although the three Atlantic species generally occur shallower to the north of their ranges. The Pacific wolffishes, *Anarhichas orientalis* and *Anarrhichthys ocellatus*, also occur in rocky reefs usually not deeper than 164 yd (150 m). All the wolffishes seem to lurk about in crevices and under overhangs when not foraging, sometimes in more open areas. The prowfish mostly occurs over soft bottoms of the outer shelf, but juveniles have been collected as shallow as 11 yd (10 m). One prowfish was caught in a trawl at 738 yd (675 m). Juvenile prowfish often occur in midwater where they may linger. The adult quillfish's habitat is probably free-swimming just above the bottom in outer shelf waters by day then migrating into surface waters at night to feed. One researcher speculated that it may burrow in mud, but without direct evidence.

Behavior

Most zoarcoids are cryptic and solitary throughout their lives, but may congregate in shelters or around food sources temporarily. Some deep-sea eelpouts have been photographed gathering around mammal carcasses and baited traps to feed on the amphipods that devour the carcass or bait. Some gunnels are known to share rock crevices for shelter. Social organization is not well documented in these hard-to-observe fishes. What behavioral patterns are known exist mostly as notes in the scientific literature from aquarium or scuba observations focusing mainly on reproduction. Nest guardianship has been noted for a few ronquils, pricklebacks, gunnels, eelpouts, and the wolf-eel, and this is discussed in the next section. During the winter in higher latitudes intertidal zoarcoids migrate into deeper water to avoid freezing, but populations are very localized and no great migrations occur. Thus, population movements are solely through larval dispersal by nearshore currents. Territoriality, as far as known, is probably weak among established adults of nearshore zoarcoids, since many share close spaces, but nothing is known of this in the deeper-living, cryptic species.

Feeding ecology and diet

The vast majority of zoarcoids are grazing predators of small invertebrates such as worms, crustaceans, molluscs, and echinoderms. Many switch prey items by season or as they age, but a significant number are dietary specialists, a situation often localized for a particular species. No in-depth study of the biology of the ronquils has been undertaken. Postlarvae (0.2–0.4 in [5–11 mm] in length) of *Ronquilus jordani* in southern British Columbia had been feeding near the surface on crustaceans (Cladocera, Copepoda, Cirrepedia larvae), clam larvae, and polychaete worms. The stripedfin ronquil, *Rathbunella hypoplecta*, is known to feed on the bottom primarily on crustaceans, but fish in one sample had fed exclusively on sea slugs (nudibranchs). A nest-guarding male *R. hypoplecta* had cannibalized part of his brood. Adult ronquils, then, are probably all opportunistic bottom grazers. The diets and feeding habits of a few eelpouts have been studied. Bottom-living species rely on crustaceans, particularly amphipods, but also eat smaller fishes, sea snails, clams, other kinds of crustaceans and, to a lesser extent, brittle stars and polychaete worms. The diets of the mesopelagic eelpouts *Lycodapus mandibularis* and *Melanostigma pammelas* off California diversify with increasing size but mainly consist of a few kinds of crustaceans. Young pricklebacks of several species in the water column off British Columbia feed on copepods and clam larvae. As they grow and settle into bottom habitats, a dietary shift occurs and most adult pricklebacks then rely on polychaetes, amphipods, sea snails, nudibranchs, various kinds of algae, bryozoans, shrimps, crabs, and sponges. Although this appears to be a rather diverse diet when several species are examined from different areas, many of the xiphisterine pricklebacks are herbivores, eating only a few kinds of red and green algae. Atlantic pricklebacks have been recorded with similar diets: amphipods, copepods, ostracods, brittle stars, clams, sea cucumbers (Holothuria) and polychaetes. Gunnels, living among rocks and seaweeds with pricklebacks, feed on the same prey. The Atlantic butterfish, *Pholis gunnellus*, eats various mollusks,

A large male wolf-eel (*Anarrhichthys ocellatus*) eats a sea urchin. (Photo by Brandon D. Cole/Corbis. Reproduced by permission.)

especially sea snails, crustaceans, and polychaetes, and has been known to raid other fishes' nests to consume their eggs. In the Canadian Pacific, *Pholis ornata* and *Pholis laeta* are known to switch diets with age. Both mostly rely on harpacticoid copepods when young, but large fish shift mostly to caprellid amphipods (*P. laeta*) or clam siphons (*P. ornata*). Young wrymouths in the Canadian Pacific also rely on copepods when feeding in midwater. The diets of adults are not well recorded, but food recorded for the Atlantic wrymouth consists of various crustaceans, mollusks, and fish. The wolffishes all have canine teeth in the front of their jaws and viselike molars in the back, dental adaptations for dealing with hard-shelled prey. The Atlantic wolffishes prey on mollusks like whelks, cockles, clams, and mussels as well as large crabs, hermit crabs, starfish, and sea urchins, the latter rather formidable as food. The Pacific wolf-eel, *Anarrhichthys ocellatus*, eats mostly commercial crabs (*Cancer magister*) and other hard-shelled invertebrates. Fish rarely enter the diets of wolffishes. Not much is known of the food habits of the prowfish, graveldiver, or quillfish. Young prowfish are free-swimming and are known to associate with jellyfish. Remnants of amphipods that also associate with jellyfish, as well as jelly fragments, have been found in stomachs of large prowfish. Thus they may utilize various jellyfishes as a food source, at least in part, throughout their life. Graveldivers are burrowers and probably hunt tiny interstitial invertebrates. The quillfish, with its small, upturned mouth and free-swimming habits, probably preys on a variety of small planktonic crustaceans and mollusc larvae.

Reproductive biology

Because nearshore zoarcoids are generally small, cryptic fishes and offshore species usually live beyond scuba depths, almost all observations on the reproductive biology of these fishes come from aquarium observations on a few nearshore species. A few zoarcoids captured from scientific research vessels have been examined in some detail, but generally little about their reproduction is known.

Direct observations on courtship are few, and fertilization is internal in all known bottom-living species. Males of the elongate zoarcoids (eelpouts, gunnels, pricklebacks, wolf-eel) wrap around receptive females snake-like and fertilization is internal, or eggs are fertilized as they are laid in clusters. The pectoral fins in courting male viviparous eelpouts (*Zoarces viviparus*) turn bright red and males assume a characteristic coiled position with the fins outstretched. When females are receptive males then assume a transverse position under and alongside the female for mating. Pelagic eelpouts like *Lycodapus* may have little formal courtship behavior and probably just pair up during a protracted spawning season and release eggs and milt into the water column. The northern ronquil (*Ronquilus jordani*) has a rather long courtship with the male displaying to the female with body and fin quivering and fanning. It is generally assumed that most of the nearshore zoarcoids spawn during daylight hours when they can see one another for courting, but the Japanese tidepool gunnel (*Pholis nebulosa*) courts and spawns from about midnight to dawn.

Egg size and number vary greatly in zoarcoids, owing to the variation in adult body size. Reported sizes range from about 0.1–0.3 in (1–7 mm), but the eggs of the quillfish, graveldiver, and prowfish are unknown. Fecundity ranges from less than 20 eggs per female in some primitive eelpouts (Gymnelinae) to about 50,000 in one wolffish. Eggs can be clear, white, yellowish, or orange with oil droplets usually yellow, orange, or red, and most are adhesive since they are laid in nests. Incubation times range from about two weeks to three months. Yolk sac larvae of bottom-living zoarcoids generally stay near the bottom, but some become planktonic for anywhere from a few seconds to minutes (*Zoarces americanus*), or up to two years (*Anarrhichthys ocellatus*).

Nests are generally simple hideaways among most zoarcoids. Ronquils lay loose egg masses on flat surfaces, which are fanned and usually guarded by the male. In the stripedfin ronquil (*Rathbunella hypoplecta*), there is an aquarium observation of a female laying eggs six times every two weeks for a total of about 10,000 eggs during the three months. A new clutch was laid just days after the most recent spawn hatched, and the male guarded the egg mass. Bottom-living eelpouts excavate a crater in muddy bottoms and coil around a clump of eggs. However, the Atlantic midwater eelpout (*Melanostigma atlanticum*) is known to burrow tunnels into soft muddy bottoms with a male and female spawning, but sometimes several apparently non-spawning individuals may occupy the tunnels to help circulate water in this anoxic environment. Gunnels and pricklebacks lay eggs on seaweeds (*Zostera*, some algae) or in nests in rock crevices and holes; some gunnels are known to use empty oyster or mussel shells. Wolffishes build nests in rocky shelters or lay their eggs among seaweeds and stones.

Parental care is fairly limited in most zoarcoids that are known for this behavior. The male of the smooth ronquil noted above sometimes ate some of the eggs he was guarding. Guardianship of nests may be limited to fanning with charging displays against intruders (ronquils) or of parents holing up with the eggs with no overt signs of care simply because the nest area provides good cover (gunnels and pricklebacks).

The spawning season of zoarcoids can be short, just a few months between courtship and egg hatching, or protracted into the better part of a year. Some deep-sea eelpouts spawn only once or twice at the end of their lives, and seasonality may not exist for these. Ronquils and many gunnels and pricklebacks spawn from the spring to summer; other gunnels and pricklebacks are autumn to winter spawners. The North Atlantic wolffishes spawn from spring to winter, and the more northerly populations start earlier in the year than southern populations. The northwestern Atlantic wrymouth (*Cryptacanthodes maculatus*) is a winter spawner, but the Japanese wrymouth (*Cryptacanthodes bergi*) spawns from spring to summer. On the basis of larval captures in scientific collecting nets, quillfish spawn at least in spring but probably have a longer season. Nothing is known of the spawning habits of the prowfish or graveldiver.

Conservation status

Most nearshore zoarcoids are fairly common in their preferred habitats. This includes most ronquils, gunnels, pricklebacks, shallow-dwelling eelpouts, and wolffishes where they're not fished. Other species are rare, and this is a result of either naturally low population densities or that they are rarely caught or seen, such as the graveldiver, prowfish, and quillfish. No zoarcoids have been considered endangered or threatened, mostly owing to these fishes' usually cryptic habits and limited commercial use.

Significance to humans

The zoarcoids are a fairly insignificant group of fishes in human history. No mythology or significant art or literature exists, but it is interesting to note that in the Middle Ages it was commonly held that the European viviparous eelpout (*Zoarces viviparus*) was said to birth all eels (order Anguilliformes), owing to its live-bearing mode of reproduction. Significant fisheries do not exist for any zoarcoid today. However, after World War II a minor longline fishery for wolffishes (*Anarhichas*) developed across the North Atlantic, but catches have declined dramatically since. An unsuccessful attempt was made to develop a fishery for the American ocean pout (*Zoarces americanus*) after World War II, but natural parasitic infections in the flesh quickly killed the plan. No zoarcoid poses any real danger to humans except wolffishes. Trawl-caught *Anarhichas* species have been known to snap at fishermen and a large, speared wolf-eel (*Anarrhichthys ocellatus*), because of its strength and imposing teeth, may inflict serious bites to careless scuba divers.

1. Ocean pout (*Zoarces americanus*); 2. Saddleback gunnel (*Pholis ornata*); 3. Wolf-eel (*Anarrhichthys ocellatus*); 4. Black prickleback (*Xiphister atropurpureus*); 5. Wrymouth (*Cryptacanthodes maculatus*). (Illustration by Marguette Dongvillo)

Species accounts

Wolf-eel
Anarrhichthys ocellatus

FAMILY
Anarhichadidae

TAXONOMY
Anarrhichthys ocellatus Ayres, 1855, San Francisco, California.

OTHER COMMON NAMES
English: Pacific wolfeel.

PHYSICAL CHARACTERISTICS
Body eel-like. Dorsal and anal fins long and low, continuous with a small caudal fin. Pectoral fins large and fan-like. Pelvic fins absent. Background color blue, greenish, brown, or gray-brown. The body and head are covered with white-lined black spots. Scales are small and rounded and embedded in the skin. There is no lateral line canal. The teeth in front are canine-like, and the rear teeth are molar-like. Juveniles are orangish brown, with spots more numerous and larger than in adults; the spots sometimes form into stripes. This is the largest zoarcoid, reaching 2.2 yd (203 cm).

DISTRIBUTION
Southeastern Bering Sea to off Imperial Beach, California. Records for the Sea of Japan or Kamchatka are erroneous.

HABITAT
Deep rocky reefs in caves or crevices. Juveniles are free-swimming for an extended period.

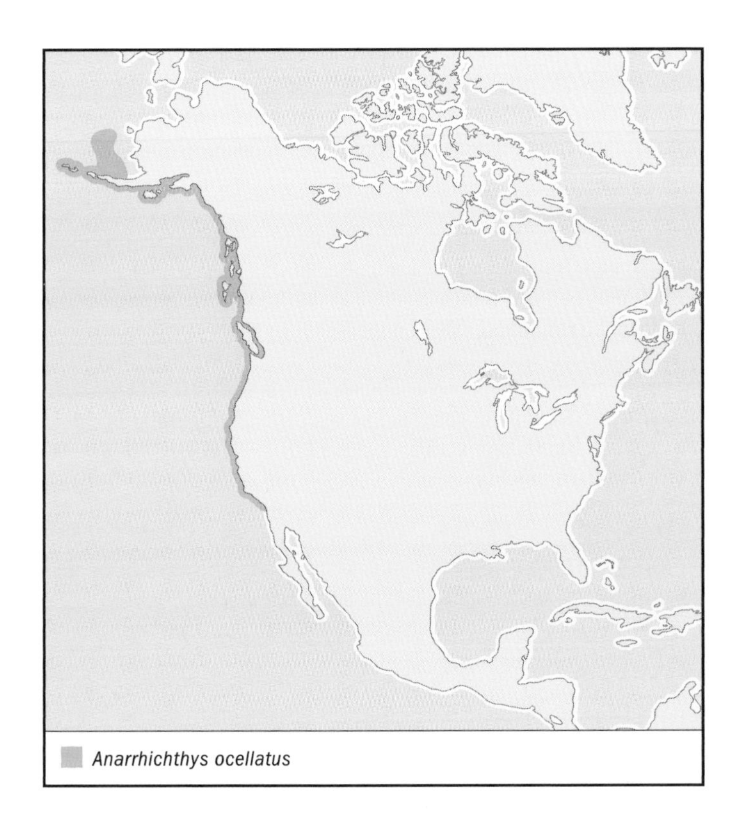

BEHAVIOR
Cryptic and solitary or lives with an apparent life-long mate in a den. It is a dusk and dawn predator but also feeds during the day. Territorial disputes occur with large individuals displacing mated pairs occasionally.

FEEDING ECOLOGY AND DIET
Stalking predator. Feeds mainly on hard-shelled invertebrates and occasionally fish. Wolf-eels grab their prey with their enlarged canines and crush it with their rear molars. Known prey consists of crabs, clams, mussels, sea urchins, sand dollars, snails, and abalone. A population of wolf eels at the head of the Monterey submarine canyon in California has relied on sand dollars at least seasonally, a food presumably rather low in nutrition for its bulk. Free-swimming juveniles eat plankton and small fish.

REPRODUCTIVE BIOLOGY
Pairs form at about four years of age (length about 3.3 ft [1 m]), and eggs are first laid at about seven years. Spawning occurs from October to February. Courtship consists of the male repeatedly bumping the female's abdomen; when she is receptive he coils around her snake-like. Eggs are fertilized as they are laid in clumps of about 7,000–10,000, and the female gathers these up into a ball and wraps around them, occasionally turning them for aeration. Both parents guard the nest, and only one at a time leaves to forage. Hatching occurs at 13–16 weeks, depending on water temperature. Juveniles are free-swimming for up to two years, then settle on open bottom until they take up their sedentary den life.

CONSERVATION STATUS
Not threatened. This species is common and widespread along its range. A minor fishery exists among scuba divers, skiff fishermen, and bottom trawlers.

SIGNIFICANCE TO HUMANS
The wolf-eel is a good eating fish, and the scuba and small boat fishery has been significant and sustained in some areas of central California and Puget Sound, Washington. Wolf-eel teeth have been found in a native American village site in central California, indicating a fishery at least 9,000 years old. Tribes in the Pacific Northwest reserved this delicacy for their shamans. ◆

Wrymouth
Cryptacanthodes maculatus

FAMILY
Cryptacanthodidae

TAXONOMY
Cryptacanthodes maculates Storer, 1839, off Massachusetts, United States.

OTHER COMMON NAMES
English: Congo eel, ghostfish; French: Terrassier tacheté.

PHYSICAL CHARACTERISTICS
Body eel-like. Head flattened above. Eyes small. Mouth large and up-turned. Pectoral fins small. Pelvic fins absent.

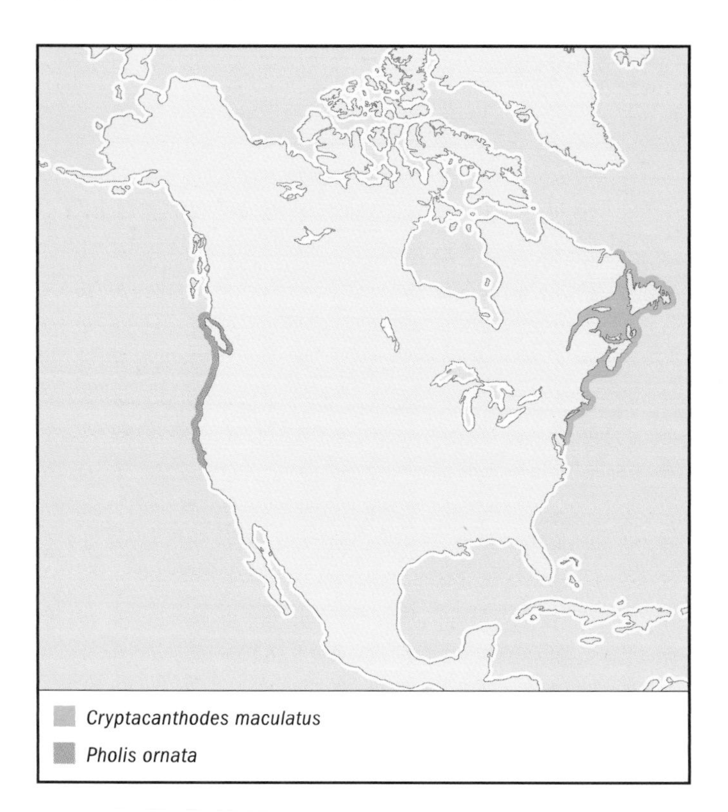

Dorsal and anal fins long, low, and continuous with caudal, which is well developed. Background color brown to reddish brown, with 2–3 irregular rows of dark brown spots or blotches running from head to tail. The dorsal and anal fins have rows of smaller brown spots. Abdomen pale gray. Scales absent.

DISTRIBUTION
Southern Labrador south to New Jersey.

HABITAT
Found from just above the low tide mark (the intertidal zone) to almost 656 yd (600 m). Constructs branching tunnels in riverine mudflats, where it lives singly. May live in burrows or hide in crevices or under rocks in deeper water.

BEHAVIOR
Due to their secretive burrowing habits, there have been no observations on natural behavior. One wrymouth was kept in an aquarium, where it could not burrow. It lived instead in a piece of rubber tubing. Thus it is likely that they hide almost all the time and venture forth only to forage.

FEEDING ECOLOGY AND DIET
Grazing or ambush predator. Feeds on small crustaceans (amphipods, shrimps, and hermit crabs), mollusks (limpets, sea snails, clams, and mussels), and occasionally fishes.

REPRODUCTIVE BIOLOGY
Eggs and larvae are unknown, but wrymouths undoubtedly spawn on the bottom and probably guard nests. Fry have been collected in scientific sampling nets in early spring, indicating winter spawning.

CONSERVATION STATUS
Rarely seen. Not threatened.

SIGNIFICANCE TO HUMANS
None known. ◆

Saddleback gunnel
Pholis ornata

FAMILY
Pholidae

TAXONOMY
Gunnellus ornatus Girard, 1854, San Francisco, California.

OTHER COMMON NAMES
None known.

PHYSICAL CHARACTERISTICS
Body eel-like. Pectoral fin small and half as long as the head. Pelvic fins are minute, mere splints. Dorsal and anal fins continuous with a caudal, which is well developed. Color can change from olive-green to brown dorsally; abdomen yellow to reddish. There is a dark bar under the eye and two above the eye across the head. There are several (about 13–15) U-shaped or V-shaped markings ("saddles") along the base of the dorsal fin reaching onto it. No lateral line canal. Scales minute, rounded, and embedded in the skin.

DISTRIBUTION
Northern Vancouver Island, British Columbia south to Carmel, California.

HABITAT
Migrates seasonally from brackish, estuarine mudflats and subtidal sea grass beds under rocks in autumn to deeper water out to about 44 yd (40 m), where its major competitor, *Pholis laeta*, occurs. Where that species does not co-occur with the saddleback, some remain in the intertidal zone year-round. The young settle in the rocky intertidal or shallow channels in winter and move into more brackish water later.

BEHAVIOR
Solitary and cryptic. Aggregates for spawning. Probably not territorial.

FEEDING ECOLOGY AND DIET
Grazer. Dietary shifts correlated with age. In British Columbia the young feed mainly on small crustaceans (copepods) and polychaete worms, while older fish switch to clam siphons, relying less on small crustaceans (copepods and tanaids).

REPRODUCTIVE BIOLOGY
Spawns in rocky intertidal or deeper sea grass beds under rocks in winter. There is probably a single spawning of one male and one female; both parents guard the eggs. Larvae are pelagic for a brief period and settle in rocky intertidal areas then move into brackish mudflats with seasonal growth of seaweeds, at least in bays and estuaries.

CONSERVATION STATUS
Not threatened. Common and widespread along its range. It is cryptic and rarely encountered by humans.

SIGNIFICANCE TO HUMANS
Because of its secretive habits and small size, there has never been a fishery for the saddleback gunnel, and it does not make a good aquarium fish. Thus the species has been of little significance to humans. ◆

Black prickleback
Xiphister atropurpureus

FAMILY
Stichaeidae

TAXONOMY
Ophidium atropurpureum Kittlitz, 1858, Alaska (no specifics). Correct generic placement should probably be in *Xiphidion* Girard, 1858 (or 1859), pending further research.

OTHER COMMON NAMES
None known.

PHYSICAL CHARACTERISTICS
Body eel-like. Pectoral fins are minute, of only 11 or 12 rays. Pelvic fins absent. Dorsal and anal fins continuous with caudal, which is well developed and has a whitish band at its base. Color reddish brown to black; abdomen is lighter. Head has three broad, black eye bars with whitish borders. Scales are minute, rounded, and covered with skin.

DISTRIBUTION
Kodiak Island, Alaska south to Baja California Norte, Mexico.

HABITAT
Intertidal zone in rock pools among seaweeds and in crevices out to about 32.8 ft (10 m). Also found under wharf pilings and in boat harbors where shelter (human trash usually) can be found.

BEHAVIOR
Solitary and cryptic without territoriality. Parental care of the eggs occurs (see below).

FEEDING ECOLOGY AND DIET
Omnivore. Feeds on seaweeds and invertebrates on or associated with the bottom, primarily crustaceans, worms, and sea snails. Hatchlings caught in surface-towed nets in British Columbia had been feeding on copepod crustaceans and clam larvae but probably eat any small planktonic animals.

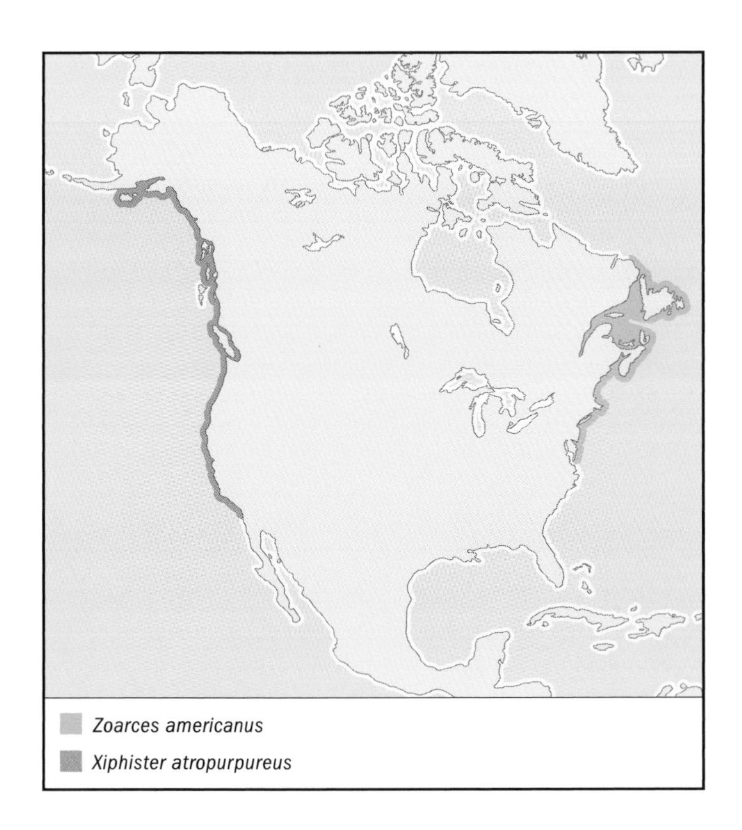

REPRODUCTIVE BIOLOGY
Spawning occurs from winter to spring throughout the range, under rocks along protected pebbly or shelly beaches in winter and shifting to other beaches that are more exposed in spring. Females lay about 900–1,700 eggs about 0.1 in (2 mm) in diameter, and males guard the site. Males may spawn with more than one female. Territoriality is non-existent in that several males may congregate under the same rock, and sometimes other species, such as clingfishes, are found nearby. Hatchlings are about 0.3 in (8.5 mm) long, and metamorphosis occurs at about 0.7 in (18 mm) when they become free-swimming and feed on small plankton.

CONSERVATION STATUS
Not threatened. Common and widespread along its extensive range. It is cryptic and rarely encountered by either humans or predators.

SIGNIFICANCE TO HUMANS
Owing to its secretive habits and small size there has never been a fishery for the black prickleback, and it does not make a good aquarium fish; thus the species has been of little significance to humans. ◆

Ocean pout
Zoarces americanus

FAMILY
Zoarcidae

TAXONOMY
Blennius americanus Bloch and Schneider, 1801, American seas.

OTHER COMMON NAMES
English: Muttonfish, yowler; French: Loquette d'Amérique.

PHYSICAL CHARACTERISTICS
Body eel-like but rather stout in adults. Pectoral fins large and fan-like. Pelvic fins have short splints. The dorsal and anal fins are continuous with the caudal fin; the dorsal fin has about 15–25 tiny spines at its rear. Background color usually muddy yellowish, tinged with brown above and becoming darker with age. Belly usually yellowish but can be olive-green. Mottling on the sides is brown, but the pattern is individually variable. Teeth are green in northern populations owing to predation on sea urchins. Scales are minute, round, and not overlapping.

DISTRIBUTION
Southern Labrador south to Virginia.

HABITAT
Adults found on the outer shelf to about 219 yd (200 m) on sandy or muddy bottoms. Young may come into intertidal areas among seaweed and rocks. Free-swimming hatchlings sometimes found in estuaries of large rivers to the north.

BEHAVIOR
Very little is known of behavioral traits in ocean pout, mostly owing to the difficulty of observing them in their usual offshore habitat. Based on aquarium observations, they probably live passive, solitary lives without territories and seem only to

congregate for spawning. Spawning consists of several copulations over many hours. Large spawning males are aggressive to smaller males at this time, as are females with non-spawning females. Parental care of the eggs occurs (see below).

FEEDING ECOLOGY AND DIET
Grazer. Feeds on invertebrates on or within the sea floor, mostly on crustaceans (amphipods, *Cancer* crabs, and hermit crabs especially), sea urchins, worms, bivalves (clams and scallops), sea snails, and brittle stars. Fish found in a few stomachs occasionally are probably scavenged.

REPRODUCTIVE BIOLOGY
Spawning occurs in early to mid-Autumn and consists of males approaching ripe females and rolling on their sides or even upside down under the female. Fertilization is internal and several copulations occur for 2–3 minutes each over many hours, perhaps up to half a day. Egg laying occurs around 6–17 hours after the last copulation. Large females lay a clutch of eggs in rocky areas more numerous than smaller females. A female at nearly 35.4 in (90 cm), near the maximum size, contained over 4,100 ripe eggs; another at 21.6 in (55 cm), and probably spawning for the first or second time, had 1,300 eggs. The eggs are pale yellow and measure 0.24–0.28 in (6–7 mm) in diameter. Upon laying her egg mass the female fans and wipes her skin over the eggs for around 30 minutes. This coats the mass in an antibiotic mucus. Then the female wraps herself tightly around the mass (now white in color), which helps stick it together into an egg ball. Females remain passive while guarding their eggs except for intermittent swimming in circles while fanning the eggs with their pectoral fins. Incubation lasts for three months, during which females probably do not feed much. Fry hatch in mid-winter, and yolk sac resorption occurs in seconds. Fry have a very short planktonic phase while working their way inshore, where they develop over the first few years of their lives.

CONSERVATION STATUS
Not threatened. Common in nearshore environments as young where predation is probably low. Also common offshore even on the fishing banks off New England and Nova Scotia.

SIGNIFICANCE TO HUMANS
A minor fishery for ocean pout began in Massachusetts in the 1930s, with fish being sold "round" in Boston markets. In 1943, as a war effort, a concerted attempt was made to sell whole fillets, and landings reached almost 4.4 million lb (2 million kg) in 1944. However, ocean pout are afflicted with a microsporidian (Protozoa) parasite that produces unsightly lesions in the flesh. Landings dropped to under 6,100 lb (2,767 kg) by 1948, and the fishery subsequently failed. Attempts to revive the fishery in the 1970s also failed. ◆

Resources

Books

Andriashev, A. P. *Fishes of the Northern Seas of the USSR.* Moscow: Academy of Sciences, USSR, 1954.

Bigelow, H. B. and W. C. Schroeder. *Fishes of the Gulf of Maine.* Washington, DC: U. S. Fish and Wildlife Service, 1953.

Breder, C. M., Jr. and D. E. Rosen. *Modes of Reproduction in Fishes.* Jersey City: T.F.H. Publications, 1966.

Burgess, W. E. and H. R. Axelrod. *Fishes of California and Western Mexico.* Neptune City, NJ: T.F.H. Publications, 1984.

Eschmeyer, W. N. and E. S. Herald. *A Field Guide to Pacific Coast Fishes of North America.* Boston: Houghton Mifflin, 1983.

Fitch, J. E. and R. J. Lavenberg. *Tidepool and Nearshore Fishes of California.* Berkeley: University of California Press, 1975.

Goodson, G. *Fishes of the Pacific Coast.* Stanford: Stanford University Press, 1988.

Hart, J. L. *Pacific Fishes of Canada.* Ottawa: Fisheries Research Board of Canada, 1973.

Kiernan, A. M. *Systematics and Zoogeography of the Ronquils, Family Bathymasteridae (Teleostei: Perciformes).* Seattle: University of Washington, Ph.D. Dissertation, 1990.

Leim, A. H. and W. B. Scott. *Fishes of the Atlantic Coast of Canada.* Ottawa: Fisheries Research Board of Canada, 1966.

Masuda, H. et al., eds. *The Fishes of the Japanese Archipelago.* Tokyo: Tokai University Press, 1984.

Mecklenburg, C. W., T. A. Mecklenburg, and L. K. Thorsteinson. *Fishes of Alaska.* Bethesda: American Fisheries Society, 2002.

Nawojchik, R. *A Systematic Revision of Zoarcoid Fishes of the Family Cryptacanthodidae (Teleostei: Perciformes).* Seattle: University of Washington, M.S. Thesis, 1986.

Whitehead, P. J. P. et al., eds. *Fishes of the North-eastern Atlantic and the Mediterranean.* Vol. 3. Paris: United Nations Educational, Scientific, and Cultural Organization, 1986.

Wingert, R. C. *Comparative Reproductive Cycles and Growth in Two Species of* Xiphister *(Pisces, Stichaeidae) from San Simeon, California.* Fullerton: California State University, M. A. Thesis, 1974.

Periodicals

Anderson, M. E. "Systematics and Osteology of the Zoarcidae (Teleostei: Perciformes)." *Smith Institute of Ichthyology, Bulletin* 60 (1994): 1–120.

Barsukov, V. V. "Fauna of the USSR. Fishes. Family Zubatok (Anarhichadidae)." *Zoological Institute, Academy of Sciences, USSR* 5, no. 5 (1959): 1–171.

Barton, M. "Comparative Distribution and Habitat Preferences of Two Species of Stichaeoid Fishes in Yaquina Bay, Oregon." *Journal of Experimental Marine Biology and Ecology* 59 (1982): 77–87.

Hughes, G. W. "The Comparative Ecology and Evidence for Resource Partitioning in Two Pholidid Fishes (Pisces: Pholididae) from Southern British Columbia Eelgrass Beds." *Canadian Journal of Zoology* 64, no. 1 (1985): 76–85.

Marliave, J. B. "Seasonal Shifts in the Spawning Site of a Northeast Pacific Intertidal Fish." *Journal of the Fisheries Research Board of Canada* 32, no. 10 (1975): 1687–1691.

———. "The Life History and Captive Reproduction of the Wolf-Eel at the Vancouver Public Aquarium." *International Zoo Yearbook* 26 (1987): 70–81.

Resources

Marliave, J. B., and E. E. DeMartini. "Parental Behavior of Intertidal Fishes of the Stichaeid Genus *Xiphister*." *Canadian Journal of Zoology* 55, no. 1 (1977): 60–3.

Matarese, A. C., A. W. Kendall, Jr., D. M. Blood, and B. M. Vinter. "Laboratory Guide to Early Life History Stages of Northeast Pacific Fishes." *NOAA Technical Report NMFS* 80 (October 1989): 496–531.

Peden, A. E. and G. W. Hughes. "Distribution, Morphological Variation, and Systematic Relationship of *Pholis laeta* and *P. ornate* (Pisces: Pholididae) with a Description of the Related Form P. nea n. sp." *Canadian Journal of Zoology* 62, no. 2 (1984): 291–305.

Richardson, S. L. and D. A. DeHart. "Records of Larval, Transforming and Adult Specimens of the Quillfish, *Ptilichthys goodei*, from Waters off Oregon." *Fishery Bulletin* 73, no. 3 (1975): 681–5.

Yao, Z. and L. W. Crim. "Copulation, Spawning and Parental Care in Captive Ocean Pout." *Journal of Fish Biology* 47 (1995): 171–3.

Yatsu, A. "A Revision of the Gunnel Family Pholididae (Pisces, Blennioidei)." *Bulletin of the National Science Museum* 7, no. 4 (December 1981): 165–190.

M. Eric Anderson, PhD

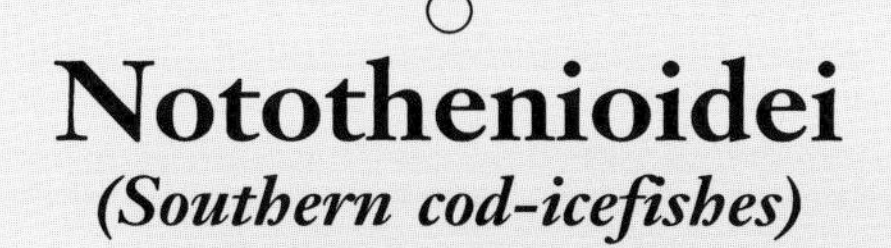

Notothenioidei

(Southern cod-icefishes)

Class Actinopterygii

Order Perciformes

Suborder Notothenioidei

Number of families 8

Photo: A scuba diver observes a blackfin icefish (*Chaenocephalus aceratus*) as it swims off the coast of Signy Island, one of the South Orkney Islands near the Antarctic Peninsula. (Photo by Rick Price/Corbis. Reproduced by permission.)

Evolution and systematics

Over the past 40 million years, the suborder Notothenioidei has evolved in high latitudes of the Southern Hemisphere from generalized blennioid ancestors, producing a remarkable variety of ecological, morphological, physiological, and biochemical specializations. This diversity has led to their recognition as the only known example of a marine species swarm, rivaling the freshwater swarms of cichlid fishes in the African rift lakes and the landlocked cottoids of Lake Baikal. Over the past three decades, intensive study of the group has been driven by the increase in internationally funded research in Antarctica that followed the Antarctic Treaty of 1959.

The basic taxonomic framework for the suborder, established in the early 1900s, recognized five families; as of 2002, the Notothenioidei included 122 species in 43 genera, divided into eight families:

- Bovichtidae (thornfishes, 10 species in two genera); the most primitive family
- Pseudaphritidae (one catadromous species, *Pseudaphritis urvillii*, the Australian congolli or tupong)
- Eleginopidae (one species, *Eleginops maclovinus*)
- Harpagiferidae (spiny plunderfishes, six species in one genus)
- Artedidraconidae (plunderfishes, 25 species in four genera)
- Nototheniidae (notothens, or "Antarctic cods," 49 species in 12 genera)
- Bathydraconidae (dragonfishes; 15 species in 11 genera)
- Channichthyidae (icefishes, 15 species in 11 genera)

There are no unequivocal notothenioid fossils; thus, little is known of the origins of the suborder. Earlier claims of Eocene and Miocene notothenioid fossils from the Antarctic Peninsula and New Zealand have been discounted. The only substantial Antarctic fish fossils are found in Eocene deposits on Seymour Island, near the Antarctic Peninsula. These beds contain a cosmopolitan shallow-water fossil fish fauna that is completely different from the modern Antarctic fauna, presently dominated by notothenioids. As Antarctica became glaciated, the hardy notothenioids replaced the earlier fishes and then diversified under the impetus of habitat destruction associated with cyclical glacial advances.

Physical characteristics

Notothenioids have a diversity of sizes and body forms. Most are small fishes, on the order of 1 ft (30 cm) in length, but some species may be as small as 4.3 in (11 cm) or as long as 6 ft (1.8 m). None is highly colored; most have black, brown, or gray mottling on paler backgrounds. They have two to three lateral lines on the trunk and well-developed sensory canals on the head, jaw, and pre-operculum. Three plate-like radials characterize the pectoral girdle. All notothenioids lack a swim bladder.

Many of the 96 species of Antarctic notothenioids coexist with sea ice, near the freezing point of seawater (28.6°F, or −1.87°C). They are protected from freezing by an antifreeze glycopeptide (AFGP), consisting of repeating groups of three amino acids (-alanine-alanine-threonine-)$_n$, with a disaccharide (galactose-N-acetylgalactosamine) attached to each threonine. AFGP works differently from normal colligative antifreezes such as salt or glycol, which only depend on the number of ions or molecules to lower freezing temperature. Instead, AFGP molecules bind to mircroscopic ice crystals and interfere with the attachement of additional water molecules, preventing ice from growing to a size that would damage living cells. Thus a very small number of AFPG molecules can have a disproportionate effect on freezing. The gene encoding AFGP has evolved from part of a gene for the digestive enzyme trypsin.

As well as being important for survival, AFGPs are metabolically expensive. In the high-latitude notothenioids that produce AFGPs, changes in the kidney reduce their loss. Instead of forming urine by pressure filtration in capillary bundles (glomeruli), Antarctic notothenioids eliminate wastes by selective secretion into the urine, and AFGPs are retained. Their kidneys lack the glomeruli characteristic of most other vertebrates. In contrast, the 26 species of notothenioids living outside Antarctic waters rarely encounter freezing seawater, lack AFGPs, and have glomeruli.

Distribution

Three of the eight notothenioid families are primarily non-Antarctic:

- Bovichtidae: These are found mainly north of the Antarctic convergence, in Australia/Tasmania, southern New Zealand, and southern South America. One species occurs along the Antarctic Peninsula.
- Pseudaphritidae: The family inhabits streams and estuaries of southeastern Australia and Tasmania.
- Eleginopidae: These fishes are confined to southern South America.

The remaining five families have distributions that are primarily Antarctic or subantarctic:

- Harpagiferidae: The family is found mainly in sub-Antarctic regions, from Patagonia, the Falkland Islands, South Georgia, and the Scotia Arc to the tip of the Antarctic Peninsula and in Kerguelen, Crozet, and Heard Islands.
- Artedidraconidae: These fishes are exclusively Antarctic, from South Georgia along the Scotia Arc to the Antarctic Peninsula and around continental Antarctica.
- Nototheniidae: Thirty-four species are truly Antarctic, and the remainder occur in sub-Antarctica, Patagonia, and southern New Zealand.
- Bathydraconidae: The family occurs in the Antarctic, from South Georgia along the Scotia Arc and around the Antarctic continent.
- Channichthyidae: The family is confined primarily to the Antarctic but has one species in Patagonia/Falkland Islands. The other 14 species are found from South Georgia through the Scotia Arc and around the Antarctic continent; they also occur in the Bouvet and Kerguelen Islands but not as far as the Heard, Campbell, or Auckland Islands.

Habitat

Notothenioids are widespread in coastal waters of the Southern Ocean, on or over the continental shelf. Their depth range is from the surface to 1,370 fathoms (2,500 m). Aside from their dispersive larval stages, most avoid the open ocean, living on or near the bottom as adults. Several species of Nototheniidae and Channichthyidae have abandoned their ancestral benthic existence, however, to move up into the water column, becoming pelagic. Lacking swim bladders, these midwater forms have increased buoyancy in other ways. Their bones are less mineralized than those of demersal species, and they have high oil contents. One widespread pelagic species, the sardine-like nototheniid *Pleuragramma antarcticum*, retains larval features, including a persistent notochord surrounded by greatly reduced vertebral centra.

Behavior

Notothenioids tend to be sedentary bottom species that swim infrequently. Where swimming has been observed, it is labriform, propelled by undulatory sculling of the pectoral fins. The trunk and tail are used only for short, fast dashes. Even the pelagic forms are relatively inactive.

Feeding ecology and diet

Demersal notothenioids are generalized predators. Some forage nonselectively; others are ambush predators. Pelagic species, such as *P. antarcticum*, are more selective, feeding on copepods, larger midwater crustacea (e.g., krill, such as *Euphausia superba*), and pteropods (e.g., *Limacina*). In turn, these smaller midwater fishes are eaten, along with krill and squid, by the few large mesopelagic fish (such as the toothfish, *Dissostichus mawsoni*), by birds (penguins, skuas, and petrels), and seals. *D. mawsoni* is eaten by Weddell seals (*Leptonychotes weddelli*) and by killer whales (*Orcinus orca*).

Reproductive biology

Many Antarctic and subantarctic notothenioids breed biennially, with oocytes taking two years to ripen. Eggs are large and yolky and usually are spawned on or near the bottom. Spawning times differ with latitude and between species, ranging from early spring to early winter. Embryonic development and growth are slow, and most larvae hatch 6–12 months after spawning. The larval stages are pelagic, settling to the bottom after feeding in the plankton for six to nine months.

Conservation status

None of the notothenoids are on the IUCN Red List. Although the majority of notothenioids are not exploited and have widespread and abundant populations, several species have been the target of fisheries. The marbled rockcod, *Notothenia rossii*, has been seriously overfished but is no longer targeted, and it is recovering, albeit slowly. The Patagonian toothfish, *Dissostichus eleginoides*, has been exploited since the 1980s and has been given the designation of data deficient in a report on the conservation status of Australian fishes complied by the Threatened Fishes Committee of the Australian Society for Fish Biology.

After depleting the South Georgia/Falkland Islands toothfish fishery, attention shifted to Macquarie and Kerguelen Islands and then further west, to the southern Indian Ocean. In the mid-1990s the estimated illegal catch exceeded the legal catch, but in 1999 the Commission for the Conservation of Antarctic Marine Resources implemented a strict catch-reporting regime for toothfish, reducing the 1999/2000 illegal catch to an estimated 8,418 tons (7,637 tonnes), of a total 33,660 tons (30,536 tonnes). Since 1996 there has been a limited exploratory fishery for the closely related Antarctic toothfish, *D. mawsoni*, in the Ross Sea, where a catch of 1,000 tons (907 tonnes) was reported in 2001. Because most notothenioids share characteristics of low fecundity and slow growth, their fisheries are highly susceptible to overfishing and should be monitored closely.

Significance to humans

Some notothenioids have been harvested for fish meal and oil (e.g., *P. antarcticum*), but the group is exploited mainly for human consumption. The mackerel icefish, *Champsocephalus gunnari*, is the basis of small, but sustainable fisheries near South Georgia and the Kerguelen Islands. There is a ready market for the two toothfish species, *D. eleginoides* and *D. mawsoni*, which are highly palatable.

1. Mackerel icefish (*Champsocephalus gunnari*); 2. Emerald notothen (*Trematomus bernacchii*); 3. Naked dragonfish (*Gymnodraco acuticeps*); 4. Sailfin plunderfish (*Histiodraco velifer*); 5. Maori chief (*Notothenia angustata*). (Illustration by Michelle Meneghini)

Species accounts

Sailfin plunderfish
Histiodraco velifer

FAMILY
Artedidraconidae

TAXONOMY
Dolloidraco velifer Regan, 1914, McMurdo Sound, Antarctica.

OTHER COMMON NAMES
None known.

PHYSICAL CHARACTERISTICS
Attains a length of up to 7.5 in (19.2 cm) (length/depth ratio, 5:1). It has a large, depressed head with a conspicuous barbel on the chin; this barbel is one of the characteristic features of the family. The first dorsal fin is tall and narrow, consisting of only three flexible spines, and is directly above the operculum. The second dorsal fin is disproportionately tall and sail-like. The second dorsal, caudal, pectoral, and pelvic fins have brown and yellow striations. The body color is a light tan, with irregular dark blotches. Scales are absent, except for parts of the two lateral lines on the trunk.

DISTRIBUTION
Coastal waters of Antarctica, from the Weddell Sea clockwise to the Ross Sea.

HABITAT
This is a bottom-dwelling fish, found at depths of 48–2,190 ft (15–667 m). It is taken occasionally by scuba divers on the mud or gravel bottoms in McMurdo Sound and Terra Nova Bay.

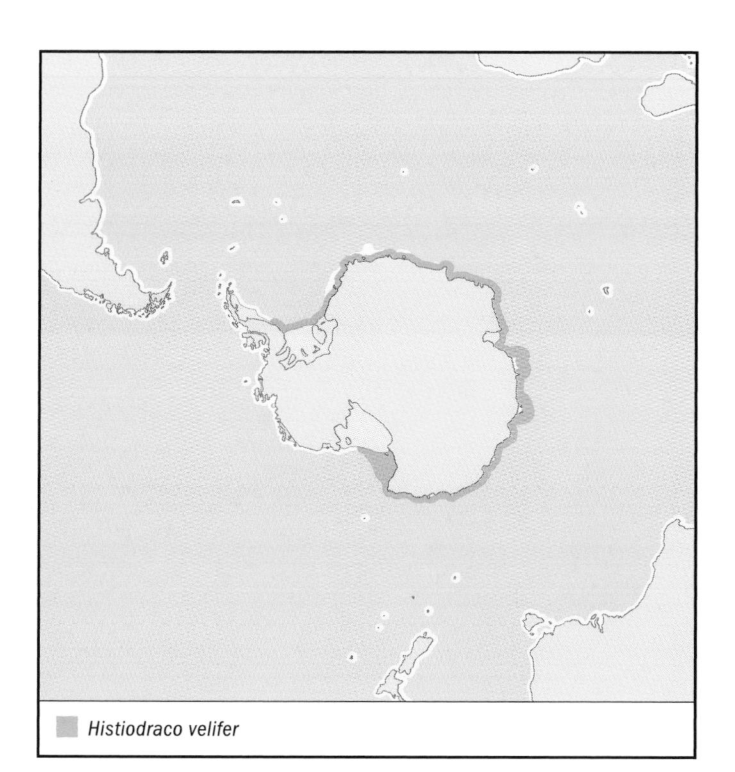

BEHAVIOR
In the aquarium, the sailfin plunderfish sits quietly on the bottom. The barbel is extended and occasionally twitched, mimicking a small worm. Histological examination shows that the barbel is highly innervated, with many tiny capsules resembling Pacinian corpuscles (a comomn type of pressure-sensitive sense organ amongst the vertebrates). Touching the barbel with forceps elicits a feeding lunge. This behavior implies that natural feeding may target mobile foragers, such as fish and amphipods.

FEEDING ECOLOGY AND DIET
Very little is known about the diet of the sailfin plunderfish, other than what can be inferred from behavioral observations in the aquarium. Where they have been examined, gut contents include krill and polychaete worms. Like most of the benthic notothenioids, it is probably an opportunistic feeder.

REPRODUCTIVE BIOLOGY
Nothing is known of the reproductive biology, as very few specimens have been collected.

CONSERVATION STATUS
Although this is an uncommon species, poorly represented in museum collections, it is not threatened.

SIGNIFICANCE TO HUMANS
None known. ◆

Naked dragonfish
Gymnodraco acuticeps

FAMILY
Bathydraconidae

TAXONOMY
Gymnodraco acuticeps Boulenger, 1902, Cape Adare, North Victoria Land, Antarctica.

OTHER COMMON NAMES
None known.

PHYSICAL CHARACTERISTICS
Grows to 13.5 in (34 cm) in length. It has an elongated, scaleless body (length/depth ratio, 9:0) with a long, pointed head. The coloring is yellowish to olive-brown, with darker blotches on the sides; it is pale on the ventral surface. The lower jaw extends well beyond the upper and bears two prominent fangs that are slanted backwards to prevent the escape of prey. In common with all members of the dragonfish family, it lacks the first spinous dorsal fin. This species produces abundant mucus when it is freshly caught.

DISTRIBUTION
South Shetland, the Antarctic Peninsula, and the eastern Antarctic continental shelf from the Weddell Sea to the Ross Sea.

HABITAT
Found mainly in shallow inshore waters, to a depth of 162 ft (50 m), although it has been caught as deep as 1,800 ft (550 m). It commonly resides in crevices or under rock ledges or anchor ice.

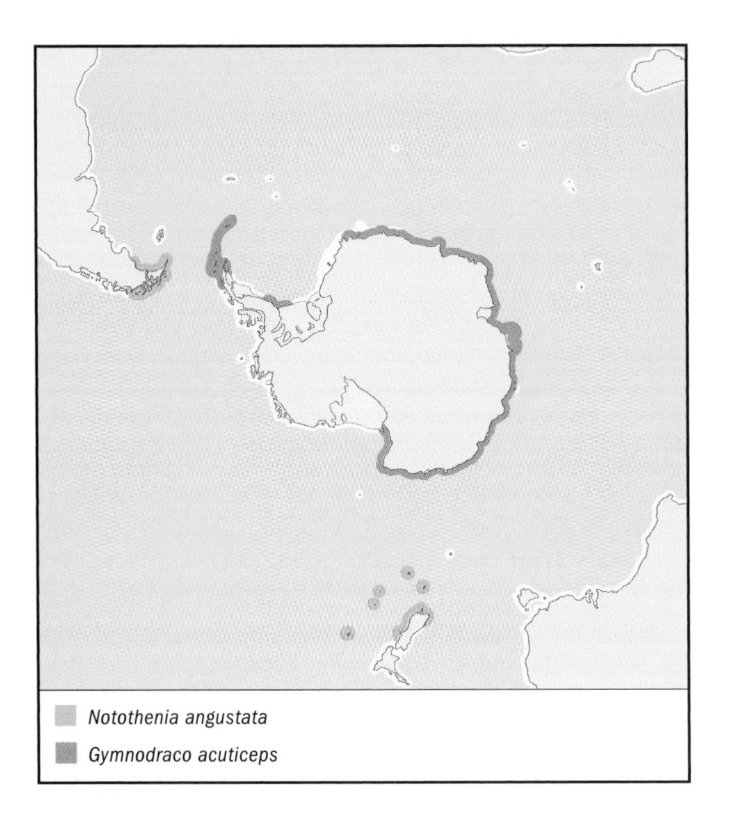

BEHAVIOR
This is an aggressive fish that rapidly consumes other aquarium inhabitants.

FEEDING ECOLOGY AND DIET
The diet varies with location. Near the Antarctic Peninsula, the dragon fish feeds on krill; in McMurdo Sound, fish, amphipods, fish eggs, and polychaetes are taken.

REPRODUCTIVE BIOLOGY
Spawning occurs in September; 0.12-in (3-mm) eggs are attached to a stone as a flattened patch of about a thousand and guarded. In the wild, adults tend egg masses until the larvae hatch and disperse. Development takes about a year. Free-swimming larvae, about 0.6 in (15 mm) long, hatch out in early summer and remain in the zooplankton for about six months.

CONSERVATION STATUS
Not threatened.

SIGNIFICANCE TO HUMANS
This species is proving to be a valuable research subject. Beginning in 2001 clusters of developing embryos near McMurdo Station have been harvested for investigation into the ontogeny of AFGPs. ◆

Mackerel icefish
Champsocephalus gunnari

FAMILY
Channichthyidae

TAXONOMY
Champsocephalus gunnari Lönnberg, 1905, Cumberland Bay, South Georgia.

OTHER COMMON NAMES
English: Crocodile icefish, pike glassfish; French: Poisson des glaces; Spanish: Draco rayado; Russian: Ledyanaya ryba.

PHYSICAL CHARACTERISTICS
Elongated, scaleless, pikelike body (length/depth ratio, about 7:2) with an elongated snout. The jaws do not protract. In coloring it is silvery gray, darker on the back, and silvery on the belly, with dark vertical stipes on the sides, reminiscent of those seen in mackerel. The gills are a pale yellowish; the blood is colorless, completely lacking hemoglobin. The absence of hemoglobin is a remarkable feature of all 15 icefish species, apparently caused by a single massive mutation that deleted the β-globin gene. Far from being an adaptation to Antarctic conditions, this has been termed a "disaptation." Several factors have enabled channichthyids to survive this evolutionary catastrophe: frigid seawater and blood plasma can carry more dissolved oxygen, and low body temperatures limit metabolic requirements. Subsequent evolution has compensated for the loss of an oxygen-binding pigment: blood volume, heart, vessel, and gill size and the perfusion rate of gills and blood vessels all have increased. As a consequence, channichthyids are surprisingly active, competing successfully with other notothenioids.

DISTRIBUTION
South Georgia and the islands of the Scotia Arc, southward to the northern Antarctic Peninsula; it also inhabits Bouvet, Kerguelen, and Heard Islands. A closely related species, *C. esox*, occurs in the Falklands Islands and Patagonia and is the only channichthyid to occur outside Antarctic/subantarctic waters.

HABITAT
This is a coastal species, found mainly between 330 and 1,140 ft (100–350 m). Mature adults are found offshore in summer, moving inshore to spawn in the fall (March–May).

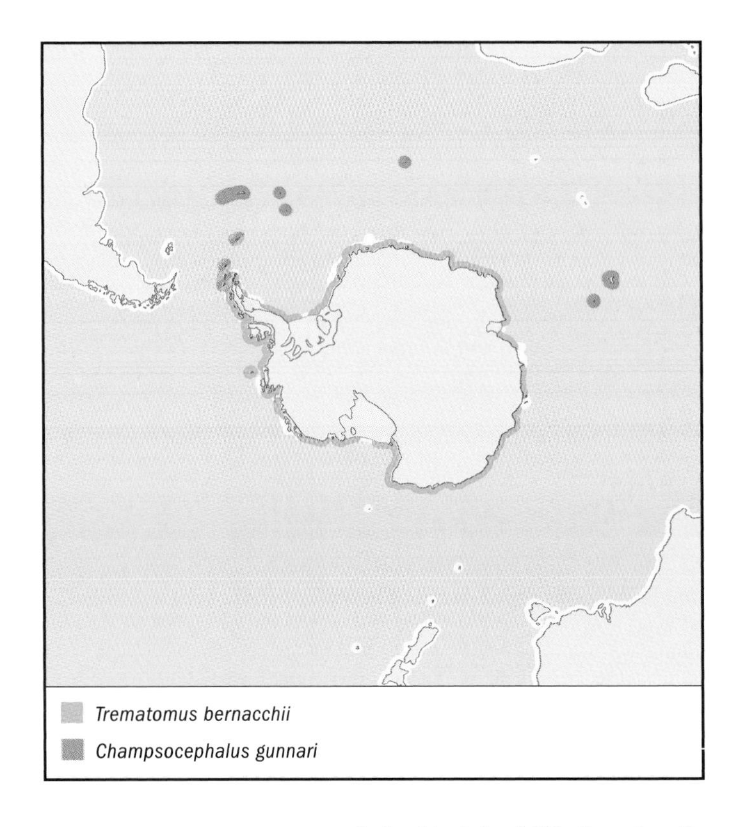

BEHAVIOR
The species aggregates in locations with dense krill populations, remaining near the bottom during the day and migrating upward with the krill at night.

FEEDING ECOLOGY AND DIET
The mackerel icefish feeds mainly on krill (*Euphausia superba*); it takes other euphausiids, mysids, and hyperiid amphipods (*Themisto gaudichaudii*) when krill numbers are low.

REPRODUCTIVE BIOLOGY
Reproductive parameters vary considerably between the different stocks. Maturation of oocytes takes less than a year, producing 1,300–31,000 eggs that are 0.12–0.16 in (3–4 mm) in diameter. When krill are scarce, oocytes may be resorbed, and as many as 60% of mature females may be nonreproductive. A marked three-year periodicity in reproduction is reported from the Kerguelen stock. Near South Georgia, spawning occurs from March to May; it takes place later in South Orkney and South Shetland and as late as July to August in the Kerguelen stock. Eggs are deposited in the depth range of 330–660 ft (100–200 m) and hatch after 30–180 days, depending on the stock. Newly hatched larvae are 0.05–0.07 in (12–17 mm) long and feed on copepods within 12 mi (20 km) offshore; larger juvenile fish move further offshore but remain over the continental shelf. Juveniles grow at 2.5–3.5 in (6.4–8.9 cm) per year, reaching sexual maturity at three years at a length of 10 in (25 cm).

CONSERVATION STATUS
Not listed by the IUCN. The various stocks of mackerel icefish have been targeted by minor trawl fisheries since 1974. Catches fluctuate from year to year, but they appear to be sustainable for the South Georgia and Kerguelen stocks. In South Orkney and South Shetland, the stock was depleted rapidly below sustainable levels and is no longer being fished.

SIGNIFICANCE TO HUMANS
A total catch of 4,295 tons (3,896 tonnes) was reported from the Commission for the Conservation of Antarctic Marine Resources management area in 1999–2000. ◆

Maori chief
Notothenia [Parenotothenia] angustata

FAMILY
Nototheniidae

TAXONOMY
Notothenia angustata Hutton, 1875, Dunedin Harbour, New Zealand.

OTHER COMMON NAMES
English: Black cod.

PHYSICAL CHARACTERISTICS
Grows to 2 ft (60 cm) in length. Mature specimens have a massive body (length/depth ratio, 5:1) with a large, slightly flattened head; heavy bony ridges over the eyes; and large scales. The coloring is dark olive-gray to blue-black, with varying light spots and lines; there are vague dark vertical bands on the trunk. Juveniles are less flattened, with deeper bodies, deard reddish bronze coloring, and prominent light spots.

DISTRIBUTION
Southern waters of New Zealand, from Stewart Island to Kaikoura. It also is found in Chatham, Snares, Auckland, and Campbell Islands and in Punta Arenas, Chile, and adjacent water.

HABITAT
Adults are demersal on rocky bottoms, from near shore to 330 ft (100 m) depth. Juveniles are found in tide pools and among kelp near the shore.

BEHAVIOR
Like most other notothenioids, the Maori chief is solitary and never occurs in schools. Adults are shy and sedentary, foraging among rocks and seaweed mainly at night. During the day, they hide in crevices or under seaweed, occasionally lunging for passing prey. Smaller fish are sucked into their mouth by the sudden expansion of the jaws and opercula. Juveniles are more active than adults, often leaving the bottom to forage in the upper layers of kelp beds.

FEEDING ECOLOGY AND DIET
The Maori chief feeds on a wide range of bottom invertebrates, crabs, and smaller fishes. Occasionally, it ingests seaweed.

REPRODUCTIVE BIOLOGY
The life cycle of the Maori chief is not well known. Eggs ripen in November and presumably are deposited on the bottom. Juveniles appear in rock pools in the summer (January) after an incubation period of about 30 days and a brief pelagic dispersal phase.

CONSERVATION STATUS
Not listed by the IUCN. This species was caught for the local market up to the 1950s, when stocks were reported to be overfished. There is currently no commercial fishery for this species, nor has it been assigned a quota under New Zealand's quota management system.

SIGNIFICANCE TO HUMANS
This is said to be a palatable species, but its fearsome appearance and coarse texture relegate its use mainly to bait in lobster pots. ◆

Emerald notothen
Trematomus bernacchii

FAMILY
Nototheniidae

TAXONOMY
Trematomus bernacchii Boulenger, 1902, Cape Adare and Duke of York Island, North Victoria Land, Antarctica.

OTHER COMMON NAMES
English: Emerald rockcod, bernak; French: Bocasson émeraude; Russian: Trematom-pestryak; Spanish: Austrobacalao esmerelda.

PHYSICAL CHARACTERISTICS
Reaches a maximum length of 12 in (30 cm) and a maximum weight of 12.3 oz (350 g). It is thick-bodied (length/depth ratio, about 3:9) and light brown to pinkish brown on the sides, with irregular darker blotches; the belly is silvery gray. It is covered with ctenoid scales, except for a patch between the eyes, where there may be a single row of scales. Three distinct color varieties are known:

- continuous white band across the nape and opercula
- an interrupted band, consisting of a narrow medial stripe on the nape and on each operculum
- no white marking

DISTRIBUTION
Coastal waters of Antarctica, including the Antarctic Peninsula and South Shetland and South Orkney Islands.

HABITAT
It is most common on boulder, rock, or gravel bottoms or in sponge beds at depths of 18–1,200 ft (5–370 m), but it has been caught as deep as 2,280 ft (700 m). Smaller individuals are found in shallow water, whereas mature fish are more common in deeper water.

BEHAVIOR
The emerald notothen is a sedentary demersal species, spending most of its time immobile on the bottom. It swims infrequently and slowly, sculling with the pectoral fins and making short excursions across the bottom or ascending about 3 ft (1 m) into the water column. In the aquarium, dominant individuals establish and defend territories.

FEEDING ECOLOGY AND DIET
This is an opportunistic feeder and scavenger, primarily on benthic invertebrates. Small crustaceans are the most common dietary item, followed by polychaetes, molluscs, fish, and fish eggs. It ingests seaweed, but its nutritional contribution is unknown.

REPRODUCTIVE BIOLOGY
The life span of the emerald notothen is 8–10 years; in McMurdo Sound, first spawning occurs at five years. Females spawn biennially; initial oocyte development takes one year and active vitellogenesis (deposition of yolk within the growing egg) another year. Spawning takes place over a brief period, from October to November in Adélie Land and December to January in McMurdo Sound. Fecundity is on the order of 1,500–3,000 eggs, which are attached to rocks, algae, or sponges; mature eggs are 0.16 in (4 mm) in diameter. There are reports that the emerald notothen tends egg masses in sponges.

CONSERVATION STATUS
Not threatened. The emerald notothen is widespread and common; considering its small size and high latitude, it is not likely to come under any significant pressure from fisheries. Its limited mobility probably would lead to extremely rapid depletion of any local populations that might be subjected to commercial fishing.

SIGNIFICANCE TO HUMANS
This species has played an important scientific role in the study of anatomical, physiological, and biochemical adaptations to the Antarctic environment. Most significantly, it has featured prominently in the elucidation of biological antifreeze compounds. ◆

Resources

Books

di Prisco, G., B. Maresca, and B. Tota, eds. *Biology of Antarctic Fish.* Berlin: Springer-Verlag, 1991.

di Prisco, G., E. Pisano, and A. Clarke, eds. *Fishes of Antarctica: A Biological Overview.* Milan: Springer-Verlag, 1998.

Eastman, J. T. *Antarctic Fish Biology: Evolution in a Unique Environment.* San Diego: Academic Press, 1993.

Gon, O, and P. C. Heemstra, eds. *Fishes of the Southern Ocean.* Grahamstown, South Africa: J. L. B. Smith Institute of Ichthyology, 1990.

Kock, Karl-Hermann. *Antarctic Fish and Fisheries.* Cambridge: Cambridge University Press, 1992.

Miller, Richard Gordon. *History and Atlas of the Fishes of the Antarctic Ocean.* Carson City, NV: Foresta Institute for Ocean and Mountain Studies, 1993.

Pisano, E., C. Ozouf-Costaz, C. Bonillo, and F. Mazzei "Pathways of Chromosomal Change During Evolution of Notothenioid Fishes." In *Antarctic Ecosystems: Models for Wider Ecological Understanding*, edited by W. Davison, C. Howard-Williams, and P. Broady. Christchurch: Caxton Press, 2000.

Periodicals

Chen, L., A. L. DeVries, and C. H. C. Cheng. "Evolution of Antifreeze Glycoprotein Gene from a Trypsinogen Gene in Antarctic Notothenioid Fish." *Proceedings of the National Academy of the USA* 94 (April 1997): 3811–3816.

Eastman, J. T. "Antarctic Notothenioid Fishes as Subjects for Research in Evolutionary Biology." *Antarctic Science* 12 (2000): 276–287.

Eastman, J. T., and R. R. Eakin. "An Updated Species List for Notothenioid Fish (Perciformes: Notothenioidei), with Comments on Antarctic Species." *Archive of Fishery and Marine Research* 48 (2000): 11–20.

Eastman, J. T., and A. R. McCune. "Fishes on the Antarctic Continental Shelf: Evolution of a Marine Species Flock?" *Journal of Fish Biology* 57, suppl. A (2000): 84–102.

Kock, K.-H., and I. Everson. "Biology and Ecology of Mackerel Icefish, *Champsocephalus gunnari*: An Antarctic Fish Lacking Hemoglobin." *Comparative Biochemistry and Physiology* 118A, no. 4 (1997): 1067–1077.

Lack, M. "Antarctic Toothfish: An Analysis of Management, Catch and Trade." *TRAFFIC Oceania* (2001): 1–27.

Lack, M., and G. Sant. "Patagonian Toothfish: Are Conservation and Trade Measures Working?" *TRAFFIC Bulletin* 19, no. 1 (2001): 1–18.

Montgomery, J., and K. Clements. "Disaptation and Recovery in the Evolution of Antarctic Fishes." *Trends in Ecology and Evolution* 15 (2000): 267–271.

Stankovic, A., K. Spalik, E. Kamler, et al. "Recent Origin of Sub-Antarctic Notothenioids." *Polar Biology* 25 (2002): 203–205.

Zhao, Y., M. Ratnayake-Lecamwasam, S. K. Parker, et al. "The Major Adult α-Globin Gene of Antarctic Teleosts and Its Remnants in the Hemoglobinless Icefishes: Calibration of the Mutational Clock for Nuclear Genes." *Journal of Biological Chemistry* 273, no. 24 (1998): 14745–14752.

Resources

Organizations

Australian Society for Fish Biology. 123 Brown Street (PO Box 137), Heidelberg, Victoria 3084 Australia. Phone: 61 (3) 9450 8669. Fax: 61 (3) 9450 8730. E-mail: john.koehn@nre.vic.gov.au Web site: <http://www.asfb.org.au>

Commission for the Conservation of Antarctic Marine Living Resources (CCAMLR). PO Box 213, North Hobart, Tasmania 7002 Australia. Phone: 61 (3) 6231 0366. Fax: 61 (3) 6234 9965. E-mail: ccamlr@ccamlr.org Web site: <http:/www.ccamlr.org>

TRAFFIC International. 219c Huntington Road, Cambridge, CB3 0DL United Kingdom. Phone: 44 (0) 1223 277427. Fax: 44 (0) 1223 277237. E-mail: traffic@trafficint.org Web site: <http://www.traffic.org>

Other

Prof J. T. Eastman, Ohio University. "Antarctic Fishes." (March 27, 2003) <http://www.oucom.ohiou.edu/dbms-eastman/index.htm>

John A. Macdonald, PhD

Trachinoidei

(Weeverfishes and relatives)

Class Actinopterygii
Order Perciformes
Suborder Trachinoidei
Number of families 13

Photo: A four-spined sandperch (*Parapercis tetracantha*) in the Balayan Bay of Luzon Island in the Philippines. (Photo by Fred McMonnaughey/Photo Researchers. Reproduced by permission.)

Evolution and systematics

A group of fishes referred to as the "trachinoids" was recognized in 1860 and has undergone significant modification over the years, with several groups variously added or subtracted and then added again. The current notion of the Trachinoidei actually is quite similar to that of 150 years ago, although a few groups have been put in the group that were not described at that time. The Trachinoidei is almost certainly an unnatural grouping of fishes, meaning that there is little evidence to suggest that its included families are related more closely to each other than they might be to any perciform family outside the trachinoids. Pietsch, and then Pietsch and Zabetian, offered the first comprehensive modern phylogenetic studies of relationships within the suborder. They did not include Pholidichthyidae and Trichodontidae in their surveys, considering the examined taxa to represent "the core of, but not necessarily to delimit," the suborder. Their results, listed sequentially in phylogenetic order (with each group being the sister to all remaining taxa) were as follows: i) Cheimarrichthyidae; ii) Pinguipedidae; iii) Percophidae, Trichonotidae, and Creediidae; iv) Champsodontidae and Chiasmodontidae; v) Leptoscopidae; vi) Ammodytidae; vii) Trachinidae; and viii) Uranoscopidae.

Johnson, and later Mooi and Johnson, reexamined much of the data and determined that many of the features used to define trachinoid relationships by Pietsch and Zabetian are distributed widely among many perciforms and do not necessarily indicate close affinity. The families themselves seem to be natural groups, except the Percophidae. In this grouping, the members of the percophid subfamily Hemerocoetinae exhibit derived features of the suspensorium, indicating a phylogenetic relationship with the Creediidae and Trichonotidae to the exclusion of the percophid subfamilies Percophinae and Bembropinae. Mooi and Johnson refuted any close relationship between Champsodontidae and Chiasmodontidae, suggesting that the former more likely has relatives among the Scorpaeniformes and that relatives of the remaining trachinoid families might be searched for more fruitfully among other perciforms rather than among themselves. Mooi and Johnson hinted that trichodontids might be scorpaeniforms, but Nazarkin and Voskoboinikova suggested a new suborder for the family, emphasizing its differences from other trachinoids but not resolving its enigmatic phylogenetic position. McDowall maintained that the monotypic Cheimarrichthyidae should be considered a member of the Pinguipedidae. The relationships of the Pholidichthyidae also are unclear.

The fossil record of trachinoids is scant. A member of the Trachinidae is described from the base of the Middle Eocene from Monte Bolca, Italy (about 49 million years ago, or mya). Reported from the Agnev Formation (Upper Miocene, about 10 mya) of Sakhalin Island, Russia, are a species of Trichodontidae and a species put in a new family, the Trispinacidae, which is considered to be related to the Pinguipedidae.

There are 13 families with 235+ species in 53 genera: Ammodytidae (seven genera, 27 species), Champsodontidae (one genus, 13 species), Cheimarrichthyidae (monotypic), Chiasmodontidae (four genera, 15+ species), Creediidae (seven genera, 16+ species), Leptoscopidae (three genera, five species), Percophidae (11 genera, 44 species), Pholidichthyidae (one genus, two species), Pinguipedidae (five genera, 50 species), Trachinidae (two genera, four species), Trichodontidae (two genera, two species), Trichonotidae (one genus, six species), and Uranoscopidae (eight genera, 50 species).

Physical characteristics

Not surprisingly, given that most families are not related closely, physical characteristics vary considerably among trachinoids. Ammodytids, the sand lances, are small to moderate sized, growing to about 16 in (40 cm). They have narrow, elongate bodies with small heads. The lower jaw protrudes beyond the upper, and there are no teeth in the jaws. These fishes have a single long dorsal fin without spines and reduced or absent pelvic fins. The body has a series of oblique folds of skin called plicae. Ammodytids swim with a distinctive eel-like undulation that makes them easy to identify in the water. Champsodontids, the gapers, are small, growing only to 6 in (15 cm), but they are large-headed, with a large mouth armed with long teeth. There is a large spine on the preopercle and two dorsal fins, the first with four to six spines. The pectoral fin is very small, but the pelvic fin is quite large. The body is covered with spinoid, nonoverlapping scales that are raised above the body surface by a small pedicle; a series of sensory papillae can be found between the scales.

Cheimarrichthyids, the New Zealand torrentfishes, are smaller than 8 in (20 cm), with a broad, somewhat flattened, wedge-shaped head; subterminal mouth; a robust, scaled body with a long, soft dorsal fin preceded by three to five short spines; and large pectoral and pelvic fins. Chiasmodontids, the deep-sea swallowers, are moderately sized bathypelagic fishes, growing to 10 in (26 cm). They are elongate, with a large head, a large mouth with enormous teeth, a rugose head pitted with sensory pores, a scaleless body that sometimes has prickles, a large-pored lateral line, and two separate dorsal fins, the first with seven to 13 flexible spines. These fishes have a hugely distensible gut that permits them to eat prey twice their own size, giving rise to their common name. Some species bear photophores. Creediids, the sandburrowers, are the smallest trachinoids, at 1.5–3.0 in (3.5–8 cm). They are elongate, with slightly protruding eyes, a fleshy snout projecting beyond the lower jaw, and a scaled body with a single dorsal fin and reduced pelvic fins. Leptoscopids, the southern sandfishes, reach about 16 in (40 cm). They are elongate, with a broad, blunt head and small, dorsally situated eyes; a dense fringe of cirri bordering the lips; a scaled body; a single dorsal fin without spines; very deep pectoral fins, and widely separated pelvic fins.

A spotted sanddiver (*Trichonotus setiger*) dives head first under the sand at the first sign of danger. (Photo by David Hall/Photo Researchers, Inc. Reproduced by permission.)

A blue cod (*Parapercis colias*) near Fiordland, New Zealand. (Photo by Michael Patrick O'Neill/Photo Researchers, Inc. Reproduced by permission.)

Percophids, the duckbills or flatheads, are small, reaching 12 in (30 cm), and elongate; most have a depressed head, large eyes, a scaled body, and two dorsal fins, the first with two to six spines that can be very elongate in some species. Pholidichthyids, the convict blennies, are eel-shaped and grow to 18 in (45 cm), with a rounded head, a single nostril, a scaleless body, spineless long dorsal and anal fins that are continuous with the caudal fin, and reduced pelvic fins. Pinguipedids, the sandperches, are mostly elongate and grow to about 12 in (30 cm). They have an almost cylindrical and scaled body; a single dorsal fin with a shorter, spinous anterior portion; and large pelvic fins. Trachinids, the weeverfishes, are elongate, growing to 17 in (42 cm), with a large oblique mouth (fringed in one species), eyes situated dorsally, and a strong opercular spine with a venom gland. There are two dorsal fins; the first has five to eight strong spines, each bearing a venom gland, and the second is long and directly opposed to the equally long anal fin. The pectoral fins are large, and there are small ctenoid scales in oblique rows along the body.

Trichodontids, the sandfishes, are superficially similar to trachinids. They reach 11.8 in (30 cm), but they are scaleless and have spines on the preopercle rather than the opercle and a longer first dorsal fin with 10–15 spines. Their lips are fringed. Trichonotids, the sanddivers, have a scientific name similar to that of the previous family, which might cause confusion. They are very different fishes, in that they are very elongate and cylindrical, growing to only about 6 in (16 cm). They have a dorsal iris flap with narrow extensions resembling eyelashes, a lower jaw with a fleshy extension beyond the upper jaw, a single long dorsal fin with three to eight anterior spines that are elongate and filamentous in males of some species, fanlike pelvic fins that are larger than the pectoral fins, and a scaled body.

The protruding lower jaws or snouts of this and several other of the sanddiving families presumably facilitate substrate penetration. Uranoscopids, the stargazers, are so named because the eyes are on the top of the head and directed dorsally, looking skyward in typical species. Stargazers, which reach 30 in (75 cm) have a squarish, bulldog-like head encased in sculptured bones, with an almost vertical mouth having fringed lips, a heavy body with or without scales, and a prominent cleithral spine (putatively with a venom gland) just dorsal to the pectoral fin. These fishes have a single dorsal fin with up to four rudimentary spines or two dorsal fins, the first having four to five weak spines, and very large rectangular and fleshy pectoral fins. Some species have a wormlike appendage extending from the respiratory valve of the lower jaw, which is used as "bait" and some have electric organs derived from modified eye muscles, which are reported to be capable of producing 50 volts.

The color pattern varies considerably among trachinoids. Often they are pale or silvery, and they may have bars, saddles, and spots, sometimes with dark markings on the dorsal and caudal fins. Species of some families (e.g., pinguipedids, percophids, and trichonotids) exhibit sexual dichromatism as well as dimorphism. Chiasmodontids, like most bathypelagic fishes, are dark brown or black.

Distribution

Ammodytidae are cold to tropical marine fish found in the Arctic, Atlantic, Indian, and Pacific Oceans and the Black, Mediterranean, and Red Seas. Champsodontidae is a marine family of the Indo-Pacific. Cheimarrichthyidae live in freshwater streams of New Zealand; the larvae are marine. Chiasmodontidae are oceanic marine fishes found in the Atlantic, Indian, and Pacific Oceans. Creediidae is an Indo-West Pacific marine fish. Leptoscopidae are marine fishes found near coastal Australia and New Zealand. Percophidae are marine fishes of the Atlantic, Indian, and Pacific Oceans and the Red Sea. Pholidichthyidae are marine fishes of the Indo-West Pacific, found from coastal Philippines to northern Australia. Pinguipedidae inhabit the Atlantic, Indian, and Pacific Oceans. Trachinidae are marine fishes distributed in the northeastern Atlantic Ocean and the Mediterranean and Black Seas. Trichodontidae are found in the northern Pacific Ocean. Trichonotidae is a marine family of the Indo-West Pacific. Uranoscopidae inhabit the Atlantic, Indian, and Pacific Oceans and the Black, Mediterranean, and Red Seas. Within many of these families, particular genera and species have much more restricted distributions.

Habitat

Most species are marine inshore fishes of tropical to cold temperate regions and are associated in some way with sandy to muddy substrate—sitting on it, in it, or hovering near it to dive for protection when threatened. Exceptions include the Pholidichthyidae, which live in burrows excavated under coral; the Cheimarrichthyidae, which hug the bottom of fast-flowing portions of freshwater streams; the Champsodontidae, which can be found as deep as 3,600 ft (1,100 m) and form large shoals that rise to the surface at night; and the Chiasmodontidae, which are oceanic bathypelagic fishes found almost to 9,500 ft (2,900 m). Even uranoscopids can be found as deep as 2,300 ft (700 m) and percophids to at least 1,970 ft (600 m). Leptoscopids and uranoscopids are the only families with species that normally are found in estuaries as adults.

Behavior

Most trachinoids are solitary or form loose aggregations in appropriate habitat. Ammodytids are an exception, often forming dense schools ranging from hundreds to several thousand individuals. Champsodontids also occur in large shoals. Only pinguipedids are known to be territorial. For example, *Parapercis cylindrica* is a protogynous hermaphrodite and polygynous, defending a territory of about 180 ft^2 (17 m^2), within which two to five females defend smaller areas. Male trichonotids have been observed displaying to each other using their long, filamentous dorsal fin spines. It is likely that hemerocoetine percophids, which also have elongated anterior dorsal fin spines, display in a similar manner. Several families have species that exhibit dichromatism and dimorphism, which suggests that display plays a role in communication between or among the sexes.

The activity of ammodytids is related to tidal currents and light levels; they avoid strong tidal currents and are diurnal feeders. At night they hide in sandy substrate and form large shoals during daylight hours and low current and tidal periods. European species are known to bury in the bottom during the low light intensity of northern winters. In contrast, uranoscopids and trachinids apparently move about more at night. Champsodontids exhibit diel movements, moving from the depths to the surface at night. Cheimarrichthyids may migrate to spawn, with females moving from the upper reaches to the lower reaches, where the males are, in summer or autumn. It has been suggested that some species of ammodytids migrate inshore for summer and offshore for winter months. Trichodontids move to shallow rocky shores to spawn, returning to deeper waters after the eggs are laid.

Feeding ecology and diet

All trachinoids are carnivorous, and most are piscivorous, or at least include fishes in the diet. Ammodytids and some of the smaller trachinoid families, such as Trichonotidae, feed on zooplankton or small crustaceans. Pinguipedids feed principally on small benthic crustaceans, and leptoscopids include marine worms in the diet. Uranoscopids, leptoscopids, creediids,

A well-camouflaged marbled stargazer (*Uranoscopus bicinctus*) lies in wait for small fish. (Photo by David Hall/Photo Researchers. Reproduced by permission.)

Two male trichonotids display to each other with their fins erect. (Illustration by Barbara Duperron)

trichonotids, and perhaps trachinids and trichodontids are sit-and-wait predators, hiding in the substrate with only the lips, the top of the head, and the eyes protruding, sucking in unsuspecting fishes and other prey.

Several families that burrow in sand have vertical mouths to facilitate upward attack and fringed lips and gill openings that are thought to prevent sand from entering their mouths and gills while they wait for prey. Uranoscopids also have nasal passages opening into the mouth cavity (which is highly unusual in fishes), permitting them to breathe without opening the mouth while buried. Some have a worm-like appendage inside the mouth that they use as a lure. One genus, *Astroscopus*, has modified eye muscles that can produce an electric shock that might be used to stun prey, though it might instead ward off predators. Pinguipedids and percophids rest on the bottom and chase down prey that is on or near the substrate. Ammodytids forage in the water column. Juvenile pholidichthyids leave their burrows in swarms to feed on plankton during the daytime, streaming back into their tunnels each night to join their parents. The adults have never been observed to leave the burrows, and their diet remains unknown. Chiasmodontids alternate from a chase-and-grab to a floating-trap mode of feeding. They have very large teeth and an extensible stomach to hold on to and eat almost any prey item that they might run across in the relatively food-scarce bathypelagic habitat.

Reproductive biology

Courtship and mating information is scant for most trachinoids. All studied members of the pinguipedid genus *Parapercis* are protogynous hermaphrodites, and at least some are territorial and have harems. Males exhibit head bobbing and pectoral fanning of harem females beginning about 40 minutes before sunset. This behavior culminates in a short ascent from the bottom (less than about 3 ft, or 1 m) for spawning within half an hour after sunset. The genus spawns year-round, with peak activity in summer, a pattern characteristic of tropical species in general and likely the case for tropical representatives of other trachinoid families.

Ammodytids spawn demersal, adhesive eggs, forming clumps on sand or gravel substrate in shallow water. Eggs are not quite spherical and are small, about 0.04 in (1 mm). Females produce 1,800–22,000 eggs, which take from two to 13 weeks to hatch pelagic larvae that are 0.12–0.18 in (3–4.5 mm) long. Most ammodytid species spawn in fall and winter, though some European species spawn in spring. Trichodontids spawn in winter, producing 600–2,300 eggs in a gelatinous mass attached to either rocks or seaweed, depending on the species. Eggs are relatively large, about 0.14 in (3.5 mm), and yellow, taking from two months to one year to hatch large, pelagic larvae, about 0.6 in (15 mm). Cheimarrichthyids produce up to 30,000 small (about 0.02 in, or 0.6 mm), demersal eggs in rivers during late summer and fall; upon hatching larvae go to sea as part of a pelagic stage.

Champsodontids, chiasmodontids, creediids, leptoscopids, pinguipedids, and trachinids probably are broadcast spawners and produce unornamented, small (< 0.06 in, or 1.5 mm), pelagic eggs, with most shore species hatching in two to six days. The newly hatched pelagic larvae are 0.08–0.18 in (2–4.5 mm) long. Uranoscopids also have pelagic eggs, but they are unusually large (up to 0.1 in, or 2.5 mm); at least some are ornamented with a polygonal network, perhaps helping to maintain the eggs in the water column. Both trachinids and northern representatives of uranoscopids spawn in the spring and summer. Eggs are unknown for the Percophidae, Pholidichthyidae, and Trichonotidae.

Trachinoid larvae can be quite distinctive. Most chiasmodontid larvae bear tiny spicules, whereas one genus, *Kali*, has a unique "gargaropteron" stage, with a large head, oversized pectoral and pelvic fins, and no spicules. Champsodontid larvae have extensive head spination and bear an elongate and narrow opercular appendage that can be as much as 40% of the body length, until it gradually shortens and is resorbed by the time the fish grows to about 0.4 in (10 mm). At least one species of trachinids, *Echiichthys vipera*, has large preopercular spines and large, heavily pigmented pelvic fins. Pholidichthyid larvae and small juveniles (less than 12 in, or 30 cm) are strikingly colored with black and white stripes. They live in burrows with the parents at night, where they attach to the ceiling and hang motionless from mucous strands secreted by the four glands on top of their heads.

Conservation status

Several species and even families have restricted ranges and exhibit considerable endemism. Many species within families are considered common, in that they are encountered frequently, but population estimates are unknown, except for some species of ammodytids. Although several species are known only from a few specimens, this is generally thought to be due to collecting artifact (difficulty of surveying habitats and so forth) rather than necessarily reflecting actual rarity. No trachinoids are considered endangered, although populations of ammodytids and the trichodontid *Arctoscopus* are monitored because of their commercial importance. The latter experienced overfishing in parts of Japan and had not recovered by 2000.

Significance to humans

Only a few trachinoids are targets for fisheries. Ammodytids are the basis of an important fishery in the North Sea and Japan. Total catch for the multinational North Sea fishery averaged almost 937,000 tons (850,000 metric tons) per year over the years 1996–2000, but this figure fluctuated quite widely and has been more than one million tonnes in a single season. This is an industrial fishery producing fishmeal and oil. Ammodytids are a food fish in Japan, marketed fresh or dried, averaging about 110,000 tons (100,000 metric tons) per year from 1995 to 1999. There is no major fishery for ammodytids in North America, though a minor bait fishery exists in New England, landing only about 22 tons (20 metric tons).

In addition, ammodytids have an indirect impact on humans, as they are important forage fish for seabirds, marine mammals, and other fishes, acting as intermediaries in the food chain by feeding on zooplankton and providing energy to higher predators. Fluctuations in seabird populations in the North Sea appear to be correlated with changes in ammodytid availability, and the abundance of sand lances as a forage fish has implications for such fishery species as mackerel and yellowtail flounder. Similarly, champsodontids are thought to be important forage fishes for commercial species. The dry, but tasty flesh of the greater weever (*Trachinus draco*) is esteemed highly in southern Europe, where it is taken commercially in small quantities; the venomous glands of its spines apparently have discouraged a large fishery. The trichodontid *Arctoscopus japonicus* supports an important commercial fishery as a food fish in Japan. Colloquially, the Japanese name is *kaminarino*, the "thunderfish," referring to its appearance inshore for spawning in late November, when thunderstorms are a common phenomenon in the northern areas where it occurs.

It has been suggested that the common name of the Trachinidae, weeverfishes, derives from the Anglo-Saxon word *wivere*, meaning "viper." Their severe toxicity has been known for at least 2,400 years, since the time of Apollodorus of Alexandria, who is said to have written a book on stinging and biting animals. Modern study has shown that trachinids have grooved opercular and dorsal fin spines bearing venom glands that cause severe pain and occasionally fever, vomiting, and heart failure. Shrimpers sometimes trap trachinids as by-catch and need to handle them with care. Uranoscopids also should be avoided, as they have stout cleithral spines used with good effect for protection. These spines, too, have been reported to be associated with venom glands, but some modern researchers suggest that this association needs to be reappraised. Some uranoscopids also have electric organs and so should be treated with respect. Because members of these two families bury themselves in the sand in shallow water, they are of concern for beachgoers and swimmers, who might tread on them.

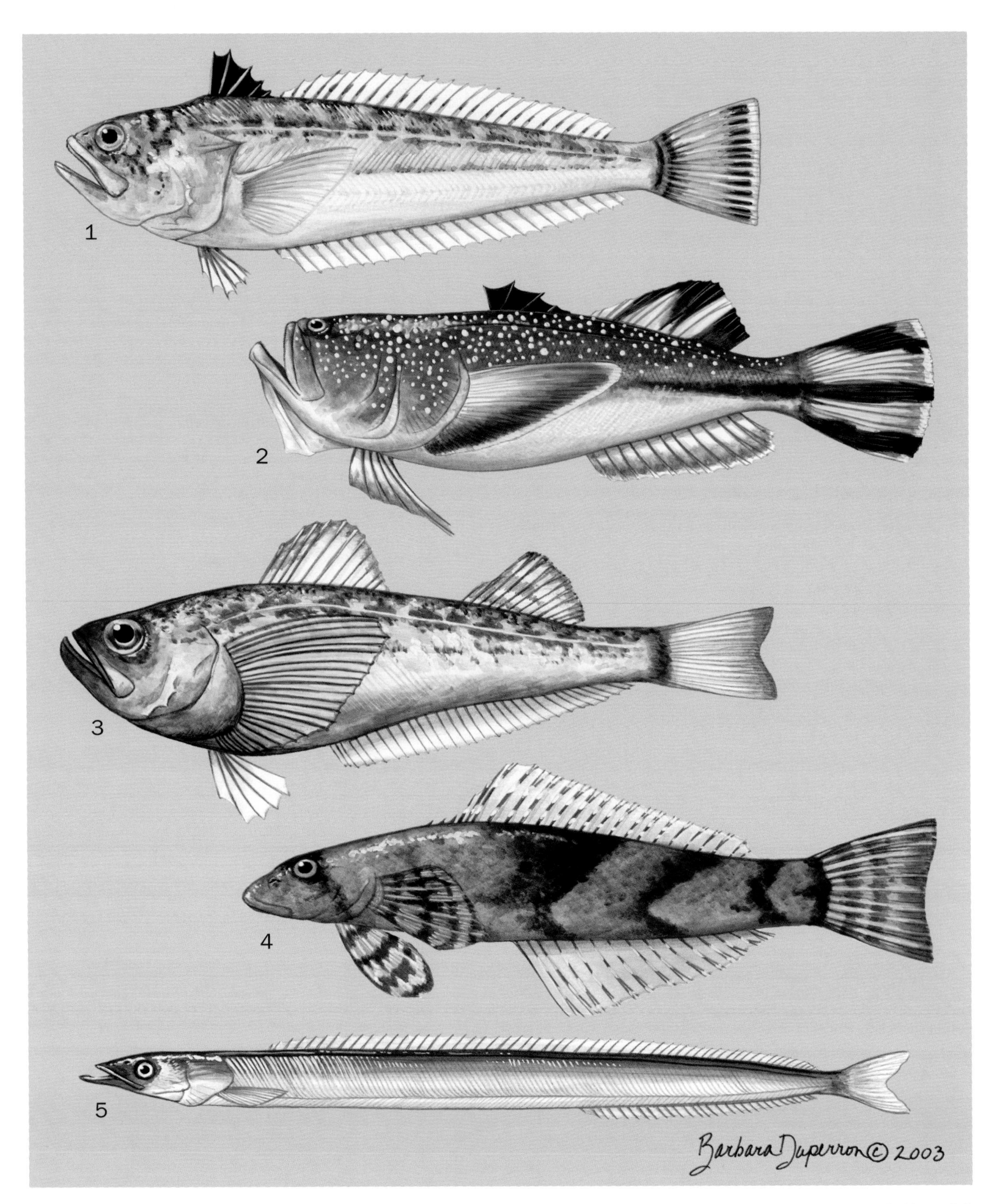

1. Lesser weever (*Echiichthys vipera*); 2. Northern stargazer (*Astroscopus guttatus*); 3. Sailfin sandfish (*Arctoscopus japonicus*); 4. Torrentfish (*Cheimarrichthys fosteri*); 5. Inshore sand lance (*Ammodytes americanus*). (Illustration by Barbara Duperron)

Species accounts

Inshore sand lance
Ammodytes americanus

FAMILY
Ammodytidae

TAXONOMY
Ammodytes americanus DeKay, 1842, Stratford, Connecticut, United States.

OTHER COMMON NAMES
English: American sand lance, sand eel, lance.

PHYSICAL CHARACTERISTICS
Slender and elongate, with a long head and sharply pointed snout and a large, toothless mouth, with the lower jaw projecting far beyond the upper. Long and spineless dorsal fin with 52–61 segmented rays, long and spineless anal fin with 26–33 segmented rays, and forked caudal fin. Has 106–126 (usually 112–124) oblique folds of skin called plicae with cycloid scales underlying them and 63–71 vertebrae (usually 65–70). Coloring is olive, brownish, or bluish green above, with silvery sides and a white belly; some have a longitudinal stripe of iridescent steel blue along each side. Grows to 6.3 in (16 cm).

DISTRIBUTION
Atlantic coast of North America from as far south as Chesapeake Bay (perhaps Cape Hatteras) to Newfoundland and northern Labrador.

HABITAT
Usually shallow water (< 6.5 ft, or 2 m) in estuaries or along coasts over sand or fine gravel substrates used for burrowing.

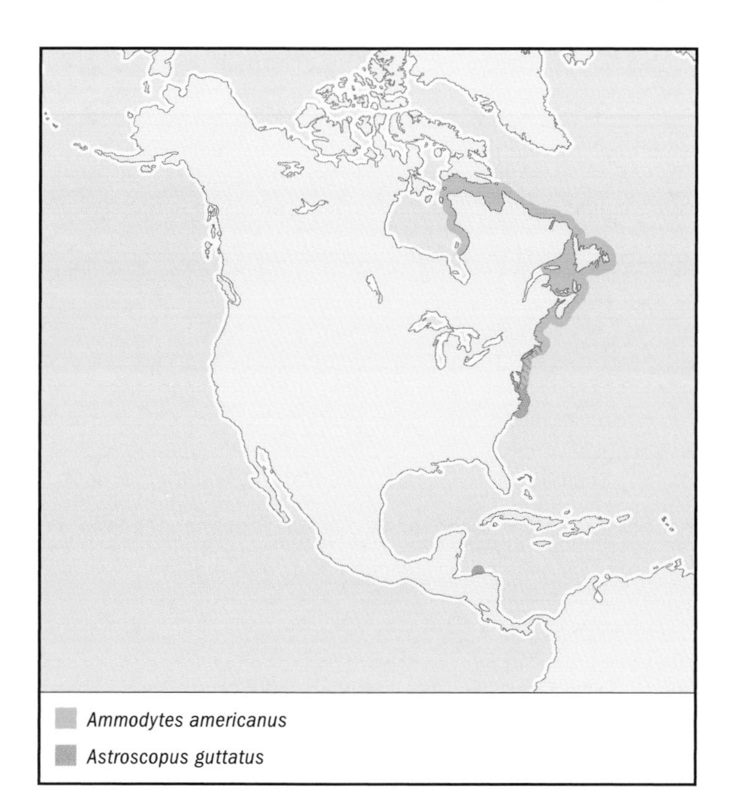

BEHAVIOR
Form schools of up to several thousand individuals, usually of similar-sized fishes. At high tide they may burrow into the sand and remain on exposed flats until the next tide. Daily movements are not known, and burying behavior, rather than offshore/inshore movements, might explain their sudden appearances and disappearances. It is thought that they spend a good deal of their time buried in the substrate, particularly at night and in the winter.

FEEDING ECOLOGY AND DIET
Feeds on zooplankton, especially copepods but also mysids, euphausids, chaetognaths, salps, urochordates, eggs, dinoflagellates, diatoms, and fish fry. Little is known about feeding ecology; even the times and places they feed are the subject of controversy. Inshore sand lances as well as other sand lance species are an important forage species for larger fishes, marine birds, and mammals; they act as agents of energy transfer in the food chain, from zooplankton to higher level predators.

REPRODUCTIVE BIOLOGY
Spawning has not been observed, but it occurs during fall and winter, peaking in December and January and ending in March, probably near shore, where current speeds are low. Most reach reproductive age at the end of their second year, with females' egg production estimated at 1,800–5,200 eggs each season. (Related species have been estimated to produce more than 20,000 eggs.) Eggs are demersal and hatch pelagic larvae after 30–74 days, depending on water temperature. Maximum life span is about 12 years.

CONSERVATION STATUS
Not listed by the IUCN.

SIGNIFICANCE TO HUMANS
There is little direct significance to humans, with only occasional use in the bait-fish industry. Historical landings are up to 75 metric tonnes per year, though usually well below this amount. They are of great ecological importance and play a significant role as forage fish for at least 20 commercial species including mackerel, herring, cod, hake, pollock, Atlantic salmon, and several flatfish species. Sand lances are also substantial components in the diet of some seabirds (such as terns and cormorants) and marine mammals (such as fin and humpback whales, porpoises, and seals) whose presence impacts tourism. ◆

Torrentfish
Cheimarrichthys fosteri

FAMILY
Cheimarrichthyidae

TAXONOMY
Cheimarrichthys fosteri Haast, 1874, Otira River, New Zealand. Placed in its own family or considered a member of the Pinguipedidae. The latter placement is based on general similarity; the only cladistic studies provide no evidence of a close relationship of *Cheimarrichthys* and pinguipedids.

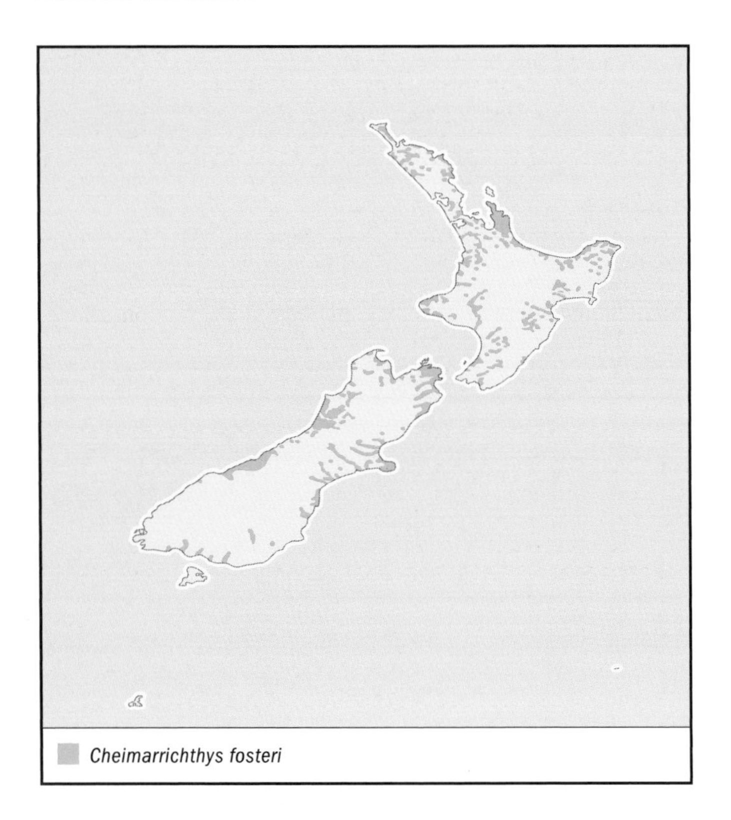

Cheimarrichthys fosteri

OTHER COMMON NAMES
Maori: Papanoko.

PHYSICAL CHARACTERISTICS
A broad, somewhat flattened, wedge-shaped head; subterminal mouth; and robust, scaled body. The dorsal fin has three to five short spines and 18–21 segmented rays; the anal fin has one or two spines and 14–16 segmented rays. The pectoral and pelvic fins are large, with about 50 lateral line scales. Grows to 8 in (20 cm).

DISTRIBUTION
Endemic to New Zealand.

HABITAT
Gravel-bottomed rivers in swift riffles from sea level to 2,300 ft (700 m) in elevation and almost 180 mi (300 km) from the sea.

BEHAVIOR
Females occupy areas upstream and males the lower reaches. Diadromous, with juveniles returning from the sea in spring and summer, when they have grown to 0.6–0.8 in (16–20 mm) and are already fully benthic.

FEEDING ECOLOGY AND DIET
Eats slow-moving benthic aquatic insects, especially midges, beetles, and caddis flies, which it probably grazes from cobble/boulder substrates. Thought to move at night from riffles to pools to feed.

REPRODUCTIVE BIOLOGY
It is suggested that females migrate downstream for spawning in summer/autumn, although spawning sites and behavior are unknown. Estimated to lay up to 30,000 small eggs, 0.02 in (0.6 mm) in diameter, that probably sink into the substrate, with larvae developing at sea.

CONSERVATION STATUS
Not threatened, though upstream migration is disrupted easily by man-made structures such as dams.

SIGNIFICANCE TO HUMANS
Of traditional importance to the Maori, but subject to no present utilization. ◆

Lesser weever
Echiichthys vipera

FAMILY
Trachinidae

TAXONOMY
Echiichthys vipera Cuvier, 1829, northern England. Originally described as *Trachinus vipera* but given its own genus in 1861 by Bleeker.

OTHER COMMON NAMES
English: Lesser weeverfish, adder pike, black fin, little weever.

PHYSICAL CHARACTERISTICS
Small, elongate, and compressed, with a large, oblique mouth; small, dorsally placed eyes; and a strong, grooved opercular spine bearing a venom gland. Two dorsal fins, the first with five to eight strong spines with anterolateral grooves bearing venom glands and the second long and directly opposed to the anal fin. Distinguished from other family members by fringed lips, a second dorsal fin with 21–24 segmented rays, an anal fin with one spine and 24–26 segmented rays, and rounded pectoral fins. Has about 60 lateral line scales. Coloring is brown dorsally, mottled with darker spots, and pale ventrally. The first dorsal fin is black; the caudal fin is pale with a dusky tip. Grows to 6 in (15 cm).

DISTRIBUTION
Eastern North Atlantic Ocean, including the Azores, and the Mediterranean Sea.

HABITAT
Inshore on sand, mud, or gravel, to 500 ft (150 m), in winter.

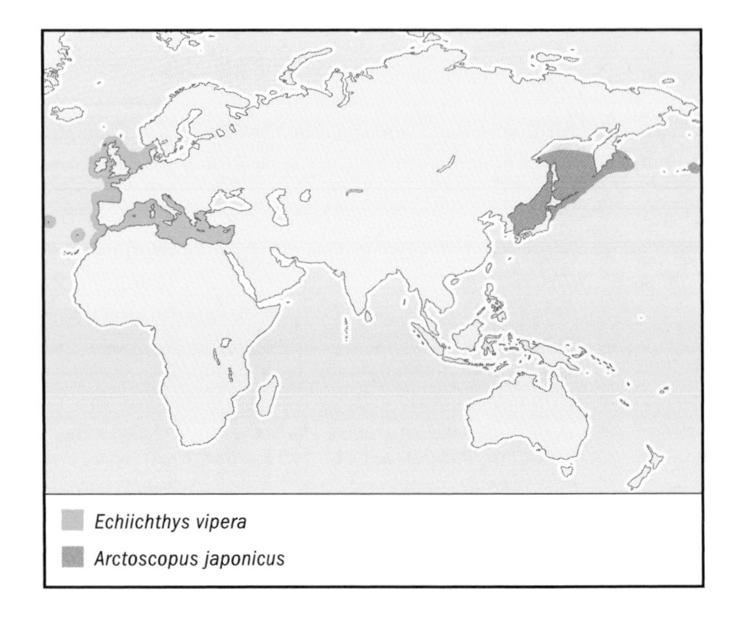

Echiichthys vipera
Arctoscopus japonicus

BEHAVIOR
During the day hides buried in substrate, with only the eyes and first dorsal fin exposed; becomes more active at night, when it feeds.

FEEDING ECOLOGY AND DIET
Feeds chiefly on crustaceans and fishes.

REPRODUCTIVE BIOLOGY
Breeds June to August, spawning planktonic eggs 0.04–0.06 in (1–1.4 mm) in diameter.

CONSERVATION STATUS
Not listed by the IUCN.

SIGNIFICANCE TO HUMANS
Because of the venomous spines and its inshore, bottom-dwelling habits, it poses a hazard to swimmers. Shrimp fishermen sometimes capture them in their nets and need to handle them carefully. This species is too small for commercial fishery and is taken mostly as incidental catch. ◆

Sailfin sandfish
Arctoscopus japonicus

FAMILY
Trichodontidae

TAXONOMY
Arctoscopus japonicus Steindachner, 1881, Strietok, Sea of Japan, and, questionably, Sitka, Alaska. Originally described as *Trichodon japonicus* by Steindachner, but placed in its own genus by Jordan and Evermann (1896) to emphasize the differences from the only other species in the family, *T. trichodon.*

OTHER COMMON NAMES
None known.

PHYSICAL CHARACTERISTICS
Compressed, wedge-shaped fishes, with an upturned mouth, fringed lips, spines on the preopercle, and a scaleless body. Two widely spaced dorsal fins, the first having eight to 14 spines and the second having 12–15 segmented rays. Spineless anal fin, with 29–32 segmented rays, and large pectoral fins. Brown mottling dorsally, pale or silvery ventrally, dark bands on both dorsal fins. Grows to 7 in (17 cm).

DISTRIBUTION
Korea to the Sea of Japan, Sea of Okhotsk, and the Bering Sea to Alaska.

HABITAT
Sandy-mud bottom at 650–1,300 ft (200–400 m), sitting on or in the substrate.

BEHAVIOR
Nothing is known.

FEEDING ECOLOGY AND DIET
Not known; perhaps a sit-and-wait predator or, similarly to trachinids, hides in substrate during the day and is active at night. Feeds primarily on mysids, crangonids, and small fishes.

REPRODUCTIVE BIOLOGY
During November and December there is a conspicuous spawning migration from deeper water to areas of seaweed at 6–33 ft (2–10 m). Eggs, which are about 0.14 in (3.5 mm), are stuck onto *Sargassum* species in spherical masses of about 600–2,300; they hatch asynchronously in about two months. Juveniles school and spend about three months in shallow water before moving into deeper water.

CONSERVATION STATUS
Fisheries in parts of northern Japan collapsed in the early 1980s. Record catches in the late 1960s reached 20,000 tons and held at about 15,000 tons until the late 1970s. By 1984 the catch had plummeted to just 74 tons. A moratorium on fishing was enforced from September 1992 to September 1995, and fishing began again during the spawning season of 1995, but catches remained below 1000 tons up to 1999.

SIGNIFICANCE TO HUMANS
An important food fish in northern Japan, caught by trawl net, set net, and dragnet as the fish come inshore to breed. They are eaten fresh or stored pickled in a mixture of salt and yeast for later consumption. The eggs, called *buriko,* also are eaten, particularly at the New Year. When catches were plentiful, excess fish were dried and used as fertilizer. ◆

Northern stargazer
Astroscopus guttatus

FAMILY
Uranoscopidae

TAXONOMY
Astroscopus guttatus Abbott, 1860, Cape May, New Jersey.

OTHER COMMON NAMES
None known.

PHYSICAL CHARACTERISTICS
Squarish head with flattened dorsal surface and large, vertical mouth with fringed lips lacking the worm-like appendage characteristic of some other species. Eyes situated dorsally and robust and scaled body. Two dorsal fins, the first with four to five spines and the second with 13–15 segmented rays; anal fin with one spine and 12 segmented rays. Large pectoral fins and cleithral spine just dorsal of the pectoral fin. Small, irregular, white spots on dark background dorsally and gray ventrally with obscure blotches. First dorsal fin dark, second dorsal fin with several distinct oblique bars, caudal fin with alternating black and white stripes, and pectoral fin dark with a pale margin. Grows to 22 in (56 cm) and 20 lb (9.1 kg). Has electric organs modified from the eye muscles in pouches behind the eyes reported to produce up to 50 volts.

DISTRIBUTION
Eastern coast of North America from New York south to North Carolina. One report as far south as Honduras.

HABITAT
Sandy substrate in coastal waters to 130 ft (40 m).

BEHAVIOR
Benthic, spending most of its time buried in the substrate.

FEEDING ECOLOGY AND DIET
A sit-and-wait predator that lies buried on the bottom, with only the top of the head, eyes, and mouth exposed, waiting for small fishes or crustaceans. Lunges at prey aggressively and sucks prey into the large mouth.

REPRODUCTIVE BIOLOGY

Spawns in spring and summer on the bottom, producing pelagic eggs. Larvae are pelagic, settling on sandy bottoms of inshore bays at about 0.6 in (15 mm), remaining there until they grow to 8–12 in (20–30 cm), when they move further offshore.

CONSERVATION STATUS

Not listed by the IUCN.

SIGNIFICANCE TO HUMANS

Of no commercial or recreational value. When caught, it should be handled with care, owing to the sharp, possibly venomous cleithral spine and the electric organs. ◆

Resources

Books

Carpenter, K. E., and V. H. Niem, eds. *FAO Species Identification Guide for Fishery Purposes.* Vol. 6, *The Living Marine Resources of the Western Central Pacific.* Part 4, *Bony Fishes: (Labridae to Latimeridae), Estuarine Crocodiles, Sea Turtles, Sea Snakes and Marine Mammals.* Rome: FAO, 2001.

Collette, Bruce B., and Grace Klein-MacPhee, eds. *Bigelow and Schroeder's Fishes of the Gulf of Maine.* 3rd edition. Washington, DC: Smithsonian Institution Press, 2002.

Halstead, Bruce W. *Poisonous and Venomous Marine Animals of the World.* Vol. 3, *Vertebrates.* Washington, DC: U.S. Government Printing Office, 1970.

McDowall, R. M. *Freshwater Fishes of New Zealand.* Auckland: Reed, 2001.

Mecklenburg, Catherine W., T. Anthony Mecklenburg, and Lyman K. Thorsteinson. *Fishes of Alaska.* Bethesda, MD: American Fisheries Society, 2002.

Murdy, Edward O., Ray S. Birdsong, and John A. Musick. *Fishes of Chesapeake Bay.* Washington, DC: Smithsonian Institution Press, 2002.

Nelson, J. S. *Fishes of the World.* 3rd edition. New York: John Wiley and Sons, 1994.

Tortonese, E. "Trachinidae." In *Fishes of the North-eastern Atlantic and the Mediterranean*, vol. 2., edited by P. J. P. Whitehead, M.-L. Bauchot, J.-C. Hureau, J. Nielsen, and E. Tortonese. Paris: UNESCO, 1986.

Watson, W., A. C. Materese, and E. G. Stevens. "Trachinoidea: Development and Relationships." In *Ontogeny and Systematics of Fishes*, edited by H. G. Moser, W. J. Richards, D. M. Cohen, M. P. Fahay, A. W. Kendall Jr., and S. L. Richardson. Special Publication no. 1. Lawrence, KS: American Society of Ichthyologists and Herpetologists, 1984.

Wheeler, Alwyn. *The Fishes of the British Isles and North-West Europe.* East Lansing: Michigan State University Press, 1969.

Periodicals

Ida, H., P. Sirimontaporn, and S. Monkolprasit. "Comparative Morphology of the Fishes of the Family Ammodytidae, with a Description of Two New Genera and Two New Species." *Zoological Studies* 33, no. 4 (1994): 251–277.

Johnson, G. D. "Percomorph Phylogeny: Progress and Problems." *Bulletin of Marine Science* 52, no. 1 (1993): 3–28.

McDowall, R. M. "Biogeography of the New Zealand Torrentfish, *Cheimarrichthys fosteri* (Teleostei: Pinguipedidae): A Distribution Driven Mostly by Ecology and Behaviour." *Environmental Biology of Fishes* 58 (2000): 119–131.

———. "Relationships and Taxonomy of the New Zealand Torrent Fish, *Cheimarrichthys fosteri* Haast (Pisces: Mugiloididae)." *Journal of the Royal Society of New Zealand* 3 (1973): 199–217.

Mooi, R. D., and G. D. Johnson. "Dismantling the Trachinoidei: Evidence of a Scorpaenid Relationship for the Champsodontidae." *Ichthyological Research* 44 (1997): 143–176.

Nazarkin, M. V. "*Trispinax ladae* gen. et sp. nov.: A Species of the New Family of Trachinoid Fishes Trispinacidae (Perciformes, Trachinoidei) from the Miocene of Sakhalin Island." *Journal of Ichthyology* 42, no. 6 (2002): 419–426.

Nazarkin, M. V., and O. S. Voskoboinikova. "New Fossil Genus and Species of Trichodontidae and the Position of This Family in the Order Perciformes." *Journal of Ichthyology* 40, no. 9 (2000): 687–703.

Okiyama, M. "Contrast in Reproductive Style Between Two Species of Sandfishes (Family Trichodontidae)." *Fishery Bulletin* 88 (1990): 543–549.

Pietsch, T. W. "Phylogenetic Relationships of Trachinoid Fishes of the Family Uranoscopidae." *Copeia* 1989, no. 2 (1989): 253–303.

Pietsch, T. W., and C. P. Zabetian. "Osteology and Interrelationships of the Sand Lances (Percifomres: Ammodytidae)." *Copeia* 1990, no. 1 (1990): 78–100.

Other

Clark, E., S. Kogge, D. Nelson, and A. Thomas. "Burrow Distribution, Diel Activity, and Behavior of *Pholidichthys leucotaenia* (Pholidichthyidae)." *Joint Meeting of Ichthyologists and Herpetologists.* <http://www.asih .org/meetings/2002/ Abstracts.pdf>

Clifton, K. E., and L. M. Clifton. "A Survey of Fishes from Various Coral Reef Habitats Within the Cayos Cochinos Marine Reserve, Honduras." (28 Dec. 2002). <http//rbt.ots .ac.cr/revistas/suplemen/honduras/10cli1.htm>

International Council for the Exploration of the Sea, Reports. <http://www.ices.dk/reports/acfm/2001/wgnssk/ S-13%20-%20San4.pdf>

Randall D. Mooi, PhD
G. David Johnson, PhD

Blennioidei

(Blennies)

Class Actinopterygii
Order Perciformes
Suborder Blennioidei
Number of families 6

Photo: A mimic blenny (*Aspidontus taeniatus*) peers out from a wormhole in coral near the Florida Islands, Solomon Islands. (Photo by Fred McConnaughey/Photo Researchers, Inc. Reproduced by permission.)

Evolution and systematics

The suborder Blennioidei has a long and convoluted history of classification, with a variety of families historically being moved into and out of the suborder. Stability finally was established for the suborder when Springer defined the Blennioidei as a monophyletic suborder (i.e., a taxon that includes all the evolutionary descendents, and only those descendents, of a common ancestor). Although the composition of the suborder has been established, the phylogenetic relationships to other families within the order Perciformes remain elusive.

The members of the six families (Blenniidae, Chaenopsidae, Clinidae, Dactyloscopidae, Labrisomidae, and Tripterygiidae) included in the Blennioidei uniquely share a distinctive shape of the joined pelvic bones; another important character is the absence of the second infrapharyngobranchial bone (PB2) from the dorsal gill arch. Springer documented the absence of the PB2 and described the shape of the joined pelvic bones as "a somewhat bean- or nut-like pod, which is open ventrally, and has a dorsally extending flange anteriorly on each side." In addition to this unusual pelvic shape and the loss of the PB2, several other characters that are rarely found in non-blennioid taxa define the Blennioidei. For example, there are cirri on the eye and often on the nape (absent in Dactyloscopidae), the bases of the pelvic fins are positioned anterior to the bases of the pectoral fins, several hypural bones supporting the caudal fin are fused, the anal fin has fewer than three spines, and all segmented rays are unbranched. Johnson described another character, the lack of a neural spine on the first vertebra, further corroborating that the Blennioidei make up a monophyletic group. Although one or a few of these individual characters occur variously in other families, the combination of more than a few of the many shared specialized characters is not found in non-blennioid families.

Few blennioid fossils have been discovered. The combination of their small size and a predominant occurrence along coastlines exposed to wave action and surge probably contributed to the rarity of blenny fossils. The oldest reported blennioid fossil is a late Eocene blennioid otolith described by Nolf.

As of 2002, there are 127 genera and about 750 recognized species in the Blennioidei. Being small, cryptic fishes, blennioids are frequently difficult to collect and identify. The use of rotenone (a natural plant substance found in the roots of a variety of leguminous plants, which aids in the collection of cryptic fishes) to conduct scientific biodiversity surveys of fishes is resulting in the continuous discovery of previously unknown new species of blennioids. If discoveries of new blennioid taxa continue at the current pace, there ultimately may be 1,000 or more valid species of blennioids.

Physical characteristics

Blennioids exhibit amazing diversity in body shapes. They are typically small fishes, usually less than 4 in (15 cm) in length, with many species growing no longer than 2 in (5 cm). Their slender, elongate bodies are very flexible. Most blennioids (not dactyloscopids) have cirri on the head, particularly over the eyes and often on the nape. The only members of the suborder attaining lengths of more than about 10 in (25 cm) are specimens belonging to one Indo-Pacific genus, *Xiphasia*, which have extremely elongate bodies with a depth of less than 0.8 in (2 cm) but reaching lengths of 21.3 in (54 cm). Body shapes range from short and stout and completely or almost completely scaled to elongate and blunt-headed and entirely lacking scales. Several species in the tube blenny genus *Acanthemblemaria* have enlarged spines adorning their heads. Some of the sand stargazers have flattened heads with eyes

A barnacle blenny (*Acanthemblemaria* sp.) from the eastern Pacific near Costa Rica. (Photo by Mark Smith/Photo Researchers, Inc. Reproduced by permission.)

set on protruding stalks. These fish burrow into the sand with only the stalked eyes sticking out as they lie in wait for unsuspecting prey.

Color patterns vary widely, ranging from drab, camouflaging mottled patterns of brown and tan to brilliant reds, yellows, and iridescent blues. Males and females of many blennioid taxa have extremely different colors and patterns. Historically, these differences sometimes have resulted in males and females of the same species being described as two different species.

Distribution

Blennioids are found in shallow marine, estuarine waters of every continent but rarely in freshwaters. The highest diversity of blennioids is in shallow tropical and temperate waters of the Caribbean, western Atlantic, Indian, and Pacific Oceans. Blennioids have not been found in the Arctic Ocean, but they are present in the shallow coastal waters of every other ocean. There is even one triplefin species known from the coast of Antarctica.

Habitat

Blennioids reside in almost every underwater marine habitat imaginable, and some even climb out of the water onto rocks. They are predominately benthic (a few are secondarily free-swimming) and found on or near coral and rocky reefs, where they live in holes and crevices. Other blennioids may live on the bottom among mangrove roots, in sea grass beds, on oyster reefs, or in sandy areas. Although most species live at depths shallower than 66 ft (20 m) and typically less than 33 ft (10 m), *Bathyblennius antholops* is known only from a specimen taken at a depth between 330 and 420 ft (101–128 m). The freshwater blenny, *Salaria fluviatilis*, of northern Africa and southern Europe is one of the few blennioids that lives in freshwater streams and rivers. (Some Omobranchini and *Phenablennius* live in freshwater habitats as well.)

Behavior

Blennioids usually are seen in shallow water, sitting on rocks or coral rubble and using their pelvic fins as props to lift the head off the bottom. They quickly dart into a hole or crevice when approached by a diver. The Indo-Pacific rockhopper blennies of the genera *Alticus* and *Andamia* are found on rocks in high-energy surf zones. The rockhoppers sit just above the waterline and, when threatened, leap from the rock and use quick flicks of the anal and tail fins to skip across the surface of the water to the next emergent rock.

Mimetic associations have been documented for many of the saber-toothed blennies. Mimicry is the process of one species or specimen gaining survival value by changing superficially to resemble another species or specimen. Saber-toothed blennies display previously known types of mimicry, as well as several newly discovered mimetic associations. Batesian mimicry (the most common form) involves a palatable species adapting to resemble an unpalatable species. Müllerian mimicry is defined as two unpalatable species resembling each other. When a predatory and aggressive species resembles a nonagressive species, this is referred to as aggressive mimicry. Social mimicry is seen when two or more species with a similar color pattern swim together in a group for mutual protection.

Feeding ecology and diet

Most blennioids are predatory, eating small benthic invertebrates. The members of the family Blenniidae exhibit a broad range of feeding habits. Some blennies have a row of flexible, comblike teeth used to scrape algae and any associated organisms in the algal mat. Other blennies range in feed-

A large-banded blenny (*Ophioblennius steindachneri*) between two pencil urchins near Caldwell Rock, Galápagos Islands. (Photo by Fred McConnaughey/Photo Researchers, Inc. Reproduced by permission.)

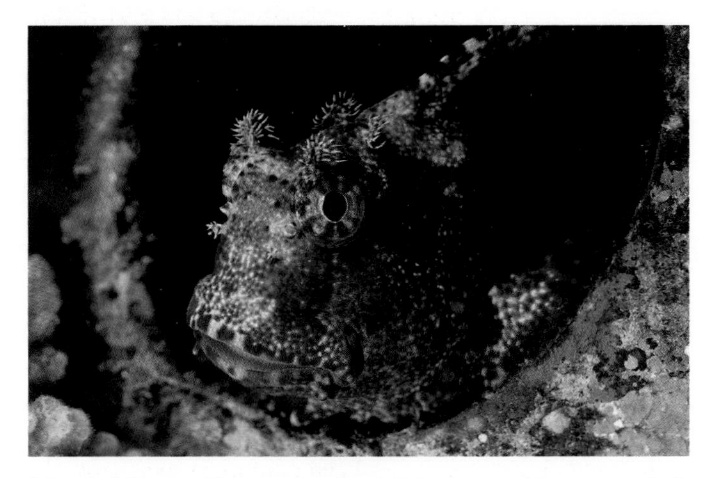

A jeweled blenny (*Salarias fasciatus*) living in an old coconut shell. (Photo by David Hall/Photo Researchers. Reproduced by permission.)

ing habits from fin ray–eating and membrane-eating aggressive mimics to tube-dwelling plankton pickers. Blennioids are also subject to predation; while the exact predators have not been identified, blennioids are probably eaten by groupers, snappers, and other large piscivores.

Reproductive biology

Reproductive biology is unknown for all but a few blennioids. In the family Blenniidae, males of some species guard and display in front of a small territory, often an empty shell or a rock crevice. The female enters the male's territory; lays large, demersal eggs that adhere to the surface of the shell or rock as they are fertilized by the male; and then leaves. The female sometimes visits two or more males in sequence, repeating the spawning process with each one. The male then guards the eggs and fans water over them until they hatch. Newly hatched fry swim toward the surface and feed on plankton until they settle. The duration of the larval stage varies widely—some blennies settle within a few days of hatching, whereas others may have an extended pelagic larval stage, called the ophioblennius stage, lasting up to one or two months.

The South African, Australian, and Mediterranean members of the Clinidae are ovoviviparous, but the South American members of the family lack an intromittent organ and are oviparous. The tropical eastern Pacific labrisomid genus *Xenomedea* is also ovoviviparous. Males of the labrisomid genus *Starksia* have a distinctive tubular intromittent organ, but live embryos have been found only in a few eastern Pacific *Starksia* species and have not been recorded for any of the Atlantic species.

Conservation status

As of 2002 most populations of marine blennioids were stable, but increasing destruction of coral reefs and coastal pollution are potential threats. The greatest conservation threat to blennioids comes from habitat destruction and pollution. Four species in three families appear on the IUCN Red List—*Entomacrodus cadenati* (Blenniidae) listed as Data Deficient; *Coralliozetus tayrona* (Chaenopsidae) listed as Vulnerable; *Protemblemaria punctata* (Chaenopsidae) listed as Vulnerable; and *Clinus spatulatus* (Clinidae) listed as Endangered. There are no conservation or preservation efforts underway.

Significance to humans

Although blennioids are eaten and can be seen regularly in fish markets from the Philippines to Indonesia and in Peru and Chile in the eastern Pacific, they are not considered commercially important. The primary use of blennioids by humans is as aquarium fish. Specimens of the eastern Pacific blenniid genus *Scartichthys* are consumed in Peru, where eating the flesh supposedly produces a drunken effect. The common name for these fish in Peru is *borracho*, which means "drunk." On Easter Island the Rapa Nui people consider the small patuki blenny to have transformational qualities and associate it with fertility.

1. Blackcheek blenny (*Starksia lepicoelia*); 2. Rosy weedfish (*Heteroclinus roseus*); 3. White-lined comb-tooth blenny (*Ecsenius pictus*); 4. Secretary blenny (*Acanthemblemaria maria*); 5. Sand stargazer (*Dactyloscopus tridigitatus*); 6. Striped poison-fang blenny (*Meiacanthus grammistes*); 7. Hairtail blenny (*Xiphasia setifer*); 8. Miracle triplefin (*Enneapterygius mirabilis*). (Illustration by Patricia Ferrer)

Species accounts

Hairtail blenny
Xiphasia setifer

FAMILY
Blenniidae

TAXONOMY
Xiphasia setifer Swainson, 1839, Vizagapatam, Philippines.

OTHER COMMON NAMES
Japanese: Unagiginpo; South Africa: Slangblennie; Oman and Micronesia: Snakeblenny.

PHYSICAL CHARACTERISTICS
May reach lengths of 23.6 in (60 cm). Large fangs in each jaw but not associated with a toxin gland. Body with alternating bands of dark brown and pale brown. Caudal fin with middle two rays very elongate and forming a hairlike filament and dorsal fin originating above the eye and continuing to the caudal fin. Extremely long and slender body.

DISTRIBUTION
Widely distributed from the western Indian Ocean to the western central Pacific.

HABITAT
Usually found in tubes or burrows in mud or sandy bottoms but may be seen swimming above the bottom or at the surface at night while feeding; shelters in tubes or burrows during the day. Specimens have been captured by trawl at 164–177 ft (50–54 m) off western India.

BEHAVIOR
Seldom seen underwater but sometimes attracted to night lights, probably in search of small invertebrates to eat.

FEEDING ECOLOGY AND DIET
Appears to be nocturnal and to forage on small invertebrates.

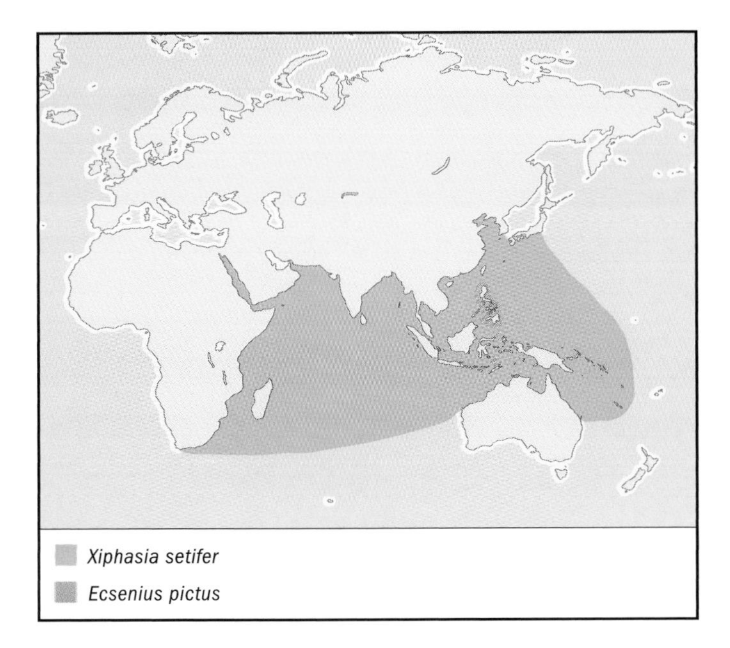

REPRODUCTIVE BIOLOGY
Not known.

CONSERVATION STATUS
Not listed by the IUCN.

SIGNIFICANCE TO HUMANS
None known. ◆

Striped poison-fang blenny
Meiacanthus grammistes

FAMILY
Blenniidae

TAXONOMY
Meiacanthus grammistes Valenciennes in Cuvier and Valenciennes, 1836, Java, Indonesia.

OTHER COMMON NAMES
English: Grammistes blenny; Japanese: Hige-nijiginpo.

PHYSICAL CHARACTERISTICS
May reach lengths of 3.5 in (90 mm). Has a toxin-producing gland enclosed in the dentary at the base of each enlarged, grooved lower-jaw fang. Body has alternating black and white stripes. White stripes are yellowish anteriorly; black stripes are narrow as they approach the base of the caudal fin, where

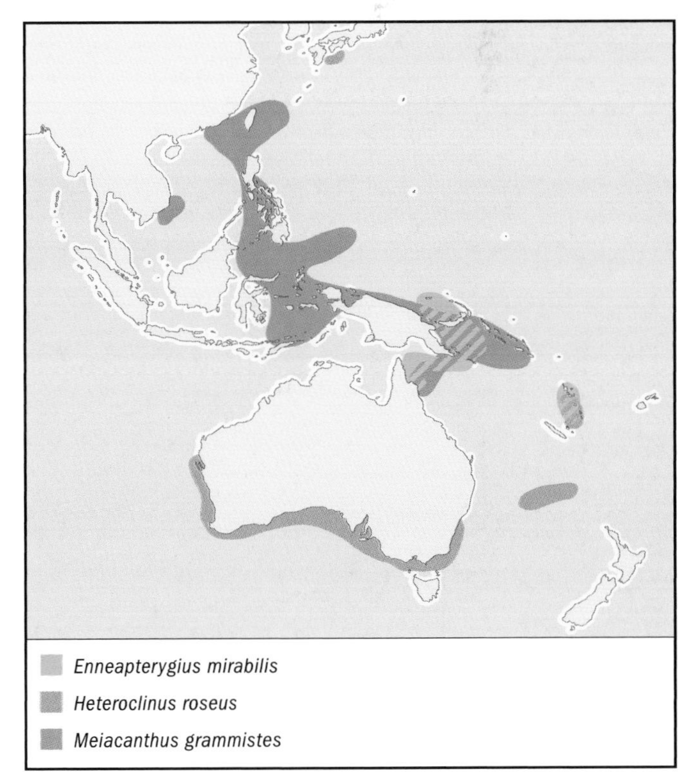

they break up into small black spots. Caudal fin is translucent, with small black spots; dorsal and anal fins have a black submarginal stripe.

DISTRIBUTION
Widely distributed throughout the western central Pacific from the Ryukyu Islands in the north to the Great Barrier Reef in the south and eastward to the Solomon Islands.

HABITAT
Usually found on or near coral reefs and may be seen swimming above the reef; shelters in tubes or holes in the reef.

BEHAVIOR
This is a venomous species that is able to inject venom using modified fangs in the lower jaw. Unlike most blennies, these fishes have a well-developed swim bladder that allows them to swim easily above the reef. The combination of the swim bladder and their ability to inject venom into a potential predator allows them to swim without restriction above the reef and to forage over a relatively wide area. Like most species of *Meiacanthus*, the striped poison-fang blenny has mimics. Two species are known to be Batesian mimics, a blenniid (*Petroscirtes breviceps*) and a cardinalfish (*Cheilodipterus nigrotaeniatus*). Both of these species swim with the striped poison-fang blenny and avoid predation as a result of potential predators' aversion to the bite of the striped poison-fang blenny.

FEEDING ECOLOGY AND DIET
Forages on small benthic invertebrates on or near the reef.

REPRODUCTIVE BIOLOGY
Unknown but probably similar to that of other blennies. Among other blennies, the male displays to entice one or more females to lay demersal eggs in a hole in the reef, and the male probably guards the eggs until they hatch.

CONSERVATION STATUS
Not listed by the IUCN.

SIGNIFICANCE TO HUMANS
Sold as aquarium fish. Potential medicinal uses for substances produced by the venom glands have not been explored. ◆

White-lined comb-tooth blenny
Ecsenius pictus

FAMILY
Blenniidae

TAXONOMY
Ecsenius pictus McKinney and Springer, 1976, Molucca Islands, Indonesia.

OTHER COMMON NAMES
Micronesia: Pictus blenny.

PHYSICAL CHARACTERISTICS
May reach lengths of 2 in (50 mm). Has anterior and posterior canines on each side of the lower jaw and lacks orbital and nape cirri. Body is brown with a series of narrow white stripes ending at the caudal peduncle. The caudal peduncle is yellowish with several brown bands.

DISTRIBUTION
Known to occur only in the western central Pacific in the Philippines, Indonesia, and the Solomon Islands.

HABITAT
Usually found on or near coral reefs at depths of 36–131 ft (11–40 m).

BEHAVIOR
A benthic fish, generally seen perched on coral.

FEEDING ECOLOGY AND DIET
Forages on small benthic invertebrates on or near the reef.

REPRODUCTIVE BIOLOGY
Not known.

CONSERVATION STATUS
Not listed by the IUCN.

SIGNIFICANCE TO HUMANS
None known. ◆

Secretary blenny
Acanthemblemaria maria

FAMILY
Chaenopsidae

TAXONOMY
Acanthemblemaria maria Böhlke, 1961, Treasure Island, Bahamas.

OTHER COMMON NAMES
None known.

PHYSICAL CHARACTERISTICS
May reach lengths of 2 in (50 mm). Head has well-developed bony spines covering the top almost to the dorsal

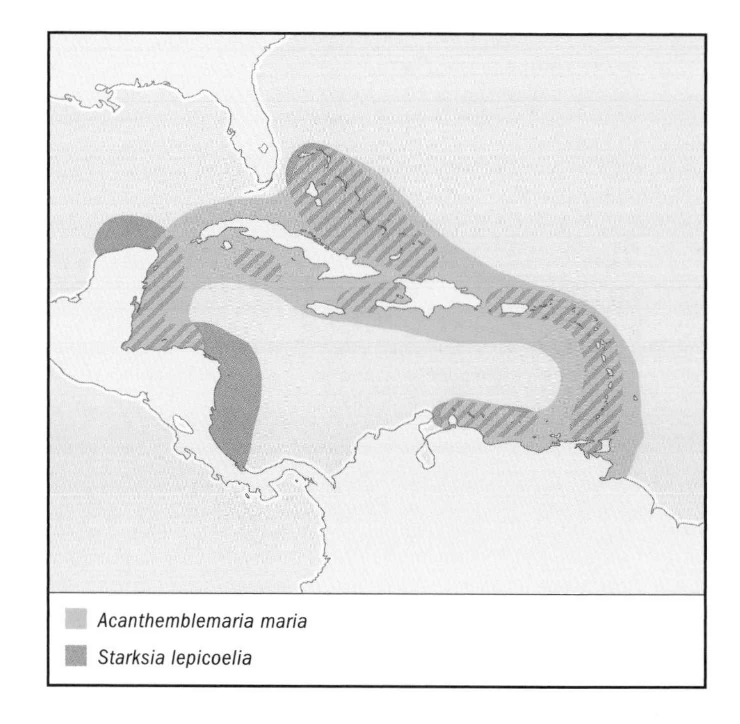

fin origin. Body has broad brown bands separated by narrow pale areas; dark bands sometimes form large oval spots midlaterally.

DISTRIBUTION
Bahamas eastward to Belize and southward to Tobago.

HABITAT
Inhabits small tubes or holes abandoned by tube worms or other invertebrates in rock or limestone slopes and usually is found at depths shallower than 33 ft (10 m).

BEHAVIOR
Generally seen with only the head projecting from a hole, retreating into the hole on the approach of a diver or a large fish.

FEEDING ECOLOGY AND DIET
Feeds on small benthic invertebrates.

REPRODUCTIVE BIOLOGY
Oviparous. Spawning has not been observed but is probably similar to that of other egg-laying blennies.

CONSERVATION STATUS
Not listed by the IUCN. Healthy numbers exist in the western Atlantic and Caribbean. Local populations could be threatened by habitat destruction or pollution.

SIGNIFICANCE TO HUMANS
Potential significance as aquarium fish. ◆

Rosy weedfish
Heteroclinus roseus

FAMILY
Clinidae

TAXONOMY
Heteroclinus roseus Günther, 1861, Freycinets Harbour, Western Australia.

OTHER COMMON NAMES
Japanese: Beni-asahiginpo.

PHYSICAL CHARACTERISTICS
May reach maximum lengths of 5.7 in (145 mm). The rosy weedfish is distinguished from other *Heteroclinus* species by having a black spot, bordered above and below by a white stripe, on the pectoral fin-base.

DISTRIBUTION
Primarily anti-subtropical in distribution, occurring north of approximately 20° N latitude and south of roughly 20° S latitude, and absent in tropical equatorial waters. A northern population occurs in Japan and a disjunct southern population living around the southern half of Australia, Lord Howe Island, Kermadec Islands, New Caledonia, and Vanuatu.

HABITAT
The rosy weedfish lives on rocky reefs from the intertidal zone to depths of 115 ft (35 m).

BEHAVIOR
Rarely seen underwater owing to its very cryptic coloration. Sits on the bottom using its pelvic fins as supports to lift the head.

FEEDING ECOLOGY AND DIET
Predatory, probably feeding on small benthic invertebrates.

REPRODUCTIVE BIOLOGY
The presence of an intromittent organ in males suggests that it (like all other clinids in the Indo-Pacific) is ovoviviparous.

CONSERVATION STATUS
Not listed by the IUCN.

SIGNIFICANCE TO HUMANS
None known. ◆

Sand stargazer
Dactyloscopus tridigitatus

FAMILY
Dactyloscopidae

TAXONOMY
Dactyloscopus tridigitatus Gill, 1859, Barbados.

OTHER COMMON NAMES
Spanish: Mirón ojilargo.

PHYSICAL CHARACTERISTICS
May reach lengths of 3 in (75 mm). A distinctive Atlantic sand stargazer with the eye perched on a slender stalk on top of the head. It differs from other Atlantic sand stargazers in that it has a tube in the last lateral line scale at the base of the caudal fin.

DISTRIBUTION
Bermuda, northeastern Gulf of Mexico, Bahamas, and the Caribbean as far as Brazil.

HABITAT
Soft, sandy bottoms from shallow water off beaches to depths of 98 ft (30 m).

BEHAVIOR
Burrows into the sand with only the stalked eyes on the top of the head exposed.

FEEDING ECOLOGY AND DIET
It is predatory on small prey, presumably invertebrates, but the sand stargazer has not been observed feeding.

REPRODUCTIVE BIOLOGY
Oviparous. Males and females mature at about 1.8 in (45 mm) in length. Spawning had not been observed as of 2002.

CONSERVATION STATUS
Not listed by the IUCN. Pollution is the biggest potential threat.

SIGNIFICANCE TO HUMANS
None known. ◆

Blackcheek blenny
Starksia lepicoelia

FAMILY
Labrisomidae

TAXONOMY
Starksia lepicoelia Böhlke and Springer, 1961, Green Cay, Bahamas.

OTHER COMMON NAMES
Spanish: Sapito carinegro.

PHYSICAL CHARACTERISTICS
May reach a maximum length of 1.4 in (35 mm). The blackcheek blenny can be distinguished from other Atlantic *Starksia* by the presence of a black spot covering most of the cheek in mature males, a completely scaled belly, and male genital papilla entirely separate from the first anal fin spine.

DISTRIBUTION
May be found on coral reefs from the Bahamas and throughout the Caribbean.

HABITAT
Coral reefs at depths of 26–66 ft (8–20 m).

BEHAVIOR
Secretive and rarely seen underwater.

FEEDING ECOLOGY AND DIET
Predatory on small benthic invertebrates.

REPRODUCTIVE BIOLOGY
Mature females have well-developed ova in the ovaries, and the males have an intromittent organ, but the presence of embryos in a female (viviparity) has not been documented. It seems likely that the species is oviparous, as no gravid female Atlantic blackcheek blenny has ever been found.

CONSERVATION STATUS
Not listed by the IUCN.

SIGNIFICANCE TO HUMANS
None known. ◆

Miracle triplefin
Enneapterygius mirabilis

FAMILY
Tripterygiidae

TAXONOMY
Enneapterygius mirabilis Fricke, 1994, Escape Reef, Australia.

OTHER COMMON NAMES
None known.

PHYSICAL CHARACTERISTICS
May reach lengths of 1.2 in (30 mm). Distinguished from other *Enneapterygius* species by having 12 or fewer tubed scales in the anterior portion of the lateral line and by its long, filamentous spines in the first dorsal fin.

DISTRIBUTION
Known from Papua New Guinea, northern Queensland, Australia, and Vanuatu.

HABITAT
Has been found at depths of 26–121 ft (8–37 m) on coral reefs and adjacent reef slopes.

BEHAVIOR
The miracle triplefin, like so many of the blennioids, has yet to be observed alive underwater. It has been seen and collected only after the application of rotenone, without which this magnificent blenny's existence would be unknown.

FEEDING ECOLOGY AND DIET
The diet is unknown, but the presence of small, sharp, conical teeth in the jaws suggests that the miracle triplefin feeds on small benthic invertebrates.

REPRODUCTIVE BIOLOGY
Oviparous, based on the presence of mature ova in females, but breeding individuals have not been observed, and young are unknown.

CONSERVATION STATUS
Not listed by the IUCN.

SIGNIFICANCE TO HUMANS
Not currently used by humans, but its red color and high first dorsal fin make it an attractive species. ◆

Resources

Books

Gomon, M. F., J. C. M. Glover, and R. H. Kuiter. *The Fishes of Australia's South Coast.* Adelaide, Australia: State Print, 1994.

Periodicals

Johnson, G. D. "Percomorph Phylogeny: Progress and Problems." *Bulletin of Marine Science* 52, no. 1 (1993): 3–28.

Nolf, D. "Les Otolithes de Teleosteens Eocenes d'Aquitaine (Sud-Ouest de la France) et Leur Interet Stratigraphique." *Academie Royale de Belgique, Memoires de la Classe des Sciences* 19, no. 2 (1988): 7–147.

Smith, C. L., J. C. Tyler, H. Andreyko, and D. M. Tyler. "Behavioral Ecology of the Sailfin Blenny, *Emblemariopsis pandionis* (Pisces: Chaenopsidae) in the Caribbean off Belize." *American Museum Novitates* 3232 (1998): 1–40.

Smith-Vaniz, William F., Ukkrit Satapoomin, and Gerald R. Allen. "*Meiacanthus urostigma,* a New Fangblenny from the Northeastern Indian Ocean, with Discussion and Examples of Mimicry in Species of *Meiacanthus* (Teleostei: Blenniidae: Nemophini)." *Aqua, Journal of Ichthyology and Aquatic Biology* 5, no. 1 (2001): 25–43.

Springer, V. G. "Definition of the Suborder Blennioidei and Its Included Families (Pisces: Perciformes)." *Bulletin of Marine Sciences* 52, no. 1 (1993): 472–495.

———. "The Indo-Pacific Blenniid Fish Genus *Ecsenius.*" *Smithsonian Contributions to Zoology,* no. 465 (1988): 1–134.

Jeffrey Taylor Williams, PhD

Icosteoidei

(Ragfish)

Class Actinopterygii
Order Perciformes
Suborder Icosteoidei
Number of families 1

Photo: A ragfish (*Icosteus aenigmaticus*) juvenile. The ragfish transforms as it ages. The juvenile has a spotted color pattern, small pelvic fins, and has spines along its lateral line. As it grows, it changes to a chocolate brown and loses its pelvic fins. (Photo by K. Gus Thiesfeld. Reproduced by permission.)

Evolution and systematics

When the juvenile ragfish was first described by Lockington in 1880, there was apparently a premonition of the difficulty in placing this species into any existing taxonomic groupings. Indeed, the name of the single icosteoid representative, *Icosteus aenigmaticus*, means "puzzling fish with yielding bones." Lockington hypothesized the ragfish's probable relationship to the blennioid fishes. He questionably assigned it to the family Blenniidae (the blennies) but stated that it probably should be in its own family.

In the next twenty years, Lockington's contemporaries offered other taxonomic schemes. In 1881 Steindachner also examined a juvenile ragfish and concluded that it was not a blennioid and should be positioned with *Icichthys* (the medusafish). In 1887 Günther examined two juveniles and proposed a coryphaenid (dolphin) affinity. In 1887 Tarleton H. Bean described the first adult ragfish; being of such dissimilar appearance to the juvenile form, it was recognized as its own species, *Acrotus willoughbyi*, and placed in the family Icosteidae, along with *Icosteus* and *Icichthys*. Several years later *Acrotus* was elevated to its own family, Acrotidae.

The next 70 years yielded few studies of ragfish (actually, ragfishes at this time) systematics. During this period, the ragfish received little attention, doubtless because of its rare occurrence. Most ichthyologists still agreed that it was a percomorph (typical spiny-rayed fishes), though degenerate and specialized. The majority of researchers recognized a stromateoid (medusafish, butterfishes, and squaretails) affinity. A few, such as Berg in 1940, believed that the ragfish should be elevated to its own order, Icosteiformes.

In 1961 Clemens and Wilby observed that *Icosteus* (the fantailed ragfish) was the juvenile form of the adult *Acrotus* (the brown ragfish). At that time, interest in the ragfish was renewed, and since then the science of systematics has advanced in the methods of interpreting characteristics that elucidate affinities between groups. Studies on internal anatomy, external morphological features, and early life history have been adding to our knowledge of ragfish. With this information, investigators may be able to piece together the "puzzling" systematic relationships of the ragfish.

The taxonomy for this species is *Icosteus aenigmaticus* Lockington, 1880.

Other common names include ragfish, spotted ragfish, speckled ragfish, fan-tailed ragfish, giant ragfish, and brown ragfish.

Physical characteristics

Juvenile ragfish have limp bodies that are easily doubled over like a wet rag (hence the name ragfish). They are deep-bodied and laterally compressed (especially the region along the dorsal and anal fin base), with a slender caudal peduncle. The body is smooth-skinned and scaleless, except along the lateral line, where the scales produce groups of numerous small spines. The lateral line is minimally arched anteriorlyand then straight. The head is short, with an abrupt, high predorsal profile. The eyes are small. The mouth is terminal, with limited protrusibility. The teeth are minute and comblike. The snout is blunt and has been described as calflike. The dorsal fin has one spine and 52–56 soft rays, and the anal fin has 37–40 soft rays. The caudal fin is rounded and

Ragfish (*Icosteus aenigmaticus*) have cartilage rather than bones, so are limp and "rag-like" when out of the water. They live deep in the water, so it is unusual to see them. The female fish on the left measures 5.05 ft (1.54 m). The male on the right is 5 ft (1.525 m). (Photo by K. Gus Thiesfeld. Reproduced by permission.)

fanlike. The pelvic fins have one spine and four soft rays. All fins have small spinules on the surface. Smaller juveniles (less than 4.7 in, or 120 mm, in length) have mottled dark blotches over a light grayish purple background. Juveniles longer than 4.7 in (120 mm) have numerous irregular spots over a light grayish purple background, which are less prominent along the ventral region of the body.

At 14.2–16.5 in (360–420 mm), ragfish undergo a remarkable transformation into an adult form that is very different from the juvenile form. The body becomes more elongated, and a fleshy keel becomes pronounced along the ventral midline. The lateral line scale spines disappear, and the number of dorsal and anal fin rays decrease as the skin along the anterior origins encroaches on the leading spine and rays. The pelvic fins disappear. The caudal fin margin broadens and becomes emarginate to slightly forked. The spinules on the fins are reabsorbed. Finally, the coloration turns to a uniform dark brown over the entire body and fins.

Ragfish have been reported to reach a length of more than 6.9 ft (2.1 m). Adult males seem to attain the greatest length, while most females grow to about 5 ft (1.6 m). Adult females appear to be deeper bodied and weigh nearly twice as much as a male of similar size.

Distribution

Ragfish are distributed from San Onofre, California, to immediately north of the Aleutian Islands and across the northern Pacific to Kochi City, southeastern Japan. They are considered rare but are less so north of Cape Mendocino, California.

A ragfish (*Icosteus aenigmaticus*) female that measures 48 in (1.2 m). Note the absence of spotted coloration, pelvic fins, and lateral line spines. (Photo by K. Gus Thiesfeld. Reproduced by permission.)

Habitat

Ragfish are captured from the surface to depths of more than 1,640 ft (500 m). They usually are offshore over deeper water, which suggests an epipelagic or mesopelagic habitat. The juvenile's spotted coloration, however, might imply a benthic habitat.

Behavior

There is little information on ragfish behavior. They most likely are not schooling fish, because they always are captured individually. Rarely, large ragfish are found stranded on the shore. Ragfish have never been observed in their natural habitat.

Feeding ecology and diet

Little is known of the feeding ecology of the ragfish. Those that are collected rarely have identifiable food items in the stomach. In 1968 Fitch and Lavenberg reported small fishes, squid, and octopods as possible prey items. There also is evidence of gelativory in ragfish. Gelativory is the consumption of jellyfishes and their relatives. Preliminary investigations of the internal anatomy suggest morphological specializations to accommodate a gelativorous diet.

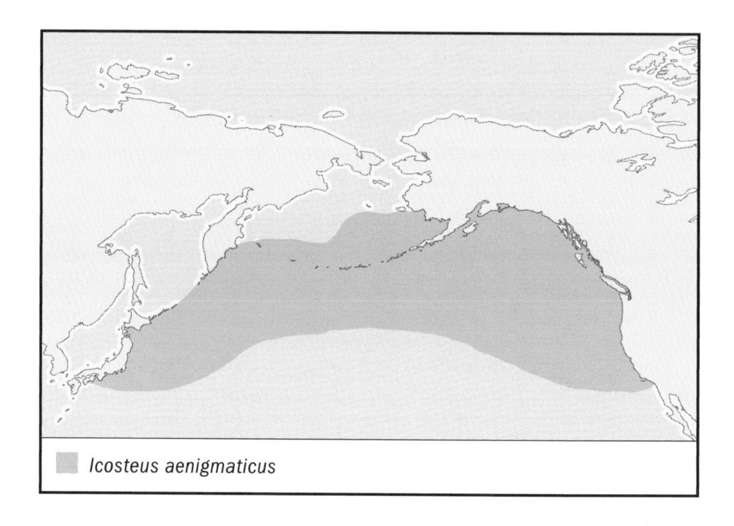

Ragfish (*Icosteus aenigmaticus*). (Illustration by Michelle Meneghini)

Ragfish probably are preyed upon by piscivorous (fish-eating) fishes. There is one record from Washington of a small juvenile ragfish found in the stomach of a tuna. In 1938 Cowan reported that a colleague identified the adult ragfish as the whaler's "bastard halibut," which forms a large part of the diet of sperm whales taken in deep water off Queen Charlotte Island, British Columbia.

Reproductive biology

Ragfish are oviparous and ready to spawn at seven to nine years of age, according to Fitch and Lavenberg, and they spawn in summer. In 1968 Allen studied four adult females and postulated that spawning occurs in late winter and early spring. Allen observed that eggs were largest (to 0.12 in, or 3.0 mm, in diameter) at that time and estimated fecundity to be 430,000 eggs for a 59-in (1.5-m) female. It is not known whether ragfish undergo spawning migrations or become territorial. Eggs are nonadhesive and planktonic, with a single oil globule. The larvae are distinctly pigmented and planktonic.

Conservation status

This species is not listed by the IUCN.

Significance to humans

Ragfish have no sport or commercial fishing value. They occur rarely in the bycatch of commercial shrimp and bottom fish trawlers.

Resources

Books

Fitch, J. E., and R. J. Lavenberg. *Deep-Water Teleostean Fishes of California.* Berkeley: University of California Press, 1968.

Periodicals

Allen, G. H. "Fecundity of the Brown Ragfish, *Icosteus aenigmaticus* Lockington, from Northern California." *California Fish and Game* 54, no. 3 (1968): 207–214.

Bean, T. H. "Description of a New Genus and Species of Fish, *Acrotus willoughbyi*, from Washington Territory." *Proceedings of the U.S. National Museum* 10 (1887): 631–632.

Clemens, W. A., and G. V. Wilby. "Fishes of the Pacific Coast of Canada." *Bulletin (Fisheries Research Board of Canada)* 68 (1961): 1–443.

Cowan, I. M. "Some Fish Records from the Coast of British Columbia." *Copeia* 1938, no. 2 (1938): 97.

Günther, A. "Report on the Deep-Sea Fishes Collected by the H.M.S. *Challenger* during the years 1873–1876." *Zoology* 22 (1887): 46–47.

Lockington, W. N. "Description of a New Genus and Some New Species of California Fishes (*Icosteus aenigmaticus* and *Osmerus attenuatus*)." *Proceedings of the U.S. National Museum* 3 (1880): 63–68.

Steindachner, F. "Ichthyologische Beitrage (11)." *Sitzungsber. Kais. Akad. d. Wiss. Wien. Math. Naturw. Kl. Bd.* 83 (1881): 396–397.

K. Gus Thiesfeld, BS

○

Gobiesocoidei

(Clingfishes and singleslits)

Class Actinopterygii
Order Perciformes
Suborder Gobiesocoidei
Number of families 1

Photo: Clingfish (*Lepadichthys lineatus*) use their sucking disc (modified pelvic fins) to cling to rocks, marine plants, or even sea urchins. (Photo by Animals Animals ©W. Gregory Brown. Reproduced by permission.)

Evolution and systematics

No clingfishes and singleslits have been identified from the fossil record, and the evolutionary history remains to be examined with molecular methods. The examination of morphological relationships has led to the formulation of competing hypotheses to describe the history and systematic position of clingfishes and singleslits, and these questions are far from resolved. Some workers place these fishes in the order Gobiesociformes along with the dragonets (Callionymidae) and draconetts (Draconettidae). Others place the Gobiesocidae in the suborder Gobiesocoidei within the Perciformes, next to the suborder Callionymoidei (Callionymidae and Draconettidae); such is the treatment in this chapter.

The family Gobiesocidae is partitioned into two subfamilies, the Gobiesocinae (clingfishes) and the Cheilobranchinae (singleslits) within the Gobiesocoidei. Both subfamilies are united by the presence of a joint that connects the supracleithrum and cleithrum bones within the pelvic girdle. Two other specializations link these two subfamilies: a joint connecting the interopercle and epihyal and a heart with a structure unique to both subfamilies. The singleslits are sometimes recognized as a separate family within the suborder Gobiesocoidei; the family consists of a single genus, *Alabes*, with four species. Briggs recognized the following subfamilies (denoted here as tribes), which now are placed under the Gobiesocinae: the primitive Trachelochismini, the monotypic Haplocylicini, the Lepadogastrini, the monotypic Chorisochismini, the Diplocrepini, the Gobiesocini, and the highly specialized Diademichthyini, which are mostly commensal with sea urchins and crinoids. Briggs concluded that more specialized taxa are found in the tropical Indo-West Pacific region; intermediate taxa are found in the eastern Pacific, western Atlantic, and eastern Atlantic; and relict Southern Hemisphere taxa are confined to the cool temperate waters of southern Africa, Australia, and New Zealand. There are at least 120 species of clingfishes (two with two subspecies each and one with three subspecies) in 44 genera plus undescribed genera and species.

The known genera include *Acyrtops* (two species), *Acyrtus* (two species), *Alabes* (four species), *Apletodon* (four species, one with two subspecies), *Arcos* (five species), *Aspasma* (one species), *Aspasmichthys* (one species), *Aspasmodes* (one species), *Aspasmogaster* (four species), *Chorisochismus* (one species), *Cochleoceps* (four species), *Conidens* (two species), *Creocele* (one species), *Dellichthys* (one species), *Derilissus* (two species), *Diademichthys* (one species), *Diplecogaster* (three species, one with three subspecies), *Discotrema* (one species), *Ecklonaichthys* (one species), *Gastrocyathus* (one species), *Gastrocymba* (one species), *Gastroscyphus* (one species), *Gobiesox* (26 species), *Gouania* (one species), *Haplocylix* (one species), *Kopua* (one species), *Lecanogaster* (one species), *Lepadichthys* (nine species), *Lepadogaster* (three species, one with two subspecies), *Liobranchia* (one species), *Lissonanchus* (one species), *Modicus* (two species), *Opeatogenys* (two species), *Parvicrepis* (one species), *Pherallodichthys* (one species), *Pherallodiscus* (one species), *Pherallodus* (two species), *Posidonichthys* (one species), *Propherallodus* (one species), *Rimicola* (five species), *Sicyases* (two species), *Tomicodon* (13 species), and *Trachelochismus* (two species).

Physical characteristics

The clingfishes are denoted mainly by the presence of pelvic fins modified into a sucking disk. Unlike the gobies (Gobiidae, Perciformes), which possess pelvic fins modified into a suction cup, the clingfishes lack a spinous dorsal fin. With some exceptions, clingfishes are tadpole-like, with broad, depressed heads and flattened bodies. Some species, however, are elongate with pointed snouts. Clingfishes lack scales but shield themselves with a mucus that forms a protective coat around their bodies. The lateral line is well developed anteriorly but is either small or missing posteriorly.

A crinoid clingfish (*Discotrema crinophila*) from Myanmar. (Photo by Animals Animals © Don Brown. Reproduced by permission.)

There is a well-developed network of sensory pores on the head. There is one dorsal and one anal fin.

The sucking disc is of two types: a single disc, in which the anterior and posterior components form a single continuous structure, and a double disc, in which the larger anterior and smaller posterior components form two discs separated by a single wall or edge. The pectoral fins generally are small and are positioned posterior to the sucking disk. The dorsal nostril and, usually, the posterior nostril are both tubular. The swim bladder is absent. Color patterns are variable and range from black to orange, brown, green, or red. Additionally, there may also be contrasting stripes, bars, or spots of yellow, blue, green, brown, gray, or white. The larvae are elongate, torpedo-shaped, and laterally compressed, and they become more rounded with growth. The gas bladder, located above the gut, is present but is lost after settlement. Body sizes typically are small, less than 1.9–2.4 in (5–6 cm) in total length, but at least one temperate species from southern Africa, the rocksucker (*Chorisochismus dentex*), reaches 11.8 in (30 cm). Clingfishes are sexually dimorphic, with differences in the size of the urogenital papillae and also in body size—males generally are larger than females. Two species reportedly produce a skin toxin.

The singleslits are eel-like in shape, with diminutive pelvic fins and a vestigial sucking disc. Their dorsal and anal fins are modified to resemble those of true eels. Soft rays are present only in the caudal fin. The posterior nostril is absent in one species. Color patterns are also variable and range from translucent pink or green to light green, brown, gray, or black. Spots or bars of black, gray, brown, or yellow may also be present. Usually they are less than 1.9 in (5 cm) in length, but one species, the common shore-eel (*Alabes dorsalis*) of Australia, reaches 4.7 in (12 cm).

Distribution

Clingfishes occur in marine and brackish waters in tropical, warm temperate, and temperate zones of the Atlantic, Indian and Pacific Oceans. General distribution patterns may be described best by relying upon Briggs' classification scheme, as modified here. Thus, the Trachelochismini consists of Southern Hemisphere relict species restricted, except for two Northern Hemisphere species, to temperate Australia and New Zealand. The monotypic Haplocylicini is endemic to New Zealand. Members of the Lepadogastrini occur in the eastern Atlantic and Mediterranean Sea, with one species found in deeper waters of southern Africa. The Chorisochismini is monotypic and also is found in southern Africa, but in shallower, warm temperate waters. The Diplocrepini is distributed across the Indo-West Pacific, with one deepwater relict species occurring in the Caribbean. The Gobiesocini occurs in the western Atlantic and eastern Pacific; some also are found in freshwater streams of the Caribbean, Central America, and Cocos Island in the eastern Pacific. One monotypic genus occurs in southern Africa. The Diademichthyini also is found in tropical waters of the Indo-West Pacific region. Singleslits are restricted to temperate inshore waters of the southern Australian mainland, Tasmania, and Norfolk Island.

Habitat

Clingfishes are largely inshore fishes adapted to clinging to the substrate. Thus, they are able to colonize high-energy habitats and withstand breaking waves and surging waters. Many species occur on or under boulders and rocks, in crevices, and on rocky slopes and rock faces. Species adapted to living in tide pools are capable of remaining out of water for a number of days, if they are kept moist and out of sunlight. Other species are found in close association with corals, soft corals, sponges, and ascidians. Some species live among crinoids or sea urchins. The crinoid clingfish, *Discotrema crinophila*, lives secluded among the arms of crinoids. *Diademichthys lineatus* has strayed from the general clingfish body plan by evolving an elongated shape that nearly mimics the long spines of *Diadema* sea urchins. Other species cling to the blades of sea grasses and to seaweed and kelp. Some members of the genus *Gobiesox* occur in freshwater streams. Singleslits also live in inshore marine waters, usually in shallow tide pools and often under stones or in close association with seaweed, the stems or branches of which they may cling to or wrap around with their prehensile caudal fins and flexible bodies.

Behavior

Clingfishes and singleslits tend to be secretive, although the reef-associated species commensal with sea urchins and other organisms are more mobile and swim in the water column. Territoriality has been reported for numerous species and is suspected for the rest. This trait is more pronounced in males. Most observations of other behavioral activity have been confined to aquarium studies; field investigations, where physically possible, are desirable.

Feeding ecology and diet

The clingfishes generally feed on smaller benthic invertebrates, but there are significant exceptions to this rule. Clingfish diets, depending upon species, body size, and habitat, include various algae, crustaceans (including amphipods, copepods, small crabs and shrimps), polychaete worms, small bivalves, limpets and other gastropods, chitons, body parts of host sea urchins, and small fishes. Members of the genus *Cochleoceps* are cleaner fishes that remove ectoparasites from boxfishes, porcupinefishes, morwongs, and other species. The singleslits feed mainly on small benthic invertebrates. Clingfishes are preyed upon by other fishes, shorebirds, and possibly crabs living in the intertidal zone.

Reproductive biology

Reproductive behavior and ecology have been described for a number of species, mainly from aquarium studies, and, as expected, there is species-specific variation around a common theme. Courtship is paired and initiated by the male within his territory. Generally, the male nudges the female's abodomen; if the female responds positively, he moves parallel to her flank and undulates or quivers. After some time, the female will undulate her body and lay demersal eggs singly until a small clutch is produced. (Demersal eggs are those that are deposited upon the substrate, such as the sea floor.) Eggs may be laid on stones, algae, or other substrates. Depending upon the species, egg laying may last from several minutes to a few hours. The male fertilizes the eggs during this process, and, upon completion, the clutch either is guarded by the male or is abandoned by the pair. There is some evidence of mating by internal fertilization, with female parental care, in the weedsucker (*Eckloniaichthys scylliorhiniceps*). This species is a monotypic clingfish endemic to South Africa that may share this trait with a few other species. Spawning or mating may be seasonal, during warmer months, at higher latitudes. At tropical or subtropical latitudes, spawning or mating may take place year-round or seasonally if water temperatures become either too warm or too cold. Larvae of all species are probably planktonic.

Conservation status

No species is currently on the IUCN Red List. Marine and freshwater species with limited distributions may be at risk from habitat destruction or pollution.

Significance to humans

Some species are collected for the aquarium trade.

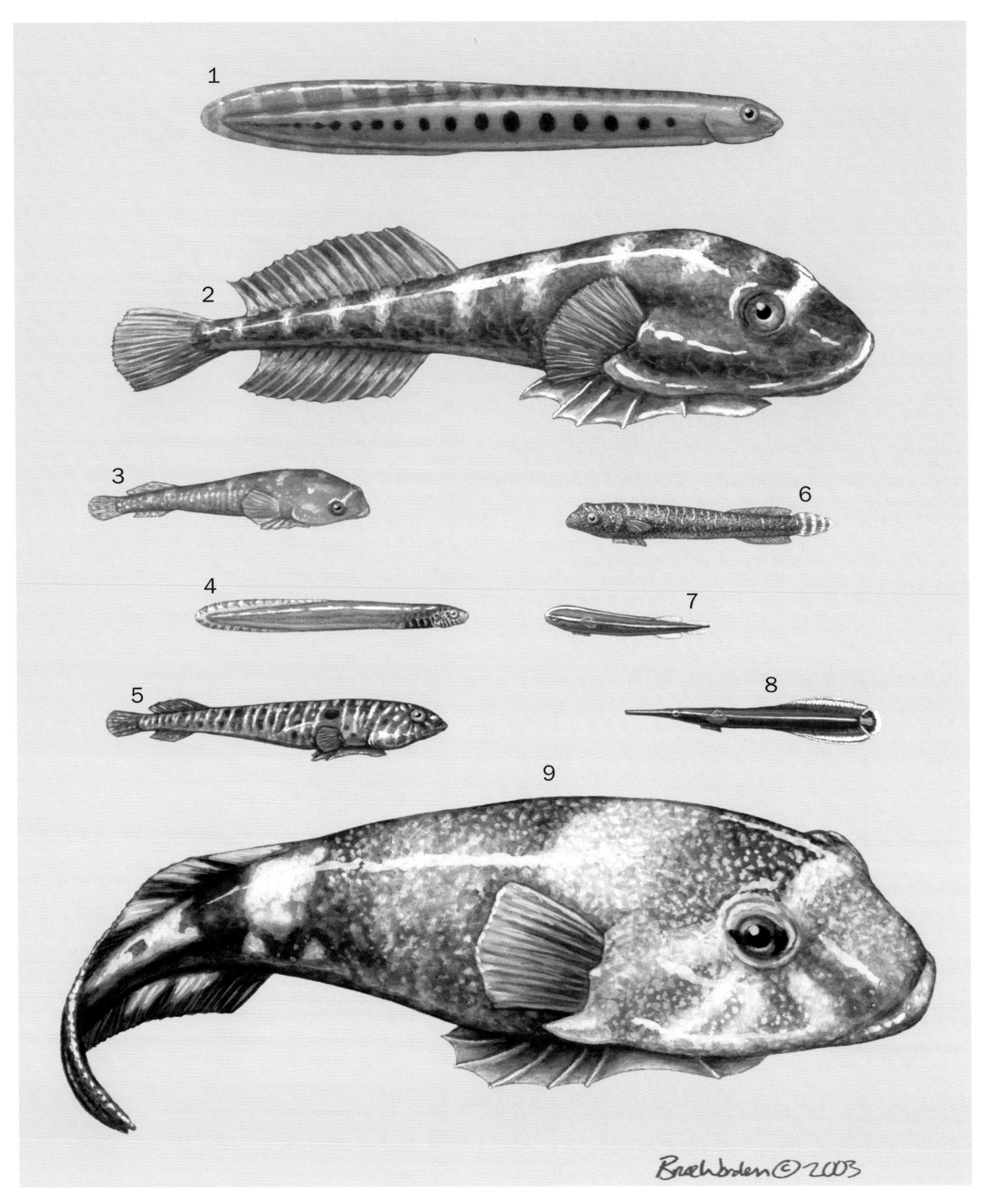

1. Common shore-eel (*Alabes dorsalis*); 2. Northern clingfish (*Gobiesox maeandricus*); 3. Two-spotted clingfish (*Diplecogaster bimaculata bimaculata*); 4. Pygmy shore-eel (*Alabes parvulus*); 5. Sonora clingfish (*Tomicodon humeralis*); 6. Eastern cleaner-clingfish (*Cochleoceps orientalis*); 7. Crinoid clingfish (*Discotrema crinophila*); 8. Urchin clingfish (*Diademichthys lineatus*); 9. Rocksucker (*Chorisochismus dentex*). (Illustration by Bruce Worden)

Species accounts

Common shore-eel
Alabes dorsalis

FAMILY
Gobiesocidae

TAXONOMY
Alabes dorsalis Richardson, 1845, attributed to the northwest coast of Australia, but this is likely a type locality error.

OTHER COMMON NAMES
English: Shore eel.

PHYSICAL CHARACTERISTICS
Body small and elongate (eel-like) with reduced fins. Vestigial sucking disc. Color variously gray to black, with prominent green spots along the flank and gray to black blotches regularly spaced along the dorsal fin. Grows to 4.7 in (12 cm) in length.

DISTRIBUTION
Australia from southern New South Wales south to Tasmania and west to central South Australia.

HABITAT
Intertidal zone under rocks.

BEHAVIOR
Poorly known. This species is secretive and seldom seen. The common shore-eel moves rather like a true eel among and under rocks, into holes, and along the bottom of tide pools. Nothing is known about its social behavior, although it may be territorial, like others in its family. May tolerate exposure to air during exceptionally low tides.

FEEDING ECOLOGY AND DIET
Feeds upon small benthic invertebrates.

REPRODUCTIVE BIOLOGY
Unknown. Spawning probably is seasonal, during warmer months, and paired with the deposition of demersal eggs. Larvae are probably planktonic.

CONSERVATION STATUS
Not listed by the IUCN. May be at risk from habitat destruction or pollution.

SIGNIFICANCE TO HUMANS
May be collected infrequently for aquaria. ◆

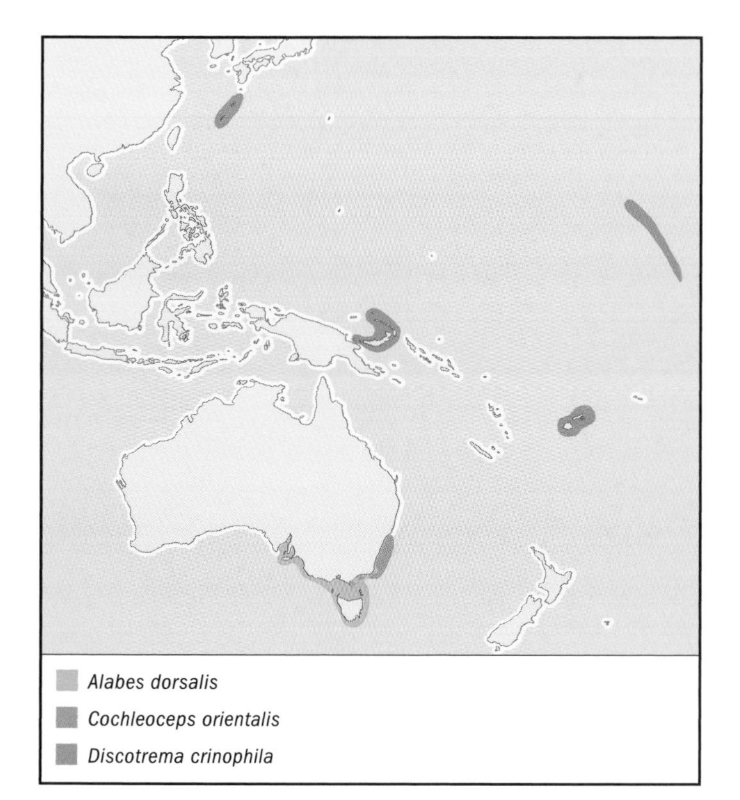

Crinoid clingfish
Discotrema crinophila

FAMILY
Gobiesocidae

TAXONOMY
Discotrema crinophila Briggs, 1976, between Stuart and Yanuca islands, Benga Lagoon, Benga, Fiji Islands.

OTHER COMMON NAMES
None known.

PHYSICAL CHARACTERISTICS
Elongate body, with a rounded snout. There are 8–9 dorsal fin soft rays and 7–8 anal fin soft rays. Color is black, with a yellow stripe that runs along each flank from the snout to the base of the caudal fin and a mid-dorsal yellow stripe that joins the lateral stripes at the snout.

DISTRIBUTION
Western Pacific, from the Ryukyu Islands of southern Japan south to Taiwan, southeast to the Bismarck Archipelago in Papua New Guinea, and east to Fiji and Christmas Island, Line Islands (Kiribati). Specimens reported from the Great Barrier Reef, and possibly the Bismarck Archipelago, represent an undescribed species.

HABITAT
Lives among the arms of crinoids on coral reefs at a depth range of 26–66 ft (8–20 m).

BEHAVIOR
Poorly known. Swims about the arms of crinoids.

FEEDING ECOLOGY AND DIET
Feeds upon small invertebrates.

REPRODUCTIVE BIOLOGY
Poorly known but likely lays demersal eggs, perhaps at the base of a crinoid.

CONSERVATION STATUS
Not listed by the IUCN.

SIGNIFICANCE TO HUMANS
May be collected infrequently for aquaria. ◆

Eastern cleaner-clingfish
Cochleoceps orientalis

FAMILY
Gobiesocidae

TAXONOMY
Cochleoceps orientalis Hutchins, 1991, Big Island, off Wollongong, New South Wales, Australia.

OTHER COMMON NAMES
None known.

PHYSICAL CHARACTERISTICS
Small, tadpole-shaped body with rounded caudal fins and a moderately sized sucker disc. There are 5–6 dorsal fin soft rays and 4–6 anal fin soft rays. Color pattern is tiny brown to red spots on an orange to yellow or greenish-yellow background; irridescent blue dashes, lines, or spots occur on the dorsum. Grows to 2.2 in (5.5 cm) in length.

DISTRIBUTION
Southwest Pacific from central New South Wales south to eastern Victoria, Australia.

HABITAT
Occurs on sponges or ascidians on deeper temperate rocky reefs and on kelp or seaweed in shallower subtidal habitats. Found as deep as 33 ft (10 m).

BEHAVIOR
In addition to its cleaning behavior, this species probably is territorial.

FEEDING ECOLOGY AND DIET
A cleanerfish, it feeds upon ectoparasites plucked from various species of fishes.

REPRODUCTIVE BIOLOGY
Poorly known, but females deposit eggs on algae within a male's territory, and both parents provide care.

CONSERVATION STATUS
Not listed by the IUCN.

SIGNIFICANCE TO HUMANS
May be collected infrequently for aquaria. ◆

Northern clingfish
Gobiesox maeandricus

FAMILY
Gobiesocidae

TAXONOMY
Gobiesox maeandricus Girard, 1858, San Luis Obispo, California, United States.

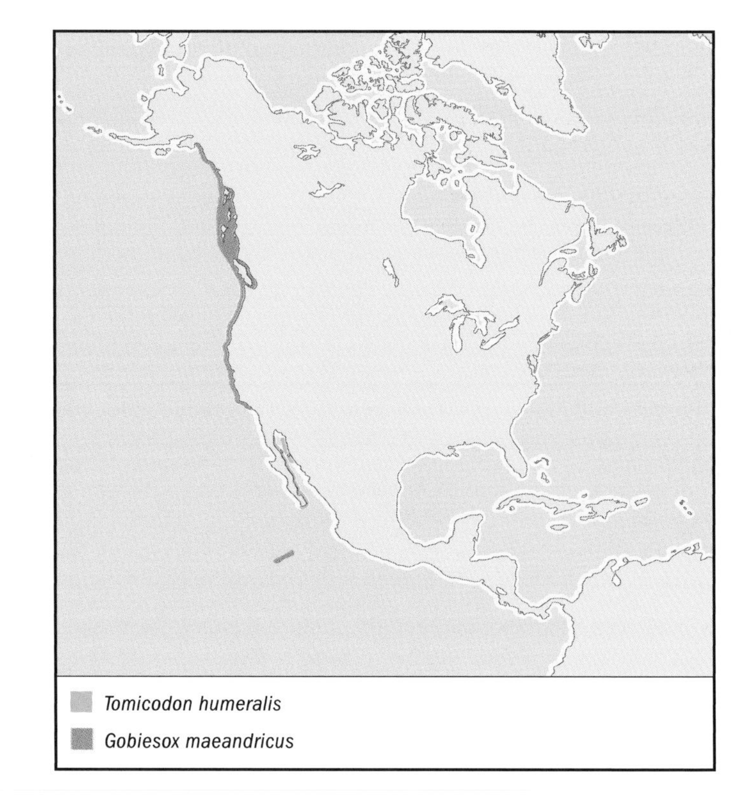

OTHER COMMON NAMES
None known.

PHYSICAL CHARACTERISTICS
There are 13–16 dorsal soft rays; 12–14 anal soft fin rays; a rounded caudal fin; and short, rather broad pectoral fins with a fleshy palp that bridges between the base of the fin and the gill opening. The color pattern consists of a light olive-brown or cherry-red background with either dark reticulations or mottled shades of lighter color and a series of white bars between the eyes. In juveniles the white bars also run along the back and on the edge of the caudal fin. Grows to about 6.3 in (16 cm) in length. Males tend to be larger than females.

DISTRIBUTION
Eastern Pacific from southeastern Alaska south to southern California; also found at Revillagigedo Island off Baja California, Mexico.

HABITAT
Lives among algae or under rocks in the intertidal zone and in the upper canopy of kelp forests, to a depth of 26 ft (8 m). The ability to breathe atmospheric air with its gills allows it to tolerate exposure, if it is kept moist under rocks or algae.

BEHAVIOR
Males maintain territories. Both males and females tend to be somewhat secretive, clinging to surfaces and moving only to feed or interact with others.

FEEDING ECOLOGY AND DIET
Feeds upon small mollusks and crustaceans found on rocks, algae, and kelp.

REPRODUCTIVE BIOLOGY
As with others in this genus, this species likely spawns demersal eggs in a nest prepared by a male within his territory.

Spawning may take several hours. Males provide parental care. The larvae are planktonic.

CONSERVATION STATUS
Not listed by the IUCN.

SIGNIFICANCE TO HUMANS
Collected mainly for public aquaria. ◆

Pygmy shore-eel
Alabes parvulus

FAMILY
Gobiesocidae

TAXONOMY
Alabes parvulus McCulloch, 1909, rock pools near Sydney, New South Wales, Australia.

OTHER COMMON NAMES
None known.

PHYSICAL CHARACTERISTICS
Elongate and eel-like body with reduced fins. Vestigial sucking disc. Grows to 1.9 in (5 cm) in length.

DISTRIBUTION
Temperate waters of southern Australia and Norfolk Island.

HABITAT
Inhabits the intertidal zone. Found in tide pools, in brown seaweed beds and under rocks, to a depth of 20 ft (6 m).

BEHAVIOR
Poorly known. Has been found in small groups of three to six individuals, and social interactions may be expected to take place regularly.

FEEDING ECOLOGY AND DIET
Feeds upon small benthic invertebrates.

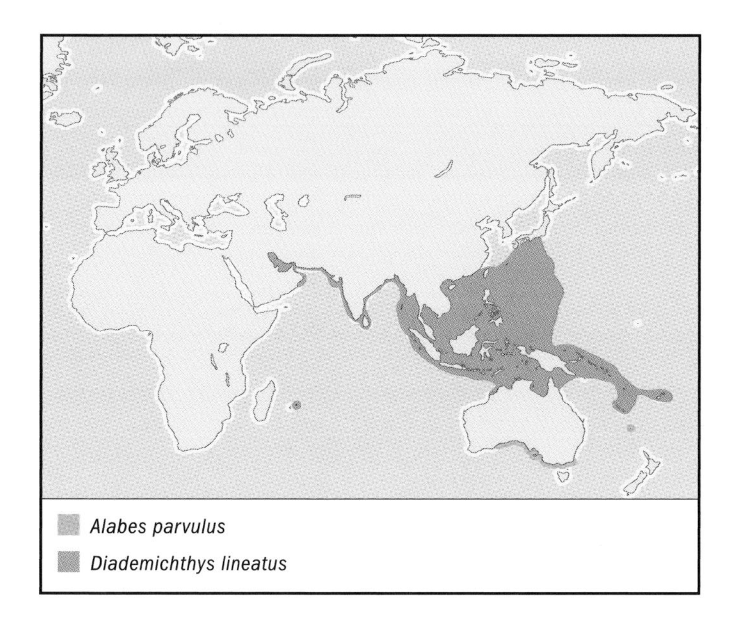

REPRODUCTIVE BIOLOGY
Largely unknown. Spawning is probably seasonal, during warmer months, and is paired with the deposition of demersal eggs. Larvae are probably planktonic.

CONSERVATION STATUS
Not listed by the IUCN.

SIGNIFICANCE TO HUMANS
May be collected infrequently for aquaria. ◆

Rocksucker
Chorisochismus dentex

FAMILY
Gobiesocidae

TAXONOMY
Chorisochismus dentex Pallas, 1769, type locality not specified.

OTHER COMMON NAMES
None known.

PHYSICAL CHARACTERISTICS
Large, tadpole-like body with a very broad head. There are 7–9 dorsal fin soft rays, 6–7 anal fin soft rays, 21–24 pectoral fin soft rays, and 8–10 caudal fin rays. The teeth are large and conical in shape. Perhaps the largest of all clingfishes, this species reaches 11.8 in (30 cm) in length.

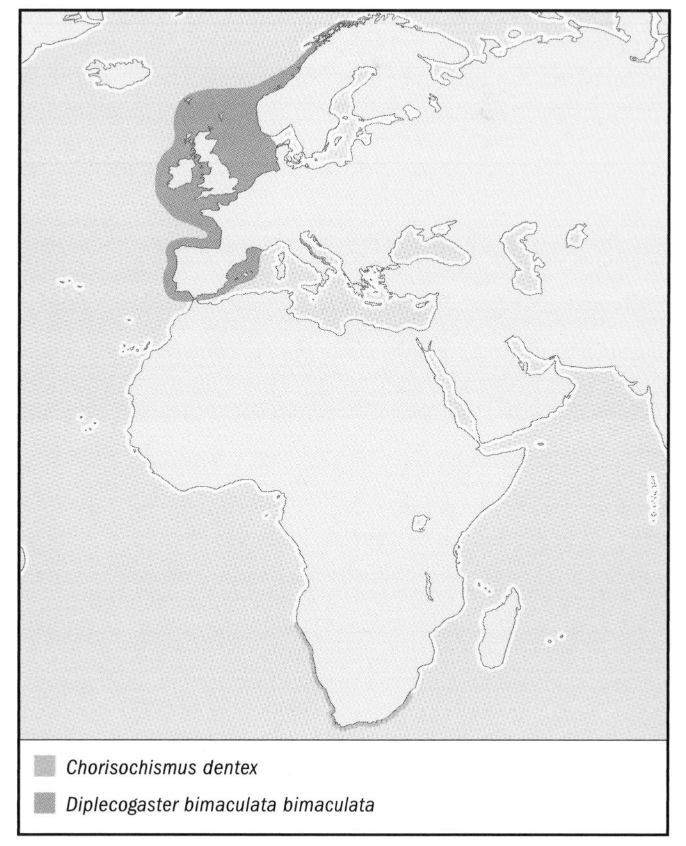

DISTRIBUTION
Southeast Atlantic, from Namibia south to northern Natal in South Africa.

HABITAT
Tide pools and rocks and boulders in the intertidal zone.

BEHAVIOR
Poorly known. Males likely are territorial. Owing to its large size, rocksuckers are probably more mobile than other demersal clingfishes.

FEEDING ECOLOGY AND DIET
A carnivore that preys upon limpets and sea urchins. This species uses its large upper incisiform teeth to pry limpets off rocks. The undigested shell fragments are passed through the gut and emerge in mucous capsules.

REPRODUCTIVE BIOLOGY
Poorly known. Females probably lay demersal eggs in a male's territory, and the male performs parental care. The larvae are probably planktonic.

CONSERVATION STATUS
Not listed by the IUCN.

SIGNIFICANCE TO HUMANS
May be collected infrequently for aquaria. ◆

Sonora clingfish
Tomicodon humeralis

FAMILY
Gobiesocidae

TAXONOMY
Tomicodon humeralis Gilbert, 1890, Puerto Refugio (Angel Island) and La Paz, Mexico.

OTHER COMMON NAMES
Spanish: Chupapiedra de Sonora.

PHYSICAL CHARACTERISTICS
Compressed, elongate body with a broad head and a large sucking disc. There are 8–9 dorsal fin soft rays, 6–7 anal soft fin rays, 17–19 pectoral fin soft rays, and 9–10 caudal fin soft rays. The color pattern consists of light diagonal stripes along the entire body, with a pair of dorsal spots positioned just behind the head, about even with the pectoral fins. Grows to 3.3 in (8.5 cm) in length. Sexually dimorphic, with males larger than females.

DISTRIBUTION
Endemic to the Gulf of California from Punta Borrascoso to Guaymas in Sonora and from San Felipe to Cabo San Lucas in Baja California.

HABITAT
A common clingfish found mainly in the upper and mid-intertidal zones, usually under rocks, to which they cling, and in little or no water. If kept moist, this fish is able to withstand extreme temperatures when exposed to air.

BEHAVIOR
This species is secretive, clinging to the undersides of rocks and moving over rocky surfaces to feed. Their movements are related to tidal movements. Activity is greater at high tide, and there is little or no activity at low tide. Males are territorial. Both males and females secrete large amounts of mucus, which coats their bodies and protects them from desiccation.

FEEDING ECOLOGY AND DIET
A diurnal predator that feeds upon small crustaceans, including barnacles and limpids.

REPRODUCTIVE BIOLOGY
Spawning begins in late spring or early summer and is paired with demersal courtship. A single male may mate with more than one female. Eggs are adhesive and laid on the underside of rocks, where they are guarded by the male, sometimes with the aid of one or more females. During low tide, the parents secrete mucus that protects the eggs from exposure. The larvae are planktonic.

CONSERVATION STATUS
Not listed by the IUCN.

SIGNIFICANCE TO HUMANS
May be collected infrequently for aquaria. ◆

Two-spotted clingfish
Diplecogaster bimaculata bimaculata

FAMILY
Gobiesocidae

TAXONOMY
Diplecogaster bimaculata bimaculata Bonnaterre, 1788, seas of England, although no types were designated.

OTHER COMMON NAMES
English: Two-spotted sucker.

PHYSICAL CHARACTERISTICS
Flattened and elongated body, with a somewhat triangular snout. The dorsal fin has 4–7 soft rays, and the anal fin has 4–7 soft rays. The gill opening is minute. Color varies but usually is red with yellow on the ventral surface and a pattern of blue and brown spots over the body surface. Males are distinguished by a purple or red spot, ringed in yellow, on each flank, behind the pectoral fins. Grows to 2.4 in (6 cm) in length.

DISTRIBUTION
Eastern Atlantic Ocean, from Norway and the Faroe Islands south to Gibraltar and the western Mediterranean and Adriatic. The subspecies *Diplecogaster bimaculata euxinica* is found in the Black Sea and *D. bimaculata pectoralis* from the offshore Canary and Cape Verde Islands and the Azores.

HABITAT
A temperate species found on sea grass beds, rocky bottoms, bivalve banks, and mud bottoms. Reported to favor shells. Depths range between 59 and 180 ft (18–55 m) in cooler waters and down to 328 ft (100 m) in the warmer Mediterranean.

BEHAVIOR
Not well known. Males likely are territorial.

FEEDING ECOLOGY AND DIET
Feeds on small benthic invertebrates.

REPRODUCTIVE BIOLOGY
Demersal courtship and spawning occur in the spring and summer months, with golden-colored eggs laid in masses under shells or stones. Males provide most of the parental care. Larvae are planktonic.

CONSERVATION STATUS
Not listed by the IUCN.

SIGNIFICANCE TO HUMANS
Occasionally recovered from trawls working soft mud bottoms or bivalve banks. ◆

Urchin clingfish
Diademichthys lineatus

FAMILY
Gobiesocidae

TAXONOMY
Diademichthys lineatus Sauvage, 1883, New Caledonia.

OTHER COMMON NAMES
None known.

PHYSICAL CHARACTERISTICS
Elongate and slender body, with spatulate snout and small fins. There are 13–15 dorsal fin soft rays, 12–14 anal fin soft rays, and 25–26 pectoral fin soft rays. Color is reddish or reddish brown, with paired yellow stripes and a yellow blotch on the caudal peduncle that extends to the caudal fin. Grows to 1.9 in (5 cm) in length.

DISTRIBUTION
From Oman and Mauritius in the western Indian Ocean east to Indonesia and Fiji, north to southern Japan, and south to northern Australia.

HABITAT
Inhabits shallow coral reefs among the long-spined sea urchins and branching corals, usually in holes or protected areas.

BEHAVIOR
This clingfish swims, often in a dancing or undulating motion, between the spines of sea urchins, within coral heads, and in holes or small caves where sea urchins might be found. Their behavior is not known in any great detail and remains to be studied.

FEEDING ECOLOGY AND DIET
Takes burrowing bivalves on corals, eggs of shrimp commensal with sea urchins, and the tube feet of sea urchin hosts. Juveniles feed on the pedicellariae and sphaeridia of host urchins and on copepods and the eggs of shrimp that are commensal with sea urchins.

REPRODUCTIVE BIOLOGY
This species probably courts and spawns in a manner similar to that of others in its family, except that courtship bouts most likely take place in the water column. The eggs are small and demersal, and the larvae are pelagic.

CONSERVATION STATUS
Not listed by the IUCN. May be at risk from the loss of sea urchins because of overfishing, disease, or other factors and from habitat destruction and the effects of pollution on reefs.

SIGNIFICANCE TO HUMANS
May be collected for the aquarium trade. ◆

Resources

Books

Bohlke, James E., and Charles C. G. Chaplin. *Fishes of the Bahamas and Adjacent Tropical Waters.* 2nd edition. Austin: University of Texas Press, 1993.

Briggs, J. C. "Clingfishes." In *Encyclopedia of Fishes*, edited by J. R. Paxton and W. N. Eschmeyer. San Diego: Academic Press, 1995.

Briggs, J. C., and J. B. Hutchins. "Clingfishes and Their Allies." In *Encyclopedia of Fishes*, 2nd edition, edited by J. R. Paxton and W. N. Eschmeyer. San Diego: Academic Press, 1998.

Eschmeyer, W. N., ed. *Catalog of Fishes.* 3 vols. San Francisco: California Academy of Sciences, 1998.

Helfman, G. S., B. B. Collette, and D. E. Facey. *The Diversity of Fishes.* Malden, MA: Blackwell Science, 1997.

Kuiter, R. H. *Guide to Sea Fishes of Australia.* Sydney: New Holland, 1997

Leis, Jeffrey M., and Brooke M. Carson-Ewart, eds. *The Larvae of Indo-Pacific Coastal Fishes.* Boston: Brill, 2000.

Lythgoe, J., and G. Lythgoe. *Fishes of the Sea: The North Atlantic and Mediterranean.* London: Blandford Press, 1991.

Masuda, H., K. Amaoka, C. Araga, T. Uyeno, and T. Yoshino, eds. *The Fishes of the Japanese Archipelago.* Tokyo: Tokai University Press, 1984.

Myers, R. F. *Micronesian Reef Fishes: A Field Guide for Divers and Aquarists.* 3rd edition. Barrigada, Guam: Coral Graphics, 1999.

Nelson, J. S. *Fishes of the World.* 3rd edition. New York: John Wiley and Sons, 1994.

Randall, J. E., G. R. Allen, and R. C. Steene. *Fishes of the Great Barrier Reef and Coral Sea.* Honolulu: University of Hawaii Press, 1996.

Smith, M. M., and P.C. Heemstra, eds. *Smiths' Sea Fishes.* Berlin: Springer-Verlag, 1986.

Thomson, Donald A., Lloyd T. Findley, and Alex N. Kerstitch. *Reef Fishes of the Sea of Cortez.* 2nd edition. Tucson: University of Arizona Press, 1987.

Thresher, R. E. *Reproduction in Reef Fishes.* Neptune City, NJ: T.F.H. Publications, 1984.

Periodicals

Briggs, J. C. "New Species of *Rimicola* from California." *Copeia* 2002 (2002): 441–444.

Gosline, W. A. "A Reinterpretation of the Teleostean Fish Order Gobiosociformes." *Proceedings of the California Academy of Sciences* series 4, 37 (1970): 363–382.

Organizations

IUCN/SSC Coral Reef Fishes Specialist Group. International Marinelife Alliance-University of Guam Marine Laboratory, UOG Station, Mangilao, Guam 96913 USA. Phone: (671) 735-2187. Fax: (671) 734-6767. E-mail: donaldsn@uog9.uog.edu Web site: <http://www.iucn.org/themes/ssc/sgs/sgs.htm>

Terry J. Donaldson, PhD

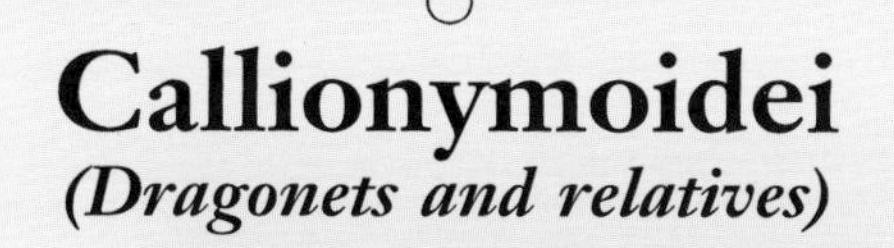

Callionymoidei

(Dragonets and relatives)

Class Actinopterygii
Order Perciformes
Suborder Callionymoidei
Number of families 2

Photo: A young fingered dragonet (*Dactylopus dactylopus*) on lava sand near the Sulawesi Island of Indonesia. (Photo by Fred McConnaughey/Photo Researchers. Reproduced by permission.)

Evolution and systematics

Members of this suborder include the Callionymidae (the dragonets) and the Draconettidae (deepwater dragonets). As with many other perciform fishes, members of this suborder likely radiated in the Eocene period. The systematic placement of this suborder begs further consideration as it has been placed under the nonperciform order Gobiesociformes (Gosline, 1970; Leis and Carson-Ewart, 2000), or linked to the family Gobiesocidae in a closely related suborder, the Gobiesocoidei, within the Perciformes (Helfman et al., 1997).

The Callionymidae consists of at least 17 genera and 156 species, although exact numbers vary because of two different and competing classifications proposed by T. Nakabo (1982 and 1983) and R. Fricke (1983). The Callionymidae genera include:

- *Anaora* (one species)
- *Bathycallionymus* (two subspecies)
- *Callionymus* (82 subspecies)
- *Calliurichthys* (one species)
- *Dactylopus* (one species)
- *Diplogrammus* (six subspecies)
- *Draculo* (five subspecies)
- *Eleutherochir* (one subspecies)
- *Foetorepus* (eight subspecies)
- *Neocynchiropus* (two subspecies)
- *Paracallionymus* (one subspecies)
- *Paradiplogrammus* (two subspecies)
- *Progogrammus* (one species)
- *Pseudocalliurichthys* (three subspecies)
- *Repomucenus* (10 subspecies)
- *Spinicapitichthys* (one species)
- *Synchiropus* (29 subspecies)

The Draconettidae consists of only two genera, *Centrodraco* (10 subspecies, but one with 2 subspecies) and *Draconetta* (one species).

Physical characteristics

The Callionymidae are relatively small, sometimes diminutive, usually elongate fishes, having both a depressed trunk and broad or depressed head. Most of these fishes are sexually dimorphic. Males tend to have larger body sizes and dorsal fins, greater fin ornamentation, and more distinctive color patterns. Dragonets usually have 2 dorsal fins, the first containing 1–4 spines, the second has only 6–11 rays; the last ray is divided at the base, as in the anal fin. Dorsal fins of males may also be large or high, and also have elongated rays. Dragonets also have relatively large pelvic and pectoral fins, with the former positioned forward of the latter. The anal fin has 4–10 rays. The shape of the caudal fin varies, but may be elongate or spade-like and include elongated rays. The preoperculum has well-developed spines and bars along the margin, but no opercular or subopercular spines are present. Dragonets are further distinguished by the presence of small gill openings, generally pore-like, positioned at the top of the head or along the upper flanks, with gill membranes that are united at the isthmus that separates the gills. Their bodies lack scales, but their lateral lines are quite well developed. Many species have color patterns well suited as camouflage, but others are brilliantly, if not spectacularly, colored. Some species, such as members of the genus *Foetorepus*, produce an acidic, bitter-tasting slime that covers their bodies and may serve as a toxic deterrent to predators.

Dragonet larvae develop quite rapidly, and most species are able to settle with a full compliment of fins at a small size. The larvae are denoted mainly by having a relatively large head, a short snout that lengthens with growth, a small, protrusible mouth placed terminally on the head, the presence of teeth only after settlement, a large eye that migrates dorsally with growth, and a gas bladder located anterior to the gut, which disappears in adults. The gills are free from the isthmus at first but become attached after the larvae undergo flexion. The bodies of dragonet larvae are heavily pigmented, with most of the pigment concentrated on the ventral, rather than dorsal, surface of the body.

The Draconettidae have small, elongated, and rounded bodies, pointed snouts, relatively large pectoral fins, and elongated pelvic fins. The gill openings are relatively broad. There are two nostrils on each side of the head. Scales are absent, but the body is distinguished by the presence of a grooved or vestigial lateral line that is well developed on the head. These fishes are also distinguished by having large eyes. There are two dorsal fins, with three spines on the first and 12–15 rays on the second. There are 12–13 soft rays on the anal fin. The operculum and suboperculum have single strong spines but these are absent on the preopercle. Their larvae are very poorly known. The body is elongate and compressed, the head is also compressed and is of moderate size, the snout is pointed, and the mouth is small and reaches as far as the anterior margins of the large, round eyes. The gas bladder is absent, and the fin elements are present in larvae at least 0.31 in (0.8 cm) long.

Distribution

The Callionymidae are found in tropical and warm temperate waters of the Atlantic, Indian, and Pacific Oceans. Species are found at either continental or insular localities, however most species occur in the Indo-Western Pacific. The Draconettidae share approximately the same broad distribution, but are limited to seamounts or deeper edges of continental shelves, and thus live at cooler water temperatures.

Habitat

Most members of the Callionymidae are found on coral pavement, rubble, sand, mud, or on other soft bottoms. Some species occur exclusively in corals, tide pools, or among rocks; however, others may be found among algae-covered rocks on flats or shorelines, or mangrove roots in brackish water. The two species of the genus *Bathycallionymus* occur on deep flats. Members of the family occur within a depth range of less than 39 in (1 m) at low tide to over 1,312 ft (400 m). Little is known about the habitats of the Draconettidae. They may likely be found in rubble and sand or amongst rocks along deep-slope edges or on top of seamounts.

Behavior

The behavior of dragonets has been studied for only a few species and, for these, most effort has been directed toward reproductive behavior. Males utilize their longer first dorsal fins for displaying in both male-male aggressive encounters, and male-to-female social and courtship interactions. Male body color may also be utilized, especially during courtship.

Virtually nothing is known about the behavior of members of the Draconettidae. Their large eyes likely aid them in movement, feeding, social interactions, and reproductive behavior in deep demersal habitats. The dorsal fins of males may be utilized for signaling, but their effectiveness under dimly lit conditions remains to be demonstrated.

Feeding ecology and diet

Callionymids and draconettids feed on small benthic invertebrates. Larger species, such as members of the callionymid genera *Callionymus*, *Foetorepus*, *Repomucenus*, and *Synchiropus*), feed upon correspondingly larger prey. Predators of dragonets and draconets are not well known, but it is likely that they are preyed upon by larger benthic-feeding or opportunistic predatory fishes.

Reproductive biology

Dragonets appear to be gonochoristic, with no evidence of sex change reported. Males maintain relative large home ranges compared to females and these may be defended territorially during periods of reproduction. Mating systems may consist of pairs or single-male dominated mating groups of two or more females. Some smaller, cryptically colored males may also be in these mating groups and will attempt to court and spawn with females, but will often fail because of aggressive interactions with the dominant male. Courtship and spawning typically commences prior to sunset for most species or possibly just after dawn in some species. A male will approach a female and display his erect fins to her, and, as if carrying her, will ascend with her a short distance into the water column, where pelagic eggs are released and fertilized. These eggs are spherical in the genera *Calliurichthys* and *Repomucenus*, and in *Callionymus* from the Atlantic Ocean. The mangrove dragonet (*Paradiplogrammus enneactis*), and perhaps others in this genus, produces a buoyant egg mass that later breaks up prior to hatching. Spawning may be seasonal, depending upon latitude and water temperature, and may demonstrate one or more peaks during an annual cycle. Serial spawning, with the daily production of eggs during the season, has been suggested for some species. Smaller females of some species with bimodal spawning seasons may defer reproduction until the second season (late summer or autumn).

Next to nothing is known about the reproductive biology of the Draconettidae. Both the spawning mode and eggs are unknown. Owing to their close relationship with the Callionymidae, they are sexual dimorphic, with males being larger than females. They are not likely to be hermaphroditic. Spawning is probably paired and pelagic, with a rapid, short ascent into the water column where eggs and sperm are released.

Conservation status

One species of callionymid, the St. Helena dragonet (*Callionymus sanctaehelenae*), which is endemic to the island of St. Helena in the South Atlantic, is currently listed by the IUCN as Critically Endangered. Some callionymids important to the aquarium trade, such as the mandarinfish (*Synchiropus splendidus*) and *Repomucenus* spp. that are important as food fishes, may also be at risk from overfishing, habitat destruction, and recruitment failure as a consequence of pollution or the creation of anoxic conditions that lead to hypoxia among larvae. No members of the Draconettidae are listed by the IUCN, but these fishes may be at risk from the effects of the deep trawling of their habitats.

Significance to humans

A number of dragonet species are important in the aquarium trade, such as the mandarinfish, and some species, such as *Repomucenus* spp., are taken directly or as bycatch in commercial or subsistence food fisheries, or for the production of fish meal. The Draconettidae may be taken incidentally by deep-trawling fishing vessels, but appear not to have any commercial significance to humans.

1. Draconett (*Centrodraco insolitus*); 2. Draconetta (*Draconetta xenica*); 3. Richardson's dragonet (*Repomucenus richardsonii*); 4. Mandarinfish (*Synchiropus splendidus*); 5. Lancer dragonet (*Paradiplogrammus bairdi*). (Illustration by Marguette Dongvillo)

Species accounts

Lancer dragonet
Paradiplogrammus bairdi

FAMILY
Callionymidae

TAXONOMY
Paradiplogrammus bairdi Jordan, 1888, off Pensacola, Florida, United States.

OTHER COMMON NAMES
English: Coral dragonet; Spanish: Dragoncillo coralino.

PHYSICAL CHARACTERISTICS
Total length 4.5 (11.4 cm). Sexually dimorphic; males are larger, have larger fins, extended dorsal fin, and more distinctive color patterns on body and fins. Body is elongate and scaleless in both sexes. Have 4 dorsal spines on first dorsal fin, 9 soft rays on second dorsal fin; 8 anal fin rays, but no spines. Color pattern is complex and varies between sexes. Dorsal half of body has marbling and mottling in various shades of brown, black, and white; white, roundish blotches between narrow brown bars on ventral half of body. First dorsal fin of male marked by yellow swirl and second dorsal fin with downward-pointing pattern of dark bands. Males also have blue concentric lines and rows of spots on first dorsal fin, blue bars with narrow orange margins on preopercle, orange spots, small blue lines and crescentlike markings on dorsal half of body, and blue spots around ventral half of eyes.

DISTRIBUTION
Western Atlantic, from Bermuda and southern Florida, United States, east and south to the Bahamas, south to the Lesser Antilles and northern South America, and west into the Gulf of Mexico.

HABITAT
Benthic; frequents sand patches on shallow reefs, rocky shorelines, and sea-grass flats at depths of 3.3–298.5 ft (1–91 m).

BEHAVIOR
As with other dragonets, utilizes a relatively cryptic coloration to provide camouflage as it forages along the bottom. Social interactions, especially between males, can include the defense of territories or females, and are distinguished by erect fin displays.

FEEDING ECOLOGY AND DIET
Feeds upon small benthic invertebrates.

REPRODUCTIVE BIOLOGY
Hermaphroditism has not been reported. Sexually dimorphic characters in males, mainly fin size, are used in reproductive behavior. Males initiate courtship by displaying erect fins towards females. Spawning is paired and occurs after short ascent into water column, where the eggs are fertilized. Eggs are pelagic, egg mass is likely buoyant and breaks up prior to hatching. Larvae generally typical of other dragonets.

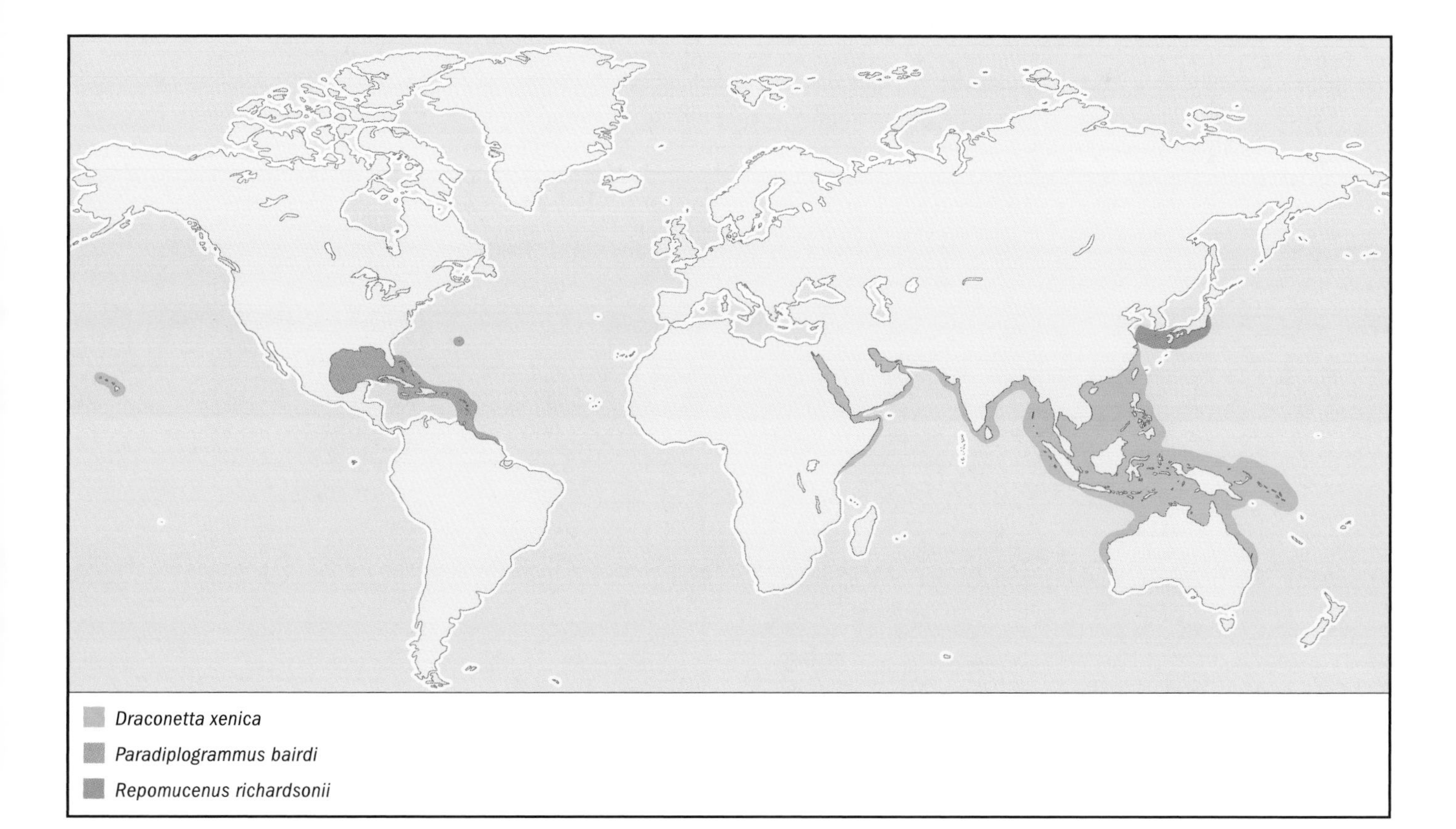

CONSERVATION STATUS
Not listed by the IUCN.

SIGNIFICANCE TO HUMANS
Collected for the aquarium trade. May be taken incidentally in bottom trawls by fishing vessels operating on offshore banks, but of no value as a food fish. ◆

Richardson's dragonet
Repomucenus richardsonii

FAMILY
Callionymidae

TAXONOMY
Repomucenus richardsonii Bleeker, 1854, Nagasaki, Japan.

OTHER COMMON NAMES
Japanese: Nezumi-gochi.

PHYSICAL CHARACTERISTICS
Total length 6.7 (17 cm). Sexually dimorphic in character development. Body elongate and strongly depressed with broad, flattened head. Scales lacking. First dorsal fin with 4 spines; 9–10 soft rays on second, elongated dorsal fin. Anal fin has 9–10 soft rays. Relatively large pectoral and pelvic fins. Caudal fin elongate. Color pattern on body is light to medium brown. Males have three dark blotches on first dorsal fin and oblique dark lines along lower half of flank. Faint dark mottling on anal fin and ventral portion of caudal fin. Females and young males have large black mark on first dorsal fin; females also have three faint white spots on second dorsal fin. May live only as long as two years.

DISTRIBUTION
Northwestern Pacific, from central Japan west into the East China Sea.

HABITAT
Coastal bottoms of sand and mud.

BEHAVIOR
Behavior not well known. Species is suited for movement, foraging, and crypsis on sandy or muddy bottoms. Social interactions include aggressive behavior between rival males, usually during courtship periods, with larger males dominating smaller ones. Aggressive behavior is characterized by display of erect fins during interactions.

FEEDING ECOLOGY AND DIET
Feeds mainly upon benthic invertebrates.

REPRODUCTIVE BIOLOGY
Pronounced sexual dimorphism in color, body size, and fin development; males have greater body size as well as elongated fin spines and rays. These differences contribute toward reproductive success because of their value in defending against rival males or their signal function in attracting females. Courtship and spawning is paired, and commences with male displays of erect fins and results in pelagic spawning in water column. Eggs produced serially, likely spawned on daily basis with one or two spawning seasons (spring and early autumn). No evidence of hermaphroditism. Recruitment of larvae is also seasonal.

CONSERVATION STATUS
Not listed by the IUCN. May be vulnerable or threatened by overfishing, habitat destruction, and pollution that causes anoxic conditions.

SIGNIFICANCE TO HUMANS
Occurs in both subsistence and commercial fisheries. Flesh favored in Japan. Also collected for the aquarium trade. ◆

Mandarinfish
Synchiropus splendidus

FAMILY
Callionymidae

TAXONOMY
Synchiropus splendidus Herre, 1927, Bungau, Philippines.

OTHER COMMON NAMES
Japanese: Nishiki-teguri.

PHYSICAL CHARACTERISTICS
Total length about 3.1 in (8 cm). Sexually dimorphic; males larger than females and have larger dorsal fins. Small, somewhat elongated body lacks scales but has well-developed lateral line. Two dorsal fins, first with 4 spines and second with 8 dorsal rays. There are 6–8 soft rays on the anal fin. The preopercle has strong spine, but no spines present on operculum or suboperculum. Body coloration of vivid green or blue markings on ground color of orange, or rarely red.

DISTRIBUTION
Western Pacific and eastern Indian Ocean from the Philippines and Indonesia (Java), east to Pohnpei (Micronesia) and New Caledonia, south to northwestern Australia (Rowley Shoals) and to the southern Great Barrier Reef, and north to

the Ryukyu Islands of southern Japan. May also occur in Tonga.

HABITAT
Coral reefs on inshore reefs and in protected lagoons. Microhabitat mainly coral heads, silty rubble, and even leaf litter. Ranges in depth 3.3–59 ft (1–18 m).

BEHAVIOR
Generally solitary and cryptic within a home range. Emerges to feed during early morning, just before dusk, or in cloudy weather. Social interactions are relatively few and usually not aggressive.

FEEDING ECOLOGY AND DIET
Feeds upon small benthic invertebrates.

REPRODUCTIVE BIOLOGY
Gonochoristic and polygynous, with a single male mating with more than one female daily. Courtship and spawning occurs during short period at dusk throughout most of the year. Females move to one or more specific areas before the onset of courtship. Males also move to and between these areas seeking females to court. Larger males dominate smaller males and prevent them from courting with females. Males use elaborate fin displays and circle females repeatedly during courtship. More than one bout of courtship may be necessary before a female is ready to spawn. Then, the pair rises slowly into the water column to spawn and fertilize a small batch (12–205) of pelagic eggs. After spawning, female returns to her home area to sleep, and male moves on to attempt to court and spawn with other females. Reported to have been bred in captivity. Larvae are pelagic.

CONSERVATION STATUS
Not listed by the IUCN. May be vulnerable or threatened by overfishing, destructive fishing (including the use of cyanide), and destruction of habitat.

SIGNIFICANCE TO HUMANS
Important and highly prized aquarium fish harvested mainly in Southeast Asia and imported to the United States as part of the aquarium trade. ◆

Draconett
Centrodraco insolitus

FAMILY
Draconettidae

TAXONOMY
Centrodraco insolitus McKay, 1971, off northwest Australia (17°17′ S, 119°51′ E, at a depth of 1,148 ft [350 m]).

OTHER COMMON NAMES
None known.

PHYSICAL CHARACTERISTICS
Total length about 5.1 (13 cm). Small, elongate, rounded body lacking scales. Lateral line grooved. Large eye and pointed snout. Two dorsal fins with three dorsal spines present on the first; the first dorsal spine is the longest. Both operculum and preoperculum have a single strong spine. Last dorsal and anal rays are branched, others are simple. Pectoral fin large and rounded. Pelvic fin elongate. Body pink with yellow blotches, and dark spots near pectoral fin and caudal peduncle.

DISTRIBUTION
Eastern Indian Ocean off coast of northwest Australia.

HABITAT
Deep-bottom rubble and sand to 1,040–1,148 ft (317–350 m) deep.

BEHAVIOR
Nothing known.

FEEDING ECOLOGY AND DIET
Likely feeds upon small benthic invertebrates.

REPRODUCTIVE BIOLOGY
Nothing known, but likely a pelagic spawning fish with paired courtship. Sexually dimorphic; males larger than females.

CONSERVATION STATUS
Not listed by the IUCN. May be vulnerable to damage caused by deep trawling of habitats.

SIGNIFICANCE TO HUMANS
No direct significance. May be taken incidentally in deep-trawling fisheries. ◆

Draconetta
Draconetta xenica

FAMILY
Draconettidae

TAXONOMY
Draconetta xenica Jordan and Fowler, 1903, Suruga Bay, Namazu, Japan.

OTHER COMMON NAMES
Japanese: Inaka-numeri.

PHYSICAL CHARACTERISTICS
Standard length 3.5 in (9 cm). Small, elongate, rounded body lacks scales but has grooved lateral line. Eye is large and snout pointed. There are two dorsal fins with three dorsal spines present on the first; first dorsal spine is the longest. Operculum and preoperculum have single strong spine. Last dorsal and anal rays branched, others are simple. Pectoral fins large and rounded, pelvic fins elongate. Body pink to whitish pink, with pink markings on dorsal, anal, and caudal fins.

DISTRIBUTION
Continental edges from East Africa to the Hawaiian Islands.

HABITAT
Deep-bottom rubble or sand.

BEHAVIOR
Nothing known.

FEEDING ECOLOGY AND DIET
Probably feeds upon small benthic invertebrates.

REPRODUCTIVE BIOLOGY
Largely unknown, but likely a pelagic spawning fish with paired courtship. Sexually dimorphic, males larger than females.

CONSERVATION STATUS
Not listed by the IUCN. May be vulnerable to damage caused by deep-trawling of habitats.

SIGNIFICANCE TO HUMANS
No direct significance. May be taken incidentally in deep-trawling fisheries. ◆

Resources

Books

Bohlke, J. E., and C. C. G. Chaplin. *Fishes of the Bahamas and Adjacent Tropical Waters*, 2nd edition. Austin: University of Texas Press, 1993.

Eschmeyer, W. N., ed. *Catalog of Fishes*, 3 vols. San Francisco: California Academy of Sciences, 1998.

Fricke, R. *Revision of the Genus Synchiropus (Teleostei: Callionymidae)*. Braunschweig, Germany: J. Cramer, 1981.

———. *Revision of the Indo-Pacific Genera and Species of the Dragonet Family Callionymidae (Teleostei)*. Braunschweig, Germany: J. Cramer, 1983.

Gloerfelt-Tarp, T., and P. J. Kailola. *Trawled Fishes of Southern Indonesia and Northwestern Australia*. Jakarta: Directorate General of Fisheries (Indonesia), German Agency for Technical Cooperation, Australian Development Assistance Bureau, 1984.

Helfman, G. S., B. B. Collette, and D. E. Facey. *The Diversity of Fishes*. Oxford: Blackwell Science, 1997.

Leis, J. M., and B. M. Carson-Ewart, eds. *The Larvae of Indo-Pacific Coastal Fishes*. Boston: Brill, 2000.

Masuda, H., K. Amaoka, C. Araga, T. Uyeno, and T. Yoshino, eds. *The Fishes of the Japanese Archipelago*. Tokyo: Tokai University Press, 1984.

Myers, R. F. *Micronesian Reef Fishes*. 3rd edition. Barrigada, Guam: Coral Graphics, 1999.

Nelson, J. S. *Fishes of the World*. 3rd edition. New York: John Wiley & Sons, 1994.

Randall, J. E., G. R. Allen, and R. C. Steene. *Fishes of the Great Barrier Reef and Coral Sea*, Revised and Expanded Edition. Honolulu: University of Hawaii Press, 1996.

Thresher, R. E. *Reproduction in Reef Fishes*. Neptune City, NJ: T.F.H. Publications, 1984.

Periodicals

Fricke, R., "Revision of the Family Draconettidae (Teleostei), with Descriptions of Two New Species and a New Subspecies." *Journal of Natural History* 26 (1992): 165–195.

Gosline, W. A. "A Reinterpretation of the Teleostean Fish Order Gobiosociformes." *Proceedings of the California Academy of Sciences*, 4, no. 37 (1970): 363–382.

Ikejima, K., and M. Shimizu. "Annual Reproductive Cycle and Sexual Dimorphism in the Dragonet, *Repomucenus valenciennei*, in Tokyo Bay, Japan." *Ichthyological Research* 45 (1998): 157–164.

———. "Disappearance of a Spring Cohort in a Population of the Dragonet, *Repomucenus valenciennei*, with Spring and Autumn Spawning Peaks in Tokyo Bay, Japan." *Ichthyological Research* 46 (1999): 331–339.

———. "Sex Ratio in the Dragonet *Repomucenus valenciennei*." *Ichthyological Research* 46 (1999): 426–428.

Nakabo, T. "Comparative Osteology and Phylogenetic Relationships of the Dragonets (Pisces: Callionymidae) with Some Thoughts of Their Evolutionary History." *Publications of the Seto Marine Biological Laboratory* 28 (1983): 1–73.

———. "A New Species of the Genus *Foetorepus* (Callionymidae) from Southern Japan with a Revised Key to the Japanese Species of the Genus." *Japanese Journal of Ichthyology* 33 (1987): 335–341.

———. "Revision of the Genera of the Dragonets (Pisces: Callionymidae)." *Publications of the Seto Marine Biological Laboratory* 27 (1982): 77–131.

Randall, J. E. "Review of the Dragonets (Pisces: Callionymidae) of the Hawaiian Islands, with Descriptions of Two New Species." *Pacific Science* 53 (1999): 185–207.

Sadovy, Y., G. Mitcheson, and M. B. Rasotto. "Early Development of the Mandarinfish, *Synchiropus splendidus* (Callionymidae), with Notes on its Fishery and Potential for Culture." *Aquarium Sciences and Conservation* 3 (2001): 253–263.

Organizations

IUCN/SSC Coral Reef Fishes Specialist Group. International Marinelife Alliance-University of Guam Marine Laboratory, UOG Station, Mangilao, Guam 96913 USA. Phone: (671) 735-2187. Fax: (671) 734-6767. E-mail: donaldsn@uog9.uog.edu Web site: <http://www.iucn.org/themes/ssc/sgs/sgs.htm>

Terry J. Donaldson, PhD

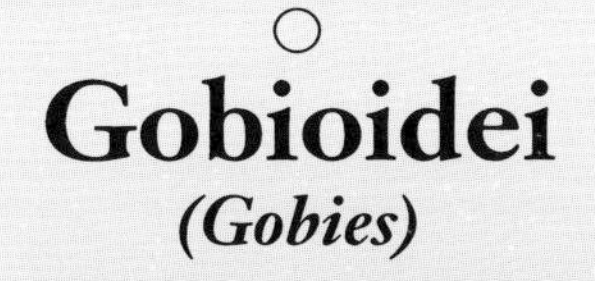

Gobioidei

(Gobies)

Class Actinopterygii
Order Perciformes
Suborder Gobioidei
Number of families 2–9

Photo: The twinspot goby (*Signigobius biocellatus*) has false eyespots on its dorsal fin that provide protective mimicry to help ward off predators. (Photo by David Hall/Photo Researchers. Reproduced by permission.)

Evolution and systematics

The suborder Gobioidei contains about 2,121 species in about 268 genera, representing about 9% of living teleostean species. Several different groups of acanthomorphs (spiny-rayed teleosts) have been suggested as the group to which gobioids are most closely related, but the proposed relationships are usually based on relatively little evidence. A morphological study in 1993 suggested that gobioid fishes might be related to one of the following: the gobiesocids (clingfishes) with the callionymoids (dragonets); the trachionoids (weeverfishes and their relatives); or the hoplichthyids (ghost flatheads).

The taxonomy and phylogenetic (evolutionary) relationships within the Gobioidei are also poorly known. It is unsurprising, therefore, that the classification of gobioid fishes has been described as "chaotic." Since 1973 they have been classified in as few as two families, Rhyacichthyidae and Gobiidae; or as many as nine families: Rhyacichthyidae, Odontobutidae, Eleotridae (called Eleotrididae by some authors), Xenisthmidae, Microdesmidae, Ptereleotridae, Kraemeriidae, Schindleriidae, and Gobiidae (including the subfamilies Gobiinae, Amblyopinae, Oxudercinae, Sicydiinae, and Gobionellinae).

Gobioid fishes are characterized by a lack of several bones in the head and axial skeleton, the presence of free sensory neuromasts borne on raised papillae on the head, and sperm-duct glands associated with the testes. The Rhyacichthyidae, or loach gobies, comprising just two species, is accepted as the most primitive gobioid group. The remaining gobioid groups share several features not seen in the Rhyacichthyidae; for example, the modification of the bones in the head, the reduction of the lateral line canal on parts of the head, and the complete loss of the lateral line canal on the body.

Several features of the skeleton and scales unite the seven families of higher gobioids (Eleotridae, Xenisthmidae, Microdesmidae, Ptereleotridae, Kraemeriidae, Schindleriidae, and Gobiidae) into one large group. Some authors apply the name Gobiidae to this whole group, which does not include the Odontobutidae. However, molecular analyses of the cytochrome b genes in gobioid fishes (made in 2000), and morphological analyses (made in 2002) indicate that the Odontobutidae, as distinct from the other higher gobioids, is an artificial assemblage (i.e., the characters that apparently unite species in the Odontobutidae are not genuinely comparable across all the species, or the characters are not restricted only to those species but may also be found in other, unrelated species).

Since 1998 two new gobioid genera, *Protogobius* and *Terateleotris*, have been described with lateral line canals extending onto the body (a primitive feature that is otherwise restricted to the Rhyacichthyidae). The affinities of these two genera to other gobioid fishes are problematic, and neither has been formally assigned to a family. Morphological evidence suggests similarities between *Protogobius* and *Rhyacichthys*, but this is based mainly on shared primitive features. Analysis of mitochondrial cytochrome b genes also showed that *Protogobius* and *Rhyacichthys* appear to be closely related. *Terateleotris* lacks several of the derived features that are shared by the seven families of higher gobioids, so this group may be related to either of the more basal groups, Rhycichthyidae or Odontobutidae.

A late Cretaceous or early Tertiary origin of gobioid groups has been proposed. Fossil otoliths that might have come from gobioid fishes have been found in the Harudi Formation of Katchchh, India (Lutetian age; 43.6–52 million years old) and the Naggulan Formation of Java (Bartonian

Neon gobies (*Gobiosoma oceanops*) on star coral (*Montastrea annularis*) in Glover's Reef, Belize. (Photo by John Lidington/Photo Researchers, Inc. Reproduced by permission.)

age; 40–43.6 million years ago). The oldest reliably identified skeletal remains are of *Pomatoschistus* (?) cf. *bleicheri* (family Gobiidae), from the Headon Hill Formation, Isle of Wight, Great Britain, and date to the Priabonian age (36.6–40 million years old). The oldest fossil material of any eleotrid gobies are otoliths from Faluns de Gaas, Landes, France, and date to the Rupelian age (30–36.6 million years old).

Gobioids play a particularly important role in the freshwater fish communities of oceanic islands, as well as long-isolated fragments of Gondwana such as Madagascar and Australia.

Physical characteristics

Gobioids are small to medium size fishes, with a cylindrical or laterally compressed body. Most species are between about 1.6 and 3.9 in (40–100 mm) in size but several species are smaller. Approximately 40% of gobioid genera include species under 1.6 in (40 mm) maximum body length, and 22% include species under 1.2 in (30 mm). *Trimmatom nanus* (family Gobiidae) is the shortest known vertebrate at sexual maturity, with females ripening at 0.3 in (8 mm). *Schindleria praematura* (family Schindleriidae) is the lightest known vertebrate at sexual maturity, not weighing more than 0.03 oz (0.8 g). There are some examples of large gobioid fishes, although these are rare and they are usually freshwater species. For example, the fat sleeper (*Dormitator maculatus*) and the marble sleeper goby (*Oxyeleotris marmorata*) in the family Eleotridae are reported to reach over 23.6 in (60 cm) total length.

The gobioid head is usually short and broad, and may be depressed (flattened dorsally). The eyes are positioned on the dorsolateral part of the head. There are usually two separate dorsal fins, the first with five or more weak spines (typically six), and the second dorsal fin with one weak spine and five or more soft, segmented rays. The anal fin also usually has one weak spine and five or more soft, segmented rays. The pectoral fins are generally broad and rounded, but may appear pointed in some species. The paired pelvic fins are positioned thorasically, just below the pectoral fins. Each pelvic fin has one spine and five, soft, segmented rays. The caudal fin is usually broad and rounded, but may appear long and lanceolate in some species. Most species have scales covering the body, although some species lack scales or have them restricted to the posterior part of the body. The size of the scales varies between species, from small and sometimes well embedded in the skin to relatively large. Coloration is variable between species; some are very brightly colored, whereas many others are shades of brown or off white. Both sexes possess a well-differentiated urogenital papilla, posterior to the anus. The urogenital papilla is short and rounded in mature female fish, whereas it is longer and more pointed in mature males.

Gobioid fishes have free, sensory neuromast organs that are distributed on the head in rows of raised papillae. The rows are arranged in diagnostic patterns that are variations of two basic types; a longitudinal arrangement (where all the rows on the cheek run horizontally along the cheek); and a transverse arrangement (where several of the rows run vertically down the cheek). Precise similarities in the arrangement of the papillae rows have been used to infer phylogenetic relationships between groups and hence to classify species. Many gobioid fishes have the pelvic fins united to form a cup-shaped, weakly suctorial disc; this is typical of most, but not all of the Gobiidae, and some of the Microdesmidae and Kraemeriidae. The pelvic fins are separate in the Rhyacichthyidae, Odontobutidae, Xenisthmidae, Eleotrididae, and Ptereleotridae, and are absent in Schindleriidae.

Distribution

Gobioid fishes are common in the tropics, but may also be found in subtropical and temperate regions. They are absent from polar regions. Most of the species are found in marine coastal waters and in estuaries. They are usually in shallow waters, although some species have been found at depths of about 2,625 ft (800 m). About one-tenth of the known species inhabit fresh waters, including the Orinoco and Amazon River basins in South America, and the Nile, Niger, and Congo River basins in Africa. Some species have very widespread distributions. For example, the dusky sleeper (*Eleotris fusca*) is found in fresh, brackish, and marine waters of the Indo-Pacific, from East Africa through to French Polynesia in the South Pacific. In contrast, some other species have very restricted distributions.

Habitat

Gobioid fishes are found in several diverse marine and brackish-water habitats. About half the known species of gobioid fishes are found in coral reef habitats. Gobiid genera such as *Gobiodon* and *Tenacigobius* spend much of their life among the branches of corals such as the black or thorny corals (Antipatharia) and the sea fans and sea whips (Gorgonacea). Gobiosomine gobies (family Gobiidae) such as *Evermannichthys*, *Risor*, and *Gobiosoma*, include tiny species that live exclusively within sponges.

Many gobiid species live epibenthically (on the surface) over sandy or muddy substrates. European species of *Gobius* and *Pomatoschistus* are very common in these habitats. Mudskippers (for example, some species of the gobiid genus *Periophthalmus*) are benthic inhabitants of mangroves. They are highly amphibious, crawling out onto the mud or onto the mangrove roots, and some species also construct burrows, into which they retreat at high tide. Representatives of several other groups of gobioid fishes exhibit a burrowing lifestyle. For example, species of Kraemeriidae (or sand gobies) and the eleotrid genus *Calumia* burrow in sand; the gobiid *Luciogobius* burrows in gravel, and species of amblyopine gobiids burrow in mud. Some species, such as the blind goby (*Typhlogobius californiensis*), from the Eastern Pacific, and the arrow goby (*Clevelandia ios*), live inside the burrows of marine invertebrates.

There are a few examples of blind eleotrid and gobiid species that live in the subterranean fresh waters of caves and sinkholes; for example, *Oxyeleotris caeca* from Papua New Guinea, *Glossogobius ankaranensis* from Madagascar, and the blind cave gudgeon (*Milyeringa veritas*) from Australia. The two species of loach goby (*Rhyacichthys aspro* and *Rhyacichthys guilberti*) show several anatomical specializations for life in torrential rivers and streams. Although the Rhyacichthyidae represents the most basal gobioid lineage, this is not taken as evidence that ancestral gobies also lived in fast-flowing rivers. Instead, it is assumed that this habitat is a specialization of the Rhyacichthyidae. Many species of the Sicydiinae (a subfamily of Gobiidae), including o'opo alamo'o (*Lentipes concolor*), also inhabit torrential hill streams. Their pelvic fins form a strong suctorial disc that allows them to attach firmly to rocks, preventing the fish from being swept away by the swift currents. Some gobies, such as species of *Chlamydogobius*, live in freshwater desert habitats. The Dalhousie goby (*Chlamydogobius gloveri*) can tolerate temperatures up to 111°F (43.9°C); the desert goby (*Chlamydogobius eremius*) can tolerate temperatures between 41 and 106°F (5–41°C), salinities in the range 0–60 parts per thousands and oxygen levels as low as 0.8 parts per million.

Some gobiid species are nektonic, living in the midwater region of shallow coastal waters and estuaries. Cool-temperate midwater dwellers include small aphyiine gobies (family Gobiidae) such as the transparent goby (*Aphia minuta*) and the crystal goby (*Crystallogobius linearis*). Some small nektonic species are also found in warm-temperate and tropical regions; for example, species of *Parioglossus* are found around mangrove roots.

Behavior

Some nektonic species form schools. For example, species of *Parioglossus* or *Gobiopterus* may form small schools around mangrove roots. Epibenthic species are less likely to form schools; however, large populations (in terms of number of individuals) of epibenthic species may be found in some ecosystems. For example, the common goby is densely distributed through some English estuaries. Some species, such as sicydiines, have a larval phase that migrates upriver in schools of very large numbers. The adults of many species are territorial, especially during the breeding season. Some species live in a commensal relationship with invertebrate species. For example, the gorgeous prawn goby (*Amblyeleotris wheeleri*) lives in the burrows of alpheid shrimps. Other species, such as the neon goby, act as "cleaner" fish, picking parasites off large fishes. Mudskippers (e.g., *Periophthalmus barbarus*) are best known for their amphibious lifestyle.

A pair of fire gobies (*Nemateleotris magnifica*) swim over sand and rubble near the islands of Vanuatu Republic. (Photo by Fred McConnaughey/Photo Researchers. Reproduced by permission.)

Feeding ecology and diet

Gobioid fishes commonly feed on invertebrates. Epibenthic species feed on crustaceans, small worms, and insect larvae associated with the benthos. Some coral-dwelling gobies, such as gobiodontines, feed on the polyps of the corals on which they live. Nektonic species living in the water column, such as the glass goby (*Gobiopterus chuno*) from fresh and brackish waters of Asia, feed on plankton. Large eleotrid gobies may feed on other, smaller fish. Various gobioid fish species are specialized for grazing on algae. Sicydiine gobies, such as oopo alamoo have rows of fine teeth that are ideal for scraping algae off rock surfaces. Some algal feeders have

Fighting blue-spotted mudskippers (*Boleophthalmus boddarti*) on a mudflat in Pulau Ubin, Singapore. (Photo ©Tony Wu/www.silent-symphony.com. Reproduced by permission.)

elongate, elaborately coiled guts to assist in the digestion of the tough algal material, e.g., species of *Chlamydogobius* from Australia and *Kelloggella* from the Western Pacific. "Cleaner" gobies, such as the seven-spine, gobiosomine gobies (e.g., *Gobiosoma genie* and the neon goby *Elacatinus oceanops*) are specialized feeders that pick parasites off the scales and skin of other fish, sometimes entering the mouth and gills to extract food items. Another seven-spine goby, the nineline goby (*Ginsburgellus novemlineatus*) shelters on rocks beneath the test, or outer skeleton, of the sea urchin (*Echinometra locunter*) and feeds on the tube feet of the urchin.

The small size of many gobioid species allows them to exploit meiobenthic food resources that are too small for other species. The small size of the gobies might also put them below the threshold for attack by some large, predatory fishes. However, the small size also places the fish at risk from a variety of other predators. For example, coral-dwelling gobiid species of *Eviota* less than 1.2 in (30 mm) long may be eaten by slightly larger coral-dwelling cardinalfishes 1.8–2.1 in (45–53 mm) long. The nineline goby, which hides under sea urchins, can fall prey to long-snouted predators like trumpetfishes, that can delve between the spines of the urchin. Large invertebrates, such as shrimps, can overpower small fishes, and there is a reported case of a small goby being ingested by the nemertean worm *Lineus longissimus*. Small reef-dwelling gobies regularly fall prey to piscivorous mollusks of the genus *Conus*. Freshwater gobioids are regularly preyed upon by water snakes and fish-eating birds.

Reproductive biology

The male may defend a small territory and nest site that, frequently, is a small space under some stones. Often the male is responsible for caring for the developing eggs. The female will lay a few to several hundred small eggs, attaching them to the underside of rocks, vegetation, or onto corals. The eggs usually hatch in a few days, and the young may be dispersed by water currents. In freshwater species such as sicydiine gobies (e.g., o'opo alamo'o), the larvae are probably swept downstream by the water current; they spend a few weeks to months at sea before returning to fresh water. Frequency of spawning is very variable between different species of gobies. Some species, such as the Japanese ice goby (*Leucopsarion petersi*), are typically semelparous species, spawning once in a single spawning season, before dying. Other species are iteroparous, relying on more than one spawning event during their lifespan to achieve a satisfactory reproductive output. This "repeat-spawning" behavior is often seen in small species that face reasonably high predation and a short lifespan. The common goby (*Pomatoschistus microps*), which grows up to 3.5 in (9 cm) total length and can convert 40% of its food into gonad energy, may spawn up to six times over a three-month period, or as many as nine times during its 18-month life cycle. In contrast, the burrowing species Fries' goby (*Lesuerigobius friesii*) reaches only 1.6 in (4 cm) longer than *P. microps*, but it does not mature until it is two years old, spawns just twice per season, and may live for 11 years. Egg size and number can also vary widely between species of gobies. The smallest eggs are 0.01 in (0.26 mm) in diameter, produced by the empire gudgeon (*Hypseleotris compressa*) from Australia and New Guinea; the largest are 0.1 in (2.4 mm) in diameter produced by the knout goby (*Mesogobius batrachocephalus*) from the Black and Caspian Seas. Some small gobies that live in or around corals and sponges may change sex, which ensures that there are always enough individuals of both sexes to ensure reproduction and survival of the population.

Conservation status

The 2002 IUCN Red List of Threatened Species includes five gobioid species ranked as Critically Endangered and 26 ranked as Vulnerable. Another 21 species are ranked under Lower Risk/Conservation Dependent or Lower Risk/Near Threatened, and 27 are listed as Data Deficient (i.e., there is inadequate information to make a direct, or indirect, assessment of the risk of extinction).

The critically endangered species are threatened by the degradation of their habitat and the introduction of invasive species. For example, *Chlamydogobius micropterus* is from small springs, bores, and drains in Queensland, Australia, that are threatened by water extraction. The Edgbaston goby (*Chlamydogobius squamigenus*) is from similar Queensland habitats that are being degraded by trampling from humans and domestic livestock. The introduction of the mosquitofish (*Gambusia holbrooki*) presents another threat to *C. squamigenus*. The poso bungu (*Weberogobious amadi*), which is restricted to Lake Poso in Sulawesi, was probably last observed in 1985 and might be extinct. It has been threatened by pollution, the introduction of alien fish species to the lake and by diseases carried by those introduced species. The sinarapan (*Mistichthys luzonensis*) from the Philippines was previously thought to be possibly extinct because of the introduction of tilapia to Philippine fresh waters. However, it is still found in lakes of Luzon and is ranked as Lower Risk/Conservation Dependent by the IUCN.

The 26 species listed as Vulnerable include several that have restricted ranges. For example, four of the eight eleotrid species listed as Vulnerable are species of *Mogurnda* that are restricted to a single lake (Lake Kutubu) in Papua New Guinea. The gobiid *Economidichthys trichonis*, which is endemic to Lake Trichonis in Greece, is threatened by agricultural activities that may cause eutrophication of the lake. *Knipowitschia punctatissima* from Italy is threatened by industrial and agricultural activity, and *Knipowitschia thessala* from Greece is threatened because some of its native watercourses are now dry.

Significance to humans

Gobioid fishes do not support important commercial fisheries, but subsistence fisheries exist for some species in the tropical Atlantic and Indo-Pacific. The sinarapan is a delicacy around the Philippine lakes where it is found, and fry of sicydiine gobies are a delicacy in parts of the Caribbean and Indo-Pacific, where they may be made into a paste. According to FAO statistics, the 2000 global capture for all gobioid fishes was 51,199 tons (52,021 tonnes); 63% of this was caught in the northwest Pacific by countries of the Russian Federation, and 15% was caught in the western Central Pacific by the Philippines. Aquaculture produced 376 tons (382 tonnes) of gobioid fishes in 2000, with a value of $3.38 million. Many gobioid species probably represent important food resources to larger fish species that are commercially fished.

Gobioid fishes form a small part of the ornamental fish trade which, by 1992, had a global retail trade value of $3,000 million. In Hong Kong, a major distributor of ornamental fishes, gobies represented only about 1% of the total number of fishes observed in a market survey in 1996 and 1997. Gobies are an important complement of the biological diversity of coral reefs, which are popular sites for tourism. Therefore, these reef-dwelling gobies represent an indirect source of sustainable income to tropical countries that are using their reefs as attractions for properly managed ecotourism.

Introduced species of gobies include the round goby (*Neogobius melanostomus*) in the Great Lakes region of North America; the Japanese goby (*Tridentiger trigonocephalus*), introduced to North American and Australian coastal waters; and the Indo-Pacific *Butis koilomatodon*, introduced into coastal waters of Nigeria, Panama, and Brazil. Translocated *Glossogobius giuris* and *Hypseleotris agilis* appear to have played a role in the extirpation of the endemic cyprinid species flock of Lake Lanao in the Philippines. Native fish species may suffer from competition with these alien species, and the economic cost of monitoring the spread of exotics, and attempting to control them, may be significant.

1. Violet goby (*Gobioides broussoneti*); 2. Large-scale spinycheek sleeper (*Eleotris amblyopsis*); 3. O'opo alamo'o (*Lentipes concolor*); 4. Neon goby (*Gobiosoma oceanops*); 5. Fire goby (*Nemateleotris magnifica*); 6. Dwarf pygmy goby (*Pandaka pygmaea*); 7. Atlantic mudskipper (*Periophthalmus barbarus*). (Illustration by Amanda Humphrey)

1. Whip coral goby (*Bryaninops yongei*); 2. Gorgeous prawn-goby (*Amblyeleotris wheeleri*); 3. Arno goby (*Padogobius nigricans*); 4. Samoan sand dart (*Kraemeria samoensis*); 5. Blind cave gudgeon (*Milyeringa veritas*); 6. Marble sleeper (*Oxyeleotris marmorata*); 7. Loach goby (*Rhyacichthys aspro*). (Illustration by Amanda Humphrey)

Species accounts

Large-scale spinycheek sleeper
Eleotris amblyopsis

FAMILY
Eleotridae

TAXONOMY
Culius amblyopsis Cope, 1871, Suriname.

OTHER COMMON NAMES
English: Sleeper.

PHYSICAL CHARACTERISTICS
Reaches about 3.9 in (10 cm) total length over most of range, largest individual observed from North America about 5.5 in (14 cm). Two dorsal fins, pelvic fins separated, and rounded caudal fin. Torpedo-like body shape with flattened head. Tan to dark brown, sometimes changes to dark trunk laterally with light tan dorsum, especially in juveniles. May have rows of small dark spots on sides. Two dark streaks radiating rearward from eyes on cheeks. A large dark spot on the upper pectoral fin base and some specimens with two elongate spots extending from pectoral fin base onto pectoral fin rays. Stout spine on lower cheek faces forward.

DISTRIBUTION
Common from North Carolina to French Guiana, and in Cuba and Hispaniola. Less frequently known from Brazil and the Antilles.

HABITAT
Euryhaline (able to live in waters of a wide range of salinities); adults and juveniles inhabit continental estuaries, but are also known from fresh water in Central America and occasionally in the Antilles. Associates with floating, emergent, and marginal vegetation, such as mangroves, river cane, and water hyacinths. Pelagic marine larval stage.

BEHAVIOR
A lethargic "sit-and-wait" predator. Postlarvae join other species to form large schools which periodically move into estuaries and fresh waters after a marine larval period.

FEEDING ECOLOGY AND DIET
Predaceous; feeds on shrimps and small fishes.

REPRODUCTIVE BIOLOGY
Presumably amphidromous, with adults living and reproducing in fresh water or estuaries and a marine larval period.

CONSERVATION STATUS
Considered vulnerable in North Carolina at the edge of its range, but not listed by the IUCN.

SIGNIFICANCE TO HUMANS
None known. ◆

Blind cave gudgeon

Milyeringa veritas

FAMILY
Eleotridae

TAXONOMY
Milyeringa veritas Whitley, 1945, Milyering, Yardie, 20 mi (32 km) southwest of Vlamingh Head, North-West Cape, Western Australia. Populations show a low degree of heterozygosity, which is typical of cave-dwelling fishes. This species is possibly closely related to the Indo-Pacific eleotrid genus *Butis.*

OTHER COMMON NAMES
English: Cave gudgeon.

PHYSICAL CHARACTERISTICS
Reaches about 2 in (5 cm) total length. The head is broad and concave over the snout. The eyes are not visible. Small first dorsal fin with four very short spines. Second dorsal fin with nine longer rays. Pelvic fins separate. The caudal fin is slightly pointed. The head is naked but the body is covered with small scales (about 28 in series along the flanks). The fish is very poorly pigmented. The top of the head is pale yellow, with a dark gray triangular mark over the brain. The posterior nostrils are pink. There is a purplish spot over the operculum. The body is pale gray and fins are flesh colored.

DISTRIBUTION
Western and northeastern coastal plains of the Cape Range Peninsula and Barrow Island, Western Australia.

HABITAT
This is a troglobitic species; i.e., an obligatory cave dweller, found in shallow wells (at depths of up to 1.6 ft/0.5m), sinkholes, and deeper cave systems (at depths of up to 105 ft/32 m) that are freshwater or anchialine (located near the sea and flooded with sea water). In anchialine systems the fishes may live in freshwater layers that overlie the salt water. They are also found in brackish water, tolerating salinities from 0.27–34 parts per thousand and temperatures of 70–83°F (21–28°C).

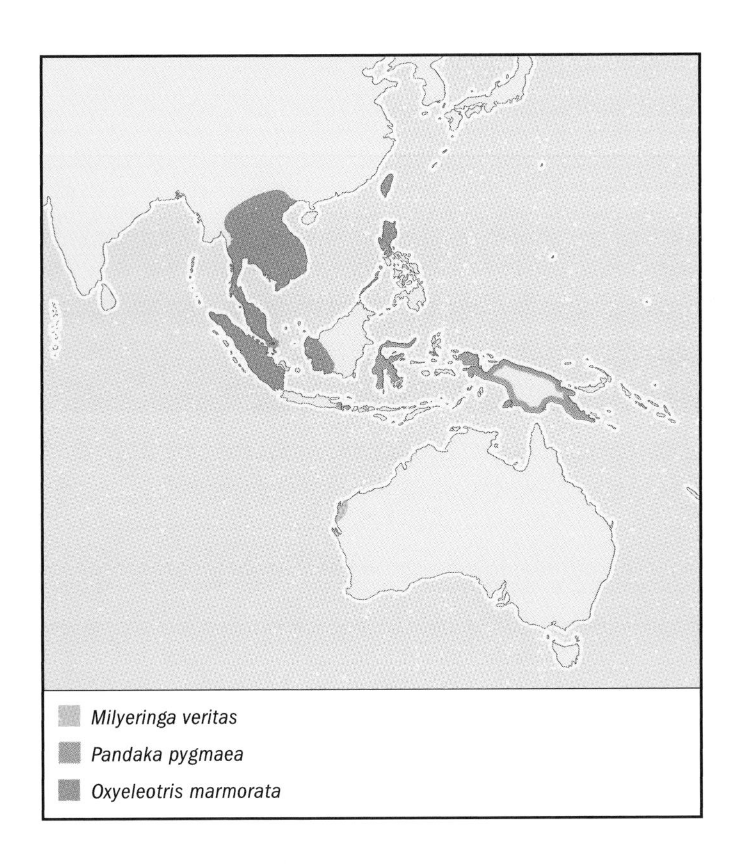

BEHAVIOR
Swims slowly, with the pectorals pointed out to the sides and the pelvics extended slightly forward.

FEEDING ECOLOGY AND DIET
Possibly feeds on insects or other small animals that drop into the cave waters.

REPRODUCTIVE BIOLOGY
Random mating apparently occurs between subpopulations throughout the range.

CONSERVATION STATUS
Listed as Data Deficient in the IUCN 2002 Red List, but as threatened by the Western Australian Conservation Act. Major threats come from habitat disturbance due to urban and agricultural development, and water extraction causing a lowering of the water table. Water quality in the underground systems is being compromised by pollution from nutrients and other chemicals.

SIGNIFICANCE TO HUMANS
None known. ◆

Marble sleeper

Oxyeleotris marmorata

FAMILY
Eleotridae

TAXONOMY
Eleotris marmorata Bleeker, 1852, Bandjarmasin, Borneo, and Palembang, southeast Sumatra, Indonesia.

OTHER COMMON NAMES
English: Marble goby, marbled sand goby; German: Marmorgrundel; Cantonese: Soon hock; Japanese: Kawaanago; Khmer: Trey damrei; Laotian: Pa bou; Malay: Bakutut, belantuk; Tagalog: Bia; Thai: Pla bu jak; Vietnamese: Cá bong.

PHYSICAL CHARACTERISTICS
The largest gobioid fish, this species may reach 35.4 in (90 cm) total length. Torpedo-like body shape with flattened head. Oblique, terminal mouth. Two dorsal fins and rounded caudal fin. Pelvic fins separate. Body brown with dark mottling giving a marbled appearance.

DISTRIBUTION
Southeast Asia in Laos, Cambodia, Vietnam, Thailand, Malay Peninsula, Sumatra and western Borneo. Also reported from Luzon (Philippines), and introduced into Taiwan for aquaculture.

HABITAT
Found in rivers, lakes, swamps, ditches, and ponds; water may be muddy or clear and over muddy, sandy, or gravel bottoms. This species may also be found in brackish waters around the mouths of rivers and canals.

BEHAVIOR
Solitary, nocturnal hunter that prowls slow-moving streams, lakes, and swamps. During the daytime it rests at the bottom, taking cover among rocks, woody debris, or vegetation. This may be the typical behavior for most sleepers, hence the common name.

FEEDING ECOLOGY AND DIET
Predaceous; primarily eats small fishes, but also takes crustaceans, insects, and mollusks. Larval fish in culture ponds feed on cladocerans, rotifers, chironomids, and brachiopods.

REPRODUCTIVE BIOLOGY
Reach sexual maturity at approximately 3.9–4.7 in (10–12 cm) standard length. Spawns January through October. The eggs are about 0.09 in (2.3 mm) long and 0.02 in (0.63 mm) wide. Hatching starts at about 41 hours after fertilization, reaching its peak about 60–70 hours after fertilization. The males care for the eggs and guard the newly hatched fry. The larvae are initially pelagic but change to a more benthic habitat after about 25–30 days, becoming quite sedentary after about 40 days.

CONSERVATION STATUS
Not listed by the IUCN, but the species may be declining in some parts of its native range due to overharvesting.

SIGNIFICANCE TO HUMANS
A highly prized food fish in Southeast Asia, where it is also raised in ponds. In 2000, 277 tons (282 t) of marble sleeper were produced by aquaculture in Malaysia, Singapore, and Thailand. This quantity represents about 74% of the total global aquaculture of gobioid fishes for 2000, and had a value of about $1.2 million. In Borneo, marble sleepers captured for markets in Singapore and Japan may bring a fisherman $20 per lb ($10 per kg). ◆

Gorgeous prawn-goby
Amblyeleotris wheeleri

FAMILY
Gobiidae

TAXONOMY
Cryptocentrus wheeleri Polunin and Lubbock, 1977, Aldabra Atoll, Seychelles.

OTHER COMMON NAMES
English: Gorgeous goby, Wheeler's shrimp goby; Afrikaans: Mooidikkop; Gela: Iga taotao; Tagalog: Bia.

PHYSICAL CHARACTERISTICS
Reaches about 3.2 in (8 cm) total length. Body with scales (62–68 in longitudinal series along the flanks). Second dorsal and anal fins, each with 12 soft fin rays. Pelvic fins not forming a strong sucking disk; fins connected between innermost rays but not between spines of each fin. Head and body a light yellowish green or brownish dorsally, with six reddish, vertical, or oblique bands on flanks; numerous small, bluish spots, or reddish spots on head. A vertical red stripe runs from eye to corner of mouth. Fins transparent, and all except pectorals with numerous blue spots and often with reddish bases, and dorsal fins with orange spots near edge. Caudal fin has oblique red stripe.

DISTRIBUTION
Throughout much of the Indo-Pacific region, from East and South Africa and the southern Red Sea to Fiji; in the Pacific, north to Taiwan, and the Ryukyu Islands of southern Japan, and south to the Great Barrier Reef

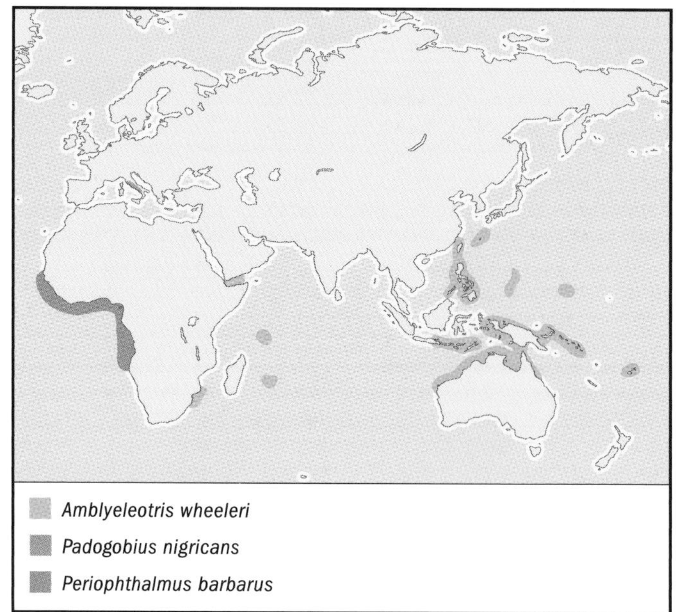

HABITAT
Lives on sand and rubble patches of reef flats between 16 and 98 ft (5–30 m) deep. Found in association with alpheid prawns (usually *Alpheus ochrostriatus*) that live in burrows.

BEHAVIOR
Typical of several species of gobies that live symbiotically with prawns. The goby rests just outside the prawn's burrow, and retreats into the burrow at the approach of danger. The goby and prawn may stay in reasonably close contact with each other when either leaves the burrow. The goby benefits from the shelter afforded by the burrow, and might also be "cleaned" by the prawn. The goby acts as a "watchman" for the prawn. The goby can signal the presence or absence of danger by flicking movements of its tail, and the prawn feels these signals through its antennae, which it places on the goby's tail. If the prawn cleans the goby of small particles attached to its body, then this may provide a useful food source for the prawn.

FEEDING ECOLOGY AND DIET
Nothing known.

REPRODUCTIVE BIOLOGY
Nothing known.

CONSERVATION STATUS
Not threatened.

SIGNIFICANCE TO HUMANS
Commercial importance in the aquarium trade. ◆

Whip coral goby
Bryaninops yongei

FAMILY
Gobiidae

TAXONOMY
Cottogobius yongei Davis and Cohen, 1969, Darvel Bay, west of Tatagan Island, Borneo.

OTHER COMMON NAMES
Japanese: Garasuhaze; Samoan: Mano'o.

PHYSICAL CHARACTERISTICS
Small, reaches about 1.2 in (3 cm) standard length. The preoperculum has a slightly scalloped edge, visible just behind the mouth. Body with scales (26–58 in longitudinal series along the flanks). The pelvics form a small suctorial disk. The head is transparent except for some brownish patches on the sides, a distinct reddish gold stripe that extends dorsally around the snout, and some reddish spots and a band over the top of the head. The eyes are bright golden with reddish rims. The body is transparent, with about six darkish or reddish brown bands on flanks. The ventral part of the body is mottled brown, and there is a large brown or reddish patch at the base of the pectoral fin. The dorsal, anal, and caudal fins may be reddish.

DISTRIBUTION
From the Red Sea, to the Seychelles in the Indian Ocean. More common in the Pacific Ocean from Indonesia east to French Polynesia, north to Hawaii and southern Japan, and south to the Great Barrier Reef.

HABITAT
Apparently restricted to small patches that are widely dispersed. The fish live only on the sea whip (*Cirripathes anguina*), found on drop offs in strong currents and on sheltered back reefs, from 9.9 to 148 ft (3–45 m) deep.

BEHAVIOR
Small sea whips are occupied by a single immature fish. Larger sea whips are occupied by a male-female pair. Juveniles and a second small female may also be present.

FEEDING ECOLOGY AND DIET
Nothing known.

REPRODUCTIVE BIOLOGY
The male-female pair on the sea whip are probably monogamous. About 6–8 in (15–20 cm) in from the tip of the sea whip, the fish clear living tissue from the whip in a narrow band about 0.9 in (23 mm) wide, exposing the underlying skeleton. The eggs are attached in a wide band over this "nest." Individual fish can change sex, apparently either way. The largest female to settle on a whip might become a male and the next largest fish remain a female, thus forming a pair. Fish that lose their mates might revert to a bisexual state, allowing them to mate with any fish of either sex that subsequently settles on the sea whip.

CONSERVATION STATUS
Not threatened.

SIGNIFICANCE TO HUMANS
Found in the aquarium trade, but not common. ◆

Violet goby

Gobioides broussoneti

FAMILY
Gobiidae

TAXONOMY
Gobioides broussonnetii Lacepede, 1800, Suriname.

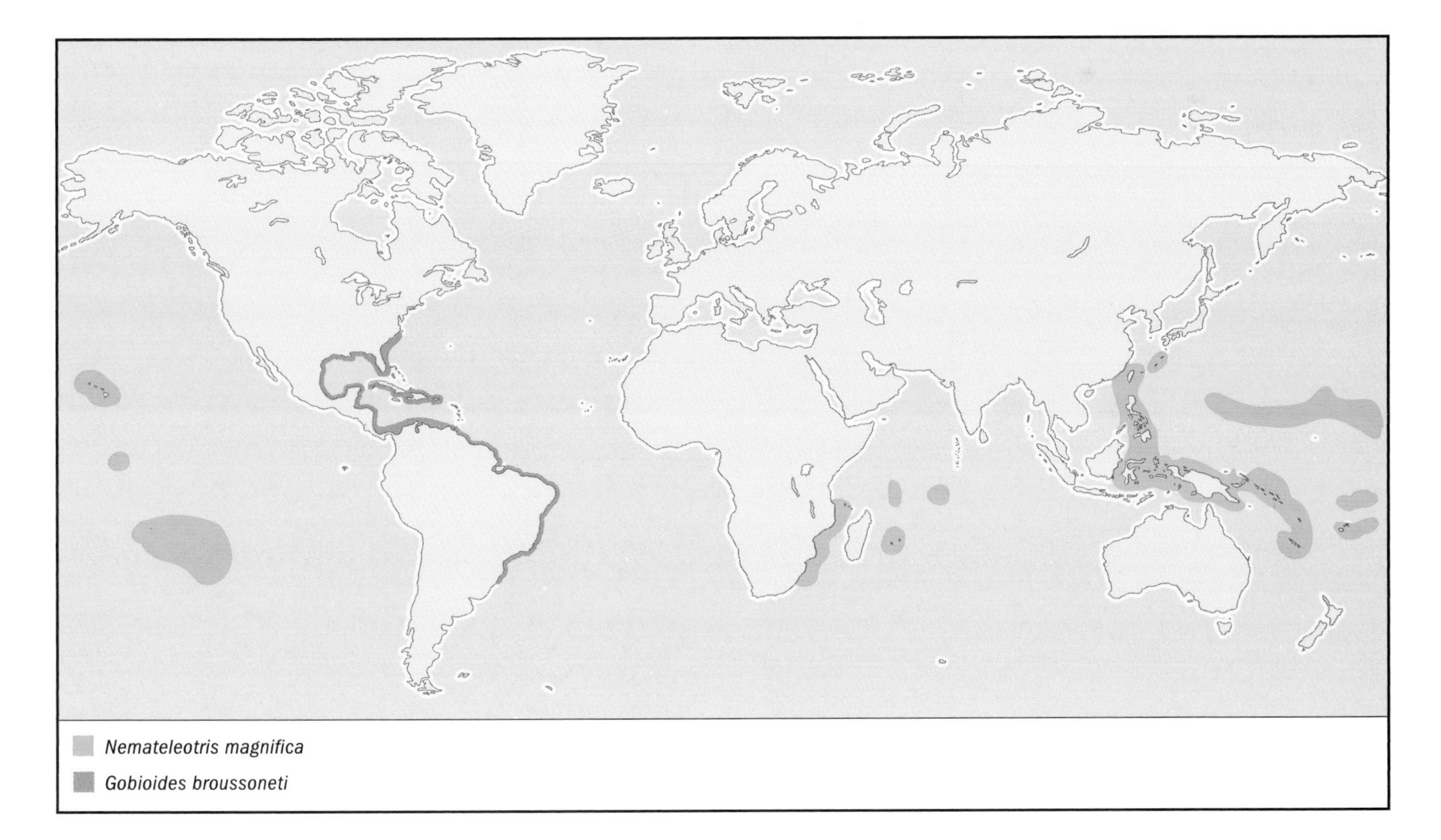

OTHER COMMON NAMES
German: Lila Aalgrundel; Spanish: Lamprea; Portuguese: Aimore.

PHYSICAL CHARACTERISTICS
To about 21.6 in (55 cm) total length. The body is elongate and eel-like. Dorsal fins united to form a long continuous fin confluent with the caudal fin. Anal fin also connected to caudal fin. The caudal fin is long and pointed. Pelvic fins are united to form a concave disk. The mouth is large and oblique. The eyes are small and high on the head. The body and head are purple brown, the abdomen and underside of the head are white. A series of dark chevrons, angled forward, extends along the sides.

DISTRIBUTION
Western Atlantic, Gulf and Caribbean coasts of the Americas, extending from Brunswick County, North Carolina, to Rio Grande do Sul, Brazil. Also on coasts of Cuba, Hispaniola, and Puerto Rico in the Caribbean.

HABITAT
Lives on muddy bottoms and in burrows; inhabits tidal freshwater and low salinity estuaries offshore from large rivers; one individual was found 298.5 ft (91 m) off the Mississippi River in Louisiana.

BEHAVIOR
No published information is available, but some is forthcoming from the aquarium trade. Males display for females before spawning by raising and lowering the dorsal fin, and swimming in and around the potential nest site. Males become territorial after spawning and guard the nest.

FEEDING ECOLOGY AND DIET
The long coiled gut suggests that it feeds on diatoms, algae, and tiny invertebrates sifted from sediments.

REPRODUCTIVE BIOLOGY
In the aquarium, spawning may last 12–24 hours. Eggs are deposited in nests under an object, presumably within a burrow in the wild. Eggs hatch in 36–38 hours. Males usually guard the fry until they swim freely.

CONSERVATION STATUS
Not threatened.

SIGNIFICANCE TO HUMANS
No commercial fishing values, but becoming an increasing part of the aquarium trade because of its interesting appearance. ◆

Neon goby
Gobiosoma oceanops

FAMILY
Gobiidae

TAXONOMY
Elacatinus oceanops Jordan, 1904, Garden Key, Tortugas, Florida.

OTHER COMMON NAMES
English: Blue neon goby.

PHYSICAL CHARACTERISTICS
Small, slender goby with an pointed, bulbous snout. Maximum length about 2.4 in (6 cm). Has two dorsal fins and a square-shaped caudal fin about the same height as the body. The pelvic fins are joined to form a cuplike sucking disk. Dorsally the fish is dark, with an iridescent pale blue stripe running laterally along the lower edge of the dorsum, extending from the snout to the caudal fin. This lateral streak in fishes from Belize is blue in the middle bordered by white. The underside of the head and the abdomen are pale.

DISTRIBUTION
Southern Florida, the Flower Garden coral reefs 100 mi (161 km) off the Texas-Louisiana border, Alacran Reef of northern Yucatán, Quintana Roo state (Yucatán peninsula of Mexico), and Belize.

HABITAT
Inhabits coral reefs and tropical rocky substrates at depths from 3.3 to 131 ft (1–40 m). Has also been observed associated with large sponges.

BEHAVIOR
Rests on coral and withdraws into crevices when threatened. A cleaner goby that picks parasites off other fishes, the male cleans the undersides of rocks, corals, or shells to prepare the surface for eggs.

FEEDING ECOLOGY AND DIET
This is a cleaner goby that picks parasites off other fishes. As with other cleaner gobies, this species waits for other fishes at stations on top of coral heads. It will then swim along the fish and remove ectoparasites from its skin. Its diet includes parasites, fish scales (perhaps accidentally removed), and benthic invertebrates. Larvae feed on small plankton.

REPRODUCTIVE BIOLOGY
Males clean the undersides of rocks, corals, or shells to prepare the surface for eggs. The male swims in front of the female and entices her into the shell or crevice. Spawning occurs with the male and female quivering side-by-side, depositing the eggs on the ceiling of the nest. Individuals can spawn multiple clutches, several times within a month. The eggs are about 0.08 in (2 mm) long, and 0.04 in (1 mm) wide. About 300–450 eggs are deposited in a nest about 0.5 sq in (3 sq cm). The pair remains together while caring for the eggs. The male guards the nest and presumably circulates oxygen-rich water over the eggs, using his pectoral and caudal fins. Eggs hatch in 7–10 days. Larvae are 0.16 in (4 mm) at hatching, and the parents do not care for the fry.

CONSERVATION STATUS
Not threatened.

SIGNIFICANCE TO HUMANS
This colorful fish is a popular, nonaggressive aquarium species that does not require much space. It is commercially important in the saltwater aquarium trade. ◆

O'opo alamo'o
Lentipes concolor

FAMILY
Gobiidae

TAXONOMY
Sicyogaster concolor Gill, 1860, Hilo, Hawaii. Part of the subfamily Sicydiinae.

OTHER COMMON NAMES
Hawaiian: 'alamo'o.

PHYSICAL CHARACTERISTICS
To 5.3 in (13.4 cm) total length. A tubular species with a blunt head, subterminal mouth, and rounded caudal fin. Pelvic fins are united to form a cup and adhere to the body. Gray or brownish trunk with irregular lateral markings that sometimes form bands. Breeding males transform from the drab color to black on the anterior half of the body, and red or white tinged with red posteriorly, with white median fins.

DISTRIBUTION
Hawaiian Islands.

HABITAT
Adults and juveniles ascend waterfalls to inhabit headwaters and upper reaches of perennial rain-forest streams on the high islands of the Hawaiian archipelago. It has been reported from sites as high as 2,947 ft (900 m) above sea level. This species has planktonic marine larvae.

BEHAVIOR
Larvae of this amphidromous species wash out to sea where they may remain for a period of three months. Post-larvae return to fresh water, and ascend waterfalls to colonize the upper reaches of streams; sicydiine gobies probably use their pelvic fins and their teeth to gain purchase on the rock surface as they climb waterfalls. Genetic evidence suggests that they do not return to the natal stream.

FEEDING ECOLOGY AND DIET
Predominantly herbivorous, has rows of fine teeth that are ideal for scraping algae from the surface of rocks in the stream bed. As the old teeth break or wear down, they are replaced by fresh teeth. Freshwater shrimps are also taken in small numbers.

REPRODUCTIVE BIOLOGY
Breeds from October through June. Males may establish and patrol reproductive territories, but evidence for this is inconclusive. Males defend nest cavities prior to breeding and pair with females prior to spawning for two to six days. Spawning may last several hours, during which females deposit eggs on the underside of the roof of the nest cavity before being chased out by the male. Individual females may spawn four clutches a season, with an average of nearly 14,000 eggs. After spawning, males apparently guard the nests. Although not directly observed, males may also clear detritus from nest cavities. Eggs hatch in three to six days.

CONSERVATION STATUS
Not currently threatened; classified as Data Deficient by the IUCN. This species was listed as a Category 1 candidate for Endangered Species classification by the U.S. Fish and Wildlife Service from 1989 to 1996. It was removed from the list when it was discovered to be more widely abundant than previously believed. Many populations were not identified prior to listing, in part because upstream sites were undersampled.

SIGNIFICANCE TO HUMANS
No commercial value. ◆

Arno goby
Padogobius nigricans

FAMILY
Gobiidae

TAXONOMY
Gobius fluviatilis var. *nigricans* Canestrini, 1867, Arno River, Italy.

OTHER COMMON NAMES
Italian: Ghiozzo di ruscello.

PHYSICAL CHARACTERISTICS
Males reach slightly beyond 4.9 in (12.5 cm) total length. Tubular species with a blunt head and terminal mouth. Two dorsal fins, second with 12–14 fin rays. Rounded caudal fin. Pelvic fins form a sucking disc, which does not extend as far as the anus. There are 44–49 ctenoid scales along the midline of the flanks. Brownish or light brown dorsally, and whitish ventrally. Five characteristic brown bands over the back, the first at the base of the pectoral fin and the last at the caudal peduncle. Dorsal fins with alternating light and dark gray bands; the first dorsal fin has a yellowish orange border.

DISTRIBUTION
Formerly widespread through fresh waters of central Italy. Currently found in the Serchio, Arno, Ombrone, and Tiber River basins. Populations have become restricted because of environmental degradation and pollution.

HABITAT
Small to medium-sized streams with moderate current. Usually found in clear waters, over a stony substrate.

BEHAVIOR
Usually stationary, positioned under stones, occasionally moves quickly from one stone to another.

FEEDING ECOLOGY AND DIET
Predaceous; feeds on oligochaetes, crustaceans, and insect larvae.

REPRODUCTIVE BIOLOGY
Probably reaches sexual maturity during the first year, at 1.6–2 in (4–5 cm). Males become very dark during the reproductive period, which usually occurs between May and June. The male is territorial and guards a nesting cavity under a stone, attracting females with a courtship display and making noises as he does. The male carries small stones out from the nest in his mouth and deposits them near the female, and then quickly returns to the nest. He repeats this behavior until the female enters the nest, and then swims several times upside down over the roof of the nest. The female attaches herself to the roof of the nest using the pelvic disk and spawns 100–200 eggs, which are stuck to the roof. Several females can spawn in the same nest, which may hold over 1,500 eggs at various stages of development. The male cares for the eggs until they hatch, after about 17 days.

CONSERVATION STATUS
Listed as Vulnerable by the IUCN, with a suspected population reduction of at least 20% over the last 10 years, or three generations. Populations are threatened by environmental degradation and, in some areas, by the introduction of *Padogobius martensii*, which probably competes for similar ecological niches.

SIGNIFICANCE TO HUMANS
There is no commercial or sports fishery for this species. However, the flesh tastes good and fishes are collected illegally and eaten fried. ◆

Dwarf pygmy goby
Pandaka pygmaea

FAMILY
Gobiidae

TAXONOMY
Pandaka pygmaea Herre, 1927, Philippines (possibly Malabon).

OTHER COMMON NAMES
German: Zwerggrundel; Tagalog: Bia.

PHYSICAL CHARACTERISTICS
A very small fish; females reach 0.6 in (1.5 cm) total length. Body is covered with scales, up to 25 in a longitudinal series along the flanks. The second dorsal and anal fins have six or seven soft rays. The pelvic fins are united into a small sucking disk. The body has a conspicuous pigmentation, with about four cross bands of dark spots on the flanks, and heavy pigmentation on the bases of all the fins except the pelvics. In males, the first dorsal fin is tipped with blue and the lower third is yellowish.

DISTRIBUTION
First reported as possibly from Malabon, a suburb of Manila in the Philippines. The presence of the species in the Malabon River was confirmed in the 1950s. Subsequently, it has been found at an island off Palawan (Philippines), the east coast of peninsular Thailand, Singapore, Bali, Sulawesi, and New Guinea.

HABITAT
Found along river banks in the Philippines from January to June, when the water is clear and the tide is rising. During the rainy season the fishes apparently move into ponds and canals. Also found in brackish waters and around mangroves. Some specimens were apparently caught in marine waters adjacent to Palawan. Fishes kept in aquaria can acclimate to marine, brackish, or freshwater conditions.

BEHAVIOR
Usually stationary, on surfaces of rocks or plants, or hanging in midwater. Swimming occurs in an erratic series of starts.

FEEDING ECOLOGY AND DIET
Aquarium-kept specimens are voracious, feeding on brine shrimps and other small or chopped invertebrate material.

REPRODUCTIVE BIOLOGY
Nothing known.

CONSERVATION STATUS
Listed as Critically Endangered by the IUCN, with an observed or suspected population reduction of 90% or more over the last 10 years, or three generations. Populations are threatened by introduced species, hybridization, pathogens, pollutants, competitors, or parasites.

SIGNIFICANCE TO HUMANS
None known. ◆

Atlantic mudskipper
Periophthalmus barbarus

FAMILY
Gobiidae

TAXONOMY
Gobius barbarus Linnaeus, 1766, Liberia. Based on a neotype specimen designated in 1989.

OTHER COMMON NAMES
French: Sauteur de vase atlantique; German: Schlammspringer; Spanish: Saltafango atlantico; Akan: Soetsi; Portuguese: Saltao-da-vasa.

PHYSICAL CHARACTERISTICS
To about 6 in (15 cm) total length. Prominent eyes with ventral eyelids perched on top of large head. Two dorsal fins. Muscular pectoral fin and pelvic fin bases used for crawling and climbing. Pelvic fins separate. Body tan but light ventrally, and with black diagonal bars dorsally on flanks; opalescent spots on head and anterior trunk; light margin on first dorsal fin may be tinged with light blue, lower portion dark; second dorsal fin with dark submarginal band over a light band.

DISTRIBUTION
West African coast from Senegal to Angola, including most islands.

HABITAT
Inhabits the littoral zone of mangrove estuaries and muddy intertidal flats where it lives in burrows.

BEHAVIOR
Amphibious; lives in burrows in the mangal zone. Territorial, forages for food on the mudflats and around the mangroves themselves. They can flee predators by skipping or hopping across the flats and into the mangrove forests or into their burrows. When on land, mudskippers keep a mouthful of water for extracting oxygen via the gills, and they can breathe through the skin which is well served with blood vessels.

FEEDING ECOLOGY AND DIET
Eats crustaceans, worms, and insects from the intertidal zone.

REPRODUCTIVE BIOLOGY
Reproduction occurs in burrows.

CONSERVATION STATUS
Not threatened.

SIGNIFICANCE TO HUMANS
No commercial fishing value; plays a minor role in the aquarium trade. ◆

Samoan sand dart
Kraemeria samoensis

FAMILY
Kraemeriidae

TAXONOMY
Kraemeria samoensis Steindachner, 1906, Samoan Islands.

OTHER COMMON NAMES
English: Sand dart; Afrikaans: sandspies.

PHYSICAL CHARACTERISTICS
Small, elongate fish, reaches 1.4 in (3.5 cm) total length. Lacks scales. Head with minute eyes, and a "chinlike" forward projection of the lower jaw. Six or seven small flaps project from the lower edge of the preoperculum, just behind the mouth, and five or six flaps project from the lower edge of the operculum covering the gills. There is a single, long, dorsal fin, and a long anal fin. Specimens from the Indian Ocean have fewer opercular flaps than those from the Pacific, and might represent a different species.

DISTRIBUTION
The Indo-Pacific region, from East Africa to the Society Islands.

HABITAT
Buries into loose coral sand in inshore areas where there is strong wave action.

BEHAVIOR
Nothing known.

FEEDING ECOLOGY AND DIET
Feeds on polychaetes.

REPRODUCTIVE BIOLOGY
Nothing known.

CONSERVATION STATUS
Not threatened.

SIGNIFICANCE TO HUMANS
None known. ◆

Fire goby
Nemateleotris magnifica

FAMILY
Ptereleotridae

TAXONOMY
Nemateleotris magnificus Fowler, 1938, Buka Buka Island, Gulf of Tomini, Sulawesi, Indonesia.

OTHER COMMON NAMES
English: Fire dartfish, firefish; French: Gobie de feu; Afrikaans: Vuur-dikkop; Japanese: Hatatatehaze; Malay: Roket antene; Samoan: Mano'o-sugale.

PHYSICAL CHARACTERISTICS
Maximum size 3.5 in (9 cm) total length. This is one of the "hovering" gobies, a species that actually hovers in the water column instead of hugging the bottom like most of its kin. The body form is more compressed laterally (deeper and thinner) instead of the more typical tubular goby shape. The anterior portion of the first dorsal fin is greatly elongate, almost as long as the fish itself in some specimens. The pelvic fins are separate. The second dorsal and anal fins are long and, when extended, have the appearance of a feather, or the tail end of an arrow or a dart. The fish has a rounded caudal fin. The head has a yellow mask over the snout and eyes. The front half of the fish (other than the mask) is white, the back half grades from orange to red, with green streaks in the median fins converging posteriorly.

DISTRIBUTION
Tropical Indian and Pacific Oceans, from South Africa to the Marquesas. In the Pacific, extends north to the Hawaiian and Ryukyu Islands, and south to New Caledonia.

HABITAT
Coral-reef species found on the upper slopes of the outer reef. Although observed at depths from 20–200 ft (6–61 m), it is usually found at less than 92 ft (28 m). Creates burrows to which it can retreat when threatened.

BEHAVIOR
Occurring singly or in small group, this species hovers in the water column, a few centimeters above sand or rubble, selectively feeding on drifting zooplankton. Usually has a small territory around a hole, cave, or burrow into which it retreats at

the threat of danger. The burrow may be shared. Has a habit of flicking the first dorsal spine up and down, erecting the dorsal fin when threatened or defending territory.

FEEDING ECOLOGY AND DIET
Feeds on zooplankton, including crustacean larvae and copepods.

REPRODUCTIVE BIOLOGY
Spawning occurs in burrows.

CONSERVATION STATUS
Not threatened.

SIGNIFICANCE TO HUMANS
This beautiful fish is an important saltwater aquarium species. ◆

Loach goby
Rhyacichthys aspro

FAMILY
Rhyacichthyidae

TAXONOMY
Platyptera aspro Valenciennes, 1837, west Java and Sulawesi. Although *Rhyacichthys aspro* has several specializations of its own, it is considered to be a primitive representative of the Gobioidei.

OTHER COMMON NAMES
Ilokano: Kampa; Japanese: Tsubasahaze; Visayan: Dalapakan.

PHYSICAL CHARACTERISTICS
Reaches 12.6 in (32 cm) total length. Head depressed dorsally, with a wedgelike snout, and with small eyes. Mouth ventrally placed on head with a fleshy upper lip. Body laterally compressed toward tail, which is slightly forked; body with 27–40 scales in longitudinal series on flanks. Lateral line system well developed on head and body. Pelvic fins separate, each with a patch of enlarged musculature on the anterior border. Body light brown with several darker, longitudinal stripes on flanks; dark bands running from eye toward anterior tip of snout. Dorsal fins with alternating light and dark longitudinal stripes; caudal and pectoral fins pale with several dark vertical bands.

DISTRIBUTION
Islands of the western Pacific rim, extending in an arc from the Ryukyu Islands of southern Japan, through Taiwan, Indonesia, and the Philippines, to New Guinea and the Solomon Islands.

HABITAT
Clings to rocks and boulders in fresh waters of torrential hill streams. The streamlined head and body shape makes this species well suited to living in these fast-flowing waters. The fish may use the ventral part of head, thoracic region of the body, and the pectoral and pelvic fins to attach to rocks. This might be done by friction between the rock surface and rough surfaces on the underside of the body, or by suction, in which a low-pressure water flow between the body and the rock "sticks" the fish to the substrate.

BEHAVIOR
Swims in a series of rapid darts and, at the approach of danger, quickly moves into crevices under large rocks or boulders.

FEEDING ECOLOGY AND DIET
Herbivorous, browses on algae on the surfaces of stones.

REPRODUCTIVE BIOLOGY
Nothing known.

CONSERVATION STATUS
Not threatened.

SIGNIFICANCE TO HUMANS
None known. ◆

Resources

Books

Akihito, Prince. "Some Morphological Characters Considered to Be Important in Gobiid Phylogeny." In *Indo-Pacific Fish Biology: Proceedings of the Second International Conference on Indo-Pacific Fishes*, edited by T. Uyeno, R. Arai, T. Taniuchi, and K. Matsuura. Tokyo: Ichthyological Society of Japan, 1986.

Akihito, Prince, M. Hayashi, T. Yoshino, K. Shimada, H. Senou, and T. Yamamoto. "Suborder Gobioidea." In *The Fishes of the Japanese Archipelago*, edited by H. Masuda, K. Amoaka, C. Araga, T. Uyeno, and T. Yoshino. Tokyo: Tokai University Press, 1984.

Colin, P. *The Neon Gobies; the Comparative Biology of the Gobies of the Genus* Gobiosoma, *Subgenus* Elacatinus, *(Pisces: Gobiidae) in the Tropical Western North Atlantic Ocean.* Neptune City, NJ: T. F. H. Publications, 1975.

Harrison, I. J., and M. L. J. Stiassny. "The Quiet Crisis: A Preliminary Listing of the Freshwater Fishes of the World That Are Extinct or 'Missing in Action.'" In *Extinctions in Near Time: Causes, Contexts, and Consequences*, edited by R. D. E. MacPhee. New York: Kluwer Academic/Plenum Publishers, 1999.

Hoese, D. F. "Gobies." In *Encylopedia of Fishes*, edited by J. R. Paxton and W. N. Eschmeyer. San Diego: Academic Press, 1995.

———. "Gobioidei: Relationships." In *Ontogeny and Systematics of Fishes*, edited by H. G. Moser. Gainesville, FL: American Society of Ichthyologists and Herpetologists, 1984.

Miller, P. J. "The Tokology of Gobioid Fishes." In *Fish Reproduction: Strategies and Tactics*, edited by G.W. Potts and R. J. Wooton. London: Academic Press, 1984.

——— "Reproductive Biology and Systematic Problems in Gobioid Fishes." In *Indo-Pacific Fish Biology: Proceedings of the Second International Conference on Indo-Pacific Fishes*, edited by T. Uyeno, R. Arai, T. Taniuchi, and K. Matsuura. Tokyo: Ichthyological Society of Japan, 1986.

Myers, R. F. "Suborder Gobioidei." In *Micronesian Reef Fishes: A Comprehensive Guide to the Coral Reef Fishes of Micronesia*, 3rd edition. Guam: Coral Graphics, 1999.

Nelson, J. S. *Fishes of the World*, 3rd edition. New York: John Wiley & Sons, 1994.

Nishimoto, R. T., and J. M. Fitzsimons. "Courtship, Territoriality, and Coloration in the Endemic Hawaiian Freshwater Goby, *Lentipes concolor*." In *Indo-Pacific Fish Biology: Proceedings of the Second International Conference on Indo-Pacific Fishes*, edited by T. Uyeno, R. Arai, T. Taniuchi, and K. Matsuura. Tokyo: Ichthyological Society of Japan, 1986.

Patterson, C. "Osteichthyes: Teleostei." In *The Fossil Record*, Vol. 2, edited by M. J. Benton. London: Chapman and Hall, 1993.

Torricelli, P. "Gobiidae." In *I pesci delle acque interne italiane*, edited by G. Gandolfi, S. Zerunian, P. Torricelli, and A. Marconato. Instituto Poligrafico e Zecca dello Stato, Italy: Ministero dell'Ambiente e Unione Zoologica Italiana, 1991.

Periodicals

Akihito, A. Iwata, T. Kobayashi, K. Ikeo, T. Imanishi, H. Ono, Y. Umehara, C. Hamamatsu, K. Sugiyama, Y. Ikeda, K. Sakamoto, A. Fumihito, S. Ohno, and T. Gojobori. "Evolutionary Aspects of Gobioid Fishes Based upon a Phylogenetic Analysis of Mitochondrial Cytochrome B Genes." *Gene* 259, nos. 1 and 2 (2000): 5–15.

Allen, G. R., and D. F. Hoese. "A Review of the Genus *Mogurnda* (Pisces: Eleotrididae) from New Guinea with Descriptions of Three New Species." *Ichthyological Exploration of Freshwaters* 2, no. 1 (1991): 31–46.

Erdman, D. S. "The Green Stream Goby, *Sicydium plumieri*, in Puerto Rico." *Tropical Fish Hobbyist* (Oct. 1986): 70–74.

Hoese, D. F., and A. C. Gill. "Phylogenetic Relationships of Eleotridid Fishes (Perciformes: Gobioidei)." *Bulletin of Marine Science* 52, no. 1 (1993): 415–440.

Larson, H. K. "A Review of the Australian Endemic Gobiid Fish Genus *Chlamydogobius*, with a Description of Five New Species." *The Beagle, Records of the Museums and Art Galleries of the Northern Territory* 12 (1995): 19–51.

———. "A Revision of the Gobiid Genus *Bryaninops* (Pisces), with a Description of Six New Species." *Beagle, Occasional Papers of the Northern Territory Museum of Arts and Sciences* 2 (1985): 57–93.

Miller, P. J. "The Adaptiveness and Implications of Small Size in Teleosts." *Symposia of the Zoological Society of London* 44 (1979): 263–306.

———. "The Endurance of Endemism: the Mediterranean Freshwater Gobies and Their Prospects for Survival." *Journal of Fish Biology* 37, supp. A (1990): 145–156.

———. "The Functional Ecology of Small Fish: Some Opportunities and Consequences." *Symposia of the Zoological Society of London* 69 (1996): 175–199.

———. "Grading of Gobies and Disturbing of Sleepers." *NERC News* (Oct. 1993): 16–19.

———. "The Osteology and Adaptive Features of *Rhyacichthys aspro* (Teleostei: Gobioidei) and the Classification of Gobioid Fishes." *Journal of Zoology (London)* 171 (1973): 397–434.

Munday, P. L., S. J. Pierce, G. P. Jones, and H. K. Larson. "Habitat Use, Social Organization and Reproductive Biology of the Seawhip Goby, *Bryaninops yongei*." *Marine and Freshwater Research* 53 (2002): 769–775.

Murdy, E. O. "A Review of the Gobioid Fish Genus *Gobioides*." *Ichthyological Research* 45, no. 2 (1998): 121–133.

———. "A Taxonomic Revision and Cladistic Analysis of the Oxudercine Gobies (Gobiidae: Oxudercinae)." *Records of the Australian Museum* 11 (1989): 1–93.

Norman, A. "The Smallest Fish There Is." *Freshwater and Marine Aquarium* 4, no. 1 (1981): 26–27, 82.

Polunin, N. V. C., and R. Lubbock. "Prawn-Associated Gobies (Teleostei: Gobiidae) from the Seychelles, Western Indian Ocean: Systematics and Ecology." *Journal of Zoology (London)* 183 (1977): 63–101.

Shibukawa, K., A. Iwata, and S. Viravong. "*Terateleotris*, a New Gobioid Fish Genus from Laos (Teleostei, Perciformes), with Comments on Its Relationships." *Bulletin of the National Science Museum, Tokyo* Series A 27, no. 4 (2001): 229–257.

Thacker, C. "Phylogeny of the Wormfishes (Teleostei: Gobioidei: Microdesmidae)." *Copeia* 2000, no. 4 (2000): 940–95.7

Tyler, J. C. and J. E. Bohlke. "Records of Sponge Dwelling Fishes, Primarily of the Caribbean." *Bulletin of Marine Science* 22 (1972): 601–642.

Winterbottom, R. "Search for the Gobioid Sister Group (Actinopterygii: Percomorpha)." *Bulletin of Marine Science* 52, no. 1 (1993): 395–414.

Other

"Taxonomy Browser." National Center for Biotechnology Information Databases [cited January 14, 2003]. <http://www.ncbi.nlm.nih.gov/Taxonomy/taxonomyhome.html>

FishBase [cited January 14, 2003]. <http://www.fishbase.org/search.cfm>

"FISHSTAT Plus: Universal Software for Fishery Statistical Time Series." Food and Agriculture Organization of the United Nations, Fisheries Department, Fishery Information, Data and Statistics Unit [cited January 14, 2003]. <http://www.fao.org/fi/statist/FISOFT/FISHPLUS.asp>

Harrison Ian, J., and Melanie L. J. Stiassny. "List of Fish Extinctions Since A.D. 1500." Committee on Recently Extinct Organisms (CREO) [cited January 14, 2003]. <http://creo.amnh.org/pdi.html>

Ian J. Harrison, PhD
Frank Pezold, PhD

Acanthuroidei
(Surgeonfishes and relatives)

Class Actinopterygii
Order Perciformes
Suborder Acanthuroidei
Number of families 6

Photo: A yellowtailed surgeonfish (*Prionurus laticlavius*) school swims over the lava bottom of the Roca Redonda, Galápagos Islands. (Photo by Fred McConnaughey/Photo Researchers, Inc. Reproduced by permsiion.)

Evolution and systematics

The suborder Acanthuroidei comprises six families and 128 species. The families are Ephippidae (spadefishes, or batfishes, with eight genera and 15 species), Scatophagidae (scats, with two genera and four species), Siganidae (rabbitfishes, or spinefoots, with one genus, two subgenera, and 27 species), Luvaridae (louvar, with one genus and one species), Zanclidae (Moorish idol, with one genus and one species), and Acanthuridae (surgeonfishes and tangs, with six genera and 80 species). Some researchers also place the Chaetodontidae (butterflyfishes), Pomacanthidae (angelfishes), and Drepaneidae (sicklefishes) in this suborder on the basis that these fishes are sister taxa to the acanthuroid fishes. This view is not accepted universally, and so these three families are treated elsewhere in this volume.

The Ephippidae and Scatophagidae date from the Lower Eocene, while the Siganidae, Zanclidae, and Acanthuridae are from the Middle Eocene. The origin of the Luvaridae is uncertain. The suborder is considered to be monophyletic, with or without the inclusion of the Chaetodontidae, Pomacanthidae, and Drepaneidae. One feature common to half the members of this suborder is the presence of sharp, venomous spines (Scatophagidae and Siganidae) or sharp peduncular spines or keels (Acanthuridae), which also may be venomous.

Physical characteristics

All but one species in this suborder have small to medium-size bodies that are compressed and may be deep, disc-like, ovate, or slightly elongated. The spadefishes, or batfishes (Ephippidae), have deep and highly compressed bodies, continuous dorsal fins, and highly elevated dorsal, anal, and pelvic fins. Their scales are small and ctenoid, and their mouths are small and terminal. The scats (Scatophagidae) have deep, compressed bodies, a pronounced forehead, a mouth that is nonprotrusible, and a deeply notched dorsal fin. Fin spines are sharp and reportedly venomous. The rabbitfishes, or spinefoots (Siganidae), have compressed, somewhat elongated bodies; some in the subgenus *Lo* have a pronounced snout. Siganids are distinguished by the presence of venomous fin spines. Color patterns vary, and some species are quite distinctive. The monotypic louvar (Luvaridae) has a large (up to 78.7 in, or 200 cm, in length) fusiform body that is slightly compressed and quite streamlined. The head is blunt, with a projecting forehead; the mouth is small and somewhat protrusible; and the opercular spines are flattened. There is a groove positioned just above the eye. There are no pelvic fins, and the caudal peduncle has a large keel and small accessory keels. The scales are dentroid and quite small. Maximum body sizes range from 6–36 in (15–91 cm) for spadefishes and batfishes, 3.5–16 in (9–40 cm) for scats, 8.5–21 in (22–53 cm) for rabbitfishes and spinefoots, and 6.3–39.5 in (16–100 cm) for surgeonfishes. Moorish idols reach about 9 in (23 cm), and louvars reach about 79 in (200 cm) in length.

The Moorish idol (Zanclidae) is also monotypic within its family. This distinctive species has a disc-like body that is strongly compressed and dorsal spines that form an elongated whiplike streamer or filament. The color pattern is a striking arrangement of yellow, black, and white. Surgeonfishes (Acanthuridae) have compressed and disc-like bodies and possess one or more scalpel-like caudal spines or keeled peduncular plates located on each side of the caudal peduncle, which vary in size with species, may be venomous, and are capable of inflicting a painful, if not serious, wound. The caudal fin is strongly lunate, emarginate, or truncate. Many species are quite colorful, whereas others are seemingly drab until close inspection reveals often minute but exquisite details.

Detail of sharp scapel (modified spine) on caudal peduncle of a powder blue tang (*Acanthurus leucosternon*). (Illustration by Joseph E. Trumpey)

Distribution

Five of the six families are distributed in continental and insular inshore habitats of tropical, subtropical, and warm temperate seas. The exception is the louvar, a species that is pelagic throughout its rather wide range. Spadefishes, or batfishes, and surgeonfishes occur in the Atlantic, Indian, and Pacific Oceans. Scats are limited to the Indo-West Pacific, whereas the Moorish idol is found there and in the eastern Pacific. Rabbitfishes occur in the Indo-West Pacific but also have colonized the eastern Mediterranean, via the Suez Canal.

Habitat

All but the louvar, which is pelagic, inhabit shallow coral and rocky reefs, although some species inhabit estuaries or even freshwater habitats. Spadefishes and batfishes favor marine coral and rocky reefs as well as brackish water habitats that include mangroves, sea grasses, or sand and rubble flats. The scats have the greatest flexibility, moving from coral and rocky reefs into brackish water estuaries and freshwater streams. Rabbitfishes also are found on coral (some exclusively) or rocky reefs, on algal or sea grass beds, or around mangroves, but some enter brackish waters. Louvars move up and down in the pelagic water column. The Moorish idol is found on coral or rocky reefs, as are the surgeonfishes, although the latter also may occur on sea grass and algae beds and around mangroves.

Behavior

All members of this suborder except one may be found singly, in small groups, or in aggregations, schools, or mixed-species schools. The exception is the louvar, a solitary fish that probably aggregates only to breed. Many of these fishes, including some rabbitfishes and surgeonfishes, are territorial. Others, such as some rabbitfishes, patrol home ranges in pairs. Numerous species of rabbitfishes also have post-larvae that often form large schools when they settle onto reefs as post-larvae and recruit into (join) the population as juveniles. Batfishes tend to hover singly or in small groups near structures but can form large schools in deeper water as well. Their juveniles mimic fallen leaves. Scats also move about

A school of blue tangs (*Acanthurus coeruleus*) eating the eggs of a sergeant major near the Saba Island of the Netherland Antilles. (Photo by Andrew J. Martinez/Photo Researchers, Inc. Reproduced by permission.)

near shelter but form larger aggregations in estuaries, bays, or open water.

Feeding ecology and diet

Spadefishes, or batfishes, are omnivorous and feed on benthic algae and invertebrates. Some species are capable of crushing bivalve or gastropod shells and crustaceans. Although scats have markedly herivorous tendencies, they are also known to be omnivorous; they even feed on detrital materials and have been observed to feed on feces discharged by ships or sewage systems. Aquatic macrophytes consitute the bulk of the stomach contents of both *S. argus* and *S. tetracanthus* in freshwater habitats. Rabbitfishes are herbivorous or omnivorous, and schools of juveniles are efficient, if not rapid, grazers. The louvar feeds on gelatinous mesoplankton in the water column. The Moorish idol feeds primarily on sponges but also plucks other benthic invertebrates. Surgeonfishes are largely herbivorous and browse or graze on benthic algae. Some species are adapted to feeding on plankton in the water column.

Reproductive biology

Almost all members of this suborder spawn pelagic eggs. The exceptions appear to be the scats, which spawn demersal eggs and engage in parental care, and some of the rabbitfishes, which spawn eggs that sink to the bottom. Mating strategies vary within and between families. Courtship and spawning are accomplished in pairs, in pairs within spawning aggregations, or in groups of varying sizes at spawning aggregation sites. Thus, spadefishes, or batfishes, spawn in pairs and possibly in aggregations. Scats spawn demersal eggs, and at least one species practices parental care. Rabbitfishes of the subgenus *Lo* form relatively long-term monogamous pairs. Others in the family, however, spawn in aggregations, in either short-term pairs or groups. Rabbitfishes have been spawned in captivity, and some species are now important in mariculture. Louvars probably spawn in aggregations in the open sea. Moorish idols spawn in pairs or in pairs within aggregations. Surgeonfishes spawn in pairs, as pairs within aggregations, and in groups. The larvae of many of these fishes are well adapted to a prolonged existence in the open sea.

Conservation status

No species is listed by the IUCN, but some species collected for the aquarium trade, including certain surgeonfishes and batfishes, may be at risk locally because of overfishing.

Significance to humans

Rabbitfishes and surgeonfishes are important in subsistence and artisanal fisheries. The louvar is taken as a minor commercial food fish, as are some of the spadefishes, or batfishes. Scats are grown for the live reef food fish trade or taken in subsistence fisheries. Fishes important to the aquarium trade include scats, batfishes, and certain rabbitfishes and surgeonfishes. The Moorish idol also is collected for aquaria, but they do not do well in captivity.

The unusual shape, coloration, and movement of the pinnate batfish (*Platax pinnatus*) probably serve to mimic a toxic flatworm. This is a juvenile batfish swimming near Ambon, Indonesia. (Photo by David Hall/Photo Researchers, Inc. Reproduced by permission.)

1. Yellow tang (*Zebrasoma flavescens*); 2. Striped bristletooth (*Ctenochaetus striatus*); 3. Yellowspotted sawtail (*Prionurus maculatus*); 4. Orangespine unicornfish (*Naso lituratus*); 5. Palette tang (*Paracanthurus hepatus*); 6. Lined surgeonfish (*Acanthurus lineatus*). (Illustration by Joseph E. Trumpey)

1. Louvar (*Luvarus imperialis*); 2. Foxface rabbitfish (*Siganus vulpinus*); 3. Intermediate stage of longfin spadefish (*Platax teira*); 4. Moorish idol (*Zanclus cornutus*); 5. Spotted scat (*Scatophagus argus*). (Illustration by Joseph E. Trumpey)

Species accounts

Lined surgeonfish

Acanthurus lineatus

FAMILY
Acanthuridae

TAXONOMY
Acanthurus lineatus Linnaeus, 1758, "Indies." No type specimens are known.

OTHER COMMON NAMES
French: Chirurgien zébré; German: Blaustreifen-Doktosfisch; Japanese: Nijihagi.

PHYSICAL CHARACTERISTICS
The body is compressed and disc-like and possesses a large, venomous, scalpel-like caudal spine on each side of the caudal peduncle. The caudal fin is strongly lunate. There are nine spines and 27–30 soft rays in the dorsal fin, three spines and 25–28 soft rays in the anal fin, and 16 soft rays in the pectoral fin. Body coloration is striking and consists of a yellowish green background, bright blue stripes (narrow and oblique on the head) edged with black on the flank but lacking the black edge on the head, and lavender blue to bluish white on the belly. The pelvic fins are bright orange. The caudal fin has two vertical lines of dark blue set against a background of purplish gray, replaced by yellow at the base of the fin. The remaining fins are purplish gray yielding to greenish yellow at the base. Larvae are acronurus, in that they are orbicular in shape; transparent, with a silvery sheen over the head, chest, and abdomen; and scaleless and have narrow, vertically oriented ridges along the body. (Acronurus refers to the last, well-developed but seemingly different, post-larval stage of surgeonfishes in the genera *Acanthurus*, *Ctenochaetus*, and *Zebrasoma*. Originally this stage was proposed to be a separate genus of surgeonfishes until closer examination revealed otherwise.) The second anal, second dorsal, and pelvic spines are elongated and venomous. Larvae are well adapted to the pelagic realm and may remain there for more than 39 days. Adults may grow to 15 in (38 cm) in length and can live for as long as 30–45 years.

DISTRIBUTION
Indo-West Pacific, from East Africa east to the Marquesas and the Tuamotu Archipelago, north to southern Japan, and south to New South Wales, Australia; strays rarely to the Hawaiian Islands. Replaced by the very similar species *A. sohal* in the vicinity of the Arabian Peninsula (Red Sea and Persian Gulf).

HABITAT
Usually found on exposed reef fronts at depths of 3.3–9.8 ft (1–3 m).

BEHAVIOR
Highly territorial. Patrols territories along the reef front and aggressively attacks conspecifics, other surgeonfishes, and other herbivores. Temporarily assumes a darker coloration of the head during territorial displays and encounters. Wields its caudal spines as effective weapons against attack.

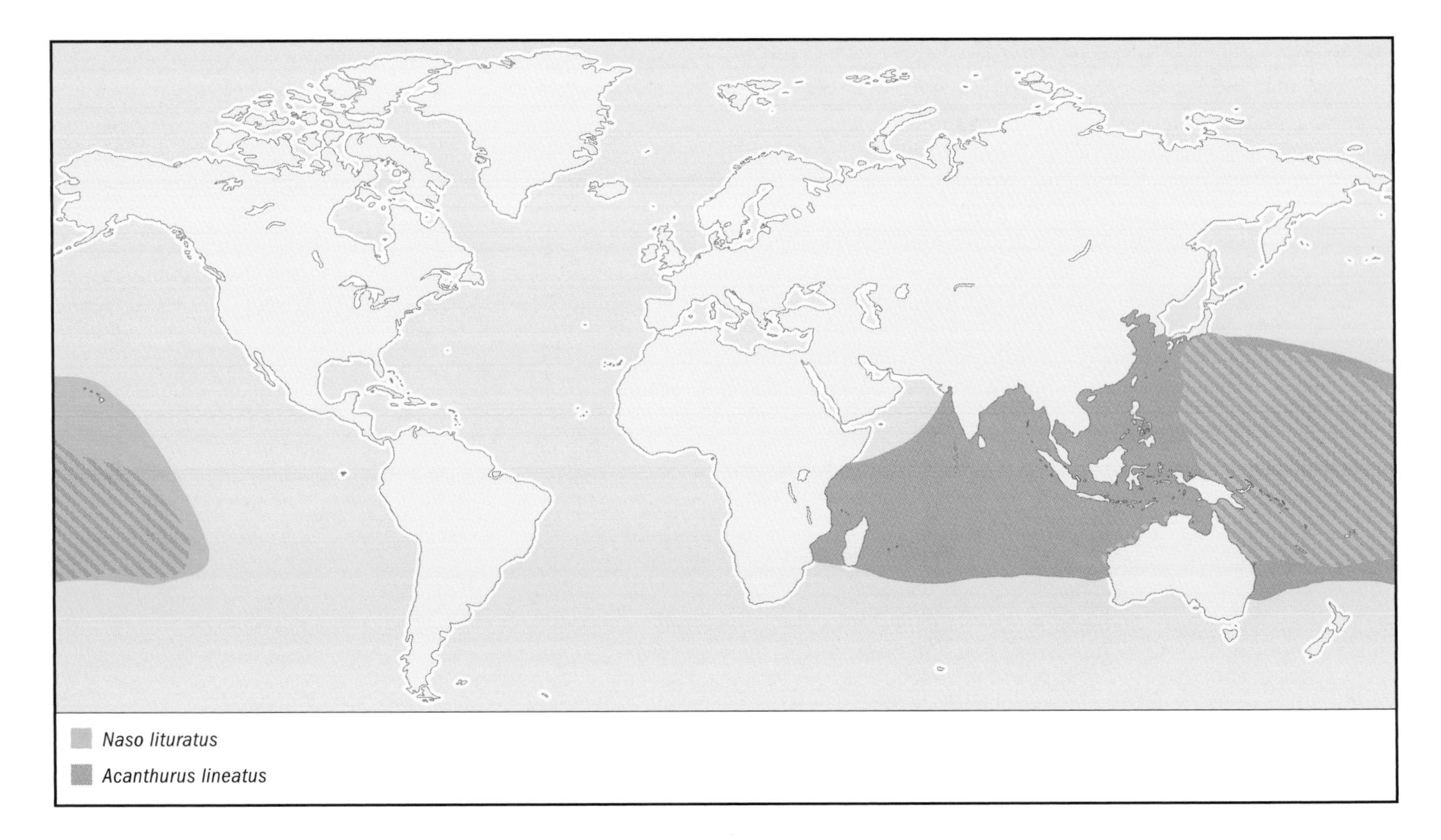

FEEDING ECOLOGY AND DIET
Herbivorous. Browses on filamentous and fleshy algae using specialized teeth.

REPRODUCTIVE BIOLOGY
Migrates to and spawns in aggregations at specific sites, although it sometimes spawns in pairs. Reported to spawn in early morning during the first 1–3 hours of a falling tide just before the full moon. Has been observed spawning at sunset as well. Tidal state, rather than time of day, more likely accounts for the temporal pattern of spawning. Spawning occurs year-round at lower latitudes but may be seasonal at higher latitudes. Eggs and larvae are pelagic.

CONSERVATION STATUS
Not listed by the IUCN.

SIGNIFICANCE TO HUMANS
An important species in subsistence and artisanal food fisheries. Also collected for larger aquaria. ◆

Orangespine unicornfish
Naso lituratus

FAMILY
Acanthuridae

TAXONOMY
Naso lituratus Forster in Bloch and Schneider, 1801, probably Tahiti, French Polynesia.

OTHER COMMON NAMES
French: Nason a éperon orange; German: Gelbklingen-Nasendoktor; Japanese: Miyako-tenguhagi.

PHYSICAL CHARACTERISTICS
The body is compressed and disc-like, although it elongates with growth. The snout is pronounced, and the forehead slants at a 45-degree angle from just forward of the dorsal fin. The caudal peduncle consists of two peduncular plates, each featuring a large keel facing forward. The caudal fin is emarginate and sexually dimorphic; the males have filaments that trail from each corner of the fin. There are six spines and 28–31 soft rays in the dorsal fin, two spines and 29–31 soft rays in the anal fin, 17–18 soft rays in the pectoral fin, and one spine and three soft rays in the pelvic fin. The body is dark grayish brown with a yellowish patch on the nape, a yellow margin along either side of the snout from the eye to just behind the mouth, a black snout, and an orange mouth. The peduncular plates and keels are bright orange. The dorsal fin is black with a thin blue margin at the base and a broader blue margin along the outer edge. The caudal fin is also black, with a yellow submarginal band on the edge. The anal fin is yellow at the base, followed by an orange band, a narrow submargin of light blue, and a narrow outer margin of black. The pelvic fin is yellow, and the pectoral fin is a faint black. Grows to about 18 in (46 cm) in length.

DISTRIBUTION
Western Pacific, from Surugu Bay, Honshu, Japan, south to the Great Barrier Reef and New Caledonia, and east to the Hawaiian and Pitcairn Islands; recently reported from Clipperton Island in the Eastern Pacific. Also found in the eastern Indian Ocean from the coast of Western Australia south to Ningaloo Reef. Replaced by *N. elegans* in the western Indian Ocean, including southern Indonesia, and the Red Sea.

HABITAT
Coral and rocky reefs in tropical and subtropical waters, generally above 98 ft (30 m) but occasionally as deep as 295 ft (90 m).

BEHAVIOR
Occurs singly or in small groups.

FEEDING ECOLOGY AND DIET
Herbivorous. Browses on benthic algae but favors tough or leafy species from the genera *Sargassum*, *Dictyota*, and *Pocockiella*.

REPRODUCTIVE BIOLOGY
Reported to spawn in pairs but also may spawn in aggregations. Eggs and larvae are pelagic. Larval life is lengthy, in excess of 69 days.

CONSERVATION STATUS
Not listed by the IUCN.

SIGNIFICANCE TO HUMANS
Taken in subsistence and artisanal fisheries and a popular target of spearfishers. Juveniles and subadults also are collected for the aquarium trade. ◆

Palette tang
Paracanthurus hepatus

FAMILY
Acanthuridae

TAXONOMY
Paracanthurus hepatus Linnaeus, 1766, Ambon and Molucca islands, Indonesia. Western Atlantic localities are in error.

OTHER COMMON NAMES
English: Blue tang, hippo; German: Paletten-Doktorfisch.

PHYSICAL CHARACTERISTICS
The body is compressed and somewhat disc-like; it elongates with growth. The caudal fin is truncated in adults, but slightly rounded in juveniles. There are nine spines and 19–20 soft rays in the dorsal fin, three spines and 18–19 soft rays in the anal fin, 16 soft rays in the pectoral fin, and one spine and three soft rays in the pelvic fin. The caudal peduncle has a single folding spine on each side. The body is a vivid bright blue, and the belly is a paler shade of blue or, in Indian Ocean specimens, yellowish blue. A black band curves backward from the eye to the caudal peduncle. A second black band runs back along the middle of the body from just behind the pectoral fin and joins the upper band just before the caudal peduncle. The caudal fin is mainly yellow, the yellow arising just ahead of the caudal spine, with black margins that arise at the caudal peduncle. Grows to 10.2 in (26 cm) in length.

DISTRIBUTION
Tropical and subtropical waters of the Indo-West Pacific, from East Africa east to Micronesia, the Line Islands, and the Samoa Islands, north to Kochi Prefecture in Japan, and south to northern New South Wales, Australia. Observations of this species in Hawaiian Islands are attributed to releases of aquarium stock.

HABITAT
Outer coral and rocky reefs and channels in clear water areas with strong or moderate current. Juveniles and subadults prefer

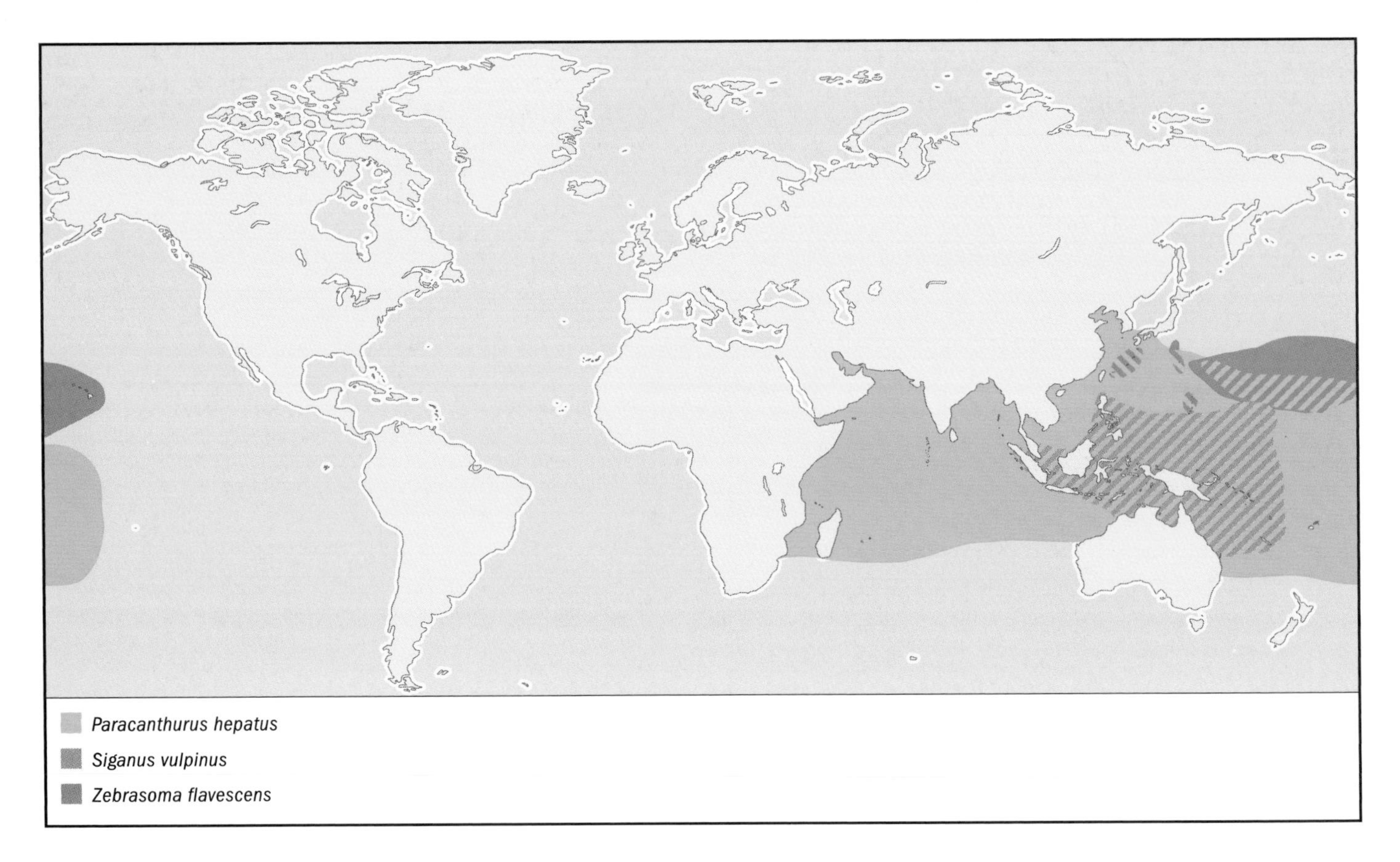

to shelter in shrublike corals. Not common and patchy in distribution wherever it occurs. Depth range of 6.6–131 ft (2–40 m).

BEHAVIOR
Occurs in small aggregations that hover 3.3–6.6 ft (1–2 m) above the substrate.

FEEDING ECOLOGY AND DIET
Omnivorous. Feeds on zooplankton and benthic algae.

REPRODUCTIVE BIOLOGY
Reproductive biology is not well known. Courtship probably is paired or paired within a spawning aggregation. Eggs and larvae are pelagic, with a larval life in excess of 37 days.

CONSERVATION STATUS
Not listed by the IUCN. Localized populations are vulnerable to overfishing of juveniles for the aquarium trade.

SIGNIFICANCE TO HUMANS
Collected for the aquarium trade, in which it is highly prized. Juveniles and subadults that shelter in corals are relatively easy to collect. ◆

Striped bristletooth
Ctenochaetus striatus

FAMILY
Acanthuridae

TAXONOMY
Ctenochaetus striatus Quoy and Gaimard, 1825, Guam, Mariana Islands.

OTHER COMMON NAMES
German: Brauner Borstenzahndoktor; Japanese: Sazanamihagi.

PHYSICAL CHARACTERISTICS
The body is compressed and disc-like, with a venomous, scalpel-like caudal spine on each side of the caudal peduncle. The caudal fin is lunate. There are eight spines and 27–31 soft rays in the dorsal fin, three spines and 24–28 soft rays in the anal fin, and 17 (sometimes 16) soft rays in the pectoral fin. Adult coloration varies from dark gray to orangish brown. Can change color temporarily from dark brown to light tan. Melanistic, albinistic, and xanthic color patterns have been reported as well. There are several fine pale blue lines along the flank. Fine orange spots appear on the upper head and back to the anterior base of the dorsal fin. Both the dorsal and anal fins have narrow dark brown and pale blue bands, running longitudinally in alternation. Juveniles and subadults are especially colorful, with a small black spot located at the posterior base of the dorsal fin. Grows to 10.2 in (26 cm) in length.

DISTRIBUTION
Indo-West Pacific, from the Red Sea south to East Africa, east to Pitcairn Island and French Polynesia (absent from the Marquesas Islands), north to southern Japan, and south to the Great Barrier Reef and New Caledonia.

HABITAT
Found on coral, rock, pavement, and rubble of exposed reefs, reef flats, and lagoons in tropical waters. Depth range from 3.3 to 98 ft (1–30 m).

BEHAVIOR
Found singly but also in small and very large aggregations. Also participates in mixed-species aggregations with other surgeonfishes and parrotfishes, among others.

FEEDING ECOLOGY AND DIET
Herbivorous. Feeds upon benthic algae, blue-green algae, and diatoms. Reported to cause ciguatera poisoning because of their diet. Ciguatera poisoning is caused by the cumulative deposition of a class of polyether toxins within the tissues of fishes. The toxins are produced by certain microscopic dinoflagellate organisms of the genus *Gambierdiscus*, which are transmitted by the food chain, increasing in intensity by a factor of ten in each successive level within the chain. The striped bristletooth ingests these organisms while feeding upon algae. If a contaminated fish is consumed by man, the concentrated poison contained within its tissues causes neurological damage that can be fatal.

REPRODUCTIVE BIOLOGY
Migrates to outer channels or exposed reefs to spawn in aggregations on an outgoing tide. Eggs and larvae are pelagic.

CONSERVATION STATUS
Not listed by the IUCN.

SIGNIFICANCE TO HUMANS
Taken for subsistence and artisanal food fisheries but may cause ciguatera poisoning at some localities. Also collected infrequently for the aquarium trade but often they do not do very well in captivity; juveniles are more highly prized than adults. ◆

Yellowspotted sawtail
Prionurus maculatus

FAMILY
Acanthuridae

TAXONOMY
Prionurus maculatus Ogilby, 1887, Port Jackson, New South Wales, Australia.

OTHER COMMON NAMES
German: Gelbfleken-Sägedoktor; Japanese: Nizadai.

PHYSICAL CHARACTERISTICS
Body is compressed and slightly elongated, with a protruding mouth and a snout with a dorsal profile that is slightly concave. there are nine spines and 24–26 soft rays in the dorsal fin, three spines and 23–25 soft rays in the anal fin, 17–18 soft rays in the pectoral fin, and one spine and five rays in the pelvic fin. the caudal fin is slightly emarginate to truncate in shape. as with other members of this genus, this species is distinguished by the presence of three peduncular plates, each having a lateral keel on either side of the caudal peduncle. the body coloration is bluish gray with several small yellow spots. the peduncular plates are black, and the keels are blue, usually with a white mark on the last plate. the caudal fin is gray with yellow spots and has a deep blue margin, as does the anal fin. juveniles have a series of narrow yellow bars that eventually, with age, become vertical rows of yellow spots. grows to about 17.3 in (44 cm) in length.

DISTRIBUTION
Western Pacific and restricted to the Southern Hemisphere, from southern Queensland on the Great Barrier Reef south to New South Wales and east to Lord Howe Island, Norfolk Island, and the Kermadec Islands. Waifs recruit seasonally to New Zealand during warmer summer months but probably do not reproduce successfully.

HABITAT
Tropical and warm temperate waters in protected areas of coral reefs, rocky reefs, and bays but also on outer coral and rocky reefs. Depth range is 6.6–98 ft (2–30 m).

BEHAVIOR
This species usually is found in aggregations.

FEEDING ECOLOGY AND DIET
Herbivorous. Feeds on benthic algae that grows mainly on rocks or coral substrates.

REPRODUCTIVE BIOLOGY
Probably pair-spawns but also may form spawning aggregations in which pair spawning occurs. Males probably change color pattern to attract a female from the aggregation, and the two touch their ventral surfaces together before the release of gametes, without an upward dash into the water column. Eggs and larvae are pelagic.

CONSERVATION STATUS
Not listed by the IUCN.

SIGNIFICANCE TO HUMANS
May be taken incidentally as a food fish. ◆

Yellow tang

Zebrasoma flavescens

FAMILY
Acanthuridae

TAXONOMY
Zebrasoma flavescens Bennett, 1828, Hawaiian Islands.

OTHER COMMON NAMES
German: Gelber Segelflosser; Japanese: Kiirohagi.

PHYSICAL CHARACTERISTICS
Body is compressed and disc-like, with a concave head, a pronounced snout, and a protruding mouth. The dorsal fin is highly elevated and has four to five spines and 23–26 soft rays. The anal fin has three spines and 19–22 soft rays, and the pectoral fin has 14–16 soft rays. The body is a bright yellow, but the caudal spine sheath is white. Known to hybridize with *Z. scopas.* Grows to more than 7.9 in (20 cm) in length.

DISTRIBUTION
Tropical waters of the northern Pacific Ocean, from Minami-tori-shima (Marcus Island) east to Wake Island, the Marshall Islands (uncommon), the Hawaiian Islands, and Johnston Island. Also in the Ogasawara, Ryukyu, and Mariana Islands. A waif was reported from Hong Kong, and other records, both in Polynesia and in the Indian Ocean, are probably xanthic morphs of *Z. scopas.*

HABITAT
Coral and rocky reefs, either exposed reef slopes to 265 ft (81 m) or in bays and lagoons as shallow as 3.3 ft (1 m).

BEHAVIOR
May be found in small groups or singly. Groups often move from point to point to browse on algae. Sometimes observed in mixed-species schools.

FEEDING ECOLOGY AND DIET
Herbivorous. Browses on filamentous algae on coral reefs.

REPRODUCTIVE BIOLOGY
This species has two mating systems. Single males may defend territories, court passing females, and engage in pair spawning in the water column. Alternatively, group spawning at spawning aggregation sites occurs. The spawning season is limited by the effects of seasonally cooler water temperatures at higher latitudes. Eggs and larvae are pelagic.

CONSERVATION STATUS
Not listed by the IUCN but subject to overfishing, particularly in parts of the Hawaiian Islands, for the aquarium trade.

SIGNIFICANCE TO HUMANS
An important species in the aquarium trade and the number one exported species from the Hawaiian Islands. ◆

Longfin spadefish

Platax teira

FAMILY
Ephippidae

TAXONOMY
Platax teira Forsskal, 1775, Luhaiya, Yemen, Red Sea.

OTHER COMMON NAMES
English: Teira batfish; French: Poisson lune; German: Langflossen-Fledermausfisch; Japanese: Mikazuki-tsubame-uo.

PHYSICAL CHARACTERISTICS
Body is deep and highly compressed. The dorsal fin is continuous, and the anal, dorsal, and pelvic fins are greatly elevated. Scales are ctenoid and small. The mouth is terminal, with tricuspid teeth. Juveniles also have deep bodies and elevated anal, dorsal, and pelvic fins (the latter extremely long), with the dorsal and anal fins extending backward beyond the caudal fin. These fins are reduced in size somewhat with age (i.e. the fins do not grow with age relative to the rest of the body). Larvae have a deep or moderately deep body and a relatively compressed caudal fin; the head and trunk are combined into a broad, ball-like structure that eventually deepens with growth. There are no scales present on the body of the larvae at settlement. Settlement refers to the transformation of a pelagic-dwelling larva to a Reef-dwelling post-larva or juvenile. Grows to more than 17.7 in (45 cm) in length.

DISTRIBUTION
Tropical and subtropical waters from the Red Sea and Indian Ocean east to Fiji, north to southern Japan, and south to Australia's Great Barrier Reef and Lord Howe Island.

HABITAT
Coral and rocky reefs, usually drop-offs, deep lagoons, and pinnacles along the outer reef to a depth of at least 66 ft (20 m). Juveniles usually are found in protected areas of shallow reefs and bays.

BEHAVIOR
Occurs singly or in small aggregations as adults and juveniles. Juveniles often mimic floating leaves in shallow waters.

FEEDING ECOLOGY AND DIET
Omnivorous, feeding on a variety of small benthic invertebrates and algae.

REPRODUCTIVE BIOLOGY
Probably pair-spawns within aggregations or smaller groups. Eggs and larvae are pelagic.

CONSERVATION STATUS
Not listed by the IUCN.

SIGNIFICANCE TO HUMANS
Collected for the aquarium trade. ◆

Louvar

Luvarus imperialis

FAMILY
Luvaridae

TAXONOMY
Luvarus imperialis Rafinesque, 1810, Sicily, Italy.

OTHER COMMON NAMES
Japanese: Amashiira.

PHYSICAL CHARACTERISTICS
Fusiform body, which is compressed slightly and streamlined. Pelvic fins are absent. The head is blunt, with a projecting forehead, and there is a groove above the eye. Opercular spines are flattened. Scales are dentroid and minute. The caudal peduncle has small accessory keels that flank the large caudal keel. The mouth is small and slightly protrusible. Metamorphosis from larvae to adults involves considerable change. Numerous juvenile characters are absent in adults. For example, juveniles possess jaw teeth, fin spines, and a rudimentary locking mechanism. Grows to 78.7 in (200 cm) in length.

DISTRIBUTION
Tropical, subtropical, and temperate waters of the Atlantic (including the Mediterranean Sea), Indian, and Pacific Oceans.

HABITAT
Epipelagic, from the surface to relatively deeper waters.

BEHAVIOR
Poorly known. This species generally is solitary in the open ocean but probably aggregates for courtship and spawning.

FEEDING ECOLOGY AND DIET
Feeds on gelatinous mesoplankton, primarily jellyfishes and ctenophores.

REPRODUCTIVE BIOLOGY
Not well known. Spawning commences in late spring and runs throughout the summer months. Species is reportedly quite fecund; one female examined had an estimated 47.5 million eggs. Pelagic spawning with pelagic larvae.

CONSERVATION STATUS
Not listed by the IUCN.

SIGNIFICANCE TO HUMANS
May be taken incidentally in commercial fisheries. ◆

Spotted scat

Scatophagus argus

FAMILY
Scatophagidae

TAXONOMY
Scatophagus argus Linnaeus, 1758, India. No type specimens are known.

OTHER COMMON NAMES
Japanese: Kurohoshimanjûdai.

PHYSICAL CHARACTERISTICS
Body is deep and compressed, with a deeply notched dorsal fin, a pronounced forehead, and a nonprotrusible mouth. There

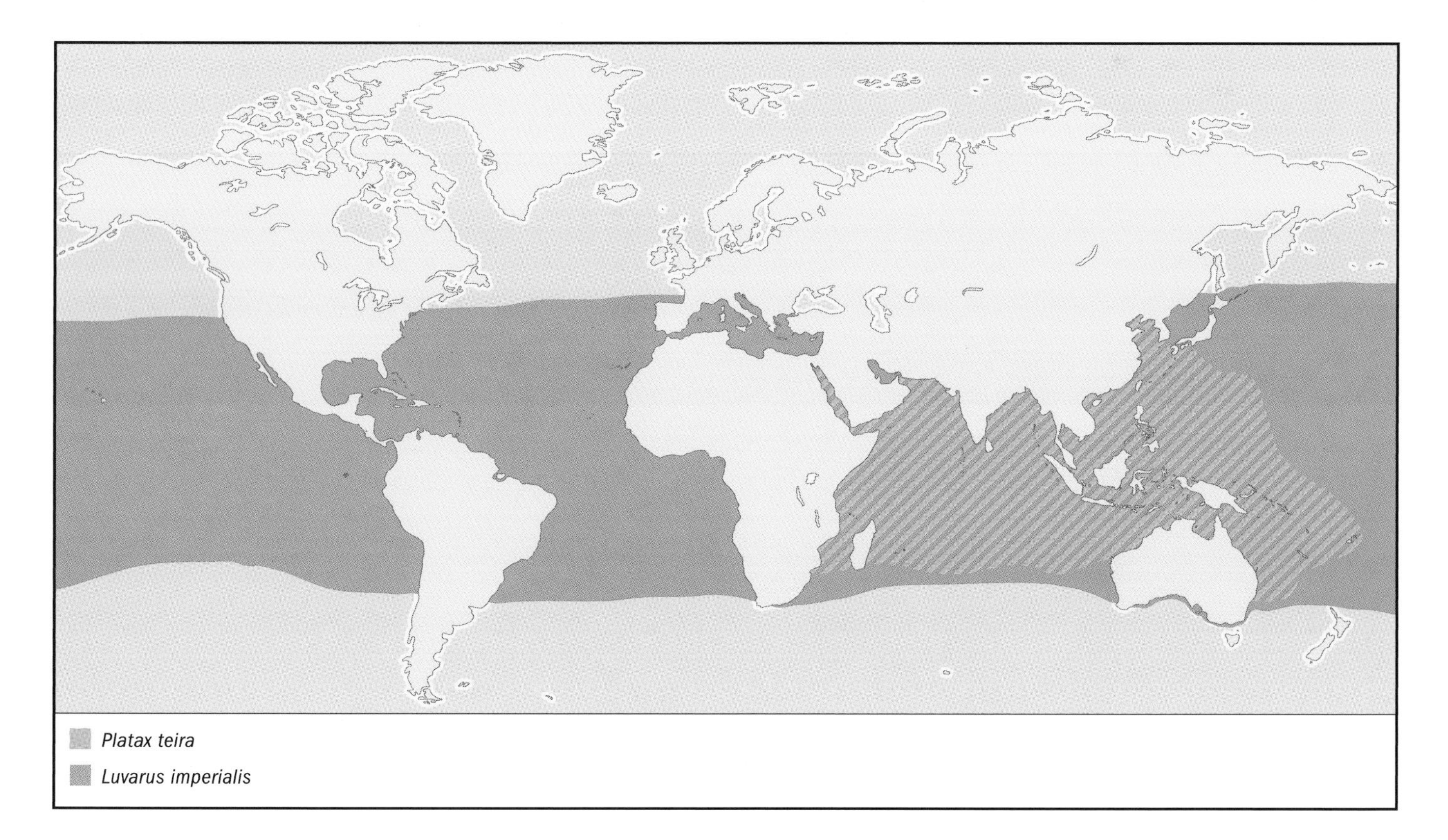

are 10–11 spines and 16–18 soft rays in the dorsal fin, four spines and 13–15 soft rays in the anal fin, and 16 branched rays in the caudal fin. The spines are sharp and are reported to be venomous. The pelvic fins are attached by an axillary process. Body color is silvery green to orange, with numerous small dark spots on the flank and on the base of the dorsal, caudal, and anal fins. Larvae have a specialized stage similar to the tholichthys of larval butterflyfishes (Chaetodontidae). Grows to about 15 in (38 cm) in length.

DISTRIBUTION
Indo-West Pacific region from Kuwait east to Vanuatu and New Caledonia, south to Australia, north to southern Japan, and east to Pohnpei (Micronesia). Also reported from Samoa and the Society Islands.

HABITAT
Inshore marine, brackish, and freshwaters, usually in turbid estuaries, harbors, back bays, mangrove stands, and reefs.

BEHAVIOR
Not well known. Occurs in small groups or aggregations.

FEEDING ECOLOGY AND DIET
Omnivorous, feeding on small invertebrates, including worms, crustaceans, and insects, and on benthic algae. Genus name was coined from observation of members of this genus feeding on feces and sewage discharged by ships into harbors and bays.

REPRODUCTIVE BIOLOGY
Not well known, except from aquarium observations. Reported to spawn demersal eggs and engage in parental care, traits inconsistent with most other species in this suborder, though somewhat similar to the Siganidae.

CONSERVATION STATUS
Not listed by the IUCN.

SIGNIFICANCE TO HUMANS
Collected for the aquarium trade for display in marine and freshwater aquaria. Also taken for subsistence or artisanal food fisheries and has been grown by commercial aquaculture for the live reef food fish trade. ◆

Foxface rabbitfish
Siganus vulpinus

FAMILY
Siganidae

TAXONOMY
Siganus vulpinus Schlegel and Muller, 1844, Ternate Island, Moluccas Islands, Indonesia.

OTHER COMMON NAMES
English: Foxface; German: Gelbes Dachsgesicht; Japanese: Hifuki-aigo.

PHYSICAL CHARACTERISTICS
Body is compressed and elongate, with minute cycloid scales on the trunk. There is a single dorsal fin with 13 spines and 10 soft rays. The pelvic fins are positioned at the thorax. The pectoral fin usually has 16 rays. Fin spines are stout and venomous. The snout is long and tubelike. Body color is predominately yellow, with a chocolate-brown forehead and snout and white spotted with brown on the preopercle and opercle and on the flank directly behind the gills. The thorax

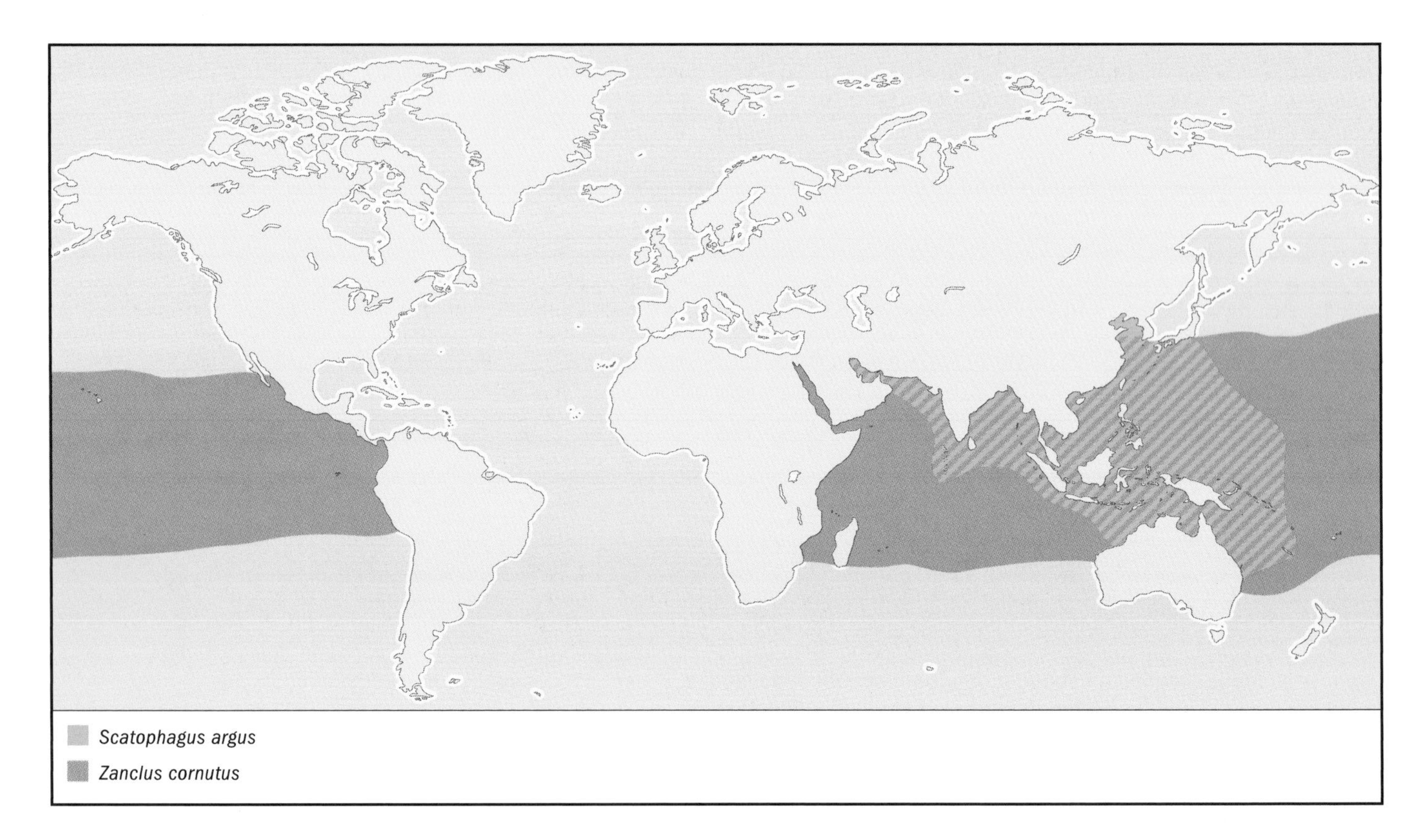

is chocolate brown, and this color extends down to the first two pelvic fin spines. The dorsal, anal, and caudal fins are yellow; the pectoral and pelvic fins are white. There is some variation between individuals. Grows to at least 9.8 in (25 cm) in length.

DISTRIBUTION
Indo-West Pacific region from Sumatra east to the Marshall Islands and south to the Great Barrier Reef and New Caledonia.

HABITAT
Seaward reefs and outer slopes, lagoons, and deeper reef flats to a depth of 98 ft (30 m). Prefers lush coral growth. Juveniles and subadults take refuge in staghorn corals.

BEHAVIOR
Larvae, juveniles, and subadults occur in large schools. Pair formation, possibly monogamous, occurs when the species reaches 3.9 in (10 cm) in length. Pairs patrol a restricted home range and at times may defend a territory.

FEEDING ECOLOGY AND DIET
A browsing herbivore that feeds upon benthic algae found on the substrate or upon dead corals.

REPRODUCTIVE BIOLOGY
Descriptions of reproductive biology of this and some other siganid species are based largely upon observations of captive individuals augmented by some field observations. This species spawns in pairs at dusk or dawn in relation to moon phase. Eggs are spherical and adhesive, measuring 0.02–0.025 in (0.55–0.66 mm), and are scattered on substrate (some siganid species have pelagic eggs, however). Larvae hatch out and measure about 0.08 in (2 mm) in length. They have a large yolk, an unformed mouth, a straight gut, unpigmented eyes, and a pattern of body pigmentation that is relatively stable with growth. Body shape is elongate, but this becomes deeper and laterally compressed with growth. Fin development begins before the yolk is absorbed and spines appear at about 0.11 in (3 mm) in length. Fin rays form at about 0.24 in (6 mm), but scales do not form until larval settlement. The larvae have the following specializations for pelagic life: long dorsal and pelvic fins that form early in development, fin spines that are serrate in shape, and the formation of spines upon the head.

CONSERVATION STATUS
Not listed by the IUCN.

SIGNIFICANCE TO HUMANS
May be taken by subsistence and artisanal fisheries for food. Also collected for the aquarium trade. ◆

Moorish idol
Zanclus cornutus

FAMILY
Zanclidae

TAXONOMY
Zanclus cornutus Linnaeus, 1758, Indian seas. No type specimens are known.

OTHER COMMON NAMES
French: Cocher jaune; German: Halterfisch; Japanese: Tsunodashi.

PHYSICAL CHARACTERISTICS
Superficially, this species resembles the butterflyfishes (Chaetodontidae). The body is strongly compressed and disc-like, the snout is tubular, and the mouth is small, with many elongated, bristle-like teeth. The third dorsal spine is long and whiplike (the "streamer"). Coloring consists primarily of three vertical bands of white and yellow in alternation with two bands of black. The caudal fin is also black and fringed with white or yellowish white. A small patch of yellow and a horizontal band of white occur on the snout. Larvae are distinguished by a third spine that is elongate and more than twice the length of the larva's body, which develops into the "streamer" of adults. Also has a spine positioned prominently above the corner of the mouth. Body size is large, up to about 3 in (7.5 cm) in length, at metamorphosis from postlarvae to juveniles. Adults grow to at least 5.9 in (15 cm) in length.

DISTRIBUTION
Tropical and subtropical waters of the Indo-Pacific region from East Africa in the Indian Ocean east to the Hawaiian Islands, Rapa Island, and Ducie Atoll in the east-central Pacific and in southern Gulf of California south to Peru in the eastern Pacific. In the western Pacific, north to southern Japan, south to Australia and Lord Howe Island, and throughout Micronesia, Polynesia, and Melanesia.

HABITAT
Clear seaward coral and rocky reefs, reef flats, and turbid inner lagoons. Depth range from less than 9.8 ft (3 m) to more than 591 ft (180 m).

BEHAVIOR
Found singly, in pairs, or in small groups of three or more. Occasionally, large aggregations in excess of 150 fishes are seen; such aggregations may be for spawning.

FEEDING ECOLOGY AND DIET
Feeds on benthic encrusting invertebrates, especially sponges, that it plucks from the substrate.

REPRODUCTIVE BIOLOGY
Still poorly known. Pair-spawns at dusk on seaward reefs within small groups or, possibly, in larger aggregations. Eggs and larvae are pelagic, and larval life can be relatively long, thus potentially explaining their broad pattern of distribution as adults.

CONSERVATION STATUS
Not listed by the IUCN. Aquarium trade is restricted in Germany.

SIGNIFICANCE TO HUMANS
Collected for the aquarium trade but does not do well in most aquaria. Also taken in subsistence food fisheries. ◆

Resources

Eschmeyer, W. N., ed. *Catalog of Fishes.* 3 vols. San Francisco: California Academy of Sciences, 1998.

Helfman, Gene S., Bruce B. Collette, and Doug E. Facey. *The Diversity of Fishes.* Malden, MA: Blackwell Science, 1997.

Leis, J. M., and B. M. Carson-Ewart, eds. *The Larvae of Indo-Pacific Coastal Fishes: An Identification Guide to Marine Fish Larvae.* Boston: Brill, 2000.

Masuda, H., K. Amaoka, C. Araga, T. Uyeno, and T. Yoshino, eds. *The Fishes of the Japanese Archipelago.* Tokyo: Tokai University Press, 1984.

Myers, R. F. *Micronesian Reef Fishes: A Field Guide for Divers and Aquarists.* 3rd edition. Barrigada, Guam: Coral Graphics, 1999.

Nelson, J. S. *Fishes of the World.* 3rd edition. New York: John Wiley and Sons, 1994.

Randall, John E. *Surgeonfishes of the World.* Honolulu: Bishop Museum Press/Mutual Publishing, 2001.

Randall, John E., Gerald R. Allen, and Roger C. Steene. *Fishes of the Great Barrier Reef and Coral Sea.* Honolulu: Bishop Museum Press, 1996.

Smith, M. M., and P. C. Heemstra, eds. *Smiths' Sea Fishes.* Berlin: Springer-Verlag,1986.

Thresher, R. E. *Reproduction in Reef Fishes.* Neptune City, NJ: T. F. H. Publications, 1984.

Periodicals

Craig, P. C., J. H. Choat, L. M. Axe, and S. Saucerman. "Population Biology and Harvest of the Coral Reef Surgeonfish *Acanthurus lineatus* in American Samoa." *Fishery Bulletin* 95 (1997): 680–693.

Johnson, G. D. "Percomorph Phylogeny: Progress and Problems." *Bulletin of Marine Science* 52, no. 1 (1993): 3–28.

Woodland, D. J. "Revision of the Fish Family Siganidae with Descriptions of Two New Species and Comments on Distribution and Biology." *Indo-Pacific Fishes* 19 (1990): 1–136.

Organizations

IUCN/SSC Coral Reef Fishes Specialist Group. International Marinelife Alliance-University of Guam Marine Laboratory, UOG Station, Mangilao, Guam 96913 USA. Phone: (671) 735-2187. Fax: (671) 734-6767. E-mail: donaldsn@uog9.uog.edu Web site: <http://www.iucn.org/themes/ssc/sgs/sgs.htm>

Terry J. Donaldson, PhD

Scombroidei

(Barracudas, tunas, marlins, and relatives)

Class Actinopterygeii
Order Perciformes
Suborder Scomboidei
Number of families 6

Photo: A marlin leaps out of the water. (Photo by Scott Kerrigan/Corbis. Reproduced by permission.)

Evolution and systematics

The first modern definition of the scombroid fishes as a suborder was by Regan in 1909. He clearly separated scombroids from percoid families, such as the jacks (Carangidae) and dolphinfishes (Coryphaenidae). Regan recognized four divisions within the Scombroidei: Trichiuriformes (Gempylidae and Trichiuridae), Scombriformes (Scombridae), Luvariformes (Luvaridae), and Xiphiiformes (Xiphiidae, Istiophoridae, and three fossil families). The suborder was redefined by Collette et al. (1984), and the Luvaridae were shown to be a highly specialized oceanic member of the typically reef-associated surgeonfishes (Acanthuroidei) by Tyler et al. (1989). Collette et al. (1984) included the Scombrolabricidae in the Scombroidei as its most primitive member. An alternative definition of the Scombroidei proposed by Johnson in 1986 excludes the Scombrolabracoidei but includes the barracudas (Sphyraenidae) as the most primitive member of the Scombroidei.

The 6 families are Sphyraenidae (1 genus, 20 species), Trichiuridae (9 genera, 32 species), Gempylidae (16 genera, 23 species), Scombridae (15 genera, 53 species), Xiphiidae (1 genus, 1 species), and Istiophoridae (3 genera, 9 species).

A school of barracuda in defensive behavior surround a diver in the Bismarck Sea near Papua New Guinea. (Photo by Jeff Rotman/Photo Researchers, Inc. Reproduced by permission.)

Physical characteristics

Perciform fishes with epiotic bones of the skull separated from each other by the supraoccipital bone; gill membranes

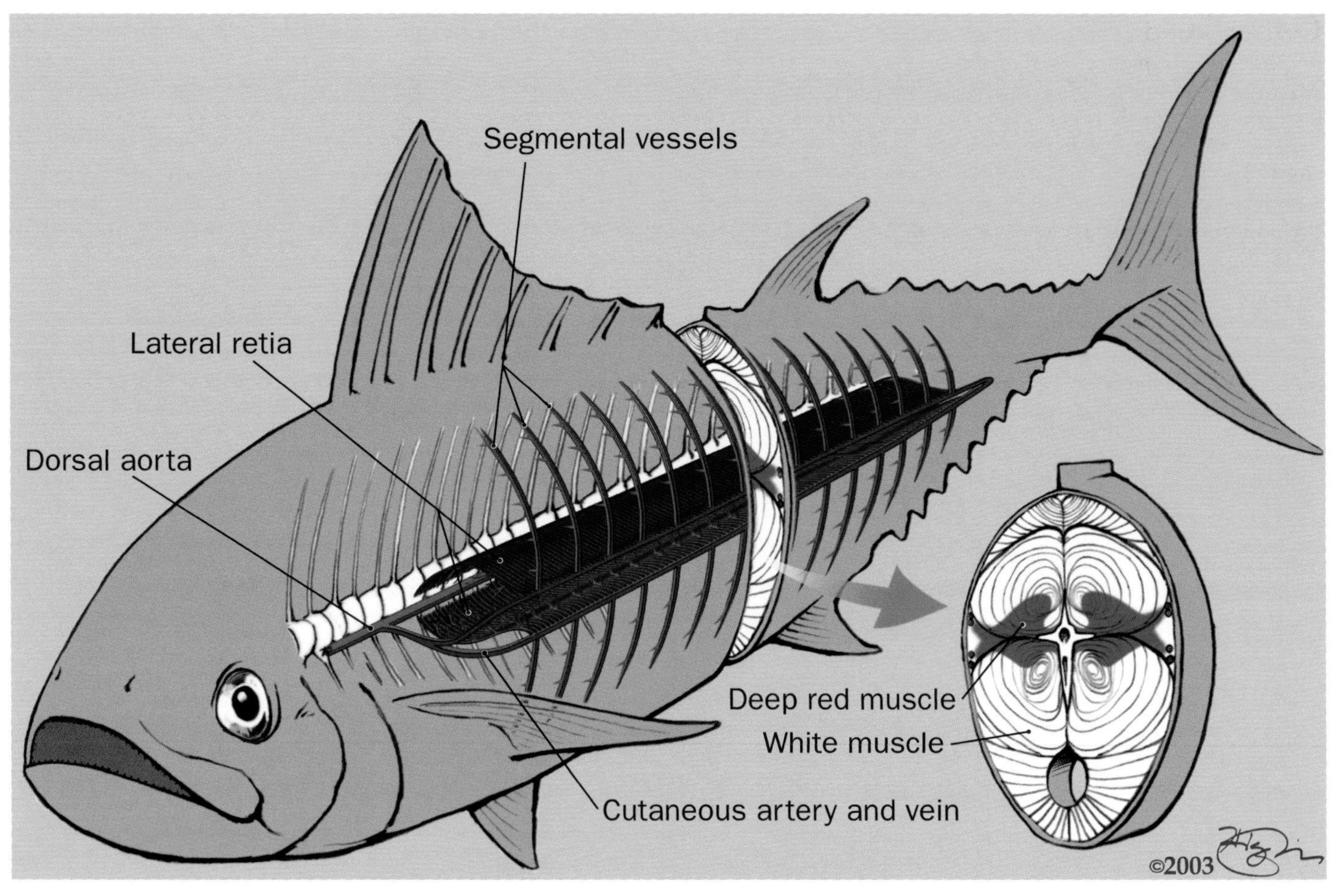

Tuna can regulate their temperature through blood flow. Increased blood flow in vessels near the skin allows heat to dissipate into the environment, cooling the fish. (Illustration by Jonathan Higgins)

free from the isthmus; premaxillae beak like, upper jaw not protrusile, predorsal bones lost (except for a small one in *Ruvettus*, *Thyrsites*, and *Tongaichthys*, and three well-developed ones in *Gasterochisma*); second epibranchial bone of pharyngeal arch extending over top of the third infrapharyngobranchial bone (except in *Gasterochisma*); 24 or more vertebrae; interorbital commissure of the supraorbital lateral line canals incomplete or absent

Distribution

These fishes are found in marine and estuarine waters of tropical and temperate oceans of the world. One species of Spanish mackerel moves long distances up the Mekong River.

Habitat

Most scombroid fishes are epipelagic; some are mesopelagic.

Behavior

Some scombroids (such as mackerels) form very large schools while others (such as wahoo and snake mackerels) are essentially solitary.

Feeding ecology and diet

Most scombroids are active predators, feeding on a wide variety of fishes, squids, and crustaceans, but some (such as the mackerels) filter small planktonic organisms out of the water with their long gillrakers.

Smaller species of scombroids, such as mackerels, fall prey to all larger predacious sea animals. Whales, porpoises, sharks, tunas, bonito, bluefish, and striped bass take a heavy toll. Cod often eat small mackerel; squids destroy great numbers of young fish; and seabirds of various kinds follow and prey upon the schools when these are at the surface.

Medium-sized scombroids such as bonitos and skipjack are preyed upon by tunas, billfishes, and sharks. Predators of large tunas like the bluefin include killer whales and such sharks as the white shark and the mako shark. Large billfishes are so active and powerful that they have few enemies but are preyed upon by killer and sperm whales and mako sharks.

Reproductive biology

Scombroids are dioecious (separate sexes) and most display little or no sexual dimorphism in structure or color pattern. Batch spawning of most species takes place in tropical and subtropical waters, frequently inshore. The eggs are pelagic and hatch into planktonic larvae.

Conservation status

Populations of three species, swordfish (*Xiphias gladius*), Atlantic bluefin tuna (*Thunnus thynnus*), and Atlantic albacore (*Thunnus alalunga*), have been greatly reduced by fishing. All three are listed by the IUCN as Data Deficient. In addition, the IUCN categorizes the Monterrey Spanish mackerel (*Scomberomorus concolor*) as Endangered, the southern bluefin tuna (*Thunnus maccoyii*) as Critically Endangered, and the bigeye tuna (*Thunnus obesus*) as Vulnerable.

Significance to humans

Many species of scombroid fishes are of great importance as food fishes—mackerel, Spanish mackerel, bonito, and tuna—and as sport fishes—Spanish mackerel, tuna, swordfish, and marlin.

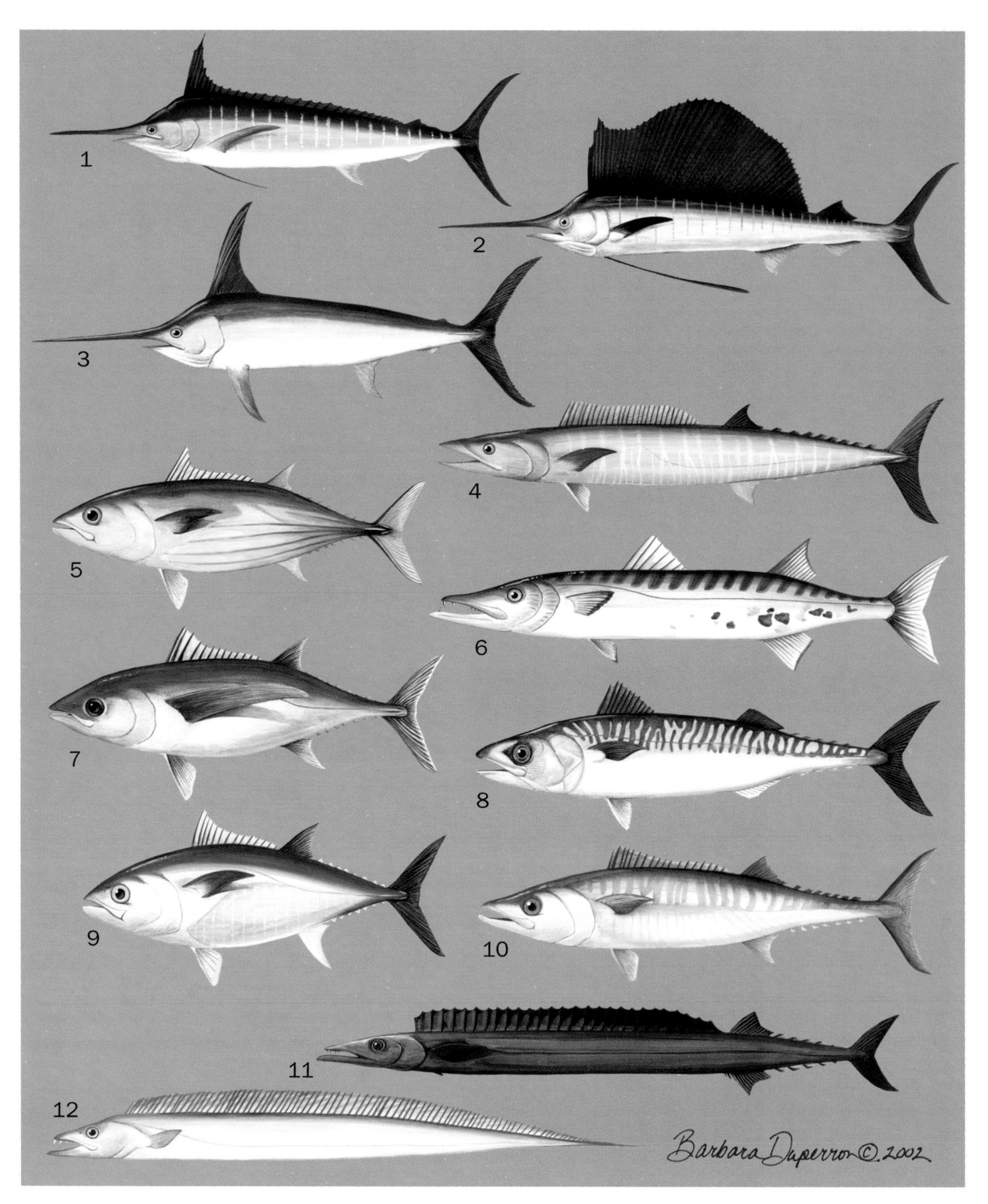

1. Blue marlin (*Makaira nigricans*); 2. Sailfish (*Istiophorus platypterus*); 3. Swordfish (*Xiphias gladius*); 4. Wahoo (*Acanthocybium solandri*); 5. Skipjack tuna (*Katsuwonus pelamis*); 6. Great barracuda (*Sphyraena barracuda*); 7. Albacore (*Thunnus alalunga*); 8. Atlantic mackerel (*Scomber scombrus*); 9. Atlantic bluefin tuna (*Thunnus thynnus*); 10. King mackerel (*Scomberomorus cavalla*); 11. Snake mackerel (*Gempylus serpens*); 12. Largehead hairtail (*Trichiurus lepturus*). (Illustration by Barbara Duperron)

Species accounts

Snake mackerel
Gempylus serpens

FAMILY
Gempylidae

TAXONOMY
Gempylus serpens Cuvier, 1829, Tropic of Cancer.

OTHER COMMON NAMES
French: Escolier serpent; Spanish: Escolar de canal.

PHYSICAL CHARACTERISTICS
Maximum length approximately 40 in (1 m), commonly to 24 in (60 cm). Body greatly elongate and strongly compressed. Tips of upper and lower jaws with cartilaginous processes. Three immovable and zero to three movable fangs anteriorly in upper jaw. First dorsal fin long with 26–32 spines, second dorsal fin with 11–14 soft rays followed by five or six finlets. Caudal fin well developed. Scales absent except on posterior part of body.

DISTRIBUTION
Worldwide in tropical and subtropical seas; adults also caught in temperate waters.

HABITAT
Strictly oceanic, epipelagic and mesopelagic from the surface to depths of 656 ft (200 m).

BEHAVIOR
Usually solitary. Adults migrate to the surface at night; larvae and juveniles stay near the surface only during the day.

FEEDING ECOLOGY AND DIET
Feed on fishes such as lanternfishes, flyingfishes, sauries, and scombrids and on squids and crustaceans.

REPRODUCTIVE BIOLOGY
Males mature at approximately 17 in (43 cm) standard length, females at 20 in (50 cm). Spawn in tropical waters throughout the year. Fecundity is estimated at 300,000 to one million eggs.

CONSERVATION STATUS
Not threatened.

SIGNIFICANCE TO HUMANS
There is no directed fishery for snake mackerel, but it sometimes appears as bycatch in the tuna long-line fishery. ◆

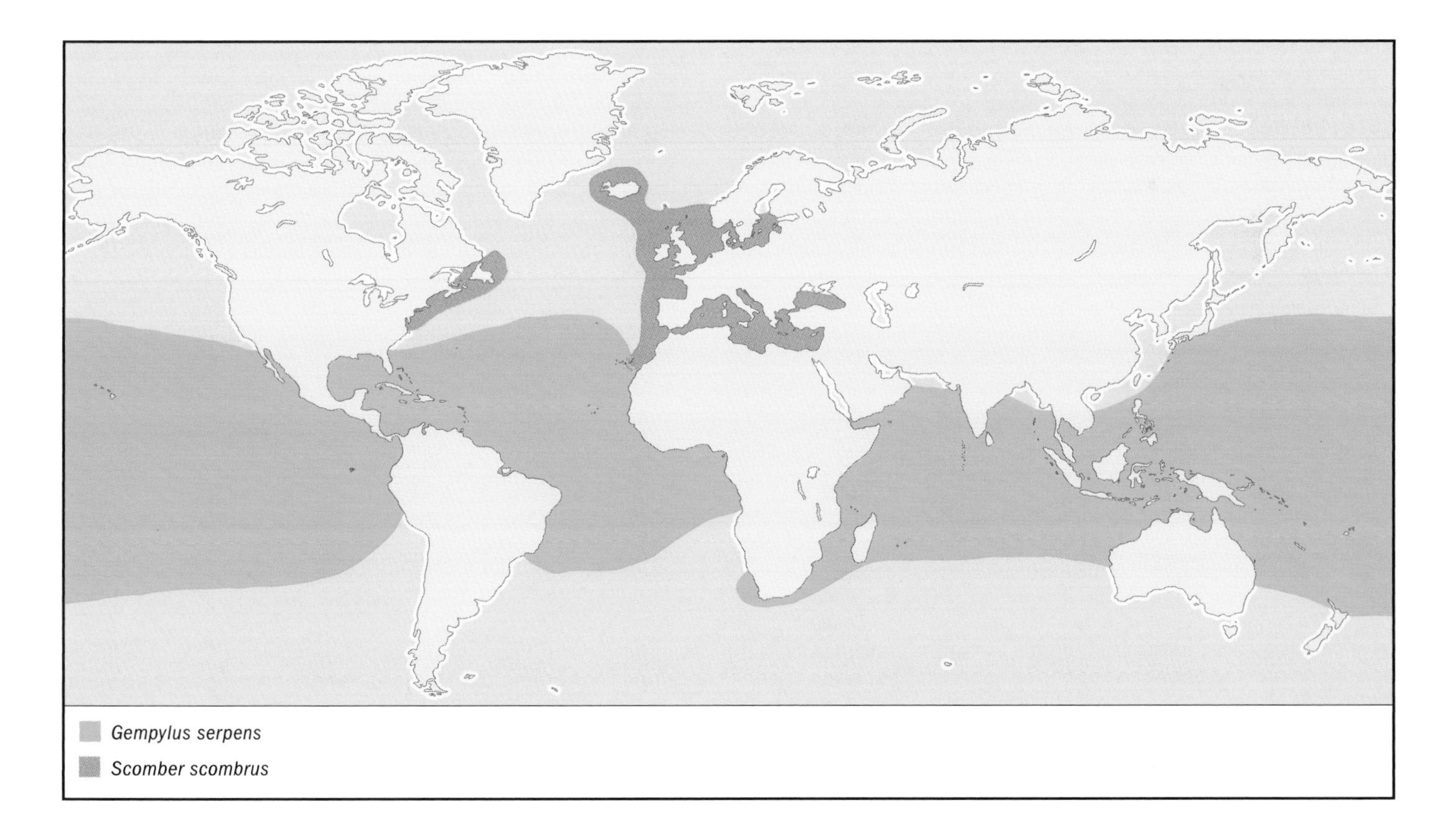

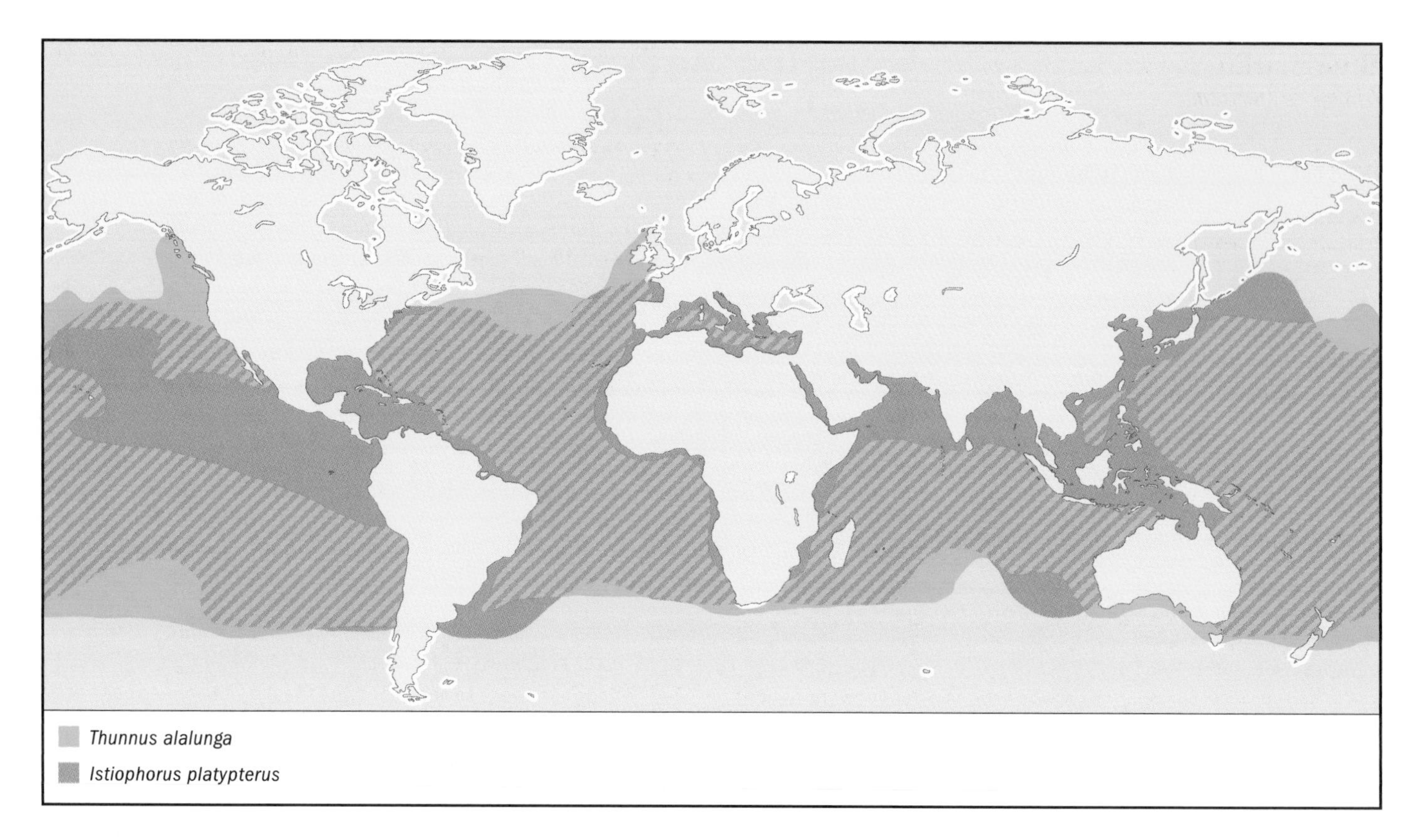

Sailfish

Istiophorus platypterus

FAMILY
Istiophoridae

TAXONOMY
Istiophorus platypterus (Shaw, 1792), Indian Ocean. Some authors, such as Nakamura (1985), differentiate the Atlantic sailfish, *I. albicans* (Latreille, 1804), as a separate species.

OTHER COMMON NAMES
French: Voilier; Spanish: Pez vela.

PHYSICAL CHARACTERISTICS
Sailfishes reach a maximum size of approximately 11 ft (3.5 m) total length and 220 lb (100 kg). Body fairly compressed. Two dorsal and two anal fins, the first dorsal fin sail-like and remarkably higher than greatest body depth, with 42–49 rays. Second dorsal fin with six or seven rays, slightly posterior to second anal fin which also has six or seven rays. Pectoral fins moderate, 18–20 rays. Pelvic fins extremely long, almost reaching the anus, depressable into a groove, with one spine and several rays tightly fused together. Jaws and palatine bones with small, file-like teeth. Gill rakers absent. Left and right branchiostegal membranes broadly united. Vertebrae, 24. Swim bladder made up of many small bubble-shaped chambers.

DISTRIBUTION
Most researchers consider the sailfish to be a single pantropical species occurring in all three major oceans.

HABITAT
An epipelagic and oceanic species, usually found above the thermocline. Sailfishes have a strong tendency to approach continental coasts, islands, and reefs.

BEHAVIOR
Sailfishes occasionally form schools or smaller groups of 3–30 individuals, but more often occur in loose aggregations over a wide area.

FEEDING ECOLOGY AND DIET
Feeding behavior has been observed by fishermen: One or several sailfish locate a school of prey fish (such as sardines, anchovies, mackerel, or jack mackerel) and pursue the school at half speed with their fins half-folded back into the grooves. They then drive at the prey at full speed with fins completely folded back and make sharp turns with fins expanded to confront part of the school and strike the prey with their bills. They feed on the killed or stunned fish, usually head first as well as on a variety of fishes, crustaceans, and squid.

REPRODUCTIVE BIOLOGY
Spawning occurs with males and females swimming in pairs or with two or three males chasing one female. Sailfishes spawn throughout the year in tropical and subtropical waters with peak spawning in local summer seasons. Ripe ovarian eggs are approximately 0.03 in (0.85 mm) in diameter and have a single oil globule. Eggs shed from a captured female averaged 0.05 in (1.30 mm) in diameter.

CONSERVATION STATUS
Not threatened.

SIGNIFICANCE TO HUMANS
Sailfishes are often taken as bycatch by the commercial surface tuna long liners. United Nations Food and Agriculture Organization (FAO) catch statistics for 1991–2000 show catches of 11.1–23.7 thousand tons (10.1–21.5 thousand metric tons) per year by 42 countries. Sailfishes are primarily important as a sportsfish taken by trolling at the surface. The all-tackle gamefish record is a 221-lb (100.2-kg) fish taken off Santa Cruz Island, Ecuador. ◆

Blue marlin
Makaira nigricans

FAMILY
Istiophoridae

TAXONOMY
Makaira nigricans Lacepede, 1802, Bay of Biscay. Some authors, such as Nakamura (1985), differentiate the Indo-Pacific blue marlin, *M. mazara* (Jordan and Snyder, 1901), as a separate species.

OTHER COMMON NAMES
French: Makaire bleu; Spanish: Aguja azul.

PHYSICAL CHARACTERISTICS
Blue marlin reach approximately 16 ft (5 m) in total length and weigh more than 1,984 lb (900 kg). Body not very compressed, nape highly elevated, body deepest at level of pectoral fins. Two dorsal and two anal fins. First dorsal fin with 39–43 rays; height of anterior lobe less than greatest body depth, not sail-like. Second dorsal fin with six or seven rays, slightly posterior to second anal fin. Pectoral fins with 20–23 rays, depressible against sides of the body. Pelvic fins shorter than pectoral fins with one spine and two rays. Lateral line looped. Body covered with densely imbedded scales, each with one or two long, acute spines. Left and right branchiostegal membranes broadly united. Vertebrae, 24. Swim bladder made up of many small bubble-shaped chambers. Body blue dorsally, silvery white ventrally, first dorsal fin membrane blue-black, unspotted; body has approximately 15 obscure vertical light bars.

DISTRIBUTION
Most researchers consider the blue marlin to be a single pantropical species occurring in all three major oceans. Blue marlin are the most tropical of the billfishes, chiefly distributed in equatorial areas.

HABITAT
Epipelagic zone of oceans, usually waters with surface temperatures of 71.6–87.8°F (22–31°C).

BEHAVIOR
Observations suggest that marlins use their bills to stun their prey.

FEEDING ECOLOGY AND DIET
Feed mostly in near-surface waters but sometimes make trips to relatively deep waters for feeding. Prey includes dolphinfishes, tuna-like fishes, particularly frigate tunas, and squids.

REPRODUCTIVE BIOLOGY
Little is known about spawning grounds or seasons. The eggs are very small, approximately 0.04 in (1 mm) in diameter, and pelagic, hatching into planktonic larvae.

CONSERVATION STATUS
Not threatened.

SIGNIFICANCE TO HUMANS
Being excellent foodfishes, all species of marlin are of some importance to fisheries, particularly in Japan. They are mostly caught on long lines incidental to the target tuna species. FAO catch statistics for 1991–2000 show catches of 26.7–37.3 thousand tons (24.2–33.8 thousand metric tons) per year by 34 countries. Blue marlin are especially important to sportfishermen and are much sought after off Cuba and on the Bahamas side of the Straits of Florida. The all-tackle gamefish record is a 1,402-lb (636-kg) fish taken off Vitória, Brazil. ◆

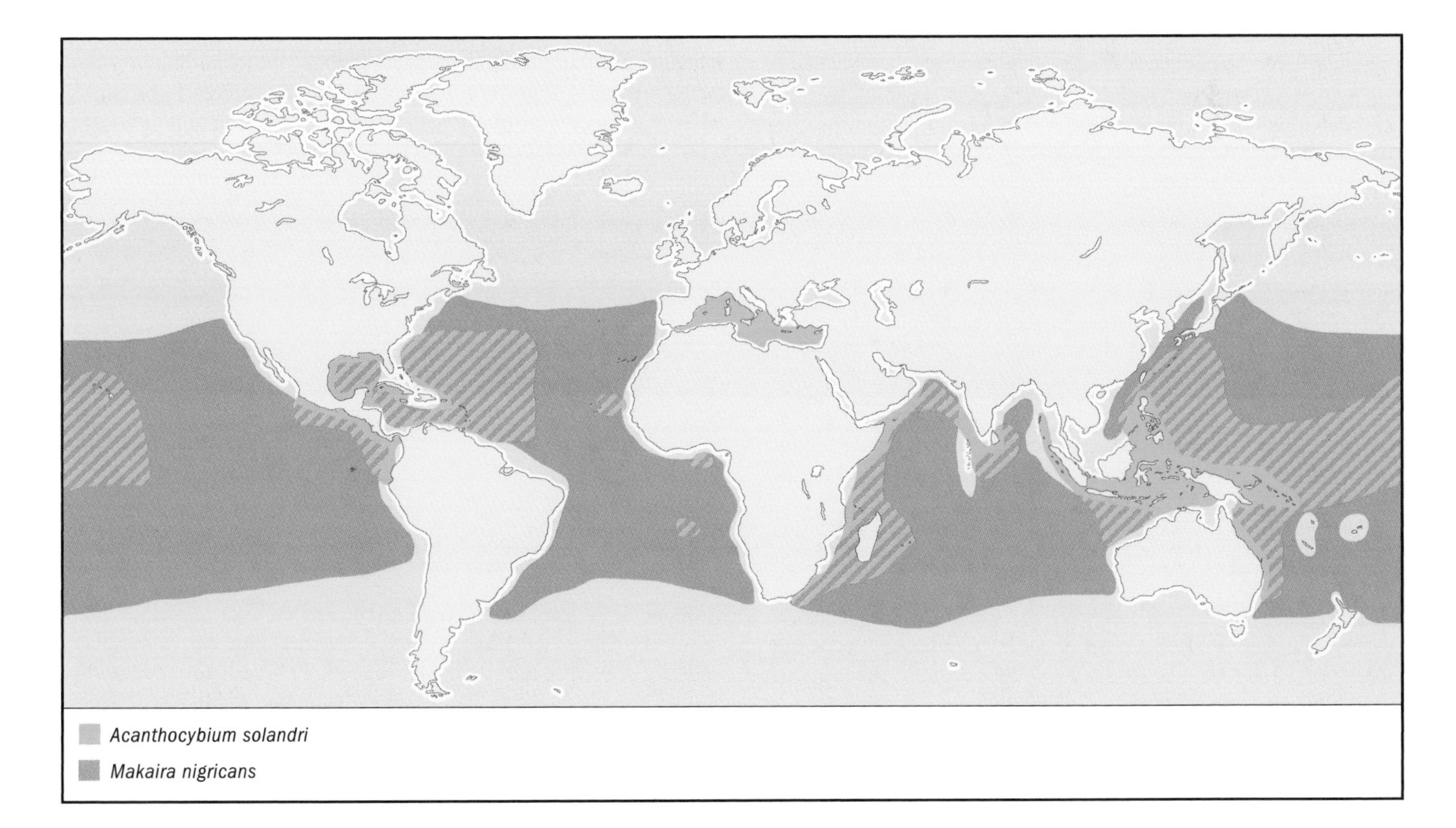

Wahoo

Acanthocybium solandri

FAMILY
Scombridae

TAXONOMY
Acanthocybium solandri (Cuvier, 1832), type locality unknown.

OTHER COMMON NAMES
French: Thazard-bâtard; Spanish: Peto.

PHYSICAL CHARACTERISTICS
Maximum size 83 in (210 cm) fork length weighing 183 lb (83 kg) or more. Body very elongate, fusiform, and only slightly compressed. Mouth large with strong, triangular, compressed, and finely serrate teeth closely set in a single series. Snout approximately as long as rest of head. Gill rakers absent. Two dorsal fins, the first consisting of 23–27 spines, the second with 12–16 rays followed by eight or nine finlets. Anal fin of 12–14 rays followed by nine finlets. Body covered with small scales. Swim bladder present. Vertebrae, 62–64. Back iridescent bluish green, sides silvery with 24–30 cobalt-blue vertical bars.

DISTRIBUTION
Tropical and subtropical waters of the Atlantic, Pacific, and Indian oceans including the Caribbean and Mediterranean seas.

HABITAT
Epipelagic zone ocean.

BEHAVIOR
Frequently solitary or forming small, loose aggregations rather than compact schools.

FEEDING ECOLOGY AND DIET
Piscivorous, preying on pelagic fishes such as scombrids, porcupinefishes, flyingfishes, herrings, scads, and lanternfishes and on squids.

REPRODUCTIVE BIOLOGY
Spawning seems to extend over a long period. Fish in different maturity stages are frequently caught at the same time. Fecundity is believed to be quite high; 6 million eggs were estimated for a 52-in (131-cm) female.

CONSERVATION STATUS
Not threatened.

SIGNIFICANCE TO HUMANS
There do not appear to be any organized fisheries for wahoo, but it is greatly appreciated when caught. FAO catch statistics for 1991–2000 show catches of 2,121–3,460 tons (1,924–3,139 metric tons) per year by 30 countries. In many areas (Caribbean, Hawaii, Great Barrier Reef), wahoo is more important as a gamefish taken on light to heavy tackle through surface trolling with spoons, feather lures, or strip bait. The all-tackle gamefish record is a 159-lb (71.9-kg) fish taken off Baja California. ◆

Skipjack tuna

Katsuwonus pelamis

FAMILY
Scombridae

TAXONOMY
Katsuwonus pelamis (Linnaeus, 1758), "pelagic, between the tropics."

OTHER COMMON NAMES
French: Bonite à ventre rayé; Spanish: Listado.

PHYSICAL CHARACTERISTICS
Maximum fork length approximately 43 in (108 cm) corresponding to a weight of 72–76 lb (32.5–34.5 kg), commonly to 31 in (80 cm) and 18–22 lb (8–10 kg). Body fusiform, elongate, and rounded. Two dorsal fins separated by a narrow interspace, the first with 14–16 spines, the second dorsal and anal fins followed by seven to nine finlets. Pectoral fins short, with 26 or 27 rays. Body naked except for anterior corselet and lateral line. Caudal peduncle very slender with a strong lateral keel between two smaller keels. Swim bladder absent. Gill rakers numerous, 53–63 on first gill arch. Back dark purplish blue, lower sides and belly silvery, with four to six very conspicuous longitudinal dark bands.

DISTRIBUTION
Cosmopolitan in tropical and warm-temperate seas but absent from the Black Sea.

HABITAT
An epipelagic oceanic species with adults distributed within the 59°F (15°C) isotherm. Aggregations of this species tend to be associated with convergences, boundaries between cold and warm water masses. Depth distribution ranges from the surface to about 853 ft (260 m) during the day.

BEHAVIOR
Skipjack show a strong tendency to school in surface waters. Schools are associated with birds, drifting objects, sharks, whales, and other tuna species.

FEEDING ECOLOGY AND DIET
Skipjack are opportunistic feeders preying on any forage available. Feeding activity peaks in early morning and late afternoon. Food items include fishes, crustaceans, and mollusks.

REPRODUCTIVE BIOLOGY
Skipjack spawn in batches throughout the year in equatorial waters and from spring to early fall in subtropical waters. The spawning season becomes shorter as distance from the equator increases. Fecundity increases with size but is highly variable. The number of eggs per season in females 16–24 in (41–87 cm) fork length ranges from 80,000 to two million.

CONSERVATION STATUS
Not threatened.

SIGNIFICANCE TO HUMANS
Skipjack make up approximately 40% of the world's total tuna catch and have come to replace yellowfin as the dominant commercial species of tuna. Catches of skipjack were reported to FAO by 94 countries for 1991–2000, 1,584–2,191 thousand tons (1,437–1,988 thousand metric tons) per year. The highest catches reported for 2000 were by Japan, 376 thousand tons (341 thousand metric tons), and Indonesia, 298 thousand tons (270 thousand metric tons). Skipjack are taken at the surface, mostly with purse seines and pole-and-line gear, occasionally with long lines. They are marketed fresh, frozen, and canned (as light-meat tuna). They are also a game fish, the all-tackle gamefish record is a 45-lb (20.5-kg) fish caught on Flathead Bank, Baja California. ◆

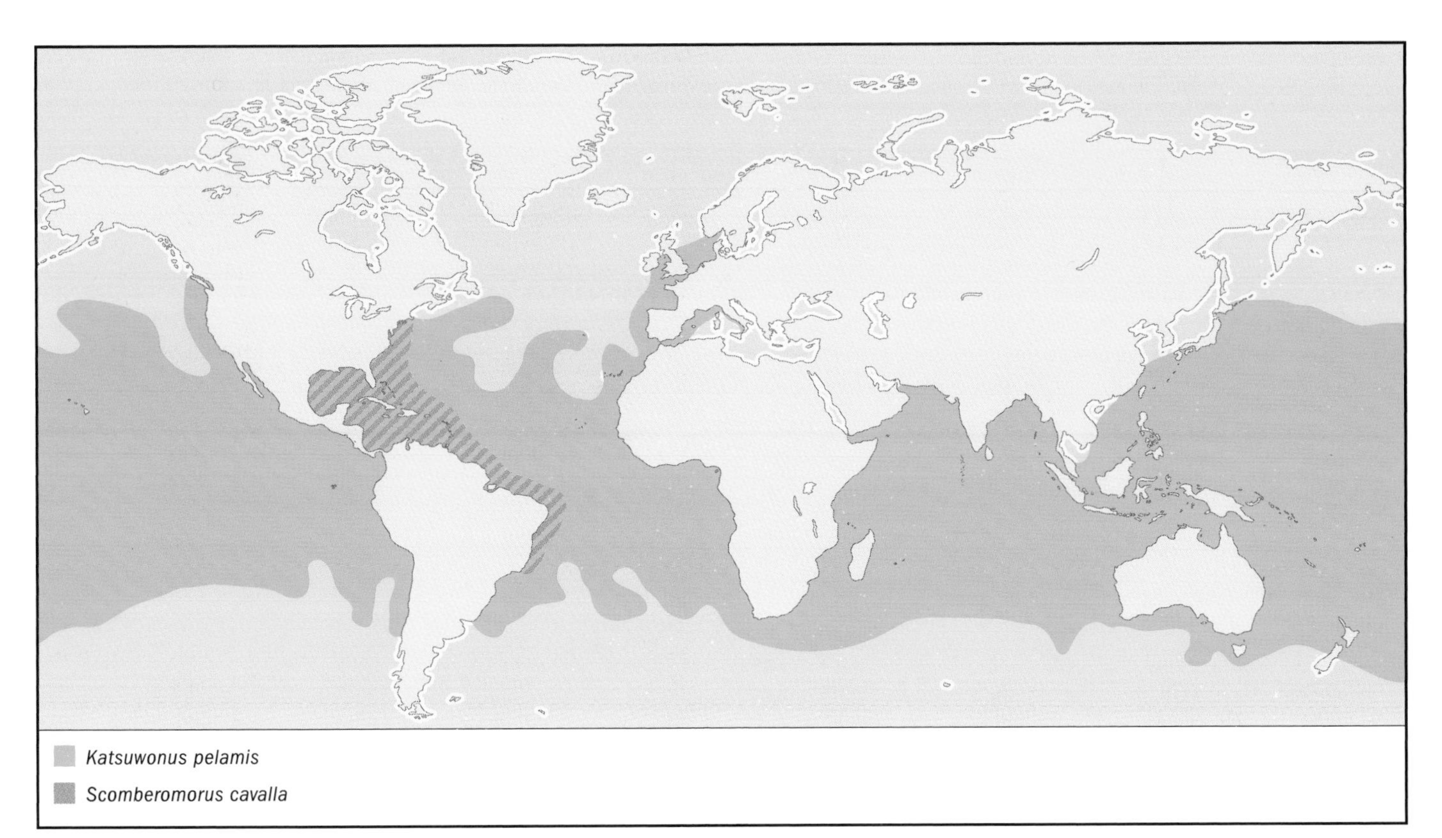

Atlantic mackerel

Scomber scombrus

FAMILY
Scombridae

TAXONOMY
Scomber scombrus Linnaeus, 1758, Atlantic Ocean.

OTHER COMMON NAMES
French: Maquereau commun; Spanish: Caballa del Atlantico.

PHYSICAL CHARACTERISTICS
Maximum fork length is 22 in (56 cm), commonly to 19 in (30 cm). Body fusiform, tapering rearward to a very slim caudal peduncle and anteriorly to a pointed snout. Eyes large with so-called adipose eyelids covering front and hind margins of the eye except for a slit over the pupil. Mouth large, filled with small, sharp, conical teeth in upper and lower jaws. Gill rakers, 30–36 on lower limb of first gill arch. Two widely separated dorsal fins, the first with 11–13 spines, the second with 11 or 12 rays. Both second dorsal and anal fins followed by five finlets. Entire body covered with small scales. Swim bladder usually absent. Vertebrae, 31. Upper surface dark steely to greenish blue. Body barred with 23–33 dark transverse bands; belly unmarked.

DISTRIBUTION
North Atlantic Ocean, including the Baltic, Mediterranean, and Black seas. Replaced by other species of *Scomber* in other oceans.

HABITAT
An epipelagic and mesodemersal species, most abundant in cold and temperate shelf areas.

BEHAVIOR
Mackerel are swift-moving, swimming with very short sidewise movements of the rear part of the body and the powerful caudal fin. They need so much oxygen that they must swim constantly to bring sufficient water across their gill filaments. Mackerel gather in dense schools of fish that are approximately the same size and age. They overwinter in moderately deep water along the continental shelf and move inshore and northward in the spring.

FEEDING ECOLOGY AND DIET
Atlantic mackerel are opportunistic carnivores that swallow their food whole. Food is captured by active pursuit or by passive filtration with the gill rakers. Juveniles feed on zooplankton. Adults also eat crustaceans but add squids and small fishes to their diet.

REPRODUCTIVE BIOLOGY
In the western North Atlantic, spawning takes place from Chesapeake Bay to Newfoundland, beginning in the south in spring and progressively extending northward into the summer. Most spawning occurs within 10–30 mi (16–48 km) of shore. Mackerel do not begin spawning until the water has warmed to approximately 46.4°F (8°C). The chief production of eggs takes place at temperatures of 48.2–57.2°F (9–14°C). Maturity is attained at 2–3 years of age. Estimates of fecundity range from 285,000 to 1.98 million eggs for females 12–17 in (307–438 mm) fork length. The eggs are 0.04–0.05 in (1.09–1.39 mm) in diameter, have one oil globule, and generally float in the surface water layer above the thermocline.

CONSERVATION STATUS
Not threatened.

SIGNIFICANCE TO HUMANS
Mackerel are delicious fish but do not keep as well as fishes that have less oil in their tissues. There are important fisheries in the northwest Atlantic, northeast Atlantic, and Mediterranean and Black seas. FAO catch statistics for 1991–2000 show catches of 6.16–9.42 thousand tons (5.59–8.55 thousand metric tons) per year by 36 countries. Atlantic mackerel are caught mainly with purse seines. Mackerel also are taken by anglers; the all-tackle world record is a 3-lb (1.2-kg) fish taken off Norway. ◆

King mackerel
Scomberomorus cavalla

FAMILY
Scombridae

TAXONOMY
Scomberomorus cavalla (Cuvier, 1829). Based on Marcgrave's guarapucu, Brazil.

OTHER COMMON NAMES
French: Thazard barre; Spanish: Carite lucio.

PHYSICAL CHARACTERISTICS
Maximum size 98 in (250 cm) fork length and 79–99 lb (36–45 kg), commonly to 28 in (70 cm). Body elongate, strongly compressed. Snout much shorter than rest of head. Posterior portion of maxilla exposed, reaching to a vertical with hind margin of eye. Gill rakers on first arch few, 6–11, usually 8–10. Two dorsal fins, scarcely separated, the first with 14–16 spines. Eight or nine dorsal finlets; 9 or 10 anal finlets. Lateral line abruptly curving downward below second dorsal fin. Body completely covered with small scales; no anterior corselet developed. Vertebrae, 41–43. Back iridescent bluish green, sides silvery, adults do not have bars or spots, but juveniles may have spots on sides. Unlike many other species of Spanish mackerels, adults have no black area in the first dorsal fin.

DISTRIBUTION
Confined to the western Atlantic Ocean from Massachusetts to Rio de Janeiro.

HABITAT
Epipelagic and neritic, often found in outer reef areas.

BEHAVIOR
King mackerel are present throughout the year off Louisiana in the Gulf of Mexico, southern Florida, and the state of Ceará in northeastern Brazil. Large schools of king mackerel migrate north along the coast of the United States in spring and return south during the fall.

FEEDING ECOLOGY AND DIET
As for other members of the genus, food consists largely of fishes and small quantities of penaeid shrimps and squids. Various herring-like fishes are particularly important components of the diet.

REPRODUCTIVE BIOLOGY
Spawning takes place from May through September in the western Gulf of Mexico, particularly in September at depths between 115 and 591 ft (35 and 180 m) over the middle and outer continental shelf. In Brazil, fecundity of females 25–48 in (63–123 cm) fork length ranges from 345,000 to 2,280,000 eggs.

CONSERVATION STATUS
Not threatened.

SIGNIFICANCE TO HUMANS
King mackerel is an important species for recreational, commercial, and artisanal fisheries throughout their range. Eight nations reported to FAO catches of 6.0–13.1 thousand tons (5.4–11.9 thousand metric tons) per year from 1991 to 2000, but several other countries combine king mackerel with other species of Spanish mackerel, so the total catch and the number of countries fishing for this species is higher. Commercial fisheries in the southeastern United States use hook-and-line, snapper hooks, gill nets, and trolled lure or small cut bait. In Brazil, gill nets and trolling are major ways of catching king mackerel. In the United States, sport fishing with hook-and-line is practiced in North Carolina, Florida, Mississippi, and Louisiana. The all-tackle world game fish record is a 94-lb (42.18-kg) fish caught off San Juan, Puerto Rico. ◆

Albacore
Thunnus alalunga

FAMILY
Scombridae

TAXONOMY
Thunnus alalunga (Bonnaterre, 1788), Sardinia.

OTHER COMMON NAMES
French: Germon; Spanish: Atún blanco.

PHYSICAL CHARACTERISTICS
Maximum fork length 50 in (127 cm). A large tuna, deepest at a more posterior point than in other tuna species, at or only slightly anterior to second dorsal fin. Two dorsal fins, separated by only a narrow interspace, the first with 11–14 spines, the second with 12–16 rays; anal fin with 11–16 rays; both second dorsal and anal fins followed by seven to nine finlets. Pectoral fins remarkably long, usually 30% of fork length or longer, reaching posteriorly well beyond origin of second dorsal fin. Teeth small and conical, in a single series. Gill rakers, 25–31. Caudal peduncle very slender with a strong lateral keel between two smaller keels. Corselet of large scales anteriorly; rest of body covered with small scales. Swim bladder present. Ventral surface of liver striated. Back metallic dark blue; lower sides and belly silvery white; second dorsal and anal fins yellow; anal finlets dark; white margin on posterior margin of caudal fin.

DISTRIBUTION
Cosmopolitan in tropical and temperate waters of all oceans and Mediterranean Sea, north to 45–50°N, south to 30–40°S. Offshore often extend into cooler waters.

HABITAT
Epipelagic and mesopelagic zones of ocean; abundant in surface waters of 60.1–66.9°F (15.6–19.4°C). Deeper-swimming large albacore are found in waters of 55.9–59.4°F (13.3–15.2°C).

BEHAVIOR
Throughout their range, albacore migrate over great distances and appear to form separate groups at different stages of the life cycle. At least two stocks, northern and southern, are believed to exist in both the Atlantic and Pacific oceans.

FEEDING ECOLOGY AND DIET
Feeds on crustaceans, fishes, and squids.

REPRODUCTIVE BIOLOGY
Albacore tend to spawn in subtropical waters, although they do spawn in tropical waters in some places. Albacore mature at approximately 5 years of age and 35 in (90 cm) fork length in the Atlantic Ocean. Fecundity increases with size generally. A 44-lb (20-kg) female may produce 2–3 million eggs per season released in at least two batches.

CONSERVATION STATUS
Listed as Data Deficient by IUCN.

SIGNIFICANCE TO HUMANS
There are important fisheries for albacore in the Atlantic and Pacific oceans. FAO catch statistics for 1991–2000 show catches of 185–280 thousand tons(168–254 thousand metric tons) per year by 59 countries. The highest landings reported for 2000 were by Japan, 69 thousand tons (62.6 thousand metric tons), and Taiwan, 57.0 thousand tons (51.7 thousand metric tons). With increasing effort in surface fisheries, the world catch has been gradually declining, particularly in the Atlantic Ocean. Albacore are caught by four types of fishing operations: long lining, live-bait fishing, trolling, and purse seining. Albacore is packed as "white-meat" tuna. The all-tackle game fish record is an 88-lb (40.0-kg) fish caught in the Canary Islands. ◆

Atlantic bluefin tuna
Thunnus thynnus

FAMILY
Scombridae

TAXONOMY
Thunnus thynnus (Linnaeus, 1758). Pacific bluefin were considered a subspecies of *T. thynnus* for many years but have recently been raised to species level (Collette et al., 2001) as *T. orientalis* Temminck and Schlegel, 1844.

OTHER COMMON NAMES
English: Northern bluefin tuna; French: Thon rouge; Spanish: Atún.

PHYSICAL CHARACTERISTICS
Maximum fork length more than 118 in (300 cm), commonly to 79 in (200 cm). A very large tuna, deepest near middle of first dorsal fin base. Two dorsal fins, separated by only a narrow interspace, the first with 11–14 spines, the second with 12–16 rays; anal fin with 11–16 rays, both second dorsal and anal fins followed by 7–10 finlets. Pectoral fins very short, less than 80% of head length, never reaching the interspace between the dorsal fins. Teeth small and conical in a single series. Gill rakers, 34–41. Caudal peduncle very slender with a strong lateral keel between two smaller keels. Corselet of large scales anteriorly; rest of body covered with small scales. Swim bladder large. Ventral surface of liver striated. Back metallic dark blue, lower sides and belly silvery white; first dorsal fin yellow or bluish, second dorsal reddish brown, anal fin silvery gray, anal finlets dusky yellow edged with black; no white margin on posterior margin of caudal fin.

DISTRIBUTION
North Atlantic Ocean from Labrador and Newfoundland south into Gulf of Mexico and Caribbean Sea. Replaced by the closely related Pacific bluefin tuna in the north Pacific.

HABITAT
Epipelagic, usually oceanic but seasonally coming very close to shore. Bluefin are found in moderately warm seas but are more tolerant of cold water than are most of their relatives. Offshore in the northwest Atlantic, large bluefin are taken at surface temperatures of 43.5–83.8°F (6.4–28.8°C). Tunas have evolved elaborate rete mirabilia, "wonder nets," of capillaries that act as countercurrent heat exchangers. These heat exchangers form thermal barriers that prevent metabolic heat loss and enable bluefin to maintain a high internal temperature, as high as 83.8°F (28.8°C) for a bluefin taken in 45.1°F (7.3°C) water.

BEHAVIOR
Atlantic bluefin migrate long distances from their spawning grounds off Florida in the western Atlantic and in the Mediterranean Sea in the eastern Atlantic. Tag returns show there is some mixing between eastern and western Atlantic but there is ongoing debate about the proportion of individuals that cross the ocean.

FEEDING ECOLOGY AND DIET
Feed on a variety of fishes, crustaceans, and squids.

REPRODUCTIVE BIOLOGY
Onset of maturity is at approximately 4–5 years. Large adults (10 years and older) spawn in the Gulf of Mexico and in the Mediterranean Sea. Females weighing 592–661 lb (270–300 kg) may produce as many as 10 million eggs per spawning season.

CONSERVATION STATUS
Listed by IUCN as Data Deficient. Western Atlantic bluefin were fished intensively in the 1960s by purse seiners targeting small fish for canneries (Safina, 2001). Obvious depletion led to reduction in the east coast purse seine fleet. Western Atlantic bluefin catches averaged approximately 8,818 tons (8,000 metric tons) during the 1960s and 6,062 tons (5,500 metric tons) during the 1970s. During the 1970s, commercial targeting switched to large fish for export to Japan for sashimi. The offer to buy giant bluefin at $1.45 per pound (0.5 kg) (instead of the previous $0.20 to $0.50 per pound in autumn 1972) greatly increased U.S. fishing pressure on giant bluefin.

SIGNIFICANCE TO HUMANS
Important as both a food fish and a sport fish. FAO catch statistics for 1991–2000 show catches of 29.7–60.2 thousand tons (26.9–54.6 thousand metric tons) per year by 44 countries. The belly meat of bluefin, when containing much fat, reaches astronomical prices in the Japanese market for sashimi. Individual bluefin in prime condition have sold for as much as $68,000, approximately $45 per pound (0.5 kg), but a new record price of $173,600 (¥20 million) was reached for a 444-lb (201-kg) bluefin sold in Tokyo's Tsukiji Central Fish Market in January 2001. The all-tackle game fish record for a "giant" bluefin is a 1,497-lb (679-kg) fish taken off Nova Scotia. ◆

Great barracuda
Sphyraena barracuda

FAMILY
Sphyraenidae

TAXONOMY
Sphyraena barracuda (Walbaum, 1792), West Indies.

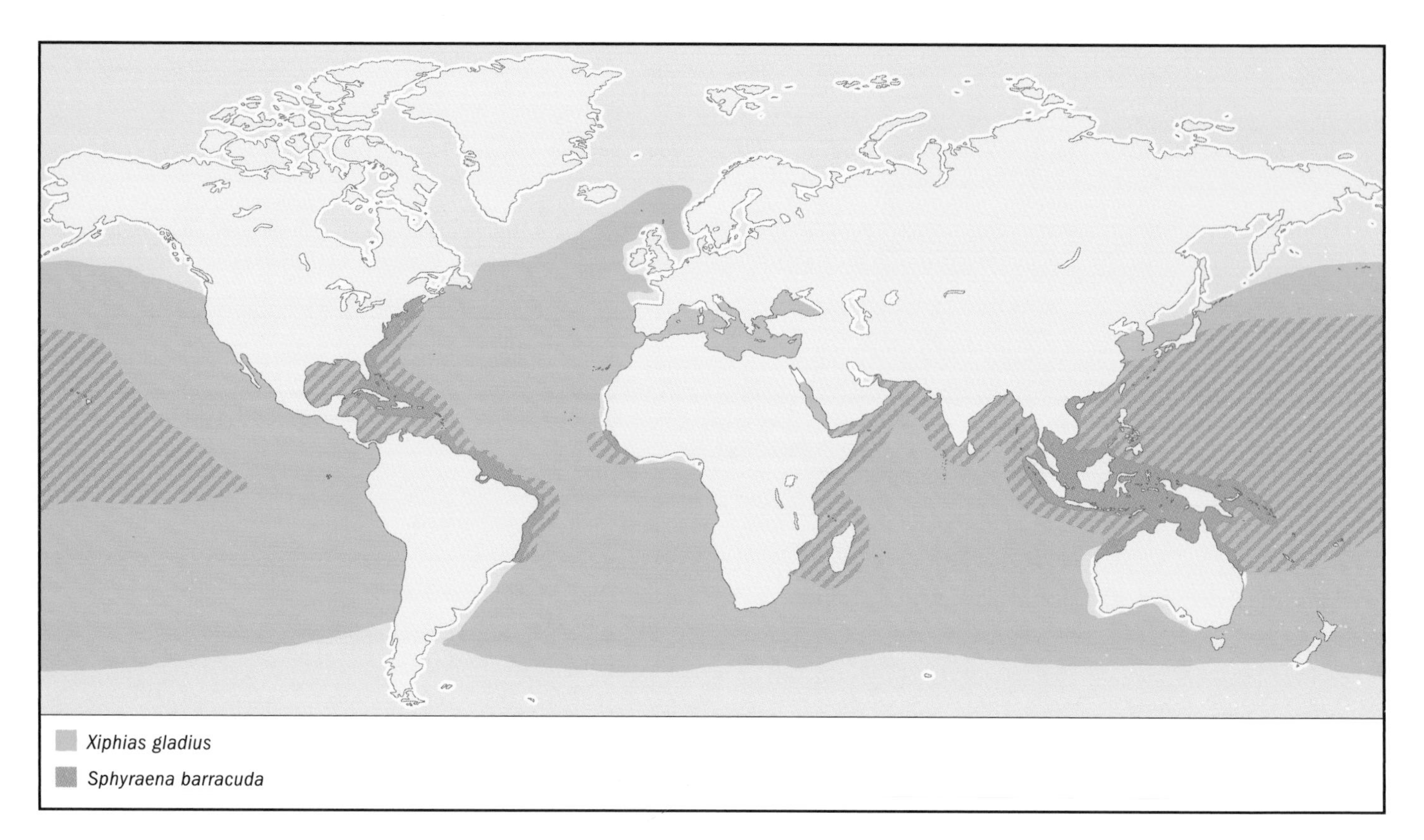

OTHER COMMON NAMES
French: Barracuda; Spanish: Picuda barracuda.

PHYSICAL CHARACTERISTICS
Twenty-four vertebrae. Reaches 79 in (200 cm), commonly to 51 (130 cm). Body elongate and slightly compressed. Head large with a long, pointed snout. Mouth large, tip of maxilla reaching to or extending beyond anterior margin of the eye in adults. Lower jaw projecting beyond upper jaw without a fleshy tip. Strong, pointed vertical teeth of unequal sizes in both jaws and in roof of mouth. Two dorsal fins, far apart, the first with five strong spines, its origin slightly behind pelvic fin origin. Tip of adpressed pectoral fin reaching to or extending beyond pelvic fin origin. Gill rakers are absent. Lateral line well developed, straight, with 80–90 scales. Deep green to steel gray above, sides silvery, abruptly becoming white on ventral surface. Adults have several oblique dark bars, usually 18–22, on upper sides and usually have several to many scattered inky blotches on posterior part of lower sides.

DISTRIBUTION
Worldwide in tropical and subtropical waters except eastern Pacific Ocean.

HABITAT
Small individuals are mostly found in shallow waters over sandy and weedy bottoms, often forming schools. Adults usually are solitary dwellers of reef areas and offshore areas.

BEHAVIOR
The fearsome appearance of barracudas combined with voracious feeding behavior and a notable curiosity toward divers has made them one of the most familiar families of marine fishes.

FEEDING ECOLOGY AND DIET
Voracious predators of small fishes, squid, and crustaceans.

REPRODUCTIVE BIOLOGY
Most males mature at two years and all are mature by three years. Some females mature at three years and all are mature at four. The spawning season is from April through October off southern Florida. Females contain 500,000 to 700,000 mature eggs at one time but spawn several times in a season.

CONSERVATION STATUS
Not threatened.

SIGNIFICANCE TO HUMANS
Unprovoked attacks on humans have been documented, but they are very rare, and most have occurred because the swimmer was carrying or wearing a silvery, bright object, which a barracuda misidentifies as prey. Human consumption of large individuals may cause ciguatera, a kind of fish poisoning. The toxicity of barracuda flesh is related to the food habits of larger barracuda, which accumulate in their flesh toxin from their prey. There is good evidence that the source of the toxin is a dinoflagellate. Barracuda are not targets of a directed fishery but are caught with hand lines, trolling gear, bottom trawls, gill nets, and trammel nets. Catch data are not reported to the FAO by species; all species of the genus are combined. Marketed fresh and salted. There is a tie for all-tackle game fish record between an 85-lb (38.5-kg) fish from Christmas Island, Kiribati, and one from the Philippines. ◆

Largehead hairtail
Trichiurus lepturus

FAMILY
Trichiuridae

TAXONOMY
Trichiurus lepturus Linnaeus, 1758, South Carolina.

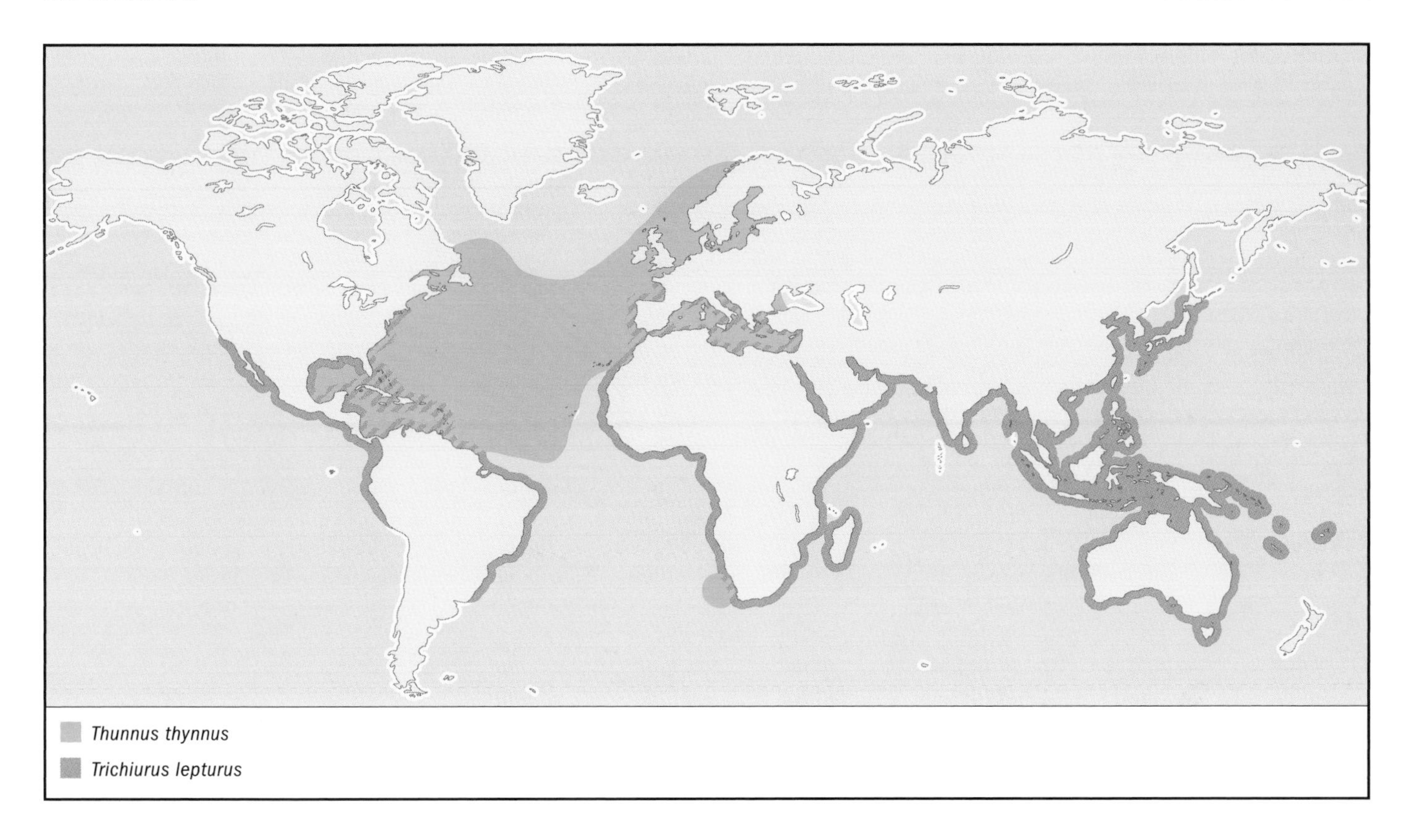

OTHER COMMON NAMES
English: Atlantic cutlassfish; French: Poisson sabre commun; Spanish: Pez sable.

PHYSICAL CHARACTERISTICS
Maximum length 47 in (120 cm) total length, commonly 20–39 in (50–100 cm). Body extremely elongate and strongly compressed, ribbon-like tapering to a point. No caudal fin. Eye large, contained 5–7 times in head length (head length divided by eye diameter equals 5–7). Mouth large with a cartilaginous process at tip of upper and lower jaws. Two or three pairs of enlarged fangs with barbs near tip of upper jaw and another pair near tip of lower jaw. Dorsal fin high and long, no notch between spinous and soft portions, three spines and 130–135 soft rays. Anal fin reduced to 100–105 minute spinules, usually embedded in the skin. Pectoral fins short; pelvic fins absent. Vertebrae numerous, 162–168. Fresh specimens steel blue with silvery reflection, color becoming silvery gray after death.

DISTRIBUTION
Tropical and temperate marine waters of the world if the eastern Pacific *Trichiurus nitens* is regarded as a synonym of *T. lepturus* Nakamura and Parin, 1993.

HABITAT
Benthopelagic, continental shelf to 1,148 ft (350 m), occasionally in shallow waters and at the surface at night.

BEHAVIOR
Juveniles and small adults form schools about 100 m above the bottom during the daytime and form loose feeding aggregations at night-time near the surface. Large adults feed on pelagic prey near the surface during the daytime and migrate to the bottom at night.

FEEDING ECOLOGY AND DIET
Young and immature fish feed mostly on krill, small planktonic crustaceans, and small fishes. Adults become more piscivorous and feed on a wide variety of fishes plus squids and crustaceans.

REPRODUCTIVE BIOLOGY
In the Gulf of Mexico, spawning occurs offshore at depths greater than 151 ft (46 m). The fish mature at 2 years of age and a size of approximately 12 in (30 cm) preanal length for females and 11 in (28 cm) for males. Adult females produce 33,000–85,000 eggs. Eggs are pelagic with a diameter of 0.067–0.074 in (1.7–1.9 mm) and hatch in 3–6 days.

CONSERVATION STATUS
Not threatened.

SIGNIFICANCE TO HUMANS
An important food fish in many parts of the world. FAO catch statistics for 1991–2000 show catches of 867,145–1,631,253 tons (786,661–1,479,848 metric tons) per year by 44 countries in 11 FAO fishing areas. Caught with a variety of nets. Also taken by anglers; the all-tackle world record is an 8-lb (3.7-kg) fish taken off Rio de Janeiro. ◆

Swordfish
Xiphias gladius

FAMILY
Xiphiidae

TAXONOMY
Xiphias gladius Linnaeus, 1758, European seas.

OTHER COMMON NAMES
French: Espadon; Spanish: Pez espada.

PHYSICAL CHARACTERISTICS
Reaches a maximum size of 175 in (445 cm) total length and approximately 1,190 lb (540 kg). The body is elongate and cylindrical. Two widely separated dorsal fins in adults, the first much larger than the second, the first with 34–49 rays, the second with four to six rays. Two separate anal fins in adults, the first with 12–16 rays, the second with three or four rays. Pectoral fins falcate, located low on body sides, with 17–19 rays. Caudal fin large and lunate, with a deep notch on upper and lower profiles of caudal peduncle. Fine file-like teeth and scales with small spines are present in juveniles but become embedded in the skin with growth at approximately 3 ft (1 m) in length. Left and right branchiostegal membranes separated distally. Vertebrae, 26. Back and sides of body blackish brown, gradually fading to light brown on ventral side.

DISTRIBUTION
Cosmopolitan in tropical, temperate, and sometimes cold waters of all oceans, including the Mediterranean, Black, and Caribbean seas.

HABITAT
This is an epipelagic and mesopelagic, oceanic species, usually found in surface waters warmer than 55.4°F (13°C), the optimum temperature range being 64.4–71.6°F (18–22°C). Swordfishes have been acoustically tracked and directly observed from submersibles in the western Atlantic and eastern Pacific to depths of 2,024 ft (617 m).

BEHAVIOR
It is likely that swordfishes use the sword to stun or kill some prey, as shown by slashes on the bodies of squid and fishes found in swordfish stomachs. The brain and eyes of swordfish are warmer than the water in which they live. The tissue that heats the brain is developed from the superior rectus muscle of the eye. The brain heater protects the central nervous system from rapid cooling during vertical excursions of as much as 984 ft (300 m) that these fish may make through a temperature range as great as 34.2°F (19°C) in 2 hours.

FEEDING ECOLOGY AND DIET
Adult swordfishes are opportunistic feeders, known to forage for their food from the surface to the bottom over a wide depth range. Over deep water, they feed primarily on pelagic fishes including tuna, dolphinfishes, lancetfishes, flyingfishes, and pelagic squids. In shallow waters swordfishes often take neritic pelagic fishes such as mackerel, herring, anchovies, sardines, and sauries. Large adults may make feeding trips to the bottom for demersal fishes.

REPRODUCTIVE BIOLOGY
In the western Atlantic, spawning apparently occurs throughout the year in the Caribbean, the Gulf of Mexico, and waters off Florida. Swordfishes spawn in the upper water layers at depths of 0–246 ft (0–75 m) and temperatures of approximately 73.4°F (23°C). Swordfishes first spawn at 5 or 6 years of age and 59–67 in (150–170 cm) eye-fork length and lay 2–5 million eggs.

CONSERVATION STATUS
Listed by IUCN as Data Deficient. Populations of swordfish have been greatly reduced by fishing, particularly in the Atlantic Ocean. A consumer boycott of Atlantic swordfish organized by the United States–based Natural Resources Defense Council and Sea Web affected prices enough to gather crucial momentum toward a recovery plan for depleted swordfish in the Atlantic (Safina, 2001).

SIGNIFICANCE TO HUMANS
Appreciation of swordfish as a food fish is recent. Swordfishes brought only approximately $0.24 per pound (0.5 kg) in 1919 and $0.60 in 1946. There are important fisheries for swordfish in all three major oceans. Swordfishes are caught by long line, harpoon, drift gill net, set nets, and other fishing gear. FAO catch statistics for 1991–2000 show catches of 76–116 thousand tons (69–105 thousand metric tons) per year by 77 countries. Restrictions on the sale of swordfish containing levels of mercury greater than 0.5 ppm in Canada and the United States in the early 1970s caused collapse of the Canadian fishery and severely restricted landings in the United States. The mercury guidelines were raised to 1.0 ppm in 1979, and by 1980 catch and effort had reached a new high in the northwest Atlantic. Swordfishes are important sport fishes. The all-tackle game fish record is a 1,182-lb (536.2-kg) fish taken off Chile. ◆

Resources

Books

Collette, B. B., C. Reeb, and B. A. Block. "Systematics of the Tunas and Mackerels (Scombridae)." In *Tuna: Physiology, Ecology, and Evolution*, edited by Barbara A. Block and E. Donald Stevens. Vol. 19, *Fish Physiology*. San Diego: Academic Press, 2001, 1–33.

Safina, C. "Tuna Conservation." In *Tuna: Physiology, Ecology, and Evolution*, edited by Barbara A. Block and E. Donald Stevens. Vol. 19, *Fish Physiology*. San Diego: Academic Press, 2001, 413–466.

Periodicals

Collette, B. B., and C. E. Nauen. "FAO Species Catalogue: Scombrids of the World: An Annotated and Illustrated Catalogue of Tunas, Mackerels, Bonitos, and Related Species Known to Date." *FAO Fisheries Synopsis* 2, no. 125 (1983).

Collette, B. B., T. Potthoff, W. J. Richards, S. Ueyangi, J. L. Russo, and Y. Nishikawa. "Scombroidei: Development and Relationships." In *American Society of Ichthyologists and Herpetologists Special Publication No. 1.*, edited by H. G. Moser, W. J. Richards, D. M. Cohen, M. P. Fahy, A. W Kendall Jr., and S. L. Richardson, 591–620. Lawrence, KS: Allen Press, 1984.

de Sylva, D. P. "Systematics and Life History of the Great Barracuda, *Sphyraena barracuda* (Walbaum)." *Studies in Tropical Oceanography* 1 (1963): 1–179.

Johnson, G. D. "Scombroid Phylogeny: An Alternative Hypothesis." *Bulletin of Marine Science* 39 (1986): 1–41.

Nakamura, I. "FAO Species Catalogue: Billfishes of the World: An Annotated and Illustrated Catalogue of Marlins, Sailfishes, Spearfishes and Swordfishes Known to Date." *FAO Fisheries Synopsis* 5, no. 125 (1985).

Resources

Nakamura, I., and N. V. Parin. "FAO Species Catalogue: Snake Mackerels and Cutlassfishes of the World Families Gempylidae and Trichiuridae: An Annotated and Illustrated Catalogue of the Snake Mackerels, Snoeks, Escolars, Gemfishes, Sackfishes, Domine, Oilfish, Cutlassfishes, Scabbardfishes, Hairtails and Frostfishes Known to Date." *FAO Fisheries Synopsis* 15, no. 125 (1993).

Regan, C. T. "On the Anatomy and Classification of the Scombroid Fishes." *Annual Magazine of Natural History, Series 8* 3 (1909): 66–75.

Tyler, J. C., G. D. Johnson, I. Nakamura, and B. B. Collette. "Morphology of *Luvarus imperialis* (Luvaridae), with a phylogenetic analysis of the Acanthuroidei (Pisces)." *Smithsonian Contributions to Zoology* 1989, no. 485: 1–78.

Organizations

Inter-American Tropical Tuna Commisssion. 8604 La Jolla Shores Drive, La Jolla, CA 92037-1508 USA. Phone: (858) 546-7100. Fax: (858) 546-7133. Web site: <http://www.iattc.org>

International Commission for the Conservation of Atlantic Tunas. Calle Corazón de Maria, 8, 6th floor, Madrid, E-28002 Spain.

Bruce B. Collette, PhD

Stromateoidei

(Butterfishes and relatives)

Class Actinopterygii
Order Perciformes
Suborder Stromateoidei
Number of families 6

Photo: The butterfish (*Peprilus triacanthus*) is a small schooling fish that is found off the East Coast of the United States. They are named for their flavor, not their appearance. (Photo by Animals Animals ©Zig Leszczynski. Reproduced by permission.)

Evolution and systematics

The fossil record of the Stromateoidei is not very extensive, consisting mostly of isolated otoliths, the paired earstones present in the membranous labyrinth of the inner ear of many fishes that can aid in detecting motion. Otoliths are difficult to identify because they usually lack the diagnostic features of families of fishes. Irrespective of this, the earliest otoliths attributed to a stromateoid date from the early Tertiary period (about 50 million years ago), of France, Belgium, and England. Fossils known from more than otoliths, such as skeletons, are rare, and these are also fragmentary, such as *Psenicubiceps alatus* from Russia, currently assigned to the stromateoid family Nomeidae. The earliest stromateoid fossils are believed to be from Denmark, from deposits as old as 60 million years, but these cannot presently be assigned with confidence to the Stromateoidei.

The Stromateoidei is a suborder of the large bony fish order Perciformes. The composition of this suborder has changed only slightly in the past 35 years; it was mostly assembled in a landmark study in 1966 by Humphry Greenwood, Donn E. Rosen, Stanley H. Weitzman, and George S. Myers, in which all families were grouped together except the Amarsipidae, which was added in 1969. The Stromateoidei currently includes 6 families, 15 genera, and some 65 species. The families are: Amarsipidae (with a single genus and species, *Amarsipus carlsbergi*), Nomeidae (driftfishes; 3 genera and 15 species), Centrolophidae (medusafishes; 7 genera and 27 species), Tetragonuridae (squaretails; 1 genus, *Tetragonurus*, and 3 species), Stromateidae (butterfishes; 3 genera and 13 species), and Ariommatidae (1 genus, *Ariomma*, and 6 species).

Within the Perciformes, stromateoids are more closely related to certain generalized families that share a specific arrangement of the ramus lateralis accessorius (a facial nerve complex), including the Kyphosidae, Girellidae, Scorpididae, Arripididae, Kuhliidae, Microcanthidae, Oplegnathidae, and Terapontidae. Of these families, the stromateoids are more closely related to the Kyphosidae (chubs), sharing with them details of their tooth pattern and larval pigmentation. Evolutionary relationships within the Stromateoidei indicate that the Amarsipidae are the most basal group, that the Centrolophidae may not have unique features, and that the butterfishes (Stromateidae) are the most derived stromateoids.

Physical characteristics

Stromateoids are moderate-sized fishes up to 47.2 in (1.2 m) in length, generally with round and somewhat large eyes (with associated adipose tissue in ariommas), small mouths, forked caudal fins (with 15 branched rays), and elongated dorsal fins that may be continuous or subdivided. The body is slender to deep, and is either compressed laterally or rounded in cross-section. The dorsal fins have spines (weak in some species); an anal fin with one to three spines; the dorsal and anal fins generally terminate at the same level; pelvic fins are absent in adults of some species (e.g., butterfishes, although the pelvic bones are present); pelvic fins are small in others, and very large in the man-of-war fish (*Nomeus gronovii*). The jaw teeth are small, usually arranged in a single series; the nostrils are double; a lateral line is present. The gill rakers range from 10–20 on the first gill arch. Scales are usually cycloid. Coloration may vary from silvery to dark brown in adults, but is usually mottled in juveniles. Some species have more elaborate color patterns, with blotches and bands.

Stromateoids, with the exception of the Amarsipidae, are unique in having a specialized pharyngeal organ (the pharyngeal sac) just anterior to the esophagus and following the last gill arch, which is specialized for further breaking down food items. The pharyngeal sacs are coated internally by small projections (papillae) that contain minute "teeth"; the structure and arrangement of the papillae vary significantly among species. All stromateoids share a specific configuration of the internal caudal fin skeleton.

Distribution

Stromateoids are found in all major oceans (except the Arctic, Baltic, Okhotsk, and Black Seas, and the Antarctic Ocean) from the high seas well offshore, to the continental shelf regions, and in large bays. They are more common in temperate and tropical waters, but a few species also occur in colder areas. Many offshore pelagic species are very widely distributed.

Habitat

All stromateoid species are marine. Some eight genera are oceanic, six genera are mostly coastal, and two are both coastal and oceanic. As a general rule, young individuals are usually epipelagic, sometimes living in association with jellyfishes, but adults occur mostly in deeper waters, either demersally (inshore or offshore down to about 1,640 ft/500 m); more in some species, but usually in shallower waters) or mesopelagically.

Behavior

The behavior of individual stromateoids, other than in relation to their association with jellyfishes, is virtually unknown. Many species form schools of moderate size, as individuals are frequently captured together. The association between stromateoids (usually juvenile) and jellyfishes requires further study, but it appears to form very early in the fish's development. Stromateoids are more resistant to jellyfish toxins than other fishes. They may also associate with salps (semitransparent barrel-shaped marine invertebrate animals), where juveniles of squaretails may seek refuge. They usually hover underneath the bell of the jellyfish, but may swim in and out of its tentacles to snatch prey items (mostly zooplankton and other invertebrates). They may also feed occasionally on the tentacles and gonads of the jellyfish, and are also subjected to predation by them. Juveniles in association with jellyfishes are usually more colorful or display more complex color patterns than adults of the same species, which are usually demersal or pelagic. Individuals of some species (e.g., the barrelfish, *Hyperoglyphe perciformis*) may congregate under floating wreckage, oceanic flotsam, planks, buoys, or other sheltered mobile habitats. Others may gather around vessels.

Feeding ecology and diet

Food items consist mostly of invertebrates (numerous crustaceans, such as barnacles, crabs, shrimps, and euphausiids (small shrimp-like crustaceans), in addition to squids, various other mollusks, and zooplankton), but many species have a predilection for jellyfishes. Small fishes may be taken occasionally, as well as urochordates. Stromateoids are preyed upon by larger fishes, including sharks, and also by their jellyfish hosts.

Reproductive biology

As with many pelagic fishes, stromateoids are broadcast spawners, releasing large quantities of eggs into the pelagic realm upon spawning. Spawning may occur on a yearly basis during the summer months in some species, but spawning periods are unrecorded for many stromateoids. The eggs are small, ranging from 0.03–0.07 in (about 0.07–0.18 cm) in diameter, when known; they are spherical, separate, and pelagic; and usually contain a single oil droplet. Hatching usually occurs under 0.2 in (0.5 cm) in length, and flexion takes place shortly thereafter. Depending on the species, juveniles begin to resemble adults at about twice their flexion length. The eggs and larvae of about one-half of all stromateoid species have yet to be described.

Conservation status

No species are listed by the IUCN.

Significance to humans

Some stromateoid species are of commercial importance locally, especially in Japan and Southeast Asia. Among the most important species are the silver pomfret (*Pampus argenteus*), with a yearly catch ranging from 11,000 to 18,000 tons (11,177–18,289 t) between 1990 and 1995; the Indian ariomma (*Ariomma indicum*), with a yearly catch reported at 9,000 tons/9,144 t off Africa; and the American butterfish (*Peprilus triacanthus*). Most other species are of little or no importance commercially. All species are harmless.

1. Black ruff (*Centrolophus niger*); 2. Man-of-war fish (*Nomeus gronovii*); 3. Butterfish (*Peprilus triacanthus*). (Illustration by Dan Erickson)

Species accounts

Black ruff

Centrolophus niger

FAMILY
Centrolophidae

TAXONOMY
Perca nigra Gmelin, 1788, eastern Atlantic.

OTHER COMMON NAMES
English: Blackfish; French: Centrolophe noir; German: Schwarzfisch; Spanish: Romerillo.

PHYSICAL CHARACTERISTICS
Length 43.3 in (110 cm). Elongate, with a rather small head and large eyes; continuous dorsal fin extends most of the length of the body (with 4–5 weak spines and 32–38 rays); anal fin with 3 spines and 20–27 rays; pelvic fins small, under pectorals; pectorals with 19–23 rays; caudal fin weakly forked; mouth rather wide, extending posteriorly beneath eyes; coloration dark bluish gray, but sometimes darker.

DISTRIBUTION
Western (from Nova Scotia to New Jersey) and eastern North Atlantic, Mediterranean Sea, and also the Pacific Ocean off Australia, New Zealand, and South America.

HABITAT
Temperate oceanic waters. Young specimens occur near surface; larger individuals are mesopelagic. Specimens have been captured as deep as 1,968 ft (600 m) off Australia and New Zealand.

BEHAVIOR
Little is known concerning its behavior. Small specimens have been found in association with jellyfishes, but adults may form schools.

FEEDING ECOLOGY AND DIET
Feeds on jellyfishes, squids, and different kinds of crustaceans. Preyed on by hake and possibly other larger fishes.

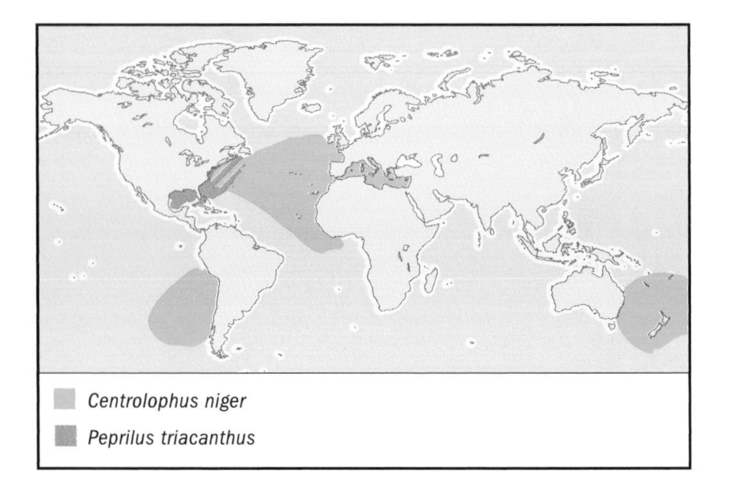

REPRODUCTIVE BIOLOGY
Eggs and larvae are pelagic; eggs are spherical, small (0.05 in/0.12 cm in diameter) and contain a single oil globule. Hatching occurs at about 0.16 in (0.4 cm) standard length, flexion at 0.2 in (0.6 cm), adult body form attained at 0.7 in (1.7 cm). Growth reported to be relatively fast. Probably a broadcast spawner.

CONSERVATION STATUS
Not threatened.

SIGNIFICANCE TO HUMANS
Not significantly consumed and therefore of minor importance. ◆

Man-of-war fish

Nomeus gronovii

FAMILY
Nomeidae

TAXONOMY
Gobius gronovii Gmelin, 1788, eastern Atlantic.

OTHER COMMON NAMES
English: Bluebottle fish, shepherd fish; Spanish: Pastorcillo, Pez azul.

PHYSICAL CHARACTERISTICS
Standard length 13.7 in (35 cm). Head somewhat tall, body tallest at level of pelvic fins, tapering posteriorly; dorsal fin continuous, originating at midbody length, with 9–12 spines and 25–27 rays (IX–XII, 24–28); pectoral fins elongate, with 21–23 rays; caudal fin forked, with some 15 rays; anal fin with two spines and 24–29 rays (I–II, 24–26); pelvic fins very long and prominent, reaching posteriorly to level of pectoral fin extremities, with one spine and five rays (I, 5); eyes large; no teeth on tongue. Juveniles have vertical, dark blue and broad bands, with blue blotches on head and fins; adults have irregular blue blotches on body, head, and fins; background color silvery, pelvic and dorsal fins black. Larger adults more uniformly dark in color.

DISTRIBUTION
Tropical and warm temperate waters of Atlantic, Pacific, and Indian Oceans.

HABITAT
Pelagic offshore, usually warm waters, commonly found in association with the siphonophore Portuguese man-of-war (*Physalia*), but larger adults may live independently in deeper waters. Juveniles may be found pelagically in waters 98 ft (30 m) deep.

BEHAVIOR
Usually found in large numbers living underneath the bell of the Portuguese man-of-war, swimming in and out among its tentacles. The mottled color pattern mimics the tentacles of *Physalia*. It lives commensally by being resistant to the venom of the siphonophore.

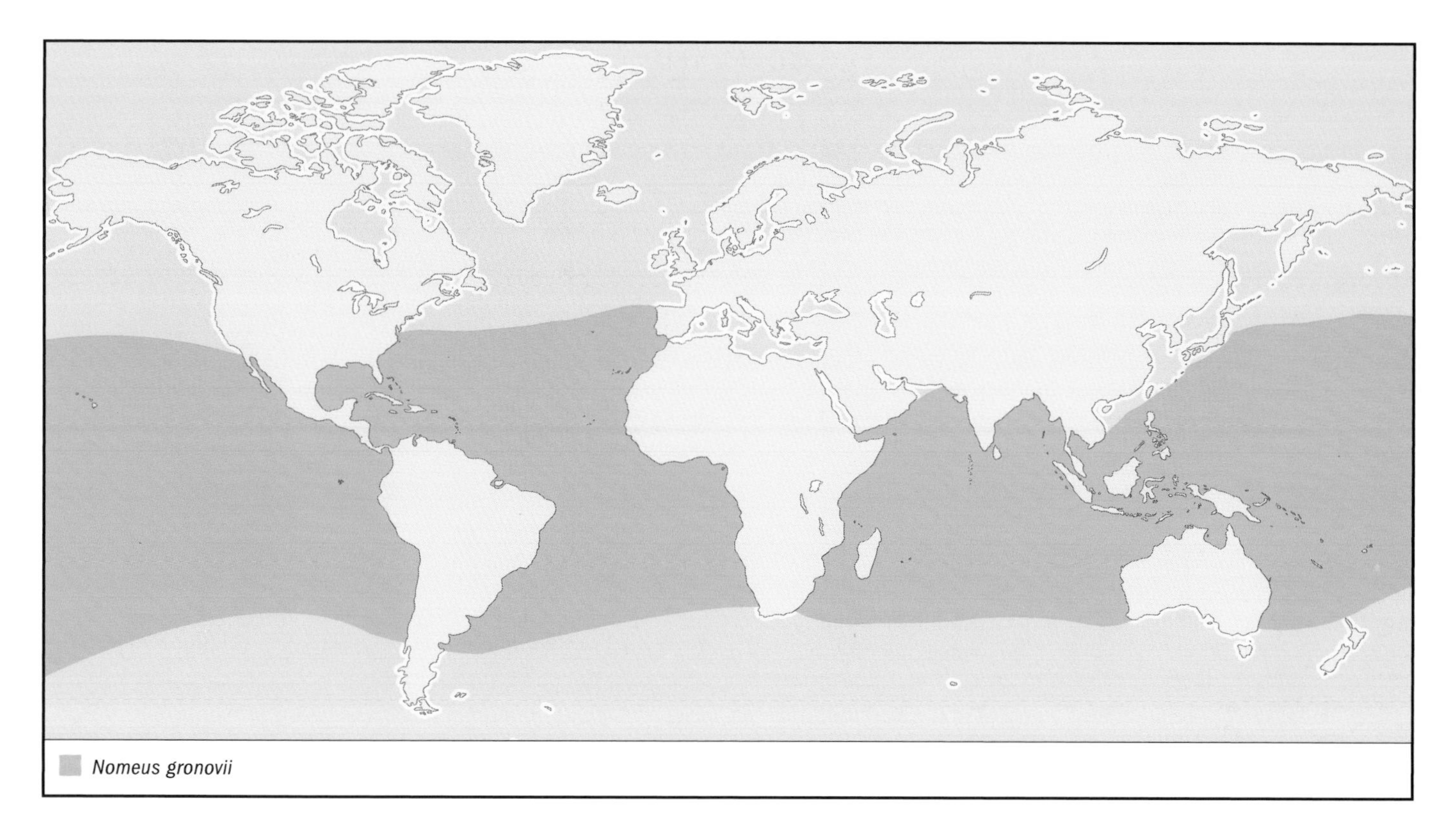

FEEDING ECOLOGY AND DIET
Feeds on the tentacles and gonads of the Portuguese man-of-war, zooplankton, and other soft-bodied jellyfishes. Sometimes eaten by its host.

REPRODUCTIVE BIOLOGY
Eggs and preflexion larvae are unknown; postflexion larvae as small as 0.27 in (0.7 cm) have been recorded, and adult form is mostly attained by 0.9 in (2.5 cm). Other aspects of its reproduction are unknown, but as with other stromateoids, eggs and larvae are pelagic and they are probably broadcast spawners. Small individuals of 0.39 in (1 cm) have been found in association with the Portuguese man-of-war, indicating that this association forms early in life.

CONSERVATION STATUS
Not threatened.

SIGNIFICANCE TO HUMANS
Of minor commercial importance, due to infrequent fishing. ◆

Butterfish

Peprilus triacanthus

FAMILY
Stromateidae

TAXONOMY
Stromateus triacanthus Peck, 1804, New Hampshire, United States.

OTHER COMMON NAMES
English: American butterfish; French: Stromaté fossette; Spanish: Palometa pintada.

PHYSICAL CHARACTERISTICS
Length about 11.8 in (30 cm). Body thin and deep, with short head and blunt snout. Single dorsal fin is taller shortly posterior to its origin, with 2–4 spines and 40–48 soft rays (II–IV, 40–48); pectoral fins moderately elongate with 1–22 fin rays; 22–25 gill rakers; lateral-line scales 96–105; anal fin almost as long as dorsal fin, with 3 spines and 37–44 rays; caudal fin deeply forked with slightly greater lower lobe. Pelvic fins absent. Color grayish blue above and silvery on sides, with many irregular dark spots laterally.

DISTRIBUTION
Western Atlantic Ocean, from off South Carolina, United States, to Nova Scotia (sometimes as a stray off the coast of Newfoundland); also further south to the coast of Florida, United States, in deeper water, as well as in the Gulf of Mexico.

HABITAT
Continental shelf; found pelagically or demersally in waters as deep as 600 ft (183 m), usually over sandy bottoms. May venture into shallow bays and estuaries.

BEHAVIOR
Little is known concerning behavior other than reproduction. During the first year may live in association with jellyfishes or freely, but forms schools as adults. Migratory patterns are common in consequence of water temperature. Appear seasonally off northeastern coast of the United States, but are generally unpredictable as to when.

FEEDING ECOLOGY AND DIET
Feeds abundantly on soft-bodied invertebrates, preferring urochordates and mollusks. Also feeds on cnidarians, ctenophores,

chaetognaths, polychaetes, and crustaceans (including amphipods, copepods, mysids, and euphausiids). Juveniles in Narraganset Bay, Rhode Island, United States, feed heavily on ctenophore *Mnemiopsis leidyi*, not by ingesting it whole but by taking small bites. Feeding occurs both day and night. Preyed on by some 30 species of fishes and squids, and are part of the diet of many commercially important fishes, such as haddocks, bluefishes, swordfishes, summer flounders, and hammerhead sharks.

REPRODUCTIVE BIOLOGY

Sexually mature between first and second years, starting about 7.1 in (18 cm) standard length (i.e., exclusive of the caudal fin). Broadcast spawners, lack specialized courtship behavior. Spawning takes place once a year during summer months. Eggs are buoyant, spherical, and transparent, measure some 0.03 in (0.08 cm) in diameter; include a single oil droplet. Larvae begin to resemble adults about 0.6 in (1.5 cm), when fin rays of dorsal, anal, and caudal fins are fully formed.

CONSERVATION STATUS

Not threatened.

SIGNIFICANCE TO HUMANS

Important commercial fishes, heavily consumed since the 1800s, and commonly caught by otter trawls, gill nets, and other means. In the early 2000s, yearly catches averaged under 4,921 tons (5,000 t), but in 1973 close to 19,684 tons (20,000 t) were landed. The flesh is considered to be very delicious and "melt-in-your-mouth," which is reflected in the common name. ◆

Resources

Books

Collette, B., and G. Klein-MacPhee, eds. *Fishes of the Gulf of Maine.* Washington, DC: Smithsonian Institution Press, 2002.

Horn, M. H. "Stromateoidei: Development and Relationships." In *Ontogeny and Systematics of Fishes,* edited by H. G. Moser, W. J. Richards, D. M. Cohen, M. P. Fahay, A. W. Kendall, Jr., and S. L. Richardson. Special Publication No. 1. American Society of Ichthyologists and Herpetologists, 1984.

Nelson, J. S. *Fishes of the World,* 3rd edition. New York: John Wiley & Sons, 1994.

Periodicals

Ahlstrom, E. H., J. L. Butler, and B. Y. Sumida. "Pelagic Stromateoid Fishes (Pisces, Perciformes) of the Eastern Pacific: Kinds, Distributions, and Early Life Histories and Observations on Five of These from the Northwest Atlantic." *Bulletin of Marine Science* 26, no. 3 (1976): 285–402.

Greenwood, P. H., D. E. Rosen, S. H. Weitzman, and G. S. Myers. "Phyletic Studies of Teleostean Fishes, with a Provisional Classification of Living Forms." *Bulletin of the American Museum of Natural History* 131 (1966): 339–456.

Haedrich, R. L. "A New Family of Stromateoid Fishes from the Equatorial Indo-Pacific." *Dana Reports* 76 (1969): 1–14.

———. "The Stromateoid Fishes: Systematics and a Classification." *Bulletin of the Museum of Comparative Zoology Harvard* 135, no. 2 (1967): 31–139.

Haedrich, R. L. and M. H. Horn. "A Key to the Stromateoid Fishes." *Woods Hole Oceanographic Institute Technical Report* 75 (1972): 45 pp.

Horn, M. H. "Swim Bladder State and Structure in Relation to Behavior and Mode of Life in Stromateoid Fishes." *U.S. Fishery Bulletin* 73: (1975) 95–109.

Mansueti, R. J. "Symbiotic Behavior Between Small Fishes and Jellyfishes, with New Data on That Between the Stromateid, *Peprilus alepidotus,* and the Scyphomedusa, *Chrysoara quinquecirrha.*" *Copeia* 1963, no. 1 (1963): 40–80.

Marcelo Carvalho, PhD

Anabantoidei

(Labyrinth fishes)

Class Actinopterygii
Order Perciformes
Suborder Anabantoidei
Number of families 3

Photo: This thick-lipped gourami (*Colisa labiosa*) variety is bred for the aquarium trade. (Photo by Mark Smith/Photo Researchers, Inc. Reproduced by permission.)

Evolution and systematics

The labyrinth fishes were first recognized as a natural assemblage by Cuvier and Valenciennes in 1831, but included the Channidae (snakeheads), in addition to the current family Anabantoidei. Bleeker (1859, 1879) added the luciocephalids (pikeheads) to this group. Jordan (1923) recognized six families, including *Luciocephalus* and Channidae. In 1963, Liem restricted the anabantoids to the families Anabantidae, Helostomatidae, Osphronemidae, and Belontiidae, thus removing *Luciocephalus* and the channids. In 1983 Lauder and Liem included *Luciocephalus*, in its own family Luciocephalidae, again in the anabantoids as the sister group to all remaining labyrinth fishes. Britz (1994, 1995), and Britz et al. (1995) demonstrated that there are no differences between Liem's families Belontiidae and Osphronemidae and that *Luciocephalus* is deeply nested within Liem's belontiids. The family name Osphronemidae applies for this monophyletic assemblage. The suborder Anabantoidei is thus divided into three families, Anabantidae, Helostomatidae, and Osphronemidae. The latter family is the sister group of the former two, and it is further subdivided into the subfamilies Belontiinae (only *Belontia*, two spp.); Osphroneminae (only *Osphronemus*, four spp.); Luciocephalinae (*Trichogaster*, four spp.; *Colisa*, four spp.; *Parasphaerichthys*, two spp.; *Ctenops*, one sp.; *Sphaerichthys*, four spp.; *Luciocephalus*, one sp., one undescribed sp.); and Macropodinae (*Macropodus*, five spp.; *Pseudosphromenus*, two spp.; *Malpulutta*, one sp.; *Parosphromenus*, 10 spp., some undescribed; *Trichopsis*, three spp.; *Betta*, 43 spp., some undescribed).

The closest relatives of anabantoids appear to be the Channidae. Based on the shared presence of parasphenoid teeth, both seem to form a larger monophyletic group with the badids and the genera *Nandus* and *Pristolepis*. The only unambiguous fossil anabantoid is a complete articulated skeleton from the Miocene epoch of Sumatra, assigned to *Osphronemus goramy*.

Physical characteristics

Anabantoids are minute (0.78 in; 20 mm) to large (23.6 in; 60 cm) percomorph fishes with a suprabranchial organ as accessory air-breathing organ. This organ consists of the suprabranchial chamber above the gill arches that houses the modified first epibranchial, termed the labyrinth. The labyrinth may have a highly complex three-dimensional shape in some species (*Anabas* and *Osphronemus*). Both the labyrinth and the wall of the suprabranchial chamber are lined with respiratory epithelium. The suprabranchial organ obtains blood from the first two afferent branchial arteries. Blood from the organ is collected in the two anterior efferent arteries that drain into the anterior cardinal vein, not into the dorsal aorta as in other teleosts. Basioccipital with paired articular processes that permit free movement with upper pharyngeal jaws. The last two characters are shared with the channids. In most anabantoids, the lacrimal and preopercular bones bear strong serrations; in anabantids, serrations occur in the subopercle, opercle bones, and sometimes the interopercle. (The name *anabantids* is vernacular for the family Anabantidae; *anabantoids* is the vernacular name for the suborder Anabantoidei.) All species except African anabantids and *Helostoma* have a exoccipital foramen medial to suprabranchial chamber, greatly enhancing hearing abilities. Most species are parasphenoid toothed, a unique derived character shared with channids, badids, *Nandus*, and *Pristolepis*. The swim bladder branches off posteriorly into two elongate diverticula that reach on either side of the hemal spines to the level of the parhypural. Many species are beautifully colored, with striking sexual dimorphism and dichromatism.

Distribution

Anabantoids occur in fresh waters of sub-Saharan Africa, and south and Southeast Asia. The genera *Ctenopoma*, *Microctenopoma*, and *Sandelia* occur only in Africa. The latter genus is confined to the Cape region and separated by a distributional gap from the other African anabantids. Most Asian species are widely distributed, but some have a greatly restricted distribution, such as *Belontia signata* and *Malpulutta kretseri*, known only from Ceylon, or *Parosphromenus deissneri*, *Betta miniopinna*, *B. schalleri*, *B. burdigala*, and *B. chloropharynx*, restricted to Banka Island in Indonesia. The

The dwarf gourami (*Colisa lalia*) is found in waters from northeastern India to Bangladesh. (Photo by Mark Smith/Photo Researchers, Inc. Reproduced be permission.)

northernmost distributed species is *Macropodus ocellatus*, from China and Korea.

Several species have been exported as food fishes or have been released accidentally from aquarium stocks to areas outside their natural ranges.

Habitat

Numerous species of anabantoids prefer still bodies of water with abundant aquatic vegetation that are exposed to the sun, but others also live in cooler, faster, mountain streams. Most anabantoids survive in oxygen-depleted waters because of their suprabranchial organ and therefore have an advantage over most other teleosts.

Behavior

In general, anabantoids differ from most other teleosts in that they rise to the water surface at intervals to exchange the air in their suprabranchial organ. This exchange is mostly achieved by flooding the suprabranchial chamber with water that enters through the gill opening. The water pushes the air out of the chamber and the mouth, either before or upon reaching the water surface. The chamber refills with air breathed in at the surface. In a second mode, air is gulped from the surface without the prior release of air from the chamber. Engulfed air is then pressed into the chamber by swallowing movements and forces some air out of the chamber, which is released from the gill opening. The latter mode is used by adult *Anabas testudineus* in the water and during the overland excursions that gave the species its vernacular name, the climbing perch. In addition to the climbing perch, at least one other anabantid species, the African *Ctenopoma multispinis* actively leaves the water and travels over land to nearby bodies of water.

Species of the genus *Colisa* show a behavior reminiscent of the spitting of the archerfishes (*Toxotes jaculatrix*). At least in captivity, representatives of *Colisa* spit a series of small droplets of water toward prey items, such as small invertebrates, above the water level to wash them down to the surface where they are taken by the fish.

Species of the genus *Trichopsis* are known for their ability to produce croaking sounds, hence their name, croaking gourami. These sounds are produced with their pectoral fins. To create the sounds, enlarged areas of tendons from the pectoral fin musculature are moved across bony knobs on some of the pectoral fin rays. The sound is enhanced by the suprabranchial chambers, which act as resonators.

All anabantoid species, except *Sandelia capensis*, show a typical, and for teleosts unusual, spawning clasp in which the male wraps around the female's body. The female is then either turned to the side or upside down when eggs are released. A similar clasp is also found in channids, badids, and *Nandus*, but not other nandids. A reduced clasp without turning of the female is shown by some mouth-brooding anabantoids.

Feeding ecology and diet

Anabantoids are diverse in regard to their feeding ecology and diet. There are extreme filter feeders, such as *Helostoma*, which feeds on small pelagic invertebrates and algae that are either filtered from the water or scraped off the substrate. Other species are omnivorous (*Anabas*, *Osphronemus*, *Trichogaster*, and *Colisa*), or have a diet with an emphasis on small invertebrates (*Microctenopoma*, *Macropodus*, *Betta*, and *Trichopsis*), but others prey on larger invertebrates and small fishes (*Ctenopoma* and *Sandelia*). *Luciocephalus* is a highly specialized predator of small fishes. *Osphronemus exodon* is an exclusively herbivorous species with external jaw teeth, which feeds on leaves of terrestrial plants, grasses, fruits, and flowers.

Reproductive biology

Although anabantoids are a fairly small percomorph group, their members exhibit a great variety of reproductive modes. The primitive mode, which occurs in *Anabas*, *Ctenopoma*, and *Helostoma*, is the absence of parental care with the release of several thousand small (ca. 0.04 in/1 mm), buoyant eggs that float due to a single large oil globule in the egg. After hatching, larvae retain the oil globule, which during development divides into two oil vesicles to the left and right of the chorda and is used as a floating organ. All species of the genus *Microctenopoma* and most osphronemids build bubble nests and guard primitively buoyant eggs and larvae. Bubble nests can consist of only a few bubbles, as in the tiny cave-brooding species of *Parosphromenus*, or be large. Mouth brooding has evolved at least twice among anabantoids, once in the lineage *Ctenops*, *Sphaerichthys*, and *Luciocephalus*, and again within the genus *Betta*.

Unusually for anabantoids, the two species of the purely South African genus *Sandelia* spawn on a substratum and have adhesive eggs. In groups with parental care, the number of eggs is usually smaller than in those without care, although several thousand eggs may be spawned in some species of *Microctenopoma* and *Trichogaster*. The number of eggs may range from 40 to several hundreds in most bubble nest builders, and from 20 to 200 in the mouth brooders. Egg size ranges from 0.03 in (0.7 mm) in *Microctenopoma* to 0.19 in (3 mm) in *Luciocephalus.*

Conservation status

Three anabantoid species are categorized as Critically Endangered by the IUCN, *Betta miniopinna*, *Betta persephone*, and *Betta spilotogena*. *Sandelia bainsii*, *Parosphromenus harveyi*, and *Betta livida* are categorized as Endangered. Another seven *Betta* species are categorized as Vulnerable, and the two Sri Lankan species *Belontia signata* and *Malpulutta kretseri* are categorized as Lower Risk.

Significance to humans

The larger species of anabantoid fishes are important as food fishes and feature in aquaculture in various parts of Asia. Many of the smaller, colorful anabantoids are very popular hardy ornamental fishes; up to several hundred U.S. dollars have been paid for a breeding pair of the conspicuously colored fighting fish *Betta macrostoma*, known as the Brunei beauty.

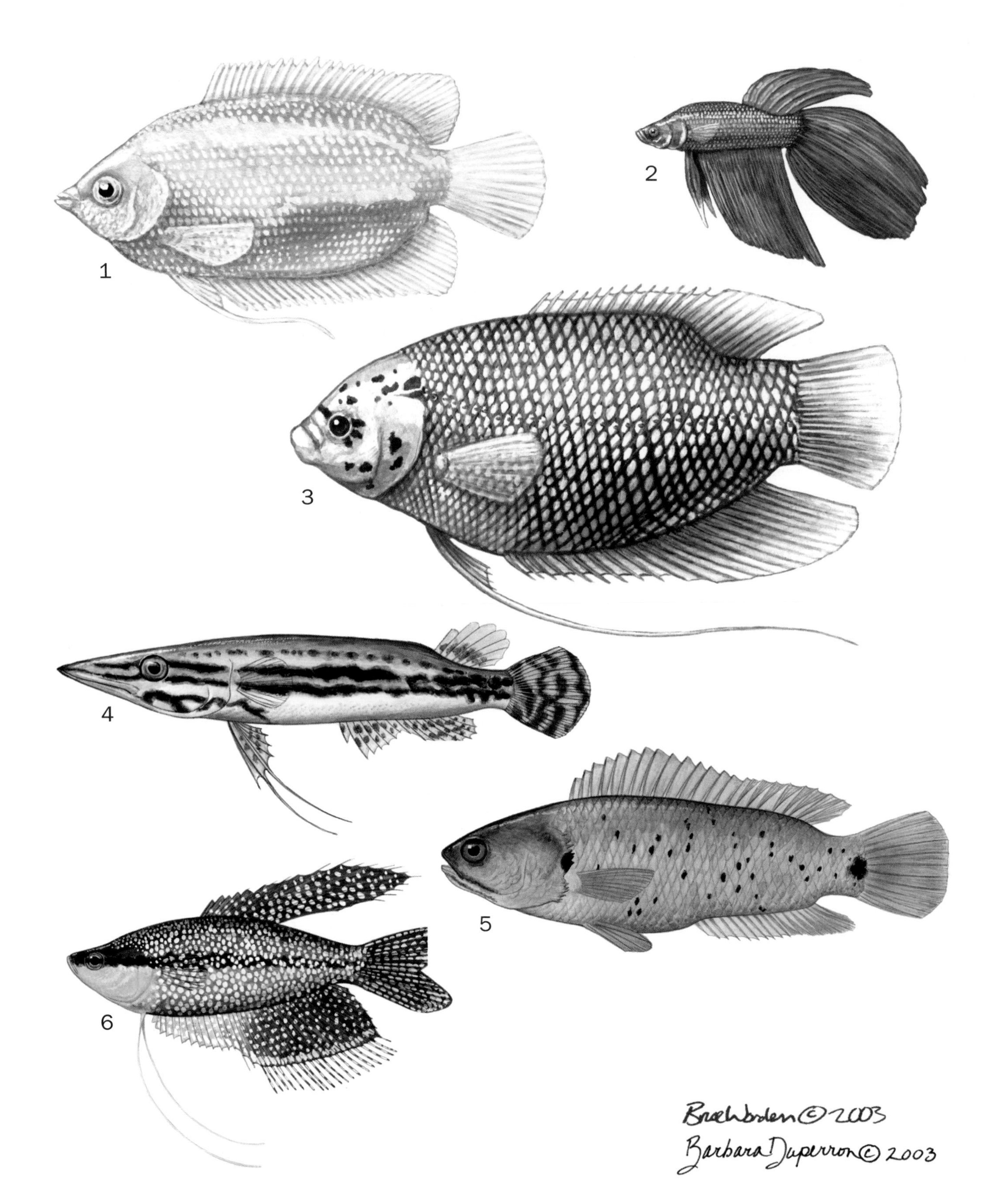

1. Kissing gourami (*Helostoma temminckii*); 2. Siamese fighting-fish (*Betta splendens*); 3. Giant gourami (*Osphronemus goramy*); 4. Pikehead (*Luciocephalus pulcher*); 5. Climbing perch (*Anabas testudineus*); 6. Pearl gourami (*Trichogaster leeri*). (Illustration by Barbara Duperron and Bruce Worden)

Species accounts

Climbing perch
Anabas testudineus

FAMILY
Anabantidae

TAXONOMY
Anthias testudineus Bloch, 1785, Japan. The diverse distribution and morphology of *A. testudineus* indicate that it may comprise more than one species.

OTHER COMMON NAMES
French: Perche grimpeuse; German: Kletterfisch; Spanish: Perca trepadora.

PHYSICAL CHARACTERISTICS
Length 9.8 in (25 cm). Robust body with wide, large head. Body shape ranges from oval and compressed to elongate and subcylindrical. Posterior edges of opercular bones, especially opercle, and subopercle, with strong spination. Without teeth on the palatine in contrast to most anabantids. Dorsal fin has 16–19 strong spines and 7–11 soft rays. Anal fin has 9–11 spines and 8–12 soft rays. Pelvic girdle without connection to pectoral girdle. Scales on the head rigidly attached to the skull bones. Scales strongly ctenoid. Lateral line interrupted at level of posterior part of spinous dorsal fin and continued two scale rows lower down to caudal peduncle. Suprabranchial organ exceptionally large and complexly folded. Coloration light beige with darker spots. A conspicuous black spot at the posterior edge of the gill cover between two prominent areas of projecting strong opercular spines and a large black ocellus on the caudal peduncle. No sexual dimorphism or dichromatism.

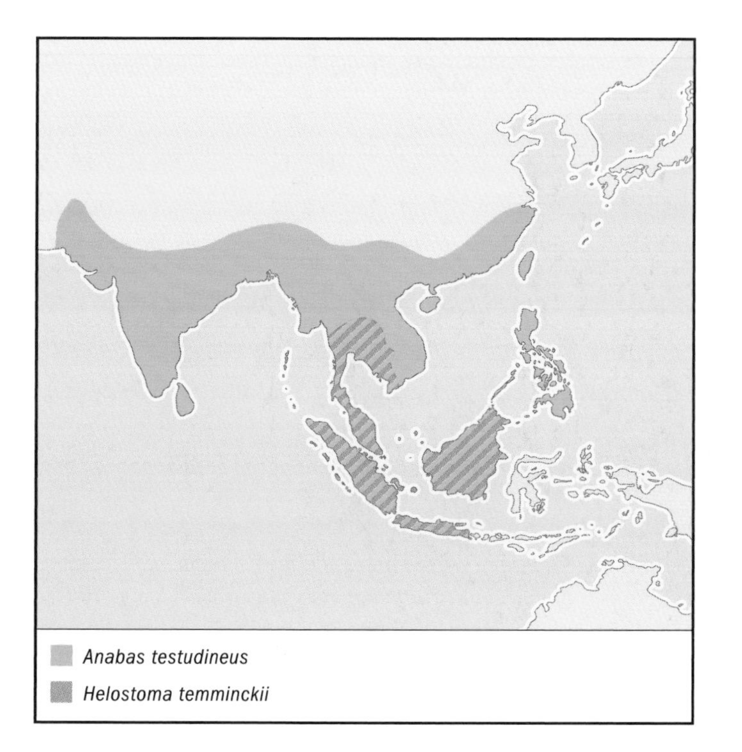

DISTRIBUTION
Widely distributed in Asia: Sri Lanka, India, Burma, Indochina, Taiwan, Sundaland (the western part of Indonesia, namely the islands of Java, Sumatra, and Kalimantan), but also introduced east of Huxley's Line (a zoogeographic distributional divide between the fauna of peninsular Southeast Asia and the Sunda islands [Sumatra, Java, and Borneo], and the fauna located on islands further to the east such as Australia, Papua New Guinea, Sulawesi, and the Philippines). This wide distributional range and the diverse physical morphology encountered indicates that more than one species is most certainly involved, but this species has not been thoroughly studied. The climbing perch has also been accidentally released in the United States.

HABITAT
Found in all types of fresh water, also survives in brackish water and tolerates water conditions unsuitable for most other fishes.

BEHAVIOR
Well known for its behavior to travel overland, first reported more than 200 years ago; uses its spiny opercular bones and a side-to-side wriggling of the body to move itself forward on land. Obligatory air breather that drowns if kept from rising to the surface to gulp air. Can survive longer periods of drought buried in the mud of the drying water bodies.

FEEDING ECOLOGY AND DIET
Omnivorous. Feeds on macrophytic vegetation, different invertebrates, and small fish.

REPRODUCTIVE BIOLOGY
No parental care. Typical spawning clasp lasts only a few seconds. Several thousand, buoyant, small (0.04 in; 1 mm) spherical eggs are spawned during one spawning phase. Eggs contain a single large oil globule. Hatching occurs after 24 hours at 82.4°F (28°C).

CONSERVATION STATUS
Not listed by the IUCN.

SIGNIFICANCE TO HUMANS
A common and popular food fish in Southeast Asia. Easily transported to the markets in buckets without water as long as the skin is kept moist, and it may survive in this condition for several days. ◆

Kissing gourami
Helostoma temminckii

FAMILY
Helostomatidae

TAXONOMY
Helostoma temminckii Cuvier, 1829, type locality not specified.

OTHER COMMON NAMES
French: Gourami embrasseur; German: Küssender Gurami; Spanish: Gurami besador, gurami besucón.

PHYSICAL CHARACTERISTICS
Length 9.8 in (25 cm). High body, laterally compressed. Head has large fleshy lips and several rows of spoon-shaped teeth unattached to the jaw bones. Wild-type coloration is greenish beige, with darker longitudinal lines along each scale row; a breed with uniform pink color is commonly cultivated in ponds. Adult fish has a highly specialized filter-feeding apparatus derived from modified gill rakers. Long dorsal fin with 16–18 spines and 13–16 soft rays; long anal fin with 13–15 spines and 17–19 soft rays.

Striking anatomical changes of the feeding system occur during development. Juveniles have normal conical teeth in jaws and on pharyngeal jaws. As the fishes mature, the jaw teeth are lost and substituted by a second type of spoon-shaped teeth in the fleshy lips. Lower pharyngeal jaws lose teeth completely, and their number is reduced in upper pharyngeal jaws, sitting on long bony bases so that the resulting structure resembles a brush.

DISTRIBUTION
Central Thailand, Malay Peninsula, Sumatra, Java, Borneo. Has been introduced to various countries as a food fish. Established populations in the Philippines, Sri Lanka, Bali, Colombia, and Florida, in the United States.

HABITAT
Occurs in sluggish streams, swamps, ponds, and lakes.

BEHAVIOR
Exhibits "kissing" behavior, in which they protrude their fleshy lips when grazing algae or during social encounters. They may also "kiss" during aggressive behavior or courtship individuals.

FEEDING ECOLOGY AND DIET
One of the most specialized filter-feeding teleosts, filters small (even unicellular) invertebrates and algae from the water. Also scrapes off algae and other *aufwuchs* (plants and animals adhering to parts of rooted aquatic plants and other open surfaces) from the substrate.

REPRODUCTIVE BIOLOGY
Exhibits typical spawning clasp. Several thousand small (0.04 in/1 mm) buoyant spherical eggs are released during one spawning phase that comprises numerous spawning bouts. Eggs with a large oil globule hatch after one day at 86°F (30°C). No parental care.

CONSERVATION STATUS
Not listed by the IUCN.

SIGNIFICANCE TO HUMANS
A valued food fish in Southeast Asia, and popular in the aquarium trade. ◆

Siamese fighting fish
Betta splendens

FAMILY
Osphronemidae

TAXONOMY
Betta splendens Regan, 1910, Menam River [= Mae Nam Chao Phraya], Thailand.

OTHER COMMON NAMES
English: Betta; French: Combatant, combattant du Siam; German: Siamesischer Kampffisch; Spanish: Combatiente siamés.

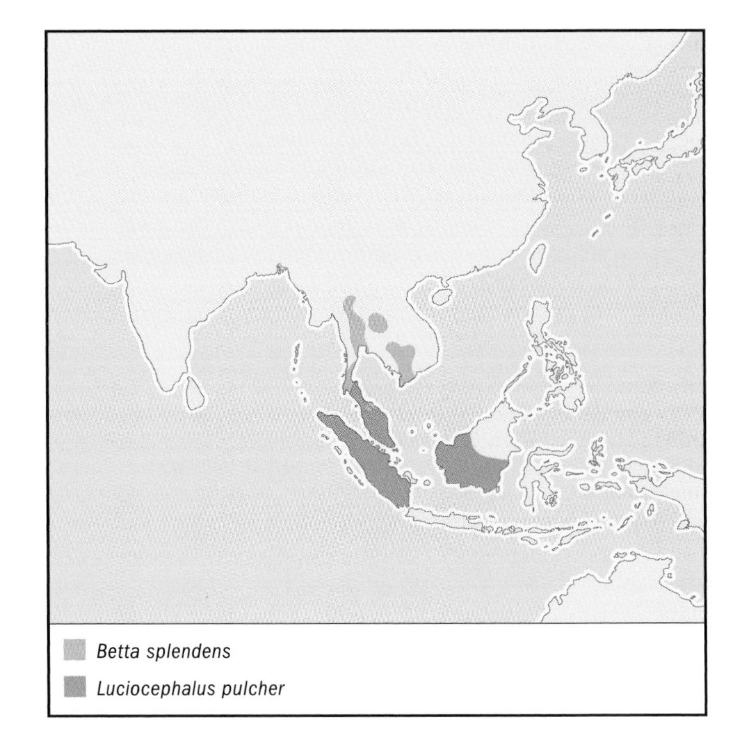

PHYSICAL CHARACTERISTICS
Up to 2.4 in (6 cm). Elongate cylindrical body, dorsal fin short with one to two spines and seven to 10 soft rays. Anal fin is long with two to five spines and 21–26 soft rays, caudal fin rounded. First soft ray of pelvic fin elongated. Sexually dimorphic; males have larger fins and a brighter coloration, females less conspicuously colored. Wild type with bluish body and blue and red fins. Two vertical iridescent marks on the opercle. Some breeds with greatly enlarged fins (sail fin) and different colors or combinations thereof, some almost completely red, blue, yellow, or black.

DISTRIBUTION
The original distributional range comprises the Chao Phraya basin in Thailand and northernmost Malay Peninsula (north of Isthmus of Kra). The species has been transported and released in various countries in Southeast Asia, and can now be found even in the Dominican Republic, Colombia, Brazil, and Florida, in the United States.

HABITAT
Tolerates a wide range of water parameters. Common in stagnant or standing water bodies with dense aquatic vegetation, especially in rice paddies and in canals. May dig into the mud when the water recedes and survive weeks in a small cocoon-like structure made of mud and probably mucus.

BEHAVIOR
This species is well known for its prominently developed aggressive behavior, especially against conspecific males. Confined to small tanks, males fight until one of them is killed. In Thailand, various breeds of *Betta splendens* are used in popular fighting matches in which people bet on the outcome.

FEEDING ECOLOGY AND DIET
Carnivorous, feeding mostly on small aquatic invertebrates, such as zooplankton and insect larvae.

REPRODUCTIVE BIOLOGY
The male constructs a bubble nest and aggressively defends the territory around it. Has typical spawning clasp. After the

spawning claps, male and female show spawning rigor, from which the male recovers earlier (after 4 s) than the female (after about 20 s). Up to 400 spherical eggs with a diameter of 0.04 in–0.05 in (1–1.4 mm) are laid per spawning sequence. They contain no oil globule and sink toward the bottom. While eggs are sinking they are collected by the male, later joined by the female, and stored in the nest. Hatching takes place after 32 to 35 hours at 84.2°F (29°C). Larvae swim free on the fourth day.

CONSERVATION STATUS
Not listed by the IUCN.

SIGNIFICANCE TO HUMANS
Very popular aquarium fish. Because of its hardy nature, often among the first species of freshwater fishes kept by beginners to the aquarium hobby. Of no interest to fisheries due to its small size. ◆

Pikehead
Luciocephalus pulcher

FAMILY
Osphronemidae

TAXONOMY
Diplopterus pulcher Gray, 1831, type locality not specified.

OTHER COMMON NAMES
German: Hechtkopf; Spanish: Cabeza de lucio.

PHYSICAL CHARACTERISTICS
Length 6 in (15 cm). Elongate, with large head and highly protrusible upper jaws. Labyrinth simple. Has separate endoskeletal ossification in front of basihyal, erroneously termed "gular element" by some authors. Caudal fin is rounded, dorsal fin is short with nine to 11 soft rays, anal fin with a median incisure and 18–19 soft rays, pelvic fin with filamentously elongate first soft ray extending to the end of the body. Color is light beige with longitudinal dark brown bands. There are several transverse stripes in the caudal fin that may be broken up into series of spots. A second, undescribed species from Sumatra and Borneo has numerous green iridescent spots along the body.

DISTRIBUTION
Malay Peninsula, Sumatra, and Borneo.

HABITAT
Smaller streams with acidic to highly acidic (down to a pH of 3.5) water; often caught among aquatic vegetation.

BEHAVIOR
The pikehead can protrude its upper jaw to about 33% of the head length, which is quite exceptional among teleosts. In captivity usually remains hidden among vegetation, from where it moves slowly toward prey. At a certain distance of around 3.9 in (10 cm), it makes a sudden rapid lunge (with a peak velocity of 150 cm/s^{-1} and a duration of 0.03 sec.) to surround the prey by protruding its upper jaws and expanding its huge mouth cavity. Suction seems to play only a minor role in capturing the prey.

FEEDING ECOLOGY AND DIET
Appears to feed exclusively on small fishes.

REPRODUCTIVE BIOLOGY
Mouth brooder with male parental care. Male defends territory around spawning site. Displays conspicuous sexually dichromatic coloration only during courtship and spawning. Mating with a reduced spawning clasp takes place at the bottom. All eggs are released during a single spawning bout, sink to the bottom, and are taken up into the male's mouth. The male mouth broods up to 150 eggs with a diameter of around 0.12 in (3 mm) for about four weeks. Eggs are pear-shaped, with a striking pattern of parallel surface ridges leading toward the micropyle (a preformed opening, the only place where sperm can enter the egg), where the ridges end in a counterclockwise spiral. This unique surface pattern also occurs in the genera *Parasphaerichthys*, *Ctenops*, and *Sphaerichthys*, demonstrating the close relationship of the four groups. Egg surface pattern may represent a sperm guiding device to enhance fertilization success. Upon release from the male's mouth, young pike-heads already measure 0.6 in (1.5 cm) long.

CONSERVATION STATUS
The species is not threatened or endangered, but may suffer in the future from habitat destruction.

SIGNIFICANCE TO HUMANS
Ornamental fish for specialized hobbyists. ◆

Giant gourami
Osphronemus goramy

FAMILY
Osphronemidae

TAXONOMY
Osphronemus goramy Lacepéde, 1801, Mauritius; China; Jakarta [Batavia], Java, Indonesia.

OTHER COMMON NAMES
French: Gourami géant; German: Riesengurami; Spanish: Gurami gigante, gurami comestible.

PHYSICAL CHARACTERISTICS
Largest species of anabantoids; up to 23.6 in (60 cm) and 19.8 lb (9 kg). High body, laterally compressed. Lateral line not interrupted and nearly straight. Dorsal fin has 11–14 spines and 12–14 soft rays. Anal fin has 10–11 spines and 20–23 soft rays. First soft ray of pelvic fin is very long, extending beyond caudal fin. Labyrinth highly complex in the adult, with the numerous folds supported by bony lamellae. Large males with prominent hump on the head. Juveniles have eight to 10 dark vertical bars and a conspicuous eyelike spot above the anal fin. Adults drab, grayish, olivaceous above and silvery or yellowish below in both sexes; no sexual dichromatism.

DISTRIBUTION
The species has been introduced in various areas outside of its natural range. Its original distribution probably comprised Thailand, the Malay Peninsula, Sumatra, Borneo, and Java. It now has established populations in India, Sri Lanka, Philippines, West Papua, Papua New Guinea, Madagascar, New Caledonia, and Colombia.

HABITAT
Occurs in swamps, lakes, and medium-to-large rivers. May also tolerate brackish water conditions.

BEHAVIOR
Nothing is known about the behavior of the giant gourami in the wild.

FEEDING ECOLOGY AND DIET
The giant gourami is omnivorous, feeding on plants, smaller vertebrates, invertebrates, and even dead animals.

REPRODUCTIVE BIOLOGY
Reaches maturity after the fourth year. The male builds a spherical to oval nest 11.8 in (30 cm) long, 7.9 in (20 cm) wide, and 3.9 in (10 cm) deep, which resembles a bird's nest, close to or below the water surface using mainly plant material. Nest building takes eight to 10 days. Eggs are deposited in the nest, and guarded by the male and female, and fanned through movements of the pectoral fins. The yellowish eggs are around 0.11 in (2.7 mm) in diameter and contain a large oil globule that makes them buoyant. Eggs may number more than 1,500 per nest. Hatching occurs after 10 days. Newly hatched fry measure 0.24–0.35 in (6–9 mm). Yolk sac is resorbed at around 15 days. After four months a length of 3.9 in (10 cm) may be reached.

CONSERVATION STATUS
Not listed by the IUCN.

SIGNIFICANCE TO HUMANS
A valued, common, and delicious food fish, eaten steamed, fried or baked. It has been introduced into more than 20 countries all over the world, and is important in aquaculture in tropical Asia. ◆

Pearl gourami
Trichogaster leeri

FAMILY
Osphronemidae

TAXONOMY
Trichopus leeri Bleeker, 1852, Sumatra.

OTHER COMMON NAMES
English: Diamond gourami, lace gourami, mosaic gourami; French: Gourami mosaïque, gourami perlé; German: Mosaikfadenfisch; Spanish: Gurami perla, Gurami mosáico.

PHYSICAL CHARACTERISTICS
Length 4.7 in (12 cm). Body laterally compressed with short dorsal fin of 5–7 fin spines and 8–10 soft rays; long anal fin with 12–14 spines and 25–30 soft rays. Pelvic fins with an extremely prolonged first soft ray behind the spine, followed by four short soft rays. This pelvic filament reaches up to two-thirds of the fish's total length, is highly movable in all three dimensions, and is used as an organ of taste because its surface is covered with numerous taste buds. A tactile function has also been demonstrated. Males can be distinguished from females by the posterior rays of the soft dorsal and soft anal fin being prolonged and projecting beyond the fin membrane. Coloration of the body consists of a grayish background, with numerous bright white spots all over the body and fins (hence the name "pearl" gourami), a black midlateral stripe that extends from the snout to the caudal peduncle, and a black spot at the base of the caudal fin. Males have bright nuptial coloration, especially when sexually active.

DISTRIBUTION
Freshwaters of Thailand, peninsular Malaysia, Sumatra, and Borneo.

HABITAT
Smaller or larger forest streams, usually with slightly acidic water.

BEHAVIOR
Nothing is known about the behavior of the pearl gourami in the wild.

FEEDING ECOLOGY AND DIET
No field data on gut contents is available, but judging from the small mouth and the numerous gill rakers, feeds on small aquatic invertebrates.

REPRODUCTIVE BIOLOGY
The onset of a reproductive period is characterized by an increasing aggressiveness of the male and the establishment of a breeding territory in which it builds a foam nest. Parts of aquatic vegetation or detritus may be incorporated into the foam mass. The foam nest is built from air gulped in at the surface and released as foam-coated bubbles below the nesting site, from either the mouth or the opercular cleft. The male's nuptial coloration is bright orange on the throat, pelvic filament, and anterior anal fin, with numerous bright white spots on the body and fins. Spawning takes place below the nest. The typical spawning clasp is performed. Eggs are usually released when the female's genital opening points to the nest. Up to 135 eggs are released during each bout of spawning, with up to 1,000 eggs per spawning sequence. A short phase (2–10 seconds) of spawning rigor follows egg release, during which the male and female remain motionless in their clasp. Eggs contain a large oil globule that renders them buoyant. Floating eggs are collected by the male and stored in the nest. Hatching occurs after 24 hours at 84.2°F (29°C). Free swimming is achieved after two to three days. The pearl gourami exhibits male parental care.

CONSERVATION STATUS
Not listed by the IUCN.

SIGNIFICANCE TO HUMANS
Consumed as a food fish and used in aquaculture. A common and popular aquarium fish. ◆

Resources

Books

Cuvier, G., and A. Valenciennes. *Histoire naturelle des poisons,* Vol. 7. Paris/Strasbourg: Levrault, 1831.

Fuller, P. L., L. G. Nico, and J. D. Williams. *Nonindigenous Fishes Introduced into Inland Waters of the United States.* Bethesda, MD: American Fisheries Society, Special Publication 27, 1999.

Kottelat, M., A. J. Whitten, S. N. Kartikasari, and S. Wirjoatmodjo. *Freshwater Fishes of Western Indonesia and Sulawesi.* Jakarta: Periplus Editions, 1993.

Vierke, J. *Labyrinthfische und verwandte Arten.* Wuppertal-Elberfeld, Germany: Engelbert Pfriem Verlag, 1978.

Periodicals

Bleeker, P. "Memoires sur les poissons á pharyngiens labyrinthiformes de l'Inde archipelagique." *Natuurk. Verh. Akad. Amsterdam.* 19 (1879): 1–56.

———. "Over de platsing in het stelsel van de Luciocephaloiden." *Natuurk. Tijdschr. Nederl. Ind.* 20 (1859): 395–397.

Britz, R. "Ablaichverhalten und Maulbrutpflege bei *Luciocephalus pulcher.*" *Aquar. Terr. Ztschr.* 47 (1994): 790–795.

———. "Egg Surface Structure and Larval Cement Glands in Nandid and Badid Fishes (Teleostei, Percomorpha), with Remarks on Phylogeny and Zoogeography." *American Museum Novitates* 3195 (1997): 1–17.

———. "The Genus *Betta*—Monophyly and Intrarelationships, with Remarks on the Subfamilies Macropodinae and Luciocephalinae (Teleostei: Osphronemidae)." *Ichthyological Exploration of Freshwaters* 12 (2001): 305–318.

———. "Ontogenetic Features of *Luciocephalus* (Perciformes, Anabantoidei) with a Revised Hypothesis of Anabantoid Intrarelationships." *Zoological Journal of the Linnean Society* 112 (1994): 491–508.

Britz, R., and J. A. Cambray. "Structure of Egg Surfaces and Attachment Organs in Anabantids." *Ichthyological Exploration of Freshwaters* 12 (2001): 267–288.

Britz, R., M. Kokoscha, and R. Riehl. "The Anabantoid Genera *Ctenops, Luciocephalus, Parasphaerichthys,* and *Sphaerichthys* (Teleostei: Perciformes) as a Monophyletic Group: Evidence from Egg Surface Structure and Reproductive Behavior." *Japanese Journal of Ichthyology* 42 (1995): 71–79.

Hall, D. D., and R. J. Miller. "A Qualitative Study of Courtship and Reproductive Behavior in the Pearl Gourami, *Trichogaster leeri.*" *Behaviour* 32 (1991): 70–84.

Jordan, D. S. "A Classification of Fishes Including Families and Genera as far as Known." *Stanford Univ. Publ., Univ. Ser., Biol. Sci.* 3 (1923): 77–243.

Kratochvil, H. "Beiträge zur Lautbiologie der Anabantoidei—Bau, Funktion und Entwicklung von lauterzeugenden Systemen." *Zool. Jb. Physiol.* 89 (1985): 203–255.

Lauder, G. V., and K. F. Liem. "The Evolution and Interrelationships of the Actinopterygian Fishes." *Bulletin of the Museum for Comparative Zoology* 150 (1983): 95–197.

———. "Prey Capture by *Luciocephalus pulcher*: Implications for Models of Jaw Protrusion in Teleost Fishes." *Environmental Biology of Fishes* 6 (1981): 257–268.

Liem, K. F. "The Comparative Osteology and Phylogeny of the Anabantoidei." *Illinois Biological Monograph* 30 (1963): 1–149.

Peters, H. M. "On the Mechanism of Air Ventilation in Anabantoids (Pisces: Teleostei)." *Zoomorphology* 89 (1978): 93–123.

Roberts, T. "*Osphronemus exodon,* a New Species of Giant Gouramy with Extraordinary Dentition from the Mekong." *Natural History Bulletin of the Siam Society* 42 (1994): 67–77.

———. "Systematic Revision of the Southeast Asian Anabantoid Fish Genus *Osphronemus,* with Descriptions of Two New Species." *Ichthyological Exploration of Freshwaters* 2 (1992): 351–360.

Roxas, H. A., and A. F. Umali. "Fresh-Water Fish Farming in the Philippines." *Philippine Journal of Science* 63 (1937): 433–468.

Sanders, M. "Die fossilen Fische der alttertiären Süsswasserablagerungen aus Mittel-Sumatra." *Verh. Geol.-Mijnbouw. Genootsch. Nderl. Kol. Geol. Ser.,* D. XI, 1ste St. (1937): 1–144.

Vierke, J. "Beiträge zur Ethologie und Phylogenie der Familie Belontiidae (Anabantoidei, Pisces)." *Zeitschrift für Tierpsychologie* 38 (1975): 163–199.

———. "Brutpflegestrategien bei Belontiiden (Pisces, Anabantoidei)." *Bonn. Zool. Beitr.* 42 (1991): 299–324.

Weber, H. "Die Sinnesfunktion der freien Bauchflossenstrahlen der Labyrinthfische (Anabantidae) und ihr Zusammenwirken mit den Augen." *Ztschr. Vergl. Physiol.* 47 (1963): 77–110.

Organizations

California Academy of Sciences. 55 Concourse Drive, Golden Gate Park, San Francisco, CA 94118-4599 USA. Phone: (415) 750-7047. Fax: (415) 750-7148. E-mail: info@calacademy.org Web site: <http://www.calacademy.org>

Food and Agriculture Organization of the United Nations (FAO) Fisheries. Viale delle Terme di Caracalla, Rome, 00100 Italy. Phone: 39 (06) 5705 1. Fax: 39 (06) 5705 3152. E-mail: FAO-HQ@fao.org Web site: <http://www.fao.org/>

Ralf Britz, PhD

Channoidei
(Snakeheads)

Class Actinopterygii
Order Perciformes
Suborder Channoidei
Number of families 1

Photo: Snakehead fishes move along the ground using their fins at a fish farm in Singapore. They are popular in parts of Asia for medicinal benefits, are aggressive, and can live out of water for days. (Photo by Jonathan Searle/Reuters New-Media inc./Corbis. Reproduced by permission.)

Evolution and systematics

The single-family (Channidae) suborder Channoidei (snakeheads) is composed of only two genera, *Channa*, which includes all Asian species, and *Parachanna*, which includes all African species. The two genera differ mainly in the morphology of the air-breathing (suprabranchial) organ; that of the latter being less developed. Currently, 25 species of *Channa* and three species of *Parachanna* have been recognized. Although snakeheads have been considered a sister group of the Anabantoidei (climbing gouramies) or the Synbranchiformes (swamp eels), the relationship of snakeheads to other fish groups still remains uncertain. The earliest fossil record of snakeheads is *Eochanna chorlakkiensis* from the Kuldana Formation, Pakistan (middle Eocene).

Physical characteristics

Snakeheads form a morphologically unique group of primarily freshwater fishes, which greatly vary in size at maturity. Some species have distinctively small pelvic fins, while a few others lack them completely. Generally, snakeheads have an elongated cylindrical body; flattened head; long, entirely soft-rayed dorsal and anal fins; a large mouth with well-developed teeth on both upper and lower jaws; tubelike anterior nostrils; a round to somewhat truncate caudal fin; cycloid or ctenoid body scales; shield-like scales on a head that superficially resemble that of a snake; a lengthy, elongated swim bladder reaching to the caudal peduncle region; and an accessory air-breathing apparatus (suprabranchial organ) in the head region. This suprabranchial organ is mainly composed of three parts: a suprabranchial chamber, epibranchial respiratory fold, and hyomandibular process.

Species of snakeheads can be distinguished based on coloration, meristics, and morphometrics, as well as the distribution of scales on the underside of the lower jaw, the shape of the head, and the morphology of the suprabranchial organs. Much taxonomic confusion has resulted from the fact that coloration in each species changes dramatically during growth, and in many cases, the color of juveniles is completely different from that of adults. One such case is the giant snakehead (*Channa micropeltes*), a popular aquarium fish that has striking black and red "racing stripes" as a juvenile, but variegated blackish markings as an adult. Many species have distinct adult coloration, such as ocellus (ocelli) on the body and/or caudal fin, vertical bands on the pectoral fins, and small spots on the body.

Distribution

Members of the genus *Channa* are widely distributed, occurring from Iran in the west, to China and southeastern Russia in the east, and throughout Southeast Asia, extending downward into the Philippines and Indonesia (Java being the southernmost location). Species of the genus *Parachanna* are restricted to central West Africa. Species are most diverse in tropical Asia. Some species are endemic to restricted areas that include special features such as tropical rainforests, such

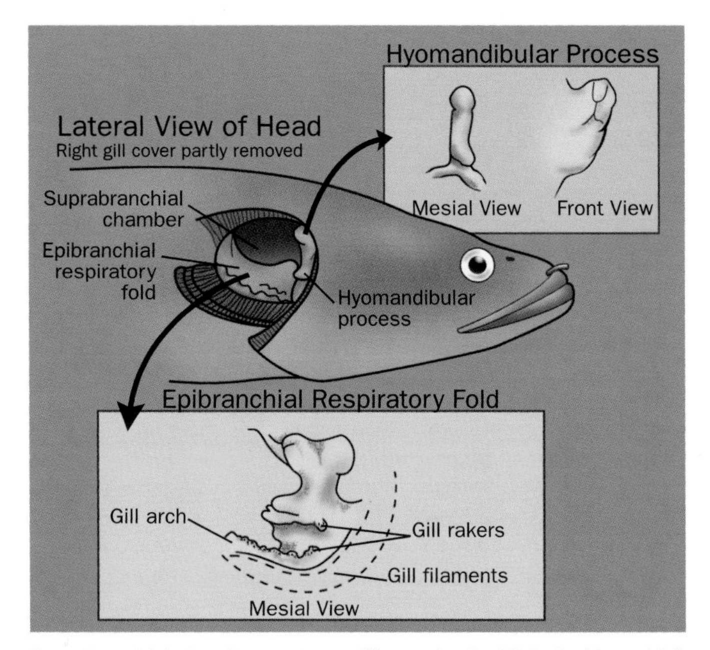

Superbranchial chamber anatomy. (Illustration by Michelle Meneghini)

as the walking snakehead (*Channa orientalis*) from southern Sri Lanka, and the orange-spotted snakehead (*Channa aurantimaculata*) and rainbow snakehead (*Channa bleheri*) from northern Assam. Some species are now found outside their natural distributions, apparently as a result of human introduction. These include established populations of the northern snakehead (*Channa argus*) in Japan, the United States, and Aral Sea basin; the blotched snakehead (*Channa maculata*) in Japan; the bullseye snakehead (*Channa marulius*) in the United States; and the striped snakehead (*Channa striata*) in the islands of Indonesia east of Wallace's line, Papua New Guinea, New Caledonia, Fiji, Hawaii, Mauritius, and Madagascar. In Madagascar the striped snakehead has had a severely detrimental effect upon the island's endemic freshwater fish fauna.

Habitat

Most snakeheads prefer stagnant or slow-running waters, usually hiding under vegetation, rocks, and sunken trees. However, large species such as the northern snakehead, bullseye snakehead, and giant snakehead can usually be found inhabiting relatively deep waters with somewhat heavy vegetation in low, open country such as large rivers, swamps, ponds, and reservoirs. One species, the banka snakehead (*Channa bankanensis*), has been found only in peat swamps, which have dark brown, highly acid water. Most small species, such as the walking snakehead, usually live in mountain streams, but can also be found in lowland habitats. The northeastern Indian barca snakehead (*Channa barca*), a large species attaining 35.4 in (90 cm), is reported to live in holes within the banks of ponds and rivers. Many snakeheads are highly adaptable, being tolerant to a wide range of environments, including polluted waters.

Behavior

Highly territorial, snakeheads usually stay hidden, and migrate only short distances. Fry of some species have been reported to be cannibalistic, opportunistically eating eggs from a later spawn. Generally, the young tend to school. Snakeheads appear to need to breach the water surface periodically to exchange the air in their suprabranchial chamber. Experiments have shown that breathing surface air is far more important to them than using their gills. This is supported by evidence from the drowning deaths of fishes that have been caught in nets under water, and could not surface to breathe. Some species of snakeheads can live out of water for several days if their bodies are wet, amazingly migrating on land during the raining season by using their bodies, pectoral fins, and caudal fins. However, no one has reported them feeding while moving on land. The striped snakehead (*Channa striata*) has been reported to survive in the bottom mud of lakes, swamps, and canals that have dried up; fishermen using long knives to cut away the mud in layers have found these fishes singly or in clusters within cavities of the mud.

Feeding ecology and diet

Active during the day, although the dwarf snakehead (*Channa gachua*) has been reported to be nocturnal, all snakeheads are predatory, ambush feeders, eating almost any animal smaller than their mouths. Usually solitary feeders as adults, as juveniles they actively migrate in schools, hunting foods such as zooplankton, small insects, and crustaceans. Adult snakeheads feed on everything from insects (both terrestrial and aquatic) to young birds, including fishes, frogs, tadpoles, lizards, geckos, mice, rats, and ducks. They primar-

Mating posture of snakeheads. (Illustration by Michelle Meneghini)

ily hunt by sight, but smell and other senses may also be involved. They sometimes jump up from the water surface to grasp their prey.

Reproductive biology

Although the reproductive biology of many species of snakeheads is still unknown, it does appear that they are monogamous, exhibit parental care, and become aggressive, especially so during breeding. Many are known to be nest breeders, the parents first clearing vegetation and then building a simple circular nest at the water surface. In these species, there is a spawning embrace with the male encircling the female, squeezing the eggs out, and fertilizing them. The eggs float upward into the nest, usually guarded. After hatching, the fry will be cared for by either parent, dependent on species. The giant snakehead, with its strong canine-like teeth, has been known to attack and seriously injure humans who disturb it, especially while guarding its brood. On the other hand, two small species, the walking and dwarf snakeheads, are known to be mouth brooders, with the male reported to keep the fertilized eggs, and later fry (for a few days), in his mouth.

Conservation status

Although generally not threatened, the status of populations of some species are poorly known. No species are included on the IUCN Red List.

Significance to humans

Most channid species are important food fishes in southern Asia and China and the flesh is considered delicious. Some medium-to-large species are cultured in ponds or in cages set in slow-running rivers. Although sold fresh and sun-dried, several species, northern snakehead (*C. argus*) and striped snakehead (*C. striata*), are known as intermediate hosts of parasites harmful to humans, including *Gnathostoma* (jaw worms), and should be cooked thoroughly at a high temperature before eating. Two species have even been used as predators to control tilapia in aquaculture ponds. Larger species are popular game fishes in Asia, and several species feature in local beliefs or myths. In October 2002, the U.S. Fish and Wildlife Service passed a rule that prohibits the importation of live snakeheads into the United States except by scientific, medical, or educational organizations, which would be required to obtain a permit to import the fishes.

1. Orange-spotted snakehead (*Channa aurantimaculata*); 2. Walking snakehead (*Channa orientalis*); 3. Rainbow snakehead (*Channa bleheri*); 4. Giant snakehead (*Channa micropeltes*); 5. Giant snakehead (*Channa micropeltes*) juvenile; 6. African snakehead (*Parachanna obscura*). (Illustration by Michelle Meneghini)

1. Striped snakehead (*Channa striata*); 2. Striped snakehead (*Channa striata*) juvenile; 3. Bullseye snakehead (*Channa marulius*); 4. Bullseye snakehead (*Channa marulius*) juvenile; 5. Ocellated snakehead (*Channa pleurophthalmus*); 6. Northern snakehead (*Channa argus*). (Illustration by Michelle Meneghini)

Species accounts

Northern snakehead

Channa argus

FAMILY
Channidae

TAXONOMY
Ophicephalus argus Cantor, 1842, Chusan Island, China.

OTHER COMMON NAMES
German: Amur-Schlangenkopf; Chinese: Hey yu; Japanese: Kamuruchi; Russian: Zmeegolov.

PHYSICAL CHARACTERISTICS
Length 44.1 in (112 cm); a large snakehead. Has relatively small scales (LL=61–72), no scales on the underside of the jaw, large canine-like teeth on the upper and lower jaws; two horizontal rows of 9–15 irregular dark brown blotches that sometimes coalesce, and brownish pectoral fins with a black mark at the base.

DISTRIBUTION
Central China (Yangtze to Luan River basins) including Korea to the Amur River basin, southern Russia; introduced and established in Japan, the United States, and republics in the former Soviet Union adjacent to Caspian Sea.

HABITAT
Lakes, swamps, marshes, reservoirs, and rivers in lowland slow-moving to stagnant temperate waters.

BEHAVIOR
Known to burrow into the mud and hibernate when the water becomes very cold, has not been reported to move over land.

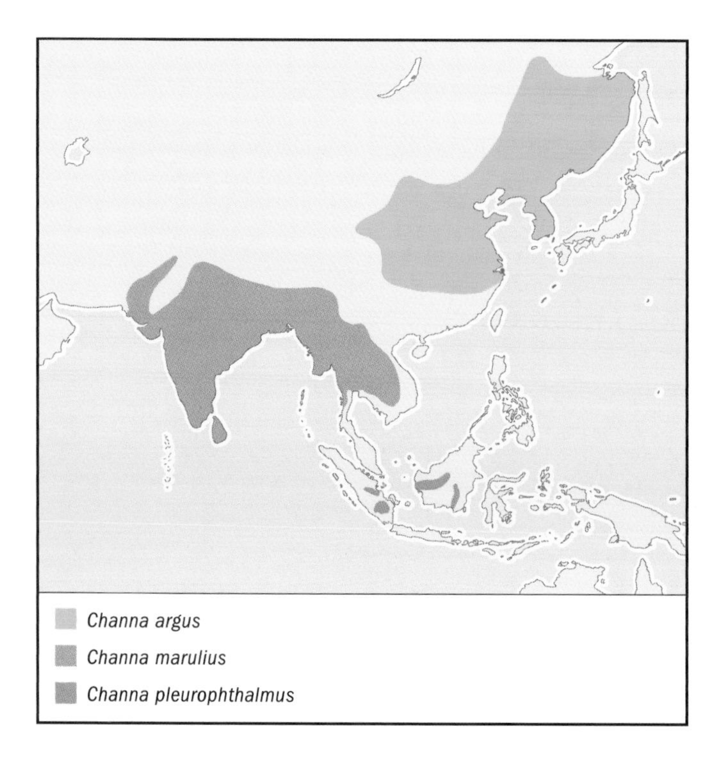

FEEDING ECOLOGY AND DIET
Feeds on fishes, frogs, prawns, worms, crayfishes, and juvenile water birds.

REPRODUCTIVE BIOLOGY
Spawns usually in early morning from spring to summer, Male and female build donut-shaped nest about 39.4 in (100 cm) in diameter using aquatic plant debris. Female deposits an average of 7,300 eggs 0.06–0.08 in (0.15–0.2 cm) in diameter; incubation takes about 45 hours at 77°F (25°C). Guarded and cared for by the parents, newly hatched fry are all black, and leave the nest when they reach 0.03 in (0.8 cm), at which size the body has become yellowish.

CONSERVATION STATUS
Not threatened; common in almost all areas within its distribution.

SIGNIFICANCE TO HUMANS
One of the most popular food fishes in China, where it is believed to be beneficial for helping in postpartum recovery. ◆

Orange-spotted snakehead

Channa aurantimaculata

FAMILY
Channidae

TAXONOMY
Channa aurantimaculata Musikasinthorn, 2000, Dibrugarh, Assam, India.

OTHER COMMON NAMES
Assamese: Naga cheng.

PHYSICAL CHARACTERISTICS
Maximum length 19.7 in (50 cm). Has two large scales on each side of lower jaw undersurface, scales relatively large (LL=51–54), black spot at the base of the pectoral fins with five vivid black vertical broad bands, and relatively small pelvic fins (less than half of pectoral fins). Very colorful, upper half of body dark brown to black, with seven or eight large irregular orange blotches, becoming yellow, golden, or orange below, and changing to blue ventrally, with many scattered small black spots.

DISTRIBUTION
Known only from Dibrugarh (Brahmaputra River basin), Northern Assam, India.

HABITAT
Streams, swamps, and ponds probably throughout the region's discontinuous patches of tropical rainforest.

BEHAVIOR
Effectively transverses wet terrestrial environments.

FEEDING ECOLOGY AND DIET
Nothing known.

REPRODUCTIVE BIOLOGY
Nothing known.

CONSERVATION STATUS
Not listed by the IUCN.

SIGNIFICANCE TO HUMANS
Sold in local markets as food fish; also known as an aquarium fish. ◆

Rainbow snakehead
Channa bleheri

FAMILY
Channidae

TAXONOMY
Channa bleheri Vierke, 1991, upper portion of Dibru River, near Guijan, Brahmaputra River basin, northeastern Assam, India.

OTHER COMMON NAMES
Assamese: Deo cheng.

PHYSICAL CHARACTERISTICS
Length 3.9 in (10 cm), one of the smallest snakeheads. Has relatively large scales (LL= 42–46), no pelvic fins, and a single large scale on each side of the underside of the lower jaw. One of the most colorful snakeheads, has very distinctive 4–11 medium-to-large irregular red or orange spots on the caudal fin which sometimes coalesce; and pectoral fins with a black spot at the base and 7–9 black concentric bands.

DISTRIBUTION
Endemic to Brahmaputra River basin, northern Assam, India.

HABITAT
Forest streams, swamps, and ponds connected with the Brahmaputra River.

BEHAVIOR
Highly capable of terrestrial movement.

FEEDING ECOLOGY AND DIET
Nothing known, but is probably a predator feeding on insects, crustaceans, and small fishes.

REPRODUCTIVE BIOLOGY
The species guards the eggs, which float at the water surface.

CONSERVATION STATUS
Not threatened, but because of its relatively small, highly restricted distribution, along with its popularity among aquarists, there is a possibility that over-collecting of natural populations may cause a serious decline in the future.

SIGNIFICANCE TO HUMANS
Due to some spiritual reasons, not considered a food fish by the local Assam populace (the Assamese common name means "ghost [or spirit] snakehead"); a popular aquarium fish. ◆

Bullseye snakehead
Channa marulius

FAMILY
Channidae

TAXONOMY
Ophiocephalus marulius Hamilton, 1822, ponds and freshwater rivers of Bengal.

OTHER COMMON NAMES
English: Great snakehead; Thai: Pla chon ngu hao, pla kalon; Burmese: Nga yan dyne; Laotian: Pa gooan; Khmer: Trey raws; Bengali (West Bengal): Sal, gajal; Sinhalese: Ara.

PHYSICAL CHARACTERISTICS
Among the largest of snakeheads (48 in or 122 cm maximum length). It has 4–6 large round black blotches on the body; some populations have white margins and small white spots on these blotches. Also has a white or orange-rimmed ocellus on the upper portion of the caudal fin base (sometimes absent or indistinct in indiviuals over 300 mm SL), moderately large scales (LL=52–70), and no scales on the underside of the lower jaw. Also, young may have an orange longitudinal band running from the tip of the head to the caudal fin that fades as the fish matures.

DISTRIBUTION
Widely distributed in tropical Asia from the Indus River basin of Pakistan and the whole Indian subcontinent, including Sri Lanka, through Myanmar and Indochina. Also introduced and established in Florida, United States.

HABITAT
Large rivers, lakes, and reservoirs; prefers clear stretches of water with a sandy and rocky bottom.

FEEDING ECOLOGY AND DIET
Known to feed mainly on fishes. It probably also eats other vertebrates and invertebrates such as frogs and crustaceans.

BEHAVIOR
The species is capable of terrestrial movement.

REPRODUCTIVE BIOLOGY
In West Bengal this species spawns from April to June. Parents use their mouths to cut elaborate passages through the weeds leading to the floating bubble nest, where 2,000–40,000 light red-yellow eggs (0.08 in [2 mm] in diameter) hatch in about 54 hours. The pair also guards the fry.

CONSERVATION STATUS
Not threatened.

SIGNIFICANCE TO HUMANS
Readily seen as a food fish in India and some parts of Southeast Asia. A number of interesting local beliefs are attributed to this fish. In central Thailand, the common name means "cobra snakehead" and comes from the superstition that its bite is very poisonous, leading to death. In Myanmar, the Karen people regard it with awe and avoid eating it, practices that arise out the belief that these fish were formerly humans who were changed into fish for their sins, as well as the belief that if a person eats one, he or she will be transformed into a lion.

Giant snakehead
Channa micropeltes

FAMILY
Channidae

TAXONOMY
Ophicephalus micropeltes Cuvier in Cuvier and Valenciennes, 1831, Java.

OTHER COMMON NAMES
English: Red snakehead; Khmer: Trey chhdaur; Laotian: Pa do; Malay: Ikan toman: Thai: Pla chado; Vietnamese: Cá bong.

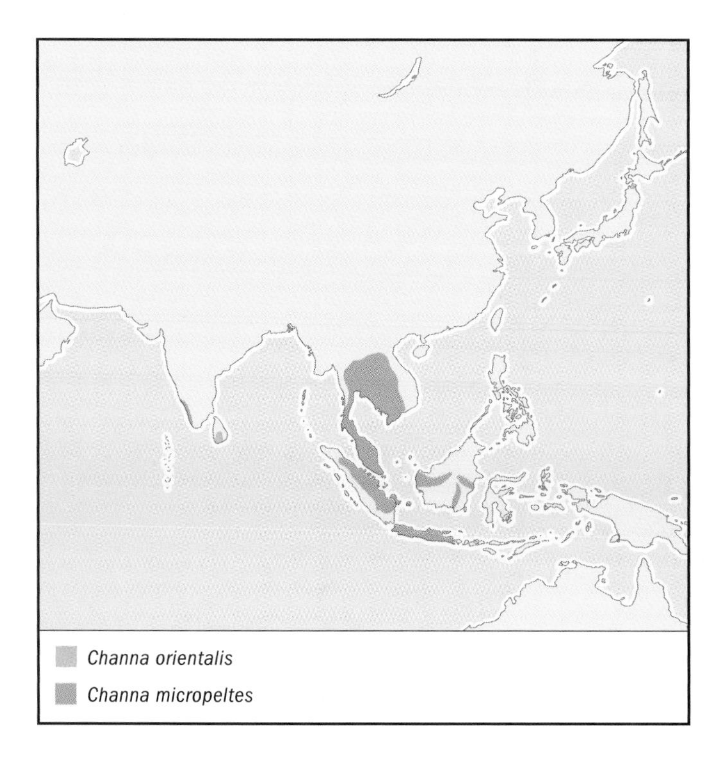

PHYSICAL CHARACTERISTICS
Length 51.2 in (130 cm); one of the largest snakeheads. Has very small scales (LL=83–106); patch of small scales near underside tip of lower jaw; large canine-like teeth on upper and lower jaws. Juveniles have vivid black horizontal stripe above and below a bright orange to red stripe running head to tail. Coloration fades into bluish black and white dappled upper body in adults.

DISTRIBUTION
Widely distributed in Southeast Asia (excluding Myanmar and northern Vietnam) and coastal region of southwestern India.

HABITAT
Large, slow-running to stagnant waters in open country.

BEHAVIOR
The most aggressive snakehead; has not been reported to travel over land.

FEEDING ECOLOGY AND DIET
Extremely voracious, consumes fishes of all kinds and sizes, frogs, and even juvenile ducks and water birds, killing far in excess of need. Young or subadults tend to school when hunting for smaller fishes.

REPRODUCTIVE BIOLOGY
Male and female build donut-shaped floating nest from surrounding aquatic vegetation into which the floating eggs are placed. Parents aggressively guard and care for eggs and fry. Once fry hatch, they soon form and forage as a dense school.

CONSERVATION STATUS
Relatively to very common in Southeast Asian range; status of population in southwestern India is unknown.

SIGNIFICANCE TO HUMANS
One of the most popularly marketed snakehead fishes in Southeast Asia. Adults are primarily game fish, juveniles are recognized worldwide as aquarium fish. ◆

Walking snakehead
Channa orientalis

FAMILY
Channidae

TAXONOMY
Channa orientalis Bloch and Schneider, 1801, India "orientali" (presumably Sri Lanka).

OTHER COMMON NAMES
English: Smooth-breasted snakehead; Sinhalese: Kola kanaya.

PHYSICAL CHARACTERISTICS
Maximum length 3.9 in (10 cm); one of the smallest snakeheads. Pelvic fins absent; body scales relatively large (LL=39–43), and a single large scale is on each side of the underside of lower jaw. Body often has bluish cast with 8–11 black descending bands on a grayish to dark brown dorsal background blending into a whitish ventral, with an orange to red outer margin on the dorsal, anal and caudal fins, a black spot at the base with 3–8 concentric black bands on the pectoral fins. Single orange or yellow-rimmed black ocellus at the back end of the dorsal fin in young usually fades away by maturity.

DISTRIBUTION
Endemic to tropical rainforest environment of southern Sri Lanka.

HABITAT
Small, shallow rivulets barely deeper than its body, and also in mountain streams, small ponds, and ditches.

BEHAVIOR
Known to effectively transverse land.

FEEDING ECOLOGY AND DIET
Feeds mostly on terrestrial and aquatic insects, caddisflies, and small fish.

REPRODUCTIVE BIOLOGY
Mouth brooders. Around 40 eggs are deposited in the floating nest, then 9–10 days after fertilization the male takes the eggs into his mouth for hatching. After hatching, either parent may hold the brood, which exits and enters through the gills, not the mouth. Fry feed on eggs laid later by their mother.

CONSERVATION STATUS
Not threatened, but a decline in the number of unpolluted streams and shrinking rainforests in Sri Lanka will probably affect future populations in the future.

SIGNIFICANCE TO HUMANS
A small number have been exported for the aquarium fish trade. ◆

Ocellated snakehead
Channa pleurophthalmus

FAMILY
Channidae

TAXONOMY
Ophicephalus pleurophthalmus Bleeker, 1851, Bandjermassing.

OTHER COMMON NAMES
Indonesian: Kerandang.

PHYSICAL CHARACTERISTICS
Total length 15.7 in (40 cm). Body is distinctively compressed laterally and quite deep compared to other snakeheads, has moderately large scales (LL= 49–55), large canine-like teeth on the upper and lower jaw, and patch of small scales near the tip of the underside of the lower jaw. Body grayish with scattered small black spots and several orange or orangey red-rimmed black ocelli, yellowish black to gray pectoral fins, and a single ocellus on the gill cover and middle of the caudal fin.

DISTRIBUTION
Islands of Sumatra and Borneo (western and southern portions), Indonesia.

HABITAT
Usually lowland, slow-moving murky rivers.

BEHAVIOR
Not known to be capable of terrestrial movement.

FEEDING ECOLOGY AND DIET
Nothing known; presumed to be a predator, feeding on fishes and other small aquatic animals.

REPRODUCTIVE BIOLOGY
Nothing known, but like other medium-to-large snakeheads, is probably a bubble nest builder.

CONSERVATION STATUS
Not threatened.

SIGNIFICANCE TO HUMANS
Marketed as a food fish within its distribution, also quite popular as an aquarium fish. ◆

Striped snakehead
Channa striata

FAMILY
Channidae

TAXONOMY
Ophicephalus striatus Bloch, 1793, Tranquebar, Malabar coast, India.

OTHER COMMON NAMES
English: Chevron snakehead, snakehead murrel; German: Quergestreifter Schlangenkopf; Burmese: Nga-yan; Khmer: Trey raws; Bengali: Shol; Laotian: Pa kho; Malay: Aruan; Indonesian: Ikan gabus; Thai: Pla chon; Vietnamese: Cá lòk.

PHYSICAL CHARACTERISTICS
Length 23.6 in (60 cm). No scales on underside of the jaw, no large canine-like teeth on the upper jaw, moderately large scales (lateral line scales 50–61). Generally dark brown above, extending into irregular blackish bands below; no bands or spots on pectoral fins. Juveniles have black spot (sometimes forming an ocellus) at posterior end of dorsal fin, but this disappears as fish reaches maturity.

DISTRIBUTION
The most widely naturally distributed snakehead, occurs from Pakistan through Southeast Asia east to Yunnan, southern China. Has also been introduced and established in tropical islands, including Madagascar, Hawaii, New Guinea, the Philippines, and Sulawesi, Indonesia.

HABITAT
Tropical stagnant to slow-running lowland waters (prefers the former) with muddy bottoms, such as ponds, swamps, and ditches.

BEHAVIOR
Can move over land during rainy season. Reported to survive in cavities in the bottom mud of lakes, swamps, and canals that have dried up.

FEEDING ECOLOGY AND DIET
In native waters, feeds on smaller fishes, frogs, prawns, and worms.

REPRODUCTIVE BIOLOGY
Spawns year round, builds a nest in shallow (11.8–39.4 in/30–100 cm), swampy stagnant areas near waters' edge. Male and female use mouth and tail to clear away dense vegetation to make donut-shaped floating nest about 11.8 in (30 cm) in diameter, into which the translucent yellow nonadhesive eggs (about 0.06 in/0.15 cm in diameter) are placed. Hatching period lasts about three days, during which male guards the nest

until a short while after the vivid reddish orange fry hatch. After hatching, fry move in a dense school while foraging, still protected by male.

CONSERVATION STATUS
Not threatened; common in almost all areas within its distribution.

SIGNIFICANCE TO HUMANS
One of the most common and important freshwater food fishes in tropical Asia, also used for control of tilapia in pond aquaculture. In Myanmar, included in a spiritual ceremony to help a sick person recover. ◆

African snakehead
Parachanna obscura

FAMILY
Channidae

TAXONOMY
Ophiocephalus obscurus Günther, 1861, West Africa.

OTHER COMMON NAMES
German: Dunkelbäuchiger Schlangenkopf; Dinka: Abioth; Ga: Hauti; Hausa: Tuhi.

PHYSICAL CHARACTERISTICS
Length 15.7 in (40 cm). Medium-sized, head somewhat concave, pointed in lateral view and depressed anteriorly; has small scales (LL= 65–78), patch of small scales near the tip of the underside of the lower jaw, and no large canine-like teeth on the upper jaw. Body brown to dark brown with several large black blotches, blackish mark at base of pectoral fins along with several rows of black spots. Young have broad blackish band on body sides and light-edged ocellar spot at the caudal fin base.

DISTRIBUTION
The White Nile and from the Senegal and Chad Rivers to the Congo River basin in western Africa; most widely distributed African species.

HABITAT
Slow-running to stagnant waters, preferably with heavy vegetation, including rivers, streams, lakes, lagoons, and marshes.

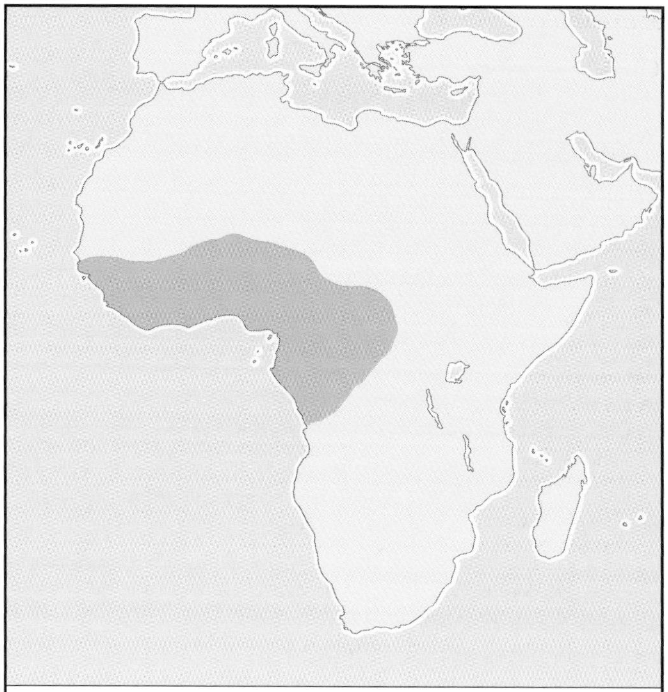

Parachanna obscura

BEHAVIOR
Adults are solitary ambush predators. Overland movement of this species has never been reported.

FEEDING ECOLOGY AND DIET
Juveniles feed on prawns, copepods, and aquatic insect larvae; adults prefer mostly fishes.

REPRODUCTIVE BIOLOGY
Female lays 2,000–3,000 eggs in October and November which are probably deposited in a nest and guarded by the male for 4–5 days after hatching.

CONSERVATION STATUS
Not threatened.

SIGNIFICANCE TO HUMANS
Of minor importance as a food fish within its distribution, cultured in ponds and also used to control tilapia in aquaculture. ◆

Resources

Books

Breder, C. M., Jr., and D. E. Rosen. *Modes of Reproduction in Fishes.* Garden City, NY: Natural History Press, 1966.

Day, F. *The Fishes of India; Being a Natural History of the Fishes Known to Inhabit the Seas and Fresh Waters of India, Burma and Ceylon,* Volume I. London: William Dawson and Sons Ltd., 1958.

———. Day, F. *The Fishes of India; Being a Natural History of the Fishes Known to Inhabit the Seas and Fresh Waters of India, Burma and Ceylon,* Part 2. London: William Dawson and Sons Ltd., 1876.

Kottelat, M., A. J. Whitten, S. N. Kartikasari, and S. Wirjoatmodjo. *Freshwater Fishes of Western Indonesia and Sulawesi.* Singapore: Periplus Editions (HK) Ltd., 1993.

Nelson, J. *Fishes of the World,* 3rd edition. New York: John Wiley & Sons, 1994.

Ng, P. K. L., and K. K. P. Lim. "Snakeheads: Natural History, Biology, and Economic Importance." In *Essays in Zoology: Papers Commemorating the 40th Anniversary of the Department of Zoology, National University of Singapore, Department of Zoology,* edited by C. L. Ming and P.K.L. Ng. Singapore: National University of Singapore, 1990.

Resources

Okada, Y. *Studies on the Freshwater Fishes of Japan.* Tsu-shi, Japan: Mie Prefecture University, 1959–1960.

Pethiyagoda, R. *Freshwater Fishes of Sri Lanka.* Sri Lanka: The Wildlife Heritage Trust of Sri Lanka, 1991.

Riehl, R., and H. A. Baensch. *Aquarium Atlas.* Melle, Germany: Mergus-Verlag, 1986.

Talwar, K. T., and A. G. Jhingran. *Inland Fishes of India and Adjacent Countries,* Vol. 2. New Delhi: Oxford & I.B.H. Publishing Co., 1991.

Periodicals

Bonou, C. A., and G. G. Teugels. "Révision systématique du genre *Parachanna* Teugels et Daget, 1984 (Pisces: Channidae)."*Revue d'Hydrobiologie Tropicale* 18, no. 4 (1985): 267–280.

Ettrich, G. "Fische voller Uberraschungen." *DATZ* 39, no. 7 (1986): 289–293.

———. "Breeding the Green Snakehead: It's a Mouthbrooder!" *Tropical Fish Hobbyist* 37, no. 10 (1989): 34–36.

Lauder, G. V., and K. F. Liem. "The Evolution and Interrelationships of the Actinopterygian Fishes." *Bulletin of the Museum of Comparative Zoology* 150 (1983): 95–197.

Liem, K. F. "The Comparative Osteology and Phylogeny of the Anabantoidei (Teleostei, Pisces)." *Illinois Biological Monograph* 30 (1963): 1–149.

Musikasinthorn, P. "*Channa aurantimaculata,* a New Channid Fish from Assam (Brahmaputra River Basin), India, with Designation of a Neotype for *C. amphibeus* (McClelland, 1845)." *Ichthyological Research* 47, no. 1 (2000): 27–37.

———. "*Channa panaw,* a New Channid Fish from the Irrawaddy and Sittangriver Basins, Myanmar." *Ichthyological Research* 45, no. 4 (1998): 355–362.

Musikasinthorn, P., and Y. Taki. "*Channa siamensis* (Günther, 1861), a Junior Synonym of *Channa lucius* (Cuvier in Cuvier and Valenciennes, 1831)." *Ichthyological Research* 48, no. 3 (2001): 319–324.

Roberts, T. R. "The Freshwater Fishes of Western Borneo (Kalimantan Barat, Indonesia)." *Memoirs of the California Academy of Science* 14 (1989): 1–210.

Roe, J. L. "Phylogenetic and Ecological Significance of Channidae (Osteichthyes, Teleostei) from the Early Eocene Kuldana Formation of Kohat, Pakistan." *Contributions from the Museum of Paleontology* 28, no. 5 (1991): 93–100.

Smith, H. M. "The Freshwater Fishes of Siam or Thailand." *Bulletin of the U.S. National Museum* 188 (1945): 1–622.

Victor, R., and B. O. Akpocha. "The Biology of Snakehead, *Channa obscura* (Gunther), in a Nigerian Pond Under Monoculture." *Aquaculture* 101 (1992): 17–24.

Vierke, J. "Ein farbenfroher neuer Schlangenkopffisch aus Assam *Channa bleheri* spec. nov." *Das Aquarium* 259 (1991): 20–24.

Zhang, C. -G., P. Musikasinthorn, and K. Watanabe. "*Channa nox,* a New Channid Fish Lacking a Pelvic Fin from Guangxi, China." *Ichthyological Research* 49, no. 2 (2002): 140–146.

Prachya Musikasinthorn, PhD

Pleuronectiformes

(Flatfishes)

Class Actinopterygii

Order Pleuronectiformes

Number of families Approximately 13

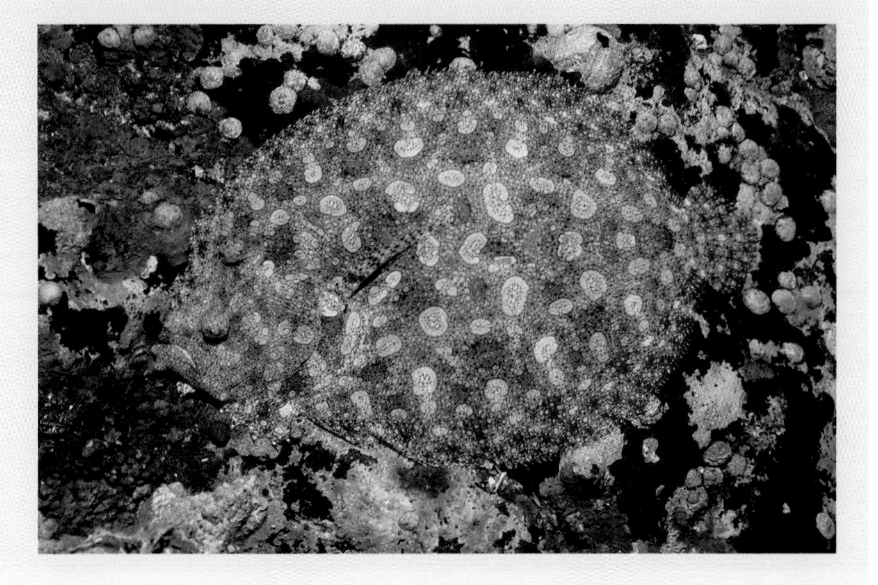

Photo: A Pacific leopard flounder (*Bothus leopardinus*) contrasting with the colorful sea ground near the Galápagos Islands. (Photo by Fred McConnaughey/Photo Researchers, Inc. Reproduced by permission.)

Evolution and systematics

The flatfish body plan, with its spectacular morphological specializations, has had a long and successful presence among marine teleost fish assemblages dating back at least to the Tertiary, more than 50 million years ago (mya). The oldest flatfish fossils are otoliths dating from early Eocene times (53–57 mya). *Eobothus minimus*, a representative of the bothoid lineage with uncertain affinities within the group, is the oldest known skeleton representative of the Pleuronectiformes, dating at least to the Lutetian (some 45 mya) in the Eocene. The oldest soles, *Eobuglossus eocenicus* and *Turahbuglossus cuvillieri*, both known from single specimens from the Upper Lutetian (Eocene) of Egypt, are early flatfish fossils that appear nearly identical to skeletons of recent soleids. The earliest bothid and pleuronectid fossils also are surprisingly "modern-looking" species dating to the Eocene. The appearance of representatives of different flatfish families in fossil deposits dating to about the same time period indicates that diversification of many of the major lineages of flatfishes took place in the distant past, earlier than 45 mya. The nearly simultaneous appearance of flatfish fossils representing different lineages and encompassing nearly all of the structural features and diversity of the order also may indicate that diversification of these lineages occurred suddenly.

It also is evident from these early fossils that anatomical specializations of flatfishes, including asymmetry of the skull, supracranial extension of the dorsal fin, and modifications of the caudal skeleton, occurred earlier than the period to which these fossil flatfishes belong. When flatfishes evolved and how rapidly they diversified are unresolved questions. Flatfish fossils are unknown from true freshwater sediments, which may indicate that the ancestor of this group was a marine fish. Because fossil flatfishes are relatively rare, our knowledge concerning their evolutionary history is still very incomplete.

Flatfishes have unique morphological specializations related to their asymmetry. Although earlier hypotheses proposed that flatfishes share a common ancestor with some as yet unidentified perciform group of symmetrical fishes, the origin and sister group of flatfishes are unknown. Interrelationships among flatfishes are not well resolved, and work continues toward understanding the evolutionary relationships of these interesting fishes. The order Pleuronectiformes is monophyletic, based on the presence of three derived characters: the ontogenetic migration of one of the eyes; the anterior position of the dorsal fin origin (overlapping the cranium); and the presence of a recessus orbitalis, a muscular, sac-like evagination in the membranous wall of the orbit that can be filled with fluid, causing protrusion of the eyes to a higher position above the surface of the head (and above the bottom when the fish is buried).

Many groups of flatfishes that traditionally were recognized as families and subfamilies do not seem to represent monophyletic groups. The lack of detailed phylogenetic studies for several pleuronectoid groups hinders understanding of the interrelationships of flatfishes even at the family level. Ongoing research using both morphological and molecular approaches is expected to provide interesting results on such interrelationships of pleuronectiform taxa. Changes in our understanding of higher relationships among these fishes can be expected as additional information is discovered.

Two major lineages of flatfishes are recognized: the Psettoidei, made up of the family Psettodidae, and the Pleuronectoidei, containing all remaining flatfish groups. The Psettodids, or spiny-flounders, are a basal group of flatfishes hypothesized to be the sister group of the Pleuronectoidei. This suborder has one family with two species of *Psettodes*. These are relatively large flatfishes that do not feature strong

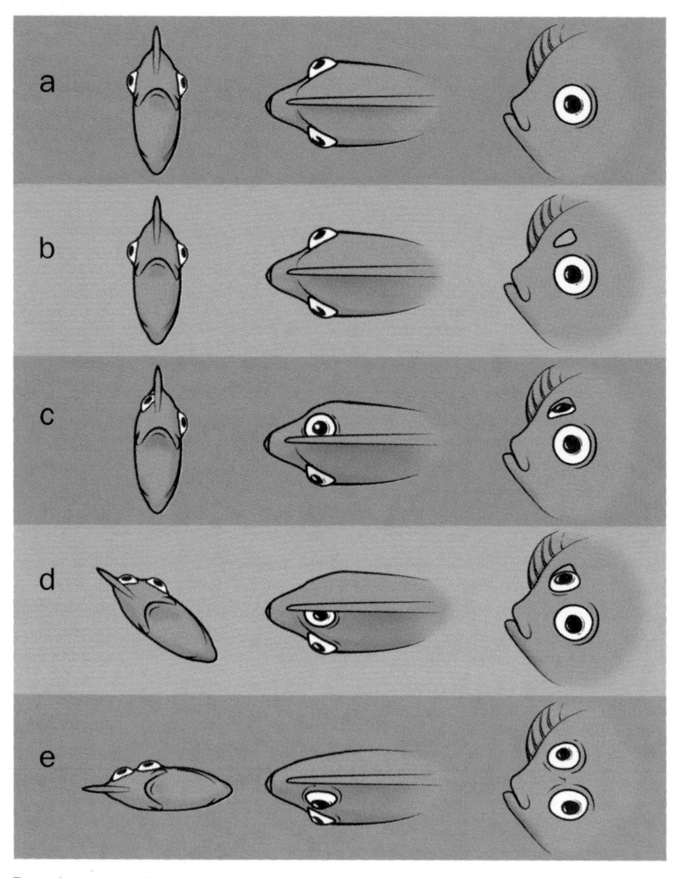

Developmental stages of eye migration in the larva from a symmetrical larva with an eye on each side of the head to the metamorphosed specimen with both eyes on same side of the head. (Illustration by Amanda Humphrey)

morphological asymmetries, as found in other flatfishes; there are both dextral and sinistral individuals in populations. These fishes are recognized easily by the posterior location of the dorsal fin, which does not advance onto the cranium anterior to the eyes; by spines in the dorsal and anal fins; by large mouths with specialized teeth; and by nearly rounded bodies without the obvious bilateral asymmetry in the lateral musculature that is evident in other flatfishes.

The Pleuronectoidei contains all of the more familiar flatfishes. At present, 13 families of pleuronectoid flatfishes are recognized, with *Tephrinectes* also representing a distinct lineage of uncertain status within the order. Phylogenetic relationships of some families and subfamilies and the monophyly of others (e.g., Paralichthyidae) are uncertain and in need of further study. Family groups within this suborder are the Citharidae, Scophthalmidae, Bothidae, Paralichthyidae, Pleuronectidae, Paralichthodidae, Poecilopsettidae, Rhombosoleidae, Achiropsettidae, Samaridae, Achiridae, Soleidae, and Cynoglossidae.

Physical characteristics

Flatfishes are deep-bodied, laterally compressed fishes that are easily and immediately recognizable anatomically, in that juveniles and adults (post-metamorphic individuals) have both eyes on the same side of the head. All flatfishes begin life as pelagic, bilaterally symmetrical fishes with an eye on each side of the head. During larval development, however, flatfishes undergo a spectacular ontogenetic metamorphosis, during which one eye migrates from one side of the head to the other, so that both eyes come to be present on the same side of the head. Depending upon the species, either the right or the left eye migrates. In relatively few species, eye migration is inde-

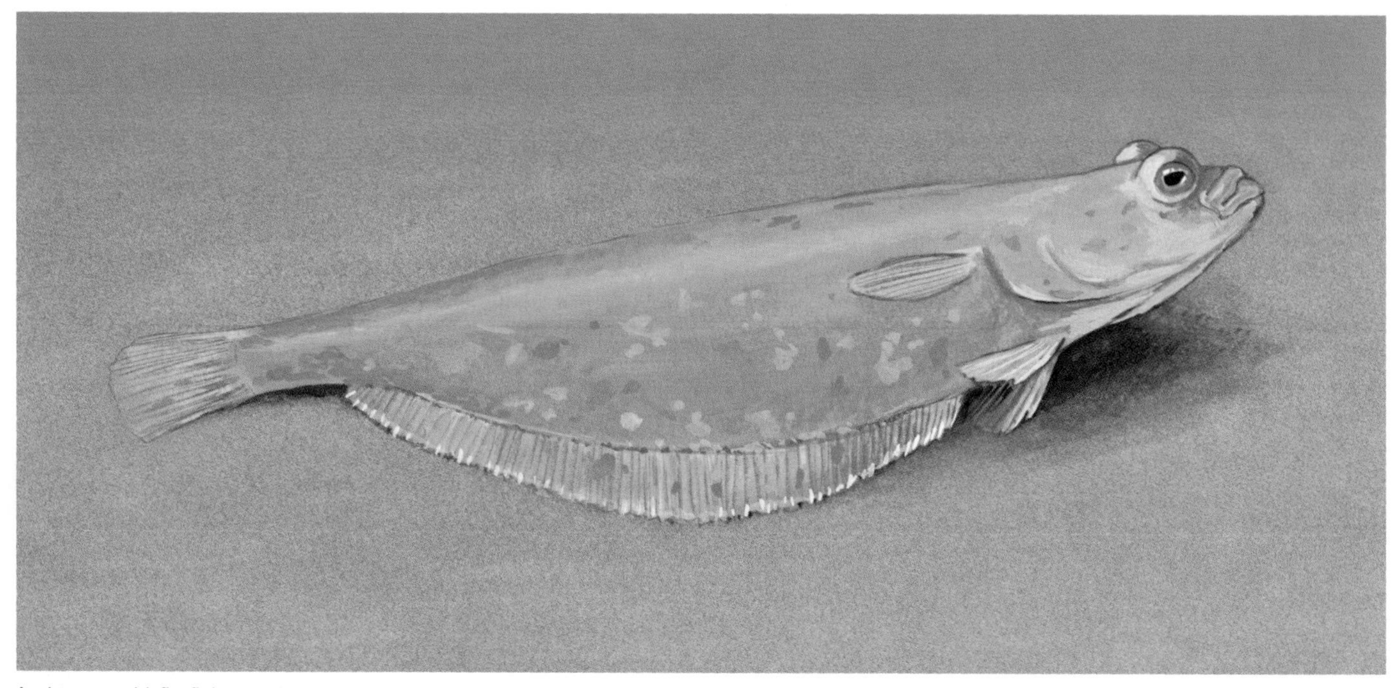

A pleuronectid flatfish with its anterior body raised off the substrate as it searches for prey. This posture represents a typical feeding behavior exhibited by benthic feeders among this important family of flatfishes. (Illustration by Amanda Humphrey)

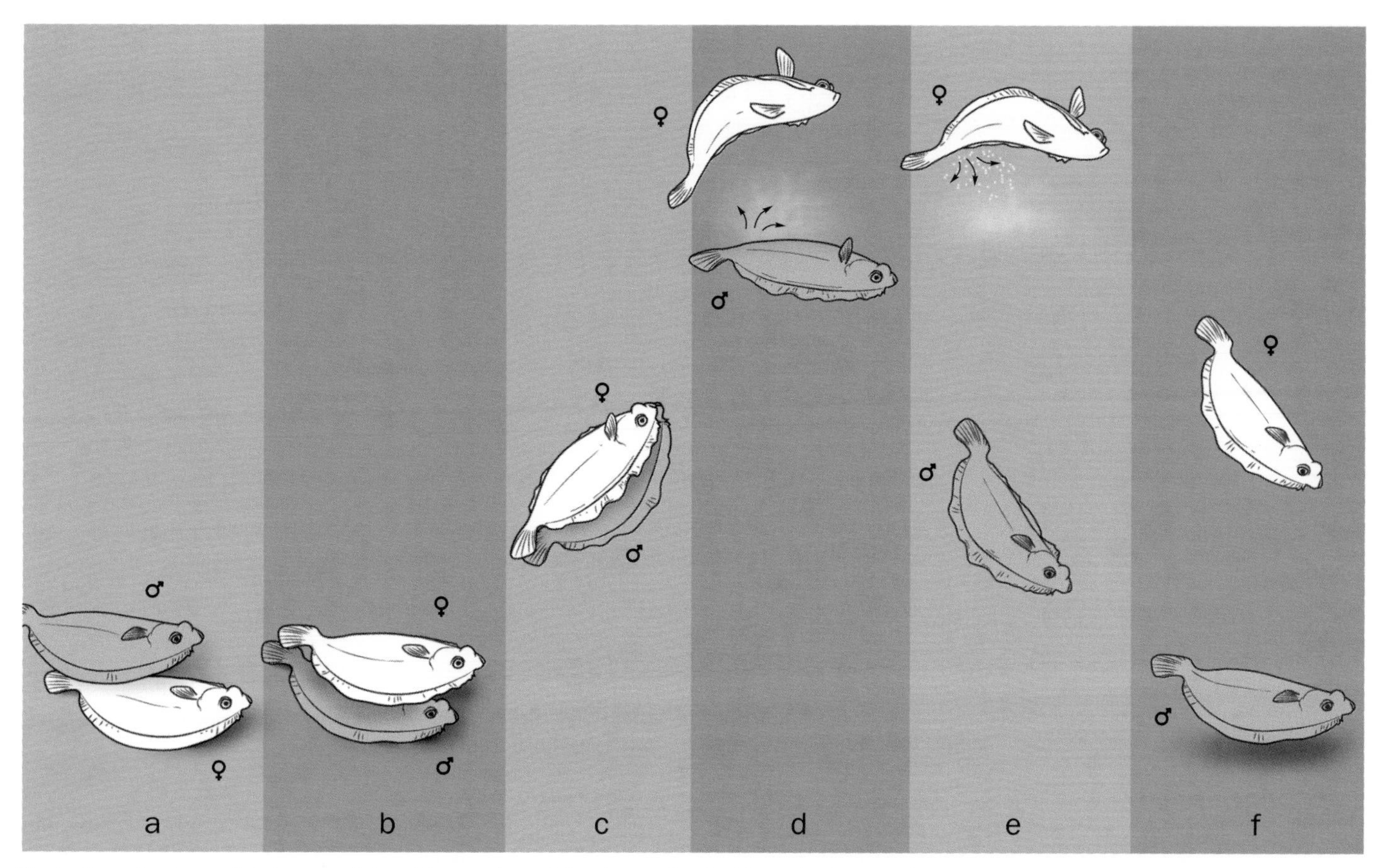

Illustration of mating of *Bothus ocellatus*. Short series of drawings reflecting the paired courtship swimming, rise off the bottom, and gamete release during reproductive events of this species. This is one of the best documented mating sequences observed for flatfishes. (Illustration by Amanda Humphrey)

terminate, but in most species eye migration is genetically fixed. The eyes may or may not come to lie in close proximity to each other when eye migration is completed.

Further deviations from a bilaterally symmetrical body plan occur in various external and internal structures, including placement of nostrils in the head, differential development of osteological features (especially bones in the anterior head skeleton), differences in jaw shape and dentition on either side of the body, degree of development of lateral body musculature, lateral line development and scale type on different sides of the body, differential coloration on ocular and blind sides, and differences in paired fin development on ocular and blind sides of the body. As a group, flatfishes are unique among fishes in their asymmetry, and they are noteworthy in that only they, among vertebrates, deviate so radically from a bilaterally symmetrical body plan.

Body shapes vary widely, ranging from nearly round, oval, and rhomboid to elongate and sometimes tapering to a sharp point. They may be either thick-bodied or thin-bodied, with or without a well-defined caudal peduncle. Flatfishes span a size range of about three orders of magnitude, from diminutive species, such as tonguefishes (*Symphurus*), which are sexually mature at a standard length (SL) of 0.98–1.6 in (2.5–4.0 cm), to giant species of halibuts (*Hippoglossus stenolepis* and *H. hippoglossus*), which reach nearly 6.6 ft (2 m) in total length and may weigh well over 661 lb (300 kg). Average total lengths of adults of most flatfish species are about 11.8 in (30 cm) or less.

Some flatfishes possess remarkable abilities to change the color and color patterns of their ocular surfaces to match the colors and patterns of the backgrounds on which they lie. Flatfishes typically exhibit distinct asymmetrical differences in pigmentation, with the ocular side of the head and body uniformly

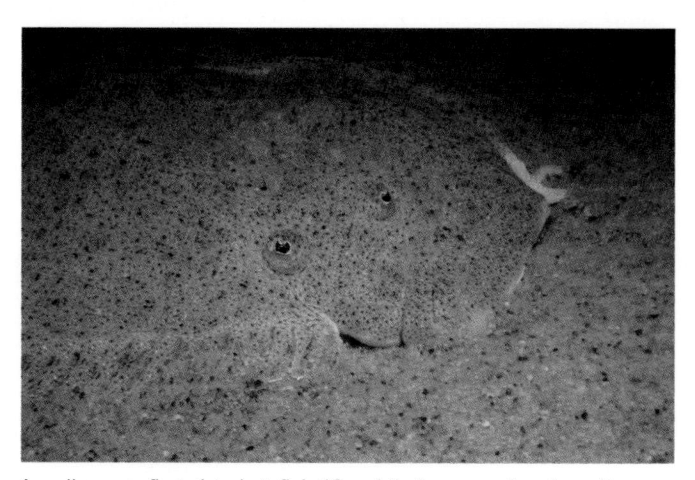
A well camouflaged turbot fish (*Scophthalmus maximus*) on the sea bed. (Photo by Lawson Wood/Corbis. Reproduced by permission.)

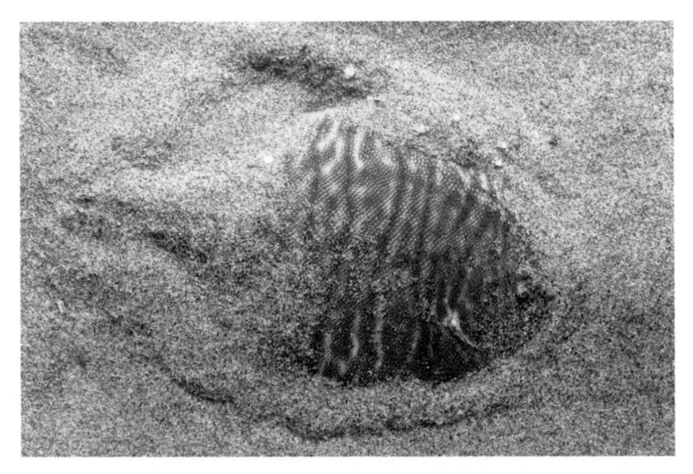

A hogchoker (*Trinectes maculatus*) burying itself into the sand. (Photo by Tom McHugh/Photo Researchers, Inc. Reproduced by permission.)

whitish to dark brown or black, upon which there may be additional markings, such as ocelli, spots, crossbanding (complete and incomplete), or longitudinal or wavy stripes. Ocelli, spots, and crossbands may be fixed in number and position and may be useful for identification of some species. In species with strong asymmetrical coloration, the blind side of the head and body is conspicuously paler than the ocular side, typically uniformly whitish to pale yellowish. Flatfishes without strong asymmetrical pigmentation usually have blind sides that are darkly pigmented, sometimes as intensely pigmented as the ocular side; in others, although the blind sides are distinctly pigmented, they are less so than the ocular surface.

Except for the spiny flounders (Psettodidae), flatfishes typically lack spines in their fins. All of the fin rays are soft. Flatfishes have a single, long dorsal fin, whose origin is located in an anterior position overlapping the cranium. The single anal fin is also long and extends along most of the ventral side of the body from a point just behind the anus nearly to, or sometimes connecting to, the caudal fin. Most flatfishes typically have paired pectoral and pelvic fins, but in some groups these fins are reduced or lost. In addition, most flatfishes have a lateral line, at least on the ocular side, and most also have a lateral line on the blind side. Adult flatfishes also lack a swim bladder (though it is present in larvae).

Distribution

Species of flatfishes have nearly global occurrence in marine habitats, ranging from Arctic and boreal marine waters to Antarctic and southern waters. They are distributed broadly throughout the world's temperate marine zones and are especially speciose in marine habitats in tropical regions. Within all regions, flatfishes are most diverse where extensive continental shelves with complex habitats are located in shallow water. The widest diversity of flatfish species occurs in the Indo-West Pacific region.

Habitat

Flatfishes occur nearly globally in marine habitats and occupy diverse bathymetric environments, ranging from shallow-water to deep-water habitats to about 6,560 ft (2,000 m). Relatively few species inhabit freshwater environments. The greatest diversity of flatfishes, about 74% of the known species, is found in habitats ranging from near shore to depths of about 328 ft (100 m) on the continental shelf.

The majority of flatfishes occur in shallow marine waters, in coastal areas and estuaries, and on the inner continental shelf, where there are soft-sediment bottom types. Flatfishes also occur on a variety of sediments on the outer continental shelf and upper continental slope. Flatfishes can be found on intertidal flats following the tide line to deeper habitats. In tropical waters, flatfishes inhabit shallow mangrove estuaries and adjacent mud flats, sea grass beds, and mud bottoms along the coast. Within reef-associated habitats, which are distributed widely across tropical oceans, flatfishes are found on reef flats, on back-reef slope areas and in lagoons associated with reefs, and around coral outcrops, as well as on sandy substrates interspersed around reef spurs. Flatfishes inhabit various sediments, including silt, mud, sand, and sand-shell mixtures, with some species also occurring on rocky or pebbly bottoms.

Behavior

Flatfishes generally lie on the bottom on their blind side. They can be found either on top of the sediments or partially buried under a fine layer of sand or silt with only their eyes protruding above the sediments. Many flatfishes are stationary for long periods of time. When swimming above the bottom, they use a "pleuronectiform" swimming mode, in which waves of muscle contraction are passed along the body, beginning in the anterior region and continuing posteriorly. Most species can utilize a more rapid escape response, where the caudal fin is brought into play, creating a powerful and speedy swimming response.

Surprisingly little is known concerning the social organization of most flatfish species. Flatfishes are non-schooling species; many occur as solitary individuals, but a few and perhaps many individuals may congregate in a general area. Males of some species may display aggressive behavior to one another during the mating season. The majority of flatfish species are diurnally active. Some species are active throughout the daytime, whereas others have peak activity at or around sunrise and sunset. Nocturnal activity is a major adaptation evident in the Soleidae, Achiridae, and Cynoglossidae.

Extensive migration patterns have been well documented for some commercially important species of northern temperate flatfishes, such as the plaice, summer flounder, and halibuts. For most flatfishes, in particular the many tropical species, little is known concerning their movements or migrations. Small, reef-associated species probably have limited home ranges and do not engage in any seasonal migrations.

Feeding ecology and diet

Flatfishes are extremely successful in conducting life on or near the bottom, where they function in pivotal ecological roles as both predator and prey. Flatfish diets include such prey as shrimps, decapod and other crustaceans, mollusks, polychaetes, and many other types of small invertebrates, as well as echinoderms, fishes, and cephalopods. Small-mouthed species, especially tonguefishes (Cynoglossidae), achirid soles (Achiridae), and true soles (Soleidae), feed on a broad spectrum of smaller epifaunal and infaunal organisms.

Halibuts, larger species of bothid and paralichthyid flounders, larger pleuronectids, and the larger scophthalmids are active predators that consume fishes, larger and more active crustaceans (shrimps, lobsters, crabs), and cephalopods (squids and octopuses). The halibuts, with their great size and swimming abilities, actively pursue and chase down their prey, whereas other large flatfishes generally are ambush predators that lie on the bottom or partially buried within the sediment, concealed by their camouflage coloration and awaiting unsuspecting prey to approach within striking distance.

All life stages of flatfishes are eaten by predators that include both invertebrates and vertebrates. While in the plankton, eggs and larvae are consumed by jellyfishes, ctenophores, arrow worms, mysid shrimps, and fishes. Young, newly settled flatfishes are attacked and consumed by crabs, shrimps, and fishes. Juvenile and adult flatfishes fall prey to a wide variety of predatory fishes, including cods, hakes, sculpins, rockfishes, striped bass, other flatfishes (sometimes their own species), monkfish, bluefish, cobia, groupers, moray eels, sea ravens, large skates, stingrays, and various sharks, as well as birds (egrets, herons, cormorants, gulls), seals, and sea lions.

Reproductive biology

The reproductive behavior of most flatfishes is not known. Direct observations of courtship and mating have not been made for the majority of flatfish species. The sexes are separate, and individuals usually do not change sex during their lifetimes. Flatfishes have external fertilization. Where reproductive behavior has been observed, individual males and females may pair up during courtship and spawning. Sometimes the mating pair is joined by other males.

Most flatfishes spawn planktonic eggs that float freely in the water column. Some pleuronectid flatfishes, such as the winter flounder, lay eggs that are demersal and adhesive, such that after the female releases them, they remain on the bottom and stick to each other and to other items. Even among species with demersal, adhesive eggs, flatfishes do not construct nests during spawning, nor do they exhibit any type of parental care. Upon hatching, flatfish larvae are planktonic and usually are found in the water column far above the bottom. Larval stages vary in duration from a few days to a couple of months; the duration of larval stages is influenced greatly by ambient water temperatures. Following eye migration and metamorphosis, young flatfish settle out of the water column and assume a benthic lifestyle, with many species utilizing shallow-water habitats as nursery areas.

A strong seasonality in reproductive period has been noted for most temperate and boreal flatfishes, with most species having one spawning season per year. The timing of spawning seasons within the year varies by species and also by latitude. Some species spawn during periods of seasonally high temperatures, whereas others spawn during wintertime. Some warm temperate species may have two spawning periods per year; for tropical and subtropical species, spawning periods may extend over several months. Spawning among temperate marine species corresponds to annual productivity cycles that are related to temperatures and photoperiods. In some tropical regions, spawning by some flatfishes also seems to correspond to seasonal monsoons, which influence productivity cycles.

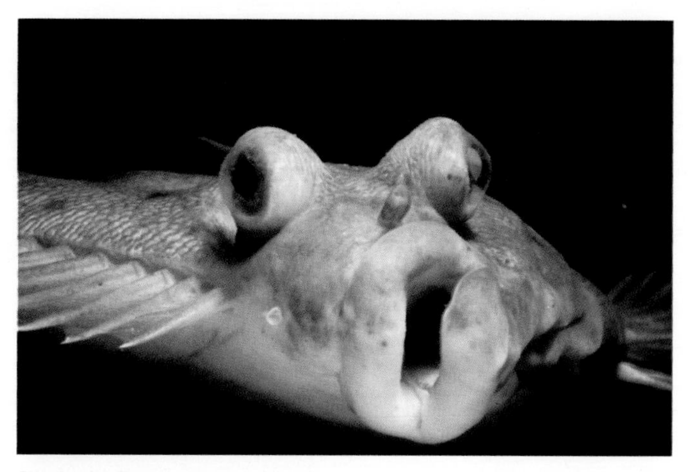

Close-up of the yellowtail flounder (*Limanda ferruginea*). (Photo by Jeffrey L. Rotman/Corbis. Reproduced by permission.)

Conservation status

Two pleuronectids are cited by the IUCN—the Atlantic halibut, which is listed as Endangered, and the yellowtail flounder, which is listed as Vulnerable. Overfishing is primarily responsible for reductions in many flatfish populations, especially for large, commercially important species. Throughout the world, stocks of commercially important flatfishes are considered to be fully exploited—for many, even overexploited. Other factors contributing to reductions in populations of flatfishes include habitat destruction and pollution, especially serious situations for flatfishes that utilize estuaries and other coastal habitats, such as sea grass meadows and mangrove forests, as nursery habitats.

Significance to humans

Flatfishes are an important group of food fishes. Medium-size and large species of most families are consumed wherever they are captured, and in some regions even the smallest flatfishes also are sold as food for people. In such regions as the North Atlantic and North Pacific Oceans and also in Southern Hemisphere regions, such as Australia, New Zealand, and South America, flatfish populations are sufficiently large to constitute major fishery resources. Some of the smaller flatfishes, especially those taken as by-catch in shrimp trawl fisheries (tonguefishes, soleids, achirids, bothids, and paralichthyids), are considered to be a nuisance by fishermen because they so firmly entangle themselves in the nets that they cannot easily be shaken out. Clearing the nets after heavy catches of these flatfishes, especially tonguefishes, soleids, and achirids, requires manually extracting fishes from the nets, a time-consuming task.

1. Peacock flounder (*Bothus lunatus*); 2. Hogchoker (*Trinectes maculatus*); 3. Windowpane flounder (*Scophthalmus aquosus*); 4. Summer flounder (*Paralichthys dentatus*); 5. Common sole (*Solea solea*); 6. Winter flounder (*Pseudopleuronectes americanus*); 7. Pacific sanddab (*Citharichthys sordidus*); 8. Plaice (*Pleuronectes platessa*); 9. Pacific halibut (*Hippoglossus stenolepis*); 10. California tonguefish (*Symphurus atricaudus*). (Illustration by Wendy Baker)

Species accounts

Hogchoker
Trinectes maculatus

FAMILY
Achiridae

TAXONOMY
Pleuronectes maculatus Bloch and Schneider, 1801, Tranquebar, India (in error).

OTHER COMMON NAMES
French: Sole bavoche; German: Amerikanische Seezunge.

PHYSICAL CHARACTERISTICS
Small, oval, dextral flatfish that have a deep and thick body without a definite caudal peduncle. The dorsal and anal fins are free from the caudal fin, and the right pelvic fin is joined to the anal fin. A relatively small head, with the snout slightly overhanging the small, subterminal mouth. The eyes are small, flat, and separated by a small space. The dorsal fin originates at the tip of the snout. No pectoral fins. The lateral line is straight. The skin is very slimy with mucus. Scales are ctenoid and very rough on both sides of the body. The ocular side is slate-olive to dark brown, with numerous conspicuous, darker transverse crossbands. There is also a longitudinal stripe along the midregion and, occasionally, a number of darker, diffuse blotches scattered over the surface. The blind side is dirty white; some specimens have numerous irregularly rounded spots, varying in both size and number, scattered over the blind side. This species reaches lengths of about 7.9 in (20 cm), with most averaging about 2.4–5.9 in (6–15 cm). They live for about seven years. Females grow larger and live longer than males.

DISTRIBUTION
Western North Atlantic in marine, estuarine, and freshwaters along the Atlantic coast of North America from Maine to the Gulf of Mexico.

HABITAT
They occur most commonly on mud, sand, or silt bottoms in coastal bays and estuaries with brackish water. In larger estuaries, young (small) fish tend to be found in upper reaches of estuaries, sometimes at considerable distances upstream into freshwater portions of coastal rivers. Fish size generally increases with increasing distances down estuary. The largest hogchokers usually are found in the lower estuary and also on the inner continental shelf to about 82 ft (25 m) and rarely to about 246 ft (75 m). Able to withstand a considerable range of temperatures of about 34–95.2°F (1.1–35.1°C). Euryhaline (able to withstand a range of salinities), ranging in salinity from freshwater to about 50 ppt. Can tolerate low oxygen conditions for periods up to 10 days. Will move out of areas with extremely low oxygen levels.

BEHAVIOR
Under laboratory conditions, hogchokers were active only during the dark period, with peak activities associated with times of slack tide in the natural habitat. Under continuous dim light, activity peaks coincided with slack tide, and fish were active in the diurnal as well as the nocturnal phase of the cycle.

FEEDING ECOLOGY AND DIET
Opportunistic, nocturnal feeders that eat a variety of small invertebrate prey, including amphipods, clam siphons, annelid worms, copepods, and small fishes. These fishes tend to macerate their food. Hogchokers are consumed by a variety of predators, including bull sharks, sandbar sharks, smooth dogfish, stingrays, striped bass, weakfish, bluefish, and cobia.

REPRODUCTIVE BIOLOGY
Mature at two to four years old and at sizes as small as about 2 in (5 cm). Probably a serial spawner. Annual fecundity has not been estimated for this species. Batch fecundity varies with fish size. Small females, about 3.5 in (9 cm), produce about 11,000 eggs, and larger females, 4.3–6.3 in SL (11–16 cm SL), produce from 23,000 to 54,000 eggs. The spawning season is April to October, but eggs have been reported as early as January, and in the southern Gulf of Mexico spawning may occur year-round. Spawning takes place in estuaries between 6 P.M. and 10 P.M., when water temperatures reach 68–77°F (20–25°C). Hogchoker eggs are pelagic in high-salinity waters and demersal in lower-salinity waters. Hatching occurs about one to two days after spawning. Eye migration begins at about 0.2 in (5 mm), 34 days after hatching, and is completed when fish are about 0.7 in (18 mm).

CONSERVATION STATUS
Not threatened.

SIGNIFICANCE TO HUMANS
Edible but noncommercial species of no interest to fisheries, owing to their small size. They sometimes are captured and sold to hobbyists in the aquarium fish trade. ◆

Peacock flounder
Bothus lunatus

FAMILY
Bothidae

TAXONOMY
Pleuronectes lunatus Linnaeus, 1758, Bahamas.

OTHER COMMON NAMES
French: Rombou lune; Spanish: Lenguado ocelado.

PHYSICAL CHARACTERISTICS
Sinistral flatfish with an oval and moderately deep body. Rounded to bluntly pointed caudal fin. The dorsal profile of the snout has a distinct notch above the nostrils, and there is a stout spine on the snout of adult males (a bony knob in females). Eyes are relatively large, with the lower eye distinctly anterior to the upper and with a broad interorbital space that is conspicuously wider in males. The moderately large and oblique mouth extends slightly beyond a vertical line through the anterior margin of the lower eye. Jaws have an irregular double row of small teeth. Ocular side upper pectoral fin rays are conspicuously elongate in males. Scales are ctenoid on the ocular side and cycloid on the blind side. The lateral line has a steep arch above the pectoral fin. Ocular side is grayish brown,

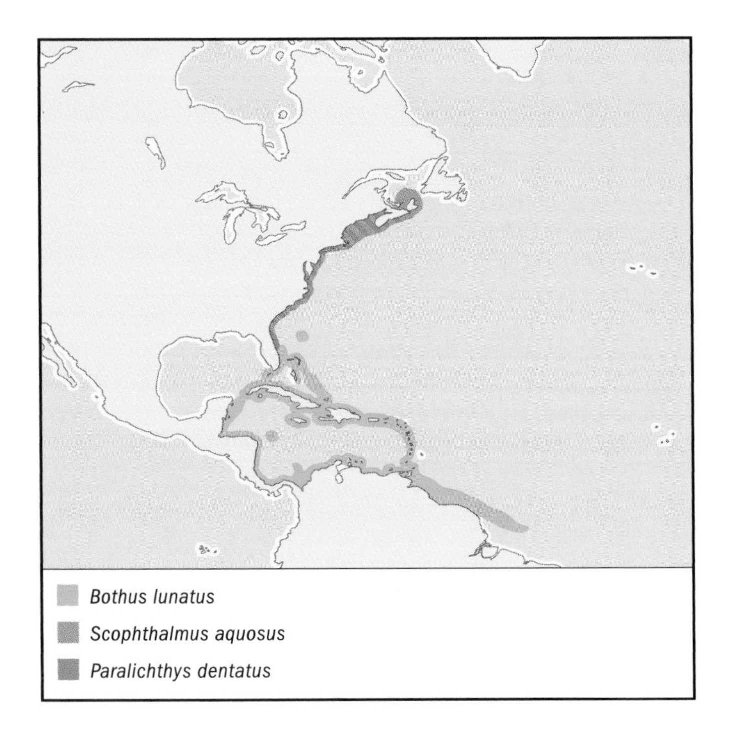

with numerous bright blue rings and rosettes covering the entire ocular side and with two to three large blackish spots on the straight portion of the lateral line. Larger individuals also have dark transverse bands on the ocular side pectoral fin. Maximum lengths to about 17.7 in (45 cm), with most individuals about 13.8 in (35 cm) long.

DISTRIBUTION
Marine coastal waters of the tropical and subtropical western Atlantic, including Bermuda, the Bahamas, Florida, throughout the Caribbean, and south to Fernando de Noronha, Brazil.

HABITAT
Shallow waters from the shore to about 213 ft (65 m). Found chiefly on sandy bottoms, often within or near coral reefs, and also in sea grass and mangrove habitats.

BEHAVIOR
Diurnally active. Often observed resting on the sandy bottom, sometimes partially buried in the sand. Occasional specimens are also observed resting on top of small coral reef tops. When swimming, they glide along just above the bottom using wave-like motions. Peacock flounder can change colors rapidly to blend in with the background.

FEEDING ECOLOGY AND DIET
Visually feeding, ambush predator that eats primarily small fishes but also consumes crustaceans and octopuses. They often lie in wait on sand patches adjacent to reef areas to intercept small fishes undertaking crepuscular migrations between reef and sea grass habitats. Lizard fishes, snappers, groupers, and various sharks and stingrays eat peacock flounders.

REPRODUCTIVE BIOLOGY
Off Bonaire in December, peacock flounder spawning took place just before sunset, with elaborate spawning behavior observed for mating pairs. Males and females would approach each other with pectoral fins held erect to initiate courtship activity. The male, with its ocular side pectoral fin held erect, first paralleled the female as they swam above the substrate. The male then positioned himself underneath the female; with their snouts touching and the male's body arched backward, the pair began a slow rise (about 15 seconds) of about 6.6 ft (2 m) off the bottom, when they simultaneously released their gametes and rapidly returned to the bottom.

CONSERVATION STATUS
Probably not threatened, but population status is unknown throughout most of its distribution. Because of its size and food qualities, this species could be susceptible to local overfishing.

SIGNIFICANCE TO HUMANS
Peacock flounder are caught incidentally in artisanal fisheries throughout their range. ◆

California tonguefish
Symphurus atricaudus

FAMILY
Cynoglossidae

TAXONOMY
Aphoristia atricauda Jordan and Gilbert, 1880, San Diego Bay, California.

OTHER COMMON NAMES
French: Langue californienne; Spanish: Lengua de perra.

PHYSICAL CHARACTERISTICS
Small, sinistral flatfish with the characteristic tonguefish teardrop-shaped body terminating posteriorly in a point without a distinct caudal fin. The small head has a pointed snout. The small eyes are set close together. A small, subterminal mouth

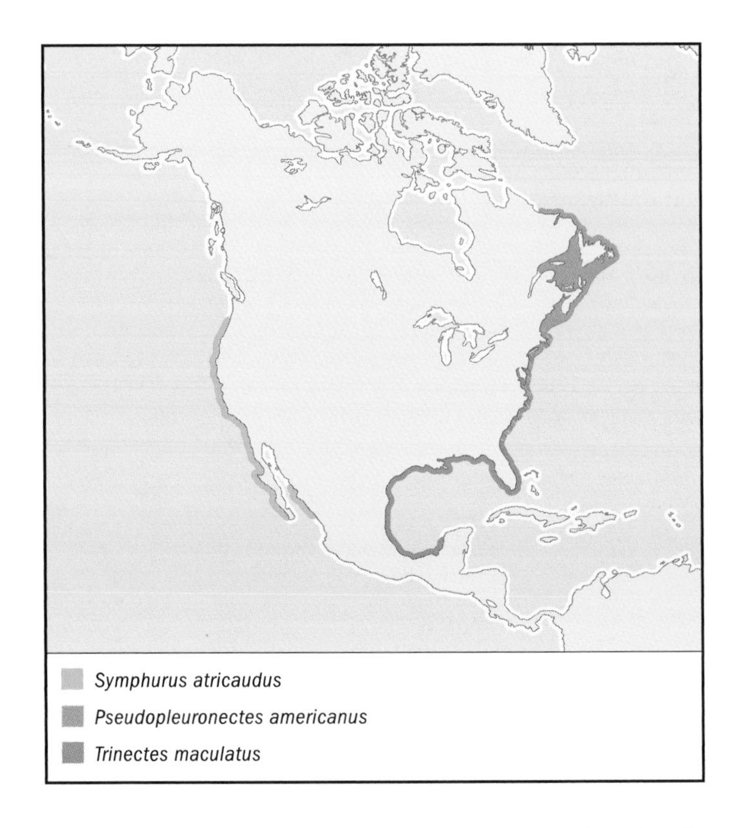

with small teeth best developed on the jaws of the blind side. Dorsal and anal fins are conjoined with the caudal fin, forming one continuous fin around nearly the entire body. Lacks a lateral line on either side of the body, lacks pectoral fins in adults, and has a pelvic fin only on the ocular side. Ocular side is uniformly medium to dark brown, with a series of complete or incomplete darker crossbands and with the posterior fifth of the body much darker than the anterior regions. Blind side is uniformly whitish or yellowish. Reach lengths to about 8.3 in (21 cm), but most are smaller, usually averaging only about 5 in (13 cm). Little is known concerning longevity, growth rates, or population structure of this species.

DISTRIBUTION
Inner continental shelf of the eastern Pacific from Washington to the Pacific side of Baja California Sur and along the western shore of Sonora and Sinaloa, Mexico.

HABITAT
Sand or mud bottoms at depths ranging from 9.8 to 328 ft (3–100 m), with most adults taken between 98.4 and 262.5 ft (30–80 m). Juveniles tend to inhabit shallower waters than do adults.

BEHAVIOR
Little is known. Probably nocturnally active and also active at other periods of low-light levels. During the daytime it remains partially or totally buried in the sediment, except for the anterior head region.

FEEDING ECOLOGY AND DIET
Consume a variety of small benthic invertebrates, including harpacticoid copepods, amphipods, ostracods, nematodes, small bivalve mollusks, and polychaetes. Predators of California tonguefish include sharks, electric rays, stingrays, and various bony fishes.

REPRODUCTIVE BIOLOGY
Little is known. They spawn planktonic eggs from June to September; larvae hatch at about 0.08 in (2 mm). Larvae transform between 0.7 and 1 in (19 and 25 mm) in length and settle to the bottom during late fall and winter. Probably a serial spawner, producing several batches of eggs during a protracted spawning season.

CONSERVATION STATUS
Not threatened.

SIGNIFICANCE TO HUMANS
Of little commercial value, owing to its small size. ◆

Pacific sanddab
Citharichthys sordidus

FAMILY
Paralichthyidae

TAXONOMY
Psettichthys sordidus Girard, 1854, San Francisco, California.

OTHER COMMON NAMES
English: Mottled sanddab, soft flounder, melgrim; Spanish: Lenguado arenero del Pacifico, lenguado.

PHYSICAL CHARACTERISTICS
Medium-sized, sinistral flatfishes with an elongate to oval body; large head with large terminal mouth; and slightly rounded, almost square caudal fin. Eyes are large, with the lower eye in advance of the upper eye and separated from it by a sharp, naked bony ridge. Pectoral fins are large and pointed. Lateral line is nearly straight. Scales are ctenoid on the ocular side of the body and cycloid on the blind side. Ocular side is a dull light brown, mottled with brown or black and sometimes yellow to orange speckles or white spots. Blind side is off-white to tan. Maximum lengths of about 16.1 in (41 cm) and weights up to about 2 lb (0.9 kg), but most are much smaller, only 4.9–5.6 oz (140–160 g). They live to be at least eight years of age.

DISTRIBUTION
Marine waters in the northern Pacific Ocean from the Sea of Japan to the Bering Sea and the Aleutian Islands south to Cape San Lucas, Baja California.

HABITAT
Adults inhabit gravel, sand, or mud-sand bottoms at 16.4–1,801 ft (5–549 m) but are most abundant at 164–492 ft (50–150 m); they rarely occur below 984 ft (300 m). Juveniles occur at shallower depths than those occupied by adults and sometimes move into tide pools.

BEHAVIOR
Diurnally active. Spend much of their time on the bottom, although occasionally they are captured at night up in the water column.

FEEDING ECOLOGY AND DIET
Opportunistic, visually oriented predators that feed principally on pelagic crustaceans, such as euphausiids, shrimps, crab larvae, calanoid copepods and amphipods, and occasionally small fishes and benthic prey, among them, annelid worms and crustaceans. Pacific sanddabs are consumed by a variety of larger predators, including blue sharks and other sharks, stingrays, and halibut, and also have appeared in the diets of Guadalupe fur seals.

REPRODUCTIVE BIOLOGY
Begin maturing between ages two and three years. Spawning takes place on or near the bottom from July to September off California. Eggs are released independently, are buoyant, and are fertilized outside the female. Females may spawn more than once during the same spawning season.

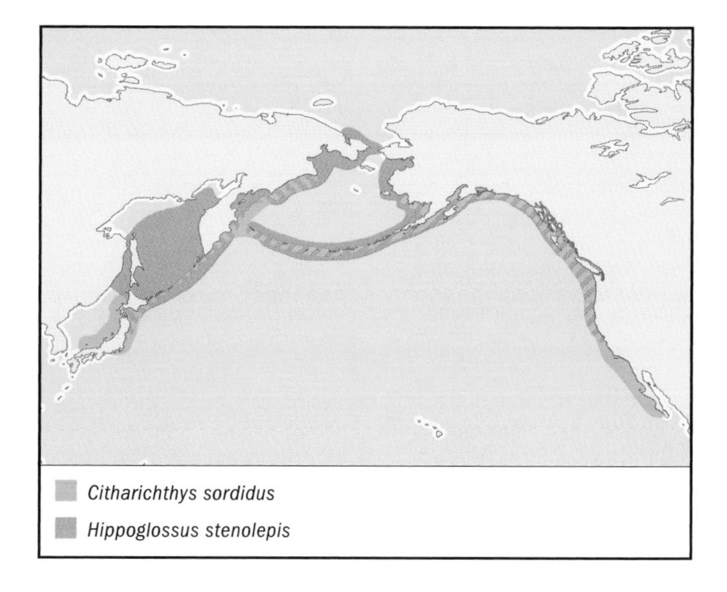

CONSERVATION STATUS
Not threatened.

SIGNIFICANCE TO HUMANS
Excellent food fish. Regarded as a delicacy in California but with less appeal elsewhere. ◆

Summer flounder
Paralichthys dentatus

FAMILY
Paralichthyidae

TAXONOMY
Pleuronectes dentatus Linnaeus, 1766, Carolina, United States.

OTHER COMMON NAMES
English: Fluke; French: Cardeau d'été; Spanish: Falso halibut del Canadá.

PHYSICAL CHARACTERISTICS
Large, sinistral flatfish with a narrow, relatively elongate and thick body. Prominent head with a large terminal mouth with a wide gape and strong canine teeth on both jaws. A rounded or doubly emarginate caudal fin. Eyes are relatively large, separated, and nearly equal in position. Lateral line is arched above the pectoral fin. Scales are small and cycloid, with secondary squamation. Varies in coloration, as individuals change color to match the background of their habitat. Ocular side coloration ranges from drab olive-green to brown to gray; the blind side usually is white. Individuals captured on white sand are nearly completely white, whereas others on dark sediments can be nearly black. Some individuals have pink, green, orange, or brown coloration on the ocular side. The ocular side is marked variously with irregular spots, often with a series of more or less distinct ocelli that are slightly darker than the background coloration, with the most posterior of these arranged in a double triangle (one above and one below the lateral line). Males grow to about 24 in (61 cm) in length, with a weight of 5.7 lb (2.6 kg), and females grow to about 37 in (94 cm) in length, with a weight of 29.5 lb (13.4 kg). Most adults are 15.7–22 in (40–56 cm) in length and weigh between 2.2 and 5.1 lb (1.0–2.3 kg).

DISTRIBUTION
Western North Atlantic in estuarine and continental shelf waters of eastern North America from Nova Scotia to Florida.

HABITAT
Inshore waters, including estuaries and even freshwater (juveniles), bays, harbors, and the inner continental shelf. Summer flounder prefer sandy or muddy bottoms, where they often are found in sand patches near and within eelgrass beds or around pier pilings. Sensitive to low oxygen concentrations and will move out of hypoxic areas (less than 3 ppm oxygen). Concentrated inshore during warmer periods of the year, with smaller fish found in very shallow water. Larger fish in the northern part of the range occur farther offshore, usually at depths of 230–509 ft (70–155 m) and deeper, even during the summer. Estuaries are important nursery areas for this species. Young summer flounder can withstand a wide range of temperatures and salinity levels and are well adapted for estuarine life. Juveniles remain in estuaries until their second year of life in the southern part of their range, but to the north they move just outside them during winter. Many young fish return to the same estuary during their second summer.

BEHAVIOR
Spend most of their lives on or close to the bottom. Occur most often on top of the sediment and do not ever bury deeply. Juveniles may be active at night, but adults appear to be most active during the daytime. Adults undertake strong seasonal inshore-offshore movements, especially in northern regions of the species' geographic range. Adults and juveniles are concentrated in shallow coastal and estuarine waters during the warmer months of the year and remain offshore in deeper waters (to about 492 ft, or 150 m) in the fall and winter, presumably to escape cold winter temperatures. Seasonal movements of summer flounder are complicated and are affected by fish size, location, and stock. Some may spend the winter in deeper bays and channels, and older fish may remain permanently on the outer shelf.

FEEDING ECOLOGY AND DIET
Diurnally active, opportunistic ambush predators. They feed while on the bottom but also rise off the bottom in pursuit of smaller fishes. The primary food of summer flounder is bony fishes, but cephalopods also are important prey of fish larger than 12.2 in (31 cm). Crustaceans, especially mysids and decapods, are important prey for smaller fishes (those fish less than 8.3 in, or 21 cm). Juveniles (3.9–7.9 in, or 10–20 cm) consume mysids, fishes, amphipods, and crabs. Feeding is most active during warmer periods and slows in winter. Spiny dogfish, blue sharks, skates, codfish goosefish, sea robins, bluefish, and winter flounder prey on summer flounder during various stages of their life history.

REPRODUCTIVE BIOLOGY
Male and female summer flounder mature at about 9.8 in (25 cm) and 11 in (28 cm), respectively, corresponding to ages two and a half years for females and two years for males. Many fish may reach maturity at one year. Spawning occurs on or near the bottom, where temperatures range from 53.6 to 66.2°F (12–19°C), and usually takes place during the autumn migration to offshore wintering grounds on the outer continental shelf. Large females (more than 26.8 in, or 68 cm) may produce in excess of four million eggs during a spawning season. Females are serial spawners, continuously producing egg batches that are shed over a period of several months (September to March). Larvae are transported to coastal areas during winter and early spring by prevailing water currents. Post-larvae and young juveniles are found in or near the mouths of estuaries. Metamorphosis is completed within bays and estuaries, where young fish settle to the bottom.

CONSERVATION STATUS
Not threatened. Exploitation of summer flounder by commercial and recreational fishers resulted in stock reductions throughout the species' range during the latter quarter of the past century. Fishery management plans have been developed to conserve and rebuild stocks by limiting commercial and recreational catches through size and season restrictions.

SIGNIFICANCE TO HUMANS
Summer flounder support the most important commercial and recreational fishery for flatfishes along the Atlantic coast of the United States. This highly prized game fish has strong fighting qualities and is excellent table fare. ◆

Pacific halibut
Hippoglossus stenolepis

FAMILY
Pleuronectidae

TAXONOMY
Hippoglossus stenolepis Schmidt, 1904, Aniva Bay, south Sakhalin Island, Sea of Okhotsk.

OTHER COMMON NAMES
English: Northern halibut, right halibut; French: Flétan du Pacifique; Spanish: Fletán del Pacifico.

PHYSICAL CHARACTERISTICS
One of the largest species of flatfishes and among the largest of bony fishes. Typically dextral flatfishes, with a thick, sturdy, elongate, and diamond-shaped body and a highly compressed caudal peduncle with a crescent-shaped caudal fin that often is indented near the edges. The head is large, with a large terminal mouth featuring a wide gape. Nearly symmetrical jaws containing two rows of prominent, conical teeth on the upper jaw and a single row of conical teeth on the lower jaw. The eyes are large, and the upper eye is slightly in advance of the lower one. The lateral line has a high arch above the pectoral fin. Small cycloid scales cover both sides of the body. Ocular side coloration is greenish brown to dark brown or black, marbled with lighter blotches; the blind side usually is white to milky white, sometimes also with blotches. Females reach lengths up to 8 ft, 9 in (267 cm), and weights of about 498 lb (226 kg); males are about 4 ft, 7 in (140 cm) and 220 lb (100 kg). Females grow considerably faster and typically live longer than do males. Almost all halibut larger than 100 lb (45.5 kg) are females. Halibut first become available to the offshore fishery at about five to seven years of age. The oldest recorded age for a male is 55 years, and the oldest recorded age for a female is 42 years.

DISTRIBUTION
Marine waters of the eastern and western North Pacific Ocean. In the west, they are found from the Sea of Japan and Okhotsk Sea north to the Gulf of Anadyr and Chukchi Sea and throughout the Bering Sea. In the eastern Pacific, this species ranges from the Gulf of Alaska southward to about Santa Barbara, California, and, rarely, southward to Point Chamalu, northern Baja California.

HABITAT
Occurs on a variety of bottom types at depths from about 19.7 to 3,609 ft (6–1,100 m) but most commonly found between 180 and 1,385 ft (55–422 m). In summer they are found between 92 and 902 ft (28–275 m) and sometimes shallower, whereas most halibut occur in deeper waters during winter. They have a preferred temperature range of 37.4–46.4°F (3–8°C).

BEHAVIOR
Diurnally active fishes found most often on the bottom. They often rise off the bottom into the water column and may at times even come close to the surface when pursuing prey. Seasonal movements of adults are associated with winter reproduction offshore and summer feeding inshore. Adult halibut move seasonally from deep water to the edge of the continental shelf and then to shallower banks and coastal waters during the summer; they move back to deep water in the winter. Migrations may be extensive—the longest migration on record was that of a fish tagged near Atka Island in the Aleutian Islands, which was recaptured at Coos Bay, Oregon, a distance of 2,500 mi (4023 km).

FEEDING ECOLOGY AND DIET
Large, powerful, opportunistic, visual feeder that consumes a wide variety of prey, including fishes, crabs, clams, squids, and other invertebrates. Small halibut eat a variety of benthic prey items and small fishes, with prey size increasing with fish length. Larger halibut consume almost anything they can catch, with fishes constituting a major portion of their diet. They also feed on squid, octopus, and diverse benthic and nektonic fishes, including cods, pollock, rockfishes, sculpins, other flatfishes, and occasionally smaller halibuts. Halibut are eaten by marine mammals, perhaps some sharks, and other halibuts, but because of their large size, they are rarely found as prey for other fish species.

REPRODUCTIVE BIOLOGY
This species spawns in deep water, 902–1,352 ft (275–412 m), at the edge of the continental shelf during winter (November to March). The Gulf of Alaska is an important spawning area. On average, females mature at 12 years of age (range, eight to 16 years), and males mature by about age eight. The number of eggs produced increases with the size of the female. Large females (those more than 250 lb, or 113.4 kg) can produce as many as two million to four million eggs annually. The eggs are buoyant and fertilized externally. Larvae hatch in about 15 days, depending on water temperature; they remain pelagic for four to five months after spawning. Eggs may be encountered anywhere between 131 and 3,068 ft (40–935 m) but are concentrated between 328 and 656 ft (100–200 m). Newly hatched larvae (0.3–0.6 in, or 8–15 mm) usually are found deeper than 656 ft (200 m). Eye migration begins at a length of about 0.7 in (18 mm). By 1.2 in (30 mm) the young fish resemble adults. With growth, young fish rise in the water and are found predominately at about 328 ft (100 m) by three to five months of age. They are transported great distances and move shoreward with currents, where they settle to the bottom at about six to seven months. After about two years, juveniles begin to move into deeper water. Fishes ages two to four years occur primarily at 361 ft (110 m) or shallower, but some at this size also have been taken at depths of 597 ft (182 m) and occasionally deeper.

CONSERVATION STATUS
Not threatened. Exploitation has resulted in stock reductions of this species throughout its range. Commercial and recreational halibut fisheries are highly regulated, with size and seasonal limitations employed to attempt to keep stocks from diminishing further or disappearing altogether.

SIGNIFICANCE TO HUMANS
Largest and most commercially important flatfish in the North Pacific Ocean. Excellent table fare highly prized by consumers. A commercial fishery for halibut has existed for longer than 100 years. ◆

Plaice
Pleuronectes platessa

FAMILY
Pleuronectidae

TAXONOMY
Pleuronectes platessa Linnaeus, 1758, European seas.

OTHER COMMON NAMES
English: European plaice; French: Plie d'Europe, carrelet; German: Scholle; Spanish: Solla europea.

PHYSICAL CHARACTERISTICS
Dextral flatfish with a deep, oval body; a relatively large head; large eyes; a small mouth; and a series of four to seven bony knobs on the head along a curved line from the eyes back to the lateral line. Teeth are best developed on the jaws of the blind side. Strong, molariform pharyngeal teeth are present on the gill arches. Lateral line is curved slightly above the pectoral fin. Scales are cycloid on both sides of the body. Distinctive ocular side coloration, consisting of a uniformly brown background with brilliant red or orange spots. Blind side usually is

uniformly white. The species can reach lengths to about 39.4 in (100 cm) and weights up to 7.9 lb (3.6 kg), but most adults average only about 13.8–19.7 in (35–50 cm) and weights of about 2.2 lb (1 kg). Individuals can reach at least 30 years of age, but most are much younger. Females grow faster and live longer than males, which rarely live longer than 10–12 years.

DISTRIBUTION
Northeastern Atlantic Ocean in marine and sometimes estuarine waters from the White and Barents Seas southward to the North Sea, including the British Isles and western Baltic Sea; off Iceland; occasionally off Greenland; and along the European coast from Germany and Denmark south to Spain and Portugal and in the western Mediterranean Sea.

HABITAT
Usually found on the inner continental shelf from shallow waters to about 656 ft (200 m) but most abundant in 33–164 ft (10–50 m); usually occur in waters of 35.6–59°F (2–15°C). Most commonly found on sandy sediments but also occur on mud or gravel bottoms. Newly settled plaice recruit to inshore waters typically between 9.8 and 88.6 ft (3–27 m), and sometimes juveniles are found in sandy intertidal pools. Plaice can tolerate reduced salinity levels but do not usually penetrate estuaries to any great degree and are not typically found in freshwater reaches within estuaries. During their first year, young plaice generally are found in shallow waters. By their second year, they begin to move to deeper waters. Older and larger plaice tend to live deeper than smaller and younger plaice.

BEHAVIOR
A diurnally active, benthic fish that lies partially buried when possible. Plaice remain stationary for long periods of time, lying partially buried in the sediment. They are often active at night, especially in shallow water, and have been reported to exhibit homing behavior, at least in near-shore environments. Where there are tidal currents, the plaice orientates itself by pointing into the current and retains its position by pressing its dorsal and anal fins against the bottom. Many individuals congregate in the same general area. Plaice, including larger individuals, sometimes move on rising tides into intertidal areas to forage, retreating with the receding tide.

FEEDING ECOLOGY AND DIET
Opportunistic, visual predators that feed mainly during daylight hours. After metamorphosis, juvenile plaice consume small polychaete worms and harpacticoid copepods, but with increasing size their diet broadens to include a wider variety of prey, such as small crustaceans, amphipods, cumaceans, and small mollusks. Larger plaice consume a greater proportion of thin-shelled bivalve mollusks, especially the siphons of burrowing species (which they nip off using the teeth of their blind-side jaws), as well as gastropod mollusks, shrimps, small crabs, and various polychaete worms. Feeding activity varies with season (temperature), with higher feeding rates occurring during warmer periods than during wintertime. The plaice takes its food in a nearly horizontal position, with its head raised slightly off the bottom. Predators that consume plaice include sculpins, lumpfish, spiny dogfish, weaver fish, seals, and cormorants. Shrimps and ctenophores prey on plaice in the early-life stages.

REPRODUCTIVE BIOLOGY
Maturity schedules vary, depending upon the area where the fish live, their food supply, and ambient temperatures. Female plaice reach sexual maturity between three and seven years of age (11.8–15.7 in, or 30–40 cm) and males at two to six years of age (7.9–11.8 in, or 20–30 cm) in the North Sea. Plaice spawn throughout their range, usually on well-defined spawning grounds. The spawning season varies with latitude and location but usually occurs in the early months of the year throughout its range (December to March in the North Sea, February to March off Denmark, and March to April off Iceland), when water temperatures are about 42.8°F (6°C). Mature fish may undertake extensive migrations from feeding grounds to discrete spawning grounds. The extent of migration depends upon the individual stocks. Spawning grounds generally are located at depths of 66–131 ft (20–40 m). Males and females may pair up and swim one above the other during spawning. Plaice do not guard a nest but rather scatter eggs, which may number up to 50,000 during a spawning event. Eggs are planktonic at first, gradually sinking as development proceeds. Hatching occurs in 18–21 days, depending on temperature. The larval stage lasts between four and six weeks, after which the fish metamorphose at about 0.4–0.7 in (10–17 mm) in length.

CONSERVATION STATUS
Not threatened, although stock sizes have declined over time as a result of overfishing, habitat destruction, and pollution.

SIGNIFICANCE TO HUMANS
One of the most familiar flatfishes in northern European waters, highly desired owing to its size, abundance, and edible qualities. It is the single most important commercial flatfish to the fisheries of Europe. The species also is targeted by a large recreational fishery. Plaice are considered a potential species for aquaculture and are kept as an aquarium species. ◆

Winter flounder
Pseudopleuronectes americanus

FAMILY
Pleuronectidae

TAXONOMY
Pleuronectes americanus Walbaum, 1792, New York.

OTHER COMMON NAMES

English: Blackback, Georges Bank flounder, lemon sole, rough flounder; French: Limande-plie rouge; Spanish: Solla roja.

PHYSICAL CHARACTERISTICS

Medium-sized, dextral flatfish with an oval and thick body with a relatively wide caudal peduncle and a broadly rounded caudal fin. The head is relatively small, with a small terminal mouth with a small gape and thick, fleshy lips. Jaws on the blind side are equipped with a series of incisor-like teeth, whereas the jaws on the ocular side are toothless or nearly so. The lateral line is nearly straight, with only a slight arch above the pectoral fin. Ocular side scales are ctenoid, whereas blind side scales are ctenoid in males and cycloid in juveniles and females. Winter flounder vary in color, depending on the bottom where they live. Larger specimens are dark muddy brown or reddish brown, olive-green, or slate-colored to almost black. On the ocular side, coloration varies from uniformly pigmented to patterned with definite flecks, spots, and darker blotches of differing hues, depending on the bottom type. The blind side usually is uniformly white and translucent with a bluish tinge toward the body margins and sometimes with yellow on the caudal peduncle. Winter flounder can attain a length of 24.8–26.4 in (63–67 cm) and weights to about 7.9 lb (3.6 kg). Winter flounder are relatively long lived, reaching a maximum age of about 15 years. After about age five, females begin to grow faster than males; they also live longer than males.

DISTRIBUTION

Western North Atlantic in estuarine and marine waters along the Atlantic coast of North America from Labrador south to Georgia and also on the offshore banks. Most abundant between the Gulf of Saint Lawrence and Chesapeake Bay.

HABITAT

Brackish waters of tidal rivers, estuaries, and river mouths to areas on the inner continental shelf. Larger and older fish tend to inhabit deeper waters than do younger, smaller fish. Typical inshore habitats consist of muddy sand, especially where the sand is broken by patches of eelgrass; clean sand; clay; and pebbly and gravelly ground. Offshore, winter flounder usually are found on hard bottom. They can survive a wide range of temperatures, from nearly the freezing point of saltwater to about 68–69.8°F (20–21°C). Their blood serum contains an antifreeze protein that helps protect them against freezing.

BEHAVIOR

Diurnally active, with activity beginning at sunrise. At night they lie flat, with heads resting on the bottom and eye turrets retracted. On muddy bottoms, winter flounder usually lie buried, all but the eyes, working themselves down into the mud soon after settling on the bottom. Fish living on tidal flats typically remain motionless during low tides but actively forage during high tides. Can change color to match background surroundings, ranging from whitish on white backgrounds to dark brown or nearly black on dark sediments. Local conditions in inshore waters appear to determine inshore distribution patterns, whereas offshore movements seem to be associated with extreme summer and winter conditions. In general, in summer months adult winter flounder stay in the shallow shore zone when the water temperature is not excessive and food availability is adequate. If these conditions are not met, they may move into deeper channels or offshore or may take evasive action. During winter in the southern parts of their range, they remain or move into shallow water to spawn, whereas in northern regions they remain inshore in protected areas and move offshore in exposed areas to avoid turbulence and drifting pack ice. Winter flounder in deeper waters, such as Georges Bank, remain there year-round. Winter flounder bury in sediments when water temperatures are below 32°F (0°C) and ice crystals are present in the water. They are active at water temperatures up to 71.6°F (22°C), beyond which they become inactive. They are sensitive to low levels (3 ppm) of dissolved oxygen.

FEEDING ECOLOGY AND DIET

Visual predators. Consume a wide variety of small invertebrates and, rarely, small fishes, such as sand lance. Principal food includes polychaetes, anthozoans, and amphipods but also shrimps, small crabs and other crustaceans, ascidians, holothurians, squids, bivalve and gastropod mollusks, and sometimes fish eggs. They often bite off clam siphons that protrude from the sand. While feeding, a winter flounder lies with its head raised off the bottom with its anterior body braced vertically against the bottom. Eye turrets are extended, and the eyes move independently of each other. After sighting its prey, the fish remains stationary, pointed toward the target, and then lunges forward and downward to seize its prey. Winter flounder are eaten by codfish, spiny dogfish, goosefish, hakes, winter skate, smooth dogfish, striped bass, bluefish, sea raven, seals, ospreys, gulls, blue herons, and cormorants.

REPRODUCTIVE BIOLOGY

In the fall, as gonads ripen, adult winter flounder remain in or move into shallow water to spawn. They spawn in winter in southern locations and in early spring (January to May) in more northern locations. Spawning in inshore waters occurs nearly at the seasonal lowest temperatures, which range from 31.1 to 35.6°F (−0.5–2.0°C), depending on latitude. Spawning on Georges Bank happens at temperatures ranging from about 37.9 to 41.9°F (3.3–5.5°C). Winter flounder spawn on sandy bottom and algal mats, often in shallow water. Spawning in estuaries occurs in areas with salinity levels as low as 11.4 ppt. Males mature at age two and females at age three off New York, when fish are 7.9–9.8 in (20–25 cm) in total length. Fish in more northern areas mature later—age three years and four months for males and three years and six months for females north of Cape Cod and age six for males and seven for females in Newfoundland. Maturity seems to be a function of size rather than age. On Georges Bank, the mean age at maturity is just under two years for both sexes.

Females produce, on average, 500,000 eggs, but larger fish can produce up to 3.5 million eggs. Winter flounder migrate into shallow water or estuaries and coastal ponds to spawn, and tagging studies show that most return repeatedly to the same spawning grounds. They are batch spawners. Females in captivity spawned up to 40 times and males up to 147 times. Males initiated all observed spawning events, which occurred throughout the night but primarily between sunset and midnight. Spawning by one pair frequently elicited sudden convergence and spawning by secondary males. Strict pair spawning is uncommon. Male and female activity patterns were almost entirely nocturnal during the reproductive season but became increasingly diurnal during the post-spawning season. Eggs are fertilized outside the body and sink to the bottom, where they stick together in clusters. Incubation takes place over 15–18 days at 37–37.9°F (2.8–3.3°C). Young larvae hatch at about 0.12–0.14 in (3.0–3.5 mm). Metamorphosis is complete when larvae are only 0.31–0.35 in (8–9 mm) long. Winter flounder exhibit little movement from areas where they settle, unless seasonal temperatures become extreme.

CONSERVATION STATUS

Not threatened, although their numbers have diminished substantially since the 1970s, and the stocks throughout most, if not all, of the species' range are considered to be overexploited. Overfishing, pollution, and habitat destruction all contribute to reductions in stock sizes of this species. Current fishery management plans are to restore stock sizes to former levels. As of 2002 little success was evident in these attempts.

SIGNIFICANCE TO HUMANS
Highly desirable commercial and recreational species. Winter flounder were the most frequently captured flatfish taken by recreational fishers along the eastern coast of the United States until about 1970. Since then, landings of fish in both commercial and recreational fisheries have declined substantially. ◆

Windowpane flounder
Scophthalmus aquosus

FAMILY
Scophthalmidae

TAXONOMY
Pleuronectes aquosus Mitchill, 1815, New York, United States.

OTHER COMMON NAMES
English: Sand dab; spotted flounder, brill; French: Turbot de sable.

PHYSICAL CHARACTERISTICS
Medium-sized, sinistral flatfish characterized by a deep, nearly round, thin, and almost translucent body with a rounded caudal fin. The head is small with a relatively large mouth having a wide gape and with a projecting lower jaw that has a knob on its ventral surface. Small teeth are present on both jaws. The first 10–12 fin rays of the dorsal fin are free from the fin membrane along the distal half and branched toward their tips, forming a conspicuous fringe. Eyes are large, separated, and nearly equal in position on the head. Pelvic fin bases are elongated and slightly asymmetrical. Scales are cycloid and smooth to the touch. The lateral line is strongly arched above the pectoral fin. The ocular side is rather translucent greenish olive or slightly reddish brown or pale slate-brown, mottled with darker and paler irregular markings and usually dotted with many small, irregularly shaped brown spots and sometimes also with white spots that vary in size. Blind side is generally whitish, occasionally with some irregular darker blotches. Can reach sizes to about 15.7 in (40 cm) and weights to about 2.2 lb (1 kg), but adults typically average only 9.8–11.8 in (25–30 cm). Can live to be 15–18 years of age, but most individuals are 11 years old or younger.

DISTRIBUTION
Western North Atlantic in lower estuarine and marine waters on the inner continental shelf of eastern North America from the Gulf of Saint Lawrence to Florida. Most abundant from Georges Bank to Chesapeake Bay.

HABITAT
Shallow, inshore waters from the high tide line down to about 656 ft (200 m), with the greatest numbers occurring at depths of less than 180 ft (55 m). They occur most often on sandy bottoms but also can be found on softer and muddier sediments.

BEHAVIOR
Diurnally active. Often lie on or within sandy sediments. Young windowpane flounder settle in shallow water inshore and tend to move into deeper offshore waters as they grow. Adults may undertake movements along the coast for considerable distances (80 mi, or 129 km, in three months) or even move across open water.

FEEDING ECOLOGY AND DIET
Visually oriented, ambush predators that forage on a variety of actively swimming prey, particularly mysids, various fishes, and decapod crustaceans, especially shrimp. They also eat chaetognaths, squids, mollusks, ascidians, polychaetes isopods, amphipods, euphausiids, and salps. Windowpane flounders, in turn, are eaten by various sharks, skates, stingrays, codfish, cobia, bluefish, and other windowpane flounders.

REPRODUCTIVE BIOLOGY
Windowpane flounder of both sexes mature at about the same size, 8.3–8.7 in (21–22 cm) between ages three and four years, with males sometimes maturing at age two. Spawning occurs from February to November, with peak spawning from May through October. There is a strong correlation between water temperatures and spawning. They spawn in the evening or at night on or near the bottom at temperatures from 42.8 to 69.8°F (6–21°C), with optimal spawning temperatures of 60.8–66.2°F (16–19°C) in the Mid-Atlantic Bight and 55.4–60.8°F (13–16°C) on Georges Bank. Eggs are transparent, have an oil globule, and are buoyant. Larvae hatch at 0.07–0.09 in (1.8–2.3 mm), usually at eight days post-spawning in temperatures of 50–55.4°F (10–13°C). Eye migration begins when they reach a length of about 0.26 in (6.5 mm), proceeds very rapidly, and typically is completed by 0.39–0.51 (10–13 mm) total length.

CONSERVATION STATUS
Not threatened.

SIGNIFICANCE TO HUMANS
Not commercially important and not targeted directly by commercial fisheries. They often are taken as by-catch during trawl fishing. ◆

Common sole
Solea solea

FAMILY
Soleidae

TAXONOMY
Pleuronectes solea Linnaeus, 1758, European ocean.

OTHER COMMON NAMES
English: Dover sole; French: Sole commune; German: Seezunge; Spanish: Lenguado commúne.

PHYSICAL CHARACTERISTICS
Dextral flatfish with an elongate but rather thick body. The dorsal fin extends to the anterior part of the head, often to a point equal to a horizontal line drawn through the upper eye. Has a smoothly rounded head, projecting snout, and a small, subterminal mouth with small teeth and with the preoperculum covered by skin. On the blind side, head and snout are covered with close-set, whitish sensory papillae. Moderately large pectoral fins. On the ocular side this fin has a distinct elliptical black patch (not surrounded by white ring) on its upper extremity. On the blind side it is only slightly smaller than the ocular side counterpart. Last rays of the dorsal and anal

fins are joined to the base of the caudal fin by a distinct membrane, and the last rays of the dorsal and anal fins overlap the bases of the caudal fin rays. Ocular side is uniformly dark brown or grayish brown, with numerous darker irregular blotches. Dorsal and anal fins are edged in white. Blind side is creamy white. Lengths reach about 27.6 in (70 cm) and weights about 6.6 lb (3 kg), but most fish are 11.8–15.7 in (30–40 cm) in length. They attain ages up to seven to eight years and perhaps older.

DISTRIBUTION
Northeastern Atlantic in estuarine and marine waters from off Norway and the western Baltic Sea and commonly off Ireland, England, and Scotland (uncommon northwest of Scotland). Also commonly along the European coast in the southern North Sea, the Mediterranean Sea, and the eastern Black Sea and southward to Senegal.

HABITAT
Occurs on soft sandy or muddy bottoms over a wide range of salinity levels and depths, ranging from shallow estuarine waters of 3.3 ft (1 m) or less down to about 656 ft (200 m). Estuaries are important nursery areas for newly settled fish and juveniles. Young-of-the-year soles appear in surf zone waters on shallow sandy beaches during summertime and can be taken even in shallow tide pools during this time. They occur at temperatures from 46.4 to 75.2°F (8–24°C). Adults and larger juveniles undertake seasonal migrations between shallower waters on the inner shelf (in warmer periods) to deeper areas (230–427 ft, or 70–130 m) on the outer shelf in winter. Larger fish generally are found in deeper waters than are smaller fish. Seasonal inshore migrations during springtime are complicated, because larger fish, besides moving inshore, also move toward definite spawning grounds. Young fish move into inshore waters earlier, usually during spring, whereas larger fishes move into shallower waters by late spring or early summer.

BEHAVIOR
Benthic fishes that live somewhat solitary lives. They spend much of the daylight hours partially buried in sediments or lying on top of them. In overcast conditions or in turbid waters, such as in estuaries, they are more active during daylight hours. Generally, they are most active at nighttime, when they sometimes are found off the bottom and up in the water column. They are found pelagically during migrations.

FEEDING ECOLOGY AND DIET
Opportunistic, nocturnal feeders that rely on chemosensory and tactile information to locate their prey. Some feeding by adults and juveniles also takes place during daylight hours when fish are active. In some estuaries, feeding activity showed a strong relationship to the tidal cycle. Juveniles may use intertidal areas as feeding grounds during flood tides. They consume a diverse array of mostly benthic prey, including amphipods, polychaetes, oligochaetes, small bivalve mollusks and siphons of bivalve mollusks, gastropod mollusks, mysids and crangonid shrimps, brittle starfish, and, to a lesser extent, small benthic fishes, such as sand eels (*Ammodytes*) and gobies. Diets often vary between habitats and in terms of size of fish, with prey sizes generally increasing with increasing fish size. Soles feed on small quantities of prey very often. Predators of sole include spiny dogfish, hakes, lizardfish, codfish, weaver fish, and cormorants.

REPRODUCTIVE BIOLOGY
Mature at three to five years of age and at sizes of about 9.1–11.8 in (23–30 cm). Spawn primarily during one well-defined season in the spring and early summer, which varies according to latitude (March to May off England and April to June farther north). Spawning grounds are located in both shallow and deep waters. Spawning takes place between 42.8 and 53.6°F (6–12°C). Females may release up to 100,000 eggs during a spawning event. Eggs are buoyant and pelagic; planktonic larvae hatch when eggs reach a length of about 0.12–0.16 in (3–4 mm). Metamorphosis occurs at about 0.47–0.59 in (12–15 mm), with settlement taking place at 0.59–0.7 in (15–18 mm) in length. Adults undertake migrations to definite spawning grounds located on the inner continental shelf at depths of 131–197 ft (40–60 m). Spawning grounds have been identified in the North Sea, the Irish Sea, and the English Channel.

CONSERVATION STATUS
Not threatened. Stock sizes have been reduced owing to overfishing. At present, efforts have focused on limiting fishing pressure and reducing mortalities, with expectations that stock sizes will not decline further.

SIGNIFICANCE TO HUMANS
Abundant and valued food species. The most valuable fishing grounds lie in the southern and central North Sea and the Bay of Biscay. Little recreational fishing occurs, primarily because of this species' nocturnal feeding habits and also because of the difficulty in catching it. ◆

Resources

Books

Able, Kenneth W., and Michael P. Fahay. *The First Year in the Life of Estuarine Fishes in the Middle Atlantic Bight.* New Brunswick, NJ: Rutgers University Press, 1998.

Bowman, Ray E., Charles E. Stillwell, William L. Michaels, and Marvin D. Grosslein. *Food of Northwest Atlantic Fishes and Two Common Species of Squid.* NOAA Technical Memorandum NMFS-NE-155. Woods Hole, MA: National Marine Fisheries Service, 2000.

Carpenter, K. E., and V. H. Niem, eds. *FAO Species Identification Guide for Fishery Purposes.* Vol. 6, *The Living Marine Resources of the Western Central Pacific.* Part 4, *Bony Fishes: (Labridae to Latimeridae), Estuarine Crocodiles, Sea Turtles, Sea Snakes and Marine Mammals.* Rome: FAO, 2001.

Chang, Sukwoo, Peter L. Berrien, Donna L. Johnson, and Wallace W. Morse. *Essential Fish Habitat Source Document: Windowpane,* Scophthalmus aquosus, *Life History and Habitat Characteristics.* NOAA Technical Memorandum NMFS-NE-137. Woods Hole, MA: National Marine Fisheries Service, 1997.

Chapleau, F., and K. Amaoka. "Flatfishes." In *Encyclopedia of Fishes,* edited by John R. Paxton and William N. Eschmeyer. 2nd edition. San Diego: Academic Press, 1998.

Collette, Bruce B., and Grace Klein-MacPhee, eds. *Bigelow and Schroeder's Fishes of the Gulf of Maine.* 3rd edition. Washington, DC: Smithsonian Institution Press, 2002.

Resources

Fischer, W., F. Krupp, W. Schneider, C. Sommer, K. E. Carpenter, and V.H. Niem, eds. *Guía FAO para la Identificación de Especes para los Fines de la Pesca: Pacífico centro–oriental.* Vols. 2 and 3, *Vertebrados—Parte 1 and Parte 2.* Rome: FAO, 1995.

Hart, J. L. *Pacific Fishes of Canada.* Bulletin 180. Ottawa: Fisheries Research Board of Canada, 1973.

Heemstra, P. C. "Achiropsettidae, Southern Flounders." In *Fishes of the Southern Ocean*, edited by O. Gon and P. C. Heemstra. Grahamstown, South Africa: J.L.B. Smith Institute of Ichthyology, 1990.

———. "Family No. 260: Pleuronectidae." In *Smiths' Sea Fishes*, edited by M. M. Smith and P. C. Heemstra. Johannesburg: Macmillan South Africa Publishers, 1986.

International Pacific Halibut Commission. *The Pacific Halibut: Biology, Fishery, and Management.* Technical Report no. 40. Seattle: International Pacific Halibut Commission, 1998.

Kramer, D. E., W. H. Barss, B. C. Paust, and B. E. Bracken. *Guide to Northeast Pacific Flatfishes: Families Bothidae, Cynoglossidae, and Pleuronectidae.* Marine Advisory Bulletin no. 47. Fairbanks: University of Alaska Fairbanks, Alaska Sea Grant College Program, 1995.

Lythgoe, J., and G. Lythgoe. *Fishes of the Sea.* Garden City, NJ: Anchor Press/Doubleday, 1975.

Mecklenburg, Catherine W., T. Anthony Mecklenburg, and Lyman K. Thorsteinson. *Fishes of Alaska.* Bethesda, MD: American Fisheries Society, 2002.

Nelson, J. S. *Fishes of the World.* 3rd edition. New York: John Wiley & Sons, 1994.

Nielsen, J. G. "Pleuronectidae." In *Fishes of the North–eastern Atlantic and the Mediterranean*, edited by P. J. P. Whitehead, M.-L. Bauchot, J.-C. Hureau, J. Nielsen, and E. Tortonese. Vol. 3. Paris: UNESCO, 1986.

O'Brien, Loretta, Jay Burnett, and Ralph K. Mayo. *Maturation of Nineteen Species of Finfish off the Northeast Coast of the U.S., 1985–1990.* NOAA Technical Report NMFS-113. Seattle: National Marine Fisheries Service, 1993.

Packer, David B., Sara J. Griesbach, Peter L. Berrien, Christine A. Zetlin, Donna L. Johnson, and Wallace W. Morse. "Essential Fish Habitat Source Document: Summer Flounder, *Paralichthys dentatus*, Life History and Habitat Characteristics." NOAA Technical Memorandum NMFS-NE-151. Woods Hole, MA: National Marine Fisheries Service, 1998.

Rackowski, J. P., and E. K. Pikitch. "Species Profiles: Life Histories and Environmental Requirements of Coastal Fishes I (Pacific Southwest). Pacific and Speckled Sanddabs." *Biological Report 82, U. S. Fish and Wildlife Service.* Washington, DC: U.S. Department of the Interior, 1989.

Scott, W. B., and M. G. Scott. *Atlantic Fishes of Canada.* Canadian Bulletin of Fisheries and Aquatic Sciences no. 219. Ottawa.

Schwarzhans, Werner. *Piscium Catalogus: Part Otolithi Piscium.* Vol. 2, *A Comparative Morphological Treatise of Recent and Fossil Otoliths of the Order Pleuronectiformes.* München: Verlag Dr. Friedrich Pfiel, 1999.

Wheeler, Alwyne. *The Fishes of the British Isles and North–West Europe.* East Lansing: Michigan State University Press, 1969.

Periodicals

Bannikov, A. F., and N. N. Parin. "The List of Marine Fishes from Cenozoic (Upper Paleocene–Middle Miocene) Localities in Southern European Russia and Adjacent Countries." *Journal of Ichthyology* 37, no. 2 (1997): 133–146.

Bengston, David A. "Aquaculture of Summer Flounder (*Paralichthys dentatus*): Status of Knowledge, Current Research and Future Research Priorities." *Aquaculture* 176 (1999): 39–49.

Berendzen, P. B., and W. W. Dimmick. "Phylogenetic Relationships of Pleuronectiformes Based on Molecular Evidence." *Copeia* 2002, no. 3 (2002): 642–652.

Brewster, B. "Eye Migration and Cranial Development During Flatfish Metamorphosis: A Reappraisal (Teleostei: Pleuronectiformes)." *Journal of Fish Biology* 31 (1987): 805–833.

Cabral, H. N. "Comparative Feeding Ecology of the Sympatric *Solea solea* and *S. senegalensis*, Within the Nursery Areas of the Tagus Estuary, Portugal." *Journal of Fish Biology* 57, no. 6 (2000): 1550–1562.

Chanet, B. "A Cladistic Reappraisal of the Fossil Flatfishes Record Consequences on the Phylogeny of the Pleuronectiformes (Osteichthyes: Teleostei)." *Annales de Sciences Naturelles, Zoologie, Paris* 18 (1997): 105–117.

———. "*Eobuglossus eocenicus* (Woodward 1910) from the Upper Lutetian of Egypt, One of the Oldest Soleids [Teleostei, Pleuronectiformi]." *Neues Jahrbuch für Paläontologie Monatehefte* 7 (1994): 391–398.

Chapleau, F. "Pleuronectiform Relationships: A Cladistic Reassessment." *Bulletin of Marine Science* 52, no. 1 (1993): 516–540.

Cooper, J. A., and F. Chapleau. "Monophyly and Intrarelationships of the Family Pleuronectidae (Pleuronectiformes), with a Revised Classification." *Fishery Bulletin* 96, no. 4 (1998): 686–726.

Hoshino, K. "Monophyly of the Citharidae (Pleuronectoidei: Pleuronectiformes: Teleostei) with Considerations on Pleuronectoid Phylogeny." *Ichthyological Research* 48, no. 4 (2001): 391–404.

Konstantinou, H., and D. C. Shen. "The Social and Reproductive Behavior of the Eyed Flounder, *Bothus ocellatus*, with Notes on the Spawning of *Bothus lunatus* and *Bothus ellipticus.*" *Environmental Biology of Fishes* 44, no. 2 (1995): 311–324.

Litvak, M. K. "The Development of Winter Flounder (*Pleuronectes americanus*) for Aquaculture in Atlantic Canada: Current Status and Future Prospects." *Aquaculture* 176 (1999): 55–64.

Pearcy, W. G., and D. Hancock. "Feeding Habits of Dover Sole, *Microstomus pacificus*; Rex Sole, *Glyptocephalus zachirus*; Slender Sole, *Lyopsetta exilis*; and Pacific Sanddab, *Citharichthys sordidus*, in a Region of Diverse Sediments and Bathymetry off Oregon." *Fishery Bulletin* 76 (1978): 641–651.

Resources

Phelan, B. A., J. P. Manderson, A. W. Stoner, and A. J. Bejda. "Size-Related Shifts in the Habitat Associations of Young-of-the-Year Winter Flounder (*Pseudopleuronectes americanus*): Field Observations and Laboratory Experiments with Sediments and Prey." *Journal of Experimental Marine Biology and Ecology* 257 (2001): 297–315.

Ramachandran, V. S., C. W. Tyler, R. L. Gregory, D. Rogers-Ramachandran, S. Duensing, C. Pillsbury, and C. Ramachandran. "Rapid Adaptive Camouflage in Tropical Flounders." *Nature (London)* 379, no. 6568 (1996): 815–818.

Stoner, A. W., and A. A. Abookire. "Sediment Preferences and Size-Specific Distribution of Young-of-the-Year Pacific Halibut in an Alaska Nursery." *Journal of Fish Biology* 61 (2002): 540–559.

Terceiro, M. "The Summer Flounder Chronicles: Science, Politics, and Litigation, 1975–2000." *Reviews in Fish Biology and Fisheries.* 11 (2002):125–168.

Other

Coughenower, D., and C. Blood. "Flatout Facts About Halibut." 1997. <http://www.uaf.edu/seagrant/Pubs_Videos/pubs/SG-ED-29.pdf>

Thomas A. Munroe, PhD

Tetraodontiformes

(Pufferfishes, triggerfishes, and relatives)

Class Actinopterygii
Order Tetraodontiformes
Number of families 9

Photo: A whitespotted filefish (*Cantherhines macrocerus*) sleeping, with dead fire coral in its mouth, at night, near the Los Rocas Islands of Venezuela. (Photo by Fred McConnaughey/Photo Researchers, Inc. Reproduced by permission.)

Evolution and systematics

Although it is remarkably derived, this order of fishes dates from the early Eocene epoch or, possibly, the late Cretaceous period. There are nine families, including the spikefishes (Triacanthodidae, 11 genera and 21 species), the triplespines (Triacanthidae, four genera and seven spp.), the boxfishes (Ostraciidae, 11 genera and 37 spp.), the filefishes and leatherjackets (Monacanthidae, 25 genera and at least 104 spp.), the triggerfishes (Balistidae, 12 genera and at least 37 spp.), the threetooth puffer (Triodontidae, one genus and one species), the pufferfishes (Tetraodontidae, 26 genera and at least 170 species, with one species that has two subspecies), the porcupinefishes (Diodontidae, seven genera and 20 spp.), and the molas or ocean sunfishes (Molidae, three genera and four spp.).

Physical characteristics

There is some variability within this order, but all are distinguished by fused teeth within their jaws. Spikefishes are compressed, high bodied (great in depth), or elongate; some have concave foreheads and others elongated, tubelike snouts. The caudal fin is rounded to truncated, and there are 12–18 soft rays in the dorsal fin and 11–16 rays in the anal fin. Coloring is orange, yellow, or pink. Triplespines are compressed and elongate, with steeply sloping foreheads, narrow caudal peduncles, and forked tails. The first dorsal spine is large. The coloring is silvery, with hints of yellow or pale olive. The boxfishes are indeed boxlike, with bodies covered in bony-like plates. They have large rounded or truncated caudal fins and the ability, in numerous species, to secrete ostracitoxin (a poison that makes the creature unpalatable) as an anti-predatation mechanism. Some species have horns that protrude from their heads. A few species are brightly colored, but most are dull or cryptic.

Filefishes and leatherjackets also vary in shape and appearance. Generally, they are compressed and somewhat elongate, and they may have steeply sloped foreheads. Others are rhomboid in shape. At least one species, *Anacanthus barbatus*, is long and slender and resembles superficially a pipefish (Sygnathidae). The first dorsal spine usually is tall and strong. The caudal fin is truncated or slightly rounded. Color patterns vary, ranging from bright colors to cryptic or nondescript shades. Triggerfishes are compressed, rhomboid, or slightly elongate, with pelvic fins that are fused into a single spine. The first dorsal spine is capable of locking by using the second spine as a trigger that locks it into place. It is used as an anti-predation mechanism. The eyes can rotate independently of each other. Color patterns also vary and may be spectacular and bright or quite dull or cryptic.

The threetooth puffer has a beak consisting of three fused teeth, with a median suture in the upper jaw. The body is

A shy toby (*Canthigaster ocellicincta*) sleeps in the Gorgonian coral at night in the South China Sea near the Sipadan Island of Borneo. (Photo by Jeff Rotman/Photo Researchers, Inc. Reproduced by permission.)

elongate, compressed, and inflatable. The caudal penduncle is long, and the caudal fin forked. Pufferfishes have four teeth fused together to form a beak, but there are medial sutures in both the upper and lower jaws. Their bodies are elongate, usually compressed or rhomboid in shape, and inflatable. There are seven to 18 soft rays in both the dorsal and anal fins, respectively. Many species have small prickles on the skin of their bellies. Their skin, entrails, and sometimes their flesh may contain tetraodotoxin, an anti-predation poison that is highly toxic. Porcupinefishes are large, elongate, robust, and somewhat box-like, and they are capable of inflation. They have numerous spines along the body. The two teeth on their jaws are fused into a beak. Molas are large, compressed, and typically high-bodied fishes, with jaws that support two fused teeth. The caudal fin may be replaced by a clavus, which is used to stablize the fish when swimming. The dorsal and anal fins are set high. Both the lateral line and swimbladder are missing from adults.

Distribution

Members of this order occur in marine, brackish, and freshwater of the tropics, subtropics, and temperate zones. The spikefishes are tropical and subtropical deepwater residents of the western Atlantic and Indo-Pacific regions. Triplespines occur in shallow marine and brackish waters of the Indo-Pacific region. Boxfishes, triggerfishes, filefishes, and leatherjackets are distributed widely, mainly in coastal waters of the tropical, subtropical, and temperate or warm temperate Atlantic, Indian, and Pacific Oceans. The monotypic threetooth puffer is a deep-slope species of the Indo-Pacific region. Marine pufferfishes occur in tropical and subtropical waters of the Atlantic, Indian, and Pacific Oceans; numerous species enter brackish water or even freshwater. Some strictly freshwater species are found in river systems of Africa, Asia, or South America. Porcupinefishes occur in tropical, subtropical, and warm temperate waters of the Atlantic and Indo-Pacific regions; at least one species is circumglobal in its distribution. Molas also are found in tropical, subtropical, and warm temperate waters of the Atlantic, Indian, and Pacific Oceans but are largely pelagic and venture inshore only occasionally.

Habitat

A birdbeak burrfish (*Cyclichthys orbicularis*) in the waters of Indonesia. (Photo by Michael Aw/The Photo Library-Sydney/Photo Researchers, Inc. Reproduced by permission.)

Habitat preferences and utilization vary widely within this order. Most species are marine, but at least 20 occur in freshwater systems. Spikefishes are bottom dwelling in relatively deep waters. Triplespines also live on the bottom but on shallow marine and brackish water sand flats and mudflats. Boxfishes frequent coral or rocky reefs in relatively shallow water; others occur on sea grass flats and algae beds. Large postlarvae or young juveniles of some species are pelagic and are part of the diet of tunas and billfishes; others enter estuaries. Triggerfishes also are associated with coral or rocky reefs but may be found on rubble and sand flats. Some species are pelagic or semipelagic, the latter dwelling in the water column but seeking shelter in holes on the bottom or along walls. Filefishes and leatherjackets are quite wide ranging in their habitat use. Many species are associated with coral or rocky reefs; others with sponge reefs, sea grass flats, algae beds, or rubble and sand flats; and still others dwell in the water column. The threetooth puffer is a benthic, deep-slope-dwelling fish. Marine and brackish water pufferfishes occur on coral and rocky reefs, on rubble, sand, mud, or sea grass flats; on algae beds; or in the water column. Freshwater species tend to be benthic. Porcupinefishes occur on coral and rocky reefs, usually to seaward, and also utilize caves and holes in relatively shallow water. They also frequent reefs dominated by sponges. The molas are pelagic fishes that prefer the upper depths of the open ocean but also venture inshore, especially near deep-slope habitats that experience upwelling of deeper, cooler water.

Behavior

To some extent, tetraodontiform fishes incorporate their unique body structure and abilities into their patterns of behavior. For defense, feeding, social interaction, or reproduction, they rely variously on body armament; their fused beaks or ability to inflate with water; or their color patterns, which may advertise toxicity. Adult boxfishes utilize their body armor and ability to secrete ostracitoxin, a compound poisonous to predators (and other fishes, including themselves, if they are kept in a confined space), as defensive mechanisms. Thus, predation risk is minimized as they swim along the bottom or up into the water column. Despite reduction in risk, however, numerous species are quite cryptic as they move about in algae or corals. Boxfishes also are territorial. A majority of triggerfishes are colorful, solitary, and quite aggressive in the defense of their territories. Those species that dwell in the water column often assemble in loose aggregations of up to hundreds of individuals as they forage off walls or along deep slopes. These species quickly retreat to individual shelters when threatened, however.

An inflated spotted puffer, or guinea fowl puffer (*Arothron meleagris*), swimming near Hawaii. (Photo byAndrew G. Wood/Photo Researchers, Inc. Reproduced by permission.)

A honeycomb cowfish (*Lactophrys polygonius*) by coral near Bonaire. (Photo by Charles V. Angelo/Photo Researchers, Inc. Reproduced by permission.)

Filefishes and leatherjackets are solitary or gather in small groups, but some form monogamous pairs that patrol a home range or territory. Others, such as *Aluterus monoceros*, form large schools. Many species are cryptic as well and take advantage of structure to mask their movements. A few species, such as juvenile *Aluterus scriptus*, swim on their sides and mimic floating leaves or other vegetation. Some swim openly in the water column, however. The color pattern of some filefishes and leatherjackets may be put to a surprising advantage. For example, the mimic filefish, *Paraluteres prionurus*, mimics the toxic saddled puffer, *Canthigaster valentini*, and thus avoids predation. Another species, the diamond filefish (*Rudarius excelsius*) mimics benthic algae.

Pufferfishes are solitary or paired or form small groups or schools. Many species utilize color pattern to advertise their toxic nature and thus avoid predation. Others are remarkably cryptic and even bury themselves in the sand. Some frequent the water column but return to the bottom for shelter or to feed. All are capable of inflating as a defensive mechanism. When they are not sheltering in holes or caves, porcupinefishes swim openly in the water column and depend on both their large spines and ability to inflate themselves to defend against predation. They reportedly have toxic flesh or organs that may contribute toward their defense as well. Some species have been observed to bury themselves in the sand. Although they are seen swimming during daylight hours, numerous species also are nocturnal. Molas swim about sideways or upright at or near the surface in the open ocean and, despite their ungainly appearance, are relatively strong and fast swimmers when necessary. They also drift on their sides in the current. When they are inshore near kelp beds, they allow themselves to be cleaned by resident cleanerfish species. The behavior of deep-dwelling spikefishes and threetooth puffers is largely unknown. Presumably, these fishes are largely solitary and move about the bottom or in the water column. The threetooth puffer is able to defend itself by inflation.

Feeding ecology and diet

Feeding patterns and dietary preferences of tetraodontiform fishes also vary. Little is known about spikefish feeding. Those species with long, slender, and highly specialized snouts likely pluck microinvertebrates from the bottom or possibly the water column. Species with relatively larger mouths probably feed on benthic invertebrates. At least one species, *Macrorhamphosodes uradoi*, feeds on the scales of other fishes. Triplespines also eat benthic invertebrates unearthed from sand or mud. Boxfishes are mainly omnivorous and feed upon small invertebrates, especially sessile species, as well as benthic algae. Triggerfishes are more wide ranging in their diet and feed upon benthic invertebrates that include gastropods, bivalves, crustaceans, and various echinoderms. Some species also eat algae, and those that dwell in the water column eat zooplankton. Similarly, filefishes and leatherjackets favor various benthic invertebrates, but some species are specialized to feed upon live corals or zooplankton. The threetooth puffer probably limits its diet to benthic invertebrates. Pufferfishes are remarkable opportunists and feed upon various invertebrates as well as algae. Larger species are quite capable of breaking open hard-shelled organisms, such as gastropods, bivalves, and crustaceans. Some species are specialized for certain invertebrates, however. Porcupinefishes prefer hard-shelled invertebrates, which they crush with their beaks. The molas utilize their parrot-like beaks to feed upon zooplankton and jellyfishes in the water column but also take fishes, mollusks, crustaceans, and brittlestars when they are inshore or near the bottom in deep-slope areas.

Two false-eye puffers (*Canthigaster papua*) courting, preparing a nest in algae on the coral reef, near Papua New Guinea. (Photo by Fred McConnaughey/Photo Researchers, Inc. Reproduced by permission.)

A smooth trunkfish (*Lactophrys triqueter*) eating sargeant major eggs, near St. Vincent, in the Caribbean. (Photo by Andrew J. Martinez/Photo Researchers, Inc. Reproduced by permission.)

Most members of this order are preyed upon while in the larva or post-larva stage. As adults, relatively few have predators. Successful predators would have to be tolerant of spines, bony plates, and ossicles formed from fused bones, and, in some families, toxins. Some toxic puffers (Tetraodontidae), however, are preyed upon by sea snakes.

Reproductive biology

Depending upon the species, tetraodontiform fishes have diverse mating systems and spawn demersal or pelagic eggs. Seasonality may be pronounced at higher latitudes or in colder, deeper waters. Reproduction also may be linked to lunar or semilunar cycles. Little is known of the reproductive biology of the deep-dwelling spikefishes, threetooth puffers, and shallow-dwelling triplespines, but their eggs are likely demersal and their larvae pelagic. Boxfishes have mainly male-dominated mating groups, court and spawn at dusk, and produce pelagic eggs and larvae. Triggerfishes have varying mating systems, including monogamy, bigamy, and polygyny. They usually spawn demersal eggs in nests that are guarded quite aggressively by females or, to a lesser extent, by males. Larvae are pelagic. The filefishes and leatherjackets studied thus far have mating systems that include monogamy, facultative monogamy, polygyny, female visiting within male territories, or promiscuity. They all appear to lay demersal eggs in nests or on algae. There may be male parental care, female parental care, biparental care. or no care of eggs. The larvae are pelagic and may be adapted for an extended pelagic existence.

Pufferfishes have diverse mating systems and patterns of courtship and spawning. The eggs generally are demersal and are laid in nests on algae or scattered on the bottom. The eggs of some species are poisonous. Porcupinefish reproductive behavior appears to vary within as well as between species. Both pair and group (single female and multiple males) spawning has been reported for *Diodon holacanthus*. Courtship begins at dusk, and spawning is pelagic in shallow water and may involve splashing at the surface as eggs and milt are released. Alternately, the eggs may sink to the bottom. The larvae are pelagic as well. Eggs of the genus *Chilomycterus* reportedly are demersal, although their larvae are pelagic. Details of courtship and spawning of molas are not well known. They are highly fecund and produce hundreds of millions of pelagic eggs. The larvae are also pelagic.

Conservation status

A few species are cited on the IUCN Red List. The queen triggerfish, *Balistes vetula* (Balistidae), the Rapa Island toby, *Canthigaster rapaensis* (Tetraodontidae), and the blunthead puffer, *Liosaccus* (= *Sphoeroides*) *pachygaster* (Tetraodontidae), are listed as Vulnerable. Five additional species in the family Tetradontidae are listed as Data Deficient. Various species may be at risk from overfishing (as a target species or as bycatch) or habitat destruction. The European Community prohibits the trade of pufferfish products.

Significance to humans

Numerous species of triggerfishes, filefishes and leatherjackets, pufferfishes, and molas are taken in subsistence and artisinal fisheries. Commercial fisheries exist for certain triplespines, triggerfishes, filefishes and leatherjackets, and pufferfishes. The latter group includes the famous fugu (*Takifugu* spp.), which is esteemed as a delicacy in Japan but poses a serious health risk from tetraodotoxin poisoning if it is not prepared properly. Boxfishes and molas also are or may be poisonous. Various species are used in Chinese medicine. Many species of boxfishes, porcupinefishes, and puffers are taken for use as decorations in the ornamental trade; for example, diodontids are often used to make lamp shades. Some triggerfishes and filefishes and leatherjackets are considered game fishes. Many species of triggerfishes, filefishes and leatherjackets, and pufferfishes (both marine and freshwater) are collected for the aquarium trade, and some are highly valued.

1. Spotted toby (*Canthigaster solandri*); 2. White-spotted puffer (*Arothron hispidus*); 3. Spot-fin porcupinefish (*Diodon hystrix*); 4. Threetooth puffer (*Triodon macropterus*); 5. Mola (*Mola mola*); 6. Yellow boxfish (*Ostracion cubicus*); 7. Fugu (*Takifugu rubripes*); 8. Scrawled filefish (*Aluterus scriptus*). (Illustration by Brian Cressman and Patricia Ferrer)

1. Spikefish (*Halimochirugus alcocki*); 2. Blackbar triggerfish (*Rhinecanthus aculeatus*); 3. Clown triggerfish (*Balistoides conspicillum*); 4. Orange-striped triggerfish (*Balistapus undulatus*); 5. Triplespine (*Triacanthus biaculeatus*); 6. Longnose filefish (*Oxymonacanthus longirostris*); 7. Longhorn cowfish (*Lactoria cornuta*). (Illustration by Bruce Worden)

Species accounts

Orange-striped triggerfish
Balistapus undulatus

FAMILY
Balistidae

TAXONOMY
Balistapus undulatus Park, 1797, Sumatra, Indonesia.

OTHER COMMON NAMES
English: Orange-lined triggerfish; French: Baliste strié; German: Orangestreifen-Drückerfisch; Afrikaans: Oranje streepsnellervis; Japanese: Kumadori.

PHYSICAL CHARACTERISTICS
Body compressed and oblong, the forehead sloping to above the eye, the caudal peduncle compressed, with a truncate caudal fin. There are three spines, 24–27 soft rays in the dorsal fin, and 20–24 soft rays in the anal fin. The first dorsal spine can be locked. Body color variably green to dark green or dark brown. There is a pattern of orange lines that curve obliquely on the posterior portion of the head and along the body. A band of orange and blue stripes runs obliquely from the mouth to below the pectoral fin. There is a large black blotch on the caudal peduncle. The caudal fin and soft fin rays of the dorsal, anal, and pectoral fins are orange. Grows to at least 11.8 in (30 cm) in total length.

DISTRIBUTION
Tropical and subtropical Indo-West Pacific from the Red Sea and East Africa south to Natal in South Africa, east through Micronesia and the Line Islands to the Marquesas and the Tuamotu Archipelago in eastern Polynesia, south to the Great Barrier Reef of Australia and New Caledonia, and north to southern Japan. New research suggests that populations in the western Indian Ocean, the Indo-Malayan region, and the Pacific may constitute three distinct species.

HABITAT
Somewhat ubiquitous on reefs; may be found on seaward or protected reefs and in lagoons at depths ranging from about 6.6 to 164 ft (2–50 m).

BEHAVIOR
Territorial and generally solitary, although individuals may live in somewhat close proximity to one another and have overlapping home ranges.

FEEDING ECOLOGY AND DIET
Omnivorous. Feeds on sponges, hydrozoans, tunicates, mollusks, echinoderms, small fishes, and algae.

REPRODUCTIVE BIOLOGY
Pair-spawns, with a single cluster of eggs deposited in a shallow nest excavated in rubble or sand. Larvae are pelagic.

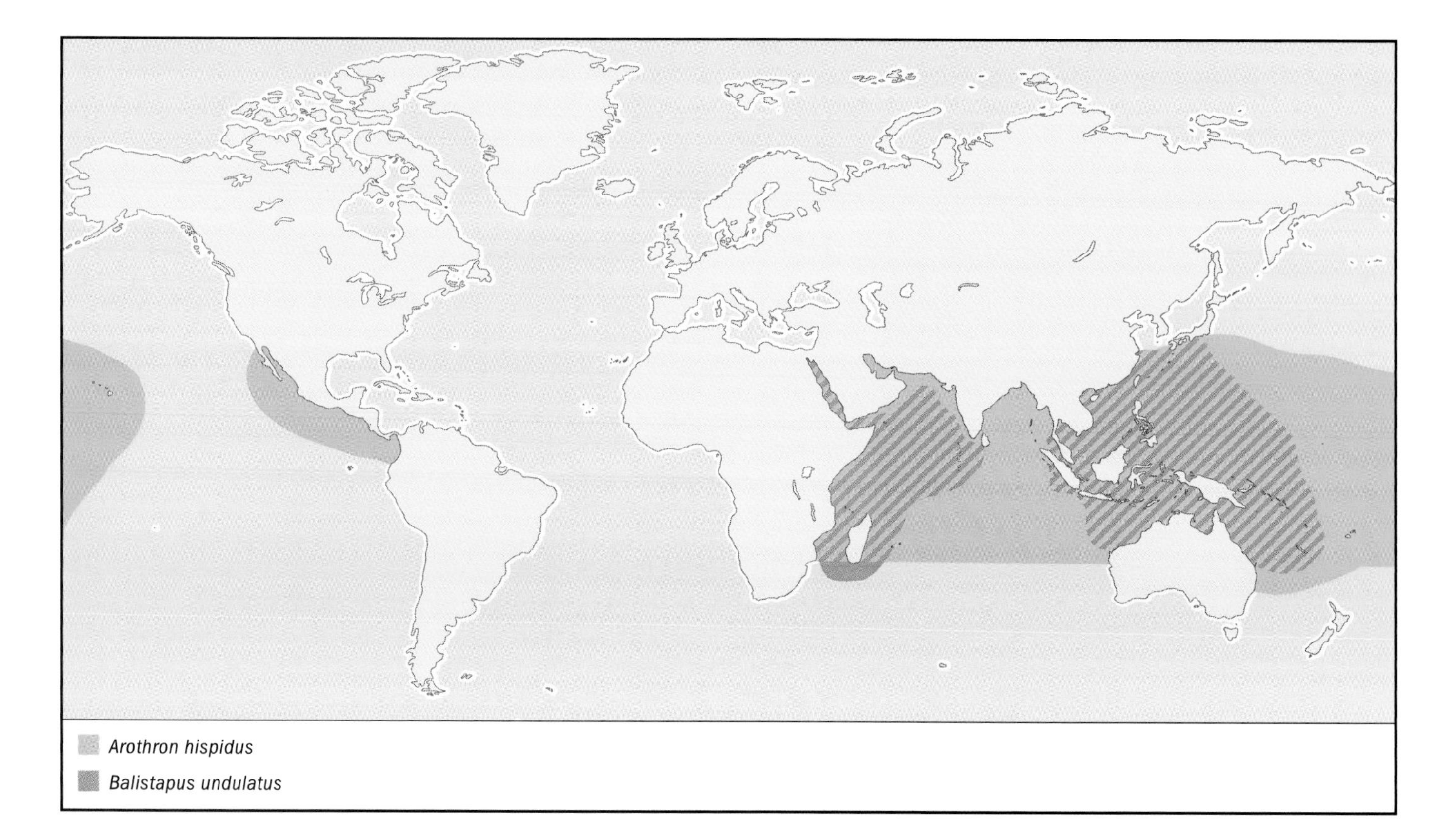

CONSERVATION STATUS
Not listed by the IUCN.

SIGNIFICANCE TO HUMANS
A minor commercial or subsistence food fish in some areas that is either dried and salted or sold fresh. Also collected for the aquarium trade. ◆

Clown triggerfish
Balistoides conspicillum

FAMILY
Balistidae

TAXONOMY
Balistoides conspicillum Bloch and Schneider, 1801, Indian and American seas.

OTHER COMMON NAMES
French: Baliste lépreaux; German: Leoparden-Drückerfisch; Afrikaans: Nas-snellervis; Japanese: Mongarakawahagi.

PHYSICAL CHARACTERISTICS
Body and caudal peduncle are compressed; the body is robust and oval in shape. There are three spines and 25–27 soft rays in the dorsal fin as well as the anal fin. The first dorsal spine can be locked into an erect position. The caudal fin is somewhat rhomboid in shape. The body's color pattern is unmistakable. The background color is black, with large white spots along the flank, a white breast and band below the eye, a white caudal peduncle, and an alternating pattern of black, white, and black on the caudal fin. The mouth is yellow or orangish yellow, with a thin yellow stripe on a black background. Faint yellow hues are found along the base of the anal fin and on the caudal peduncle. The dorsal, lower anal, pectoral, and pelvic fins are white. Grows to at least 19.7 in (50 cm) in total length.

DISTRIBUTION
Tropical and subtropical Indo-West Pacific from East Africa and South Africa east to Samoa, south to northern Australia and New Caledonia, and north to southern Japan.

HABITAT
Occurs mainly on seaward reefs or passes along or adjacent to deep slopes or walls, from just below the surge zone to at least 246 ft (75 m).

BEHAVIOR
Generally solitary, territorial, and even aggressive.

FEEDING ECOLOGY AND DIET
Feeds upon benthic invertebrates, mainly crustaceans, including crabs, as well as sea urchins, mollusks, and tunicates.

REPRODUCTIVE BIOLOGY
Reproduction is consistent with that of others in the family, but greater detail is needed. Pair-spawns. Lays and defends demersal eggs in a nest. The larvae are pelagic.

CONSERVATION STATUS
Not listed by the IUCN but relatively rare in many localities and may be vulnerable to overfishing.

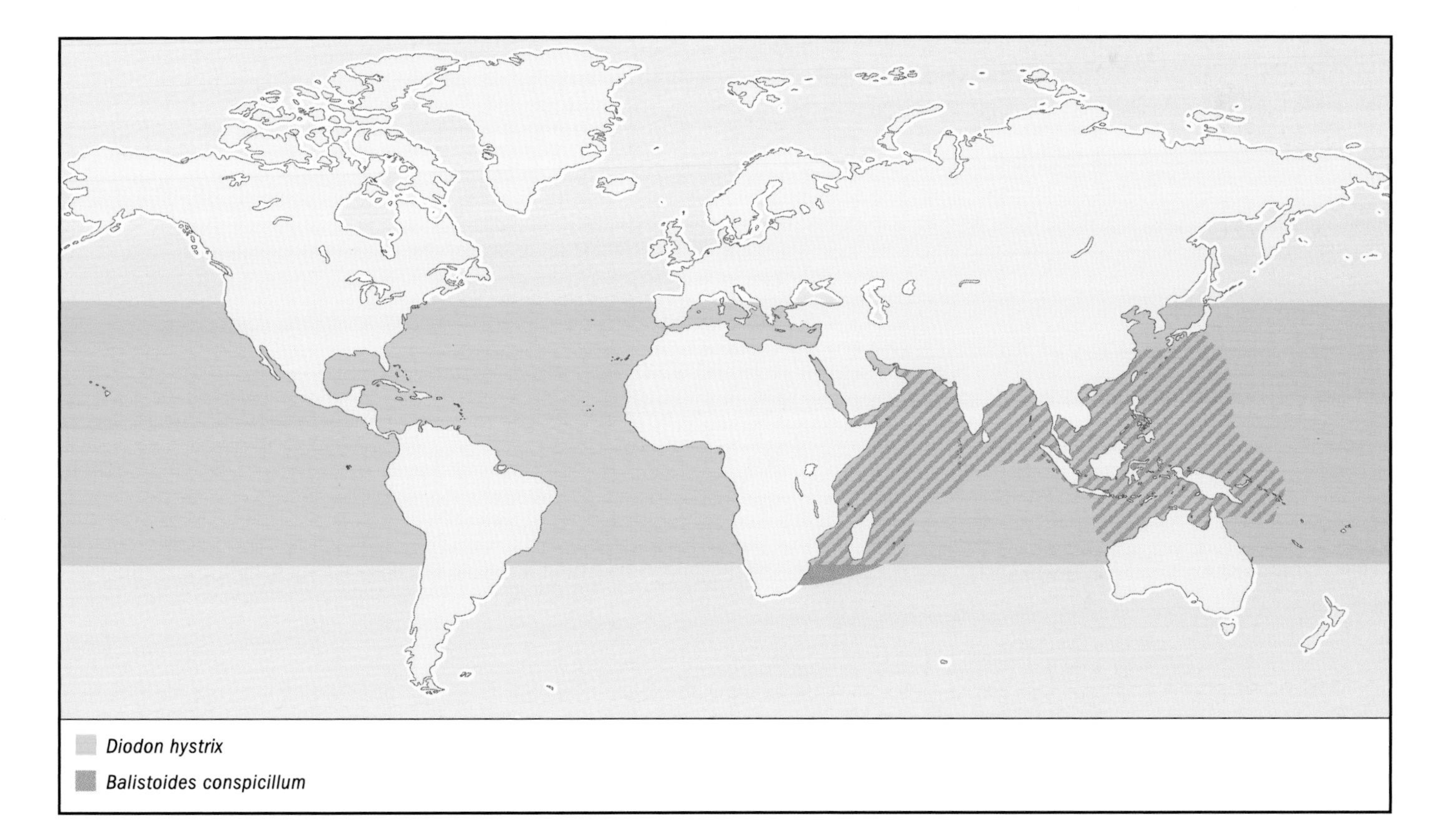

SIGNIFICANCE TO HUMANS
A highly prized aquarium fish species, often collected as juveniles or subadults but occasionally full grown for larger aquaria. Also taken in subsistence and minor commercial food fisheries but has been implicated in ciguatera poisoning in humans. ◆

Blackbar triggerfish
Rhinecanthus aculeatus

FAMILY
Balistidae

TAXONOMY
Rhinecanthus aculeatus Linnaeus, 1758, India.

OTHER COMMON NAMES
English: Picassofish; English and Hawaiian: Humu humu; French: Poisson picasso sombre; German: Gemeiner Picassodrücker.

PHYSICAL CHARACTERISTICS
Body compressed and somewhat oblong or rhomboid. The forehead slopes upward past the eye to just anterior to the dorsal fin. There are three spines and 23–26 soft rays in the dorsal fin and 21–23 soft rays in the anal fin. The color pattern is gray to grayish brown on the head and just below the dorsal fin and brown to brownish gray along the upper flank, with two thick and two thin oblique stripes of the same color extending down, in alternation, onto the lower flank and belly. An orangish brown bar extends from midflank up and back to the base of the soft rays of the dorsal fin. The lower flank and belly are white. A thin yellowish line runs from the mouth back to just past the lower part of the operculum. There is a black to brown blotch extending from the posterior portion of the flank onto the white caudal peduncle and a black blotch around the anus. The fins are pale to light gray. Reaches 11.8 in (30 cm) in total length.

DISTRIBUTION
Tropical and subtropical Indo-West Pacific from the Red Sea and East Africa; south to South Africa; east to the Tuamotu Archipelago, Marquesas, and Hawaiian Islands in Polynesia; south to northern Australia and Lord Howe Island; and north to southern Japan. Also reported from the eastern Atlantic from Senegal south to South Africa.

HABITAT
Coral and rocky reefs, mainly on reef flats or in shallow lagoons on rubble and sand. Depth range is 3.3–13 ft (1–4 m).

BEHAVIOR
Highly territorial, especially when guarding a nest. Excavates or utilizes a small hole for shelter, where it can lock itself in by extending its first dorsal spine. It sleeps on its side there. Patrols a territory and may inflict a painful bite upon intruders or passersby. Produces a whirring sound when alarmed.

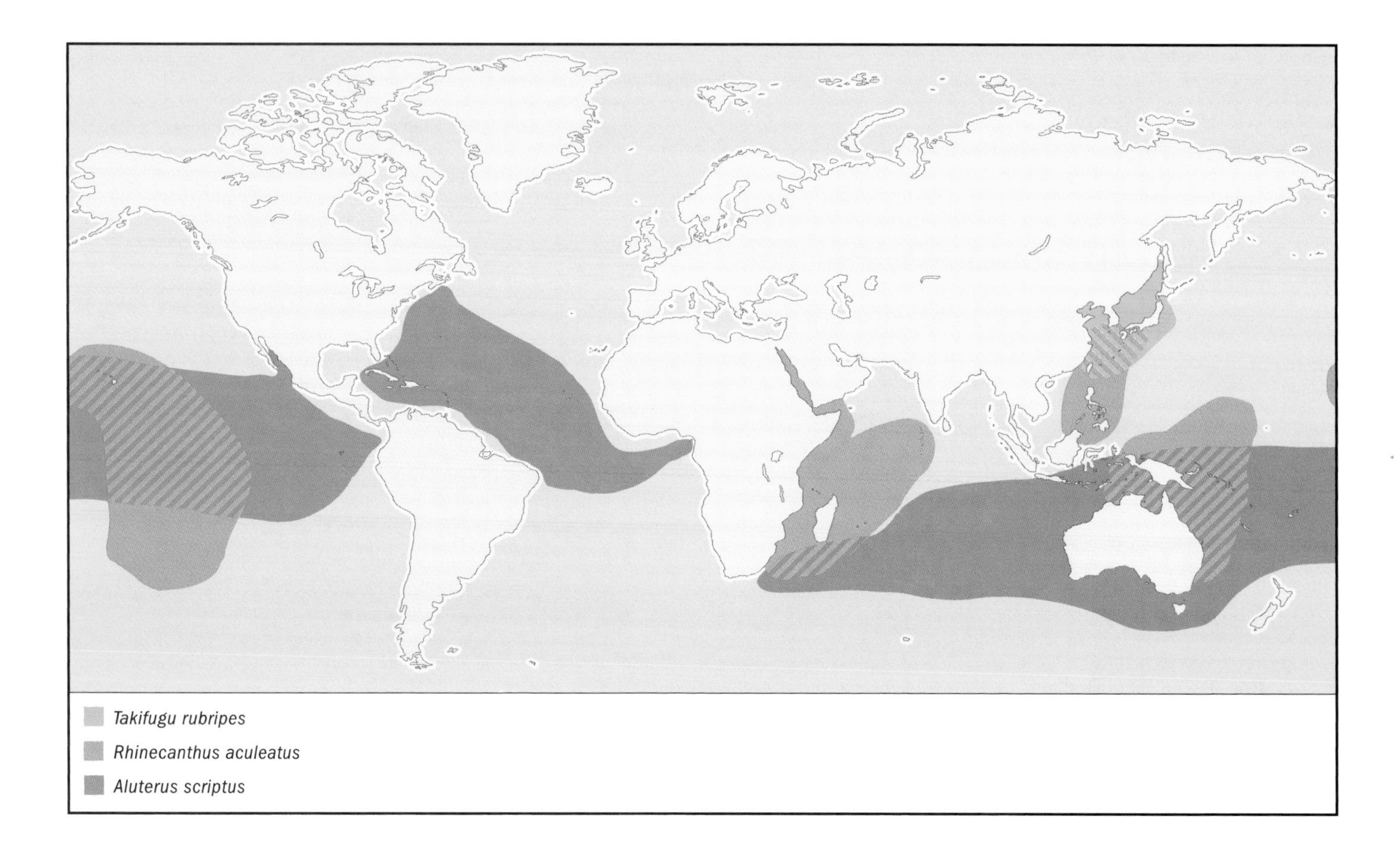

FEEDING ECOLOGY AND DIET
Omnivorous. Feeds upon mollusks, crustaceans, worms, sea urchins, corals, small fishes, tunicates, foramniferans, fish eggs, algae, and detritus.

REPRODUCTIVE BIOLOGY
The mating system is single male/multiple female polygyny, but facultative monogamy occurs occasionally. Spawning takes place on a semilunar cycle. Spawning is paired, and eggs are deposited on the substratum within a territory. The eggs are demersal and guarded. The larvae are pelagic.

CONSERVATION STATUS
Not listed by the IUCN.

SIGNIFICANCE TO HUMANS
An important fish in the aquarium trade but also taken as food in minor commercial and subsistence fisheries. ◆

Spot-fin porcupinefish
Diodon hystrix

FAMILY
Diodontidae

TAXONOMY
Diodon hystrix Linnaeus, 1758, India.

OTHER COMMON NAMES
English: Spot-fin porcupinefish; French: Diodon porc-épic; German: Gepunkter Igelfisch; Afrikaans: Penvis; Japanese: Nezumifugu.

PHYSICAL CHARACTERISTICS
Body elongate, stocky, and inflatable. Several long, strong spines cover the body, with a row of 20 stretching from the snout to the dorsal fin. The jaw has two fused teeth that are quite adept at crushing food items. There are 14–17 soft rays in the dorsal fin and 14–16 soft rays in the anal fin. The body color is grayish-tan, with small black spots and a white belly. A dusky-colored ring surrounds the belly. Grows to more than 35.4 in (90 cm) in total length.

DISTRIBUTION
Essentially circumglobal in tropical, subtropical, and warm temperate waters. From East Africa to San Diego, California, and Chile and the Galapagos Islands. Also in the Atlantic from Massachusetts and Bermuda to Brazil and from the Iberian Peninsula south to southern Africa.

HABITAT
Frequents coral and rocky reefs, usually in caves and holes to a depth of 164 ft (50 m). Swims in the water column during low-light periods.

BEHAVIOR
Generally solitary. Reportedly nocturnal but occasionally active in daylight.

FEEDING ECOLOGY AND DIET
Prefers hard-shelled benthic invertebrates that include gastropods, hermit crabs, and sea urchins.

REPRODUCTIVE BIOLOGY
Not well known, but paired courtship involving chasing has been observed just before sunset. Probably spawns demersal eggs. The larvae are pelagic.

CONSERVATION STATUS
Not listed by the IUCN. May be vulnerable to overfishing, because it is normally not common anywhere.

SIGNIFICANCE TO HUMANS
Poisonous and not taken for food but collected and dried for use as ornaments. ◆

Mola
Mola mola

FAMILY
Molidae

TAXONOMY
Mola mola Linnaeus, 1758, Mediterranean Sea.

OTHER COMMON NAMES
English: Ocean sunfish.

PHYSICAL CHARACTERISTICS
Body large, tall, and compressed. The body is scaleless, with a thick, elastic skin. The mouth is small, with a parrot-like beak formed by fused teeth. The caudal fin is replaced by a clavus, a rudder-like body-fin adaptation. Both the dorsal and anal fins have relatively short bases but are very high. These fins are flapped in a synchronous motion that allows for reasonably good speed and also for swimming sideways. The pectoral fins are small and point upward toward the dorsal fin. There are 15–18 soft rays in the dorsal fin and 14–17 soft rays in the anal fin. The swim bladder is absent in adults. The color is silvery gray. The tips of the dorsal and anal fins and the clavus are darkly colored. Grows to more than 130 in (330 cm) in total length. Flesh may be toxic.

DISTRIBUTION
Circumglobal in tropical, subtropical, and warm temperate waters.

HABITAT
Frequents the pelagic realm but will move inshore to kelp beds. May be carried inshore by upwelling to the deep slopes of coral or rocky reefs. Depth range from surface down to 985 ft (300 m).

BEHAVIOR
Solitary but occasionally found in small groups. Swims up to the surface and even exposes its dorsal fin in the air. Also lies on its side.

FEEDING ECOLOGY AND DIET
Feeds on jellyfishes, larger zooplankton, crustaceans, and fishes; also takes mollusks and brittle stars inshore.

REPRODUCTIVE BIOLOGY
Not well known, but courtship probably is paired. Adult females are very fecund. Larvae are pelagic.

CONSERVATION STATUS
Not listed by the IUCN.

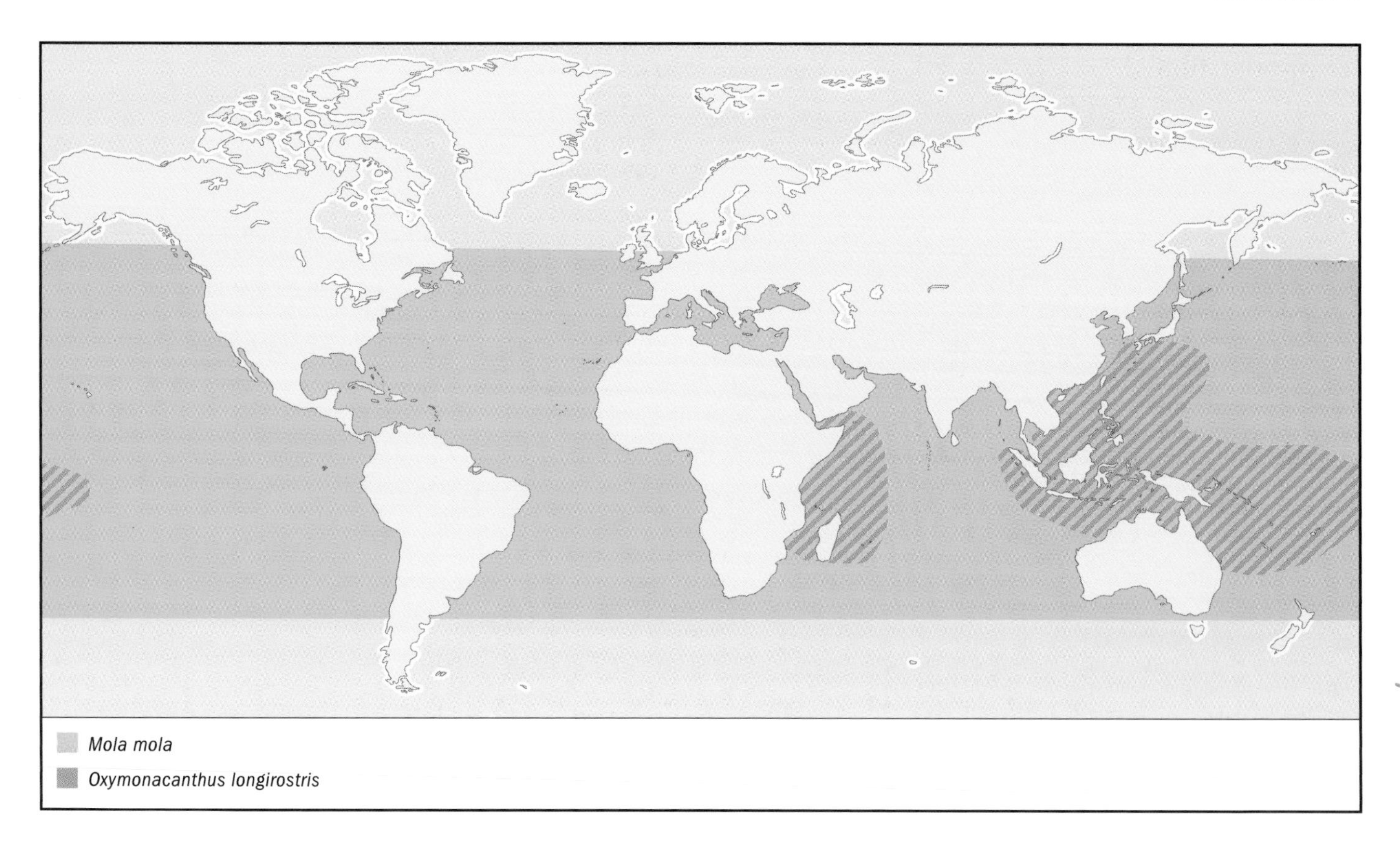

SIGNIFICANCE TO HUMANS
Prized as a delicacy in some places but avoided as poisonous elsewhere. Used in Chinese medicine. May be collected incidentally for large aquaria, but this species does not do well in captivity. ◆

Scrawled filefish
Aluterus scriptus

FAMILY
Monacanthidae

TAXONOMY
Aluterus scriptus Osbeck, 1765, China Sea.

OTHER COMMON NAMES
French: Robe de cuir; German: Schrift-Feilenfisch; Afrikaans: Bekrapte leerbaadjie; Japanese: Sôshihagi.

PHYSICAL CHARACTERISTICS
Body compressed, elongate, and somewhat rhomboid, with a concave snout and a relatively large and rounded caudal fin. There are two spines and 43–50 soft rays in the dorsal fin and 46–52 soft rays in the anal fin. The color is olive-brown to gray or tan with blue lines and spots in adults and yellowish brown with black spots in juveniles. Grows to more than 43.3 in (110 cm) in total length.

DISTRIBUTION
Tropical, subtropical, and warm temperate waters worldwide. In the Indo-Pacific from South Africa east to the Gulf of California and Colombia. In the Atlantic from Nova Scotia south to Brazil in the west and at Saint Paul's Rocks, Cape Verde, Ascension Island, and São Tomé Island in the east.

HABITAT
Found on seaward or lagoon coral reefs, rocky reefs, sea grass beds, and algae beds. Occasionally shelters under floating trees or other objects. Depth range is 13–394 ft (4–120 m).

BEHAVIOR
A solitary species that, despite its relatively large size, often moves cryptically within or between habitats.

FEEDING ECOLOGY AND DIET
Omnivorous. Feeds on hydrozoans, gorgonians, colonial anemones, tunicates, sea grass, or algae.

REPRODUCTIVE BIOLOGY
Not well known but probably lays demersal eggs. The larvae are pelagic and well equipped for a long pelagic existence, thus explaining this species' remarkable pattern of geographical distribution.

CONSERVATION STATUS
Not listed by the IUCN.

SIGNIFICANCE TO HUMANS
Collected for the aquarium trade. Also a game fish at some localities. Taken for subsistence elsewhere. This species has been reported to cause ciguatera poisoning. ◆

Longnose filefish
Oxymonacanthus longirostris

FAMILY
Monacanthidae

TAXONOMY
Oxymonacanthus longirostris Bloch and Schneider, 1801, East Indies.

OTHER COMMON NAMES
English: Harlequin filefish; French: Baliste à taches orange; German: Palettenstachier; Japanese: Tengukawahagi.

PHYSICAL CHARACTERISTICS
Body compressed and elongate with a pronounced, almost tubular snout. There are two spines and 31–35 soft rays in the dorsal fin and 29–32 soft rays in the anal fin. The color is bright green with orange spots or elongated blotches. The eyes are ringed with orange. The caudal fin is whitish with a small dark spot posteriorly. Grows to 4.7 in (12 cm) in total length.

DISTRIBUTION
Tropical and subtropical Indo-West Pacific from East Africa and Mozambique east to Samoa and Tonga, north to southern Japan, and south to the southern Great Barrier Reef and New Caledonia. The very similar congener *O. halli* replaces it in the Red Sea.

HABITAT
Found on coral-rich seaward and lagoon reefs, including reef flats, between 1.6 and 98 ft (0.5–30 m).

BEHAVIOR
Males and females most often are found in pairs that jointly patrol a shared territory. Territoriality is especially pronounced during mating season, and territorial defense is greater among males than females during this time, because females generally spend more time feeding. Aggressive behavior takes place in relation to food or mates. Males are capable of defending territories successfully alone during mating season, but females are not.

FEEDING ECOLOGY AND DIET
Feeds upon *Acropora* spp. coral polyps.

REPRODUCTIVE BIOLOGY
This species usually is monogamous but is facultatively polygynous in relation to the availability of mates locally. If males are in short supply, a male may have a two-female mating group. Eggs are demersal and laid almost daily during the season on a piece of filamentous algae, but no care is given. Algae that is toxic is preferred, because of the anti-predation advantage that it confers, in the absence of parental care, upon the eggs. The larvae are pelagic.

CONSERVATION STATUS
Not listed by the IUCN but may be vulnerable because of coral bleaching.

SIGNIFICANCE TO HUMANS
Collected for the aquarium trade but does not do well without live coral. ◆

Longhorn cowfish
Lactoria cornuta

FAMILY
Ostraciidae

TAXONOMY
Lactoria cornuta Linnaeus, 1758, India.

OTHER COMMON NAMES
French: Coffre cornu; German: Langhorn-Kofferfisch; Afrikaans: Langhoring-koeivis; Japanese: Kôngo-fugu.

PHYSICAL CHARACTERISTICS
Boxlike body, with two long "horns" projecting forward from the top of the head just in front of the eyes. The mouth is subterminal with prominent lips. The dorsal fin is relatively small and arises just forward of the caudal peduncle. The caudal fin is rounded. There are eight or nine soft rays in the dorsal fin, eight to nine soft rays in the anal fin, and nine to 10 rays in the caudal fin. Body color ranges from green and olive to light orange with blue spots and is cryptic. Grows to at least 18.1 in (46 cm) in total length.

DISTRIBUTION
Indo-Pacific in tropical, subtropical, and warm temperate waters from East Africa and the Red Sea east to the Tuamotu Archipelago and the Marquesas in eastern Polynesia, north to southern Japan and southern Korea, and south to Australia and Lord Howe Island.

HABITAT
Coral and rocky reefs, especially in algae beds or sea grasses. Juveniles enter brackish waters of estuaries. Depth range is 3.3–328 ft (1–100 m).

BEHAVIOR
Adults usually are solitary and territorial, but the juveniles form small groups.

FEEDING ECOLOGY AND DIET
Forages for benthic vertebrates by blowing water into the sand to dislodge them.

REPRODUCTIVE BIOLOGY
The mating system consists of a single male with a small group of females within its territory. Practices elaborate courtship just before or after sunset. Courtship is paired and results in the release of pelagic eggs well up into the water column above the bottom. The larvae are pelagic.

CONSERVATION STATUS
Not listed by the IUCN.

SIGNIFICANCE TO HUMANS
An interesting aquarium species but one that should be kept alone or with species with which it is not likely to interact. Also taken as a minor subsistence species that may cause ciguatera poisoning. Most are probably dried and sold as ornaments. ◆

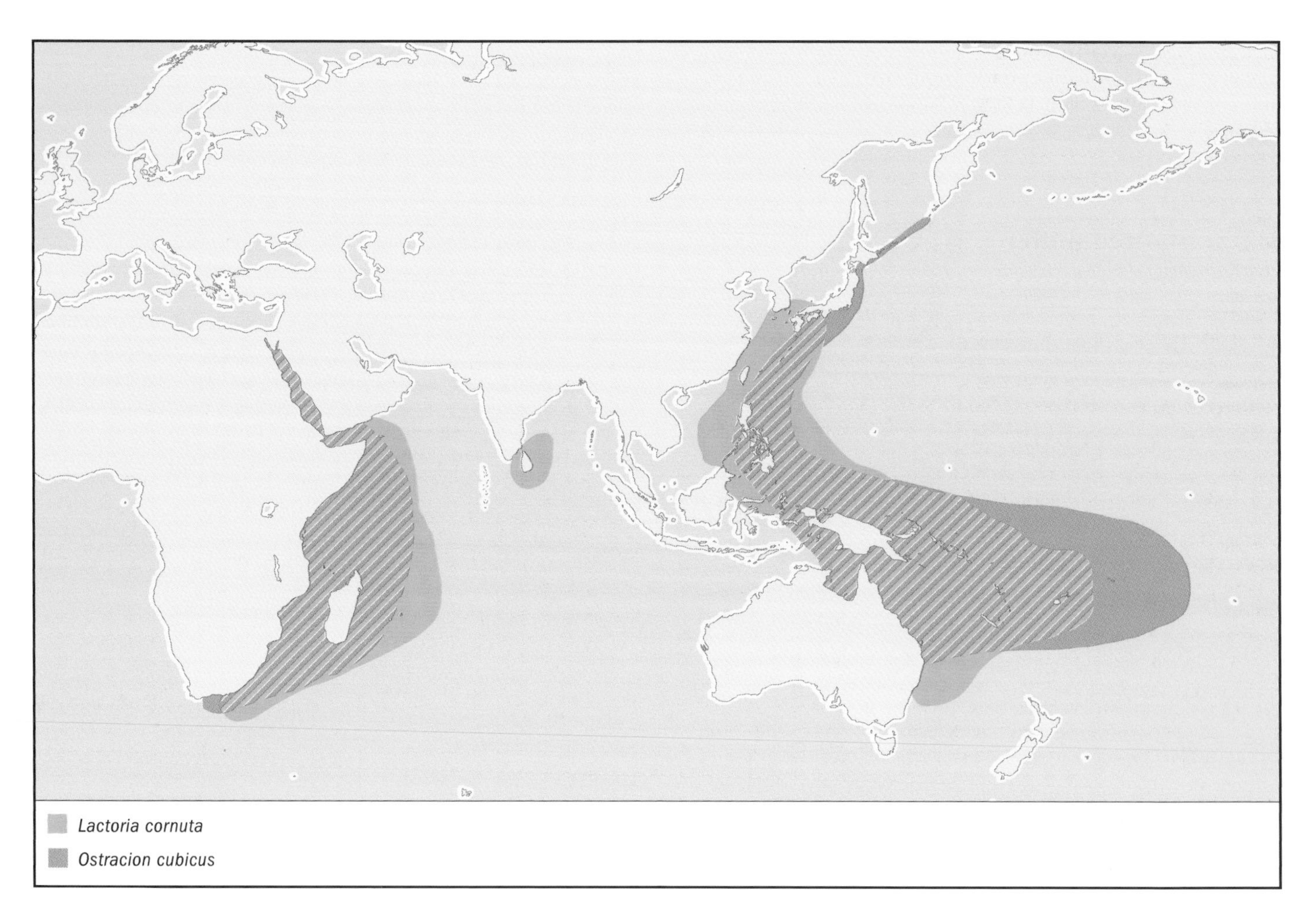

Yellow boxfish

Ostracion cubicus

FAMILY
Ostraciidae

TAXONOMY
Ostracion cubicus Linnaeus, 1758, India.

OTHER COMMON NAMES
French: Poisson cube; German: Gelbbrauner Kofferfisch; Afrikaans: Koffertije; Japanese: Minami-hakofugu.

PHYSICAL CHARACTERISTICS
Boxlike body with bony plates, a steeply sloping forehead, a subterminal mouth, a thick caudal peduncle, and a relatively large, somewhat rounded caudal fin. The dorsal and anal fins are positioned well back, just forward of the caudal peduncle. Body color of adults is a dirty yellow with a blueish hue and yellow seams between the body's bony plates. Juveniles are an attractive bright yellow with small black spots. There are eight or nine soft rays in the dorsal fin, nine soft rays in the anal fin, and 10 caudal fin rays. Total length up to 11 in (28 cm).

DISTRIBUTION
Tropical and subtropical waters of the Indo-Pacific and southeastern Atlantic, from East Africa and the Red Sea east to the Hawaiian Islands and the Tuamotu Archipelago in Polynesia, north to the Ryukyu Islands of southern Japan, south to Australia and Lord Howe Island, and off southern Africa in the Atlantic.

HABITAT
Primarily coral reefs in lagoons, on reef flats, and on protected seaward reefs. Juveniles associate with *Acropora* corals. Depth range is 3.3–148 ft (1–45 m).

BEHAVIOR
Adults are solitary and territorial. Juveniles also are solitary after settlement. Secretes ostracitoxin as a defensive mechanism when attacked or disturbed.

FEEDING ECOLOGY AND DIET
Omnivorous. Feeds upon benthic algae, various microorganisms, and foraminiferans that it strains from sediments, sponges, polychaete worms from sand flats, mollusks, small crustaceans, and small fishes. Large larvae and post-larvae often taken by pelagic predators, such as tunas.

REPRODUCTIVE BIOLOGY
Paired courtship just before or after sunset. Eggs and larvae are pelagic.

CONSERVATION STATUS
Not listed by the IUCN.

SIGNIFICANCE TO HUMANS
Collected for the aquarium trade but, due to its ability to secrete ostracitoxin, usually is a poor choice for a community aquarium. ◆

White-spotted puffer
Arothron hispidus

FAMILY
Tetraodontidae

TAXONOMY
Arothron hispidus Linnaeus, 1758, India.

OTHER COMMON NAMES
French: Tetrodon marbré; German: Weissflecken-Kugelfisch; Afrikaans: Witspikkel-blaasop; Japanese: Sazanamifugu.

PHYSICAL CHARACTERISTICS
Body elongate, robust, and capable of inflation with water. There are small spines on the body, except for the snout and caudal peduncle. The jaw teeth are fused but separated by a median suture. There are 10–11 soft rays in both the dorsal and anal fins. The caudal fin is rounded. Two fleshy tentacles emerge from each nostril. The gill opening is restricted, and the single lateral line is bent. The color is greenish brown with small white spots on the back, flanks, and caudal fin. The belly has white bars. Grows to at least 19.7 in (50 cm) in total length.

DISTRIBUTION
Tropical and subtropical waters of the Indo-Pacific from the Red Sea and East Africa to Panama and the Gulf of California; also north to Japan and throughout Micronesia and Polynesia as far as the Hawaiian Islands and Rapa Island and south to Australia and Lord Howe Island.

HABITAT
Coral and rocky reef slopes, lagoons, and inner reef flats over sand, rubble, or patch reefs. Juveniles found among weeds in estuaries or on reef flats. Depth range is 3.3–164 ft (1–50 m).

BEHAVIOR
Solitary and territorial. Individuals may bury themselves partially in the sand.

FEEDING ECOLOGY AND DIET
Omnivorous, feeding on fleshy coralline or calcareous algae, detritus, sponges, corals, anenomes, tube worms, mollusks, crabs, echinoderms, and tunicates.

REPRODUCTIVE BIOLOGY
Not well known. Probably lays demersal eggs. Larvae are pelagic.

CONSERVATION STATUS
Not listed by the IUCN.

SIGNIFICANCE TO HUMANS
Collected for the aquarium trade. ◆

Spotted toby
Canthigaster solandri

FAMILY
Tetraodontidae

TAXONOMY
Canthigaster solandri Richardson, 1845, Polynesia.

OTHER COMMON NAMES
English: Spotted sharpnose puffer.

PHYSICAL CHARACTERISTICS
Oblong and stocky body, with a pronounced snout, prickles on the belly, a rounded caudal fin, and the ability to inflate itself with water. Color is orangish red with blue spots over much of the body and on the orange caudal fin. The ventral surface is white. There is a black spot ringed in pale blue directly below the base of the dorsal fin. The dorsal fin is white, with a black base, and has eight to 10 soft rays. The anal fin also has eight to 10 soft rays. Some geographical variation in color pattern. The skin contains tetraotoxin. Grows to at least 4.3 in (11 cm) in total length.

DISTRIBUTION
Tropical and subtropical waters of the Indo-West Pacific from East Africa east to the Tuamotu Archipelago, south to the Great Barrier Reef and New Caledonia, and north to the Ryukyu Islands of southern Japan. Reportedly strays to the Hawaiian Islands.

HABITAT
Mainly coral reefs on reef flats, lagoon patch reefs, and seaward reefs. Depth range is 3.3–118 ft (1–36 m).

BEHAVIOR
Usually paired but also solitary or in groups. Swims over a home range, inspecting potential food items; also swims up in the water column. Some territorial interactions, as well.

FEEDING ECOLOGY AND DIET
Omnivorous, feeding on tunicates, bryozoans, echinoderms, crustaceans, mollusks, polychaete worms, corals, coralline red algae, green algae, and red algae.

REPRODUCTIVE BIOLOGY
Paired spawning of demersal eggs on algae growing on dead coral, rocks, or other substrata. The eggs are poisonous, and parental care is not practiced. The larvae are pelagic.

CONSERVATION STATUS
Not listed by the IUCN.

SIGNIFICANCE TO HUMANS
Collected for the aquarium trade. ◆

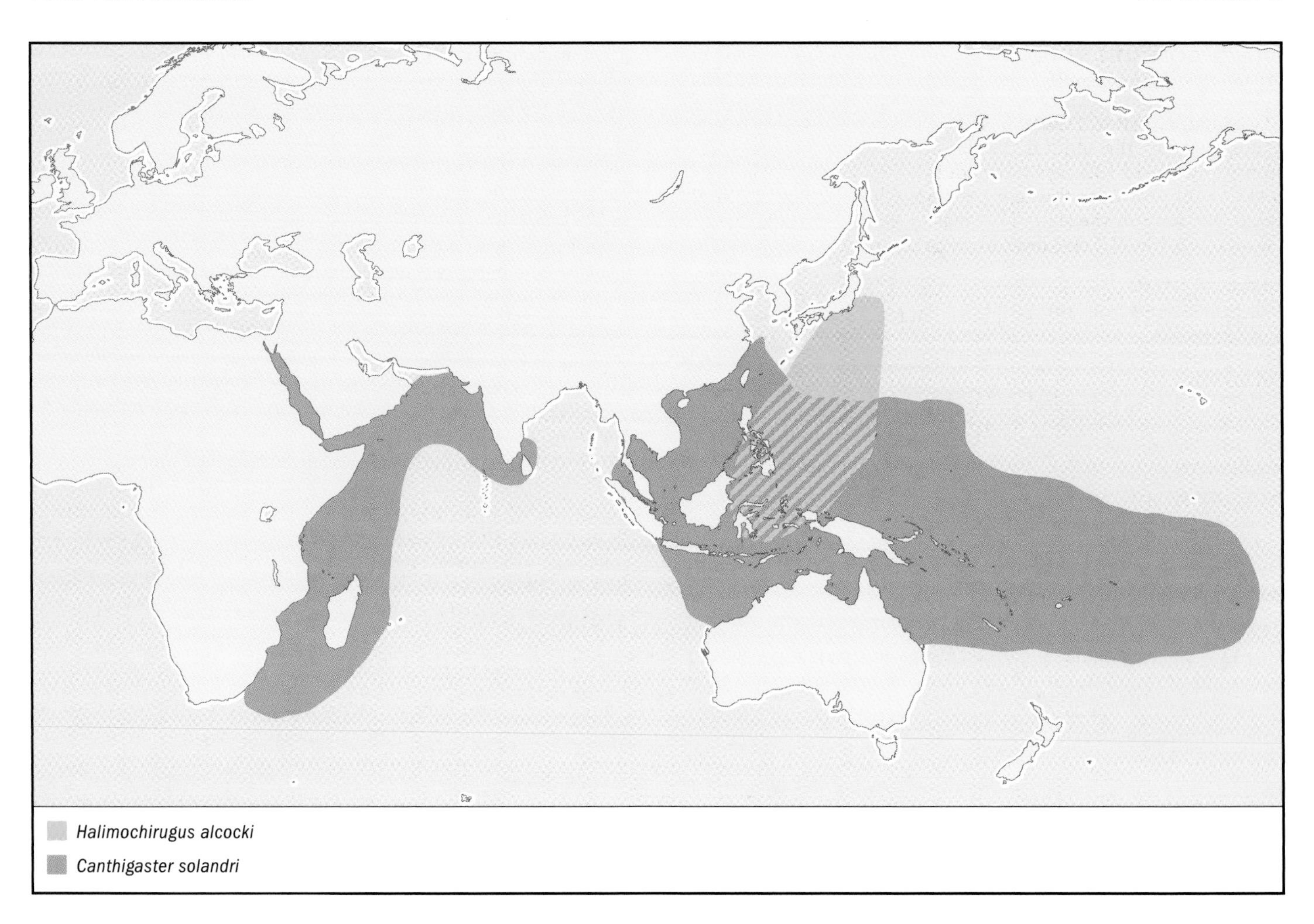

Fugu
Takifugu rubripes

FAMILY
Tetraodontidae

TAXONOMY
Takifugu rubripes Temminck and Schlegel, 1850, "Coasts of Japan."

OTHER COMMON NAMES
Japanese: Torafugu (tiger fugu).

PHYSICAL CHARACTERISTICS
Body oblong, compressed, and covered with prickles. Color is dark gray or grayish olive dorsally and white ventrally. A large black spot, ringed in white, is positioned on the flank just behind the pectoral fin. The dorsal fin has 16–19 soft rays, and the anal fin has 13–16 soft rays. The caudal fin is truncate. Poisonous, especially the ovaries, liver, and intestines. Grows to 27.6 in (70 cm) in total length.

DISTRIBUTION
Temperate waters of the northwest Pacific from the East China Sea and the Yellow Sea to the Sea of Japan as far north as Hokkaido.

HABITAT
Rocky reefs and shingle flats, usually up in the water column. Also frequents inlets and brackish waters but moves offshore with age and growth. Depth range is 3.3–656 ft (1–200 m).

BEHAVIOR
Solitary or in loose groups. This species swims about in the water column.

FEEDING ECOLOGY AND DIET
Omnivorous, feeding mainly on invertebrates.

REPRODUCTIVE BIOLOGY
The breeding season runs typically from March to May. Spawning is paired. Eggs are demersal and attached to rocks at depths of less than 65.6 in (20 m). Larvae are pelagic.

CONSERVATION STATUS
Not listed by the IUCN.

SIGNIFICANCE TO HUMANS
Highly prized in commercial fisheries of Japan. May be taken incidentally for aquaria. Also used in Chinese medicine. ◆

Spikefish
Halimochirugus alcocki

FAMILY
Triacanthodidae

TAXONOMY
Halimochirugus alcocki Weber, 1913, Arafura Sea.

OTHER COMMON NAMES
Japanese: Nagakawamuki.

PHYSICAL CHARACTERISTICS
Elongate body; the snout is tubular and elongate. There are six spines and 12–13 soft rays in the dorsal fin; the third spine is minute and protrudes through the skin, whereas the remaining spines lie beneath the skin. The anal fin has 11–12 soft rays. Grows to 6.7 in (17 cm) in total length.

DISTRIBUTION
Western Pacific from central Japan south to the Philippines and Indonesia.

HABITAT
A bottom-dwelling tropical and subtropical marine species at depths of 1,280–2,001 ft (390–610 m).

BEHAVIOR
Not well known.

FEEDING ECOLOGY AND DIET
Not well known but probably feeds on small benthic invertebrates that it takes with its tubelike snout.

REPRODUCTIVE BIOLOGY
Not well known but likely has demersal eggs and pelagic larvae.

CONSERVATION STATUS
Not listed by the IUCN.

SIGNIFICANCE TO HUMANS
Not significant as a food, sport, or aquarium species; of scientific interest, warranting further investigation. ◆

Triplespine
Triacanthus biaculeatus

FAMILY
Triacanthidae

TAXONOMY
Triacanthus biaculeatus Bloch, 1786, East Indies.

OTHER COMMON NAMES
English: Short-nosed tripodfish; Japanese: Gima.

PHYSICAL CHARACTERISTICS
Body oblong and compressed, with a long caudal peduncle, a deeply forked caudal fin with rounded lobes, a pronounced snout in the head, and a prominent dorsal spine. The dorsal fin has 20–26 soft rays, and the anal fin has 13–22 soft rays. Body color is silvery with a yellowish hue. The caudal fin is pale yellow and somewhat clear at the lobes. The fins are pale or milky to clear white, except for the dorsal spine, which is black at the base. Grows to at least 11.8 in (30 cm) in total length.

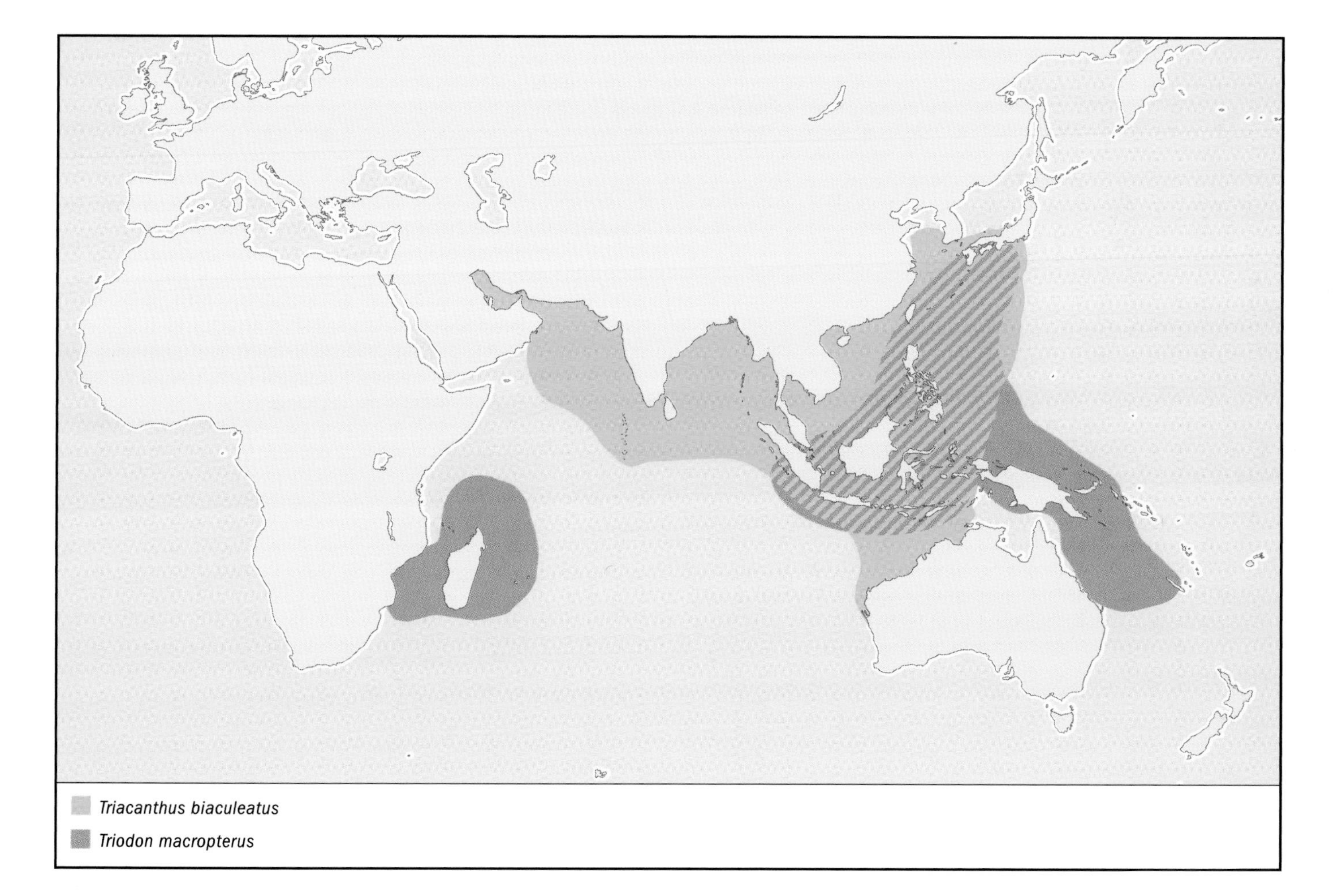

DISTRIBUTION
Indo-West Pacific region from the Persian Gulf east across the Indian Ocean to the east coast of Australia and north to China and southern Japan.

HABITAT
A shallow-water bottom-dwelling species in tropical and subtropical waters. Prefers sand or mudflats of estuaries and inshore coastal waters. Depth range to 197 ft (60 m).

BEHAVIOR
Not well known.

FEEDING ECOLOGY AND DIET
This species forages for benthic invertebrates.

REPRODUCTIVE BIOLOGY
Not well known but probably has demersal eggs, perhaps in a nest, with parental care. The larvae are pelagic.

CONSERVATION STATUS
Not listed by the IUCN.

SIGNIFICANCE TO HUMANS
A minor commercial and subsistence species that also is used in Chinese medicine. ◆

Threetooth puffer
Triodon macropterus

FAMILY
Triodontidae

TAXONOMY
Triodon macropterus Lesson, 1831, Mauritius, Indian Ocean.

OTHER COMMON NAMES
Afrikaans: Drietand-blaasop; Japanese: Uchiwafugu.

PHYSICAL CHARACTERISTICS
Body oblong and compressed, with a well-developed ventral flap. The teeth are quite large and arranged in a beaklike plate; the lower teeth are completely fused, and the upper teeth are separated by a suture. Color is golden yellow with hints of white. There is an irregular black blotch, ringed in white, located medially at the base of the ventral flap. The dorsal and anal fins are relatively small and positioned posterior to the ventral flap. The dorsal fin has two spines and 10–12 soft rays, and the anal fin has nine to 10 soft rays. The caudal fin is slightly forked. Grows to 15.7 in (40 cm) in total length.

DISTRIBUTION
Tropical and subtropical waters of the Indo-Pacific from East Africa to Indonesia and the Philippines north to Japan and south to Australia and New Caledonia.

HABITAT
A deep-slope demersal species found above the bottom at depths of 164–984 ft (50–300 m).

BEHAVIOR
Not well known. Can inflate its body with water when threatened or attacked.

FEEDING ECOLOGY AND DIET
Not well known but probably feeds on benthic invertebrates.

REPRODUCTIVE BIOLOGY
Not well known. Probably lays demersal eggs. The larvae are pelagic.

CONSERVATION STATUS
Not listed by the IUCN.

SIGNIFICANCE TO HUMANS
No commercial importance. May be taken incidentally for larger aquaria. ◆

Resources

Books

Eschmeyer, William N., ed. *Catalog of Fishes*, 3 vols. San Francisco: California Academy of Sciences, 1998.

Helfman, G. S., B. B. Collette, and D. E. Facey. *The Diversity of Fishes.* Malden, MA: Blackwell Science, 1997.

Leis, J. M., and B. M. Carson-Ewart, eds. *The Larvae of Indo-Pacific Coastal Fishes: An Identification Guide to Marine Fish Larvae.* Boston: Brill, 2000.

Masuda, H., K. Amaoka, C. Araga, T. Uyeno, and T. Yoshino, eds. *The Fishes of the Japanese Archipelago.* Tokyo: Tokai University Press, 1984.

Myers, R. F. *Micronesian Reef Fishes.* 3rd edition. Barrigada, Guam: Coral Graphics, 1999.

Nelson, J. S. *Fishes of the World.* 3rd edition. New York: John Wiley and Sons, 1994.

Randall, John E., Gerald R. Allen, and Roger C. Steene. *Fishes of the Great Barrier Reef and Coral Sea.* Honolulu: University of Hawaii Press, 1996.

Smith, M. M., and P. C. Heemstra, eds. *Smiths' Sea Fishes.* Berlin: Springer-Verlag, 1986.

Thresher, R. E. *Reproduction in Reef Fishes.* Neptune City, NJ: T.F.H. Publications, 1984.

Periodicals

Ishihara, M., and T. Kuwamura. "Bigamy or Monogamy with Maternal Egg Care in the Triggerfish, *Sufflamen chrysopterus.*" *Ichthyological Research* 43 (1996): 307–313.

Kawase, H., and A. Nakazono. "Two Alternative Female Tactics in the Polygynous Mating System of the Threadsail Filefish, *Stephanolepis cirrhifer* (Monacanthidae)." *Ichthyological Research* 43, no. 3 (1996): 315–323.

Kokita, T., and A. Nakazono. "Pair Territoriality in the Longnose Filefish, *Oxymonacanthus longirostris.*" *Ichthyological Research* 46 (1999): 297–302.

———. "Spawning Substrate Selection by Female Longnose Filefish, *Oxymonacanthus longirostris.*" *Ichthyological Research* 46, no. 4 (1999): 429–432.

Resources

Nakae, M., and K. Sasaki. "A Scale-Eating Triacanthodid, *Macrorhamphosodes uradoi*: Prey Fishes and Mouth 'Handedness' (Tetradontiformes, Triacanthoidei)." *Ichthyological Research* 49, no. 1 (2002): 7–14.

Organizations

IUCN/SSC Coral Reef Fishes Specialist Group. International Marinelife Alliance-University of Guam Marine Laboratory, UOG Station, Mangilao, Guam 96913 USA. Phone: (671) 735-2187. Fax: (671) 734-6767. E-mail: donaldsn@uog9.uog.edu Web site: <http://www.iucn.org/themes/ssc/sgs/sgs.htm>

Terry J. Donaldson, PhD

For further reading

Able, Kenneth W., and Michael P. Fahay. *The First Year in the Life of Estuarine Fishes in the Middle Atlantic Bight.* New Brunswick, NJ: Rutgers University Press, 1998.

Alexander, R. M.*Functional Design in Fishes.* London: Hutchinson, 1967.

Allan, J. David. *Stream Ecology: Structure and Function of Running Waters.* New York: Chapman & Hall, 1995.

Allen. G. R. *Freshwater Fishes of Australia.* Neptune City, NJ: T. H. F. Publications, 1989.

———. *Marine Fishes of Tropical Australia and South-east Asia.* Perth: Western Australian Museum, 1997.

Allen, G. R., and D. R. Robertson. *Fishes of the Tropical Eastern Pacific.* Honolulu: University of Hawaii Press, 1994.

Allen, G. R., and R. Swainston. *The Marine Fishes of North-Western Australia: A Field Guide for Anglers and Divers.* Perth: Western Australian Museum, 1988.

Allen, G. R., S. H. Midgley, and M. Allen. *Field Guide to the Freshwater Fishes of Australia.* Perth: Western Australian Museum, 2002.

Andriashev, A. P. *Fishes of the Northern Seas of the USSR.* Moscow: Academy of Sciences, USSR, 1954.

Banarescu, P. *Zoogeography of Fresh Waters.* 3 vols. Wiesbaden, Germany: AULA-Verlag, 1990–1995.

Beard, J. A. *James Beard's New Fish Cookery.* New York: Galahad Books, 1976.

Bemis, William E., Warren W. Burggren, and Norman E. Kemp. *The Biology and Evolution of Lungfishes.* New York: A. R. Liss, 1987.

Benton, M. J., ed. *The Fossil Record 2.* London: Chapman and Hall, 1993.

Berra, Tim M. *An Atlas of Distribution of the Freshwater Fish Families of the World.* Lincoln: University of Nebraska Press, 1981.

———. *Freshwater Fish Distribution.* San Diego: Academic Press, 2001.

Bigelow, H. B., ed. *Fishes of the Western North Atlantic.* New Haven: Sears Foundation for Marine Research, Yale University, 1963.

Bohlke, James E., and Charles C. G. Chaplin. *Fishes of the Bahamas and Adjacent Tropical Waters.* 2nd edition. Austin: University of Texas Press, 1993.

Bond, Carl E. *Biology of Fishes.* 2nd edition. New York: Harcourt Brace College Publishers, 1996.

Bone, Q., N. B. Marshall, and J. H. S. Blaxter. *Biology of Fishes.* 2nd edition. Glasgow: Blackie Academic and Professional, 1995.

Boschung, H. T., Jr., J. D. Williams, D. W. Gotshall, D. K. Caldwell, and M. C. Caldwell. *The Audubon Society Field Guide to North American Fishes, Whales, and Dolphins.* New York: Alfred A. Knopf, 1983.

Breder, C. M., Jr., and D. E. Rosen. *Modes of Reproduction in Fishes.* Garden City, NY: Natural History Press, 1966.

Briggs, John C. *Marine Zoogeography.* New York: McGraw-Hill, 1974.

———. *Global Biogeography.* Amsterdam: Elsevier, 1995.

Brooks, D. R., and D. A. McLennan. *Phylogeny, Ecology, and Behavior.* Chicago: University of Chicago Press, 1991.

Bullock, T., and W. Heiligenberg. *Electroreception.* New York: John Wiley & Sons, 1986.

Butler, Ann B., and William Hodos. *Comparative Vertebrate Neuroanatomy: Evolution and Adaptation.* New York: John Wiley and Sons, 1996.

Carwardine, Mark, and Ken Watterson. *The Shark Watcher's Handbook: A Guide to Sharks and Where to See Them.* Princeton: Princeton University Press: 2002.

Coad, B. W. *Guide to the Marine Sport Fishes of Atlantic Canada and New England.* Toronto: University of Toronto Press, 1992.

Collette, Bruce B., and Grace Klein-MacPhee, editors. *Bigelow and Schroeder's Fishes of the Gulf of Maine.* 3rd edition. Washington, D.C.: Smithsonian Institution Press, 2002.

Compagno, L. J. V. *Sharks of the World: An Annotated and Illustrated Catalogue of Shark Species Known to Date.* FAO Species Catalogue, vol. 4, part 1. Rome: Food and Agriculture Organization of the United Nations, 1984.

Cushing, Colbert E., and J. David Allen. *Streams: Their Ecology and Life.* San Diego: Academic Press, 2001.

Darlington, Philip J., Jr. *Zoogeography: The Geographical Distribution of Animals.* New York: John Wiley and Sons, 1957.

Deloach, Ned. *Reef Fish Behavior: Florida, Caribbean, Bahamas.* Jacksonville: New World Publications, Inc., 1999.

Demski, Leo S. and John P. Wourms, eds. *Reproduction and Development of Sharks, Skates, Rays and Ratfishes.* Boston, MA: Kluwer Academic Publishers, 1993.

Diana, J. S.*Biology and Ecology of Fishes.* Carmel, IN: Cooper Publishing, 1995.

di Prisco, G., B. Maresca, and B. Tota, eds. *Biology of Antarctic Fish.* Berlin: Springer-Verlag, 1991.

di Prisco, G., E. Pisano, and A. Clarke, ed. *Fishes of Antarctica: A Biological Overview.* Milan: Springer-Verlag, 1998.

Dobson, Mike, and Chris Frid. *Ecology of Aquatic Systems.* Essex, U.K.: Addison Wesley Longman, 1998.

Eastman, J. T. *Antarctic Fish Biology: Evolution in a Unique Environment.* San Diego: Academic Press, 1993.

Echelle, A. A., and I. Kornfield, eds. *Evolution of Fish Species Flocks.* Orono: University of Maine at Orono Press, 1984.

Eddy, S., and J. C. Underhill. *Northern Fishes,* 3rd edition. Minneapolis: University of Minnesota Press, 1974.

Eschmeyer, W. N. *Catalog of the Genera of Recent Fishes.* San Francisco: California Academy of Sciences, 1990.

———, ed. *Catalog of Fishes.* 3 vols. San Francisco: California Academy of Sciences, 1998.

Eschmeyer, W. N., E. S. Herald, and H. Hammann. *A Field Guide to Pacific Coast Fishes of North America.* Boston: Houghton Mifflin Company, 1983.

Etnier, David A., and Wayne C. Starnes. *The Fishes of Tennessee.* Knoxville: University of Tennessee Press, 1993.

Evans, D. H., ed.*The Physiology of Fishes,* 2nd edition. Boca Raton, FL: CRC Press, 1998.

Everhart, W. H., and W. D. Youngs.*Principles of Fishery Science,* 2nd edition. Ithaca, NY: Cornell University Press, 1981.

Fitch, J. E., and R. J. Lavenberg. *Deep-Water Teleostean Fishes of California.* Berkeley: University of California Press, 1968.

———. *Tidepool and Nearshore Fishes of California.* Berkeley: University of California Press, 1975.

Frickhinger, Karl Albert. *Fossil Atlas: Fishes.* Blacksburg, VA: Tetra Press, 1996.

Fryer, G., and T. D. Iles. *The Cichlid Fishes of the Great Lakes of Africa.* Edinburgh: Oliver and Boyd, 1972.

Fuller, P. L., L. G. Nico, and J. D. Williams. *Nonindigenous Fishes Introduced into Inland Waters of the United States.* Bethesda, MD: American Fisheries Society, Special Publication 27, 1999.

Gery, J. *Characoids of the World.* Neptune City, NJ: Tropical Fish Hobbyist Publications, Inc., 1977.

Giller, Paul S., and Bjorn Malmqvist. *The Biology of Streams and Rivers.* Oxford: Oxford University Press, 1998.

Gloerfelt-Tarp, T., and P. J. Kailola. *Trawled Fishes of Southern Indonesia and Northwestern Australia.* Jakarta: Directorate General of Fisheries (Indonesia), German Agency for Technical Cooperation, Australian Development Assistance Bureau, 1984.

Gomon, M. F., J. C. M. Glover, and R. H. Kuiter, eds. *The Fishes of Australia's South Coast.* Adelaide: State Print, 1994.

Gon, O, and P. C. Heemstra, editors. *Fishes of the Southern Ocean.* Grahamstown, South Africa: J. L. B. Smith Institute of Ichthyology, 1990.

Goodson, G. *Fishes of the Pacific Coast.* Stanford: Stanford University Press, 1988.

Gordon, B. L.*The Secret Lives of Fishes.* New York: Grosset and Dunlap, 1977.

Graham, Jeffrey B. *Air-breathing Fishes: Evolution, Diversity, and Adaptation.* San Diego, CA: Academic Press, 1997.

Groombridge, B., and M. Jenkins. *Freshwater Biodiversity: A Preliminary Global Assessment.* Cambridge, U.K.: World Conservation Monitoring Centre—World Conservation Press, 1998.

Halstead, Bruce W. *Poisonous and Venomous Marine Animals of the World.* Washington, DC: Government Printing Office, 1965–1970.

Hamlett, William C., ed. *Sharks, Skates, and Rays: The Biology of Elasmobranch Fishes.* Baltimore, MD: Johns Hopkins University Press, 1999.

Hart, J. L. *Pacific Fishes of Canada.* Ottawa: Fisheries Research Board of Canada, Bulletin 180, 1973.

Helfman, Gene S., Bruce B. Collette, and Douglas E. Facey. *The Diversity of Fishes.* Malden, MA: Blackwell Science, 1997.

Hennemann, Ralf M.*Elasmobranch Guide of the World: Sharks and Rays.* Frankfurt: Ikan, 2001.

Hoar, W. S., and D. J. Randall, eds. *Fish Physiology.* Vols. 1–20. New York: Academic Press, 1969–1993.

Hoese, H. D., and R. H. Moore. *Fishes of the Gulf of Mexico, Texas, Louisiana, and Adjacent Waters.* 2nd edition. College Station, TX: Texas A & M University Press, 1998.

Horn, Michael H., Karen L. M. Martin, and Michael A. Chotkowski, eds. *Intertidal Fishes: Life in Two Worlds*, San Diego: Academic Press, 1999.

Hoyt, E. *Creatures of the Deep: In Search of the Sea's "Monsters" and the World They Live In.* Buffalo: Firefly Books, Ltd., 2001.

Janvier, Philippe. *Early Vertebrates.* New York: Oxford University Press, 1996.

Kock, Karl-Hermann. *Antarctic Fish and Fisheries.* Cambridge: Cambridge University Press, 1992.

Kottelat, M., A. J. Whitten, S. N. Kartikasari, and S. Wirjoatmodjo. *Freshwater Fishes of Western Indonesia and Sulawesi.* Jakarta: Periplus Editions, 1993.

Kramer, D. E., W. H. Barss, B. C. Paust, and B. E. Bracken. *Guide to Northeast Pacific Flatfishes: Families Bothidae, Cynoglossidae, and Pleuronectidae.* Marine Advisory Bulletin no. 47. Fairbanks: University of Alaska Fairbanks, Alaska Sea Grant College Program, 1995.

Kuiter, Rudie H. *Guide to Sea Fishes of Australia.* London: New Holland, 1996.

Kuiter, Rudie H. *Coastal Fishes of South-Eastern Australia.* Honolulu: University of Hawaii Press, 1993.

———. *Seahorses, Pipefishes and Their Relatives: A Comprehensive Guide to Syngnathiformes.* Chorleywood, U.K.: TMC Publishing, 2000.

La Rivers, Ira. *Fish and Fisheries of Nevada.* Reno: University of Nevada Press, 1994.

Lagler, Karl F. *Ichthyology.* 2nd edition. New York: John Wiley & Sons, 1977.

Lampert, Winfried, and Ulrich Sommer. *Limnoecology: The Ecology of Lakes and Streams.* New York: Oxford University Press, 1997.

Last, P. R., and J. D. Stevens. *Sharks and Rays of Australia.* Melbourne, Australia: CSIRO, 1994.

Lee, D. S., C. R. Gilbert, C. H. Hocutt, R. E. Jenkins, D. E. McAllister, and J. R. Stauffer, Jr. *Atlas of North American Freshwater Fishes.* Raleigh: North Carolina State Museum of Natural History, 1980.

Leis, Jeffrey M., and Brooke M. Carson-Ewart, eds. *The Larvae of Indo-Pacific Coastal Fishes: An Identification Guide to Marine Fish Larvae.* Boston: Brill, 2000.

Lieske, Ewald, and Robert Myers. *Coral Reef Fishes: Caribbean, Indian Ocean and Pacific Ocean: Including the Red Sea.* London: Harper Collins, 1994.

Loiselle, Paul V. *The Cichlid Aquarium.* Melle, Germany: Tetra-Press, 1985.

Long, John A. *The Rise of Fishes: 500 Million Years of Evolution.* Baltimore, MD: Johns Hopkins University Press, 1995.

Lourie, S. A., A. C. J. Vincent, and H. J. Hall. *Seahorses: An Identification Guide to the World's Species and Their Conservation.* London: Project Seahorse, 1999.

Love, Milton S. *Probably More Than You Want to Know About the Fishes of the Pacific Coast.* 2nd edition. Perm: Izd-vo Permskogo Universiteta, 2001.

Love, Milton S., Mary Yoklavich, and Lyman Thorsteinson. *The Rockfishes of the Northeast Pacific.* Berkeley: University of California Press, 2002.

Lythgoe, J., and G. Lythgoe. *Fishes of the Sea.* Garden City, NJ: Anchor Press/Doubleday, 1975.

Mago-Leccia, F. *Electric Fishes of the Continental Waters of America.* Vol. 29, Biblioteca de la Academica de Ciencias Fisicas Matematicas y Naturales. Caracas, Venezuela: FUDECI, 1994.

Maissey, J. G. *Santana Fossils: an Illustrated Atlas.* Neptune City, NJ: T. F. H. Publishers, 1991.

Marshall, N. B. *Aspects of Deep Sea Biology.* London: Hutchinson, 1954.

Masuda, H., K. Amaoka, C. Araga, T. Uyeno, and T. Yoshino, eds. *The Fishes of the Japanese Archipelago.* Tokyo: Tokai University Press, 1984.

Matthews, William J. *Patterns in Freshwater Fish Ecology.* New York: Chapman & Hall, 1998.

Mayden, R. L., ed. *Systematics, Historical Ecology and North American Freshwater Fishes.* Stanford: Stanford University Press, 1992.

McDowall, R. M. *Freshwater Fishes of South-Eastern Australia.* Sydney: Reed Books, 1996.

———. *Diadromy in Fishes: Migrations Between Freshwater and Marine Environments.* London: Croom Helm, 1988.

———. *Freshwater Fishes of New Zealand.* Auckland: Reed, 2001.

McEachran, John D., and Janice D. Fechhelm. *Fishes of the Gulf of Mexico.* Vol. 1, Myxiniformes to Gasterosteiformes. Austin, TX: University of Texas Press, 1998.

Mecklenburg, Catherine W., T. Anthony Mecklenburg, and Lyman K. Thorsteinson. *Fishes of Alaska.* Bethesda: American Fisheries Society, 2002.

Menon, A. G. K. *The Fauna of India and the Adjacent Countries.* Madras: Amra Press, 1987.

Merrick, J. R., and G. E. Schmida. *Australian Freshwater Fishes: Biology and Management.* North Ryde, N.S.W., Australia: J. R. Merrick, 1984.

Michael, Scott W. *Reef Fishes: A Guide to Their Identification, Behavior and Captive Care.* Shelburne, VT: Microcosm Ltd., 1998.

Miller, Richard Gordon. *History and Atlas of the Fishes of the Antarctic Ocean.* Carson City, NV: Foresta Institute for Ocean and Mountain Studies, 1993.

Minckley W. L., and James E. Deacon, eds. *Battle Against Extinction, Native Fish Management in the American West.* Tucson: University of Arizona Press, 1991.

Moller, P., ed. *Electrical Fishes: History and Behavior. Fish and Fisheries Series 117.* London: Chapman & Hall, 1995.

Morrow, James E. *The Freshwater Fishes of Alaska.* Anchorage: Alaska Northwest Publishing, 1980.

Moyle, Peter B., and Joseph J. Cech, Jr. *Fishes: An Introduction to Ichthyology.* 4th edition. Upper Saddle River, NJ: Prentice Hall, 2000.

Murdy, E. O., R. Birdsong, and J. A. Musick. *Fishes of the Chesapeake Bay.* Washington, DC: Smithsonian Institution, 1997.

Myers, R. F. *Micronesian Reef Fishes: A Field Guide for Divers and Aquarists.* 3rd edition. Barrigada, Guam: Coral Graphics, 1999.

Neira, F. J., A. G. Miskiewicz, and T. Trnski, eds. *The Larvae of Temperate Australian Fishes: A Laboratory Guide for Larval Fish Identification.* Perth: University of Western Australia Press, 1998.

Nelson, J. S. *Fishes of the World.* 3rd edition. New York: John Wiley and Sons, 1994.

Page, L. M., and B. M. Burr. *A Field Guide to Freshwater Fishes of North America North of Mexico.* Boston: Houghton Mifflin Company, 1991.

Paulin, C., A. Stewart, C. Roberts, and P. McMillan. *New Zealand Fish: A Complete Guide.* National Museum of New Zealand Miscellaneous Series No. 19. Wellington: New Zealand, 1989.

Paxton, John R., and William N. Eschmeyer, eds. *Encyclopedia of Fishes.* 2nd edition. San Diego: Academic Press, 1998.

Payne, A. I. *The Ecology of Tropical Lakes and Rivers.* New York: John Wiley & Sons, 1986.

Perrine, D. *Sharks and Rays of the World.* Stillwater, MN: Voyager Press, 1999.

Pethiyagoda, R. *Freshwater Fishes of Sri Lanka.* Sri Lanka: The Wildlife Heritage Trust of Sri Lanka, 1991.

Pietsch, T. W., and D. B. Grobecker. *Frogfishes of the World: Systematics, Zoogeography, and Behavioral Ecology.* Stanford: Stanford University Press, 1987.

Pitcher, T. J., ed. *The Behaviour of Teleost Fishes.* London: Chapman and Hall, 1993.

Potts, D. T., and J. S. Ramsey. *A Preliminary Guide to Demersal Fishes of the Gulf of Mexico Continental Slope (100 to 600 fathoms).* Mobile, AL: Alabama Sea Grant Extension, 1987.

Potts, G. W., and R. J. Wootton, eds. *Fish Reproduction: Strategies and Tactics.* London: Academic Press, 1984.

Raasch, Maynard S. *Delaware's Freshwater and Brackish-Water Fishes: A Popular Account.* Neptune City, NJ: T.F.H. Publications, 1996.

Randall, John E. *Surgeonfishes of the World.* Honolulu: Bishop Museum Press/Mutual Publishing, 2001.

Randall, John E., Gerald R. Allen, and Roger C. Steene. *Fishes of the Great Barrier Reef and Coral Sea.* Honolulu: Crawford House Publishing/University of Hawaii Press, 1997.

Riehl, R., and H. A. Baensch. *Aquarium Atlas.* Melle, West-Germany: Baensch, 1986.

Robins, C. Richard, and G. Carleton Ray. *A Field Guide to Atlantic Coast Fishes of North America.* Boston: Houghton Mifflin, 1986.

Romero, Aldemaro, ed. *The Biology of Hypogean Fishes.* Dordrecht: Kluwer Academic Publishers, 2001.

Ross, Stephen T. *Inland Fishes of Mississippi.* Jackson: University Press of Mississippi, 2001.

Sadovy, Y., and A. S. Cornish. *Reef Fishes of Hong Kong.* Hong Kong: Hong Kong University Press, 2000.

Sale, Peter F., ed. *Coral Reefs Fishes: Dynamics and Diversity in a Complex Ecosystem.* San Diego: Academic Press, 2001.

Schultze, Hans-Peter, and Linda Trueb, eds. *Origins of the Higher Groups of Tetrapods: Controversy and Consensus.* Ithaca, NY: Comstock Publishing Associates, 1991.

Scott, W. B., and E. J. Crossman. *Freshwater Fishes of Canada.* Ottawa: Fisheries Resource Board of Canada, 1973.

Skelton, P. *Freshwater Fishes of Southern Africa*, 2nd edition. Cape Town: Struik, 2001.

Smith, C. Lavett. *The Inland Fishes of New York State.* Albany: New York State Department of Environmental Conservation, 1985.

———. *National Audubon Society Field Guide to Tropical Marine Fishes of the Caribbean, Gulf of Mexico, Florida, the Bahamas, and Bermuda.* New York: Knopf, 1997.

Smith, M. M., and P. C. Heemstra,eds. *Smiths' Sea Fishes.* Berlin: Springer-Verlag, 1986.

Snyderman, Marty, and Clay Wiseman. *Guide to Marine Life: Caribbean, Bahamas, Florida.* New York: Aqua Quest Publications, 1996.

Springer, Victor G., and Joy P. Gold. *Sharks in Question: The Smithsonian Answer Book.* Washington, DC: Smithsonian Institution Press, 1989.

Stiassny, Melanie L. J., Lynne R. Parenti, and G. David Johnson. *Interrelationships of Fishes.* San Diego, CA: Academic Press, 1996.

Talwar, P. K., and A. G. Jhingran.*Inland Fishes of India and Adjacent Countries.* New Delhi: Oxford & I.B.H. Publishing Co., 1991.

Thomson, Donald A., Lloyd T. Findley, and Alex N. Kerstitch. *Reef Fishes of the Sea of Cortez.* 2nd edition. Tucson: University of Arizona Press, 1987.

Thresher, Ronald E. *Reef Fishes: Behavior and Ecology on the Reef and in the Aquarium.* Saint Petersburg, FL: Palmetto Publishing Company, 1980.

———. *Reproduction in Reef Fishes.* Neptune City, NJ: T.F.H. Publications, Inc., 1984.

Tomelleri, J., and M. Eberle. *Fishes* Lawrence: University of Kansas Press, 1990.

Wheeler, A. *The Fishes of the British Isles and North-west Europe.* London: Macmillan, 1969.

———. *The World Encyclopedia of Fishes.* London: Macdonald, 1985.

Whitworth, W. R. *Freshwater Fishes of Connecticut.* State Geological and Natural History Survey of Connecticut Bulletin 114. Hartford: Connecticut Department of Environmental Protection, 1996.

Winfield, I., and J. Nelson, eds. *Cyprinid Fishes: Systematic Biology and Exploitation.* New York: Chapman & Hall, 1991.

Wischnath, Lothar. *Atlas of Livebearers of the World.* Neptune City, NJ: T.F.H. Publications, 1993.

Wootton, R. J. *Ecology of Teleost Fishes.* London: Chapman and Hall, 1990.

Organizations

American Elasmobranch Society
114 Hofstra University
Hempstead, NY 11549-1140
USA
http://www.flmnh.ufl.edu/fish/Organizations/aes/aes.htm

American Fisheries Society
5410 Grosvenor Lane
Bethesda, MD 20814
USA
Phone: 301–897–8616
Fax: 301–897–8096
main@fisheries.org
http://www.fisheries.org

American Livebearer Association
5 Zerbe Street,
Cressona, Pennsylvania 17929-1513
United States
Phone: 570-385-0573
Fax: 570-385-2781
tjbrady@uplink.net
http://livebearers.org/

American Killifish Association
280 Cold Springs Drive
Manchester, Pennsylvania 17345-1243
United States
garrybartell@sprintmailcom
http://www.aka.org

American Society of Ichthyology and Herpetology
http//www.asih.org/

American Society of Ichthyologists and Herpetologists
donnelly@fiu.edu
Phone: (305) 919-5651
http://199.245.200.110/

American Sportsfishing Association
225 Reinekers Lane, Suite 420
Alexandria, VA 22314
USA
Phone: 703–519–9691
Fax: 703–519–1872
info@asafishing.org
http://www.asafishing.org

American Zoo and Aquarium Association
8403 Colesville Road, Suite 710
Silver Spring, MD 20910
http://www.aza.org

Australian Society for Fish Biology Inc.
123 Brown Street (PO Box 137)
Heidelberg, Victoria 3084
Australia
Phone: 61 3 9450 8669
Fax: 61 3 9450 8730
john.koehn@nre.vic.gov.au
http://www.asfb.org.au

Commission for the Conservation of Antarctic Marine Living Resources (CCAMLR)
137 Harrington Street
Hobart, Tasmania 7000
Australia
Phone: 61 3 6231 0366
Fax: 61 6234 9965
webmaster@ccamlr.org
http://www.ccamlr.org

Desert Fishes Council
315 East Medlock Drive,
Phoenix, Arizona 85012
United States
602-274-5544.
stefferud@cox.net
http://www.desertfishes.org/

Food and Agriculture Organization of the United Nations (FAO) Fisheries
Viale delle Terme di Caracalla
Rome 00100
Italy
Phone: +39 06 5705 1
Fax: +39 06 5705 3152
FAO-HQ@fao.org
http://www.fao.org/fi/default.asp

Grouper and Wrasse Specialist Group, Species Survival Commission, IUCN
Department of Ecology and Biodiversity,
The University of Hong Kong
Hong Kong
China
Phone: 852 2859 8977
Fax: 852 2517 6082
yjsadovy@hkusua.hku.hk
http://www.hku.hk/ecology/GroupersWrasses/iucnsg/index.html

Inter-American Tropical Tuna Commisssion
8604 La Jolla Shores Drive
La Jolla, CA 92037-1508
United States
Phone: 619-546-7133
Fax: 619-546-7159
www.iattc.org

International Commission for the Conservation of Atlantic Tunas
Calle Corazón de Maria, 8, 6th floor
Madrid E-28002
Spain

IUCN: The World Conservation Union
Rue Mauverney 28
1196
Gland
Switzerland
Phone: 41-22-999-0000
mail@hq.iucn.org
http://www.iucn.org

IUCN/SSC Coral Reef Fishes Specialist Group
c/o IMA-Integrative Biological Research Program,
University of Guam Marine Laboratory, UOG Station
Mangilao, Guam 96913
USA
Phone: 1-671-735-2187
Fax: 1-671-734-6767
donaldsn@uog9.uog.edu
http://www.iucn.org

Menhaden Resource Council
1901 N. Fort Myer Drive, Suite 700
Arlington, VA 22209
USA
Phone: 703-796-1793
resource@menhaden.org
http://www.menhaden.org

Muskies Canada Sport Fishing and Research, Inc.
P.O. Box 814, Station C
Kitchener
Ontario N2G 4C5
Canada
http://www.trentu.ca/muskie/mc.html.

North American Native Fishes Association
1107 Argonne Dr.
Baltimore, MD 21218
USA
nanfa@att.net
http://www.nanfa.org

Salmon and Trout Association (UK)
Fishmongers' Hall, London Bridge
London EC4R 9EL
UK
Phone: 0207 283 5838
Fax: 0207 626 5137
http://www.salmon-trout.org

South African Coelacanth Conservation and Genome Resource
Somerset Street
Private Bag 1015
Grahamstown 6140
South Africa
Phone: +27 (0)46 636 1002
Fax: +27 (0)46 622 2403

TRAFFIC International
219c Huntingdon Road
Cambridge CB3 0DL
United Kingdom
Phone: 44 1223 277427
Fax: 44 1223 277237
traffic@trafficint.org
http://www.traffic.org

United States Trout Farmers Association
111 West Washington St., Suite One
Charles Town, WV 25414-1529
USA
Phone: (304) 728 2189
Fax: (304) 728 2196
http://www.ustfa.org

• • • • •

Contributors to the first edition

The following individuals contributed chapters to the original edition of Grzimek's Animal Life Encyclopedia, *which was edited by Dr. Bernhard Grzimek, Professor, Justus Liebig University of Giessen, Germany; Director, Frankfurt Zoological Garden, Germany; and Trustee, Tanzanian National Parks, Tanzania.*

Dr. Michael Abs
Curator, Ruhr University
Bochum, Germany

Dr. Salim Ali
Bombay Natural History Society
Bombay, India

Dr. Rudolph Altevogt
Professor, Zoological Institute,
University of Münster
Münster, Germany

Dr. Renate Angermann
Curator, Institute of Zoology,
Humboldt University
Berlin, Germany

Edward A. Armstrong
Cambridge University
Cambridge, England

Dr. Peter Ax
Professor, Second Zoological Institute
and Museum, University of Göttingen
Göttingen, Germany

Dr. Franz Bachmaier
Zoological Collection of the State
of Bavaria
Munich, Germany

Dr. Pedru Banarescu
Academy of the Roumanian Socialist
Republic, Trajan Savulescu Institute
of Biology
Bucharest, Romania

Dr. A. G. Bannikow
Professor,
Institute of Veterinary Medicine
Moscow, Russia

Dr. Hilde Baumgärtner
Zoological Collection of the State
of Bavaria
Munich, Germany

C. W. Benson
Department of Zoology,
Cambridge University
Cambridge, England

Dr. Andrew Berger
Chairman, Department of Zoology,
University of Hawaii
Honolulu, Hawaii, U.S.A.

Dr. J. Berlioz
National Museum of Natural History
Paris, France

Dr. Rudolf Berndt
Director,
Institute for Population Ecology,
Hiligoland Ornithological Station
Braunschweig, Germany

Dieter Blume
Instructor of Biology,
Freiherr-vom-Stein School
Gladenbach, Germany

Dr. Maximilian Boecker
Zoological Research Institute and
A. Koenig Museum
Bonn, Germany

Dr. Carl-Heinz Brandes
Curator and Director, The Aquarium,
Overseas Museum
Bremen, Germany

Dr. Donald G. Broadley
Curator, Umtali Museum
Mutare, Zimbabwe

Dr. Heinz Brüll
Director; Game, Forest, and Fields
Research Station
Hartenholm, Germany

Dr. Herbert Bruns
Director, Institute of Zoology and the
Protection of Life
Schlangenbad, Germany

Hans Bub
Heligoland Ornithological Station
Wilhelmshaven, Germany

A. H. Chrisholm
Sydney, Australia

Herbert Thomas Condon
Curator of Birds,
South Australian Museum
Adelaide, Australia

Dr. Eberhard Curio
Director,
Laboratory of Ethology,
Ruhr University
Bochum, Germany

Dr. Serge Daan
Laboratory of Animal Physiology,
University of Amsterdam
Amsterdam, The Netherlands

Dr. Heinrich Dathe
Professor and Director, Animal Park
and Zoological Research Station,
German Academy of Sciences
Berlin, Germany

Dr. Wolfgang Dierl
Zoological Collection of the State
of Bavaria
Munich, Germany

Dr. Fritz Dieterlen
Zoological Research Institute,

A. Koenig Museum
Bonn, Germany

Dr. Rolf Dircksen
Professor, Pedagogical Institute
Bielefeld, Germany

Josef Donner
Instructor of Biology
Katzelsdorf, Austria

Dr. Jean Dorst
Professor, National Museum of Natural History
Paris, France

Dr. Gerti Dücker
Professor and Chief Curator, Zoological Institute, University of Münster
Münster, Germany

Dr. Michael Dzwillo
Zoological Institute and Museum, University of Hamburg
Hamburg, Germany

Dr. Irenäus Eibl-Eibesfeldt
Professor and Director, Institute of Human Ethology, Max Planck Institute for Behavioral Physiology
Percha/Starnberg, Germany

Dr. Martin Eisentraut
Professor and Director, Zoological Research Institute and A. Koenig Museum
Bonn, Germany

Dr. Eberhard Ernst
Swiss Tropical Institute
Basel, Switzerland

R. D. Etchecopar
Director, National Museum of Natural History
Paris, France

Dr. R. A. Falla
Director, Dominion Museum
Wellington, New Zealand

Dr. Hubert Fechter
Curator, Lower Animals, Zoological Collection of the State of Bavaria
Munich, Germany

Dr. Walter Fiedler
Docent, University of Vienna, and Director, Schönbrunn Zoo
Vienna, Austria

Wolfgang Fischer
Inspector of Animals, Animal Park
Berlin, Germany

Dr. C. A. Fleming
Geological Survey Department of Scientific and Industrial Research
Lower Hutt, New Zealand

Dr. Hans Frädrich
Zoological Garden
Berlin, Germany

Dr. Hans-Albrecht Freye
Professor and Director, Biological Institute of the Medical School
Halle a.d.S., Germany

Günther E. Freytag
Former Director, Reptile and Amphibian Collection, Museum of Cultural History in Magdeburg
Berlin, Germany

Dr. Herbert Friedmann
Director, Los Angeles County Museum of Natural History
Los Angeles, California, U.S.A.

Dr. H. Friedrich
Professor, Overseas Museum
Bremen, Germany

Dr. Jan Frijlink
Zoological Laboratory, University of Amsterdam
Amsterdam, The Netherlands

Dr. H .C. Karl Von Frisch
Professor Emeritus and former Director, Zoological Institute, University of Munich
Munich, Germany

Dr. H. J. Frith
C.S.I.R.O. Research Institute
Canberra, Australia

Dr. Ion E. Fuhn
Academy of the Roumanian Socialist Republic, Trajan Savulescu Institute of Biology
Bucharest, Romania

Dr. Carl Gans
Professor, Department of Biology, State University of New York at Buffalo
Buffalo, New York, U.S.A.

Dr. Rudolf Geigy
Professor and Director, Swiss Tropical Institute
Basel, Switzerland

Dr. Jacques Gery
St. Genies, France

Dr. Wolfgang Gewalt
Director, Animal Park
Duisburg, Germany

Dr. H. C. Viktor Goerttler
Professor Emeritus, University of Jena
Jena, Germany

Dr. Friedrich Goethe
Director, Institute of Ornithology, Heligoland Ornithological Station
Wilhelmshaven, Germany

Dr. Ulrich F. Gruber
Herpetological Section, Zoological Research Institute and A. Koenig Museum
Bonn, Germany

Dr. H. R. Haefelfinger
Museum of Natural History
Basel, Switzerland

Dr. Theodor Haltenorth
Director, Mammalology, Zoological Collection of the State of Bavaria
Munich, Germany

Barbara Harrisson
Sarawak Museum, Kuching, Borneo
Ithaca, New York, U.S.A.

Dr. Francois Haverschmidt
President, High Court (retired)
Paramaribo, Suriname

Dr. Heinz Heck
Director, Catskill Game Farm
Catskill, New York, U.S.A.

Dr. Lutz Heck
Professor (retired), and Director, Zoological Garden, Berlin
Wiesbaden, Germany

Dr. Dr. H.C.Heini Hediger
Director, Zoological Garder
Zurich, Switzerland

Dr. Dietrich Heinemann
Director, Zoological Garden, Münster
Dörnigheim, Germany

Dr. Helmut Hemmer
Institute for Physiological Zoology,
University of Mainz
Mainz, Germany

Dr. W. G. Heptner
Professor, Zoological Museum,
University of Moscow
Moscow, Russia

Dr. Konrad Herter
Professor Emeritus and Director
(retired), Zoological Institute, Free
University of Berlin
Berlin, Germany

Dr. Hans Rudolf Heusser
Zoological Museum,
University of Zurich
Zurich, Switzerland

Dr. Emil Otto Höhn
Associate Professor of Physiology,
University of Alberta
Edmonton, Canada

Dr. W. Hohorst
Professor and Director,
Parasitological Institute,
Farbwerke Hoechst A.G.
Frankfurt-Höchst, Germany

Dr. Folkhart Hückinghaus
Director,
Senckenbergische Anatomy,
University of Frankfurt a.M.
Frankfurt a.M., Germany

Francois Hüe
National Museum of Natural History
Paris, France

Dr. K. Immelmann
Professor, Zoological Institute,
Technical University of Braunschweig
Braunschweig, Germany

Dr. Junichiro Itani
Kyoto University
Kyoto, Japan

Dr. Richard F. Johnston
Professor of Zoology,
University of Kansas
Lawrence, Kansas, U.S.A.

Otto Jost
Oberstudienrat,
Freiherr-vom-Stein Gymnasium
Fulda, Germany

Dr. Paul Kähsbauer
Curator, Fishes, Museum of
Natural History
Vienna, Austria

Dr. Ludwig Karbe
Zoological State Institute
and Museum
Hamburg, Germany

Dr. N. N. Kartaschew
Docent, Department of Biology,
Lomonossow State University
Moscow, Russia

Dr. Werner Kästle
Oberstudienrat, Gisela Gymnasium
Munich, Germany

Dr. Reinhard Kaufmann
Field Station of the Tropical Institute,
Justus Liebig University,
Giessen, Germany
Santa Marta, Colombia

Dr. Masao Kawai
Primate Research Institute,
Kyoto University
Kyoto, Japan

Dr. Ernst F. Kilian
Professor,
Giessen University and Catedratico
Universidad Austral,
Valdivia-Chile
Giessen, Germany

Dr. Ragnar Kinzelbach
Institute for General Zoology,
University of Mainz
Mainz, Germany

Dr. Heinrich Kirchner
Landwirtschaftsrat (retired)
Bad Oldesloe, Germany

Dr. Rosl Kirchshofer
Zoological Garden, University of
Frankfort a.M.
Frankfurt a.M., Germany

Dr. Wolfgang Klausewitz
Curator, Senckenberg Nature
Museum and Research Institute
Frankfurt a.M., Germany

Dr. Konrad Klemmer
Curator, Senckenberg Nature
Museum and Research Institute
Frankfurt a.M., Germany

Dr. Erich Klinghammer
Laboratory of Ethology,
Purdue University
Lafayette, Indiana, U.S.A.

Dr. Heinz-Georg Klös
Professor and Director,
Zoological Garden
Berlin, Germany

Ursula Klös
Zoological Garden
Berlin, Germany

Dr. Otto Koehler
Professor Emeritus,
Zoological Institute,
University of Freiburg
Freiburg i. BR., Germany

Dr. Kurt Kolar
Institute of Ethology, Austrian
Academy of Sciences
Vienna, Austria

Dr. Claus König
State Ornithological Station
of Baden-Württemberg
Ludwigsburg, Germany

Dr. Adriaan Kortlandt
Zoological Laboratory,
University of Amsterdam
Amsterdam, The Netherlands

Dr. Helmut Kraft
Professor and Scientific Councillor,
Medical Animal Clinic, University
of Munich
Munich, Germany

Dr. Helmut Kramer
Zoological Research Institute and
A. Koenig Museum
Bonn, Germany

Dr. Franz Krapp
Zoological Institute,
University of Freiburg
Freiburg, Switzerland

Dr. Otto Kraus
Professor,
University of Hamburg, and Director,
Zoological Institute
and Museum
Hamburg, Germany

Dr. Hans Krieg
Professor and First Director (retired),

Scientific Collections of the State of Bavaria
Munich, Germany

Dr. Heinrich Kühl
Federal Research Institute for Fisheries, Cuxhaven Laboratory
Cuxhaven, Germany

Dr. Oskar Kuhn
Professor, formerly University Halle/Saale
Munich, Germany

Dr. Hans Kumerloeve
First Director (retired), State Scientific Museum, Vienna
Munich, Germany

Dr. Nagamichi Kuroda
Yamashina Ornithological Institute, Shibuya-Ku
Tokyo, Japan

Dr. Fred Kurt
Zoological Museum of Zurich University, Smithsonian Elephant Survey
Colombo, Ceylon

Dr. Werner Ladiges
Professor and Chief Curator, Zoological Institute and Museum, University of Hamburg
Hamburg, Germany

Leslie Laidlaw
Department of Animal Sciences, Purdue University
Lafayette, Indiana, U.S.A.

Dr. Ernst M. Lang
Director, Zoological Garden
Basel, Switzerland

Dr. Alfredo Langguth
Department of Zoology, Faculty of Humanities and Sciences, University of the Republic
Montevideo, Uruguay

Leo Lehtonen
Science Writer
Helsinki, Finland

Bernd Leisler
Second Zoological Institute, University of Vienna
Vienna, Austria

Dr. Kurt Lillelund
Professor and Director, Institute for Hydrobiology and Fishery Sciences, University of Hamburg
Hamburg, Germany

R. Liversidge
Alexander MacGregor Memorial Museum
Kimberley, South Africa

Dr. Konrad Lorenz
Professor and Director, Max Planck Institute for Behavioral Physiology
Seewiesen/Obb., Germany

Dr. Martin Lühmann
Federal Research Institute for the Breeding of Small Animals
Celle, Germany

Dr. Johannes Lüttschwager
Oberstudienrat (retired)
Heidelberg, Germany

Dr. Wolfgang Makatsch
Bautzen, Germany

Dr. Hubert Markl
Professor and Director, Zoological Institute, Technical University of Darmstadt
Darmstadt, Germany

Basil J. Marlow, B.SC. (Hons)
Curator, Australian Museum
Sydney, Australia

Dr. Theodor Mebs
Instructor of Biology
Weissenhaus/Ostsee, Germany

Dr. Gerlof Fokko Mees
Curator of Birds, Rijks Museum of Natural History
Leiden, The Netherlands

Hermann Meinken
Director, Fish Identification Institute, V.D.A.
Bremen, Germany

Dr. Wilhelm Meise
Chief Curator, Zoological Institute and Museum, University of Hamburg
Hamburg, Germany

Dr. Joachim Messtorff
Field Station of the Federal Fisheries Research Institute
Bremerhaven, Germany

Dr. Marian Mlynarski
Professor, Polish Academy of Sciences, Institute for Systematic and Experimental Zoology
Cracow, Poland

Dr. Walburga Moeller
Nature Museum
Hamburg, Germany

Dr. H.C.Erna Mohr
Curator (retired), Zoological State Institute and Museum
Hamburg, Germany

Dr. Karl-Heinz Moll
Waren/Müritz, Germany

Dr. Detlev Müller-Using
Professor, Institute for Game Management, University of Göttingen
Hannoversch-Münden, Germany

Werner Münster
Instructor of Biology
Ebersbach, Germany

Dr. Joachim Münzing
Altona Museum
Hamburg, Germany

Dr. Wilbert Neugebauer
Wilhelma Zoo
Stuttgart-Bad Cannstatt, Germany

Dr. Ian Newton
Senior Scientific Officer, The Nature Conservancy
Edinburgh, Scotland

Dr. Jürgen Nicolai
Max Planck Institute for Behavioral Physiology
Seewiesen/Obb., Germany

Dr. Günther Niethammer
Professor, Zoological Research Institute and A. Koenig Museum
Bonn, Germany

Dr. Bernhard Nievergelt
Zoological Museum, University of Zurich
Zurich, Switzerland

Dr. C. C. Olrog
Institut Miguel Lillo San Miguel de Tucuman
Tucuman, Argentina

Alwin Pedersen
Mammal Research and
Arctic Explorer
Holte, Denmark

Dr. Dieter Stefan Peters
Nature Museum and Senckenberg
Research Institute
Frankfurt a.M., Germany

Dr. Nicolaus Peters
Scientific Councillor and Docent,
Institute of Hydrobiology and
Fisheries, University of Hamburg
Hamburg, Germany

Dr. Hans-Günter Petzold
Assistant Director,
Zoological Garden
Berlin, Germany

Dr. Rudolf Piechocki
Docent, Zoological Institute,
University of Halle
Halle a.d.S., Germany

Dr. Ivo Poglayen-Neuwall
Director, Zoological Garden
Louisville, Kentucky, U.S.A.

Dr. Egon Popp
Zoological Collection of the State
of Bavaria
Munich, Germany

Dr. H. C. Adolf Portmann
Professor Emeritus, Zoological
Institute, University of Basel
Basel, Switzerland

Hans Psenner
Professor and Director, Alpine Zoo
Innsbruck, Austria

Dr. Heinz-Siburd Raethel
Oberveterinärrat
Berlin, Germany

Dr. Urs H. Rahm
Professor, Museum of Natural History
Basel, Switzerland

Dr. Werner Rathmayer
Biology Institute,
University of Konstanz
Konstanz, Germany

Walter Reinhard
Biologist
Baden-Baden, Germany

Dr. H. H. Reinsch
Federal Fisheries Research Institute
Bremerhaven, Germany

Dr. Bernhard Rensch
Professor Emeritus, Zoological
Institute, University of Münster
Münster, Germany

Dr. Vernon Reynolds
Docent, Department of Sociology,
University of Bristol
Bristol, England

Dr. Rupert Riedl
Professor, Department of Zoology,
University of North Carolina
Chapel Hill, North Carolina, U.S.A.

Dr. Peter Rietschel
Professor (retired),
Zoological Institute,
University of Frankfurt a.M.
Frankfurt a.M., Germany

Dr. Siegfried Rietschel
Docent, University of Frankfurt;
Curator, Nature Museum and
Research Institute Senckenberg
Frankfurt a.M., Germany

Herbert Ringleben
Institute of Ornithology,
Heligoland Ornithological Station
Wilhelmshaven, Germany

Dr. K. Rohde
Institute for General Zoology,
Ruhr University
Bochum, Germany

Dr. Peter Röben
Academic Councillor, Zoological
Institute, Heidelberg University
Heidelberg, Germany

Dr. Anton E. M. De Roo
Royal Museum of Central Africa
Tervuren, South Africa

Dr. Hubert Saint Girons
Research Director,
Center for National
Scientific Research
Brunoy (Essonne), France

Dr. Luitfried Von Salvini-Plawen
First Zoological Institute,
University of Vienna
Vienna, Austria

Dr. Kurt Sanft
Oberstudienrat,
Diesterweg-Gymnasium
Berlin, Germany

Dr. E. G. Franz Sauer
Professor, Zoological Research
Institute and A. Koenig Museum,
University of Bonn
Bonn, Germany

Dr. Eleonore M. Sauer
Zoological Research Institute and
A. Koenig Museum,
University of Bonn
Bonn, Germany

Dr. Ernst Schäfer
Curator,
State Museum of Lower Saxony
Hannover, Germany

Dr. Friedrich Schaller
Professor and Chairman,
First Zoological Institute,
University of Vienna
Vienna, Austria

Dr. George B. Schaller
Serengeti Research Institute,
Michael Grzimek Laboratory
Seronera, Tanzania

Dr. Georg Scheer
Chief Curator and Director,
Zoological Institute,
State Museum of Hesse
Darmstadt, Germany

Dr. Christoph Scherpner
Zoological Garden
Frankfurt a.M., Germany

Dr. Herbert Schifter
Bird Collection,
Museum of Natural History
Vienna, Austria

Dr. Marco Schnitter
Zoological Museum,
Zurich University
Zurich, Switzerland

Dr. Kurt Schubert
Federal Fisheries Research Institute
Hamburg, Germany

Eugen Schuhmacher
Director, Animals Films, I.U.C.N.
Munich, Germany

Dr. Thomas Schultze-Westrum
Zoological Institute,
University of Munich
Munich, Germany

Dr. Ernst Schüt
Professor and Director (retired),
State Museum of Natural History
Stuttgart, Germany

Dr. Lester L. Short, Jr.
Associate Curator, American Museum
of Natural History
New York, New York, U.S.A.

Dr. Helmut Sick
National Museum
Rio de Janeiro, Brazil

Dr. Alexander F. Skutch
Professor of Ornithology,
University of Costa Rica
San Isidro del General, Costa Rica

Dr. Everhard J. Slijper
Professor, Zoological Laboratory,
University of Amsterdam
Amsterdam, The Netherlands

Bertram E. Smythies
Curator (retired),
Division of Forestry Management,
Sarawak-Malaysia
Estepona, Spain

Dr. Kenneth E. Stager
Chief Curator, Los Angeles County
Museum of Natural History
Los Angeles, California, U.S.A.

Dr. H. C. Georg H. W. Stein
Professor, Curator of Mammals,
Institute of Zoology and
Zoological Museum,
Humboldt University
Berlin, Germany

Dr. Joachim Steinbacher
Curator, Nature Museum and
Senckenberg Research Institute
Frankfurt a.M., Germany

Dr. Bernard Stonehouse
Canterbury University
Christchurch, New Zealand

Dr. Richard Zur Strassen
Curator, Nature Museum and
Senckenberg Research Institute
Frandfurt a.M., Germany

Dr. Adelheid Studer-Thiersch
Zoological Garden
Basel, Switzerland

Dr. Ernst Sutter
Museum of Natural History
Basel, Switzerland

Dr. Fritz Terofal
Director, Fish Collection, Zoological
Collection of the State of Bavaria
Munich, Germany

Dr. G. F. Van Tets
Wildlife Research
Canberra, Australia

Ellen Thaler-Kottek
Institute of Zoology,
University of Innsbruck
Innsbruck, Austria

Dr. Erich Thenius
Professor and Director,
Institute of Paleontolgy,
University of Vienna
Vienna, Austria

Dr. Niko Tinbergen
Professor of Animal Behavior,
Department of Zoology,
Oxford University
Oxford, England

Alexander Tsurikov
Lecturer, University of Munich
Munich, Germany

Dr. Wolfgang Villwock
Zoological Institute and Museum,
University of Hamburg
Hamburg, Germany

Zdenek Vogel
Director,
Suchdol Herpetological Station
Prague, Czechoslovakia

Dieter Vogt
Schorndorf, Germany

Dr. Jiri Volf
Zoological Garden
Prague, Czechoslovakia

Otto Wadewitz
Leipzig, Germany

Dr. Helmut O. Wagner
Director (retired),
Overseas Museum, Bremen
Mexico City, Mexico

Dr. Fritz Walther
Professor, Texas A & M University
College Station, Texas, U.S.A.

John Warham
Zoology Department,
Canterbury University
Christchurch, New Zealand

Dr. Sherwood L. Washburn
University of California at Berkeley
Berkeley, California, U.S.A.

Eberhard Wawra
First Zoological Institute,
University of Vienna
Vienna, Austria

Dr. Ingrid Weigel
Zoological Collection of the State
of Bavaria
Munich, Germany

Dr. B. Weischer
Institute of Nematode Research,
Federal Biological Institute
Münster/Westfalen, Germany

Herbert Wendt
Author, Natural History
Baden-Baden, Germany

Dr. Heinz Wermuth
Chief Curator,
State Nature Museum, Stuttgart
Ludwigsburg, Germany

Dr. Wolfgang Von Westernhagen
Preetz/Holstein, Germany

Dr. Alexander Wetmore
United States National Museum,
Smithsonian Institution
Washington, D.C., U.S.A.

Dr. Dietrich E. Wilcke
Röttgen, Germany

Dr. Helmut Wilkens
Professor and Director,
Institute of Anatomy,
School of Veterinary Medicine
Hannover, Germany

Dr. Michael L. Wolfe
Utah, U.S.A.

Hans Edmund Wolters
Zoological Research Institute and
A. Koenig Museum
Bonn, Germany

Dr. Arnfrid Wünschmann
Research Associate, Zoological Garden
Berlin, Germany

Dr. Walter Wüst
Instructor, Wilhelms Gymnasium
Munich, Germany

Dr. Heinz Wundt
Zoological Collection of the State
of Bavaria
Munich, Germany

Dr. Claus-Dieter Zander
Zoological Institute and Museum,
University of Hamburg
Hamburg, Germany

Dr. Dr.Fritz Zumpt
Director,
Entomology and Parasitology,
South African Institute for
Medical Research
Johannesburg, South Africa

Dr. Richard L. Zusi
Curator of Birds,
United States National Museum,
Smithsonian Institution
Washington, D.C., U.S.A.

Glossary

Adipose fin—A small, fleshy fin without rays.

Afferent—Conducting impulses toward nerve centers or blood toward the gills. Compare efferent.

Agonistic behavior—Aggressive and submissive interaction between individuals of the same species.

Albinistic—Displaying the characteristics of an albino; an organism that has deficient pigmentation and white, colorless, or translucent skin and hair.

Amphidromous—Regular migration between fresh and seawater at different stages in their development.

Ampulla—A sac- or pouch-like anatomical swelling.

Anal fin—Fin located on the undersurface of the body, behind the anus.

Andropodium—Modified anal fin exhibited by some males.

Anoxic—Extreme deficiency of oxygen.

Anthropogenic—Caused by the activities of human beings.

Antitropical—Found in both Northern and Southern Hemispheres, but not in equatorial regions.

Aplacental—Without a placenta.

Axial skeleton—Skeleton of the main body and head.

Axillary process—Modified scale present at the upper or anterior base of the pectoral or ventral fins exhibited by some fishes.

Barbel—Fleshy, tactile projection resembling tentacles located near the mouth, chin, or snout.

Basiocapital—Bone located at the back of the head or skull; the occiput.

Basioccipital—Base of the head or skull.

Bathypelagic—Living and/or feeding in open waters at depths between 3,280 and 13,125 ft (1,000 and 4,000 m).

Benthic—Relating to, living on, or occurring at the bottom of a body of water.

Benthopelagic—Relating to, living on, or occurring on the bottom or midwaters of a body of water, feeding on benthic and free swimming organisms.

Benthos—The bottom of a body of water.

Bilobed—The division of matter into two lobes.

Branchial—Relating to the gills.

Branchiostegal membrane—The gill membrane; supported by the branchiostegal rays (bones).

Buccal cavity—Mouth cavity forward of the gills.

Bycatch—Species that are not targeted as catch, but are caught along with a target species during fishing.

Carapace—A hardened shell, such as turtles or crabs have.

Catadromous—Living in freshwater, but migrating to saltwater for spawning.

Caudal fin—Fin located at the end of the body, also known as the tail fin.

Caudal keels—Ridges on either side of the caudal peduncle that often function in stabilization during fast swimming.

Caudal peduncle—A narrow part of the body located at the base of the caudal fin.

Cavernicolous—Cave dwelling.

Cecum—Cavity or pouch extending off the intestine that receives undigested food.

Cephalic—Relating or belonging to the head.

Clade—A group of biological taxa, such as species, that includes all descendants of one common ancestor.

Cladist—One who classifies organisms based on their evolutionary history.

Cleithrum—The major bone of the pectoral girdle.

Cloaca—Chamber into which the intestinal, urinary, and reproductive ducts discharge.

Confluent fins—Fins that are joined or run together, having no true separation.

Crepuscular—Active in the twilight or evening.

Crypsis—Patterns and/or coloration that make an organism more difficult for predators to detect; protective patterns or coloration.

Ctenoid scales—Scales having minute spines on exposed surface.

Cupula—A cup-shaped structure.

Cycloid scales—Scales having smooth edges, absent of spines.

Demersal—Living near, laying on, or sinking towards the bottom of the ocean.

Dentary—Lower jawbone of vertebrates.

Dentine—Material similar to but harder than bone and is the principal mass of teeth.

Dermal denticles—Teeth-like scales, also known as placoid scales, on the skin of various elasmobranchs that acts as a protective barrier and also enables faster swimming.

Dextral—Occurring on or relating to the right side of the body.

Diadromous—Regular migration between freshwater and seawater.

Dichromatism—Partial color blindness; the ability to recognize only two colors.

Diel—Involving a 24 hour period of time; occurring on a daily basis.

Diploid—Two sets of chromosomes existing in a cell or organism.

Dorsal fin—Spined or rayed fin on the dorsal surface of body.

Dorsal—Relating or belonging to the back or top surface of the body.

Dorsolateral—Belonging to, or orientated between the dorsal and lateral surfaces.

Dorsoventral—Belonging to, or orientated between the dorsal and ventral surface.

Ectodermal—Formed from the outer germ layer of an embryo.

Ectothermic—Cold-blooded animal.

Efferent—Conducting impulses away from nerve centers or blood away from the gills. Compare afferent.

Elasmobrand—Relating to the group of fishes that includes the sharks, rays, and skates.

Electric organ—Organ capable of delivering an electric shock or used to emit electrical discharges to stun prey, repel predators, or detect objects.

Endemic—The restriction of a species to a particular geographic area or continent; native.

Engybenthic—Organisms living or occurring at the bottom of a body of water.

Epibenthic—Living on the bottom of the ocean.

Epigean—Organisms that are not cave-dwellers and do not live underground.

Epipelagic—Living or feeding in the uppermost layer of water; from the surface to midwater depths of 656.17 ft (200 m).

Euryhaline—Ability to live in waters of varying salinity.

Exogenous—Introduced from, or produced outside the organism or system.

Facultative parasite—An organism that can exist off of its host.

Falcate—Having a hooked or curved shape.

Filiform—Having the shape or form of a filament.

Fin base—Portion of a fin that attaches to the body.

Fin spine—Bony structure that supports the fin in more derived fishes.

Finlet—A small, isolated fin, usually without rays, that ususally occurs dorsally or ventrally on the caudal peduncle.

Flange—A rib or rim that aids one object in attaching to another.

Flexion—The act of bending, extending or flexing; a physical structure having a bent shape.

Fry—Newly hatched juveniles, or very small adult fishes.

Fusiform—The tapering of each end.

Ganglia—Mass of nerve tissue containing nerve cells external to the brain or spinal cord.

Ganoid—Relating to, or having scales that are made of bone and an outer layer that resembles enamel.

Ganione—Substance that resembles enamel and makes up the outer layer of certain fishes' scales.

Gas bladder—Sac in the body cavity below the vertebral column; helps maintain buoyancy, may aid in respiration, and may help produce or receive sound. Also called swim bladder.

Gill—Organ for obtaining oxygen from water.

Gill cover—Flap made of bone or cartilage that covers and protects the gills. Also called operculum.

Gill rakers—Projections from the gill arch that help in retaining food particles.

Globose—Having the shape or form of a globule or ball.

Gregarious—Living in a group or colony.

Haploid—One set of chromosomes existing in a cell or organism.

Hermaphroditism—The presence of both male and female sexual organs in one individual. When both organs occur at the same time, the individual is bisexual or a simultaneous hermaphrodita; if they occur at different times, the individual is a sequential hermaphrodite.

Heterocercal—Upper lobe of the tail is larger than the lower lobe, and the vertebral column extends into the upper lobe.

Heterozygous—Having two alleles at corresponding loci on homologous chromosomes that are different.

Holarctic—Relating to, or being from the northern parts of the world.

Homocercal—A caudal fin in which all of the principal rays attach to the last vertebra.

Homologous—Structures or properties of organisms shared through common ancestry.

Hyoid—Belonging or pertaining to the tongue.

Hypogean—Lives underground.

Hypoxia—Deficiency of oxygen reaching the tissues of the body.

Hypural plate—Modified last vertebra, to which caudal fin rays attach.

Incisiform—Teeth that are flat with sharp edges.

Integument—A layer or membrane that encloses or envelopes an organism or one of its parts.

Intertidal—Shallow areas along the shore that are alternately exposed and covered by the tides.

Isosmotic—Having the same osmotic pressure on two sides of a membrane.

Isthmus—Narrow, triangular area on the underside of the body, between the gill openings.

Iteroparous—Successive production of offspring, annually or seasonal batches.

Lacustrine—Relating to, inhabiting, formed or growing in lake water.

Lamella—Thin plate or membrane; often refers to smallest divisions of gill.

Larvaceans—Small transparent animals found in marine plankton; belong to subphylum Urochordata.

Lateral line—A series of ampulla forming a sensory organ to detect movements in water. Scales are often modified with pores opening to a sensory canal on the side of a fish.

Lecithotrophic—Embryos feeding on the yolk stored in the yolk sac.

Littoral—Related to, inhabiting, or situated near a shore.

Lunate—Having the shape of a crescent.

Maxilla—Upper jaw bone.

Median fin—Fins located on the median plane.

Meiobenthic—Benthic organisms with dimensions less than 0.02 in (0.5 mm) but greater than or equal to 0.004 in (0.1 mm).

Melanistic—An organism that exhibits a high amount of melanin (black coloration) in the skin.

Melanophore—A cell containing melanin.

Mesentary—A membrane that attaches organ to the abdominal wall.

Mesopelagic—Relating to, inhabiting or feeding at midwater at depths between 656.17 ft (200 m) and 3,280.84 ft (1,000 m).

Microphthalmic—Having eyes noticeably reduced in size.

Micropredator—A predator smaller than its prey that comes into contact with its host only when needing to feed.

Midwater—The middle stratum of a body of water.

Milt—The combination of spermatozoa and seminal fluid in fishes.

Monogamy—Mating system in which a single pair joins together for spawning and may remain together for one or more seasons.

Monophyletic—Developed from or related to a single common ancestral form or stock.

Monotypic—A group containing a single representative.

Myoglobin—Protein pigment in muscles that contains iron.

Naked—A fish that has no scales.

Nares—Nostrils.

Nasohypophysical—Nostril opening.

Nektonic—Organisms that swim strongly enough to move against currents.

Neural spine—The uppermost spine of a vertebra.

Notched fin—A fin that has patterned indentation.

Obligate air breathers—An organism that must receive a certain amount of their oxygen directly from air.

Ocelli—An eye-like marking.

Ontogenetic—Changes that incur from growth or age.

Oophagy—The process of embryos feeding on eggs produced by the ovary while still inside the uterus.

Osmoregulation—The regulation of water in the body.

Otoliths—Calcareous deposit in the ear capsules of bony fishes that show daily, seasonal or annual checks, rings or layers that can be used to determine ages.

Oviduct—Duct that serves as the passage of eggs from the ovary.

Oviparous—Production of eggs that develop and hatch outside of the mother.

Ovoviviparous—Fertilized eggs are retained in the mother's body during development.

Paedomorphic—Phylogenetic retention of larval or juvenile characters in the adult stage.

Paired Fins—Fins that occur in pairs, on each side of the body.

Paleoecology—The study of ecological characteristics in ancient environments and their relationships to ancient plants and animals.

Palp (Palpus)—Segmented and tactile process on the mouth.

PCBs (polychlorinated biphenyls)—Substances used as coolants and lubricants; their manufacture was banned in the United States in 1977.

Pectoral—Relating or belonging to the forward pair of appendicular appendages.

Pectoral fins—Fins attached to the shoulder or pectoral girdle, just behind the head.

Peduncule—Narrow part by which a larger part or the whole body is attached.

Pelagic—Relating to, living or occurring in open ocean water.

Pelvic fins—Pair of fins attached to the pelvis or pelvic girdle.

Photophore—A luminous spot or light-producing organ.

Piscivorous—Diet consists solely of other fishes.

Planktivorous—Diet consists solely of passively floating or weakly swimming animal and plant life.

Polyandry—Females mate with more than one male in a season.

Polygamy—Mating system in which individuals mate with more than one partner in a season.

Polygyny—Males mate with more than one female in a season.

Polyphyletic group—An assemblage consisting of different ancestral taxa, i.e., a group based upon convergence rather than on common ancestry.

Precaudal pit—Cavity just anterior to the caudal fin.

Prehensile—Adapted specifically to enable seizing, grasping, or wrapping around.

Promiscuity—Males and females spawn together with little or no mate choice.

Protandrous—Sequential hermaphroditism in which the fish functions first as a male and then a female.

Protrusible mouth—Mouth which can project forward and out to help catch prey.

Protygynous—Sequential hermaphroditism in which the fish functions first as a female and then a male.

Ray—Segmented bony rod or element that supports a fin membrane.

Riffle stretches—Areas of rough water caused by submerged rocks or a sandbar.

Rostral—Located toward the mouth or nasal region.

Seamount—Submarine mountain rising above the deep-sea floor.

Sexual dichromism—Exhibiting both male and female forms and aspects.

Sinistral—Occurring on or relating to the left side of the body.

Sinusoidal—Relating to, or shaped like the sine curve or wave.

Sister group—Closest relative to a taxa or group. The two groups share a common ancestor.

Spinules—Minute or miniature spine.

Standard length—Standard scientific measure of a fish's length; found by measuring from the most anterior part of the snout, lip or chin to the end of the last vertebra.

Subtidal—Zone just below the low-water mark of the tide that is never exposed, even at low tide.

Swim bladder—See Gas Bladder.

Syntopic—Sharing the same habitat within the same geographical range.

Tetraploid—Four sets (two homologous pairs) of chromosomes existing in a cell or organism.

Translocated—Transferred or dislocated specimens.

Tubercle—Nodule, growth, or knob present in an organ or on the skin.

Ventral—Relating to, or located in the abdomen or belly.

Vermicular—Relating to, caused by, or resembling worms.

Vestigial—Body part that was functional in ancestral sources but has become reduced or nonfunctional descendants.

Vitellogenesis—Deposition of yolk within the growing egg.

Viviparous—Producing live young.

Fishes family list

Myxini [Class]
Myxiniformes [Order]
Myxinidae [Family]

Cephalaspidomorphi [Class]
Petromyzoniformes [Order]
Petromyzonidae [Family]
Geotriidae
Mordaciidae

Chondrichthyes [Class]
Chimaerformes [Order]
Callorhinchidae [Family]
Chimaeridae
Rhinochimaeridae

Heterodontiformes [Order]
Heterodontidae [Family]

Orectolobiformes [Order]
Parascyllidae [Family]
Brachaeluridae
Orectolobidae
Hemiscyllidae
Ginglymostomatidae
Stegostomatidae
Rhincodontidae

Carcharhiniformes [Order]
Scyliorhinidae [Family]
Proscylliidae
Pseudotriakidae
Leptochariidae
Triakidae
Hemigaleidae
Carcharhinidae
Sphyrnidae

Lamniformes [Order]
Ondontaspididae [Family]
Mitsukurinidae
Pseudocarchariidae
Megachasmidae
Alopiidae
Cetorhinidae
Lamnidae

Hexanchiformes [Order]
Chlamydoselachidae [Family]
Hexanchidae

Squaliformes [Order]
Echinorhinidae [Family]
Dalatiidae
Centrophoridae
Squalidae
Etmopteridae
Somniosidae
Oxynotidae

Squatiniformes [Order]
Squatinidae [Family]

Pristiophoriformes [Order]
Pristiophoridae [Family]

Rajiformes [Order]
Pristidae [Family]
Torpedinidae
Narcinidae
Rhinidae
Rhinobatidae
Rajidae
Plesiobatidae
Hexatrygonidae
Dasyatidae
Urolophidae
Gymnuridae
Myliobatidae
Potamotrygonidae
Mobulidae
Urotrygonidae
Platyrhinidae
Zanobatidae
Narkidae
Hypridae
Rhinopteridae

Sarcopterygii [Class]
Coelacanthiformes [Order]
Latimeriidae [Family]

Ceratodontiformes [Order]
Ceratodontidae [Family]

Lepidosireniformes [Order]
Lepidosirenidae [Family]
Protopteridae

Actinopterygii [Class]
Polypteriformes [Order]
Polypteridae [Family]

Acipenseriformes [Order]
Acipenseridae [Family]
Polyodontidae

Lepisosteiformes [Order]
Lepisosteidae [Family]

Amiiformes [Order]
Amiidae [Family]

Osteoglossiformes [Order]
Osteoglossidae [Family]
Pantodontidae
Hiodontidae
Notopteridae
Mormyridae
Gymnarchidae

Elopiformes [Order]
Elopidae [Family]
Megalopidae

Albuliformes [Order]
Albulidae [Family]
Halosauridae
Notacanthidae

Anguilliformes [Order]
Anguillidae [Family]
Heterenchelyidae
Moringuidae
Chlopsidae
Myrocongridae
Muraenidae
Synaphobranchidae
Ophichthidae
Colocongridae
Derichthydae
Muraenesocidae

Nemichthydae
Congridae
Nettastomatidae
Serrivomeridae

Saccopharyngiformes [Order]
Cyematidae [Family]
Saccopharyngidae
Eurypharyngidae
Monognathidae

Clupeiformes [Order]
Denticipitidae [Family]
Engraulidae
Pristigasteridae
Chirocentridae
Clupeidae

Gonorynchiformes [Order]
Chanidae [Family]
Gonorynchidae
Kneriidae

Cypriniformes [Order]
Cyprinidae [Family]
Gyrinocheilidae
Catostomidae
Cobitidae
Balitoridae

Characiformes [Order]
Citharinidae [Family]
Hemiodontidae
Curimatidae
Anostomidae
Erythrinidae
Lebiasinidae
Ctenoluciidae
Hepsetidae
Gasteropelecidae
Characidae
Alestiidae

Siluriformes [Order]
Diplomystidae [Family]
Icaluridae
Bagridae
Claroteidae
Australoglanidae
Siluridae
Schilbeidae
Pangasiidae
Amphiliidae
Sisoridae
Amblycipitidae
Akysidae
Chacidae
Clariidae
Malapteruridae
Ariidae
Plotosidae
Mochokidae
Doradidae
Auchenipteridae
Pimelodidae
Cetopsidae
Aspredinidae
Trichomycteridae
Callichthyidae
Scoloplacidae
Loricariidae
Astroblepidae
Nematogenyidae
Heptapteridae
Pseudopimelodidae
Auchenoglanidae
Erthistidae
Claroglanididae

Gymnotiformes [Order]
Sternopygidae [Family]
Rhamphichthyidae
Hypopomidae
Apteronotidae
Gymnotidae

Esociformes [Order]
Esocidae [Family]

Osmeriformes [Order]
Argentinidae [Family]
Microstomatidae
Opisthoproctidae
Alepocephalidae
Platytroctidae
Osmeridae
Retropinnidae
Lepidogalaxiidae
Galaxiidae

Salmoniformes [Order]
Salmonidae [Family]

Stomiiformes [Order]
Gonostomatidae [Family]
Photichthyidae
Sternoptychidae
Stomiidae

Aulopiformes [Order]
Giganturidae [Family]
Aulopodidae
Synodontidae
Chlorophthalidae
Ipnopidae
Scopelarchidae
Notosuridae
Pseudotrichonotidae
Paralepididae
Anotopteridae
Evermannellidae
Omosudidae
Alepisauridae

Myctophiformes [Order]
Neoscopelidae [Family]
Myctophidae

Lampridiformes [Order]
Veliferidae [Family]
Lampridiidae
Stylephoridae
Lophotidae
Radiicephalidae
Trachipteridae
Regalecidae

Polymixiiformes [Order]
Polymixiidae [Family]

Percopsiformes [Order]
Percopsidae [Family]
Aphredoderidae
Amblyopsidae

Ophidiiformes [Order]
Carapidae [Family]
Ophidiidae
Bythitidae
Aphyonidae
Parabrotulidae

Gadiformes [Order]
Macrouridae [Family]
Steindachneriidae
Moridae
Melanonidae
Bregmacerotidae
Muraenolepididae
Phycidae
Merluccidae
Gadidae
Lotidae
Macruronidae
Raniciptidae

Batrachoidiformes [Order]
Batrachoididae [Family]

Lophiiformes [Order]
Lophiidae [Family]
Antennariidae
Lophichthyidae
Tetrabrachiidae
Caulophrynidae
Chaunacidae
Ogcocephalidae
Brachionichthyidae
Diceratiidae
Thaumatichthyidae
Centrophrynidae
Gigantactinidae
Neoceratiidae
Melanocetidae
Ceratiidae
Himantolophidae
Oneirodidae
Linophynidae

Mugiliformes [Order]
Mugilidae [Family]

Atheriniformes [Order]
Bedotiidae [Family]
Melanotaeniidae
Pseudomugilidae
Atherinidae
Notocheiridae
Telmatherinidae
Dentatherinidae
Phallostethidae

Beloniformes [Order]
Adrianichthyidae [Family]
Belonidae
Scomberesocidae
Exocoetidae
Hemiramphidae

Cyprinodontiformes [Order]
Aplocheilidae [Family]
Profundilidae
Fundulidae
Valenciidae
Anablepidae
Poeciliidae
Goodeidae
Cyprinodontidae
Rivulidae

Stephanoberyciformes [Order]
Melamphaidae [Family]
Gibberichthyidae
Stephanoberycidae
Hispidoberycidae
Rondeletiidae
Barbourisiidae
Cetomimidae
Mirapinnidae
Megalomycteridae

Beryciformes [Order]
Anoplogasteridae (also spelled Anoplogastridae) [Family]
Diretmidae
Anomalopidae
Monocentridae
Trachichthyidae
Berycidae
Holocentridae

Zeiformes [Order]
Parazenidae [Family]
Macrurocyttidae
Zeidae
Oreosomatidae
Grammicolepidae
Caproidae
Zeniontidae

Gasterosteiformes [Order]
Hypoptychidae [Family]
Aulorhynchidae
Gasterosteidae
Pegasidae
Solenostomidae
Syngnathidae
Indostomidae
Aulostomidae
Fistulariidae
Macroramphosidae
Centriscidae

Synbranchiformes [Order]
Sychbranchidae [Family]
Chaudhuriidae
Mastacembelidae

Scorpeniformes [Order]
Dactylopteroidei [Suborder]
Dactylopteridae [Family]

Scorpaenoidei [Suborder]
Scorpaenidae [Family]
Caracanthidae
Aploactinidae
Pataecidae
Gnathanacanthidae
Congiopodidae
Peristediidae
Apistidae
Neosebastidae
Sebastidae
Setarchidae
Synanceiidae
Tetrarogidae
Triglidae

Platycephaloidei [Suborder]
Bembridae [Family]
Platycephalidae
Hoplichthyidae

Anoplopomatoidei [Suborder]
Anoplopomatidae [Family]

Hexgrammoidei [Suborder]
Hexagrammidae [Family]

Normanichthyiodei [Suborder]
Normanichthyidae [Family]

Cottoidei [Suborder]
Rhamphocottidae [Family]
Ereuniidae
Cottidae
Comephoridae
Abyssocottidae
Hemitripteridae
Agonidae
Psycholutidae
Bathylutichthyidae
Cyclopteridae
Liparidae
Cottocompheroidae

Perciformes [Order]
Percoidei [Suborder]
Centropomidae [Family]
Chandidae
Moronidae
Percichthyidae
Acropomatidae
Serranidae
Ostracoberycidae
Callanthiidae
Pseudochromidae
Grammatidae
Plesiopidae
Notograptidae
Opistognathidae
Dinopercidae
Banjosidae
Centrarchidae
Percidae
Priacanthidae
Apogonidae
Epigonidae
Sillaginidae
Malacanthidae
Lactariidae
Dinolestidae
Pomatomidae
Nematistiidae
Echeneidae
Rachycentridae
Coryphaenidae
Carangidae
Menidae
Leiognathidae
Bramidae
Caristiidae
Emmelichthyidae
Lutjanidae
Lobotidae
Gerreidae
Haemulidae
Inermiidae
Sparidae
Centracanthidae
Lethrinidae
Nemipteridae
Polynemidae
Sciaenidae
Mullidae
Pempheridae
Glaucosomatidae
Leptobramidae
Bathyclupeidae
Monodactylidae
Toxotidae
Coracinidae
Drepanidae
Chaetodontidae
Pomacanthidae
Enoplosidae
Pentacerotidae
Nandidae
Kyphosidae

Arripidae
Teraponidae
Kuhliidae
Oplegnathidae
Cirrhitidae
Chironemidae
Aplodactylidae
Cheilodactylidae
Latridae
Cepolidae
Elassomatidae
Paracorpididae
Dischistiidae
Percilidae
Nannopercidae
Gadopsidae

Labroidei [Suborder]
Cichlidae [Family]
Embiotocidae
Pomacentridae
Labridae
Odacidae
Scaridae

Zoarcoidei [Suborder]
Bathymasteridae [Family]
Zoarcidae
Stichaeidae
Cryptacanthodidae
Pholidae
Anarhichadidae
Ptilichthyidae
Zaproridae
Scytalinidae

Notothenioidei [Suborder]
Bovichthyidae [Family]
Nototheniidae
Harpagiferidae
Bathydraconidae
Channichthyidae
Pseudaphritidae
Eleginopidae
Artedidraconidae

Trachinoidei [Suborder]
Chiasmodontidae [Family]
Champsodontidae
Pholidichthyidae
Trichodontidae
Pinguipedidae
Cheimarrhichthyidae
Trichonotidae
Creediidae
Percophidae
Leptoscopidae
Ammodytidae
Trachinidae
Uranoscopidae

Blenniodei [Suborder]
Tripterygiidae [Family]
Dactyloscopidae
Labrisomidae
Clinidae
Chaenopsidae
Blenniidae

Icosteoidei [Suborder]
Icosteidae [Family]

Gobiesocoidei [Suborder]
Gobiesocidae [Family]

Callionymoidei [Suborder]
Callionymidae [Family]
Draconettidae

Gobioidei [Suborder]
Rhyacichthyidae [Family]
Odontobutidae
Eleotridae
Gobiidae
Kraemeriidae
Xenisthmidae
Microdesmidae
Ptereleotridae
Schindleriidae

Kurtoidei [Suborder]
Kurtidae [Family]

Acanthuroidei [Suborder]
Ephippidae [Family]
Scatophagidae
Siganidae
Luvaridae
Zanclidae
Acanthuridae

Scombrolabracoidei [Suborder]
Scombrolabracidae [Family]

Scombroidei [Suborder]
Sphyraenidae [Family]
Gempylidae
Trichiuridae
Scombridae
Xiphiidae
Istiophoridae

Stromateoidei [Suborder]
Amarsipidae [Family]
Centrolophidae
Nomeidae
Ariommatidae
Tetragonuridae
Stromateidae

Anabantoidei [Suborder]
Luciocephalidae [Family]
Anabantidae
Helostomatidae
Belontiidae
Osphronemidae

Channoidei [Suborder]
Channidae [Family]

Pleuronectiformes [Order]
Psettodidae [Family]
Citharidae
Bothidae
Achiropsettidae
Scophthalmidae
Paralichthyidae
Pleuronectidae
Samaridae
Achiridae
Soleidae
Cynoglossidae
Paralichthodidae
Poecilopsettidae
Rhombosleidae

Tetraodontiformes [Order]
Triacanthidae [Family]
Balistidae
Monacanthidae
Ostraciidae
Triodontidae
Tetraodontidae
Diodontidae
Molidae

• • • • •

A brief geologic history of animal life

A note about geologic time scales: A cursory look will reveal that the timing of various geological periods differs among textbooks. Is one right and the others wrong? Not necessarily. Scientists use different methods to estimate geological time—methods with a precision sometimes measured in tens of millions of years. There is, however, a general agreement on the magnitude and relative timing associated with modern time scales. The closer in geological time one comes to the present, the more accurate science can be—and sometimes the more disagreement there seems to be. The following account was compiled using the more widely accepted boundaries from a diverse selection of reputable scientific resources.

Geologic time scale

Era	Period	Epoch	Dates	Life forms
Proterozoic			2,500-544 mya*	First single-celled organisms, simple plants, and invertebrates (such as algae, amoebas, and jellyfish)
Paleozoic	Cambrian		544-490 mya	First crustaceans, mollusks, sponges, nautiloids, and annelids (worms)
	Ordovician		490-438 mya	Trilobites dominant. Also first fungi, jawless vertebrates, starfish, sea scorpions, and urchins
	Silurian		438-408 mya	First terrestrial plants, sharks, and bony fish
	Devonian		408-360 mya	First insects, arachnids (scorpions), and tetrapods
	Carboniferous	Mississippian	360-325 mya	Amphibians abundant. Also first spiders, land snails
		Pennsylvanian	325-286 mya	First reptiles and synapsids
	Permian		286-248 mya	Reptiles abundant. Extinction of trilobytes
Mesozoic	Triassic		248-205 mya	Diversification of reptiles: turtles, crocodiles, therapsids (mammal-like reptiles), first dinosaurs
	Jurassic		205-145 mya	Insects abundant, dinosaurs dominant in later stage. First mammals, lizards, frogs, and birds
	Cretaceous		145-65 mya	First snakes and modern fish. Extinction of dinosaurs, rise and fall of toothed birds
Cenozoic	Tertiary	Paleocene	65-55.5 mya	Diversification of mammals
		Eocene	55.5-33.7 mya	First horses, whales, and monkeys
		Oligocene	33.7-23.8 mya	Diversification of birds. First anthropoids (higher primates)
		Miocene	23.8-5.6 mya	First hominids
		Pliocene	5.6-1.8 mya	First australopithecines
	Quaternary	Pleistocene	1.8 mya-8,000 ya	Mammoths, mastodons, and Neanderthals
		Holocene	8,000 ya-present	First modern humans

*Millions of years ago (mya)

Index

Bold page numbers indicate the primary discussion of a topic; page numbers in italics indicate illustrations.

B

INDEX

C

D

E

F

G

INDEX

H

I

J

K

L

M

P

INDEX

Q

R

INDEX

INDEX

T

U

V

W